To my wife Selma, my children, and their families,

and

to the loving memory of my father, Joseph Haskin,
to my mother Reba, and to my wife Pam.

Volume

I

ROENTGENOLOGIC DIAGNOSIS

A Complement in Radiology to the Beeson and McDermott Textbook of Medicine

J. GEORGE TEPLICK, M.D., F.A.C.R.

Professor and Director of General Diagnosis, Department of Diagnostic Radiology,
Hahnemann Medical College and Hospital

MARVIN E. HASKIN, M.D., F.A.C.P.

Professor and Chairman, Department of Diagnostic Radiology,
Hahnemann Medical College and Hospital

Third Edition

W. B. SAUNDERS COMPANY • Philadelphia • London • Toronto

W. B. Saunders Company: West Washington Square
                          Philadelphia, PA 19105

                          1 St. Anne's Road
                          Eastbourne, East Sussex BN21 3UN, England

                          1 Goldthorne Avenue
                          Toronto, Ontario M8Z 5T9, Canada

Library of Congress Cataloging in Publication Data

Teplick, J. George.

Roentgenologic diagnosis.

Includes bibliographical references and indexes.

1. Diagnosis, Radioscopic.   I. Haskin, Marvin E., joint
   author.   II. Textbook of medicine.   III. Title.
   [DNLM: 1. Radiography — Atlases. WN17T314r]

RC78.T38 1976   616.07'572   75–10390

ISBN 0–7216–8789–X (v. 1)

ISBN 0–7216–8790–3 (v. 2)

Listed here is the latest translated edition of this book together with the
language of the translation and the publisher.

Japanese *(1st Edition)* (Volumes I and II) — Igaku Shoin, Ltd., Tokyo, Japan

Spanish *(2nd Edition)* (Volumes I and II) — NEISA, Mexico, D.F., Mexico

Roentgenologic Diagnosis   ISBN Vol I: 0-7216-8789-X
                                                      ISBN Vol II: 0-7216-8790-3

Last digit is the print number: 9   8   7   6   5   4   3   2

PREFACE
to the Third Edition

The format and organizational concepts of the Third Edition are basically unchanged, corresponding to the arrangement of diseases in the latest edition (Fourteenth) of the Beeson and McDermott: *Textbook of Medicine.* Our work continues to be a radiologic complement to the *Textbook of Medicine,* abetting correlation of clinical and physiologic alterations with the radiographic changes of the diseases and their complications.

Major revisions and additions have entered this Third Edition. Most of the text has been rewritten and updated, and over 70 new entities, 300 additional radiologic illustrations, and well over 1000 additional current references have been added. In addition, over 100 illustrations of basic radiographic abnormalities have been included in a new section at the beginning of Volume I. The updated tables of roentgen signs are located at the back of each volume.

Computerized tomography of the brain is discussed and illustrated under the relevant diseases and conditions.

The text in the section on special procedures has been augmented with radiographic illustrations. This section will appear in Volume II, following the main text.

We gratefully acknowledge the unstinting assistance of our secretarial force and of our copy editor, Karen Comerford. We especially acknowledge the constant encouragement and advice of our medical editor, John J. Hanley.

J. GEORGE TEPLICK
MARVIN E. HASKIN

PREFACE
From the Second Edition

Radiology is a specialized form of clinical examination. Radiographic findings can be completely appreciated only in light of the clinical understanding of disease. Therefore, although useful as a primary text, the relation of the Second Edition of *Roentgenologic Diagnosis* as a complement to Beeson and McDermott's *Textbook of Medicine* (Thirteenth Edition) provides an even broader understanding of disease processes and their precise clinical diagnoses.

The Second Edition maintains the basic format and organizational concepts of the First Edition. The organization corresponds to that of Beeson and McDermott's *Textbook of Medicine,* so that clinical, radiologic, and physiologic correlations can easily be made.

Major changes have been incorporated into the Second Edition. Almost the complete text has been rewritten and expanded to increase the depth of radiologic detail. More than fifty additional clinical disease entities are discussed, and some four hundred new illustrations have been added.

Traditional radiographic organization is based on roentgen signs, rather than according to disease entities as in our book. We have attempted to bridge the gap between these pedagogical methods by a new appendix in which disease entities are collated under headings of various radiologic signs, thereby providing extensive lists for differential diagnoses.

Also preceding the index in each volume is an alphabetically arranged glossary which explains the numerous special roentgenologic procedures that are frequently mentioned and illustrated throughout these volumes.

The reader will recognize disparities between the frequency and clinical importance of a disease and the number of radiograms illustrated; that is, some rare conditions are illustrated more extensively than relatively common diseases. In many cases we feel that the relative obscurity of a disease process makes an understanding of its roentgenologic changes especially important. Moreover, even when the roentgenologic changes of a disease entity are not specific in themselves, they may play a key part in diagnosis when considered in conjunction with the clinical data.

It is intentional that these volumes do not provide instruction in the fundamentals of radiology. There are no introductory sections dealing with the physical or technical principles of radiology or the fundamentals of radiologic positioning and interpretation. This simplicity of design facilitates our basic objective of demonstrating the pertinent radiographic features of the innumerable rare and common diseases that the physician encounters.

We wish to acknowledge our colleague and friend, Dr. Arnd Schimert, who participated in the preparation of the First Edition, and whose warmth and counsel have been sorely missed. Advice, cooperation, and many radiograms were generously contributed by numerous radiologists and other physicians, to whom we express our thanks. Special gratitude is due to Drs. William Likoff, Lewis C. Mills, Charles Schwartz, Michael Geduldig, Robert F. Johnston, and Paul Gonick, all of the Hahnemann Medical College and Hospital. Finally, we thank the staff of the W. B. Saunders Company. The quality of the production reflects the publisher's attitude and care.

J. George Teplick
Marvin E. Haskin

CONTENTS

v

SECTION 5 GRANULOMATOUS DISEASES OF UNPROVED ETIOLOGY

SECTION 6 MICROBIAL DISEASES

SECTION 10   CARDIOVASCULAR DISEASES

VOLUME II

SECTION 11   RENAL DISEASES

SECTION 17   DISEASES OF BONE

SECTION 18 CERTAIN CUTANEOUS DISEASES WITH SIGNIFICANT SYSTEMIC MANIFESTATIONS

SECTION 19 MISCELLANEOUS HEREDITARY DISORDERS AFFECTING MULTIPLE ORGAN SYSTEMS

BASIC ROENTGEN SIGNS

The appearance of many basic radiographic abnormalities is illustrated in this section. These abnormalities, which are the fundamental building blocks for roentgenologic diagnosis, are familiar and readily recognized by the trained radiologist. However, the medical student, the novice radiologic resident, and the nonradiologic physician, who repeatedly encounter verbal descriptions of these abnormalities or signs, more often than not are unclear about their exact radiographic appearance. It is hoped that these illustrations will acquaint the nonradiologist with these basic radiographic abnormalities.

III. GASTROINTESTINAL TRACT AND ABDOMEN

IV. JOINTS

I. CARDIOVASCULAR SYSTEM

Enlarged Pulmonary Arteries

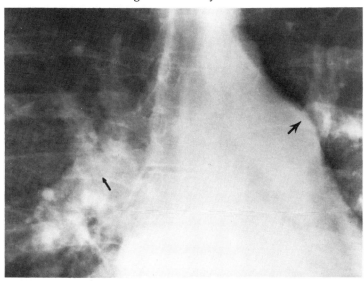

The localized bulge on the left below the aortic knob *(large arrow)* is characteristic of an enlarged main pulmonary artery. The marked enlargement of the right hilar shadow *(small arrow)*, from which large vessels emerge, is characteristic of an enlarged right pulmonary artery.

Overpenetration may be needed to evaluate the size of the left pulmonary artery, which is obscured by the enlarged main pulmonary artery shadow.

Enlarged Right Ventricle

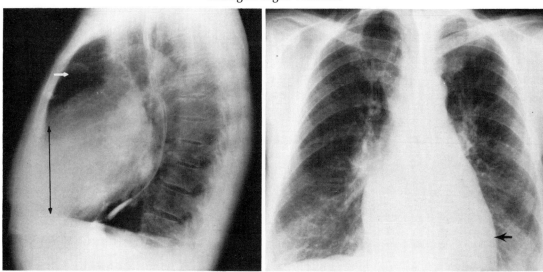

On lateral view *(left)*, an unusually large portion of the heart density *(double-headed arrow)* extends along the anterior chest wall, decreasing the amount of the anterior lung lucent space *(white arrow)*. This is quite characteristic of right ventricular enlargement.

On frontal view *(right)*, there is often maximum fullness of the left border well above the diaphragm *(arrow)*, but it is often difficult to distinguish right from left ventricular enlargement on frontal film alone.

**Decreased Vasculature:
Small Pulmonary Arteries**

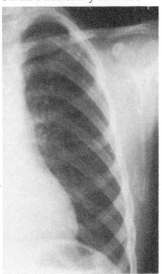

The inconspicuous left hilar shadow and small, sparse central and peripheral vessels are characteristic of markedly diminished vascularity of the lung.

Increased Vascularity (Arterial)

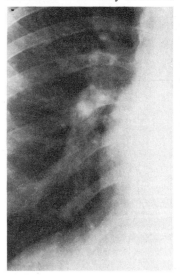

The enlarged right pulmonary artery is accompanied by unusually prominent vessels extending well into the lung periphery throughout the lung field. These findings were bilateral. This is the characteristic appearance of hypervascularity associated with increased pulmonary arterial flow.

**Vascular Redistribution
(Venous)**

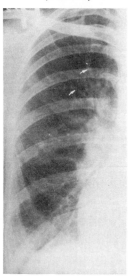

The vessels in the upper lung fields are prominent and somewhat enlarged (*arrows*), while the vascular shadows in the lower lung fields are smaller than usual.

This redistribution of blood flow to the upper lobes is significant only when present on erect films.

Enlarged Aortic Knob, Enlarged Left Ventricle

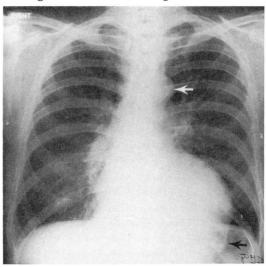

The aortic knob is more prominent (larger) than usual *(white arrow)*. The enlarged left cardiac border extends caudad and can be seen through the gastric air bubble *(black arrow)*. This extension of an enlarged left border inferiorly is characteristic of left ventricular enlargement.

Enlarged Left Ventricle

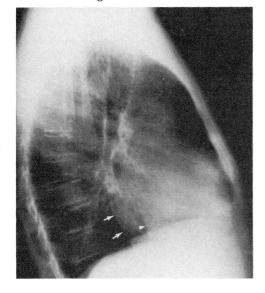

On lateral view, the posterior border of the left ventricle *(arrows)* is considerably more posteriorly located than normal and lies well behind the linear shadow of the inferior vena cava *(arrowhead)*. The latter normally lies slightly behind the left ventricular shadow.

Enlarged Left Atrium, Enlarged Main Pulmonary Artery

The left main bronchus *(small arrows)* is elevated, a characteristic finding in left atrial enlargement.

The rounded bulge *(large arrow)* below the aortic knob is typical of an enlarged main pulmonary artery.

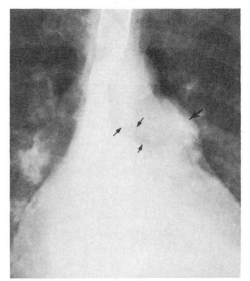

Enlarged Left Atrium

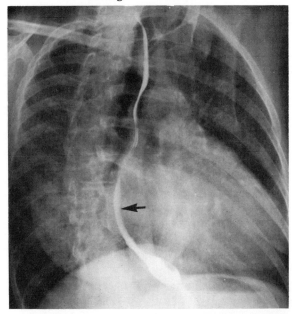

A

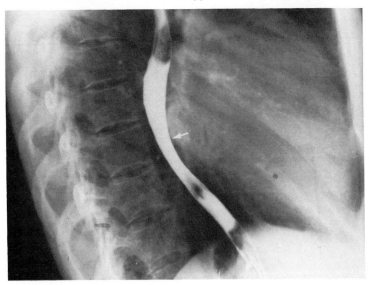

B

The smooth, concave impression on the lower portion of the barium-filled esophagus with some dorsal displacement is characteristic of left atrial enlargement. This is usually best demonstrated in the right anterior oblique projection *(A; large black arrow)* but can often be seen in the lateral projection *(B; white arrow).*

Left ventricular enlargement may also cause an esophageal indentation, but this occurs in the extreme lower portion of the esophagus, below the area of atrial impression.

Enlarged Left Atrial Appendage

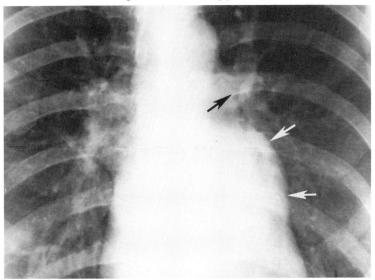

The smooth bulge *(white arrows)* on the left below the pulmonary artery segment *(black arrow)* is the characteristic appearance of enlargement or prominence of the left atrial appendage.

Pericardial Calcification

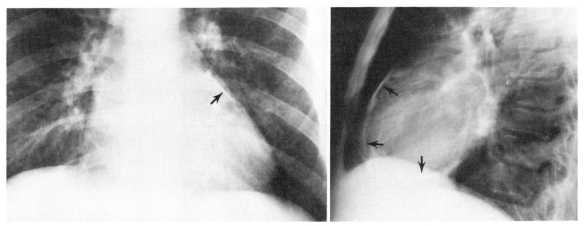

Linear plaques of calcification *(arrows)* along the cardiac borders are virtually pathognomonic of pericardial calcification.

Such calcifications are usually more readily appreciated on the lateral film.

Calcified Aortic Annulus and Valve

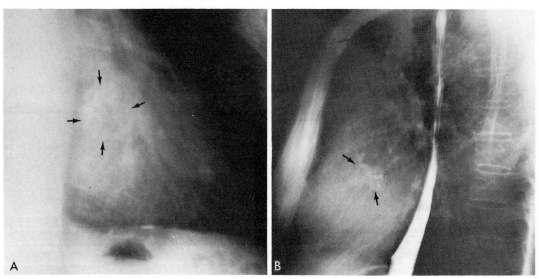

A, On frontal view, the ring-like cardiac calcification *(arrows)* just to the left of the spine is a calcified aortic annulus.

B, On a lateral view, the irregular calcifications *(arrows),* which lie in the region of the aortic root, are calcified aortic leaflets and annulus.

II. CHEST

Air Bronchograms

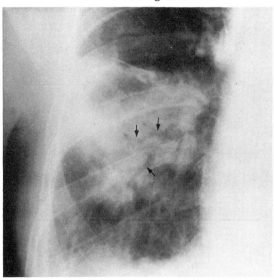

Branching linear lucencies *(arrows)* within the irregular area of density are characteristic of air bronchograms. Their presence confirms the alveolar nature of an infiltrate.

Alveolar Infiltrates with Air Bronchogram

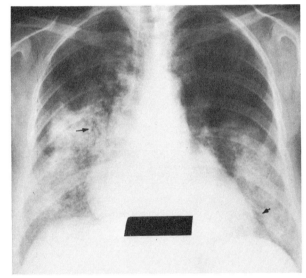

The bilateral lower lobe densities with fluffy margins are characteristic of alveolar infiltrates. Faint air bronchograms are visible *(arrows)*.

Alveologram

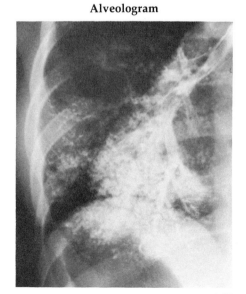

The radiographic appearance of opacification of alveoli is demonstrated by bronchography during which large numbers of alveoli were opacified by flooding of the contrast material.

Note the irregular nodule-like small densities with unsharp borders, which are characteristic of opacified primary alveolar lobules.

Pulmonary Edema: Bat-Wing Configuration

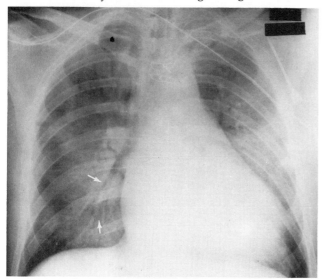

Bilateral perihilar homogeneous alveolar densities with indistinct lateral margins and usually air bronchograms *(arrows)* are characteristic of the classic picture of pulmonary edema.

Interstitial Densities

The linear densities of variable thickness and length are characteristic of diffuse interstitial infiltrates.

Kerley B Lines

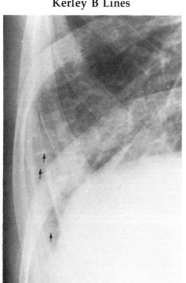

Short transverse thin lines of density in the periphery of the lower lung fields *(arrows)* are characteristic of Kerley B lines.

Interstitial Reticulonodular Pattern

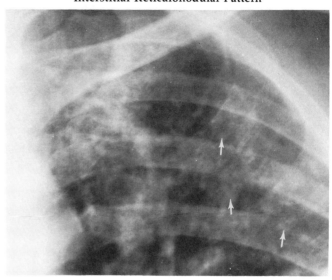

Increased interstitial densities associated with numerous small irregular nodular densities *(arrows)* are fairly characteristic of the diffuse reticulonodular pattern.

Kerley B and C Lines, Lamellar Effusion

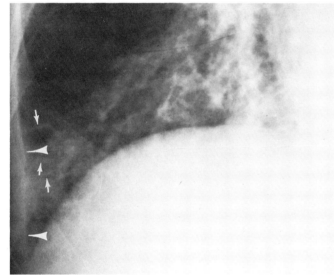

The vessels at the right base are obscured by a reticular pattern, called Kerley C lines. These are due to edema of the interstitial lymphatics. The linear transverse short densities *(arrows)* adjacent to the lower lateral chest wall are the Kerley B lines, swollen peripheral lymphatics. The linear density along the chest wall *(arrowheads)* is a lamellar effusion.

Kerley A Lines

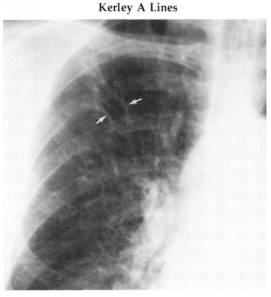

The curved linear densities *(arrows)* in the upper lung fields do not taper peripherally like the usual bronchovascular markings, and they disappeared when compensation was restored.

Their radiographic appearance and occurrence during decompensation and interstitial edema is characteristic of Kerley A lines (edema of interstitial lymphatics).

Honeycomb Lung

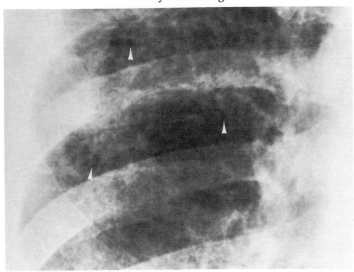

Diffuse interstitial densities interspersed with small areas of lucency *(arrowheads)* are the characteristic findings. The underlying abnormality is usually diffuse fibrosis and intervening areas of hyperinflation. Such a picture can occur in a variety of conditions.

Small Nodular Lesions

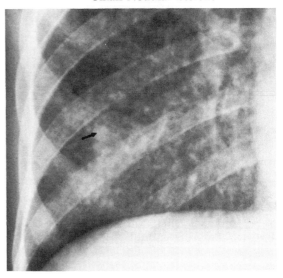

The entire lung field is filled with small, irregular nodular densities of varying size *(arrow)*. When nodules of this size or smaller are diffused throughout the entire lung, the appearance is termed *diffuse miliary disease.*

Small Nodules

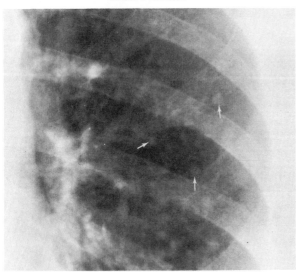

Small, rather discrete and sharp nodular lesions are seen scattered throughout the lung fields *(arrows).*

Irregular Nodular Densities

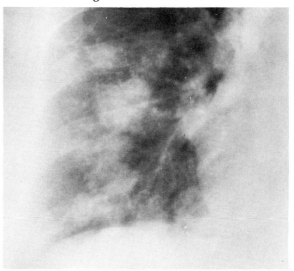

Multiple densities with ill-defined borders illustrate the appearance of large, irregular, nodular infiltrates.

Irregular Thick-Walled Cavity

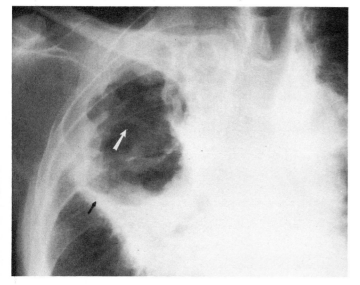

The large lucent cavity *(white arrow)* has irregular borders and is surrounded by a thick wall *(black arrow)*.

Thin-Walled Cavity

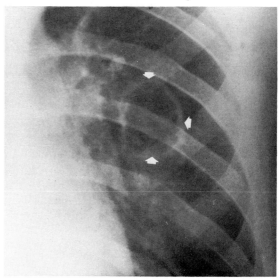

The walls surrounding the lucent cavity *(arrows)* are fairly regular and only a few millimeters thick.

Lobar Consolidation

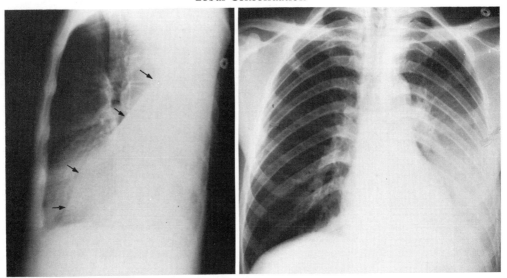

Opacification of the entire left lower lobe is best appreciated on the lateral film, in which the entire long fissure *(arrows)* is seen as the anterior boundary of the consolidation. The full extent of the density cannot be accurately appreciated from the frontal film.

Atelectasis: Left Upper Lobe

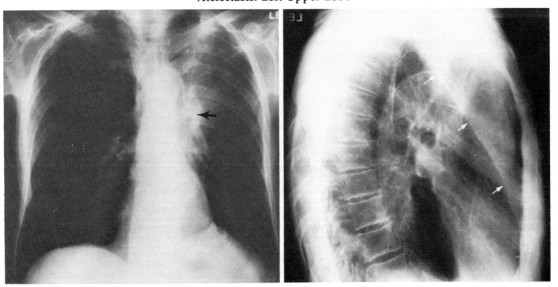

On frontal view, there is an ill-defined density in the left upper lung field, and the left hilar shadow is considerably elevated *(arrow)*. The left diaphragm is also elevated.

On lateral view, a triangular density lies in the upper anterior chest, bounded by the sharp, long fissure *(arrows)* that is considerably displaced anteriorly.

These are characteristic findings of left upper lobe atelectasis.

Atelectasis: Right Middle Lobe: Silhouette Sign

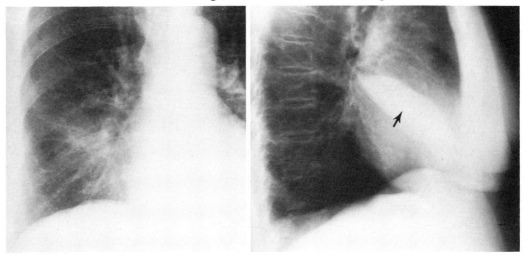

The ill-defined density in the right lower lung field obliterates the right cardiac border. On lateral view, a wedge-shaped density (*arrow*) occupies the position of the right middle lobe, which normally is considerably larger than the density. The rounded, superior border of the density is the bowed and downward displaced short fissure of the right middle lobe.

The findings are characteristic of right middle lobe atelectasis. The normally sharp outline of the right cardiac border is obscured by the middle lobe consolidation (silhouette sign).

Hyperinflation

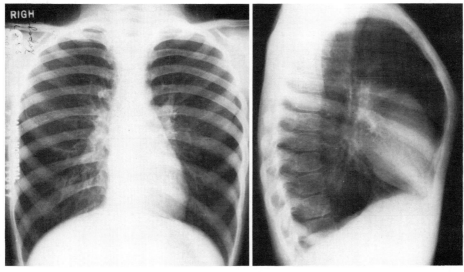

On frontal view, the low diaphragm and the diminished vascular shadows that contribute to the increased radiolucency of the lung are characteristic of hyperinflation.

On lateral view, the anteroposterior diameter of the chest is increased, the superior anterior lung is hyperlucent, and the diaphragms are low and flattened, all characteristic signs of hyperinflation.

Segmental Atelectasis: Right Lower Lobe

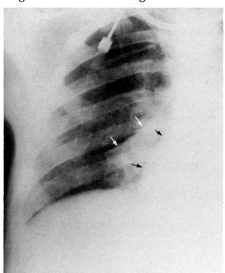

The density in the right lower lung field has a sharp, somewhat convex upper border *(white arrows)*, which represents the retracted anterior fissure of the lower lobe.

The preservation of the right cardiac border *(black arrows)* indicates that the density is posterior to the heart.

Thick Plate of Atelectasis

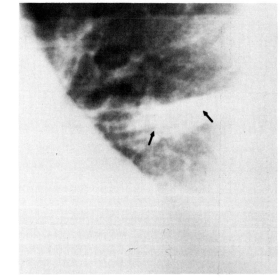

The transverse band of density *(arrows)* in the left lower lung field is fairly characteristic of a focal plate of atelectasis. Atelectatic plates can be of variable thickness and often change rapidly after adequate ventilation.

Pneumatoceles

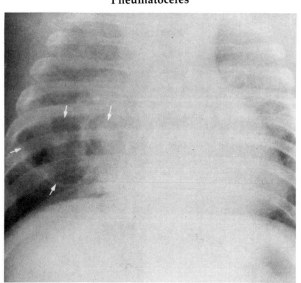

Multiple rounded lucencies with distinct thin, sharp borders *(arrows)* are characteristic of pneumatoceles.

Bullae

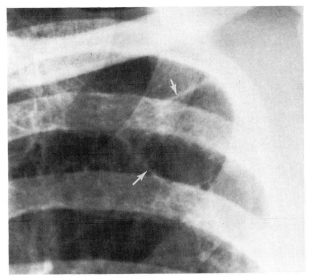

The linear strands of density *(arrows)* that represent the margins of the bullae are often the only easily identified evidence of the presence of bullae.

Bullae

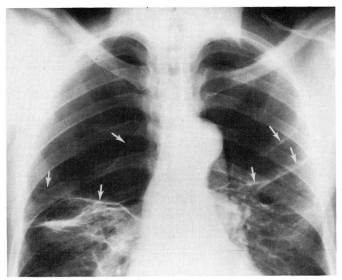

Most of both upper lung fields has been replaced by giant bullae. Their borders are the thin lines of density *(arrows)*. There is complete absence of lung markings within the bullae.

Often bullae can be identified only by the thin linear strands of their borders.

Pneumothorax

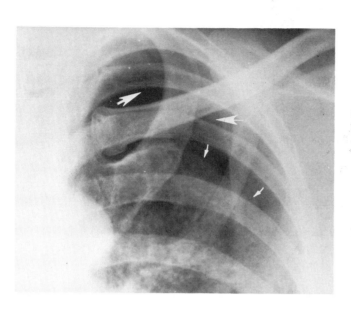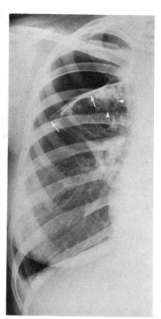

Characteristically, a pneumothorax can be identified by a sharp edge of the collapsed lung, which may be quite faint (*A; small arrows*) or quite dense, especially in a more chronic pneumothorax (*B; small arrows*). There will be a complete absence of pulmonary markings in the pneumothorax (*A; large arrows*). Careful scrutiny is often necessary to detect a small pneumothorax with a faint lung edge.

The small arrowheads in *B* outline a cavity in the collapsed upper lobe.

Pleural Effusion

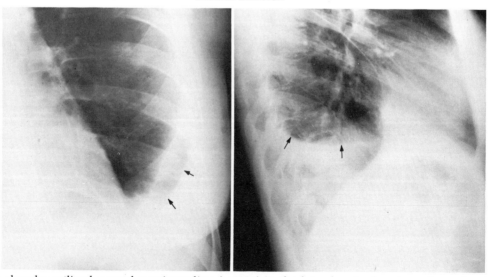

The sharply outlined curved meniscus line (*arrows*) in the lateral costophrenic angle on frontal view (left) and in the posterior angle on lateral view (right) is characteristic of a pleural effusion. The diaphragm is obliterated in the fluid area.

Pleural Effusion: Decubitus Position

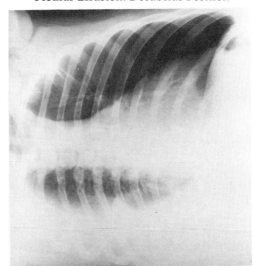

The density along the right chest wall appeared on a right side down decubitus film and represents free pleural fluid.

When nonloculated fluid is minimal, subpulmonic, or obscured by parenchymal disease, a decubitus view will reveal its presence.

Interlobar Pleural Fluid

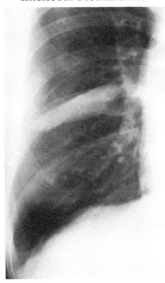

This sausage-like density in the region of the short fissure has sharp borders and represents fluid in this fissure. Its interlobar location becomes more readily apparent on a lateral film.

Lamellar Effusion, Interstitial Edema

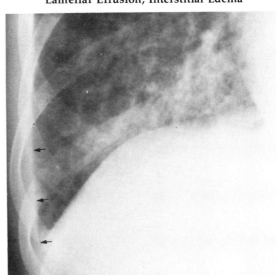

The band of density along the right lower chest wall, which has sharp straight borders (arrows), illustrates the appearance of a small lamellar effusion.

The irregular interstitial densities in the right lower lung have obliterated the normal vascular markings. This appearance is fairly typical of interstitial edema or infiltrate.

Pleural Thickening, Pleural Calcification

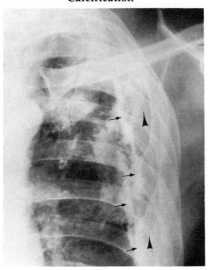

The irregular calcific densities *(arrows)* at the border of the long area of pleural thickening *(arrowheads)* are calcification of the thickened pleura.

Pleural Thickening

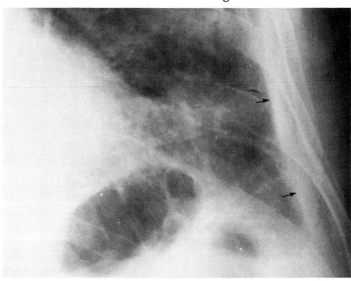

A band of increased density, with fairly sharp borders *(arrows),* lying along the chest wall is characteristic of pleural thickening. Unlike pleural fluid, this density will remain fixed in location in all positions, including decubitus or recumbent.

Pleural Nodules

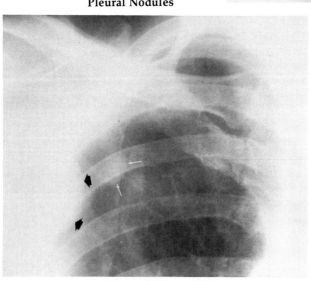

A curved density *(black arrows)* has its broad base against the chest wall. This is typical of a pleural nodule. Another nodule, not seen tangentially, has a sharp lower border *(white arrows),* but its other border is indefinite and unsharp, since it lies against the posterior pleural surface.

Hilar Adenopathy

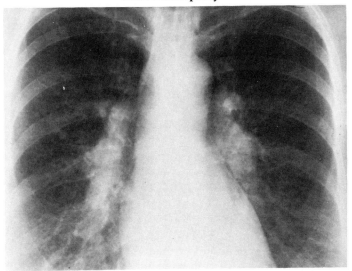

The prominent lobulated hilar shadows are characteristic of bilateral hilar adenopathy. Hilar adenopathy may sometimes be difficult to distinguish from enlarged hilar vessels, although tomography can usually resolve this problem.

Hilar Adenopathy

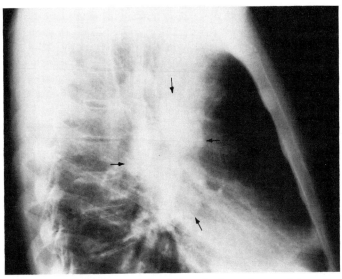

On lateral film, the lobate enlargement of the hilar shadows *(arrows)* is quite characteristic of hilar adenopathy.

Not infrequently, hilar adenopathy is more readily appreciated on the lateral film when the finding is equivocal on the frontal projection.

Mediastinal Widening

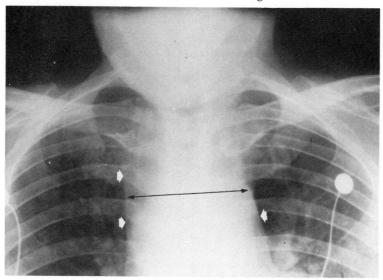

The superior mediastinum is increased in width *(double-headed arrow)* and has straight sharp borders *(white arrows)*.

Mediastinal Widening, Bilateral Masses

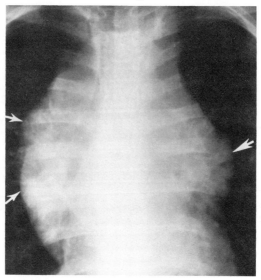

Widening of the superior mediastinum, with bilateral rounded or lobulated borders *(arrows)*, is characteristic of mediastinal masses. Such findings can also be unilateral.

Mediastinal Emphysema

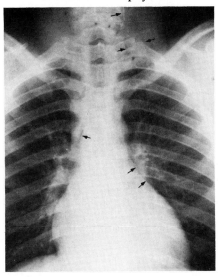

Lucent streaks (of air) medial to the mediastinal pleura, in the mediastinum around the ascending aorta, and in the neck *(arrows)* are characteristic of mediastinal emphysema.

Mediastinal Emphysema, Subcutaneous Emphysema

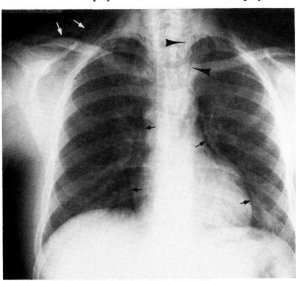

Lucent streaks of air *(black arrows)* have dissected portions of the mediastinal pleura from the heart and great vessels. Air streaks *(arrowheads)* are also seen in the superior mediastinum and have extended into the soft tissues of the neck *(white arrows)*, where they now represent subcutaneous emphysema.

These are characteristic findings in mediastinal emphysema. The air streaks along the heart borders should not be mistaken for air in the pericardium.

III. GASTROINTESTINAL TRACT AND ABDOMEN

Ulcer-Duodenal Bulb

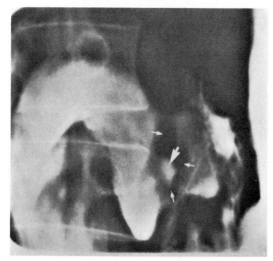

Close-up view of duodenal bulb shows an ulcer crater filled with barium *(large arrow)* surrounded by lucencies *(small arrows)* which are edematous mucosal folds.

Ulcer (Gastric) with Thickened Folds

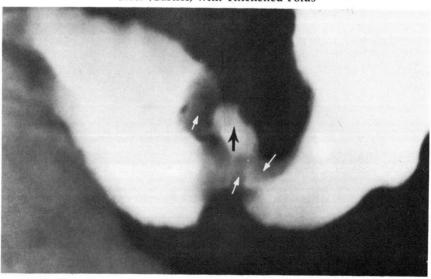

The ulcer appears as an irregular outpouching of barium *(black arrow)* arising from the pyloric area. Thickened edematous folds are the lucent areas *(white arrows)* that surround most of the ulcer.

This appearance is characteristic of an ulcer surrounded by thickened edematous folds.

Benign Gastric Ulcer with Collar

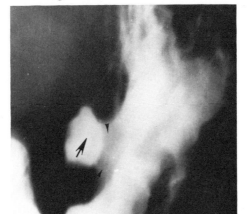

The large barium collection *(black arrow)* projecting from the gastric lesser curvature is a typical ulcer. It is surrounded by a smooth lucent zone *(arrowheads)* that is characteristic of the mucosal collar that often surrounds a benign ulcer.

Diffuse Ulceration

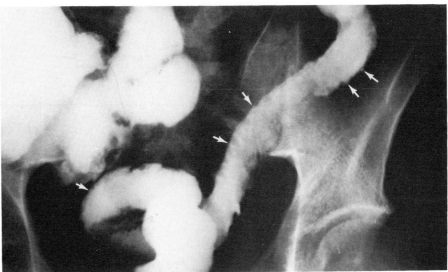

The small projections (only a few are arrowed) throughout the sigmoid are characteristic of diffuse colonic ulceration.

Deformed Duodenal Bulb

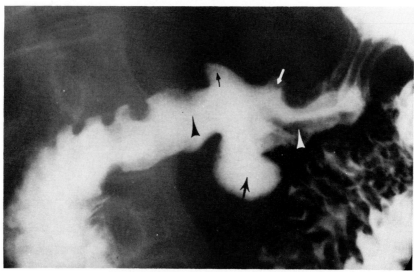

One recess of the bulb is dilated and enlarged *(large black arrow)*, while the lesser curvature recess is small and pointed *(small black arrow)*. Another irregular outpouching *(white arrow)* is just distal to the widened pyloric canal *(white arrowhead)*. The distal portion of the bulb *(black arrowhead)* cannot be distinguished from the beginning of the second portion of the duodenum.

A deformed bulb may have a variety of shapes and sizes, depending upon the severity of the inflammatory process.

Thickened Gastric Folds

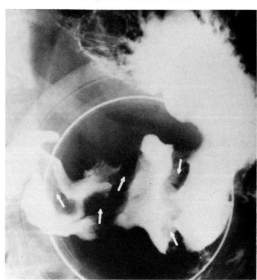

The lucent bands *(arrows)* in the lower half of the stomach are characteristic of greatly thickened gastric folds.

Thickened Folds in Small Bowel

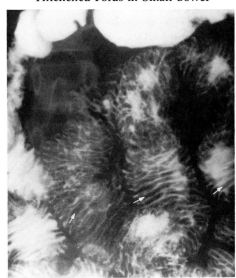

The lucent bands *(arrows)* between the barium are characteristic of thickened small bowel folds.

Thickened Walls of Small Bowel

The increased distance *(arrows)* between the barium-filled lumens of adjacent loops of small bowel is characteristic of thickening of the walls of the small bowel.

Flocculation and Moulage in Small Bowel

Small irregular collections of barium *(white arrows)* are areas of flocculation. The larger amorphous collections *(black arrows)* are isolated segments of small bowel filled with barium; these collections are called *moulage*. Both of these changes are often seen in malabsorption syndromes and in conditions in which there is hypersecretion in the small bowel.

Multiple Filling Defects of Stomach

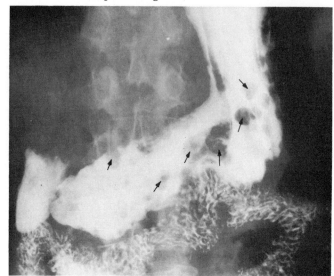

There are multiple, scattered, rounded lucencies (*arrows*) in the barium-filled colon.

Diffuse Nodularity

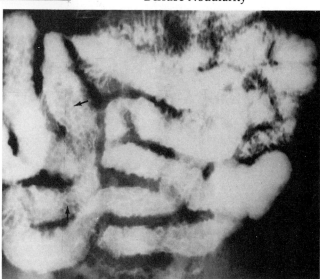

Virtually all the small bowel loops contain innumerable small rounded defects (*arrows*) indicative of diffuse nodular masses. This diffuse nodularity is sometimes referred to as "cobblestone" mucosa.

Infiltrating Lesion of Stomach

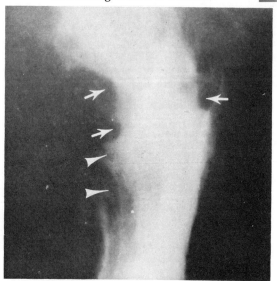

Multiple wall defects (*arrows*) and destructive changes in the gastric folds (*arrowheads*) are characteristic of an aggressive infiltrating destructive lesion.

Infiltrating and Polypoid Lesion of Stomach

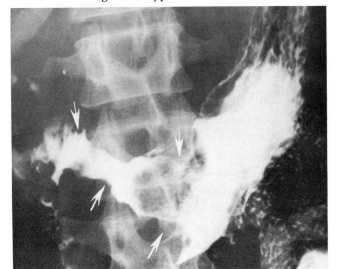

The distal half of the stomach is greatly deformed and irregular and contains many irregular filling defects *(arrows)*. No normal mucosal folds are seen in this portion.

These are characteristic findings of an infiltrating and polypoid tumor.

Infiltrating Lesion of Colon: Shoulder Defect

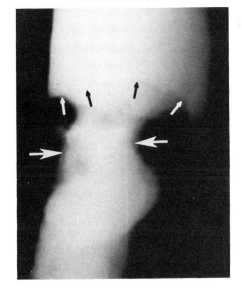

In addition to the irregularity of the walls and narrowing *(large arrows)* of the colon, there is a smooth indentation *(small black and white arrows)* on the normal segment of colon. This is a typical shoulder defect indicative of a tumor mass pressing into the normal portion of the colon.

Polypoid Defect of Stomach

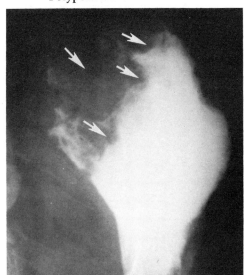

The irregular filling defect *(arrows)* in the gastric fundus is characteristic of an aggressive polypoid mass that is invading the gastric mucosa.

Annular Lesion of Colon: Shelving Defect

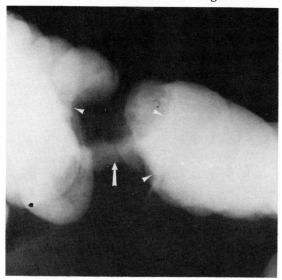

The narrowed segment *(arrow)* is irregular and devoid of mucosal pattern. Shelving or shoulder defects *(arrowheads)* are projecting into the adjacent colon, findings indicative of a mass lesion.

Annular Lesion of Colon

The segment of transverse colon is irregularly narrowed *(arrows)*, with absent mucosal pattern. These changes are characteristic of an encircling, infiltrating carcinoma.

Extramucosal Lesion of Stomach

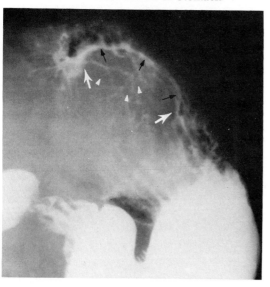

The wall of the stomach, outlined with barium *(small arrows)*, is displaced by a mass. The mucosal folds appear intact *(arrowheads)*. The lower edge of the filling defect *(lower large arrow)* has an obtuse angle, but the upper angle *(upper large arrow)* is more acute, suggesting that the lesion might be arising in the gastric wall.

Displacement of the wall or mucosa of a gastrointestinal viscus, without mucosal destruction, is characteristic of an extramucosal lesion. An acute angle at the edge of the defect is suggestive of an intramural lesion.

Intramural Extramucosal Lesion

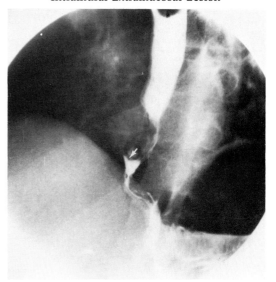

The large defect in the lower esophagus has stretched and displaced the mucosa. The barium is angled around the lower edge of the defect *(arrow)*. These findings are characteristic of an extramucosal intramural mass.

Extrinsic Mass Deformity

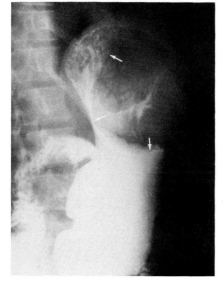

The smooth, rounded defect *(arrows)*, with well-preserved gastric mucosa, is characteristic of extrinsic mass pressure without invasion.

Mass with Pedicle (Stalk)

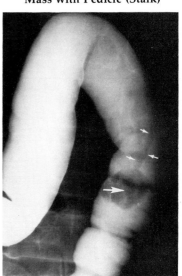

A long, linear lucency *(small arrows)* in the barium is characteristic of a pedicle. It arises from the lateral wall and terminates in the large intraluminal mass (polyp) *(large arrow)*.

Thumbprinting Defects in Colon

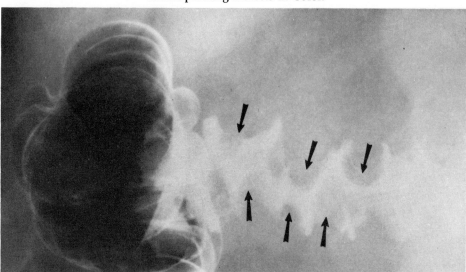

The rather rounded defects *(arrows)* are present on both walls of the colon and are characteristic of the thumbprinting defects due to submucosal hemorrhage and edema.

Liver Calcifications

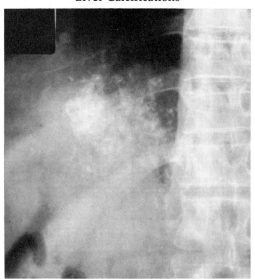

The amorphous calcifications just below the right diaphragm are intrahepatic.

Lateral films are usually necessary to rule out a thoracic location of densities that are close to the diaphragm on frontal view.

IV. JOINTS

Irregular Joint Narrowing, Articular Sclerosis

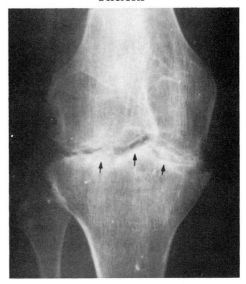

Irregularity and narrowing of both the medial and lateral aspects of the knee joint are readily apparent.

Sclerotic changes are seen in the articular end of the tibia (articular sclerosis) *(arrows)*.

Periarticular Demineralization, Periarticular Swelling

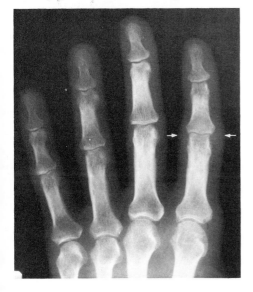

Periarticular demineralization is evidenced by the diminished bone densities adjacent to the interphalangeal joints.

The localized soft tissue bulge *(arrows)* at the interphalangeal joint illustrates characteristic periarticular swelling.

Joint Narrowing

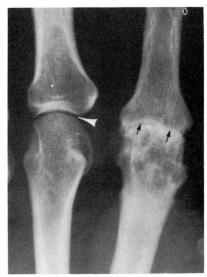

The interphalangeal joint narrowing *(arrows)* is strikingly apparent when compared with the normal joint space *(arrowhead)* in the adjacent finger.

Joint Space Narrowing, Periarticular Erosion

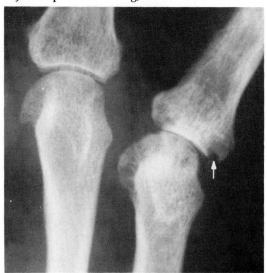

The sharp lucent defect *(arrow)* is character- istic of a periarticular erosion. These erosions are often seen in many forms of arthritis. The inter- phalangeal joint space is uniformly narrowed.

Bony Ankylosis

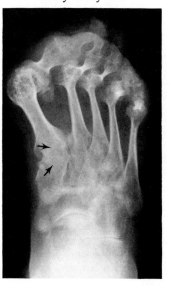

The lucent joint spaces of the anterior tarsal area have virtually disap- peared, although some are faintly visible. True bony ankylosis between the base of the first metatarsal *(upper arrow)* and the cuneiform *(lower arrow)* is apparent by continuity of the trabeculae between these bones and com- plete absence of a joint space.

Periarticular Erosion

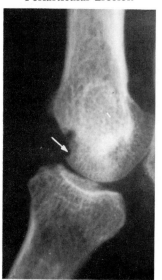

The large periarticular defect on the posterior femur has rather sharp borders and an overhanging inferior edge *(arrow)*.

Irregular Periarticular Erosions

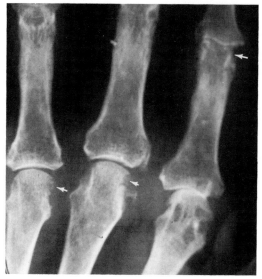

Irregular loss of bony substance at the lateral articular margins *(arrows)* is characteristic of the periarticular erosions commonly seen in the rheumatoid group of arthritides.

Chondrocalcinosis

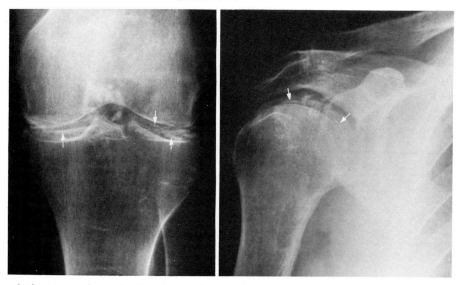

Linear calcifications of meniscal and articular cartilages *(arrows)* are visible in the knee and shoulder joints. These calcifications appear as linear densities within a joint space.

V. KIDNEYS AND URINARY TRACTS

Nephrographic Defect

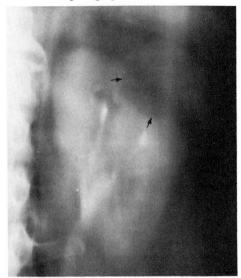

The entire kidney is opacified (nephrotomographic study), and a sharply marginated defect is apparent in the upper lateral portion (*arrows*).

Displaced and Stretched Calices

Stretching (*arrowheads*) and displacement (*arrow*) of one or more calices are indicative of a mass lesion within the kidney.

Caliceal Blunting

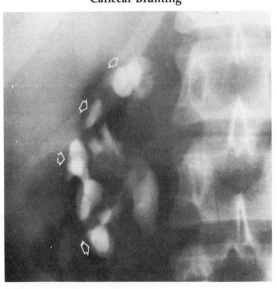

All the calices show a loss of their normal convex cupping and reveal a straight or convex border (*arrows*). These changes are due to chronic caliceal and pericaliceal inflammatory processes and are most often seen in chronic pyelonephritis.

Caliceal Irregularities and Destructive Changes

Pockets of contrast material *(arrows)* are seen adjacent to the irregular calices and demonstrate caliceal irregularities and some destructive changes in the calices and their adjacent papillae.

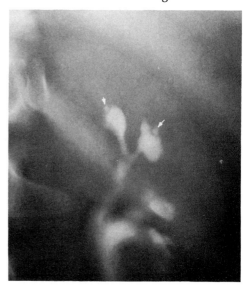

Caliceal Destruction, Filling Defects in Pelvocaliceal System

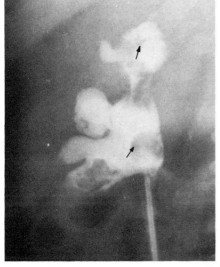

The superior caliceal borders are grossly irregular, and defects are seen within the contrast-filled collecting system, in both the calices and pelvis *(arrows)*.

Renal Cortical Thinning

The space between the blunted calices and the borders of the right kidney *(arrows)* is markedly reduced, indicative of cortical thinning and atrophy. Compare with normal left kidney.

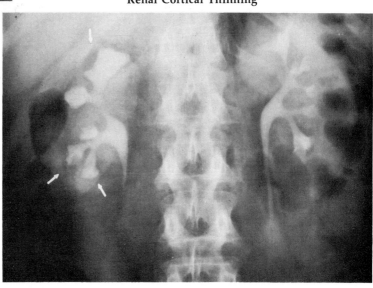

Renal Calcifications: Nephrocalcinosis

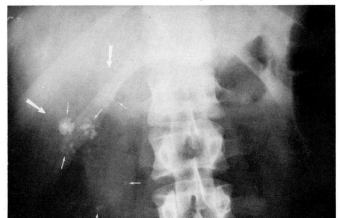

Clusters of calcification *(small arrows),* which are scattered throughout the kidney but do not involve the cortical portion, are characteristic of medullary nephrocalcinosis. Large arrows identify the margins of the kidney.

Prostatic Enlargement

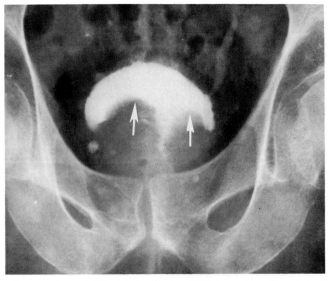

The two large, sharply marginated defects *(arrows),* which are extending into the base of the opacified bladder, are the enlarged lobes of the prostate. Note that the base of the bladder is markedly elevated from its normal position adjacent to the symphysis.

Normal Arterial Vessels: Kidney

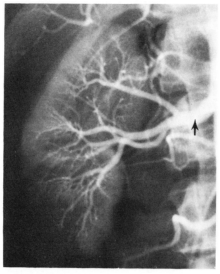

The normal intrarenal arborization of the branches of the renal artery (*arrow*) is illustrated. Note the regular branching, gradual attenuation, and the gentle undulations of the peripheral branches. Loss of these gentle undulations occurs when a vessel is stretched, and increased undulations occur when there is considerable loss of the parenchyma.

Tumor Vessels

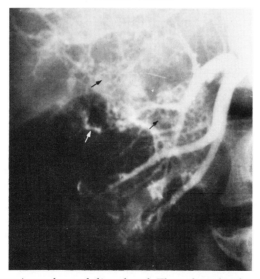

Virtually all the arteries are irregular and disordered. They show haphazard branching and some beading (*arrows*). These findings are characteristic of vessels within a malignant neoplasm.

VI. OSSEOUS SYSTEM

Osteoporosis

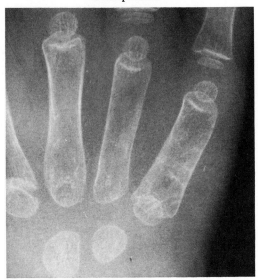

The cortices are very thin, and all the bones show a markedly diminished calcium density. There is marked absence of trabeculations, which gives the bones a ground-glass appearance. These are characteristic changes of severe osteoporosis.

Pseudofracture

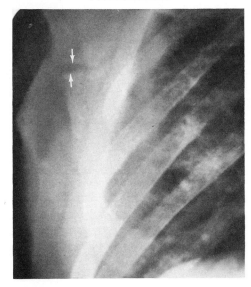

The broad lucent zone in the scapula *(arrows)* extending horizontally is characteristic of a pseudofracture. These fractures are usually symmetrically bilateral and perpendicular to the long axis of the bone and represent a healing reaction to infractions in osteomalacic bone.

Subperiosteal Resorption

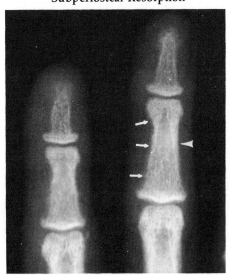

A lace-like erosive irregularity has replaced some of the normally sharp cortex *(arrows)* on the radial side of the phalanx. The opposite cortex *(arrowhead)* is virtually intact. Similar changes are present in many of the other phalanges.

This lace-like irregular erosion is typical subperiosteal resorption, characteristic of hyperparathyroidism. It is seen earliest in the phalanges.

Benign Periosteal Layering

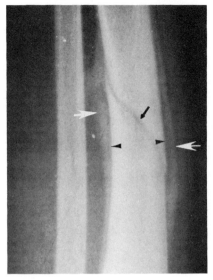

Sheet-like periosteal reactive bone *(white arrows)* is forming along the shaft in the fracture area *(black arrow)*. The thin lucent zone *(arrowheads)* between the bone and the periosteal new bone is thickened periosteum. The lucent zone is not seen in neoplastic periosteal reaction.

Periosteal Layering

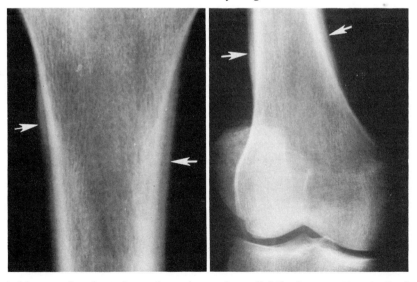

The laminated layers of periosteal new bone *(arrows)* parallel the bone cortices in the femur and upper tibia. This appearance is characteristic but nonspecific and appears in many inflammatory and other conditions. When the layers fuse, a solid mass of periosteal new bone is formed.

Codman's Triangle, Sunburst Periosteal Reaction, Soft Tissue Calcification

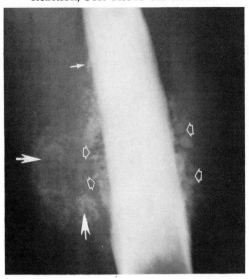

The angulated periosteal new bone *(small arrow)* at the top of the lesion is known as "Codman's triangle."

The irregular bony spicules *(open arrows)* extending horizontally from the periosteum into the soft tissues are characteristic of the "sunburst" periosteal reaction of malignant bone lesions.

The irregular soft tissue calcifications *(large arrows)* are within tumor tissue.

Endosteal Sclerosis, Osteosclerosis

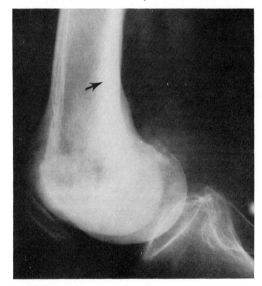

Osteosclerosis diffusely involves one femoral condyle and extends up the posterior cortex, producing endosteal sclerosis and thickening *(arrow)*.

Endosteal Sclerosis

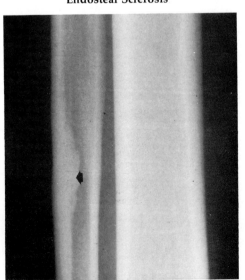

The irregular thickening *(arrow)* of the inner aspect of the cortex is characteristic of endosteal sclerosis.

Osteoblastic Bone Nodules

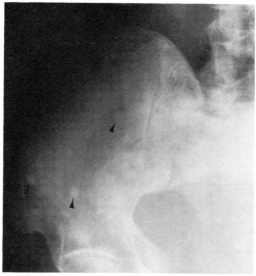

Multiple, small, rounded areas of increased density *(arrowheads)* are diffusely distributed throughout the ileum. These are typical osteoblastic nodular densities and are most often due to metastatic disease.

Osteoblastic Changes

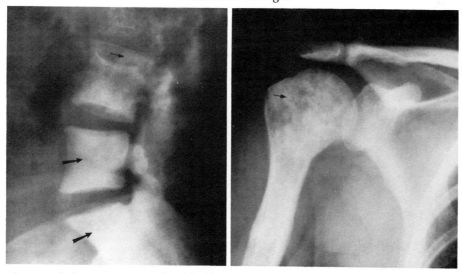

Diffuse increased density seen in the fifth lumbar body and upper sacrum *(large arrows)* is characteristic of osteoblastic change. Focal areas of osteoblastic change are also seen in the fourth lumbar body and the head of the humerus *(small arrows)*.

Medullary Calcification

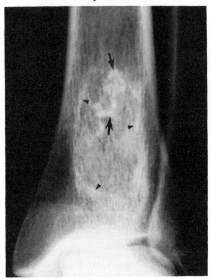

Calcific densities are present in the medullary cavity, both as irregular deposits *(arrows)* and as an oval rim *(arrowheads)*. The trabeculae are not distorted or destroyed.

Endosteal Scalloping, Lytic Lesions

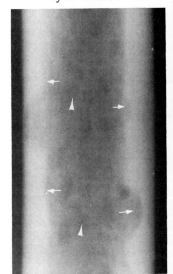

The inner layer of the cortex (endosteum) shows many areas of irregularity and erosion *(arrows)*, characteristic of endosteal scalloping.

There are also multiple small areas of medullary lysis *(arrowheads)*.

Bone Invasion from Soft Tissue Lesion

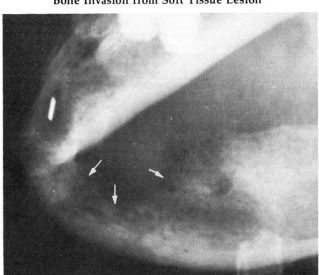

Destructive erosion of the superior bony margin *(arrows)* of the mandible is accompanied by irregular lytic lesions extending deeply into the bone.

This appearance is characteristic of bone invasion by an adjacent soft tissue malignancy.

Permeating Bone Lesion, Periosteal Reaction (Neoplastic)

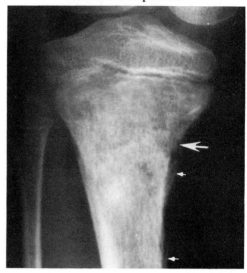

Irregular lucencies permeate the metaphysis, and there is marginal bone destruction *(large arrow)*. The involvement does not cross the epiphyseal line.

Irregular periosteal reaction is seen on the medial aspect of the tibia *(small arrows)* adjacent to the involved bone. The upper projecting edge *(upper small arrow)* of the periosteal reaction can be considered a Codman's triangle.

Sclerotic Border Around Bone Lesion

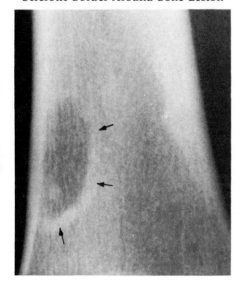

A rather thick sclerotic border *(arrows)* surrounds the lucent area lying adjacent to the cortex. The trabeculae visible within the lucent area are superimposed from the normal bone anterior to the lesion.

A sclerotic border around a bone lesion usually signifies benignity.

Sharp Lytic Bone Lesion

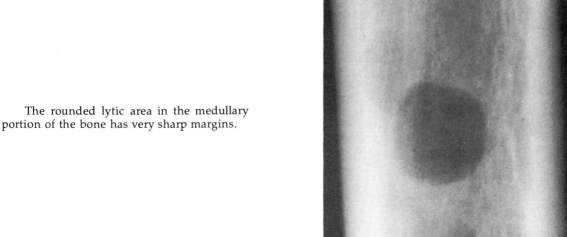

The rounded lytic area in the medullary portion of the bone has very sharp margins.

Expansile, Lytic Bone Lesion: Pathologic Fracture

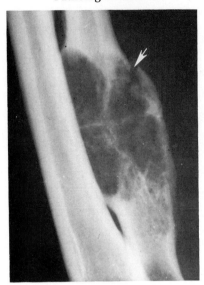

The bone is greatly expanded by the large, lucent, lobulated, intramedullary mass. A pathologic fracture of the expanded cortex is apparent *(arrow)*.

Widened Epiphyseal Line, Irregular Metaphyseal Margin

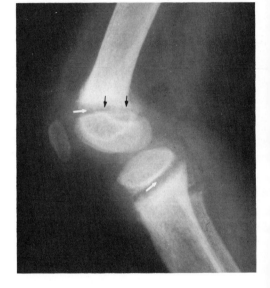

The lucent epiphyseal lines are considerably widened *(white arrows)*. The metaphyseal ends are no longer sharp but have become irregular and frayed *(black arrows)*.

Exostoses (Spurs)

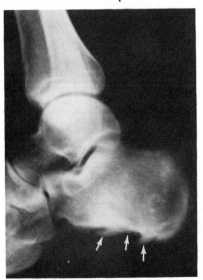

The bony projections *(arrows)* from the inferior border of the calcaneus are characteristic of reactive exostoses, or spurs. These can occur along ligamentous attachments in any bone.

Spinal Osteophytes

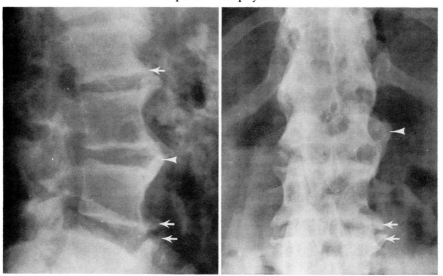

The bony outgrowths *(arrows)* extend into the anterior and lateral soft tissues and can arise from any border of a vertebral body. They do not affect the intervertebral cartilages. Adjacent osteophytes may fuse with each other *(arrowheads)*.

Wormian Bones

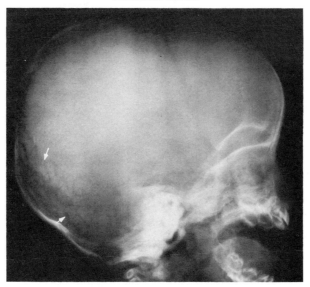

The multiple lucent lines in the posterior skull *(arrows)* are abnormal suture lines that divide the area into a large number of small, separate bones, the wormian bones.

ENVIRONMENTAL FACTORS IN DISEASE

INGESTED DRUGS AND CHEMICALS

Numerous radiographic findings are due directly or indirectly to ingested drugs and chemicals; these are therefore of considerable importance to both clinicians and radiologists.

Lead ingestion (see below) causes increased density of the metaphyseal plates in the growing skeleton. A generalized osteosclerosis can occur from *vitamin D* overdosage (p. 1333) and from chronic *fluoride* ingestion (p. 1329). *Cortisone* therapy can cause osteoporosis of the axial skeleton, aseptic necrosis of a joint, and peptic ulcer. Gastroduodenal ulceration can also result from heavy *phenylbutazone* or *salicylate* therapy.

Anticoagulant therapy is occasionally responsible for an intramural hematoma of the duodenum or the more distal small intestine, or for bleeding into the renal parenchyma (p. 858, Fig. 11–110), lesions that may radiographically be mistaken for tumors. Renal papillary necrosis (p. 825) from *phenacetin*, retroperitoneal fibrosis (p. 120) from *methysergide,* and renal calculi from prolonged *milk and antacid* therapy are all well documented.

A variety of pulmonary infiltrates can develop from administration of *salicylates, nitrofurantoin, methotrexate,* and other drugs (p. 3). In *ergotism* (p. 724) angiographic evidence of arterial spasm of the extremities may be found.

A constantly increasing number of radiologic changes due to drug therapy are being reported; the preceding group includes some of the most significant, currently.[1]

Lead Poisoning

Although chronically ingested or inhaled lead is deposited in the bones, no appreciable roentgenographic alterations are seen in the osseous system of adults. However, in children the lead is deposited in the growing ends of the bones and causes the metaphyseal border to be wider and denser — the familiar lead line. The density is caused not only by the lead but also by reactive hyperplasia of the spongiosa. When ingestion is stopped, the lead line migrates into the shaft as the bone grows, causing a transverse line of density near the metaphysis. This line eventually disappears. However, similar metaphyseal densities can occasionally occur in normal growing bones.

About half the affected children develop in later years metaphyseal dysplasias with alterations of bone modeling.

Abdominal films may demonstrate lead opacities in the gastrointestinal tract, and these may provide a diagnostic clue in infants and children who experience abdominal pain after unsuspected chronic lead ingestion. In acute cases there may be pronounced abdominal pain and central nervous system symptoms. In infants, increased intracranial pressure may lead to diastasis of the skull sutures. The lead line does not appear in acute cases.

Persistent segmental or complete dilatation of the colon often develops, and it may persist for many months following clinical recovery. It is similar to the toxic megacolon seen in many other conditions.[2-4]

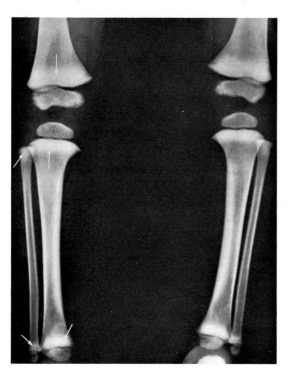

Figure 1-1 **Lead Poisoning: Metaphyseal Lead Lines.** A child had ingested lead paint over a period of many weeks. The metaphyseal ends of all the long bones *(arrows)* have a broad band of density — the radiographic lead line.

A moderate increase in density of the metaphyseal lines is difficult to evaluate, since normal variations may simulate the minimal changes caused by lead poisoning.

Figure 1–2 **Acute Lead Poisoning.** A 2 year old girl had clinical symptoms of encephalopathy. There are small irregular opacities *(small arrows)* scattered throughout the intestinal tract and rectum *(lower arrow)*. These opacities represent ingested flakes of lead paint from the child's crib. There is toxic distention of the transverse colon *(large arrows)*.

The skull films were normal, and no lead line was seen in the long bones. An abdominal film may disclose evidence of ingested opaque material before the lead line appears in the bones. In lead encephalopathy, skull films are usually negative unless the intracranial pressure increases greatly. In infants, this increase may bring about diastasis of the sutures.

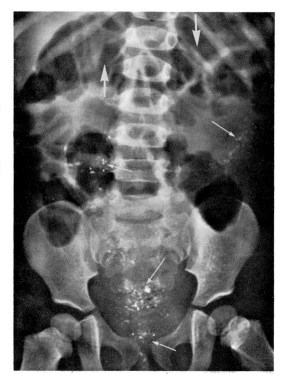

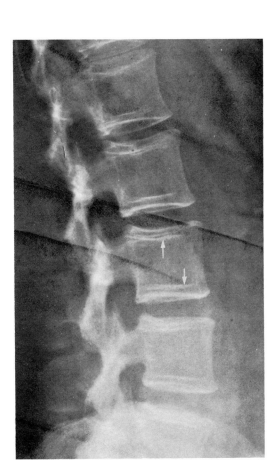

Figure 1–3 **Late Changes in Bones Due to Lead Poisoning: Striated Vertebrae.** Lines of density paralleling the superior and inferior margins *(arrows)* of each vertebra are thought to be the residuum of lead poisoning that had occurred eight years earlier. With growth, the deposits in the vertebral margins migrated to the vertebral body; the distance between the earlier lead line and the present vertebral margin represents this growth.

Pulmonary Reactions to Oral or Parenteral Drugs

Respiratory symptoms accompanied by pulmonary infiltrates can develop from drug hypersensitivity, overdose, or prolonged administration.

The pulmonary changes in acute drug hypersensitivity may take one of five patterns: (1) Acute alveolar infiltrates that are frequently bilateral may appear. These can be diffuse or basal, producing a pulmonary edema–like appearance but without cardiac enlargement. When the infiltrates are patchy, they may tend to have peripheral distribution with perihilar sparing, similar to the pulmonary infiltrates seen in eosinophilia (see p. 121). (2) Rarely, an acute diffuse interstitial pattern may be seen. (3) A chronic interstitial pattern, occasionally resulting in irreversible pulmonary fibrosis, may develop. A blood eosinophilia is sometimes associated. Nitrofurantoin and methotrexate are the most frequently reported offenders, but other antimetabolic agents, antibiotics, and antihypertensives have been implicated. (4) A lupus-like lung pattern associated with a systemic lupus syndrome can be produced by procainamide (Pronestyl). (5) Bilateral hilar adenopathy may develop. This is most frequently seen with the anticonvulsants, especially phenylhydantoin.

Discontinuation of the drugs generally leads to fairly prompt disappearance of the pulmonary changes and clinical recovery. In severe cases, steroid therapy may be lifesaving. Similar pulmonary reactions have also appeared after blood transfusions.

Acute and chronic pulmonary changes in drug addicts and drug abusers have become fairly frequent. Intravenous heroin overdose can cause pulmonary edema, coma, respiratory failure, and death. In surviving patients, clinical and radiographic recovery is rapid. Pulmonary edema may occasionally develop after oral administration of heroin or methadone. The cause of these reactions is uncertain; hypersensitivity, overdose toxicity, and contaminants have been postulated.

Other pulmonary complications from intravenous heroin and other addictive drugs include septic emboli and abscesses, pulmonary bacterial infections, and aspiration pneumonia in comatose victims. A reticulonodular fibrotic-like radiographic appearance occasionally develops, possibly from microemboli of particulate contaminants.[5-16]

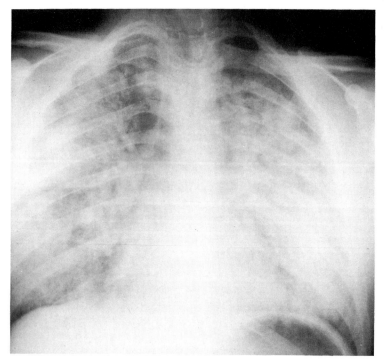

Figure 1–4 **Acute Pulmonary Edema from Heroin Overdose.** Diffuse bilateral confluent alveolar densities, characteristic of pulmonary edema, extend from the hila to the peripheries. The heart is not enlarged, and there is no pleural reaction.

The acute pulmonary edema due to heroin overdose will clear rapidly if the patient survives.

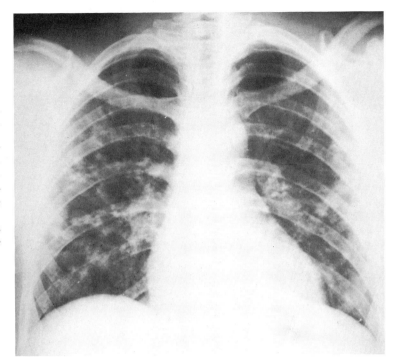

Figure 1–5 **Furodantoin Lung.** Pulmonary symptoms developed during a course of furodantoin therapy.

The lungs show nodular fluffy infiltrates bilaterally, with involvement most marked in the lung peripheries. The perihilar areas remain relatively uninvolved. There is no pleural reaction.

This distribution of lesions is suggestive of a lung reaction due to hypersensitivity.

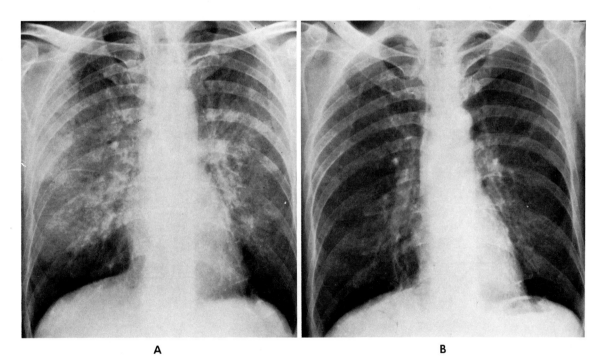

A   B

Figure 1–6 **Pulmonary Drug Reaction: Salicylate Ingestion.**

A, The patient had ingested 500 aspirin tablets in two weeks. There are mottled confluent alveolar densities in both lung fields; the bases have been spared. Although the appearance suggests pulmonary edema, there was neither cardiac enlargement nor evidence of hilar or vascular engorgement.

B, Aspirin ingestion was stopped, and four days later the chest was normal. Similar, and even more extensive, infiltrates occur more frequently after prolonged administration of nitrofurantoin and methotrexate. (Courtesy Dr. S. M. Greenstein, Newington, Connecticut.)

Thalidomide Embryopathy

Phocomelia, or short flipper-like extremities, is the principal fetal abnormality resulting from use of thalidomide during the first trimester of pregnancy.

One limb or all of the limbs, but particularly the upper extremities, may be affected. The involved bones are hypoplastic and deformed; usually the proximal bones are more severely affected. The glenoid or acetabular fossa may be missing, and sometimes an entire limb is absent.

Other congenital anomalies frequently associated with phocomelia include cardiac defects, anal and duodenal atresia, and ureteral dysplasia.[17]

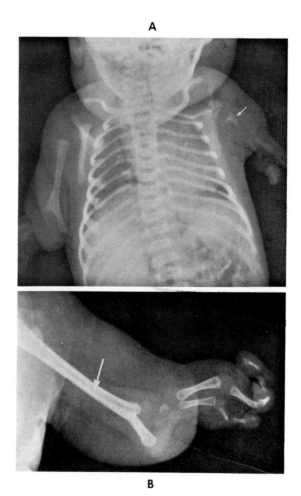

A

B

Figure 1–7 **Thalidomide Embryopathy: Phocomelia.**

A, The upper extremities, particularly the left, resemble flippers. On the left there is only a rudimentary humerus *(arrow),* and there are abnormalities of every bone in both extremities. Many bones are absent.

B, Enlargement of right upper extremity discloses forking of humerus *(arrow),* which probably represents a rudimentary radius and ulna. The bones of the wrist and hand are deranged. The lower extremities were normal.

Chemical Esophagitis

Ingestion of caustic material usually causes inflammation of the esophagus, the severity of which depends upon the amount and type of caustic or irritant.

In mild cases, spasm and irritability of the esophagus, especially the lower third, develop rapidly. As a rule, in most severe cases mucosal edema and ulceration develop, so that irregularity and mucosal distortion are evident roentgenographically. Contrast material may remain in the wall if intramural dissection occurs. Marked atony and dilatation may portend imminent perforation. Fibrosis and stricture follow healing.

Lye ingestion, most often by children, is the commonest cause of both acute chemical esophagitis and late stricture. In severe cases similar changes may involve the stomach.[18, 19]

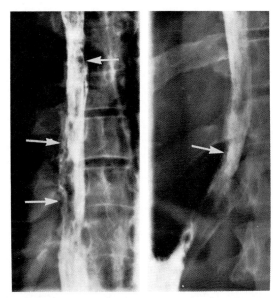

Figure 1–8 **Acute Esophagitis Following Acid Ingestion.** Esophagogram made shortly after accidental ingestion of a strong acid discloses marked irregularity of the lower half of the esophagus. Multiple scattered lucencies represent edematous mucosal folds *(arrows)* that grossly simulate varices. The lumen could not be distended further than is illustrated. Some of the barium streaks lateral to the edematous folds probably represent shallow linear ulceration. The appearance is somewhat similar to candidiasis of the esophagus. Severe chemical esophagitis eventually causes fibrotic stricture of the involved segments.

FROSTBITE

Following frostbite of the extremities, soft tissue swelling commonly develops within two to six days, but no immediate underlying bone or joint changes appear. Within one or two months, a mild to moderate demineralization may be seen in about 50 per cent of the patients; this usually disappears in a few months.

Sharply defined punched-out periarticular defects occur from six months to several years after the injury in about 10 to 20 per cent of patients. There may also be small areas of increased density in the phalangeal tufts, possibly representing bone infarcts. Marginal spurs and narrowing of the joint spaces, indistinguishable from ordinary osteoarthritis, occasionally develop in adults one year or more after the injury. The incidence and severity of the bone changes do not appear to be related to the severity of the original frostbite.

In children, late epiphyseal changes, including fragmentation, articular deformity, and premature fusion, can cause deformity or shortening of the involved phalanges.[20-23]

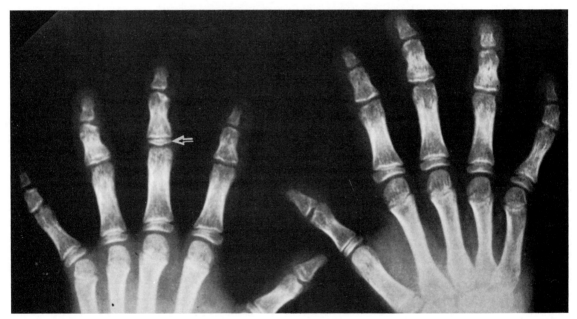

Figure 1–9 **Frostbite: Growth Disturbance.** Films of both hands of a 9 year old girl disclose that all but one *(arrow)* of the 18 epiphyses of the middle and distal phalanges have closed prematurely. These phalanges are short, stubby, and mildly deformed.

Severe frostbite of the fingers had occurred six years previously and had apparently severely damaged the epiphyseal cartilages, causing early closure and growth deformities. (Courtesy Dr. A. C. Selke, Jr., Iowa City, Iowa.)

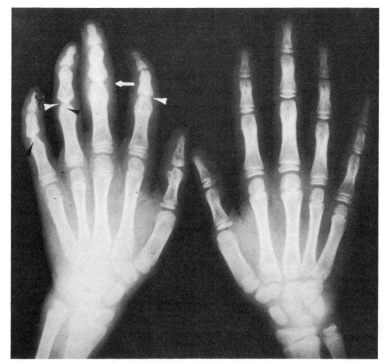

Figure 1–10 **Unilateral Frostbite: Growth Disturbances in a 10 Year Old.** In the left hand, frostbitten one year previously, there are shortened middle and distal phalanges with completely fused epiphyses, narrowed proximal interphalangeal joint spaces *(white arrowheads)*, deformed distal ends of the proximal phalanges *(black arrowheads)*, soft tissue swelling *(white arrow)*, and contracture deviations of the distal phalanges.
The unaffected normal right hand is shown for comparison.

RADIATION INJURY

The acute and chronic clinical syndromes caused by *whole body* irradiation are devoid of significant or characteristic radiographic findings.

Local radiation to tissues and organs with the large dosages often employed in radiation therapy can lead to acute and chronic tissue changes, the degree and type depending on the rate and dose and also on the specific response of the tissue or organ. A wide spectrum of pathologic alterations may occur, including acute inflammatory reaction, acute or chronic tissue necrosis, chronic fibrosis, impairment of secretory or functional activity, and even late malignant degeneration. Significant radiographic changes can develop in the lungs, kidneys, bones, and gastrointestinal tract.

Pulmonary changes generally follow intense radiation given after mastectomy or in the treatment of malignancies of the lungs, mediastinum, or esophagus. Nearly 50 per cent of patients subjected to large doses develop radiation pneumonitis, which appears within one to three months and usually progresses to fibrosis. Occasionally some degree of resolution is noted, but this rarely is complete. Pleural reaction is an infrequent occurrence.

During the stage of acute pneumonitis, a fairly homogeneous alveolar infiltrate is seen in the irradiated area. Gradually this is replaced by strands of fibrosis limited to the same area. Symptoms occur only in about half the patients in whom these pulmonary changes develop. Not infrequently, it is difficult to distinguish radiographically the radiation fibrosis from lymphangitic spread of malignancy in the same area.

Radiation injury to mature bones, if not severe, may produce local demineralization. If radiation injury is more severe, the trabecular pattern becomes coarsened and disorganized, and endosteal cortical thickening and irregularity develop. Mixed areas of dense sclerosis and focal demineralization can produce a pagetoid appearance. Pathologic fractures are not uncommon, even in bones with minimal radiographic changes. Superimposed osteomyelitis with bone destruction and soft tissue calcification may simulate metastatic or primary bone neoplasms. An associated soft tissue mass should arouse suspicion of a radiation-induced sarcoma.

In growing bones, radiation of the metaphyseal-epiphyseal area can cause marked disturbances in growth and development. Irregularity of the metaphyses and epiphyses, shortening and narrowing of bone, failure of proper tubulation, and a variety of other deformities may be late results. Scoliosis often develops in the growing spine.

Acute radiation injury to the small bowel is quite common during radiation therapy and may cause mucosal edema and ileus, with atonic dilated loops. These changes may gradually disappear, but severe injury may lead to chronic changes. The latter include thickening of the bowel wall, mucosal folds, and submucosa. Irregular submucosal thickening may cause nodular filling defects. Eventually the affected loops may become fibrotic, rigid, and stenotic, leading to obstruction. Ulceration and fistulous tracts may occasionally occur. Adhesions within the mesentery may cause a matting of many loops.

Radiation colitis is encountered most often in the pelvic colon; persistent focal spasm is the earliest finding. This may progress to a permanent stricture, usually with a smooth gradual transition to normal bowel. This process is sometimes difficult to distinguish from recurrent or primary bowel neoplasm. Ulceration and fistulous tracts may sometimes precede stricture development.

In acute radiation nephritis, an intravenous pyelogram generally shows a nonfunctioning kidney of normal size. If a retrograde pyelogram is obtained, the structure of the pelvis and calices appears normal. Recovery of function may occur if damage is not too great, but in the absence of recovery the kidney gradually decreases in size.

There are few or no significant roentgenographic changes in radiation encephalitis or myelitis. However, spinal cord arterial occlusion has been observed angiographically in cases of postradiation myelomalacia.[24-32]

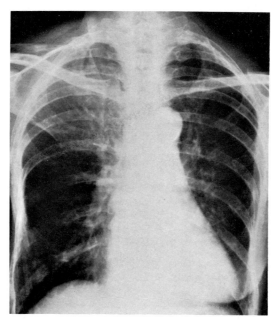

Figure 1–11 **Radiation Pneumonitis: Postmastectomy Radiation.** Four months after a course of post-operative radiation, there is density in the right upper lobe composed of fibrous strands radiating from the hilum to the periphery. There were minimal respiratory symptoms. The right breast shadow is absent.

Radiation pulmonary fibrosis appears a few months after radiation therapy and is usually permanent. Although its development depends chiefly on the intensity of radiation, other factors, including the age of the patient and individual differences in response, are significant.

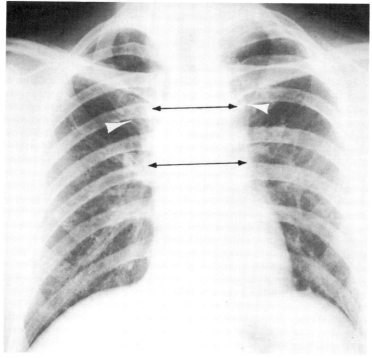

Figure 1–12 **Postirradiation Mediastinal Fibrosis.** Extensive irradiation of the hila and mediastinum two years ago has caused considerable widening of the mediastinum *(double-arrowed lines)*. The mediastinal borders are no longer sharp, but merge with longitudinal fibrotic strands *(white arrowheads)*. Both hilar shadows are almost entirely obscured by the surrounding fibrosis.

There are characteristic late changes after extensive mediastinal and hilar irradiation for lymphomatous disease.

A

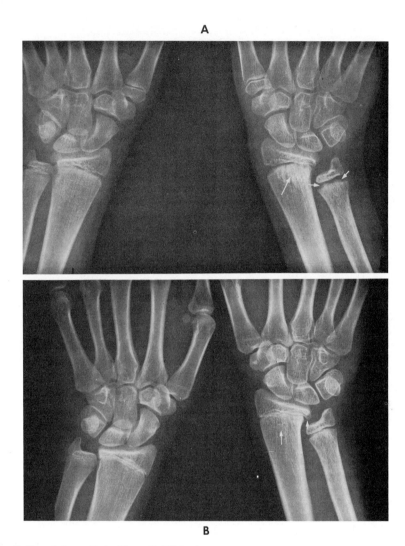

B

Figure 1–13 **Radiation Injury: Late Growth Disturbances.**

A 2 year old girl received radiation therapy for a large cavernous hemangioma on the dorsum of her right wrist.

A, At the age of 12 years she experienced pain on motion of the right wrist. There are abnormalities in the metaphyses of the right radius and ulna. There are longitudinal lines of increased trabecular density *(long arrow),* and the right radial metaphysis is thinner than its fellow. The metaphyseal margin of the ulna is somewhat irregular *(short arrows),* and the epiphyseal line is widened. Limitation of use owing to pain has resulted in slight demineralization of the bones of the right hand and wrist.

B, At 16 years of age, the edge of the right radial epiphysis overhangs *(short arrow),* and the epiphyseal line is slightly irregular. The metaphysis of the radius is narrow and contains coarsened trabeculae *(long arrow).* The right radius and ulna are 1¾ inches shorter than the left radius and ulna. The left wrist is normal. Right wrist pain disappeared when growth ceased.

Even small doses of radiation delivered to the growing ends of a bone can lead to growth disturbances and permanent deformity.

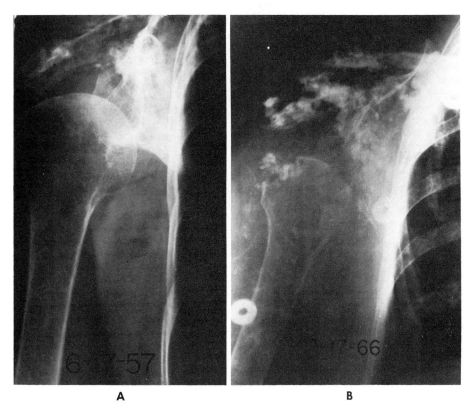

A   **B**

Figure 1–14 **Radiation Osteitis: Late Changes.**

A, Twelve years after supraclavicular radiation for breast carcinoma, there is irregular sclerosis around the glenoid of the scapula and the articular margin of the humerus.

B, Eleven years later, skin breakdown and low grade infection have produced severe destructive changes in the glenoid and humeral head, with extensive soft tissue calcification. These late results from radiation and chronic infection could easily be mistaken for malignant changes. (Courtesy Dr. D. G. Bragg, Salt Lake City, Utah.)

Figure 1–15 **Postirradiation Acute Gastroduodenitis.** Four months after an intense course of radiation to the pancreatic area, the patient developed nausea, some vomiting, and epigastric pain. The marked thickening and irregularity of the gastric *(open arrows)* and duodenal *(small arrows)* folds associated with decreased peristaltic activity are due to irradiation.

Incidentally, a calcified gallstone is seen on the right *(white arrowhead).*

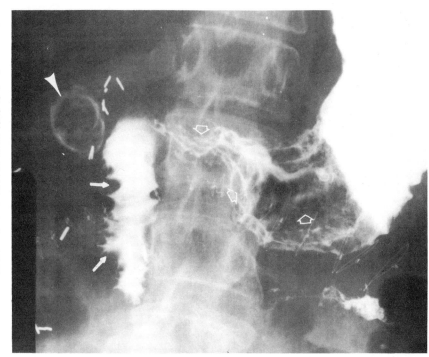

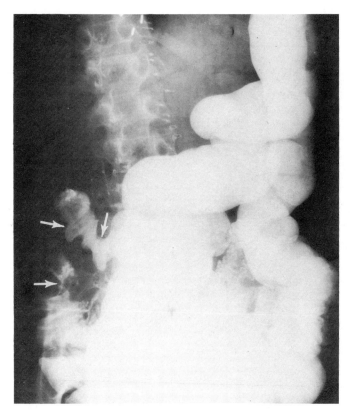

Figure 1–16 **Late Radiation Colitis.** The distal ascending and proximal transverse colon (hepatic flexure) are narrowed, irregular, and fixed, with no definite mucosal pattern *(arrows)*.

The colonic segment had been in the radiation field for right hypernephroma two and one half years previously. The involvement resembles segmental granulomatous colitis.

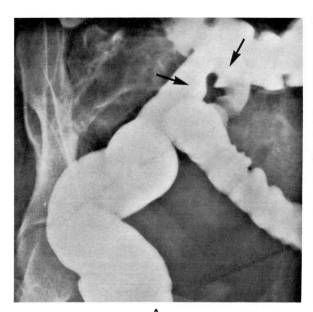

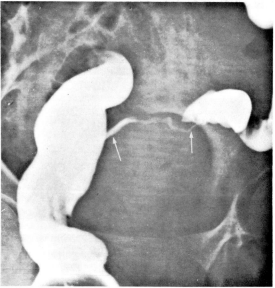

A   B

Figure 1–17 **Postirradiation Stricture of the Sigmoid.**

A, In enlarged view, the barium-filled rectum and sigmoid prior to irradiation are normal *(arrows)*.

B, Six months after radium treatment for carcinoma of the cervix, there is an area of stricture in the sigmoid *(arrows)*. The involved segment is long, with regular borders and without "shoulder" defects. These features are typical of a benign stricture and militate against the diagnosis of neoplastic involvement.

ELECTRIC INJURY

Directly following severe but nonfatal electric shock, fractures or dislocations may result from tetanic muscular spasms.

A variety of delayed bone and joint changes have been encountered, which may be due to a combination of soft tissue injury, blood vessel damage, neurologic disturbance, disuse, and superimposed infection. In some cases, direct thermal or electrical effect on a bone or joint has been postulated as the cause. The roentgen changes include bone necrosis, short fine barely visible fracture lines, periosteal reaction, demineralization, increased width of an affected bone, and discrete areas of rarefaction that may remain as permanent "holes" in the bone. Sometimes delayed ischemic necrosis can affect a bone in the path of the current but distant from the site of entry.

Because of the paucity of reports, the frequency of any of these bone changes is unknown.[33, 34]

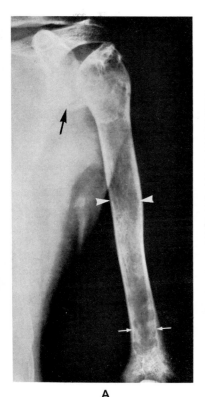 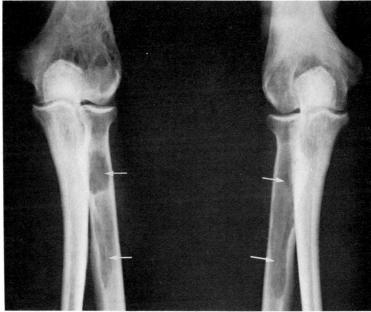

A   B

Figure 1–18 **Bone Change After Severe Electrical Injury.**

One and a half years prior to these films, the patient sustained a severe electrical injury from a 2500-volt charge passing through both upper extremities for 15 to 20 seconds.

A, The left humerus is decalcified, the medullary cavity is widened, and the cortices are thinned *(arrowheads)*. The bony fragment *(black arrow)* adjacent to the humeral neck is the deformed head from a fracture-dislocation incurred during the electrical injury. Note the sharp, circinate area of rarefaction *(white arrows)* in the lower humerus.

Similar changes, including a fractured head, were present in the right humerus.

B, Both forearms show almost identical long areas of sharply demarcated rarefaction *(arrows)* in the upper radii.

These bone changes are almost certainly the result of the electrical injury, but the underlying pathophysiologic mechanism is unknown. (Courtesy Dr. L. B. Brinn, New York City, New York.)

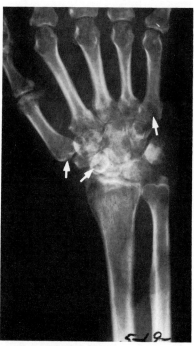

Figure 1–19 **Electrical Injury.** Four months after severe electrical burns, the bones of the wrist and forearm are severely osteoporotic from disuse. The lytic lesions in the first and fifth metacarpals and the navicular are due to the electrical effect, which may be transient intense hyperthermia causing focal bone necrosis. (Courtesy Dr. J. W. Barber, M.D., Cheyenne, Wyoming.)

HIGH ALTITUDE PULMONARY EDEMA

In susceptible individuals, rapid ascent to high altitudes (over 9000 feet) may lead to pulmonary edema that may prove fatal. This occurs within 6 to 36 hours and is usually preceded by symptoms of mountain sickness. The mechanism is unclear.

Radiographically there are characteristic alveolar densities in the region of both hila, without cardiac enlargement but with prominence of the pulmonary artery segment. Prompt clinical and radiographic recovery follows oxygen therapy or descent, but the prominent pulmonary artery segment may persist for weeks.[35-37]

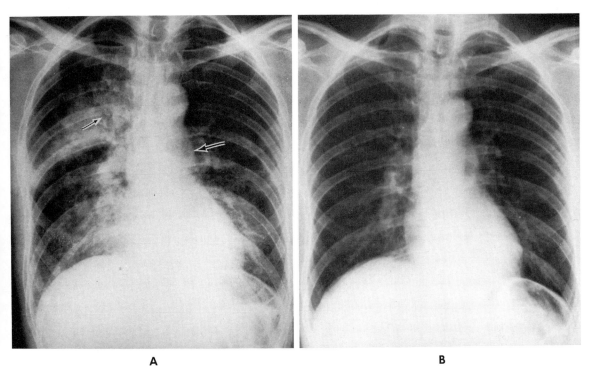

A   **B**

Figure 1–20   **Pulmonary Edema Due to High Altitude.**

A, The fluffy alveolar densities of pulmonary edema are more extensive in the right lung. The air bronchogram *(small arrow)* is characteristic of alveolar consolidation. The main pulmonary artery segment is prominent *(large arrow).*

B, Five days later, after descent and oxygen administration, the lungs are entirely clear. The pulmonary artery is still prominent. (Courtesy Dr. H. L. Fred, Houston, Texas.)

DECOMPRESSION SICKNESS (CAISSON DISEASE)

Caisson disease is due to release of dissolved nitrogen gas bubbles into the tissues following too rapid decompression from a high pressure environment. The acute stage requires prompt treatment, and radiographic studies are not performed at this time.

Radiographic changes in the bones may occur from a few months to several years after exposure to hyperbaric conditions, frequently without clinical symptoms. The lesions are forms of aseptic necrosis or bone infarcts; about two thirds are juxta-articular and one third are medullary. The distal femoral or proximal tibial diaphyses and the head and neck of the humerus or femur are the commonest sites. Bilateral symmetric lesions can occur.

The early smaller diaphyseal lesions are areas of circular or oval rarefaction; these may slowly regenerate. Larger diaphyseal lesions eventually show irregular medullary new bone in a circumscribed area, characteristic of a calcified bone infarct. There is no expansion, and the cortex is unaffected.

The juxta-articular lesions are spotty zones of subarticular rarefaction and sclerosis. Healing may produce a "snow-cap" area of osseous condensation over the articular surface. Occasionally the adjacent joint is narrowed or even obliterated; rarely, irreversible collapse of the articular cortex occurs.[38-41]

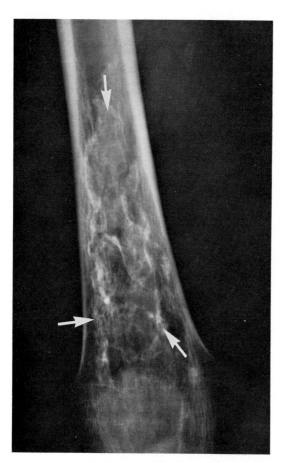

Figure 1–21 **Caisson Disease: Calcified Infarction of Bone.** The medullary portions of the metaphyseal and diaphyseal areas in the lower femur contain irregular curved linear densities *(arrows)* that represent the calcified infarct. The femur has not expanded, and there is no cortical involvement. Normal trabeculae are interspersed between the calcifications. This peculiar pattern of calcification is characteristic of bone infarction. Infarcts resulting from ischemia or sickle cell anemia have an identical appearance.

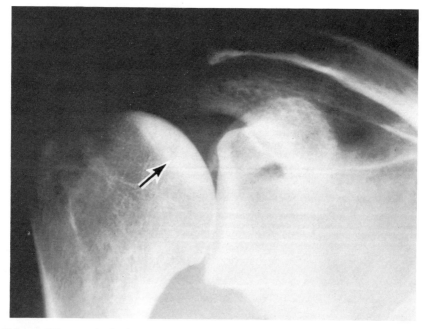

Figure 1–22 **Caisson Disease: Articular "Snow-Cap."** A sharply marginated band of sclerosis *(arrow)*, the result of a subarticular infarct, gives the characteristic "snow-cap" appearance to the humeral head. (Courtesy Dr. J. R. Nellen, Milwaukee, Wisconsin.)

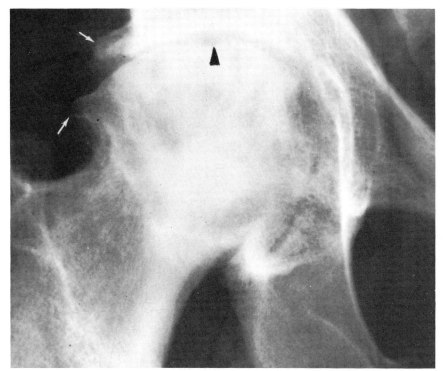

Figure 1–23 **Caisson Disease: Aseptic Necrosis.** The femoral head and neck are irregularly sclerotic, and an area of articular collapse *(arrowhead)* is apparent. Irregularity of the joint space and bony spurs *(arrows)* are due to secondary osteoarthritic changes. (Courtesy Dr. J. R. Nellen, Milwaukee, Wisconsin.)

WEIGHTLESSNESS

Weightlessness, associated with reduced physical activity, has been found to lead to some skeletal demineralization, especially in the distal extremities. This has occurred both in humans (astronauts) and primates. Restoration of the mineral content was observed after return to normal gravity and physical activity.[42]

BLAST INJURY

The pressure wave from an explosion can produce hemorrhage in the lung or gastrointestinal tract, often without externally visible injury. The pulmonary hemorrhages appear radiographically as irregular mottled alveolar densities of varying size and distribution. The densities generally disappear rapidly with clinical recovery.

In more severe cases, laceration of the lung may occur, causing pneumothorax, hemothorax, or pneumomediastinum.[43, 44]

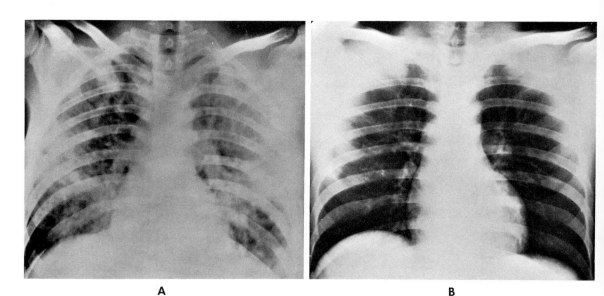

A   **B**

Figure 1–24 **Blast Injury: Pulmonary Hemorrhages.**

A, Twenty-four hours after a severe explosion, the patient developed widespread alveolar densities throughout both lung fields, with confluent lateral densities, due to intrapulmonary hemorrhage. There were no rib fractures or burns.

B, Six days later the chest is completely clear.

With cessation of active bleeding, the intra-alveolar blood is rapidly absorbed.

NEAR DROWNING

The radiologic findings in drowning individuals who are rescued and revived are similar to those seen in pulmonary edema. The pulmonary changes are due mainly to hypoxia; water aspiration plays only a minor initial role.

In milder cases, coarse to fine perihilar alveolar infiltrates are present. Severe cases show diffuse homogeneous alveolar densities with air bronchograms.

Marked clearing of the lung fields usually occurs within a few days; complete resolution may take 7 to 10 days. More prolonged infiltrates should suggest secondary infection.[45, 46]

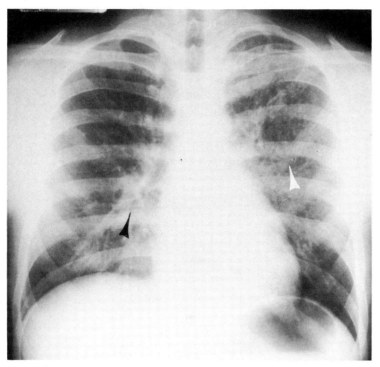

Figure 1–25 **Near Drowning.**   Six hours after revival from near drowning, a chest film of this 18 year old boy shows small, ill-defined acinar nodular densities *(white arrowhead)* in the perihilar zones, more marked on the left. More confluent densities *(black arrowhead)* are obscuring the right hilum.

These alveolar densities are from edema due to hypoxia, and not aspirated water. In more severe cases, the picture will be that of diffuse widespread pulmonary edema.

OCCUPATIONAL ACRO-OSTEOLYSIS

A syndrome that includes Raynaud-like phenomena and acro-osteolysis of the distal phalanges can develop in workers exposed to vinyl chloride in the commercial polymerization process of polyvinyl chloride.

The Raynaud phenomenon usually precedes the acro-osteolysis; occasionally the bone changes occur without prior symptoms of vasospasm.

There are several radiographic patterns. Lytic destruction of the ungual tuft is preceded by cortical loss and a half-moon cut in the cortex of the tuft. The other common pattern is a transverse band of bone dissolution through the distal half of the phalanx, proximal to the tuft. The resulting picture is striking.

The pathophysiologic mechanism is unknown. Healing may occur spontaneously or only after cessation of exposure to the vinyl chloride.

Radiographically, occupational acro-osteolysis cannot be readily distinguished from familial acro-osteolysis.[47-49]

A

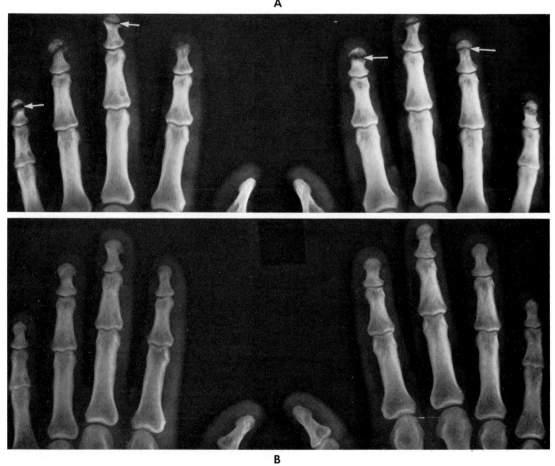

B

Figure 1–26 **Occupational (Vinyl Chloride) Acro-osteolysis, with Recovery.**

A, All the distal phalanges (except the thumbs) of both hands are fragmented by osteolytic bands *(arrows)* across the tufts.

B, One year after cessation of exposure to the chemical, complete healing has occurred.

Although the appearance of the phalanges is indistinguishable from that of congenital acro-osteolysis, in the latter condition there are usually changes in other bones, whereas in occupational acro-osteolysis only the distal phalanges are involved.

REFERENCES

1. Ansell, G.: Radiologic manifestations of drug-induced disease. Clin. Radiol., *20*:133, 1969.
2. Pease, C. N., and Newton, G. G.: Metaphyseal dysplasia due to lead poisoning in children. Radiology, *79*:233, 1962.
3. Kissel, D., et al.: Saturnine megacolon (lead poisoning). Presse Med., *68*:1739, 1960.
4. Betts, P. R., Watson, S. M., and Astley, R.: A suggested role of radiology in lead poisoning. Ann. Radiol. *16*:183, 1973.
5. Greenstein, S. M.: Pulmonary edema due to salicylate intoxication. Dis. Chest, *44*:552, 1963.
6. Nicklaus, T. M., and Snyder, A. B.: Nitrofurantoin pulmonary reaction. Arch. Intern. Med., *121*:151, 1968.
7. Rosenow, E. C., DeRemee, R. A., and Dines, D. E.: Chronic nitrofurantoin pulmonary reaction. N. Engl. J. Med., *279*:1258, 1968.
8. Clarysse, A. M., et al.: Pulmonary disease complicating intermittent therapy with methotrexate. J.A.M.A., *209*:1861, 1969.
9. Cortez, L. M., and Pankey, G. A.: Acute pulmonary hypersensitivity to furazolidine. Am. Rev. Resp. Dis., *105*:823, 1972.
10. Morrison, W. I., Wetherill, S., and Zyroff, J.: The acute pulmonary edema of heroin intoxication. Radiology, *97*:347, 1970.
11. Jaffe, R. B., and Koschmann, E. B.: Intravenous drug abuse. Am. J. Roentgenol. Radium Ther. Nucl. Med., *109*:107, 1970.
12. Thompson, J. S., Severson, C. D., et al.: Pulmonary hypersensitivity reactions induced by transfusion of non-HL-A leukoagglutinins. N. Engl. J. Med., *284*:1120, 1971.
13. Kjeldgaard, J. M., Hahn, G. W., et al.: Methadone induced pulmonary edema. J.A.M.A., *218*:882, 1971.
14. Warnock, M. L., Gharemani, G. G., et al.: Pulmonary complications of heroin intoxication. J.A.M.A., *219*:1051, 1972.
15. Horowitz, A. L., Friedman, M., et al: The pulmonary changes of bleomycin toxicity. Radiology, *106*:65, 1973.
16. Brettner, A., Heitzman, R. E., and Woodin, W. G.: Pulmonary complications of drug therapy. Radiology, *96*:31, 1970.
17. Lenz, W., and Knapf, P.: Thalidomide embryopathy. German Med. Monthly, 7:253, 1962.
18. McKibben, B. G., and Lee, S.: Stenosis of esophagus and stomach following ingestion of corrosive substances. Arch. Otolaryngol., *61*:2, 1955.
19. Martel, W.:Radiologic features of esophagogastritis secondary to extremely caustic agents. Radiology, *103*:31, 1972.
20. Vinson, H. A., and Schatzki, R.: Roentgenologic bone changes encountered in frostbite. Radiology, *63*:685, 1954.
21. Selke, A. C., Jr.: Destruction of phalangeal epiphyses by frostbite. Radiology, *93*:859, 1969.
22. Ellis, R., Short, J. G., and Simonds, B. D.: Unilateral osteoarthritis of the distal interphalangeal joints following frostbite. Radiology, *93*:857, 1969.
23. Tishler, J. M.: The soft tissue and bone changes in frostbite injuries. Radiology, *102*:511, 1972.
24. Rubin, P., et al.: Radiation induced dysplasias of bone. Am. J. Roentgenol. Radium Ther. Nucl. Med., *82*:206, 1959.
25. Freid, J. R., and Goldberg, H.: Post-irradiation changes in lungs and thorax. Am. J. Roentgenol. Radium Ther. Nucl. Med., *43*:877, 1940.
26. Devois, A., et al.: Lesions of small intestines secondary to irradiation. Ann. Radiol., *4*:185, 1961.
27. Luxton, R. W.: Radiation nephritis. Lancet, *2*:1221, 1961.
28. Pallis, C. A., et al.: Radiation myelopathy. Brain, *84*:460, 1961.
29. Bragg, D. J., Shidina, H., et al.: The clinical and radiographic aspects of radiation osteitis. Radiology *97*:103, 1970.
30. Mason, G. R., Dietrich, P., et al.: The radiologic findings in radiation-induced enteritis and colitis. Clin. Radiol. *21*:232, 1970.
31. Felson, B., et al.: Complications of radiation therapy. Semin. Roentgenol. *9*:5, 1974.
32. DiChiro, G., and Herdt, J. R.: Angiographic demonstration of spinal cord arterial occlusion in post-radiation myelomalacia. Radiology, *106*:317, 1973.
33. Brinn, L. B., and Moseley, J. E.: Bone changes following electrical injury. Am. J. Roentgenol. Radium Ther. Nucl. Med., *97*:682, 1966.
34. Barber, J. W.: Delayed bone and joint changes following electrical injury. Radiology, *99*:49, 1971.
35. Fred, H. L., et al.: Acute pulmonary edema of altitude. Circulation, *25*:929, 1962.
36. Marticorena, E., et al.: Pulmonary edema by ascending to high altitudes. Dis. Chest, *45*:273, 1964.
37. Kohli, P., and Stucki, P.: Pulmonary edema in high altitudes. Schweiz. Med. Wochenschr. *98*:845, 1968.
38. Fournier, A. M., and Jullien, G.: Radiologic aspects of caisson disease. J. Radiol. Electrol. Med. Nucl., *40*:529, 1959.
39. Rotenberg, C., and McGee, A. R.: Late bone changes in caisson disease. J. Can. Assoc. Radiol., *8*:50, 1957.

40. Poppel, M. W., and Robinson, W. T.: The roentgen manifestations of caisson disease. Am. J. Roentgenol. Radium Ther. Nucl. Med., 76:74, 1956.
41. Nellen, J. R., and Kindwall, E. P.: Occupational aseptic necrosis of bone secondary to occupational exposure to compressed air. Am. J. Roentgenol., Radium Ther. Nucl. Med., 115:512, 1972.
42. Mack, P. B., and Vogt, F. B.: Roentgenographic bone density changes in astronauts during representative Apollo space flight. Am. J. Roentgenol. Radium Ther. Nucl. Med., 113:621, 1971.
43. Zuckerman, S.: Blast injuries to the lung. Lancet 2:219, 1940.
44. Hirsch, M., and Bazini, J.: Blast injury of the chest. Clin. Radiol. 20:362, 1969.
45. Hunter, T. B., and Whitehouse, W. M.: Fresh water near drowning. Radiologic aspects. Radiology, 112:51, 1974.
46. Rosenbaum, H. T., Thompson, W. L., and Fuller, R. H.: Radiographic pulmonary changes in near drowning. Radiology, 83:306, 1964.
47. Wilson, R. H., McCormick, W. E., et al.: Occupational acro-osteolysis. J.A.M.A., 201:577, 1967.
48. Harris, D. K., and Adams, W. G. F.: Acro-osteolysis occurring in the polymerization of vinyl chloride. Br. Med. J., 3:712, 1967.
49. Chatelain, A., and Motillon, P.: A syndrome of acro-osteolysis of occupational origin and its recent observation in France. J. Radiol. Electrol. Med. Nucl., 48:277, 1967.

SECTION
2

IMMUNE DISEASE

Primary Immunodeficiency Diseases

Chronic and recurrent infection by usual and opportunistic organisms, including viruses and fungi, characterizes all the immunodeficiency diseases. There are a number of clearly defined syndromes or diseases.

The *DiGeorge syndrome,* due to a deficiency of cellular immunity, is characterized by the absence of the thymus and parathyroids. Tetany occurs in early life. Pulmonary infections, particularly with acid-fast organisms, fungi, viruses, and *Pneumocystis carinii,* are prone to occur. Absence of a thymic shadow in a neonate less than 4 days old, producing a clear lucent retrosternal space and a narrow mediastinum on frontal view, should arouse suspicion of this condition.

In *congenital hypogammaglobulinemia* or *agammaglobulinemia* (Bruton's disease), humoral immunity (immunoglobulins) is virtually absent. There is an absence of the normal nasopharyngeal adenoid tissue on the lateral film of the nasopharynx in infants and children with this condition. The thymus is normal. Sinus and pulmonary infections begin after 6 or 9 months of age; the latter are characterized by recurrent and poorly resolving pneumonias, atelectasis, bronchiectasis, and empyema, most often caused by common pyogens. Hilar adenopathy is strikingly absent with these infections. Septic arthritis is not uncommon.

About one third of patients develop a rheumatoid-like arthritis in larger joints, with synovial thickening, some effusion, but few or no bone changes.

In the *Swiss type agammaglobulinemia* (combined humoral and cellular deficiency) the thymus is small and lymphoid tissue virtually absent. Severe skin, respiratory, and gastrointestinal infections occur within 1 or 2 years of age. Candidiasis of the esophagus is common and produces characteristic edema and ulcerations radiographically. In the lungs there may be bilateral diffuse alveolar densities due to *Pneumocystis carinii* infection or diffuse interstitial infiltrates from viral agents. Gastrointestinal ulcerations and subsequent pneumatosis intestinalis can occur.[1-3]

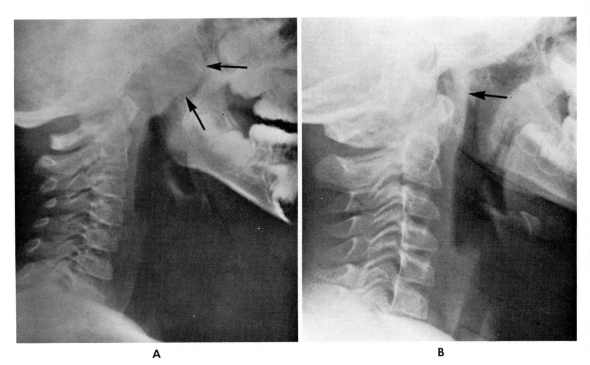

A   B

Figure 2–1 **Congenital Agammaglobulinemia, Bruton's Disease: Absence of Lymphoid Tissue.**

A, Lateral radiograph of nasopharynx of a normal 6 year old child demonstrates normal mass of adenoidal tissue *(arrows)* bulging into the air spaces of the nasopharynx.

B, In an agammaglobulinemic child of the same age, such tissue is lacking. The nasopharyngeal soft tissue line is straight *(arrow),* and the air spaces are considerably larger. If adenoidectomy has not been performed, and if there is a history of frequent severe respiratory infections, this finding strongly suggests congenital agammaglobulinemia of Bruton.

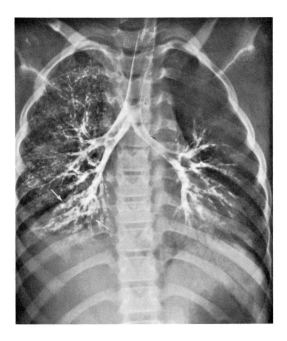

Figure 2–2 **Bruton Congenital Agammaglobulinemia: Atelectasis and Cylindrical Bronchiectasis.** Bronchogram of a child who had many pulmonary infections demonstrates crowding of the right lower lobe bronchi *(arrows),* which is indicative of segmental atelectasis. Bronchial dilatation is caused by cylindrical bronchiectasis. The hilar nodes are not enlarged. Acute and chronic respiratory infections are common complications of agammaglobulinemia and hypogammaglobulinemia.

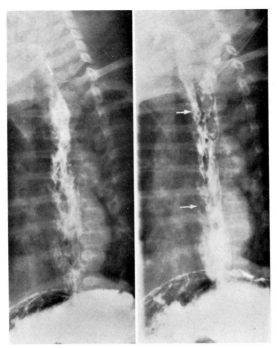

Figure 2–3 **Moniliasis of Esophagus Associated with Agammaglobulinemia.** The esophagus is filled with barium: the contour is jagged *(arrows)* and there are swollen mucosal folds throughout its length. These changes are typical. Esophageal moniliasis is rarely seen except in debilitated persons, in persons whose immunologic defenses are inadequate, as in agammaglobulinemia, or in patients undergoing immunosuppressive therapy.

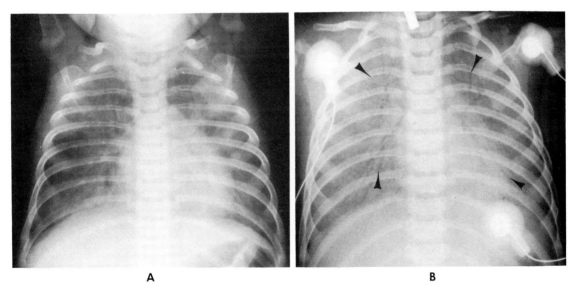

A   B

Figure 2–4 **Pneumocystis Pneumonia Complicating Agammaglobulinemia.**

A, In a 3 month old infant with agammaglobulinemia (Swiss type) and with recent fever and cough, there are diffuse densities extending from both hilar areas into the lung fields. The air bronchograms indicate that the infiltration is alveolar.

B, Eleven days later, virtually complete alveolar opacification is apparent; well-developed air bronchograms are seen through both lungs *(arrowheads)*.

Pneumocystis carinii pneumonia occurs only in individuals with immunologic deficiency or in severely debilitated infants.

Acquired Hypogammaglobulinemia

In *acquired hypogammaglobulinemia* and in *dysgammaglobulinemia,* which occur in later childhood or adult life, splenomegaly and recurrent pulmonary infections are common. Diarrhea is a frequent symptom and is often associated with giardiasis and with nodular lymphoid hyperplasia of the intestines. These nodules are found most often in the jejunum, but involvement of the duodenum, distal ileum, and colon can also occur. The diffuse evenly distributed nodules may be mistaken for intestinal folds seen on end, but the latter are more randomly distributed.

In children, diffuse lymphoid hyperplasia may have an identical appearance. Other lesions of the small or large bowel which must be differentiated include diffuse colonic polyposis, Peutz-Jeghers syndrome, and pseudopolyposis of ulcerative or granulomatous colitis.[4-6]

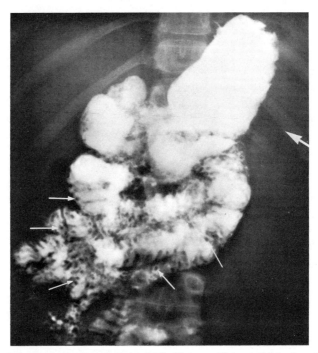

Figure 2–5 **Dysgammaglobulinemia: Small Bowel Findings.** This child had episodic respiratory infections with recurrent infiltrates in both lung fields.

Small bowel study reveals numerous small nodular lucencies *(small arrows)* in otherwise normal appearing loops. The greatest number was in the jejunum.

The spleen is considerably enlarged *(large arrow),* displacing the stomach and small bowel loops medially.

The nodular lucencies, due to islands of lymphoid hyperplasia, resemble normal mucosal folds seen on end, but the latter are never as numerous and diffuse.

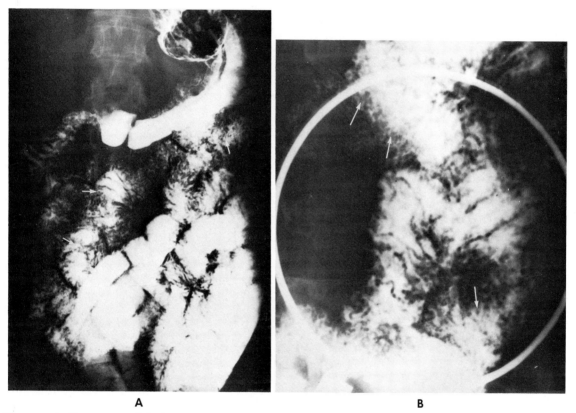

A   B

Figure 2–6 **Dysgammaglobulinemia: Typical Small Bowel Findings.**

This 42 year old woman had a history of increasingly severe and frequent upper respiratory infections during the past 15 years. Blood studies disclosed hypogammaglobulinemia and dysgammaglobulinemia.

A, The jejunal small bowel pattern is abnormal, and multiple small rounded lucencies *(arrows)* can be identified throughout the jejunum.

B, Magnified view of the jejunum demonstrates more clearly the innumerable small nodules *(arrows)* and the absence of a normal mucosal pattern. These nodules are due to lymphoid hyperplasia, a characteristic finding in dysgammaglobulinemia.

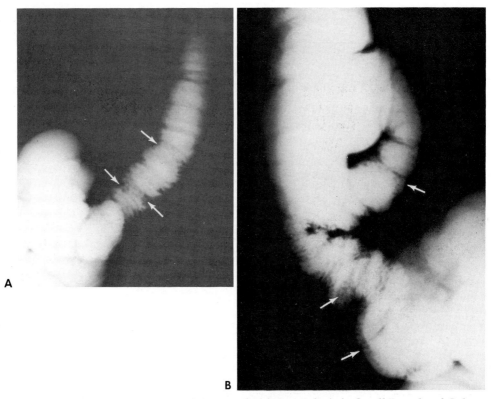

Figure 2–7 **Dysgammaglobulinemia in Adult; Lymphoid Hyperplasia in Small Bowel and Colon.**

A, There are small nodules throughout the left colon, best seen in the lower descending colon (*arrows*).
B, Similar diffuse nodularity is present in the distal small bowel (*arrows*).

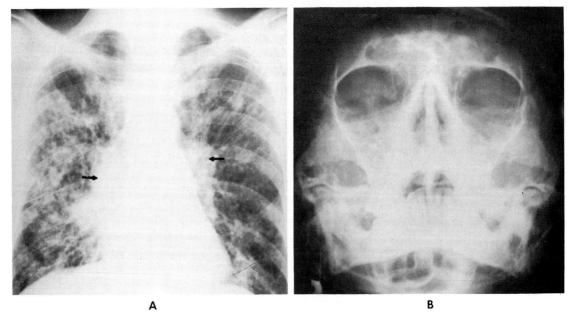

Figure 2–8 **Acquired Hypogammaglobulinemia in Adult: Diffuse Lung and Paranasal Sinus Infection.**

A, Chest film reveals diffuse reticular and small conglomerate densities. Many of these proved to be areas of bronchiectasis. The hilar nodes are enlarged (*arrows*).

The lung densities had been increasing over many years. Serum globulins were virtually completely absent.

B, All the paranasal sinuses are completely clouded by inflammatory exudates.

Chronic Granulomatous Disease (Neutrophil Dysfunction Syndrome)

This disorder is due to the inability of the neutrophilic leukocytes to kill bacteria even after phagocytosis. It is usually familial and predominantly affects males. A diversity of chronic and recurrent infections begins in childhood and may extend into adult life. The chronicity results from low-virulent intracellular bacteria within granulomatous lesions.

There may be recurring and chronic pulmonary infections, with a lobar or bronchopneumonic distribution, which are usually associated with prominent hilar adenopathy and pleural thickening. At times the infiltrates may be widespread granulomatous nodules, which may even calcify. Occasionally they simulate miliary tuberculosis. Abscesses and empyema can occur; pericarditis has been reported.

Hepatosplenomegaly, due to diffuse granulomatous abscesses, is very frequent. Punctate calcifications may develop in these organs. Less frequent findings include ureteral obstruction by intra-abdominal inflammatory masses, small bowel abnormalities, and unexplained esophageal dilatation.

Hematogenous osteomyelitis is not uncommon. There is a tendency to involve the small bones of the hands and feet. Soft tissue swelling is minimal, and unusual destructive bony changes without sequestra may be noted.[7-10]

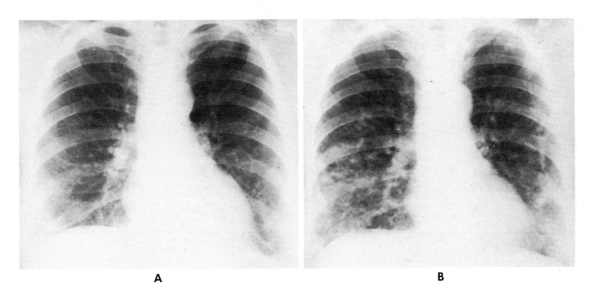

A   **B**

Figure 2–9 **Chronic Granulomatous Disease in Adult.**

A, This patient had been operated on a few years ago, when her spleen proved to be the site of diffuse granulomas and small abscesses. Current cough and low grade fever were associated with interstitial and nodular densities in the lung bases.

B, Three months later, lung involvement has increased, with many more interstitial, nodular, and confluent densities. At this time, an immunologic defect of leukocytes was uncovered.

The lung lesions were microabscesses and chronic infectious granulomas.

This condition is most often seen in children.

Allergic Rhinitis (Hay Fever), Asthma, and Gastrointestinal Allergies

During the active phase of allergic rhinitis there may be radiologic evidence of edematous thickening of the mucosa in the paranasal sinuses. A high incidence of polyps in the maxillary sinuses is found in individuals with recurrent nasal allergies.

The chest findings in bronchial asthma are discussed on page 467.

Specific gastrointestinal allergies are frequently accompanied by motility changes of the small bowel. If a small bowel study is made after adding the suspected allergen to a barium-water mixture, there may be hypermotility, segmentation, increased secretions, and even edema of the small bowel wall.

If a normal small bowel pattern reappears after withholding the suspected allergen from the diet for several days, the probability that this was the specific allergen is quite high.[1]

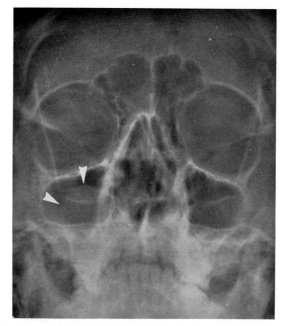

Figure 2–10 **Recurrent Allergic Rhinitis: Maxillary Sinus Polyp.** The patient experienced seasonal hay-fever. The smooth round mass *(arrowheads)* in the right maxillary antrum is a large polyp. The other paranasal sinuses are clear.

During the active phase there is often edematous thickening of the mucous membranes in the sinuses. Polyps in the nasal passages and paranasal sinuses frequently occur in patients with chronic recurrent allergic rhinitis, although polyps are also found in normal persons. Radiographically, it is difficult to distinguish a polyp from a mucocele or retention cyst.

REFERENCES

1. Margulis, A. R., et al.: Congenital agammaglobulinemia. Radiology 69:354, 1957.
2. Pernod, J., et al.: Agammaglobulinemia in the adult. J. Franc. Med. Chir. Thorac. 20:131, 1966.
3. Kirkpatrick, J. A., Capitanio, M. A., and Pereira, R. M.: Immunologic abnormalities: Roentgen observations. Radiol. Clin. North Am., 10:245, 1972.
4. Hodgson, J. R., Hoffman, H. N., and Huizenga, K. A.: Roentgenologic features of lymphoid hyperplasia of the small intestine associated with dysgammaglobulinemia. Radiology, 88:883, 1967.
5. Wolfson, J. J., Goldstein, G., et al.: Intestinal lymphoid hyperplasia of the large intestine associated with dysgammaglobulinemia. Am. J. Roentgenol. Radium Ther. Nucl. Med., 108:610, 1970.
6. Vermess, M., Waldman, T. A., and Pearson, K. D.: Radiographic manifestations of primary acquired hypogammaglobulinemia. Radiology, 107:63, 1973.
7. Ament, M. E., and Ochs, H. D.: Gastrointestinal manifestations of chronic granulomatous disease. N. Engl. J. Med., 288:382, 1973.
8. Sutcliffe, J.: Chronic granulomatous disease. Ann. Radiol., (Paris), 13:305, 1970.
9. Wolfson, J. J., et al.: Bone findings in chronic granulomatous disease of childhood. J. Bone Joint Surg., 51:1573, 1963.
10. Gold, R. H., Douglas, S. D., et al.: Roentgenographic features of the neutrophil dysfunction syndromes. Radiology, 92:1045, 1969.
11. Tiemann, F., and Lenz, H.: Roentgenologic contributions to the diagnosis of gastrointestinal allergies. Fortschr. Roentgenstr., 90:351, 1969.

3

CONNECTIVE TISSUE DISEASES ("COLLAGEN DISEASES") OTHER THAN RHEUMATOID ARTHRITIS

Systemic Sclerosis (Scleroderma)

The multiplicity of radiographic findings, involving many organs, often suggests the diagnosis in the absence of skin lesions. There are esophageal abnormalities in about half the cases. The esophagus is widened and atonic, with poor or absent peristalsis, and it may retain barium for hours if the patient remains recumbent. The mucosal pattern is poorly defined. Trapped air is often seen. The atonicity is best demonstrated by cineradiography. In advanced disease, one or more segments of rigid narrowing may develop.

Changes in the small bowel and/or colon are seen sometime in the course of disease in about three fourths of the cases. The small bowel is dilated and atonic, and motility is greatly reduced. Pouches or sacculations may occur. The valvulae conniventes appear prominent and thickened, and may have a "stacked-coin" configuration. A dilated duodenal loop is frequently seen in the early stage. Malabsorption may result from severe small bowel involvement (see Fig. 3–3). Pneumatosis cystoides intestinalis is a documented but rare complication of extensive scleroderma small bowel disease (see p. 920 and Fig. 12–47). The colon may show areas of atonicity; characteristic wide-mouthed outpouchings may project from the antimesenteric border of the colon, and these sacculations superficially resemble diverticula.

In about one fourth of cases the lungs are affected, characteristically by diffuse fibrosis that is more marked in the lower fields. As in the other collagen diseases, there may be intermittent episodes of pneumonitis and pleuritis. A honeycomb lung with small emphysematous areas scattered between the areas of fibrosis and fine nodulation is common, although the fibrosis may be confluent and limited to the bases. Aspiration pneumonitis may be a confusing complication.

Cardiomegaly resulting from myocardial or pericardial involvement, or both, occasionally with effusion, appears in about one third of cases, and cor pulmonale may supervene as a result of interstitial fibrosis.

Raynaud's phenomenon develops in about 60 per cent of cases and may lead to resorption of the ungual tufts. Atrophy of the distal soft tissues commonly occurs; often calcific deposits appear in these atrophied tissues. In longstanding cases, small articular marginal lesions may occur in the distal interphalangeal joints. Arteriographic studies disclose fairly characteristic findings: the ulnar artery is narrowed and occluded in half the cases, and stenotic or occluded digital arteries are found in 90 per cent of cases. These are permanent organic changes in contrast to the vasospastic findings in "pure" Raynaud's disease.

Joint symptoms are frequent, but usually periarticular swelling is the only finding. Rarely, however, the progressive changes of rheumatoid arthritis may develop. Osteoporosis is fairly common and may be secondary to disuse.

Films of the mandible may reveal widening of the periodontal space. Calcium deposits in soft tissues occur in about one sixth of patients, most commonly in the hands, usually in the distal soft tissues of the digits.[1-5]

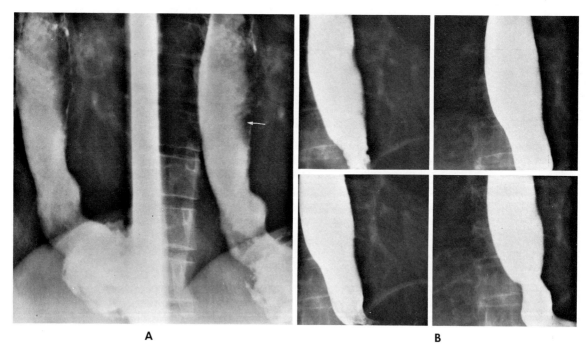

A　　　　　　　　　　　　　　　　　　　　**B**

Figure 3–1　**Scleroderma: Esophageal Changes in Two Patients.**

A, The esophagus is dilated and almost completely atonic. The mucosal pattern is ill defined. Note trapping of air *(arrow).* Gaping of the esophagogastric junction, as illustrated, is frequently seen. The esophagus may remain filled for hours when the patient is recumbent, but it empties by gravity in the erect position.

B, In another patient the esophagus has a similar appearance. There is evidence of weak peristalsis in the lower esophagus, with moderate tone in the lower esophageal sphincter.

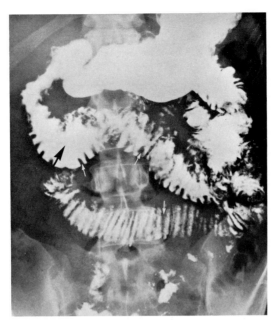

Figure 3–2 **Scleroderma: Small Bowel Changes.** The mucosal folds in the duodenum and proximal jejunum are very much thickened *(white arrows)*. The loops are moderately dilated, and peristalsis and motility are decreased.

The distal duodenum *(black arrow)* is frequently persistently dilated in scleroderma involving the small bowel.

Atonicity and duodenal dilatation are the most common small bowel findings in scleroderma. In the advanced stage, with mucosal changes, clinical malabsorption may occur.

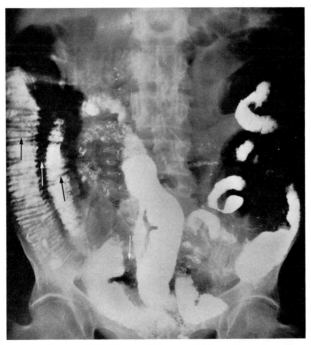

Figure 3–3 **Malabsorption in Advanced Collagen Disease (Scleroderma).** The small intestinal loops are somewhat dilated and segmented. The mucosal pattern is effaced in many of the segments. Other segments have a stacked-coin appearance *(black arrows)* due to thickening of the valvulae. The thickening of the bowel wall is indicated by the increased space *(white arrows)* between adjacent loops. Motility was greatly decreased; this film was made eight hours after oral ingestion of the barium meal.

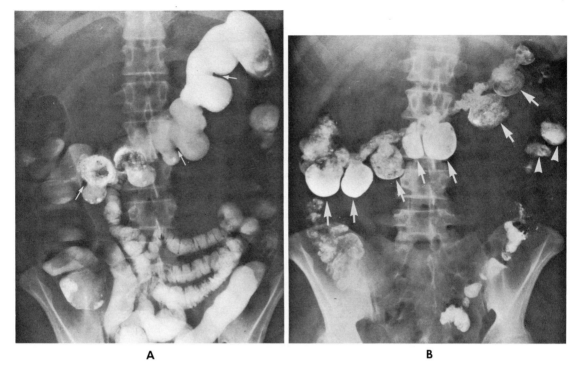

A   B

Figure 3–4 **Scleroderma of Colon: Characteristic Pseudodiverticula.**

A, In the barium-filled colon there appear to be deep asymmetric haustra in the transverse colon *(arrows)*, mainly on the inferior surface.

B, After evacuation, large wide-mouthed outpouchings *(arrows)* on the inferior (antimesenteric) border of the transverse colon remain filled with barium. A few smaller saccules, or outpouchings, are also seen in the descending colon *(arrowheads)*.

The sacculations, or outpouchings, or pseudodiverticula, with their wide-mouth and antimesenteric position, are characteristic of scleroderma and occur most frequently in the transverse colon. The pseudodiverticula are due to muscular atrophy and are best demonstrated on the postevacuation film, since they remain filled while the rest of the colon contracts and empties. Their wide-mouth and antimesenteric location in the transverse colon should readily distinguish them from ordinary colonic diverticula.

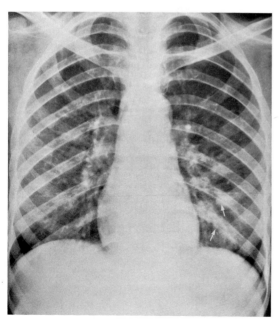

Figure 3–5 **Scleroderma: Pulmonary Findings.** There is fine reticular fibrosis in both middle and lower lung fields. Ill-defined conglomerate fibrotic densities are seen in the lower lobes *(arrows)*, but the upper lobes are relatively uninvolved.

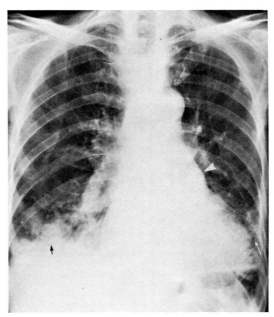

Figure 3–6 **Scleroderma: Pulmonary and Cardiac Involvement.**   There is extensive involvement at both bases, with honeycomb areas of fibrosis *(white arrow)* and ill-defined conglomerate densities. A large confluent area of fibrosis lies just above the right diaphragm *(black arrow)*. The upper lung fields appear to be unaffected.

There is pronounced cardiomegaly as a result of the myocardial involvement. The main pulmonary artery is dilated *(arrowhead)*. The basal fibrosis in particular strongly suggests scleroderma.

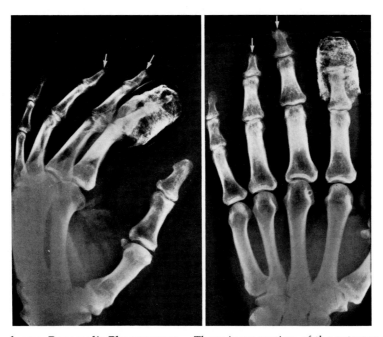

Figure 3–7 **Scleroderma: Raynaud's Phenomenon.**   There is resorption of the extreme distal edges of the third and fourth distal phalanges *(arrows)* of the left hand. (A medicated bandage is over the infected index finger.) Raynaud's phenomenon develops in 60 per cent of patients with scleroderma. Skin thickening, contractures, and soft tissue calcification may be associated.

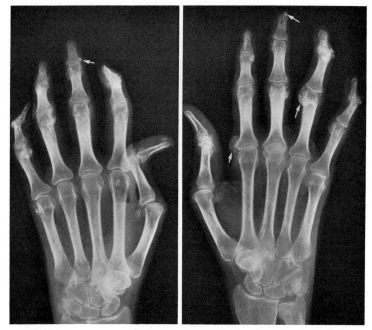

Figure 3–8 **Scleroderma; Two Patients with Calcinosis.** Amorphous calcium deposits of varying sizes *(arrows)* are seen chiefly beneath pressure points. The deformities at the distal interphalangeal joints are due to skin contracture.

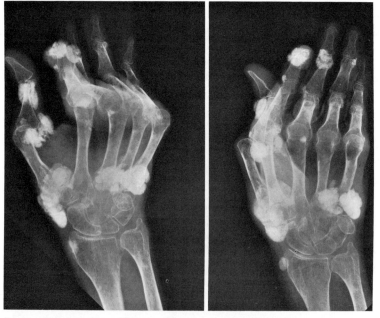

Figure 3–9 **Scleroderma: Calcium Deposition.** There are extensive calcium deposits involving the carpal areas and wrists. Flexion deformities are most marked at the metacarpophalangeal joints. Osteoporosis is due to prolonged disuse.

The tufts of the terminal phalanges are somewhat atrophic, but only minimal bone resorption has occurred. If progressive Raynaud's phenomenon is associated, resorption of tufts may be pronounced.

Polymyositis and Dermatomyositis

Roentgenologic findings in these uncommon diseases may be entirely absent, or changes may be seen in the gastrointestinal tract, lungs, and soft tissues. Dysphagia is a common symptom occurring in over four fifths of patients, but changes in the esophagus are found in only about one third of patients. Esophageal peristalsis is decreased or absent, and emptying is delayed at the esophagogastric junction and hypopharynx. These findings are similar to the esophageal changes seen in scleroderma, but they may be more pronounced in the proximal esophagus where skeletal muscle is present. Atony of the stomach and small bowel is also occasionally encountered.

Bilateral patchy pulmonary infiltrations develop that may coalesce to form larger confluent densities. Diffuse interstitial fibrosis often develops. The lung changes are similar to those seen in the other collagen diseases, but pleural effusion is not a prominent finding. Nonspecific cardiac enlargement can occur. Occasionally there is soft tissue calcification in the subcutaneous tissues and muscles underlying the skin lesions. Disuse of the stiffened joints may lead to periarticular osteoporosis.

The musculoskeletal changes are most severe in childhood dermatomyositis. There is early loss of the sharp radiographic musculosubcutaneous line, due to edema. Later, extensive subcutaneous calcifications, contractures, and osteoporosis develop.

There appears to be a high incidence of various types of malignant neoplasms associated with dermatomyositis in adults.[6-9]

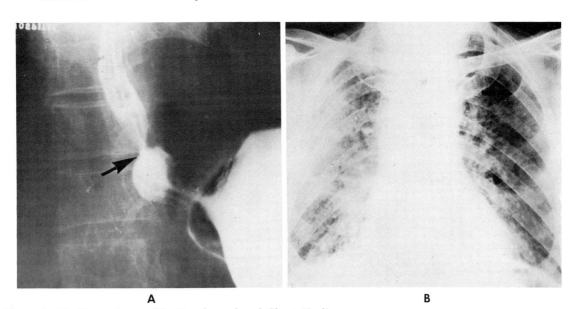

A   B

Figure 3–10   **Dermatomyositis: Esophageal and Chest Findings.**

A, In supine view the esophagus remains filled because of diminished peristalsis. The esophagogastric junction *(arrow)* is just above the hiatal hernia. Gravity emptying occurs readily when the patient is in the erect position.

The esophageal changes in scleroderma are similar but more pronounced. The esophageal symptoms and changes in dermatomyositis respond to steroids, but there generally is not a similar favorable response in scleroderma.

B, There is widespread fibrosis throughout both lung fields, and there are conglomerate areas in the right lung. Pleural reaction is usually absent in dermatomyositis. (Courtesy J. Goldfisher and E. Rubin: Ann. Intern. Med., *50*:194, 1959.)

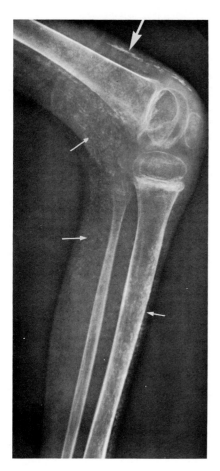

Figure 3–11 **Dermatomyositis: Soft Tissue Calcification.** Lateral view of the knee and leg in a child reveals extensive subcutaneous calcifications, most of which are short linear plaques in the muscles and subcutaneous tissues *(short arrows)*. A longer denser plaque is seen above the knee *(large arrow)*. All of the calcifications tend to follow the fascial planes.

The marked atrophy of soft tissue and the severe demineralization of bone are the result of restricted motion caused by muscular weakness and inelasticity of the skin.

Figure 3–12 **Dermatomyositis (Polymyositis): Extensive Soft Tissue Calcification in a Child.** There is widespread calcification of the subcutaneous tissues *(short arrows)* and muscles *(long arrows)*. The demarcation between the muscles and subcutaneous tissues is still well defined, but it is lost in longstanding cases. Osteoporosis is not pronounced.

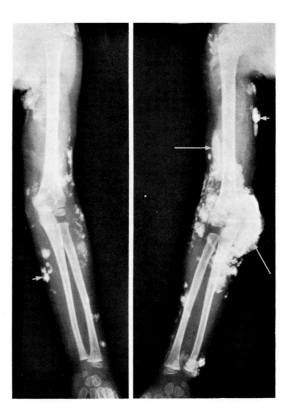

Systemic Lupus Erythematosus

Roentgen changes are extremely variable. Occasionally, patients may show few or no changes throughout the entire course of the disease or changes may appear only in the terminal stage.

Pleural disease and a variety of pulmonary lesions, often recurrent, appear during the course of the disease in a majority of patients. Pleural changes are the commonest findings, with pleural effusion occurring in about one half of cases. While a simple asymptomatic effusion can develop, most often the pleuritis is accompanied by elevation of a diaphragm and secondary plates of atelectasis at the base or bases. Much more infrequently, lesions occur that are probably due to lupoid vasculitis and are characterized by nodular miliary or fluffy infiltrates, usually rather diffuse (lupoid pneumonia). Both pleural and parenchymal lesions respond to steroids. However, probably the majority of the pulmonary lesions that are seen in lupus are pneumonias due to a variety of secondary infections. Occasionally uremic pulmonary edema may result from lupoid nephritis. If the nephrotic syndrome develops, a pleural effusion may occur.

The cardiac silhouette is often enlarged, and changes in size may be revealed in serial studies. Such alterations are usually the result of pericarditis with effusion; they can develop abruptly or rather slowly over many weeks. Endocardial or myocardial changes may also occur. All these manifestations regress during remission.

Splenomegaly is fairly common, but rarely marked. Hepatomegaly may occur, but it is difficult to evaluate by roentgenograms. Although gastrointestinal symptoms are frequent, the roentgenologic findings are minimal, consisting mainly of small bowel dilatation and poor tone. There may be hypomotility. However, in more severe cases there may be radiographic evidence of small bowel vasculitis, with spasticity, luminal narrowing, thumbprinting, and mucosal effacement. Complete clinical and radiographic recovery may result from increasing the steroid dosage.

Polyarthralgia and painful swollen joints are the most frequent clinical complaints (they appear in 90 per cent of cases). Usually no joint changes are seen radiologically, but often there are characteristic ulnar deformities at the metacarpophalangeal joints and hyperextension and hyperflexion deformities at the interphalangeal joints. Apparently, these deformities are due to mild and reversible subluxations. In a minority of patients, a typical rheumatoid arthritis develops. Avascular necrosis, most commonly of the femoral heads, is not uncommon in systemic lupus erythematosus; some cases, however, may be due to steroid therapy.

No specific urographic findings are seen in lupoid nephritis, which often progresses to small, nonfunctioning kidneys. Renal angiography may disclose abnormality of the secondary renal vessels or merely nonspecific end-stage kidneys. Renal biopsy is necessary for specific diagnosis.

Secondary infections, some of which are very severe, are frequent complications of systemic lupus erythematosus.

Unexplained pleuritis and pericarditis, particularly in younger females, should suggest systemic lupus erythematosus. The radiographic changes of pleuritis, pneumonitis, and pericarditis tend to disappear and recur.

Although this collagen disease is of unknown origin, occasionally a systemic lupus syndrome can be induced by certain drugs such as Pronestyl and isoniazid.[6, 10-15]

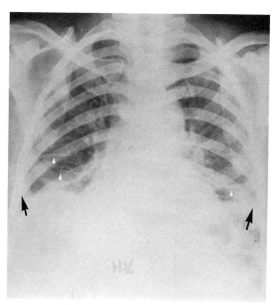

Figure 3–13 **Systemic Lupus Erythematosus: Chest Findings.** Small bilateral pleural reactions have obliterated the costophrenic angles *(arrows)*. Numerous transverse bands of density are seen in both lower lobes *(arrowheads)*, but the upper lung fields are clear. The heart shadow is moderately enlarged, probably because of pericardial effusion. These findings strongly suggest lupus erythematosus.

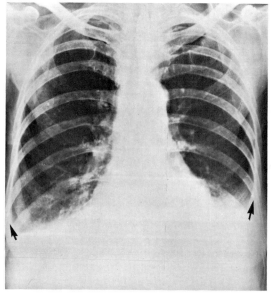

Figure 3–14 **Systemic Lupus Erythematosus: Bilateral Pleural Effusions.** Small effusions at both bases *(arrows)* have obliterated the costophrenic angles, but no other abnormalities are seen, and the heart is not enlarged.

Unexplained small pleural effusions, if recurrent or bilateral, suggest disseminated lupus erythematosus.

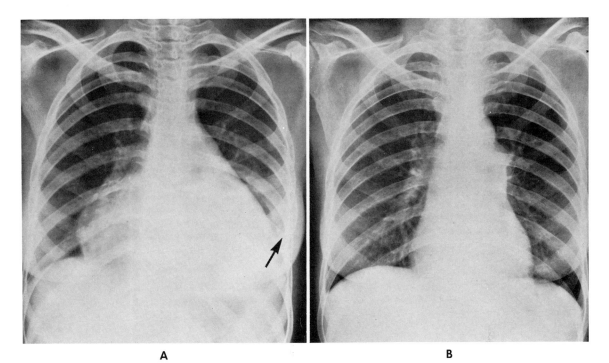

A   B

Figure 3–15 **Systemic Lupus Erythematosus: Pericardial and Pleural Effusion and Regression.**

A, The heart shadow is greatly enlarged, so that the normal contour is lost. A large pericardial effusion is seen, as well as an effusion obliterating the left costophrenic angle *(arrow)*. There are no parenchymal lesions.

B, Dramatic clinical response to steroid therapy is accompanied by return of the cardiac silhouette to normal size and shape. The pleural effusion has cleared.

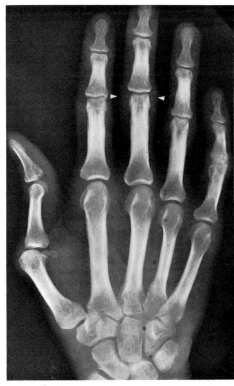

Figure 3–16 **Systemic Lupus Erythematosus: Joint Symptoms.** There is soft tissue swelling around the proximal interphalangeal joints *(arrowheads),* but no destructive synovial changes or bone alterations are apparent.

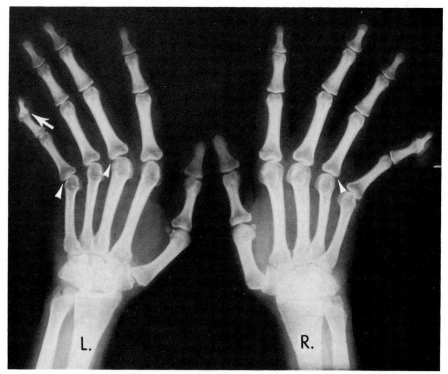

Figure 3–17 **Hands in Lupus Erythematosus.**   Both hands show marked ulnar deviation of all the phalanges *(arrowheads),* owing to ligament laxity and transient subluxations at the metacarpophalangeal joints. There is also a mild swan-neck deformity of the distal phalanges of the left fifth finger *(arrow).*
    Note the absence of radiographic changes in the joints.

Periarteritis Nodosa (Polyarteritis Nodosa)

Periarteritis may affect any tissue or organ, and the radiographic findings are quite variable and rarely specific.

Pulmonary infiltrates of various forms occur in about 25 per cent of cases. Interstitial infiltrates, especially at the bases, perivascular nodular lesions, or scattered confluent densities may occur alone or in combination. Cavitation in a confluent density is an infrequent but suggestive finding, but it must be differentiated from Wegener's granulomatosis. Frequently the infiltrates change over serial films, with areas of progression or regression. Eosinophilia and asthmatic symptoms may be associated, making the distinction from Löffler's pneumonopathy difficult. In many cases, large perihilar densities—the batwing appearance—are seen, simulating pulmonary edema, but these are probably caused by a vasculitis. Pleural fluid is uncommon. Hilar vascular enlargement is found in about one third of patients with pulmonary lesions. It may be caused by pulmonary hypertension.

Cardiac enlargement, usually due to myocardial involvement, can lead to congestive changes and episodes of true pulmonary edema.

In the gastrointestinal tract there may be areas of infarction or ulceration, and gastrointestinal bleeding may be a prominent symptom.

Multiple aneurysms of the smaller visceral arteries are virtually pathognomonic of periarteritis. The renal arteries are most frequently involved and the lesions are readily demonstrated by renal arteriography. The aneurysms are usually in the interlobar arteries and may appear as rounded densities or fusiform dilatations. Localized areas of small vessel stenosis or occlusion are also characteristic.

Although renal involvement is very common, there are no definitive pyelographic findings; excretory function may be impaired. Rupture of a renal artery aneurysm may cause a perinephric or parenchymal hematoma, radiographically simulating a renal tumor.[6, 11, 16-18]

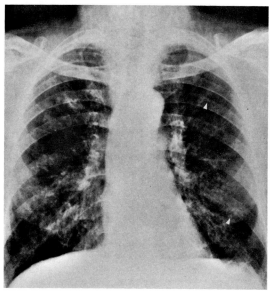

Figure 3–18 **Polyarteritis: Nodular Lung Lesions.** There are small perivascular nodular densities *(arrowheads)* throughout the lung fields, especially in the lower lobes. The pattern of these densities changed within a period of months. The hilar shadows are prominent, and there was no change in size on serial films.

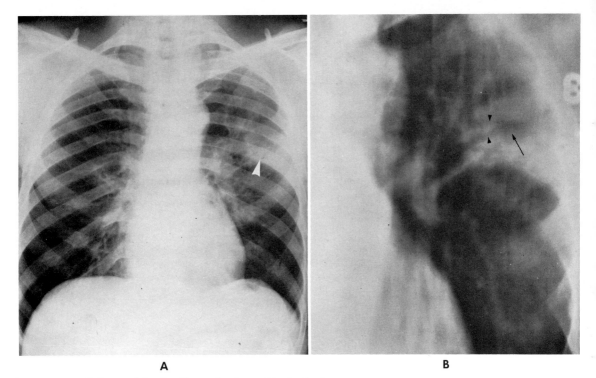

| A | B |

Figure 3–19 **Polyarteritis: Confluent Lesion with Cavitation.**

A, There is a confluent, well-circumscribed area of density *(arrowhead)* in the left upper lobe adjacent to the pleura. The left hilar shadow is prominent.

B, Planigram of the left upper lobe shows a cavity within the density *(arrow).* The cavity is connected with dilated irregular or air-filled bronchi *(arrowheads).*

This pleuropneumonitis with cavitation and bronchiectasis is most likely caused by pulmonary infarction secondary to vasculitis.

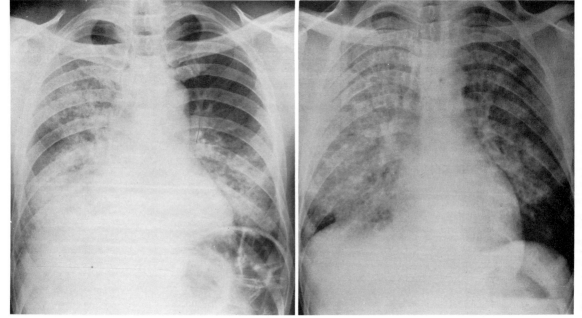

Figure 3–20 **Polyarteritis: Findings Similar to Those of Pulmonary Edema.**   In both patients there are confluent densities that extend from the hila and that are associated with cardiomegaly. The nodular pattern contrasts with the homogeneous density of uncomplicated pulmonary edema due to cardiac failure.

Although the findings are nonspecific and may be seen in a variety of other conditions, the combination of hilar prominence and pulmonary nodulation suggests the possibility of polyarteritis.

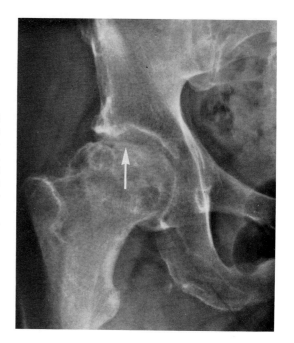

Figure 4–11 **Rheumatoid Arthritis: Aseptic Necrosis of Femoral Head Secondary to Steroid Therapy.** The upper outer aspect of the femoral head *(arrow)* is flattened and irregular. The joint space appears widened, and the acetabulum is uninvolved. Lytic cyst-like areas are seen throughout the femoral head. These changes are secondary to high steroid doses and probably represent a form of aseptic necrosis.

The hip was not involved by the rheumatoid process.

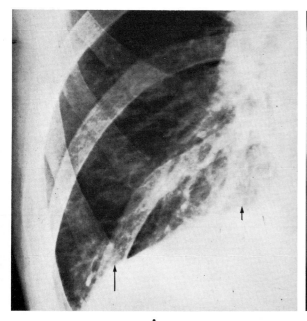

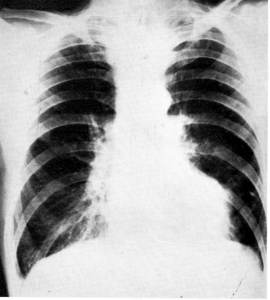

A   B

Figure 4–12   **Rheumatoid Arthritis: Lung Changes.**

A, Enlarged view of the right base discloses thickened basilar markings with some lucencies suggestive of bronchiectasis *(short arrow)*. Small nodular densities *(long arrow)* can be identified within the heavy markings.

B, In posteroanterior view the basilar markings are prominent on both sides, particularly the right. A variety of nonspecific pulmonary abnormalities, including merely accentuated markings, may occur in patients with rheumatoid arthritis. (Courtesy Dr. W. Hartl, Marburg, Germany.)

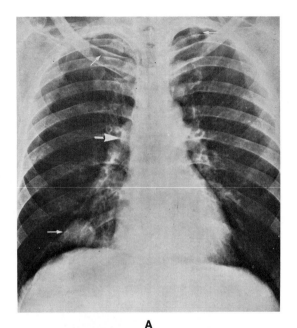

A

B   **C**

Figure 4–13   **Rheumatoid Nodules in Lung: Rapid Growth and Cavitation.**

   A, Posteroanterior view demonstrates several discrete, sharply bordered nodular densities *(small arrows)* in the lung fields, and a mediastinal density on the right *(large arrow)*. The lesions strongly suggest malignant metastases. Rheumatoid nodules were found on biopsy; the mediastinal density was probably a rheumatoid lymph node.

   B and *C,* Several months later, posteroanterior and lateral views indicate marked enlargement of the nodular masses *(small arrows)*. Most of the masses have undergone cavitation and have air-fluid levels, indicating communication with the bronchi. A section of the right sixth rib was removed during thoracotomy *(large arrow)*.

   Rheumatoid lung nodules are infrequent; rapid growth and cavitation are rare.

Sjögren's Syndrome

This syndrome of polyarthritis, dry mouth, and dry eyes is a systemic disease occurring predominantly in middle-aged women. The usual roentgenologic changes are in the joints and parotid glands. Abnormalities in the lungs and abdominal lymph nodes are also quite common; nephrocalcinosis is an occasional finding.

The arthritic changes occur in about one half of the cases and are generally indistinguishable from those of ordinary rheumatoid arthritis. In about 25 per cent of cases, however, there is juxtaarticular bone destruction in one or more metacarpals, metatarsals, or phalanges, more similar to changes seen in psoriatic arthritis.

Sialography discloses ectasia of both the ducts and the alveoli of the parotid glands in virtually every case.

Reticular and nodular lung infiltrates and often pleural thickening are found in about one third of cases. These are similar to the lesions seen in collagen vascular diseases or in rheumatoid arthritis and also respond to steroids. Complicating lipoid pneumonia may occur from repeated oil ingestion to combat the dry oral mucosa.

In many patients, lymphangiography will disclose enlarged but architecturally normal abdominal lymph nodes.

Splenomegaly and hepatomegaly occur in about 20 per cent of cases. Lymphoma has been reported in 2 to 4 per cent of patients with Sjögren's syndrome.[17-20]

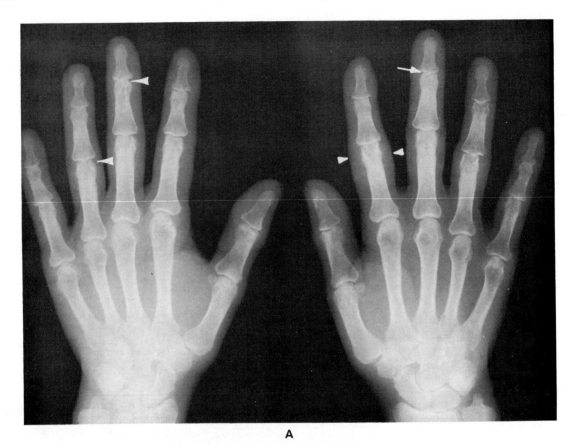

A

B

Figure 4–14 **Sjögren's Syndrome.**

A, There is soft tissue swelling around a few of the proximal interphalangeal joints *(small arrowheads),* narrowing of the proximal and distal interphalangeal joints *(arrow),* and periarticular erosions *(large arrowheads).* There is also some periarticular demineralization (fourth and fifth fingers of right hand). All these findings are fairly characteristic of rheumatoid arthritis.

B, Close-up view shows some sclerotic changes *(arrowheads)* in the heads of two middle phalanges, probably due to bone apposition. This finding is more common in psoriatic arthritis.

The usual joint changes in Sjögren's syndrome are indistinguishable from rheumatoid arthritis.

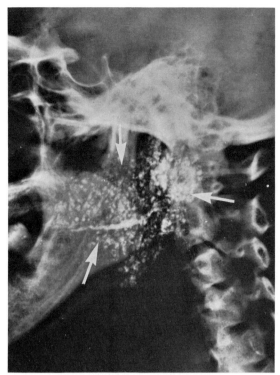

Figure 4–15 **Sjögren's Syndrome: Parotid Ectasia.**   Sialogram demonstrates absence of the normal pattern of ductile arborization. Small collections of contrast medium throughout the gland represent alveolar ectasia *(arrows).* Irregular dilatation of the proximal portion of the major duct is seen above the lower arrow.
 Ectasia of the parotid alveoli and ducts is seen in virtually every case of Sjögren's syndrome.

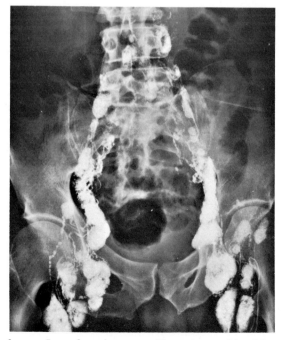

Figure 4–16 **Sjögren's Syndrome: Lymphangiogram.**   There is considerable enlargement of the inguinal, iliac, and lower abdominal nodes. Their architecture is well preserved, in contrast to the frothy appearance in Hodgkin's disease and the distorted irregular nodes of metastatic disease.

Ankylosing Spondylitis (Marie-Strümpell Spondylitis, von Bechterew's Syndrome, Rheumatoid Spondylitis)

Although once considered a variant of rheumatoid arthritis, ankylosing spondylitis is generally regarded as a distinct entity. Males are most commonly affected.

The disease almost always begins in both sacroiliac joints and progresses cephalad to the lower lumbar spine, eventually extending to the dorsal and cervical regions.

The earliest roentgenographic changes in the sacroiliac joints are blurring of the articular edges of the joints and loss of bone detail. Later, poorly marginated subchondral erosions with some periarticular sclerosis develop. The nonarticular portion superior to the joint proper also becomes involved. Bone proliferation is prominent. Ultimately, bony ankylosis ensues, generally within two to five years, and osteosclerosis becomes less marked as joint obliteration develops. Complete disappearance of the joint space eventually occurs. Erosion of the ischial tuberosity and symphysis pubis is frequent.

A similar process takes place in the apophyseal joints of the spine, although this usually is not uniform. There is ankylosis of the apophyseal and costovertebral joints; the vertebral bodies become squared, with sharp corners. Later, there is calcification in the spinal ligaments; this, plus the squared vertebral bodies, gives rise to the bamboo spine. General spinal osteoporosis is frequent.

In severe cases, involvement of the cervical spine occurs after the lumbar and dorsal spine involvement. Obliteration of the smaller joints and eventual encasement of the spine by calcified ligaments occur. Dislocation of the cervical spine, especially in the odontoaxial articulation, is not an uncommon complication. Fractures of the cervical spine constitute a rather serious complication and frequently occur through the calcified ligaments at the level of an interspace.

In 30 to 80 per cent of cases, the hips and less often the shoulders and temporomandibular joints become involved, showing radiographic changes identical to those seen in rheumatoid arthritis. In longstanding cases, additional involvement of peripheral joints, especially hands, wrists, and knees, can occur. The radiographic changes in these smaller joints resemble those of Reiter's disease or psoriasis rather than those of rheumatoid arthritis. The involvement is asymmetric, demineralization is not severe, the erosive changes are less prominent, with a sclerotic margin, and a shaggy periostitis is frequent.

An aortitis involving the ascending aorta and aortic valves can occasionally occur; it may cause aortic insufficiency.

In longstanding or burned-out spondylitis, one or more spinal arachnoid cysts or diverticula infrequently develop in the cauda equina, causing erosive bone changes in the posterior canal and producing a cauda equina syndrome.[9, 21-25]

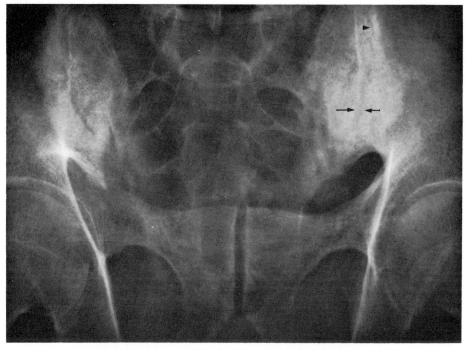

Figure 4–17 **Ankylosing Spondylitis: Sacroiliac Joint Involvement.** The sacroiliac joint is blurred, and the margins *(arrows)* are poorly defined. There are spotty areas of subchondral erosion and bilateral sclerosis. The nonarticular portion of the joint is also involved *(arrowhead)*. The process is bilateral and symmetric.

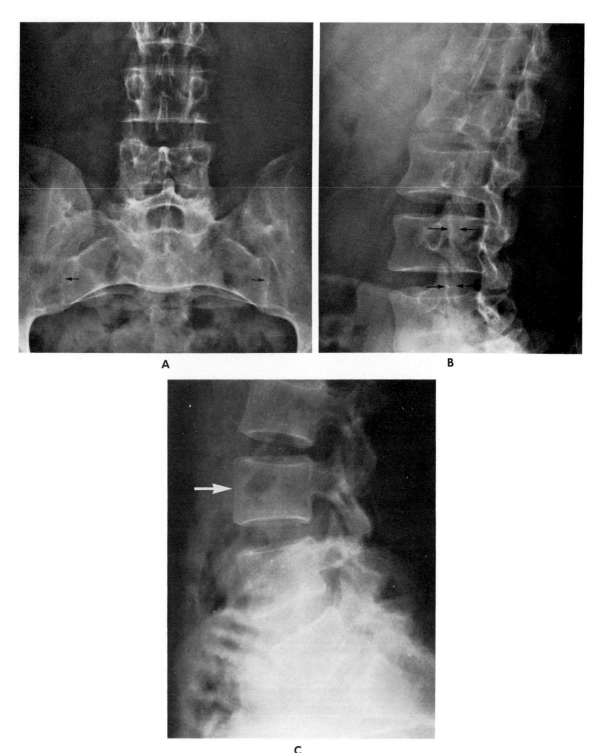

Figure 4–18 **Ankylosing Spondylitis: Sacroiliac Ankylosis: Apophyseal Erosion.**

A, There is bony ankylosis, and the sacroiliac joint spaces have been obliterated *(arrows).*

B, In the lumbar spine there is sclerosis and irregularity of the apophyseal joints *(arrows),* but ankylosis has not yet developed.

C, The lumbar vertebrae are squared off and rectangular *(arrow).* There is osteoporosis of the entire spine.

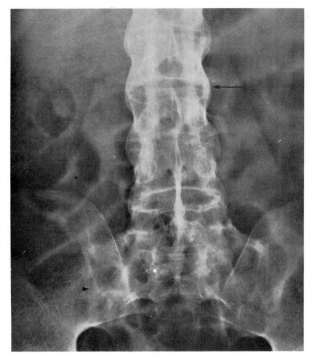
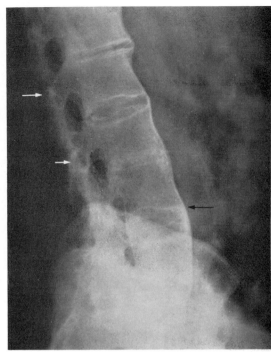

A   B

Figure 4–19 **Ankylosing Spondylitis: Bamboo Spine.**

A, Anteroposterior view discloses calcification of the paraspinous ligaments *(arrow)*, bony ankylosis of the sacroiliac joint *(arrowhead)*, and ankylosis of the apophyseal joints.

B, Lateral view demonstrates bony ankylosis of the apophyseal joints *(white arrows)*, calcification of the anterior longitudinal ligaments *(black arrow)*, and marked osteoporosis.

The picture has aptly been dubbed the bamboo spine, and it is diagnostic of ankylosing spondylitis.

Calcification of the anterior longitudinal ligament can also occur in older patients with ordinary degenerative changes of the spine.

Figure 4–20 **Apophyseal Joints in Normal Adult Spine and in Ankylosing Spondylitis.**

A, Oblique view of a normal spine discloses normal apophyseal joints *(arrows)*. The joint spaces are regular, and the articular margins are smooth and distinct.

B, Oblique view of a patient with advanced ankylosing spondylitis demonstrates virtual obliteration of the apophyseal joint spaces *(arrows)*, which have become ankylosed. The articular facets have all but disappeared. This is ligamentous calcification and bone demineralization — the bamboo spine.

The apophyseal joints become involved in ascending order following sacroiliac joint changes. The apophyseal joints are not uniformly affected, but almost all the joint spaces become hazy and ill defined, with many progressing to complete ankylosis and obliteration.

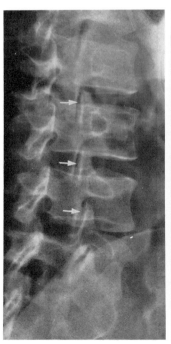
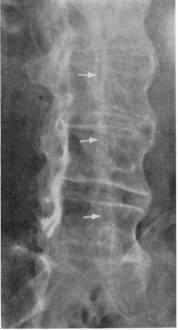

A   B

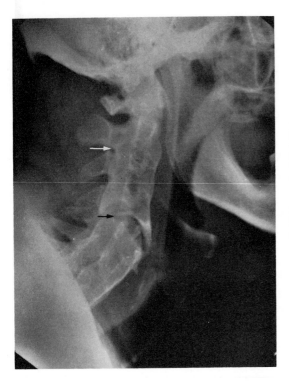

Figure 4–21 **Ankylosing Spondylitis: Cervical Spine.** Fusion of the apophyseal joints and the posterior elements produces a block appearance *(white arrow).* There is a fracture between C4 and C5 *(black arrow).* The lumbodorsal spine was also involved.

Ankylosing spondylitis of the cervical spine is always associated with involvement of the dorsal and lumbar spine. Trauma may cause fracture of the calcified ligaments without involvement of the bone itself.

Figure 4–22 **Ankylosing Spondylitis: Cauda Equina Syndrome in Two Patients: Myelograms.** In both patients the ankylosing spondylitis was of long standing. Multiple arachnoid cysts or diverticula are seen *(arrows).* In *B,* lateral view shows the posterior location of these cysts.

Bone erosion was present, but is not appreciated on the illustration. (Courtesy D. A. Gordon, Toronto, Ontario, Canada.)

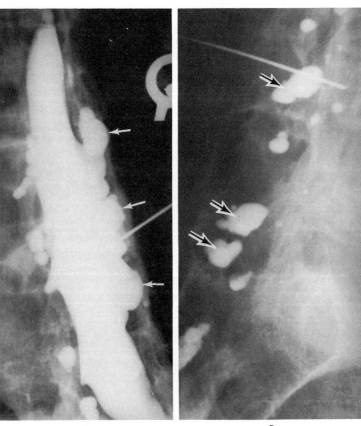

A   B

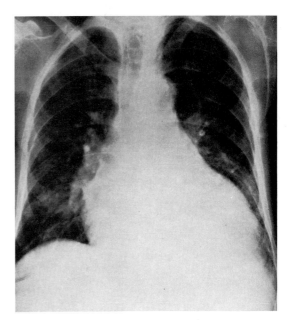

Figure 4–23 **Ankylosing Spondylitis: Aortic Insufficiency.** Cardiomegaly with enlargement of the left ventricle is due to aortic insufficiency. The lung fields are clear.

In ankylosing spondylitis, the aortic valve and the ascending aorta may be involved, and there are no roentgenologic features to distinguish this type of aortic insufficiency from aortic insufficiency due to other causes. Nevertheless, the presence of associated ankylosing spondylitis should suggest the probable cause.

Juvenile Rheumatoid Arthritis

Juvenile rheumatoid arthritis is a systemic disease primarily affecting the joints and connective tissue. It most frequently affects the areas of greatest bone growth: the knees, ankles, and wrists. The hands and feet are often spared early in the disease. Monoarticular involvement, especially of the knee, is not rare. The roentgen changes are akin to those seen in adult rheumatoid arthritis, but special and distinctive changes occur in a growing skeleton.

Periarticular soft tissue swelling and osteoporosis are the early changes. Later, cortical erosions occur, but narrowing of the joint space from cartilaginous destruction is a late finding. Periosteal new bone often appears near an involved joint, especially in the hands or feet, a finding rarely seen in adult rheumatoid arthritis. Occasionally, band-like areas of metaphyseal rarefaction similar to those seen in childhood leukemia are observed adjacent to areas of active arthritis. Synovial cysts are less frequent than in the adult disease.

Growth disturbances invariably ensue, associated with accelerated skeletal maturation and deformity of the epiphyses. The epiphyseal centers may be enlarged and ballooned, increasing the length of the bone, or they may be underdeveloped, causing shortened bones and narrowed shafts. Compression fractures of the epiphyses are not infrequent.

Cervical spondylitis affecting the small apophyseal joints is found in about two thirds of cases. These joints may ankylose, but paraspinal ossification is rare. Subluxations, especially atlantoaxial, are common. Sacroiliac involvement, extremely uncommon in adult rheumatoid arthritis, occurs in about 20 per cent of cases.

Interstitial pulmonary lesions are found in about 5 per cent of cases. Splenomegaly is more common, being detectable in about 20 per cent of cases.

The distinctive roentgen features of juvenile rheumatoid arthritis are the location of involvement, the growth disturbance, the periosteal new bone formation, and the late onset of articular erosion and narrowing of the joint space.[26–28]

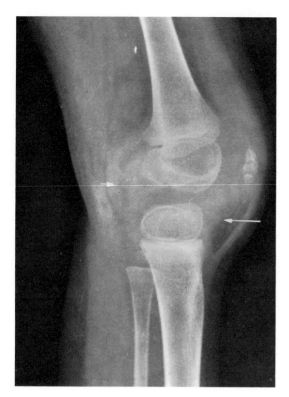

Figure 4–24 **Juvenile Rheumatoid Arthritis: The Knee.** There is marked swelling of the soft tissues around the knee. The infrapatellar fat pad *(long arrow)* is displaced anteriorly, and the posterior fat pad *(short arrow)* is displaced posteriorly, indicating the presence of fluid within the joint. The bones are diffusely porotic. The epiphyses of the femur and tibia are slightly enlarged. The joint space is still normal.

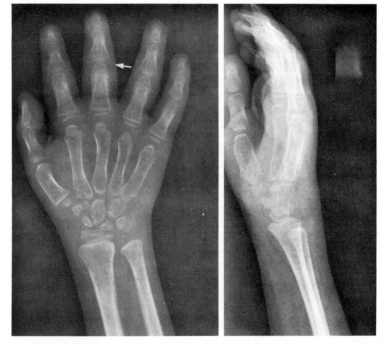

Figure 4–25 **Juvenile Rheumatoid Arthritis: The Hand and Wrist.**   Osteoporosis of all the bones of the hand and wrist, and periarticular soft tissue swelling *(arrow)* are evident. The joint spaces are preserved.

Severe clinical symptoms may be present without joint narrowing in juvenile rheumatoid arthritis; soft tissue swelling and periarticular demineralization, however, are usually seen.

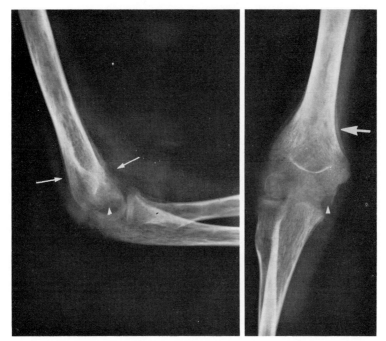

Figure 4–26 **Juvenile Rheumatoid Arthritis: The Elbow.** Displacement of the bursal fat pads *(small arrows)* around the elbow indicates the presence of fluid within the joint space. There is marked periarticular osteoporosis. Subchondral bone resorption *(arrowheads)*, periosteal reaction *(large arrow)*, and joint space narrowing are findings indicative of advanced juvenile rheumatoid arthritis.

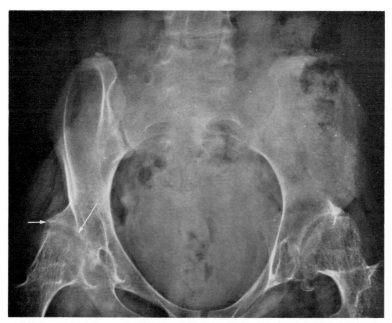

Figure 4–27 **Juvenile Rheumatoid Arthritis: The Hips.** Overgrowth of the right acetabular roof secondary to chronic subluxation *(short arrow)* has occurred; there is slight but uniform narrowing of the joint space *(long arrow)*. The cortical margins of the bone are indistinct and hazy, and osteoporosis is generalized. The entire right hemipelvis is hypoplastic. Although accelerated growth is the usual finding, growth may be retarded in some cases of juvenile rheumatoid arthritis.

The left hip is also involved, and there is osteoporosis and joint narrowing. The femoral capital epiphysis is abnormally large.

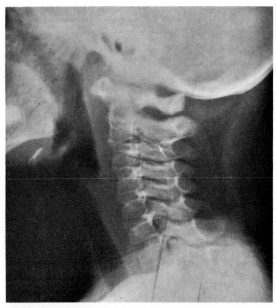

Figure 4-28 **Juvenile Rheumatoid Arthritis: Cervical Spondylitis.** There is ankylosis of the apophyseal joints of C2 and C3 *(white arrow)* and diffuse osteoporosis in all the vertebral bodies.

Psoriatic Arthritis

Psoriasis is frequently accompanied or sometimes even preceded by a form of arthritis that resembles rheumatoid arthritis both clinically and radiographically. However, there are often certain distinctive radiographic features in the arthritis seen in psoriatic patients.

The distal interphalangeal joints of the hands and feet are usually affected in psoriatic arthritis in contrast to the proximal interphalangeal joint involvement in rheumatoid arthritis. The feet usually show the most marked changes. Asymmetric joint involvement is the rule, and osteoporosis is minimal or absent. Severe destructive changes are frequent and there may be "penciled" bone ends – arthritis mutilans. A fluffy periostitis adjacent to involved joints is common, as are tuft erosions and calcaneal spurs. Otherwise, the joint changes are quite similar to those of rheumatoid arthritis; there is soft tissue swelling, narrowing of the joint spaces but with minimal periarticular demineralization, and considerable periarticular erosion and destruction.

Erosive changes in the sacroiliac joints are quite common in contrast to their rarity in rheumatoid arthritis. Sacroiliac involvement may be unilateral or asymmetric. In a high proportion of severe cases, there are spine changes akin to those seen in ankylosing spondylitis, including syndesmophytes, squared vertebras, and apophyseal fusions. Atlantoaxial subluxation is also frequent.

Distal interphalangeal joint involvement, fluffy periostitis, asymmetric disease, minimal or absent osteoporosis, tuft erosions, and sacroiliac involvement may help distinguish psoriatic arthritis from rheumatoid arthritis. Distinction from the arthritis of Reiter's disease is often more difficult.[29-31]

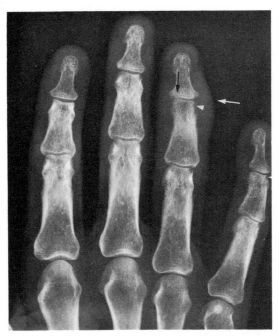

Figure 4–29 **Psoriatic Arthritis: Early Changes in Distal Interphalangeal Joints.** There is soft tissue swelling *(white arrow)*, joint space narrowing, focal loss of the articular cortex *(black arrow)*, and subcortical cystic erosion *(arrowheads)* at the distal interphalangeal joints. These changes are identical to those in rheumatoid arthritis, but in the latter they occur in the proximal interphalangeal joints.

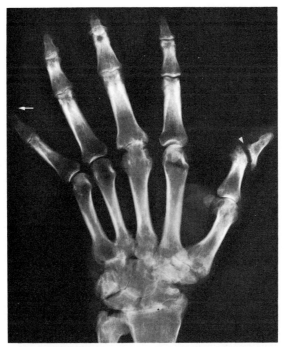

Figure 4–30 **Psoriatic Arthritis: Advanced Changes in Hand.** The joint spaces are narrowed in almost all the interphalangeal joints and in the metacarpophalangeal joints of the thumb and middle finger. Virtually all the carpal joints are ankylosed. The distal phalanx of the thumb is subluxated *(arrowhead)*, and the articular surfaces are deformed. The ungual tuft of the fifth finger has been resorbed *(arrow)*. Osteoporosis is moderate.

These changes are indistinguishable from those of rheumatoid arthritis.

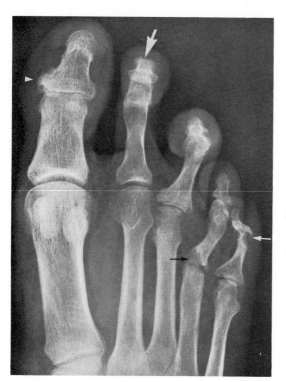

Figure 4–31 **Psoriatic Arthritis: The Foot.** Periarticular bone destruction is present in the interphalangeal joint *(small white arrow)* and the metatarsophalangeal joints *(black arrow).* Ungual tuft resorption is seen *(large white arrow).* The proliferative reaction at the base of the great toe *(arrowhead)* is secondary to the joint involvement.

Periarticular and ungual tuft bone resorption is seen in both psoriatic and rheumatoid arthritis, but is often more marked in the psoriatic form.

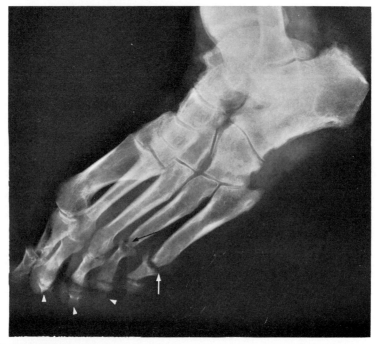

Figure 4–32 **Psoriatic Arthritis: The Foot.** There is resorption of the ungual tufts of the distal phalanges *(arrowheads)* and marked widening of the joint space in the fifth metatarsophalangeal joint, with a sharply demarcated eroded adjacent bone *(white arrow).* The metatarsal head is pointed *(black arrow).* Osteoporosis is not conspicuous. Widening of the joint space due to periarticular bone resorption is frequent in advanced psoriatic arthritis. The resorbed bones often develop tapered or pointed edges.

Reiter's Syndrome

In this triad of urethritis, arthritis, and conjunctivitis, the changes in the joints may closely resemble those of rheumatoid arthritis, but there is usually asymmetric involvement and a predilection for the joints of the lower extremities. The feet and the sacroiliac joints are favored sites. Permanent joint damage rarely develops. Osteoporosis is generally not as marked as in rheumatoid arthritis. There is exuberant periosteal new bone formation, especially in the os calcis and in the short bones near affected joints. There often is an irregular fluffy spur on the plantar aspect of the heel — a fairly specific and characteristic finding.

The earliest joint change is periarticular swelling, sometimes with slight subchondral bone resorption. There may be joint effusion, especially in the knee. Moderate osteoporosis and subchondral erosion and narrowing of the joint space may occur. Rarely, subluxation of peripheral joints develops. Periosteal new bone formation in the vicinity of affected joints occurs in about 20 per cent of cases and may parallel the shafts of the small bones of the feet, simulating hypertrophic osteoarthropathy. Changes in the sacroiliac joints consist of articular erosions, irregular widening of the joint space, and later subchondral sclerosis. One or both joints may be affected. Unlike ankylosing spondylitis, fused obliteration of the joint does not occur. In some cases, lateral bony bridging between adjacent vertebrae (syndesmophytes) is seen in the lower dorsal spine, a suggestive finding in younger individuals.

In the male, arthritis of the lower extremities, especially if limited to the feet and sacroiliac joints, if asymmetric and if associated with some periosteal reaction, should suggest Reiter's syndrome. An irregular plantar spur of the os calcis is also highly suggestive.

Rarely, nonspecific pneumonitis may be associated with this syndrome.[9, 32-35]

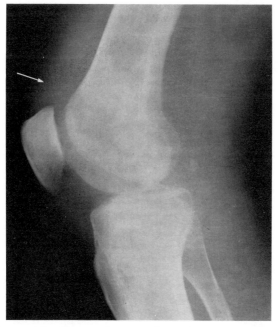

Figure 4–33 **Reiter's Syndrome: The Knee.** Periarticular soft tissue swelling with evidence of intra-articular effusion is demonstrated by swelling of the suprapatellar recess of the joint *(arrow)*. There is also some osteoporosis. The findings are nonspecific and are seen in other conditions with joint effusion.

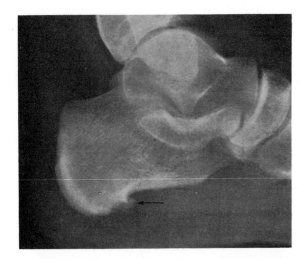

Figure 4–34 **Reiter's Syndrome: Os Calcis.** Periostitis in the plantar portion of the heel associated with an irregular fluffy calcaneal spur *(arrow)* is a characteristic finding of Reiter's syndrome. (Courtesy Drs. W. V. Weldon and R. Scalettar, Walter Reed General Hospital, Washington, D.C.)

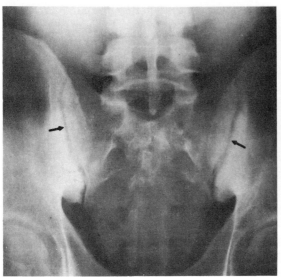

A

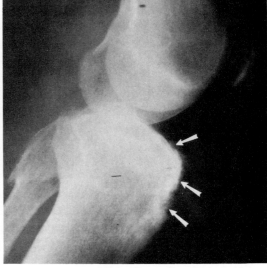

B

Figure 4–35 **Reiter's Syndrome: Classic Skeletal Changes.**

A, Both sacroiliac joints are irregular and narrowed *(arrows)*. Sclerotic changes are seen on both sides of each sacroiliac joint.

B, A fluffy irregular periosteal reaction is seen on the anterior tibia *(arrows)* adjacent to the knee joint. The healed fracture of the fibula is an incidental finding.

C, The *two* spurs on the os calcis *(thinner arrows)* have somewhat indistinct borders. Fluffy periosteal irregularities *(wider arrows)* extend throughout the posterior border of the os calcis. The other os calcis was similarly affected.

Fluffy periosteal changes near the joints of the lower extremities are highly characteristic of Reiter's arthropathy. Sacroiliac involvement is also characteristic, but radiographically it is identical to the changes of ankylosing spondylitis.

C

Caplan's Syndrome

This rather uncommon condition is characterized by the appearance of pulmonary nodules in a person with pneumoconiosis (most often anthracosilicosis) who develops rheumatoid arthritis. The pulmonary findings due to the pneumoconiosis are frequently minimal; the nodules usually appear during or following an arthritic attack.

The nodules are rounded, discrete, multiple, usually bilateral, and often peripheral. They vary in size from 0.5 to 6.0 cm. The nodules develop rapidly, and some may eventually undergo liquefaction and cavitation. The pulmonary background often discloses only minimal evidence of the underlying silicotic fibrosis. The picture may resemble that of metastatic disease.

The large conglomerate shadows of advanced silicosis can be readily distinguished from the rheumatoid pulmonary nodules of Caplan's disease: the former are single parenchymal lesions in each lung, nearly symmetric, irregular in outline, and often sharply marginated. The diffuse fibrotic background is invariably prominent in silicosis but is often minimal in Caplan's syndrome.[36-38]

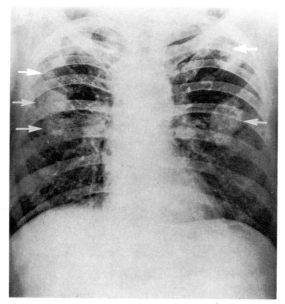

Figure 4–36 **Caplan's Syndrome.** In a 46 year old man there are multiple, rounded, sharply defined nodules of various sizes in both lung fields *(arrows)*. The nodules developed following the onset of polyarthritis ten years earlier. The diffuse fibrotic densities throughout the lungs are due to silicosis, which the patient incurred during employment as a stone crusher, prior to the onset of arthritis.

Although pulmonary nodules are infrequent complications of rheumatoid arthritis, they appear to be unusually common in patients with silicosis who develop rheumatoid arthritis. (Courtesy Dr. M. Kantor, Veterans Administration Hospital, Wilkes-Barre, Pennsylvania.)

DISEASES WITH WHICH ARTHRITIS IS FREQUENTLY ASSOCIATED

Inflammatory Bowel Disease

About half of the patients with ulcerative colitis or granulomatous enterocolitis have some arthritic complaints, but only a small percentage have clinical and radiographic features of arthritis.

Sacroiliac involvement is the most common radiographic finding and is characterized by symmetric erosive irregularity and joint widening, followed by subchondral sclerosis. Occasionally, ankylosis develops.

Peripheral joint involvement — most often in the knee and elbow — usually consists of soft tissue swelling and/or effusion. Occasionally, erosive periarticular changes are seen that are indistinguishable from those of rheumatoid arthritis. Rarely, changes identical to ankylosing spondylitis are seen.[29, 39-42]

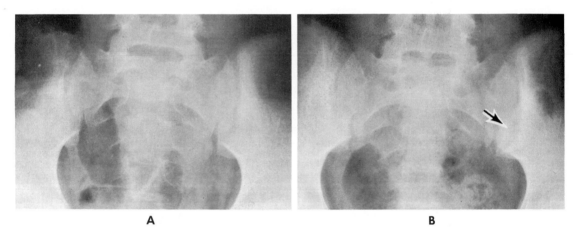

A   B

Figure 4–37   **Sacroiliac Changes in Crohn's Disease.**

A, The sacroiliac joints are normal in this young man with a several-year history of granulomatous ileocolitis.

B, Two years later, the left sacroiliac joint space is irregularly widened and hazy, with adjacent sclerosis *(arrow).* The right sacroiliac was also involved.

The appearance is quite similar to the early changes of ankylosing spondylitis.

Agammaglobulinemia Arthritis

Almost 25 per cent of patients with agammaglobulinemia have a mild clinical arthritis, which is episodic and without roentgen changes or residua. Rarely, the arthritis becomes chronic and severe, producing roentgen changes indistinguishable from those seen in rheumatoid arthritis.

A suppurative arthritis can also complicate agammaglobulinemia.[29]

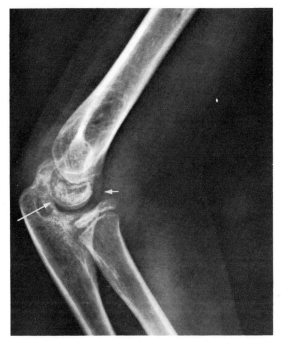

Figure 4–38 **Agammaglobulinemia: Arthritic Changes in Elbow.** Osteoporosis is evident. The cortical line of the olecranon process has disappeared *(long arrow)*. The periarticular bursal fat pads have been slightly displaced, indicating intra-articular effusion *(small arrow)*. These findings are indistinguishable from those of ordinary rheumatoid arthritis.

Polychondritis

Relapsing polychondritis — perhaps an acquired mucopolysaccharide disorder — is characterized by an inflammatory degeneration of cartilage associated with febrile episodes of respiratory and joint symptoms. Cartilaginous destruction develops in the trachea, larynx, epiglottis, bronchi, nose, and external ear. Radiographically, there may be calcifications in the ears and narrowing of the trachea. The tracheal narrowing can lead to obstructive emphysema and recurrent pneumonias.

Arthritic involvement of the hands, feet, and sacroiliac joints may lead to narrowing of the joint spaces and rather severe periarticular erosions, findings that are sometimes indistinguishable from those of rheumatoid arthritis. In some cases, dilatation of the aortic ring has been an associated finding.[43-46]

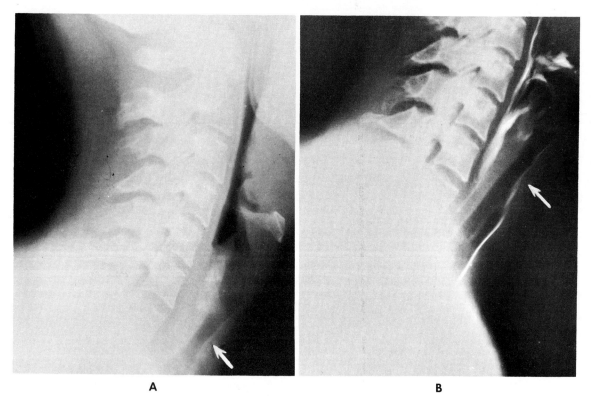

A   B

Figure 4–39 **Polychondritis with Tracheal Involvement.**

A, Lateral soft tissue film of the neck discloses narrowing of the trachea just below the larynx *(arrow).*

B, Contrast study clearly demonstrates the extensive localized narrowing *(arrow)* due to partial destruction of the cartilaginous rings of the trachea. This narrowing can lead to severe respiratory symptoms. (Courtesy Dr. C. M. Pearson, University of California Medical Center at Los Angeles.)

Acromegalic Joint Disease

Growth hormone stimulates growth of all body tissues to varying degrees. In acromegaly one sees increased thickness of the soft tissues, especially in the heels, fingers, and toes, and increased joint cartilage resulting in widening of the joint space. Associated laxness of ligaments may cause subluxation of the joints. There may be extensive spur formation, producing a radiologic picture resembling that of severe osteoarthritis but with well-preserved joint spaces.

The most distinctive features are seen in the hands and spine. The soft tissues of the fingers and toes become thickened, especially around the proximal interphalangeal joints, owing to capsular and synovial thickening. Often the normal periarticular fascial planes are obliterated. The joint space may be widened. The ungual tufts are large and the base of the distal phalanges is enlarged and broadened, creating a squared-off appearance. Exostoses are frequent; the shafts of the phalanges and metacarpals may be thickened, but this is less common.

In the spine the most frequent abnormality is increased width of the lumbar intervertebral discs. Elongation and widening of the bodies have been reported but are probably not common. There may be moderate or marked scalloping of the posterior vertebral bodies in the dorsal and lumbar spine. Osteoporosis may be a late feature.

See also *Hyperpituitarism: Acromegaly,* page 1255.[47-49]

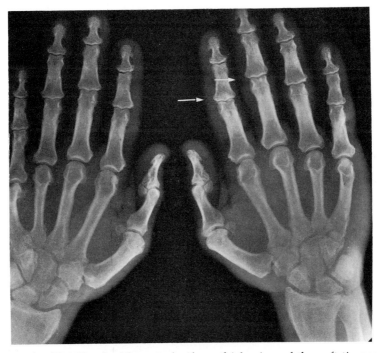

Figure 4–40 **Acromegaly: The Hand.** There is fusiform thickening of the soft tissues in the area of the proximal interphalangeal joints *(arrows)*. The ungual tufts are normal, but in more advanced cases they are enlarged. There is slight widening of the metacarpophalangeal joint spaces. The soft tissue swelling in acromegaly may suggest the early changes of rheumatoid arthritis, but there will be no evidence of subarticular bone erosion or periarticular dimineralization.

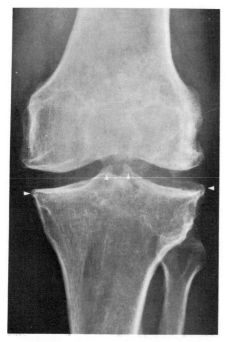 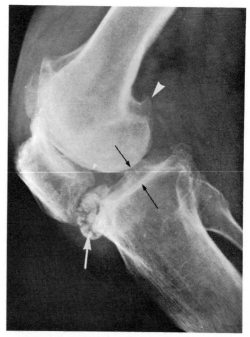

Figure 4–41 **Acromegaly: The Knee.** There are numerous changes characteristic of degenerative joint disease, including small and large bone spurs *(arrowheads),* a narrowed posterior patellar tunnel, and an intra-articular villous osteochondroma *(white arrow).* However, the actual joint spaces are widened *(black arrows),* while in ordinary degenerative arthritis of this degree, joint space narrowing is virtually always seen.

When severe degenerative changes are associated with a normal or widened joint space, the possibility of acromegaly should be considered.

Hemophilia Joint Disease

Hemophilia is often complicated by bleeding into the joints, most frequently the knees, elbows, and ankles. In acute hemarthrosis, distention of the capsule and soft tissue swelling around the joint are the principal radiographic findings. After repeated episodes of bleeding the chronic synovitis leads to thickening of the capsule and synovium. Disuse atrophy of the periarticular tissues may be apparent. With progression, the increased intra-articular pressure and formation of hematoma cause degeneration and thinning of cartilage. Erosion and cyst formation may occur in the articular ends of the bones. Intraosseous hemorrhage in subchondral bone also contributes significantly to cyst formation and subchondral bone loss. The joint margins become irregular, and the joint space is decreased. Marginal spurring and subchondral sclerosis may develop, and deposits of hemosiderin may cause increased density of the soft tissues. In young patients there is often accelerated maturation and increase in the size of the epiphyses. Osteoporosis of the epiphyses is present and may contribute to epiphyseal collapse. Avascular necrosis may also occur. In the knee, enlargement of the intercondylar notch and squaring of the patella are characteristic changes. Ankylosis and contractures frequently develop after recurrent episodes of intra-articular hemorrhage.

See also *Hemophilia,* page 1217.[48, 50–52]

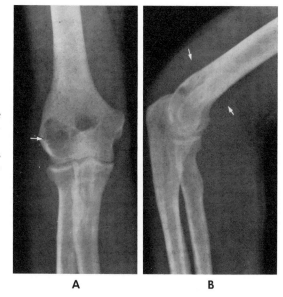

Figure 4–42 **Hemophilia: The Elbow.**

A, A large cystic area of bone erosion in the humerus *(arrow)* resulted from an intra-articular hematoma.

B, In lateral view the periarticular bursal fat pads *(arrows)* have been displaced; this indicates intra-articular effusion.

A   B

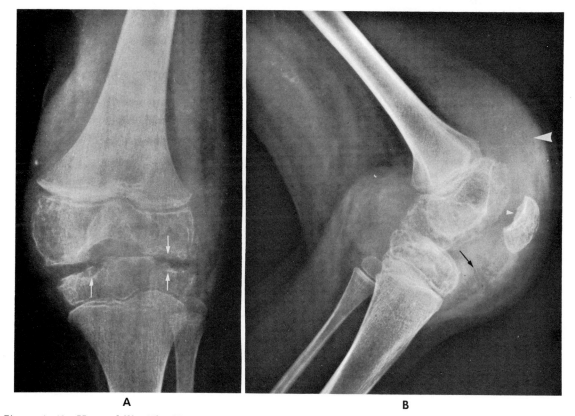

A   B

Figure 4–43 **Hemophilia: The Knee.**

A, The joint space is narrowed, and the subchondral bone is very irregular *(arrows).* The epiphysis is enlarged and demineralized, and, as is characteristic, the vertical trabeculae are prominent.

B, Lateral view indicates intra-articular effusion with displacement of the infrapatellar fat pad anteriorly *(arrow).* A hematoma *(small arrowhead)* has caused erosion of the posterior aspect of the patella. The articular bone is irregular. The increased density of the soft tissues *(large arrowhead)* is due to hemosiderin deposition from repeated hemorrhage.

Ochronosis and Joint Disease

Arthritis associated with ochronosis generally appears in the fourth decade. Deposition of homogentisic acid in cartilage is responsible for the cartilaginous degeneration and the secondary arthritic changes.

Initially the spine is affected, and roentgenograms reveal dense band-like areas of calcification within each disc. These areas are separated from the vertebral end plate by a translucent zone representing annular cartilage. Calcification of the disc is generally limited to the dorsal and lumbar spine; the cervical spine is spared. As a rule the intervertebral disc spaces are narrowed, and osteophytes are prominent.

Later, the hips, knees, and shoulders may become involved and show the progressive changes of degenerative arthritis: osteophytes, sclerotic eburnated articular margins, and irregular narrowed joint spaces. Joint effusion frequently occurs. Small joints are not affected.

See also *Ochronosis and Alcaptonuria,* page 1243.[48, 53]

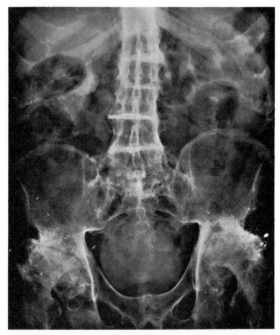

Figure 4–44 **Ochronosis: The Spine.** All the intervertebral discs are densely calcified and narrowed. A fine band of lucency at the L4 and L5 levels separates the calcified disc from the adjacent vertebral bodies. Generalized disc calcification of this kind is diagnostic of ochronosis, whereas localized calcification of one or several discs has little significance in adults.

The severe degenerative arthritis of both hips is also due to ochronosis. (Courtesy Dr. M. M. Thompson, Toledo, Ohio.)

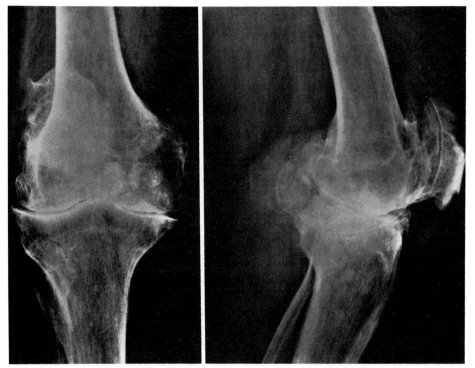

Figure 4–45 **Ochronotic Arthritis.** Anteroposterior and lateral views of the knee demonstrate narrowing of the joint space, reactive sclerosis around the joint surfaces, and much hypertrophic spurring. There is considerable irregularity of the posterior aspect of the patella, and joint debris. The picture is indistinguishable from that of ordinary advanced degenerative arthritis. (Courtesy Dr. M. M. Thompson, Toledo, Ohio.)

Other Diseases with Arthritic Symptoms

Joint symptoms occur in many conditions and diseases, but often without roentgen changes. Conditions in which symptoms are associated with some radiographic findings include familial Mediterranean fever, sarcoidosis, Behçet's syndrome, hemochromatosis, hyperparathyroidism, collagen diseases, sickle cell disease, and hypertrophic osteoarthropathy. These entities are described and discussed elsewhere.

MISCELLANEOUS FORMS OF ARTHRITIS

Neuropathic Joint Disease (Charcot Joints)

This disorder is secondary to diseases or conditions in which there is loss of pain and/or proprioceptive stimuli in a joint. The more frequent causes are tabes dorsalis, syringomyelia, diabetes (a combination of neuropathy and ischemia), and leprosy. Less common causes include the Riley-Day syndrome, paraplegia with sensory loss, and acrodystrophic neuropathy (maladie de Thévenard).

In the knee, hip, or ankle, tabes dorsalis is the usual cause; in the shoulder, syringomyelia is most often responsible. Diabetes and leprosy more often affect the joints of the feet.

The earliest roentgenologic finding is joint swelling resulting from intra-articular effusion. Subluxation and bony condensation develop, usually only in the articular end of the bone. There is generally some loss of articular bone substance, usually leaving a sharp, eburnated margin.

In the more common hypertrophic form, disorganization and fragmentation of the joint, marked irregularity of the articular surface, and new bone formation all act to produce loose fragments and osteophytes; the joint appears thoroughly disorganized. In the atrophic form the loss of articular bone with resultant smooth, even margins sometimes simulates an osteotomy. There is little or no soft tissue calcification in the atrophic form.

Usually, significant bone demineralization does not occur because the absence of pain permits continued use of the joint. The joint space ultimately becomes narrowed.

In the smaller peripheral joints, there may be pencil-like thinning of the periarticular bone, often with subluxation. A similar appearance may be seen in advanced rheumatoid or psoriatic arthritis.[54-56]

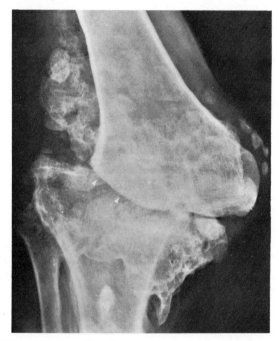

Figure 4–46 **Neuropathic (Charcot) Joint: The Knee.**   The knee joint is very much disorganized, with extensive fragmentation and debris. There has been resorption of the lateral femoral condyle, leaving a sharp eburnated margin *(arrowheads).* The bones are not demineralized. Subluxation of the tibia has occurred.

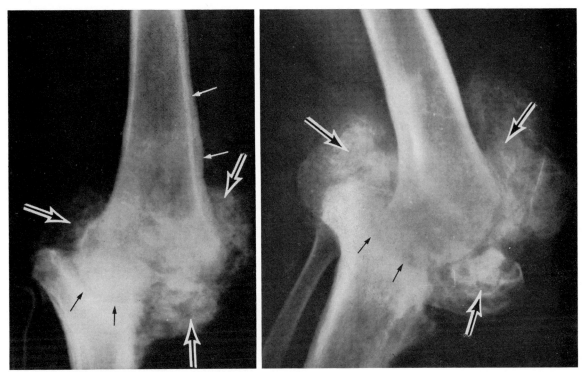

Figure 4–47 **Advanced Tabetic Charcot Joint: The Knee.** Anteroposterior and lateral views of the knee disclose anterior and medial subluxation of the femur. The medial two thirds of the articular end of the tibia has been extensively resorbed and now has a sclerotic, eburnated, and sharply marginated border *(black arrows)*.

Extensive collections of amorphous calcifications *(black-white arrows)* in the joint and surrounding soft tissues produce a bizarre but characteristic appearance. Periosteal new bone *(white arrows)* has developed along the lower femur. Note the absence of demineralization.

The resorbed, eburnated articular bone, the amorphous soft tissue calcification, and the absence of demineralization are characteristic features of the hypertrophic form of Charcot joint.

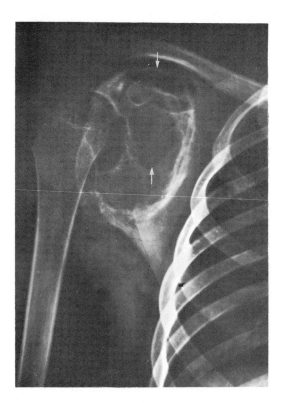

Figure 4–48 **Neuropathic Joint: Syringomyelia: The Shoulder.** There is extensive loss of scapular bone around the glenoid cavity *(arrows)*, with some sclerosis and new bone at the margins. There is subluxation of the humerus. Very little soft tissue debris or calcification is seen; this is the less florid "atrophic" form of Charcot joint.

Neurotropic changes secondary to syringomyelia may occur rather rapidly after the onset of clinically significant symptoms. (Courtesy Dr. Arlyne Shockman, Veterans Administration Hospital, Philadelphia, Pennsylvania.)

Figure 4–49 **Neuropathic Joint: Diabetic Neuropathy: The Foot.** The entire phalanx of the fourth toe, most of the proximal phalanx of the second toe, and the distal phalanx of the third toe have been destroyed. The end of the second metatarsal is pointed. The joint space is widened, and fragmentation is present.

The bone destruction and loss of bone substance suggest a neuropathy; diabetes is the most frequent cause of such changes in the foot. In diabetic vascular insufficiency without neuropathy, bone changes do not occur unless osteomyelitis develops. The sharp borders and pointed end of the resorbed metatarsal heads are characteristic features of neurotropic bone and joint disease.

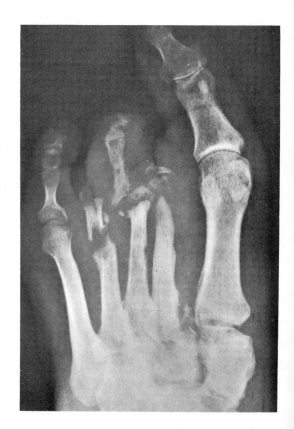

Multicentric Reticulohistiocytosis (Lipoid Dermatoarthritis)

This rare disorder of unknown origin occurs more often in females. Nodules in the skin and mucous membranes are associated with early and rapid disseminated joint involvement, most marked in the hands and feet.

Radiographically, a destructive and mutilating arthritis of the distal interphalangeal joints is common and characteristic. Resorption of the distal phalangeal tips is frequent. Periarticular erosions and resorption are also severe in the other involved joints.

Diagnosis requires biopsy of a skin lesion or synovium; histiocytes and giant cells containing PAS–positive material are diagnostic.[48, 57, 58]

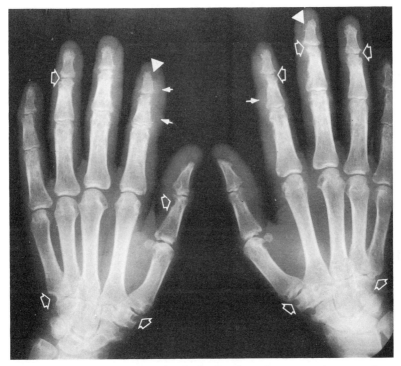

Figure 4–50 **Multicentric Reticulohistiocytosis: Both Hands.** A symmetric, severely erosive arthritis is involving all the distal interphalangeal joints and some of the carpometacarpal joints, with marked periarticular erosions *(open arrows)*. Periarticular demineralization is minimal. Soft tissue masses are apparent *(solid arrows)*. Erosive changes are seen in the distal phalangeal tufts *(arrowheads)*.

Soft tissue masses, a mutilating arthritis of the distal interphalangeal joints, and tuft erosions are the characteristic radiographic findings of this rare condition.

Chondrocalcinosis and Pseudogout

Chondrocalcinosis — calcification of the joint cartilages — can occur in a number of systemic conditions, including gout, hyperparathyroidism, ochronosis, hemachromatosis, hypervitaminosis D, diabetes, and degenerative joint disease. The cartilage calcification in these conditions may cause few or no symptoms.

In the idiopathic form, pseudogout, calcification can appear in the joint fibrocartilage or less often in hyaline cartilage. While usually bilateral, involvement may occasionally be unilateral. Pseudogout occurs in the elderly and is associated with attacks of acute joint effusion and pain, sometimes clinically simulating an acute septic arthritis. Calcium pyrophosphate crystals are always found in the synovial fluid. The knee is the most commonly affected joint; the calcifications are seen in the menisci and/or articular cartilages. The symphysis pubis, hips, wrists, elbows, or annulus fibrosus of the intervertebral discs may be involved. Often there are associated osteoarthritic changes in the affected synovial joints, frequently with discrete subchondral rarefactions.

A history of acute joint pains in an elderly individual with evidence of chondrocalcinosis in the knees and symphysis pubis is highly suggestive of pseudogout. Demonstration of calcium pyrophosphate crystals in the joint aspirate is necessary for definitive diagnosis. Asymptomatic calcification of joint cartilage is not necessarily pseudogout.[48, 59-65]

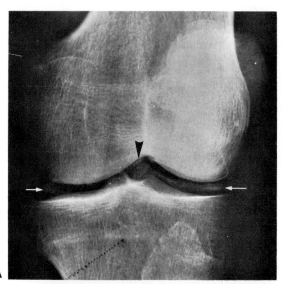

Figure 4–51 **Pseudogout: The Knee.**

A, There is calcification of both the medial and lateral menisci *(arrows),* and of the articular cartilage *(arrowhead).* Calcification was also present in the cartilage of the symphysis pubis and hip joints.

B, Enlarged view of the knee demonstrates the calcification in greater detail.

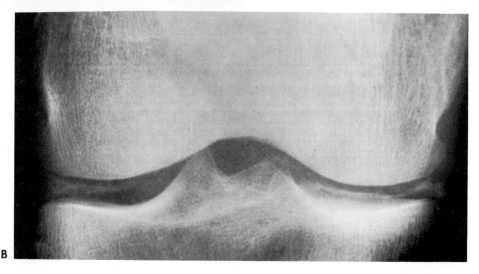

Villonodular Synovitis (Xanthomatous Giant Cell Tumor, Xanthogranuloma)

This condition occurs mostly in young adults and is characterized by single or multiple nodules of varying size within a single joint or in the surrounding soft tissues. The intra-articular form usually occurs within the knee joint, but the hip, shoulder, elbow, or ankle may be affected. The adjacent bones may be involved. Joint effusion often occurs.

Radiographically there may be soft tissue masses in the vicinity of the affected joint, often with evidence of effusion. The fat pads may be displaced. If the nodular nature cannot be discerned, the picture may be indistinguishable from a simple joint effusion. The synovial masses never calcify, a point of distinction from a synovioma. The articular cartilage is only rarely affected, and the joint space usually remains unchanged.

Periarticular demineralization does not occur unless prolonged immobility has resulted. In about one fifth of cases, circumscribed lytic areas with sclerotic margins but without periosteal reaction appear in the periarticular bones; these represent intraosseous extensions of the synovial masses.

More definitive diagnosis can be made by arthrography, preferably with air and contrast material (double contrast). The synovial cavity will be increased in size and the synovial masses will be demonstrated.

If the lesion arises from a tendon sheath (giant cell tumor of tendon), it may appear as a soft tissue mass unrelated to a joint.[69, 70]

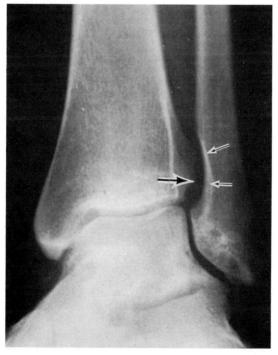

Figure 4–52 **Villonodular Synovitis: The Ankle**   A smooth erosive defect *(large arrow)* of the inner border of the fibula is delimited by cortex *(small arrows),* findings characteristic of an extrinsic pressure deformity.

A bone erosion in the vicinity of a joint is often the only positive roentgen finding in villonodular synovitis; nonspecific soft tissue swelling around the joint is also usually present. Soft tissue calcification does not occur in villonodular synovitis.

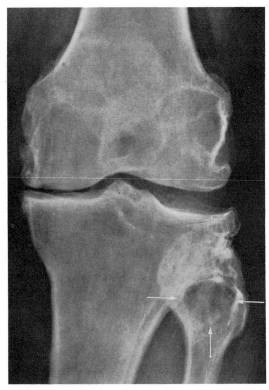

Figure 4–53 **Villonodular Synovitis of Knee.**   In the head of the fibula a cystic area with a slightly sclerotic margin *(arrows)* has resulted from growth of synovial nodules into the bone. The findings can simulate those of a primary bone lesion, especially if no joint swelling or soft tissue mass can be demonstrated. Spurring and joint space narrowing are due to unrelated osteoarthritis. (Courtesy Dr. Arlyne Shockman, Veterans Administration Hospital, Philadelphia, Pennsylvania.)

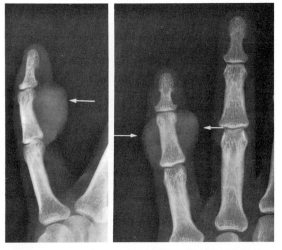

Figure 4–54 **Giant Cell Tumor of Tendon (Localized Villonodular Synovitis).**   There is a sharply demarcated, homogeneously dense, slightly lobulated mass in the volar aspect of the finger *(arrows)*. The joint space is preserved, and there is normal mineralization. Bone erosion has not occurred.

Giant cell tumors can arise from the tendon sheath in any part of the tendon. (Courtesy Dr. Arlyne Shockman, Veterans Administration Hospital, Philadelphia, Pennsylvania.)

Synovioma (Synovial Sarcoma)

This malignant tumor may arise from the joint capsule, bursa, or tendon. It occurs most frequently in the vicinity of a joint in the foot, knee, or elbow. Children and young adults are most commonly affected.

Radiographically, a fairly well-defined round or lobulated soft tissue mass is usually seen adjacent to or near a joint. The joint is almost never disturbed, and joint effusion does not occur.

Bone involvement is seen in about 25 per cent of cases. It usually occurs at some distance from the joint surface; the articular portion of the bone is rarely affected. Several types of bone changes are encountered. There may be periosteal proliferative reaction with or without erosive changes. Osteolytic erosive lesions sometimes cause progressive destruction of the bone shaft. Infrequently, a central area of osteolysis without cortical erosion is seen.

In 20 to 30 per cent of cases, amorphous deposits of calcium are seen within the mass; when present, they are an important diagnostic clue.

In villonodular synovitis (xanthomatous giant cell tumor), joint effusion is common, calcification does not occur, and erosive bone changes are usually closer to the articular surfaces and are not associated with periosteal reaction.[66-68]

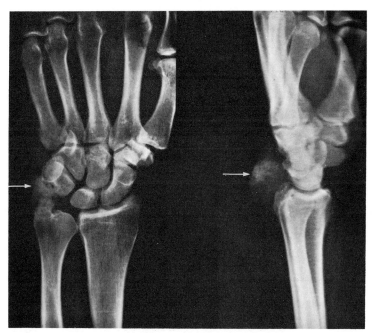

Figure 4–55 **Synovioma of Wrist: Calcification.** An irregular, poorly defined soft tissue mass in the medioposterior portion of the wrist contains punctate, ill-defined calcifications *(arrows)*. There is no bone involvement. At surgery a malignant synovioma was found.

In malignant synovioma, areas of bone destruction are often found at a distance from the joint space. Joint effusion does not occur, since the tumor does not extend into the joint.

DEGENERATIVE JOINT DISEASE (OSTEOARTHRITIS)

Degenerative Joint Disease (Osteoarthritis, Hypertrophic Arthritis)

This extremely common condition is part of the aging process but can also occur after repeated trauma to a joint. Degenerative joint changes may develop as sequelae of any other chronic joint disease.

The earliest roentgenologic changes are sharpening and thickening of the articular margins of the bones. Later, thinning of the articular cartilage causes narrowing of the joint space. These changes are most marked at points of stress or weight bearing. Marginal spurs or osteophytes subsequently develop. As thinning and erosion of the articular cartilage continue, the joint narrows irregularly, and sclerosis appears at the articular margins. The irregularity of the joint narrowing is in contrast to the smooth, even narrowing of rheumatoid arthritis.

In advanced disease, cystic rarefactions in the bone, periarticular and intra-articular calcifications, and even subluxation can appear. Local osteoporosis does not occur unless there is prolonged disuse of the joint.

Uneven joint space narrowing, marginal spurring, and periarticular sclerosis are the basic radiographic changes of hypertrophic arthritis.

The peripheral joints most often affected are those of the hands, knees, hips, and shoulders; these are discussed separately in the following pages. Spine involvement, which is extremely frequent, is considered under *Spondylosis* (p. 98).[29, 48, 71, 72]

Degenerative Joint Disease of the Knee

Spurring of the posterior aspect of the patella and increased prominence of the tibial spines are often early findings. The articular cortex becomes thickened, and additional spurring may develop on the tibial plateaus and intercondylar notches. There is subchondral osteosclerosis and concomitant narrowing of the joint space. In most instances this is most marked on the medial aspect, the area of maximum weight bearing. Cystic changes are infrequent. Loose bony or calcific fragments are frequently seen within the joint space. Occasionally, small intra-articular osteochondromas develop. These appear as irregularly calcified rounded densities that arise from and usually remain attached to the synovial villi. These villous osteochondromas can occur in an otherwise normal appearing joint.[29, 48, 71]

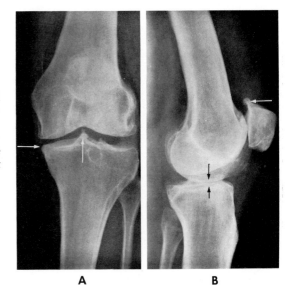

Figure 4–56 **Degenerative Joint Disease: The Knee.**

A, Small hypertrophic spurs are seen on the tibial spine and the medial edge of the tibia *(arrows)*. The medial joint space is narrowed.

B, A prominent bone spur arises from the posterosuperior surface of the patella *(white arrow)*. The joint space narrowing is apparent *(between black arrows)*.

The bones are not demineralized.

Degenerative Joint Disease of the Shoulder

Degenerative changes are rare before the fourth decade unless trauma or dislocation has occurred.

Sclerosis and thickening of the rim of the glenoid fossa are the most frequent findings. Commonly, there is little progression, and clinical symptoms are minimal or absent. In the infrequent advanced case, marginal spurring and narrowing of the joint space may occur.

Other findings include cystic changes of the tuberosity of the humerus and erosions of the humeral tuberosity and acromion, as well as spurring and sclerosis in the acromioclavicular joint.[29, 48, 71, 73]

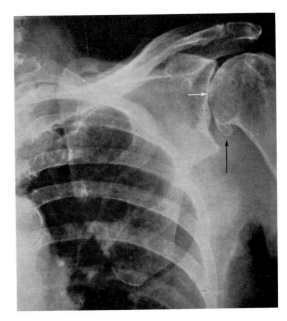

Figure 4–57 **Degenerative Joint Disease: The Shoulder.** There is sclerosis of the glenoid rim of the left shoulder; the joint space is irregularly narrowed *(white arrow),* and there is a large spur at the inferior articular surface of the humeral head *(black arrow).* The right shoulder was similarly involved.

In an older person with shoulder symptoms, marginal sclerosis frequently is the only roentgenologic finding. Extensive degenerative changes confined to one shoulder are sometimes due to earlier injuries to the rotator (musculotendinous) cuff.

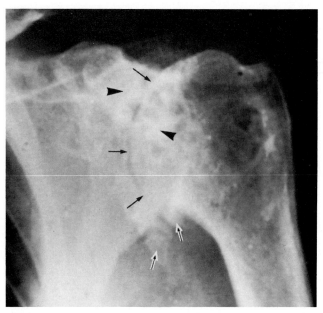

Figure 4–58 **Advanced Degenerative Joint Disease: The Shoulder.** The joint space is irregularly narrowed *(black arrows)*, and there are sclerotic changes on both articular margins *(arrowheads)*. Large spurs are seen at the inferior articular margins of the humerus and glenoid cavity *(black-white arrows)*.

The cyst-like appearance of the humeral head is mainly due to areas of nonsclerotic bone that are accentuated by adjacent sclerosis, but subarticular cystic areas are commonly found in osteoarthritis.

Degenerative Joint Disease of the Hand

In the hand, the distal interphalangeal joints are most often involved, though the proximal interphalangeal and metacarpophalangeal joints may also be affected. Early in the course of disease there are small marginal spurs or flakes at the base of the distal phalanges. As these spurs enlarge they cause bony protuberances—Heberden's nodes. The larger spurs are found on the dorsal aspect of the bone. The joint spaces are narrowed, and the articular ends become irregular. Cystic rarefactions with thin sclerotic margins may occur, but articular erosions like those seen in rheumatoid arthritis are not found. Sclerosis of the subchondral bone is an important feature. Demineralization, soft tissue atrophy, and subluxation are infrequent. This condition is seen more often in females, and there appears to be a hereditary predisposition.[29, 48, 72]

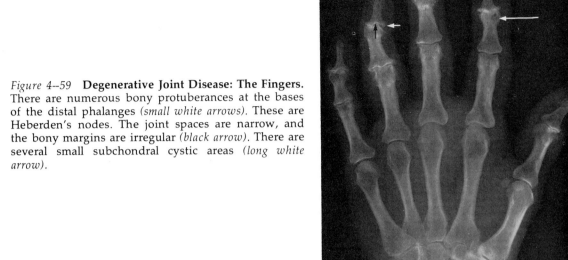

Figure 4–59 **Degenerative Joint Disease: The Fingers.** There are numerous bony protuberances at the bases of the distal phalanges *(small white arrows)*. These are Heberden's nodes. The joint spaces are narrow, and the bony margins are irregular *(black arrow)*. There are several small subchondral cystic areas *(long white arrow)*.

Degenerative Joint Disease of the Hip

Spurring, increased density of the acetabular rim, and narrowing of the joint space are the early findings. The superior and lateral portion of the joint—the point of maximum weight bearing—is affected initially. Reactive sclerosis appears in the subchondral bone on both sides of the joint, and multiple cysts frequently develop, especially in the roof of the acetabulum.

In advanced disease the femoral head becomes deformed, and there are areas of lysis and sclerosis. The joint space is greatly reduced and irregular, and may be obliterated. This stage is often termed *malum coxae senilis*.[29, 48]

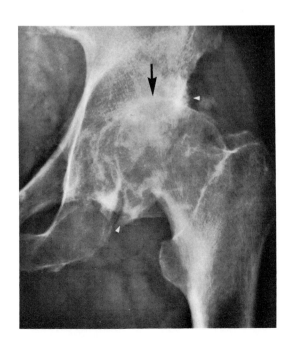

Figure 4–60 **Degenerative Joint Disease: The Hip.** The joint space is quite narrow and has nearly been obliterated at its superolateral aspect *(arrow)*; the bony density in this area is increased, and spur formation has occurred *(arrowheads)*. There is bony debris immediately lateral to the acetabular rim. Cystic changes are not a prominent feature in this case.

Spondylosis

Spondylosis is characterized by spur formation and areas of marginal sclerosis of the vertebral bodies. A narrowed intervertebral space from local disc degeneration may or may not be associated. These changes occur with advancing age and from local stress; any portion of the spine may be affected. Spondylosis is not an osteoarthritis, since the disc spaces are not true joints. The apophyseal and costovertebral articulations are the only true spinal joints.

The spurs or osteophytes can occur at any margin of stress—anteriorly, laterally, or posteriorly. A spur can become quite large and may fuse with another spur from an adjacent body. Sclerosis of the vertebral margins is also due to abnormal stress. These findings are usually asymptomatic.

Degenerative thinning of an intervertebral disc is often responsible for secondary spondylitic changes.

Spondylosis is extremely common in the cervical spine and most often involves the lower bodies (C5, C6, C7). Anterior and posterior spurs appear; the latter may project into a neural intervertebral foramen and produce symptoms of nerve root pressure. If a narrowed interspace is associated, myelography may be necessary to rule out disc herniation. Occasionally a large posterior spur can impinge upon the spinal canal and produce cord symptoms, especially if the cervical bony canal is narrow (less than 12 mm.). Frontal myelographic films may show a central defect simulating an intramedullary lesion, but on lateral view the extradural spur defect is readily apparent.

True osteoarthritis of the spine can develop in the apophyseal joints. Oblique views will demonstrate eburnated irregular articular surfaces and sharpening of the apex of the articular facet.[29, 48]

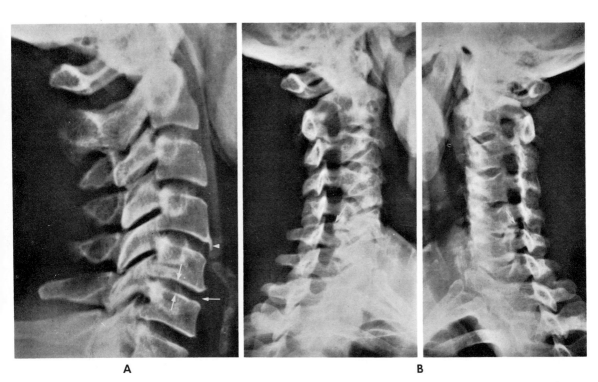

A   B

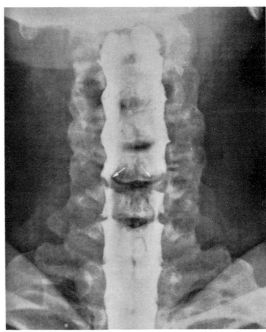

C

Figure 4–61 **Spondylosis: Cervical Spine.**

A, There is minimal narrowing of the intervertebral disc space between C5 and C6 *(arrows)* and minimal anterior spurring on the inferior aspect of C5. The anterior longitudinal ligament on the inferior aspect of C4 is calcified *(arrowhead)*. The patient had neurologic symptoms.

B, Oblique projection discloses bilateral spurs at the level of C5–C6; the spurs impinge on the intervertebral foramina *(arrows)*.

C, Anteroposterior view of myelogram demonstrates a central defect at C5–C6 *(arrows)* simulating an intramedullary lesion. Lateral view *(not shown)* revealed that the defect was due to large posterior spurs.

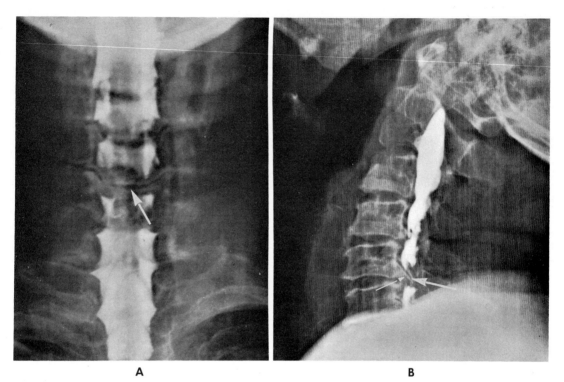

A   B

Figure 4–62 **Cervical Spondylosis: Posterior Spur: Neurologic Symptoms.**

A, An elderly male complained of neck and arm pain and had neurologic symptoms in the upper extremities. Anteroposterior view of cervical myelogram discloses a large central defect *(arrow)* at C5–C6.

B, In lateral view made with patient prone (cross-table lateral view), the defect in the oil column *(large arrow)* is caused by pressure from a large bone spur *(small arrow)* arising in the posteroinferior border of C5.

Although there are hypertrophic changes in the lower cervical spine, the intervertebral space between C5 and C6 is not significantly narrowed. In this case and the one illustrated in Figure 4–61, anteroposterior myelograms demonstrated central defects in the oil column that might have been misinterpreted as an intramedullary mass if a lateral film had not been made.

THE PAINFUL SHOULDER

Calcific Tendinitis

Calcification of a tendon sheath can occur almost anywhere in the body, but it is most common in the shoulder; the elbow, wrist, and hip are also frequent sites. When the tendon is surrounded by a bursa, the condition is often termed *calcific bursitis;* the two terms are used interchangeably.

The supraspinatus tendon near the greater tuberosity of the humerus is most often involved. There may be a few minute flecks of calcium in the tendon sheath, or large masses of calcium that fill the bursa. Calcification is noted in up to 50 per cent of patients with persistent or recurrent shoulder pain, but it is also seen in asymptomatic persons.

Fine, amorphous calcium sometimes deposited during an acute attack of shoulder tendinitis may disappear with recovery; more commonly the calcification persists, becoming more extensive and dense with additional acute attacks or during long asymptomatic periods. The deposits may ossify. If motion is severely limited, osteoporosis of disuse may develop.

Calcification may develop in other tendons of the shoulder and in the elbow, hip, and wrist. It is emphasized that clinical symptoms can occur without demonstrable calcification and, conversely, that asymptomatic tendon calcification can exist.[29, 74]

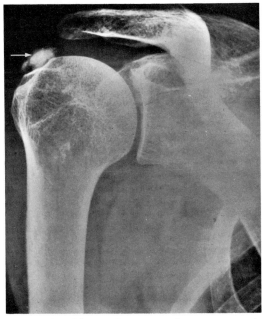

Figure 4–63 **Calcific Tendinitis of Shoulder.** There is a dense opacity *(arrow)* in the soft tissues just above the greater tuberosity. This calcification lies within the supraspinatus tendon. The patient had clinical symptoms of chronic tendinitis.

Rotator Cuff Tears

Rupture of the rotator muscle cuff of the shoulder causes a communication between the shoulder (glenohumeral) joint and the subacromial bursa. Arthrography will show contrast medium filling the bursa, a virtually pathognomonic finding of rotator cuff tear.

In chronic tears, shoulder radiographs may show changes in the acromion; its distance from the humeral head is decreased and its inferior convexity may be reversed. Subcortical cysts may develop in the acromion.

However, the arthrographic changes are the definitive findings of rotator cuff tear.[75, 76]

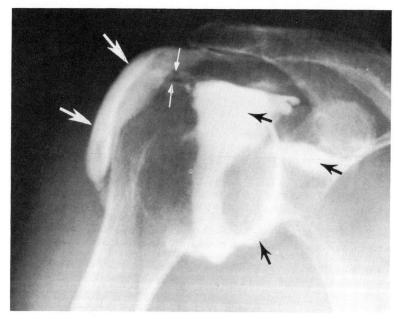

Figure 4–64 **Rotator Cuff Tear: Arthrogram.** The injected contrast material has extended from the shoulder joint spaces *(black arrows)* into the subacromion bursa *(large white arrows)*. The entry tract is also opacified *(small white arrows)*.

Opacification of the subacromial bursa during a shoulder arthrogram is virtually pathognomonic of a rotator cuff tear.

The Shoulder-Hand Syndrome (Reflex Neurovascular Dystrophy)

The roentgenographic changes in this neurovascular disturbance are nonspecific. Usually there is spotty osteoporosis that generally begins in the head of the humerus and in the wrist. In addition, soft tissue swelling on the dorsum of the wrist and hand may be seen. With progression, there is diffuse demineralization of the shoulder, wrist, and hands; contractures ultimately may develop.[29]

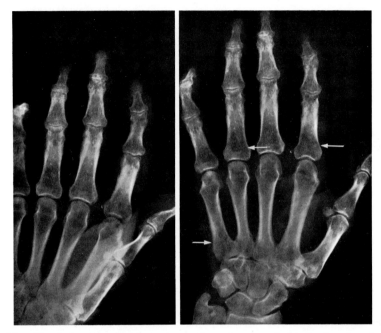

Figure 4–65 **Shoulder-Hand Syndrome.** There is spotty osteoporosis in the hand and wrist *(arrows)*. Soft tissue swelling is not demonstrated in this film. The radiographic picture is nonspecific, but it may aid in diagnosis when clinical symptoms are present.

NONARTICULAR RHEUMATISM

Tietze's Syndrome (Costal Chondritis)

This idiopathic painful swelling most frequently involves the costochondrosternal junction of the second anterior rib or less often the sternoclavicular joint. The routine chest film is most often negative, since a cartilaginous lesion does not cast a shadow on the roentgenogram. However, a soft tissue mass is usually present early and may sometimes be demonstrated on films taken tangential to the area of involvement or by tomography. The affected cartilage may eventually calcify and a periosteal reaction may develop, causing an increase in the size and density of the affected rib. The inner clavicular end may be deformed or hypertrophied if the sternoclavicular joint is involved.[77, 78]

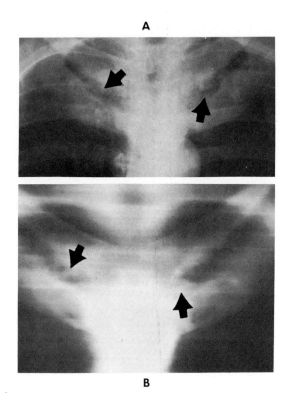

Figure 4–66 **Tietze's Syndrome.**

A, In posteroanterior view there is increased density and irregularity of the anterior end of the right first rib, and of the manubrium on the left *(arrows).*

B, Planigraphic examination better delineates the sclerosis and irregularity of the rib end and of the manubrium *(arrows).* (Courtesy Dr. I. Meschan, Winston-Salem, North Carolina.)

REFERENCES

1. Lloyd-Roberts, G. C.: Suppurative arthritis of infancy; some observations upon prognosis and management. J. Bone Joint Surg., *42B*:706, 1960.
2. Argen, S. T., et al.: Suppurative arthritis. Clinical features of 42 cases. Arch. Intern. Med., *17*:666, 1966.
3. Shawker, T. H., and Dennis, J. M.: Periarticular calcification in pyogenic arthritis. Am. J. Roentgenol. Radium Ther. Nucl. Med., *113*:650, 1971.
4. Caffey, J.: Tuberculosis of Joints. *Pediatric X-ray Diagnosis*. 6th ed. Chicago, Year Book Medical Publishers Inc., 1972, p. 1336.
5. Winston, J. M., and Hewson, S. S.: Early roentgen diagnosis of tuberculosis of the hip in children. Radiology, *79*:241, 1962.
6. Berens, D. L., Locke, L. M., Len, R., and Norcross, B. M.: Roentgen changes in early rheumatoid arthritis. Radiology, *82*:645, 1964.
7. Martel, W.: The pattern of rheumatoid arthritis in the hand and wrist. Radiol. Clin. North Am., *2*:221, 1964.
8. Fletcher, D. E., and Rowley, K. A.: The radiological features of rheumatoid arthritis. Br. J. Radiol., *25*:282, 1952.
9. Mason, R. M., Murray, R. S., Oates, J. K., and Young, A. C.: A comparative radiological study of Reiter's disease, rheumatoid arthritis and ankylosing spondylitis. J. Bone Joint Surg., *41B*:137, 1959.
10. Pratt, T. L.: Spontaneous dislocation of atlanto-axial articulation occurring in ankylosing spondylitis and rheumatoid arthritis. J. Fac. Radiol., *10*:40, 1959.
11. Hilbish, T. F., and Black, R. L.: X-ray manifestations of peptic ulceration during corticosteroid therapy for rheumatoid arthritis. Arch. Intern. Med., *101*:932, 1958.
12. Locke, B. G.: Rheumatoid lung. Clin. Radiol., *14*:43, 1963.
13. Sieniewicz, J. D., Martin, J. R., Moore, S., and Miller, A.: Rheumatoid nodules in the lung. J. Can. Assoc. Radiol., *13*:73, 1962.
14. Bland, J. N., et al.: A study of roentgenologic criteria for rheumatoid arthritis of the cervical spine. Am. J. Roentgenol. Radium Ther. Nucl. Med., *95*:949, 1965.
15. Lillington, G. A., Carr, D. T., and Mayne, J. G.: Rheumatoid pleurisy with effusion. Arch. Intern. Med. *128*:764, 1971.
16. Hall, A. P., and Scott, J. T.: Synovial cysts and rupture of the knee joint in rheumatoid arthritis. Ann. Rheum. Dis. *25*:32, 1966.
17. Silbiger, M. L., and Peterson, C. C., Jr.: Sjögren's syndrome. Am. J. Roentgenol., Radium Ther. Nucl. Med., *100*:554, 1967.
18. Bloch, K. J., Buchanan, W. W., et al.: Sjögren's syndrome. Medicine, *44*:187, 1965.
19. Castleman, B.: Case records of Massachusetts General Hospital. N. Engl. J. Med., *286*:992, 1972.
20. Whaley, K., Blair, S., et al.: Sialographic abnormalities in Sjögren's syndrome and other arthritides. Clin. Radiol., *23*:474, 1972.
21. Sharp, J.: Differential diagnosis of ankylosing spondylitis. Br. Med. J., *1*:975, 1957.
22. Rosenkranz, W.: Ankylosing spondylitis: Cauda equina syndrome with multiple spinal arachnoid cysts. J. Neurosurg. *34*:241, 1971.
23. Russell, M. L., Gordon, D. A., et al.: The cauda equina syndrome of ankylosing spondylitis. Ann. Intern. Med. *78*:551, 1973.
24. Resnick, D.: Patterns of peripheral joint disease in ankylosing spondylitis. Radiology, *110*:523, 1974.
25. Resnick, D.: Temporomandibular joint involvement in ankylosing spondylitis. Radiology, *112*:587, 1974.
26. Martel, W., Holt, J., and Cassidy, J. T.: Roentgenologic manifestations of juvenile rheumatoid arthritis. Am. J. Roentgenol. Radium Ther. Nucl. Med., *88*:420, 1964.
27. Ansell, B. M., and Bywaters, E. G.: Rheumatoid arthritis (Still's disease). Pediatr. Clin. North Am., *10*:921, 1963.
28. Barbaric, Z. L., and Young, L. W.: Synovial cysts in juvenile rheumatoid arthritis. Am. J. Roentgenol. Radium Ther. Nucl. Med., *116*:655, 1972.
29. Hollander, J. L., et al.: *Arthritis and Allied Conditions: Textbook of Rheumatology*. Philadelphia, Lea & Febiger, 1966.
30. Pugh, D. G., Slocumb, C. H., and Winkelmann, R. K.: Psoriatic arthritis: a roentgenographic study. Radiology, *75*:691, 1961.
31. Killebrew, K., Gold, R. H., and Sholkoff, S. D.: Psoriatic spondylitis. Radiology, *108*:9, 1973.
32. Weldon, W. V., and Scalettar, R. S.: Roentgen changes in Reiter's syndrome. Am. J. Roentgenol. Radium Ther. Nucl. Med., *86*:344, 1961.
33. Murray, R. S., Oates, J. K., and Young, A. C.: Reiter's syndrome. Clin. Radiol., *9*:37, 1958.
34. Sholkoff, S. D., et al.: Roentgenology of Reiter's syndrome. Radiology *97*:497, 1970.
35. Cliff, J. M.: Spinal bony bridging and carditis in Reiter's disease. Ann. Rheum. Dis., *30*:171, 1971.
36. Kantor, M., and Morrow, C. S.: Caplan's syndrome: a perplexing pneumoconiosis with rheumatoid arthritis. Am. Rev. Tuberc., *78*:274, 1958.
37. Davies, D., and Lindars, D. C.: Rheumatoid pneumoconiosis: A clinical study. Am. Rev. Resp. Dis., *97*:617, 1968.
38. Edling, N. P. G., Ohlson, L., and Swensson, A.: Rheumatoid pneumoconiosis (Caplan's disease). Acta Radiol. *8*:168, 1969.

39. Bywaters, E. G. L., and Ansell, B. M.: Arthritis associated with ulcerative colitis. A clinical and pathologic study. Ann. Rheum. Dis., *17*:169, 1958.

40. Zvaifler, N. J., and Martel, W.: Spondylitis in chronic ulcerative colitis. Arthritis Rheum., *3*:76, 1967.

41. Clark, R. L., Muhletaler, C. A., and Margulies, S. I.: Colitic arthritis. Radiology, *101*:585, 1971.

42. Mueller, C. E., Seeger, J. F., and Martel, W.: Ankylosing spondylitis and regional enteritis. Radiology, *112*:579, 1974.

43. Horns, J. W., and O'Loughlin, B. J.: Tracheal collapse in polychondritis. Am. J. Roentgenol. Radium Ther. Nucl. Med., *87*:844, 1962.

44. Dolan, D. L., et al.: Relapsing polychondritis. Analytical literature review and studies on pathogenesis. Am. J. Med., *41*:285, 1966.

45. Johnson, T. H., Jr., Mital, N., et al.: Relapsing polychondritis. Radiology, *106*:313, 1973.

46. Relapsing polychondritis. Br. Med. J., *2*:627, 1973.

47. Steinbach, H. L., Feldman, R., and Goldberg, M. B.: Acromegaly. Radiology, *72*:535, 1959.

48. Forrester, D. M., and Nesson, J. W.: *The Radiology of Joint Disease.* Philadelphia, W. B. Saunders Co., 1973.

49. Stuber, J. L., and Palacios, E.: Vertebral scalloping in acromegaly. Am. J. Roentgenol. Radium Ther. Nucl. Med., *112*:397, 1971.

50. Stiris, G.: Bone and joint changes in haemophiliacs. Acta Radiol., *49*:269, 1958.

51. Boldero, J. L., and Kemp, H. S.: The early bone and joint changes in haemophilia and similar blood dyscrasias. Br. J. Radiol., *39*:172, 1966.

52. Salerno, N. R., Menges, J. B., and Borns, P. F.: Arthrograms in hemophilia. Radiology, *102*:135, 1972.

53. Simon, G., and Zorab, P. A.: Radiographic changes in alkaptonuric arthritis; report on three cases. Br. J. Radiol., *34*:384, 1961.

54. Jacobs, J. E.: Observations of neuropathic (Charcot) joints occurring in diabetes mellitus. J. Bone Joint Surg., *40A*:1043, 1958.

55. Banna, M., and Foster, J. F.: Roentgenologic features of acrodystrophic neuropathy. Am. J. Roentgenol. Radium Ther. Nucl. Med., *115*:186, 1972.

56. Clouse, M. E., et al.: Diabetic osteoarthropathy. Am. J. Roentgenol. Radium Ther. Nucl. Med., *121*:22, 1974.

57. Buchel, E.: Multicentric reticulohistiocytosis. S. Afr. Med. J., *44*:1434, 1970.

58. Brodey, P. A.: Multicentric reticulohistiocytosis. Radiology, *114*:327, 1975.

59. McCarthy, D. J., Jr., and Haskin, M. E.: Roentgenographic aspects of pseudogout (chondro-calcinosis). Am. J. Roentgenol. Radium Ther. Nucl. Med., *90*:1248, 1963.

60. Skinner, M., and Cohen, A. S.: Calcium pyrophosphate dihydrate crystal disposition disease. Arch. Intern. Med., *123*:636, 1969.

61. Dodds, W. J., and Steinbach, H. L.: Gout associated with calcification of cartilage. N. Engl. J. Med., *275*:745, 1966.

62. Angevine, C. D., and Jacox, R. F.: Pseudogout in the elderly. Arch. Intern. Med. *131*:693, 1973.

63. Martel, W., Champion, C. K., et al.: A roentgenologically distinctive arthropathy in some patients with the pseudogout syndrome. Am. J. Roentgenol. Radium Ther. Nucl. Med., *109*:587, 1970.

64. Atkins, C. J., McIvor, J., et al.: Chondrocalcinosis and arthropathy: Studies in hemochromatosis and idiopathic chondrocalcinosis. Q. J. Med., *39*:71, 1970.

65. McIvor, J.: Idiopathic chondrocalcinosis. Clin. Radiol., *22*:370, 1971.

66. Raben, M., Calabrese, A., Higenbotham, N. L., and Phillips, R.: Malignant synovioma. Am. J. Roentgenol. Radium Ther. Nucl. Med., *93*:145, 1965.

67. Strickland, B., and Mackenzie, D. H.: Bone involvement in synovial sarcoma. Clin. Radiol., *10*:64, 1959.

68. Horowitz, A. L., Resnick, D., and Watson, R. C.: The roentgen features of synovial sarcomas. Clin. Radiol., *24*:481, 1973.

69. Breimer, C. W., and Freiberger, R. H.: Bone lesions associated with villonodular synovitis. Am. J. Roentgenol. Radium Ther. Nucl. Med., *79*:618, 1958.

70. Wolfe, R. D., and Guiliano, V. J.: Double contrast arthropathy in the diagnosis of pigmented villonodular synovitis. Am. J. Roentgenol. Radium Ther. Nucl. Med., *110*:793, 1970.

71. Traut, E. F.: Symposium on medical problems of aged. Degenerative arthritis: its causes, recognition and management. Med. Clin. North Am., *40*:63, 1956.

72. Kellgren, J. H., and Moore, R.: Generalized osteoarthritis and Heberden's nodes. Br. Med. J., *1*:181, 1952.

73. Kernwein, G. A.: Roentgenographic diagnosis of shoulder dysfunction. J.A.M.A., *194*:1081, 1965.

74. Hitchcock, E. R., and Langton, L.: Peritendinitis calcarea with special reference to the hand. J. Fac. Radiol., *10*:86, 1959.

75. Killoran, P. J., Marcove, R. C., and Freiberger, R. H.: Shoulder arthrography. Am. J. Roentgenol. Radium Ther. Nucl. Med., *103*:658, 1968.

76. Kotzen, L. M.: Roentgen diagnosis of rotator cuff tear. Am. J. Roentgenol. Radium Ther. Nucl. Med., *112*:507, 1971.

77. Skorneck, A. B.: Roentgen aspects of Tietze's syndrome; painful hypertrophy of costal cartilage and bone osteochondritis. Am. J. Roentgenol. Radium Ther. Nucl. Med., *83*:748, 1960.

78. Cardona, P., Biolcati, A. R., and Vita, G.: Gli aspetti radiologici della sindrome di Tietze. Ann. Radiol. Diagn. (Bologna), *43*:3, 1970.

SECTION

5

GRANULOMATOUS DISEASES OF UNPROVED ETIOLOGY

Sarcoidosis

Although predominantly found in adults, sarcoidosis can also occur in older children. The earliest and most frequent radiographic findings are in the chest. Intrathoracic adenopathy or parenchymal infiltrates, or both, occur in almost every case sometime during the disease.

Bilateral hilar enlargement from adenopathy is the commonest single finding. Enlarged mediastinal nodes, especially the right paratracheal group, are usually associated, but they rarely cause tracheal deviation. Unilateral hilar involvement is unusual.

Pulmonary infiltrates appear in the majority of cases. In about 80 per cent of patients with pulmonary lesions, hilar and mediastinal adenopathy are associated, and this is a helpful finding. A variety of pulmonary patterns can occur, including a fine reticulation, small miliary nodules simulating miliary tuberculosis, larger nodules, and confluent interstitial and alveolar densities. Bilateral involvement is most frequent. Rarely, cavity formation can occur; superimposed mycetoma may develop. Pleural involvement with effusion is extremely rare (one per cent), and its presence makes the diagnosis of pulmonary sarcoidosis questionable.

These pulmonary lesions may regress spontaneously or after steroid therapy. Late irreversible fibrosis occurs in about 10 per cent of cases and is characterized by nonspecific coarse and fine fibrotic strands with interspersed areas of emphysema. Cor pulmonale may develop.

Bone lesions are found in a small percentage of cases (5 to 15 per cent), usually in the phalanges of the hands or feet; they appear as punched-out lytic areas or as a lace-like coarsely trabeculated pattern. Cortical expansion and thinning may occur. Rarer instances of lytic sarcoid lesions in the long bones, skull, or vertebrae have been reported. Rarely, osteosclerotic lesions appear in the axial skeleton.

In less than 5 per cent of cases, sarcoid infiltration may enlarge one or more salivary glands, usually the parotid. Sialography will show strictured ducts and ectasia (dilated saccules) in the parenchyma.

Liver and spleen involvement is found histologically in half the cases, but splenomegaly or hepatomegaly is infrequent in adult sarcoidosis. Other infrequent findings are nephrocalcinosis secondary to the hypercalcemia, and congestive failure resulting from myocardial involvement.

A transient polyarthritis is not uncommon; periarticular soft tissue swelling is the usual finding.[1-7]

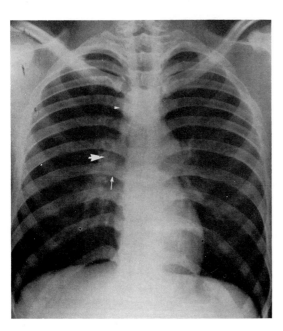

Figure 5–1 **Sarcoidosis: Adenopathy Without Pulmonary Lesions.** There is bilateral hilar adenopathy with greater enlargement of the right hilar shadow *(large arrow)*. The right paratracheal nodes are enlarged, producing a mediastinal bulge *(arrowhead)*. The lung fields are clear. Note that the "clear space" is preserved *(small arrow)* between the enlarged hilum and the heart shadow. This is frequently seen in hilar adenopathy and may help distinguish it from mediastinal adenopathy, which usually obliterates the "clear space."

Figure 5–2 **Sarcoidosis: Adenopathy with Minimal Pulmonary Lesions.** There is marked hilar enlargement bilaterally, and paratracheal and superior mediastinal adenopathy *(arrowhead)*. Ill-defined infiltrative densities extend from the lower hilar areas on both sides *(arrows)*.

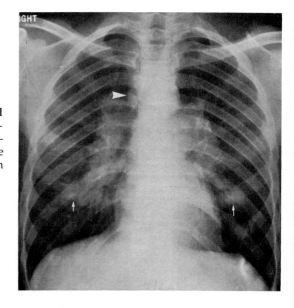

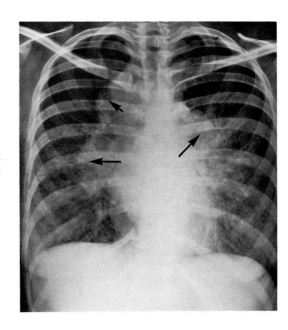

Figure 5–3 **Sarcoidosis: Massive Adenopathy with Miliary Pulmonary Lesions.** Massive enlargement of the hilar and right paratracheal nodes *(arrows)* produces a striking picture. The middle and lower lung fields are diffusely involved by miliary densities, whereas the upper lung fields are free of lesions.

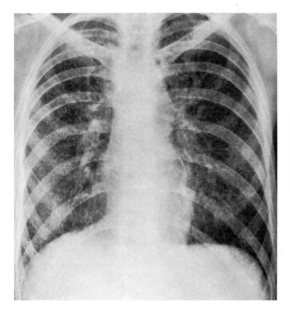

Figure 5–4 **Sarcoidosis: Diffuse Involvement Simulating Miliary Tuberculosis.** There are miliary and reticular densities uniformly scattered throughout both lung fields. No confluent lesions are evident. The hilar shadows are not enlarged, but some nodes slightly widen the superior mediastinum. Chest films in miliary tuberculosis are nearly identical.

Figure 5–5 **Sarcoidosis: Diffuse Miliary and Nodular Parenchymal Disease Without Significant Adenopathy.** Both lung fields are extensively involved by nodular densities of various sizes, many of which are ill defined. The picture resembles diffuse fibrosis. Although no hilar adenopathy is seen, there is fullness of the right upper mediastinum, probably from paratracheal adenopathy. Only about one fifth of patients with parenchymal pulmonary sarcoid lesions do not have associated hilar adenopathy.

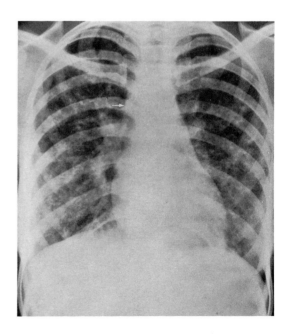

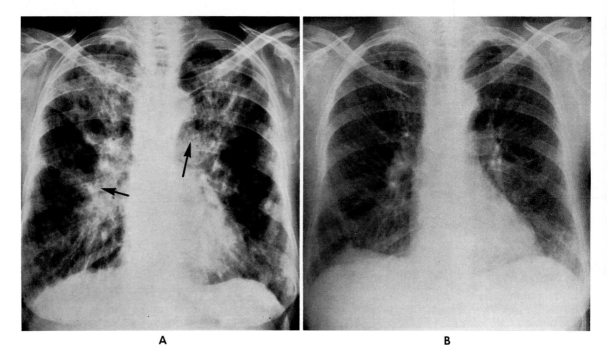

A   B

Figure 5–6 **Sarcoidosis: Extensive Pulmonary Infiltration and Hilar Adenopathy, with Recovery Following Steroid Therapy.**

A, There is extensive infiltration of both upper and lower lobes, and less involvement of the midlung fields. Hilar adenopathy *(arrows)* is pronounced, but there are no paratracheal nodes. The picture strongly resembles tuberculosis except for the hilar adenopathy.

B, Following six months of steroid therapy there is resolution of most of the infiltrates. Some apical strands remain. The hilar shadows have returned to normal size.

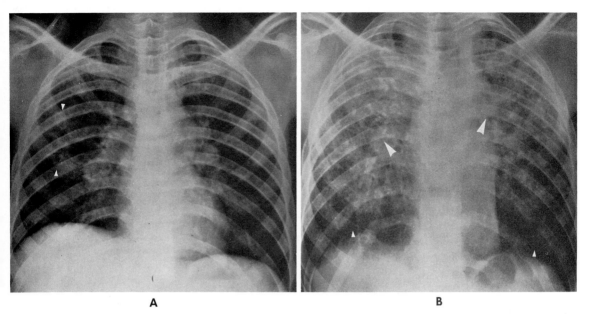

A   B

Figure 5-7 **Sarcoidosis: Progression to Irreversible Fibrosis.**

A, This patient, age 25, manifests the typical appearance: enlarged hilar nodes, enlarged right para-tracheal nodes, and localized ill-defined infiltrations in the right midlung field *(arrowheads).*

B, Three years later there is extensive fibrosis in the upper two thirds of both lung fields, with elevation of both hilar shadows *(large arrowheads).* Hilar and mediastinal adenopathy is obscured by the dense fibrosis. There is marked bilateral basal emphysema *(small arrowheads)* and numerous small areas of emphysema between the fibrotic strands. The enlarged nodes help differentiate sarcoidosis from advanced tuberculous fibrosis. Figures 5-1 to 5-7 are representative of chest sarcoidosis. All cases except the last, in which there is advanced fibrosis, are reversible either spontaneously or following steroid therapy. Note the absence of pleural reaction; had there been pleural effusion, the diagnosis of uncomplicated sarcoid could be questioned.

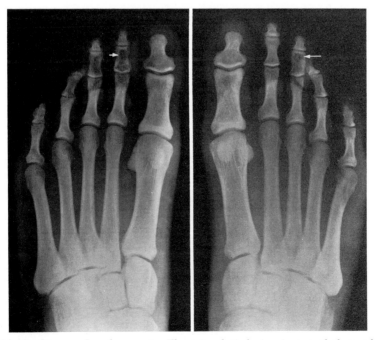

Figure 5-8 **Sarcoidosis: Osseous Involvement.**   There is a lytic lesion in one phalanx of each foot *(arrows).* The lytic area is fairly sharply delineated, and the residual trabeculae have a lace-like appearance.

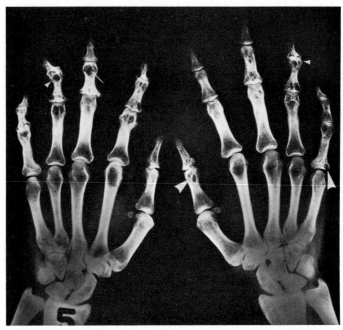

Figure 5–9 **Sarcoidosis: Multiple Bone Lesions in Hands.** All the lesions are in the peripheral portions of the phalanges. Some are of the sharp, punched-out variety *(large arrowheads),* but most have a lace-like appearance. Cortical thinning and expansion *(arrow)* is less common. Joint deformities *(small arrowheads)* can occur when articular margins are involved. Bone lesions occur in less than 10 per cent of cases; swelling and pain generally are associated.

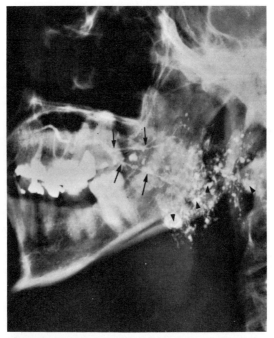

Figure 5–10 **Sarcoidosis: Sialogram in Parotid Gland Infiltration.** The normal tree-like pattern with fine terminal branches of the intraparotid ductile system has been effaced by sarcoid infiltration that has obstructed and destroyed the smaller ducts, and caused ectasia. This is seen as multiple saccular collections of opaque medium throughout the gland *(arrowheads).* Note irregularity of the principal intraglandular ducts *(arrows).* The entire gland is enlarged.

Cranial (Temporal) Arteritis

Temporal arteritis is a localized manifestation of cranial arteritis, which, in turn, is part of a generalized systemic vascular disturbance (giant cell arteritis). It occurs in older adults and appears to be closely related to polymyalgia rheumatica (senile arteritis).

Opacification of the temporal artery will demonstrate focal areas of irregularity, narrowing, and dilatation. Although these changes may mimic atherosclerotic involvement, the temporal artery in older individuals may be tortuous but virtually never shows atherosclerotic changes.

The temporal artery may be opacified by direct injection or during a cerebral (carotid) angiogram in most individuals. Sometimes similar changes are found in some of the intracranial arteries.[8-10]

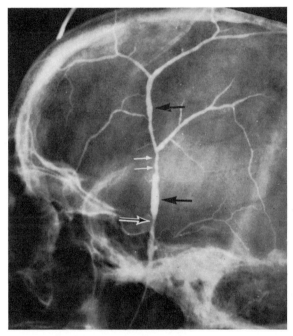

Figure 5–11 **Temporal Arteritis: Temporal Arteriogram.** The temporal artery is quite irregular and contains segments of narrowing *(black-white arrow)*, narrowing and beading *(white arrows)*, and ectasia *(black arrows)*, all of which are findings characteristic of an arteritis. (Courtesy Dr. L. A. Gillanders: Clin. Radiol., 20:149, 1969.)

Wegener's Granulomatosis

This rare condition is characterized by necrotizing granulomas of the upper respiratory tract and by a systemic vasculitis most often affecting the lungs and kidneys.

The significant radiographic findings occur in the lungs. Typically there are one or more discrete nodular or alveolar infiltrative lesions of varying size; a diameter of up to 10 cm. can occur. Extension to other portions of one or both lungs is usual. In a large percentage, thick-walled irregular cavitation develops, often with fluid levels. Enlarged hilar nodes are frequent. Pleural changes are not uncommon, but sizable effusions are quite rare.

If cavitation is absent, the radiographic picture is suggestive of polyarteritis. The cavitary lesions may resemble infectious granulomas or cavitating metastases. The variable pulmonary lesions are due to vasculitis and usually respond rapidly to steroids. However, pulmonary changes due to lung infarction or renal failure may also develop; these do not respond to steroids.

Radiographic evidence of paranasal sinus disease is found in almost every case. There may be progressive destruction of orbits, nasal bones, mastoids, and the base of the skull. Tomographic studies are helpful for demonstrating these destructive changes. The focal glomerulonephritis produces no pyelographic findings.

In lethal midline granuloma, a related condition, pulmonary and renal lesions are absent.[11-16]

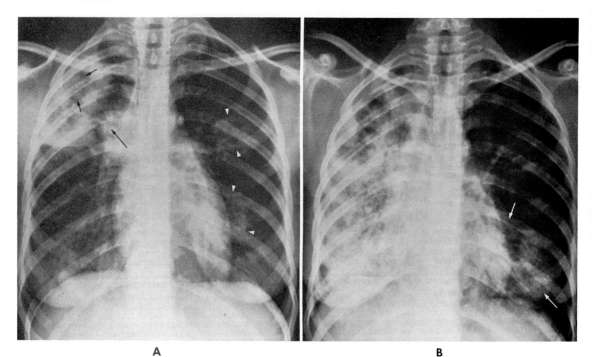

A   B

Figure 5–12 **Wegener's Granulomatosis: Rapid Extension.**

A, There is a large infiltrative density in the right upper lobe, with multiple translucencies *(short arrows)* that indicate cavitation. The medial border of the infiltrate is irregular, and the hilar nodes are considerably enlarged *(long arrow)*.

Two large round densities *(arrowheads)* in the left lung represent rapidly growing nodular lesions.

B, One month later there is extension into the entire right lower lung field. The cavities in the right upper lobe are larger. The lesion at the left base *(arrows)* has extended toward the diaphragm, and the lesions in the left upper lobe have resolved.

Although temporary regression of lesions may occur following steroid therapy, fatal progression is the usual course.

Weber-Christian Disease (Relapsing Febrile Nodular Nonsuppurative Panniculitis, Panniculitis)

Radiographic changes may sometimes appear in the soft tissues, joints, and bones. The painful tender nodules, which appear as small areas of increased density within the enlarged subcutaneous fat tissue, are best seen in the extremities.

Changes in the joints or bones are probably the result of involvement of synovial or bone marrow fat. Attacks of joint pain and swelling are common, but usually there are no accompanying radiographic findings. However, one report indicates that during acute involvement of the knee, fragmentation of the subpatellar fat pad with extension of lucent fat streaks into adjacent muscles may be a specific finding. Following recurrent acute joint attacks, narrowing of the joint space may develop.

In a small percentage of cases, lytic cyst-like defects may appear in one or more bones; if widely disseminated, multiple myeloma may be simulated.

Mesenteric panniculitis (retractile mesenteritis), in which the fatty degeneration involves only the small bowel mesentery (see below), is considered to be a variant of Weber-Christian disease.[17-21]

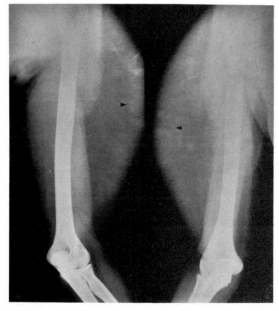

Figure 5-13 **Weber-Christian Disease.** There is pronounced thickening of the subcutaneous fat layer in both arms. Nodules of different sizes, the densities of which are greater than that of fat *(arrowheads)*, are scattered throughout the swollen, fatty panniculus. (Courtesy Dr. A. Levitan, Brooklyn, New York.)

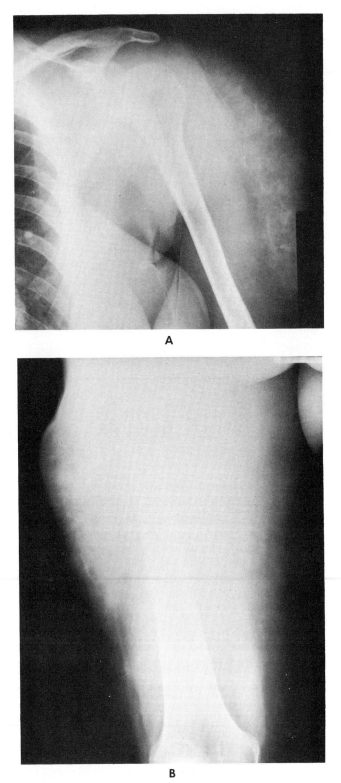

Figure 5–14 **Weber-Christian Disease.** In this 42 year old woman, there is enormous thickening of the relatively lucent subcutaneous fat tissue in the shoulder *(A)* and thigh *(B)*. Irregular bands of density are traversing the thickened fatty tissue. These are characteristic findings.

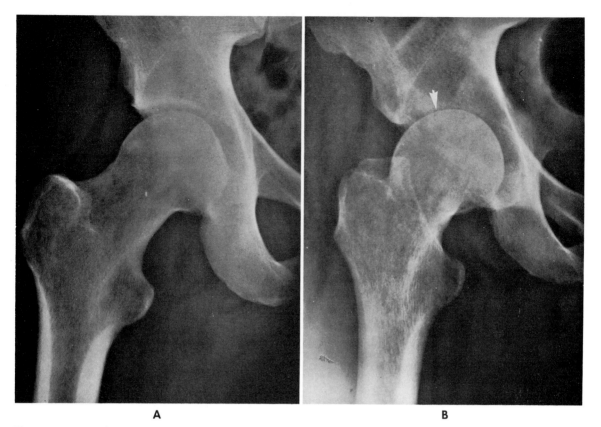

| A | B |

Figure 5–15   **Weber-Christian Disease: Progressive Degeneration of Hip Joint.**

   A, One month after onset of symptoms in a 39 year old man, there is some fuzziness of the acetabular roof, but the joint space is well preserved.

   B, Six months later, the joint space is much narrower *(arrow),* there is demineralization of the periarticular bones, and the acetabulum has deepened. A year and a half later, protrusio acetabuli had developed. (Courtesy Dr. L. Ritzman, Portland, Oregon.)

Figure 5–16   **Weber-Christian Disease: Progressive Degeneration of Knee Joint.**

   A, This is the same patient shown in Figure 5–15. During early acute attack there is osteoporosis around the joint, but the joint spaces are normal.

   B, Two years later, after a number of recurrent acute episodes, there has been considerable degeneration of the articular cartilages, with narrowing of the joint spaces *(arrows).* A coarse trabecular pattern is demonstrated in the periarticular bones. (Courtesy Dr. L. Ritzman, Portland, Oregon.)

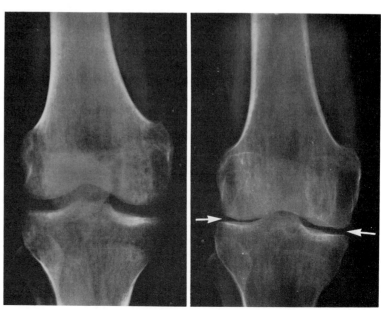

| A | B |

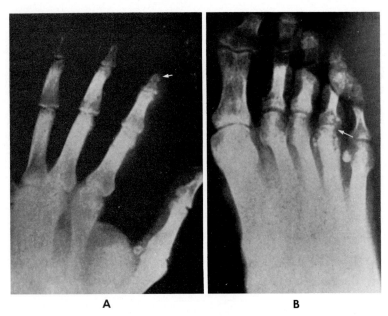

Figure 5–17 **Weber-Christian Disease: Bone and Joint Changes.**

A, In a 42 year old man there are cystic and destructive changes in the shaft and cortex of the terminal phalanx of the index finger *(arrow)*. The tuft has been destroyed.

B, The joint space of the fourth metatarsophalangeal joint *(arrow)* is narrowed; there is juxta-articular osteoporosis and some lysis of bone. These changes are similar to those described in Figures 5–15 and 5–16. (Courtesy Drs. C. DeLor and R. Martz, Columbus, Ohio.)

Retractile Mesenteritis (Mesenteric Panniculitis)

This uncommon disorder can affect individuals of either sex at any age. Chronic inflammation of the mesenteric fat leads to thickening of the mesentery, which may compress and distort adjacent small bowel loops. The clinical picture is very variable, but the most frequent symptoms are low grade fever, abdominal pain, and gastrointestinal disturbances.

Radiographically, the smaller tumefactions are undetected. Larger masses may displace the small bowel loops away from the center of the abdomen and cause compression, stretching, and distortion of the segments. These changes may suggest mesenteric cyst or neoplasm. The more diffuse mesenteric lesions cause wide separation of the bowel loops, but no intrinsic abnormalities are found in the small bowel.

Angiography will show stretching and encasement of the superior mesenteric artery around a mass and irregularity or occlusions of the distal small branches, without neovasculature.

Some authors have suggested that the condition is related to retroperitoneal fibrosis or Weber-Christian disease.[22-24]

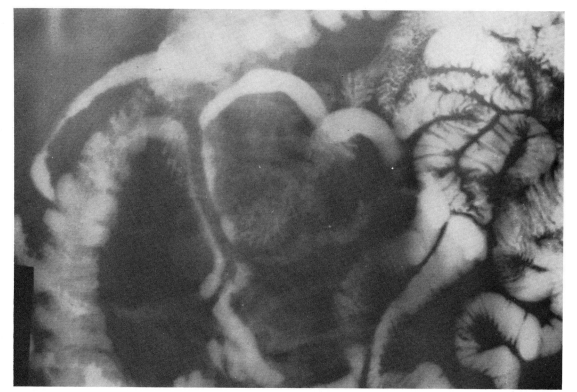

Figure 5–18 **Retractile Mesenteritis.**   The central small bowel loops are displaced, separated, stretched, and narrowed by a diffuse mesenteric mass. (Courtesy Drs. R. Marshak and A. E. Lindner: *Radiology of the Small Intestine.* 2nd ed. Philadelphia, W. B. Saunders Company, 1976.)

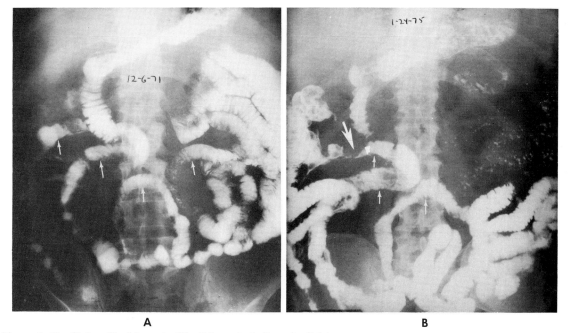

A   **B**

Figure 5–19   **Retractile Mesenteritis (Mesenteric Panniculitis).**

A, Small bowel study of a 51 year old man who had suffered a mild attack of pancreatitis discloses a number of separated and curved small bowel loops *(arrows).* No abnormalities of the small bowel were otherwise noted, and the mucosal pattern was normal. The patient was asymptomatic.

B, Three years later, the small bowel loops were almost identical in appearance *(small arrows).* Separation and curved loops were again seen. A deformity had developed in one of the loops *(large arrow).*

The appearance suggested multiple benign mesenteric masses. These proved to be masses due to retractile mesenteritis.

119

Retroperitoneal Fibrosis (Periureteral Fibrosis)

Nonmalignant retroperitoneal fibrosis is a condition of unknown origin; however, many cases have been associated with the use of methysergide for migraine. It can cause progressive obstruction of one or both ureters by a surrounding but noninvasive fibrotic plaque, most often at a level close to the pelvic brim. Inferior vena caval obstruction can also occur.

The urographic findings are almost pathognomonic. Hydronephrosis and hydroureter above the point of obstruction can be demonstrated by intravenous urography, or by retrograde studies if renal function is lost. The ureter shows medial deviation, irregularity, narrowing, and cone-like tapering at the obstructive site, a highly characteristic appearance. The constricted deviated segment is usually short, but it can be as long as 6 cm. In about one fourth of cases, involvement is bilateral; sudden loss of renal function and anuria sometimes occur.

Other roentgen signs include loss of normal retroperitoneal fat lines adjacent to the psoas shadows, and lymphangiographic evidence of lymphatic obstruction in the para-aortic nodes and channels. Associated mediastinal fibrosis has been reported in some cases. Diffuse retroperitoneal malignancy, especially lymphoma, may simulate retroperitoneal fibrosis.[25-28]

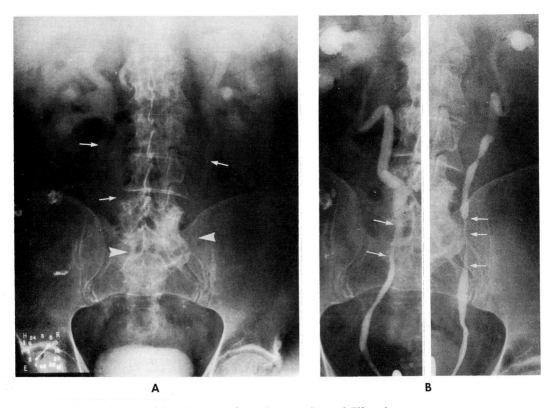

A   **B**

Figure 5–20 **Bilateral Ureteral Involvement from Retroperitoneal Fibrosis.**

A, Intravenous urogram shows dilatation of both upper urinary tracts. The dilated and tortuous upper ureters can be identified *(arrows)* up to the upper sacrum. Both ureters, especially the right one, appear to be displaced medially *(arrowheads).*

B, A composite retrograde study reveals segmental narrowing of both ureters *(arrows).* There is considerable medial displacement of the right ureter.

The medial displacement of a smooth narrowed segment is characteristic of ureteral involvement by retroperitoneal fibrosis.

Eosinophilic Pneumonia (Löffler's Syndrome, Löffler's Pneumonopathy, Pulmonary Infiltrates with Eosinophilia)

The clinical and radiographic spectrum of eosinophilic pulmonary infiltrates is broad and varied. An asthmatic or allergic background and a blood eosinophilia are usually present, but one or both may be absent.

In the milder cases there are transient, bilateral, and rapidly changing infiltrates, with minimal clinical symptoms. More chronic and severe forms show more persistent and slower changing bilateral patches of infiltrating consolidation, which clear slowly in one area while appearing in another. Cough and fever can persist for months. In the most severe forms the pneumonic infiltrates progress slowly; a debilitating and even life-threatening clinical course may occur.

Radiographically the infiltrates are nonspecific patchy conglomerate alveolar densities, appearing in any portion of the lungs but having a distinct tendency to peripheral involvement. In many chronic cases this leads to areas of pleural-like density along the lateral chest wall. Tuberculosis or granulomatous disease is usually initially suspected, but the eosinophilia and the allergic background should suggest the diagnosis. Even in the absence of eosinophilia, the densities along the peripheral lung fields in association with a chronic bilateral disease should bring to mind the possibility of eosinophilic pneumonia.

A most striking and constant feature of chronic eosinophilic pneumonia is the dramatic response to steroids, often within 48 hours.

There are several other forms of shifting pneumonopathy with eosinophilia in which more definitive etiologic agents are implicated. These include Löffler's tropical eosinophilia, which is caused by parasites, especially filariasis; occasional cases of pulmonary aspergillosis; nitrofurantoin and other drug hypersensitivity; early or prodromal lesions of some cases of periarteritis nodosa, pulmonary ascariasis and creeping eruption.[29-32]

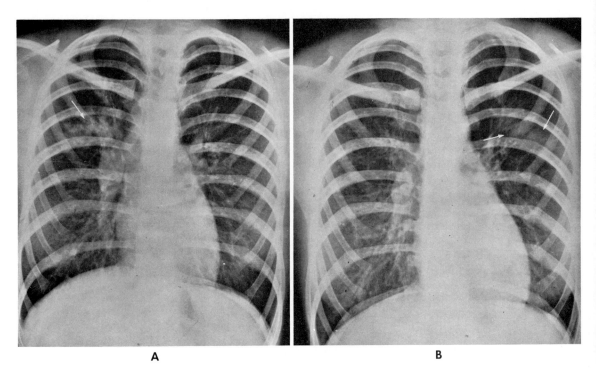

A   B

Figure 5–21 **Migratory Pneumonitis in Asthmatic (Löffler's) Pneumonia.**

A, There is a patch of infiltration *(arrow)* in the right upper lobe.

B, Four days later there is a new infiltration in the left upper lobe *(arrows).* The pneumonitis on the right has resolved completely.

This transient migratory pneumonitis (Löffler's pneumonia), seen in the asthmatic person, is also associated with eosinophilia but is apparently a different entity from the more chronic forms of eosinophilic pneumonopathy that are often also called *Löffler's pneumonia.*

Figure 5–22 **Pulmonary Infiltrates with Eosinophilia (Chronic Eosinophilic Pneumonia).**

A, A 48 year old woman had developed a chronic cough and bouts of fever for six weeks during a period of desensitization for chronic asthma. Chest film revealed several separate areas of irregular infiltration in the right middle and upper lung fields *(white arrows),* and a large confluent infiltrate in the left apex; the infiltrate extended along the outer lung border *(black arrows).* Although no cavitation was seen, a radiographic diagnosis of tuberculosis was suggested.

B, One month later, two of the right lung infiltrates had disappeared, but a large confluent lesion *(black-white arrow)* had developed in the right apex and upper lung border. On the left, the previous lesion had greatly increased, both medially *(black arrow)* and downward along the outer border of the lung *(white arrow).*

All studies for tuberculosis were negative. There was a blood eosinophilia of 30 per cent. A lung biopsy disclosed eosinophilic infiltrates without histologic evidence of polyarteritis.

C, Within a few weeks after commencement of steroid therapy, there was concomitant dramatic clinical improvement and complete clearing of the lungs. The pleural tenting at the left base *(arrow)* was the result of the surgical biopsy.

The chronicity, the changing character of the infiltrates, their tendency toward confluency along lung borders, and their dramatic response to steroids are all prominent features of chronic eosinophilic pneumonia. The blood eosinophilia and an asthmatic history are also highly characteristic.

In polyarteritis there may be a similar clinical and roentgenologic picture that also responds to steroids.

See illustration on the opposite page.

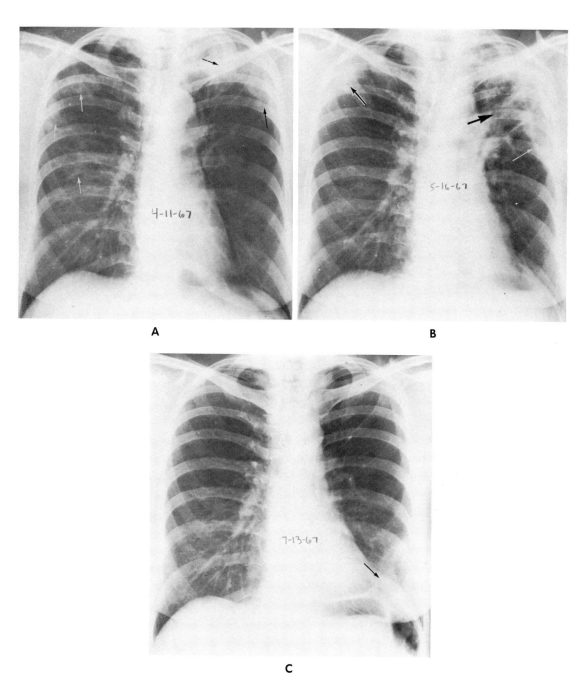

Figure 5–22 *See legend on the opposite page.*

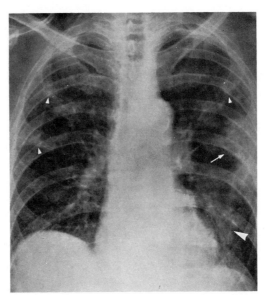

Figure 5–23 **Prolonged Eosinophilic Pneumonia in Nonasthmatic Person.** The markings are increased *(large arrowhead)*, and there are small patches of infiltration scattered throughout both lungs *(arrowheads)*. A larger irregular area of confluent infiltration lies against the left chest wall *(arrow)*.

These lesions persisted with very little change for over one month. In tropical eosinophilia the roentgenographic findings in the lung are similar. The condition is thought to be caused by filarial infestation in a hypersensitive person, and it responds to arsenical therapy.

Eosinophilic Gastroenteritis

This uncommon condition is characterized by eosinophilic infiltrates in the lower third of the stomach, especially the antrum. In about half the cases the small bowel is also involved. Clinically there is usually an allergic history, blood eosinophilia, abdominal pain, and fever.

Radiographically the distal stomach or antrum appears infiltrated and irregular, resembling carcinoma. Sometimes pylorospasm and delayed gastric emptying are the only findings. If the small bowel is involved, there may be segmental areas of thickening and rigidity of the wall, and occasional narrowing of the lumen. The appearance strongly resembles the granulomatous involvement of regional enteritis; it may also mimic an infiltrative lymphoblastoma. An increasing number of patients with only small bowel involvement have been reported.

The roentgen appearance of the stomach or small bowel, or both, in conjunction with a blood eosinophilia and prompt clinical and radiologic response to steroids, can suggest the diagnosis. This, however, can be confirmed only by biopsy.[33-35]

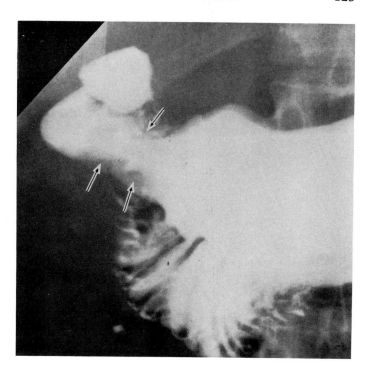

Figure 5–24 **Eosinophilic Gastroenteritis: Gastric Lesion.** This 55 year old woman with a history of chronic asthma developed abdominal pain and diarrhea. There was a blood eosinophilia of 10 per cent. An irregular infiltration *(arrows)* in the gastric antrum was thought to be carcinoma. Biopsy disclosed a benign eosinophilic infiltrate. Complete clinical and radiographic regression occurred after steroid therapy. (Courtesy Dr. H. J. Burhenne, San Francisco, California.)

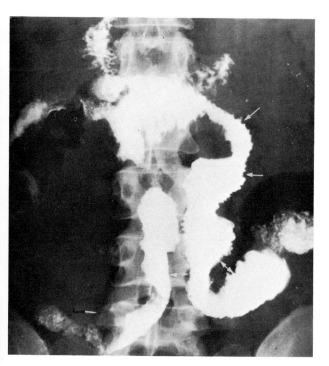

Figure 5–25 **Eosinophilic Gastroenteritis: Small Bowel Involvement.** A 31 year old woman with a history of seasonal allergic rhinitis complained of abdominal cramps, diarrhea, and episodic vomiting, all of six weeks' duration. A blood eosinophilia of 36 per cent was found.

The small bowel study showed irregular loops of small bowel with thickened walls *(double-headed arrows)* and irregular thickened mucosa *(arrows)*. Regional enteritis was the radiologic diagnosis, but biopsy disclosed extensive benign eosinophilic infiltrates of the bowel wall.

Under steroid therapy the small bowel rapidly returned to normal. Clinical recovery was complete. (Courtesy Dr. H. J. Burhenne, San Francisco, California.)

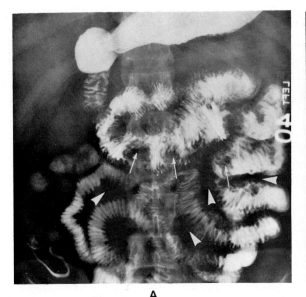 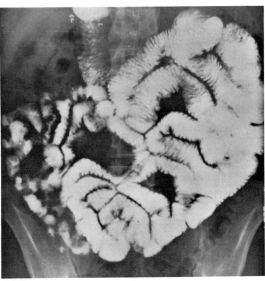

A   B

Figure 5–26   **Eosinophilic Gastroenteritis and Hodgkin's Disease in Young Male with Eosinophilia.**

A, Diffuse thickening of the small bowel walls is evidenced by the marked increase in distance between loops *(arrowheads).* Nodular masses *(arrows)* are scattered through the proximal small bowel, and there is some thickening of the valvuli, especially in the duodenum and jejunum.

A radiographic diagnosis of lymphosarcoma was suggested, but biopsy of the bowel wall and a nodule was diagnosed histologically as eosinophilic gastroenteritis.

B, After 10 days of steroid therapy the small bowel appearance has returned to normal; wall thickening, nodules, and mucosal edema have all completely disappeared.

One year later the patient developed full-blown Hodgkin's disease. It is unclear whether biopsy diagnosis was erroneous or whether there is a pathologic relationship between the two conditions.

Fibrosing Syndromes

See *Chronic Mediastinitis and Idiopathic Mediastinal Fibrosis,* p. 573, *Primary Sclerosing Cholangitis,* p. 1100, and *Retroperitoneal Fibrosis,* p. 120.

REFERENCES

1. Ellis, K., and Renthal, G.: Pulmonary sarcoidosis. Am. J. Roentgenol. Radium Ther. Nucl. Med., *88*:1070, 1962.
2. Stein, G. N., Israel, H. L., and Sones, M.: Roentgenologic study of skeletal lesions in sarcoidosis. Arch. Intern. Med., *97*:532, 1956.
3. Pfeiffer, K.: Boeck's disease of the salivary glands. Radiology, *3*:165, 1963.
4. Schmitt, E., Appelman, H., and Threatt, B.: Sarcoidosis in children. Radiology, *106*:621, 1973.
5. Kirks, D. R., McCormick, V. D., and Greenspan, R. H.: Pulmonary sarcoidosis. Am. J. Roentgenol. Radium Ther. Nucl. Med., *117*:777, 1973.
6. Young, D. A., and Laman, M. L.: Radiodense skeletal lesions in Boeck's sarcoid. Am. J. Roentgenol. Radium Ther. Nucl. Med., *114*:553, 1972.
7. Bonakdarpour, A., Levy, W., and Aegerter, E. F.: Osteosclerotic changes in sarcoidosis. Am. J. Roentgenol. Radium Ther. Nucl. Med., *113*:646, 1971.
8. Henck, V. C., Carter, C. C., and Rippey, J. G.: Giant cell (cranial) arteritis. Am. J. Roentgenol. Radium Ther. Nucl. Med., *92*:769, 1964.

9. Gillanders, Lewis A.: Temporal arteriography. Clin. Radiol., 20:149, 1969.
10. Elliott, P. D., Baker, H. L., and Brown, A. L.: The superficial temporal artery angiogram. J.A.M.A., 222:1405, 1972.
11. Felson, B., and Braunstein, H.: Non-infectious necrotizing granulomatosis. Radiology, 70:326, 1958.
12. Lynch, E. C., et al.: Pulmonary cavitation in Wegener's granulomatosis. Am. J. Roentgenol. Radium Ther. Nucl. Med., 92:521, 1964.
13. Levin, D. C.: Pulmonary abnormalities in the necrotizing vasculitides. Radiology, 97:521, 1970.
14. Hülse, R., and Jung, N.: Wegener's granulomatosis in the region of the paranasal sinuses and skull base. Fortschr. Roentgenstr., 115:561, 1971.
15. Gonzales, L., and Van Ordstrand, H. S.: Wegener's granulomatosis. Radiology, 107:295, 1973.
16. Landman, S., and Burgener, F.: Pulmonary manifestations in Wegener's granulomatosis. Am. J. Roentgenol. Radium Ther. Nucl. Med., 122:750, 1975.
17. DeLor, C. J., and Martz, R. W.: Weber-Christian disease with bone changes. Ann. Intern. Med., 43:591, 1955.
18. Goldberg, L. M., and Ritzman, L. W.: Relapsing, nodular, febrile, panniculitis. Am. J. Med., 25:788, 1958.
19. Zhentlin, N., and Golenterneck, J.: X-ray signs in Weber-Christian disease. J.A.M.A., 189:580, 1964.
20. Doel, G.: Bone involvement in Weber-Christian disease. Br. J. Radiol., 36:140, 1963.
21. Herrington, J. L. Edwards, W. H., and Grossman, L. A.: Mesenteric manifestations of Weber-Christian disease. Ann. Surg., 154:949, 1961.
22. Marshak, R. H., and Lindner, A. F.: Radiology of the Small Intestine. 2nd ed. Philadelphia, W. B. Saunders Co., 1970.
23. Carillo, F. J., Ruzicka, F. F., and Clemett, A. R.: Angiography in retractile mesenteritis. Am. J. Roentgenol. Radium Ther. Nucl. Med., 115:396, 1972.
24. Castleman, B., et al.: Retractile mesenteritis—retroperitoneal fibrosis. Case report from Massachusetts General Hospital. N. Engl. J. Med. 287:34, 1972.
25. Twigg, H.: Peri-ureteral fibrosis. Am. J. Roentgenol. Radium Ther. Nucl. Med., 84:876, 1960.
26. Jose, J. S.: Idiopathic retroperitoneal fibrosis. Br. J. Urol., 39:431, 1967.
27. Packham, D. A., and Yates-Bell, J. G.: Retroperitoneal fibrosis. Br. J. Urol., 40:207, 1968.
28. Jones, J. H., et al.: Retroperitoneal fibrosis. Am. J. Med., 48:203, 1970.
29. Incapera, F. P.: Pulmonary eosinophilia. Am. Rev. Resp. Dis., 84:730, 1961.
30. Carrington, C. B., Addington, W. W., et al.: Chronic eosinophilic pneumonia. N. Engl. J. Med., 280:787, 1969.
31. Robertson, C. L., Shackelford, G. D., and Armstrong, J. D.: Chronic eosinophilic pneumonia. Radiology, 101:57, 1971.
32. Citro, L. A., Gordon, M. E., and Miller, W. T.: Eosinophilic lung disease. Am. J. Roentgenol. Radium Ther. Nucl. Med., 117:787, 1973.
33. Edelman, M. J., and March, T. L.: Eosinophilic gastroenteritis. Am. J. Roentgenol. Radium Ther. Nucl. Med., 91:773, 1964.
34. Buchenne, J. H., and Carbone, J. V.: Eosinophilic (allergic) gastroenteritis. Am. J. Roentgenol. Radium Ther. Nucl. Med., 96:332, 1966.
35. Dalinka, M. K., and Masters, C. J.: Eosinophilic enteritis: Report of a case without gastric involvement. Radiology, 96:543, 1970.

SECTION

6

MICROBIAL DISEASES

VIRAL DISEASES

Viral Infections of the Respiratory Tract

Pneumonic infiltrates are the principal roentgenographic findings in the viral respiratory diseases. Involvement of the larynx, trachea, and bronchi generally produces no significant radiologic changes except in children, in whom edematous hypopharyngeal and perilaryngeal tissues can often be recognized radiographically.

Pneumonia of infectious origin is manifested either by interstitial or alveolar infiltrates, or by a mixture of both. Interstitial infiltrates obscure the normally sharply marginated borders and the tapered branchings of the bronchovascular markings. The markings become irregular and fuzzy in outline, lose their branching appearance, and may even appear to increase in caliber toward the lung periphery. When localized, these infiltrates are readily recognized by comparison with the unaffected portions of the lung. Diffuse fine interstitial infiltrates, however, may be extremely difficult to appreciate without comparison with earlier or later films. More extensive infiltrates may assume nodular or miliary patterns or may coalesce to produce larger, unmistakable densities.

Alveolar exudates can appear as small, fluffy, irregularly and poorly marginated densities. These often become confluent and present as larger homogeneous densities confined to one lobe. The air bronchogram, which results from visualization of air-filled smaller bronchi by contrast with surrounding consolidated alveoli, is an invaluable sign of alveolar infiltration.

In viral lung disease, interstitial infiltrates are more common than alveolar infiltrates, but alveolar densities may coexist or even dominate the picture, producing a lobular or bronchopneumonia. Sometimes miliary densities predominate.

Pulmonary infiltrates are usually quite nonspecific and are found in a huge number of diseases of both infectious and noninfectious origin. Clinical background and serial films are usually necessary to reduce the number of diagnostic possibilities. Differentiation of disease by the appearance of the infiltrates alone is only rarely possible; a patchy infiltrate in an apex may prove to be a viral pneumonia rather than tuberculosis.

The principal respiratory viral and filter-passing agents responsible for a primary pneumonia include the *Mycoplasma* group (Eaton agent), the adenoviruses, the influenza and parainfluenza viruses, and the ornithosis group (psittacosis). Epidemics of "primary atypical pneumonia" are most often caused by the Mycoplasma, but the adenoviruses are the agent in some epidemics. In children, respiratory syncytial virus is the commonest cause of serious nonbacterial pneumonia.

In adults, pleural involvement is relatively rare; in children pleural exudates are quite common.

Hilar adenopathy is an occasional finding and develops most frequently in children. The adenovirus of Enders is prone to cause hilar nodal enlargement in adults.

On a statistical basis, the radiologic findings in the lung infections of different viral groups may have certain suggestive features, but actually there is great overlap and virtually no specificity. An effort to distinguish the etiologic agent by the roentgen pattern alone is a futile academic exercise. Nevertheless, the usual and sometimes special radiographic features of each group cannot be ignored and will be discussed in the following pages.[1-3]

Respiratory Syncytial Viral Infections

In addition to causing many of the more severe upper respiratory infections of infancy and early childhood, this virus is the commonest cause of bronchitis and bronchopneumonia in infants.

Radiographically, the infiltration is most often interstitial (though it may be alveolar), and many lobes are involved; it may spread diffusely through both lungs. Segmental lobar consolidation is considerably less frequent. Pleural involvement is uncommon and minimal. Hyperinflation is usually found. The infiltration rarely lasts for more than a week.

The absence of pneumatoceles, cavitation, and significant pleural reaction will help to distinguish syncytial viral pneumonia from staphylococcus pneumonia.[1-4]

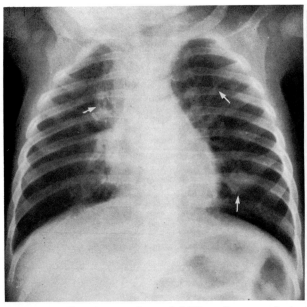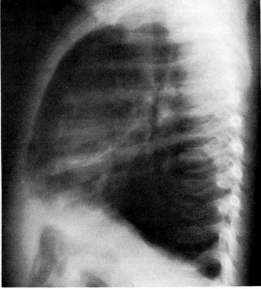

Figure 6–1 **Respiratory Syncytial Virus Pneumonia in Infant.** The diffuse, bilateral interstitial and alveolar infiltrates *(arrows)* and the hyperinflation (evidenced by low flattened position of the diaphragms) are the most common findings in this pneumonia of infancy. (Courtesy Dr. R. P. Rice, Chapel Hill, North Carolina.)

Influenza

In mild sporadic endemic influenzal pneumonia, the focal areas of patchy infiltration in the lungs cannot be distinguished from those of other viral pneumonias. In the severe epidemic form, the pulmonary infiltrates are far more extensive and variable; frequently, additional lung changes due to secondary bacterial complications are superimposed.

In primary influenzal pneumonia, there is usually a bilateral bronchopneumonia extending from the hila, but the periphery of the lungs remains clear. Massive infiltration of this type is associated with a high mortality. Segmental or pseudolobar consolidation may occur and may mimic lobar pneumonia; widespread miliary interstitial lesions have also been reported, often simulating heavy vascular markings.

Secondary bacterial complications are generally due to pneumococci, staphylococci, or streptococci and include pleural effusion, empyema, and lung abscess.[2, 3, 5-7]

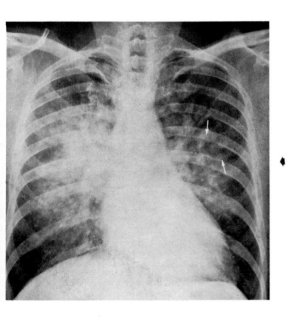

Figure 6–2 **Primary Influenzal Pneumonia: Confluent Central Bronchopneumonia.** Infiltration extends from both hila toward the periphery. On the right side, the infiltration is confluent; on the left, bronchopneumonic patches *(arrows)* can be identified. Note that the lung periphery is entirely clear. This nonspecific type of confluent bronchopneumonia, beginning centrally, is the most severe form of epidemic influenzal pneumonia.

Figure 6–3 **Primary Influenzal Pneumonia: Central Pseudolobar Consolidation:** A large area of consolidation extends outward from the right hilum. Although the homogeneous density resembles lobar pneumonia, the irregular borders *(arrow)* of the lesion may be seen; in lobar pneumonia, sharp borders of the lobar margins limit consolidation.

Areas of unilateral central consolidation advancing toward the periphery are another frequent roentgenologic finding in primary influenzal pneumonia. Additional patches of pneumonitis are seen at both bases.

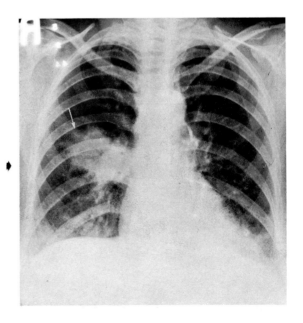

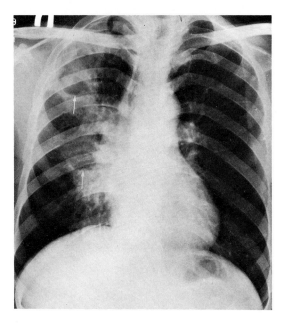

Figure 6–4 **Influenzal Pneumonia: Secondary Bacterial Infection.** There is central consolidation in the right hilar area, and consolidation in the right upper lobe. Secondary staphylococcal cavitary abscesses have developed in both areas, with air-fluid levels *(arrows).* The mortality rate with this type of infection was high during the great epidemic of 1918. Pleural effusion and empyema were also frequent secondary complications.

Adenoviral Infections

Severe upper respiratory tract disorders, including tracheobronchitis, are frequent in adenoviral infections but produce no roentgenologic changes. If pneumonia develops, patchy infiltrates that are indistinguishable from those of common viral pneumonia are usually seen; mottled peribronchial infiltrates and homogeneous densities also occur. Systemic adenopathy frequently accompanies adenoviral infections, and hilar adenopathy is often seen.

In children, there may be a diffuse bilateral bronchopneumonia with striking hyperinflation. Lobar collapse is not unusual. Chronic sequelae including bronchiectasis and unilateral hyperlucent lung can occur.[8, 9]

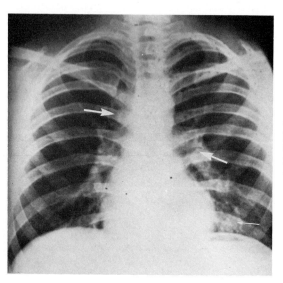

Figure 6–5 **Adenoviral Pneumonia.** There is a patchy infiltrate at the left base *(small arrow).* Note enlarged left hilar shadow and slight fullness in the right upper mediastinum *(large arrows),* both due to nodal enlargement.

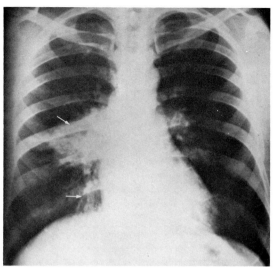

Figure 6–6 **Adenoviral Confluent Pneumonia.** A homogeneous infiltrate in the region of the right hilum extends toward the base near the cardiac border *(arrows)*. Nodal enlargement in the right hilum is somewhat obscured by the area of consolidation.

Mycoplasmal Pneumonia

A great variety of pulmonary infiltrates occur in mycoplasmal pneumonia (Eaton agent pneumonia, primary atypical pneumonia). The lesions have no radiographic features that can distinguish them from viral pneumonias. In about one half of adults with mycoplasmal pneumonia, the lesion is bilateral. A lower lobe is a favored site. The infiltrates are usually poorly marginated patchy densities, and nodular or reticular patterns are not uncommon. The perihilar lesions tend to be more dense than the peripheral infiltrates. Hilar adenopathy frequently occurs, but pleural reactions are uncommon in adults, although in some epidemics pleural exudates constantly occur.

In children, bilateral lower lobe lesions, hilar adenopathy, and pleural reactions are quite common. Mycoplasmal infiltrates tend to resolve slowly, often persisting for several weeks.[1, 10-12]

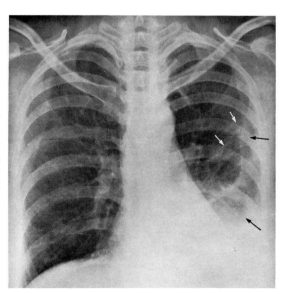

Figure 6–7 **Primary Atypical (Mycoplasmal) Pneumonia with Effusion.** Streaky infiltrative densities are seen in the left midlung *(white arrows)*. The thick pleural stripe *(upper black arrow)* and obliteration of the left costophrenic angle *(lower black arrow)* are due to a pleural exudate. Occasionally the pleural lesions are more prominent than the parenchymal ones. In most viral pneumonias, pleural reaction is uncommon.

Viral Pneumonia

A number of viral agents are causative, and, as in mycoplasmal pneumonia, a wide variety of appearances are encountered radiographically. Focal alveolar bronchopneumonic infiltrates or interstitial densities are the most frequent. Miliary, nodular, reticular, or diffuse perihilar densities can also occur. In adults involvement is usually unilateral with predilection for the bases. Pleural reaction is rare. In children, however, bilateral lower lobe bronchopneumonia, enlarged hilar nodes, and pleural reaction are more common. Chest x-rays may be normal for several days after the onset of symptoms. Roentgen changes may persist for days or even weeks after clinical recovery. The term *primary atypical pneumonia*, which has been applied to both mycoplasmal and viral pneumonias, is too nonspecific and is being employed much less often.[1-3, 11, 13]

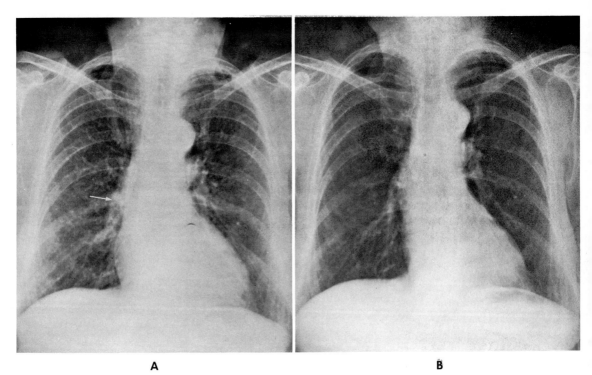

A   **B**

Figure 6–8   **Diffuse Interstitial Viral Pneumonia.**

A, The generalized increase in the interstitial pulmonary markings during an acute clinical episode is not striking. The right hilar shadow *(arrow)* did not appear unusual but definitely decreased in size after clinical recovery.

B, One week later, interstitial markings are greatly reduced, and the right hilar shadow is diminished. Had this serial film not been made, the earlier film *(A)* might have been considered normal. Subtle interstitial changes and slight hilar enlargements can often be recognized only with comparative films.

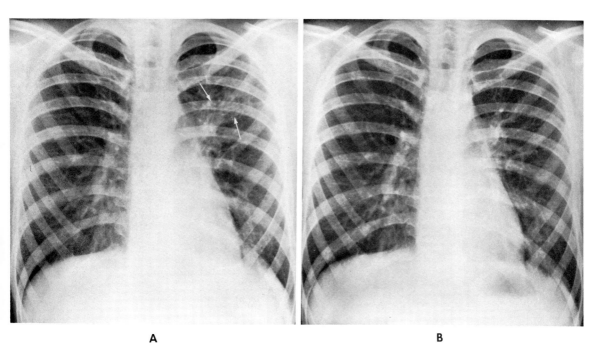

A   B

Figure 6–9 **Viral Pneumonia: Left Upper Lobe.**

A, Linear and interstitial infiltrates *(arrows)* extend from the left hilum into the left upper lobe. The posterior apical segment was involved, and the radiographic features alone are indistinguishable from those of tuberculous infiltration.

B, Eight days later, the infiltrates have disappeared, confirming the nontuberculous origin.

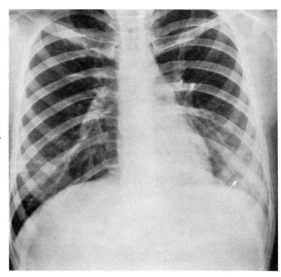

Figure 6–10 **Viral Pneumonia: Left Lower Lobe.** Interstitial and alveolar infiltrates *(arrow)* in the left lower lobe disappeared completely within one week.

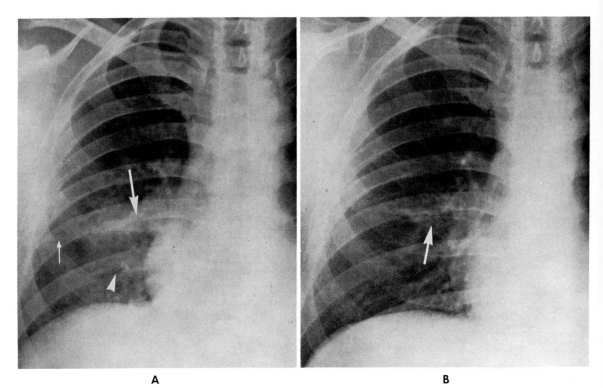

A   **B**

Figure 6–11   **Viral Pneumonia: Right Upper and Middle Lobes.**

A, An alveolar exudate is seen in the anterior inferior portion of the right upper lobe *(large arrow).* The short fissure, barely visible *(small arrow),* is the lower border of the pneumonic density. Interstitial and alveolar infiltrates are also seen in the right lower lung field *(arrowhead).* These were in the middle lobe.

B, Ten days later the upper lobe lesions have almost completely disappeared, but there are still some residual interstitial densities *(arrow)* marking the site of previous involvement.

The middle lobe has completely cleared.

Figures 6–8 to 6–11 illustrate the variable location and appearance of viral pneumonia. There are almost always interstitial infiltrates with or without alveolar consolidation; these distinguish viral from segmental lobar pneumonia. Note the absence of pleural reaction; only a small percentage of cases show pleural involvement.

The diagnosis of viral pneumonia should always be confirmed by subsequent chest films to prove resolution. Some forms of bronchogenic carcinoma may sometimes mimic viral pneumonia, but in bronchogenic carcinoma the infiltrates fail to clear completely.

Psittacosis

The initial lung lesions are one or several patchy alveolar and interstitial reticular infiltrates. These are often bilateral and frequently become confluent areas of consolidation extending from the hilum. Usually they are most marked in the lower lobes. They may assume a dense lobar distribution, simulating a bacterial pneumonia.

A distinct tendency toward migratory involvement and frequent change in appearance is commonly observed. Hilar adenopathy is not rare. Pleural reaction, often in an adjacent fissure, is frequent. The extensive pulmonary involvement is often strikingly disparate to the scant physical findings and the relatively mild clinical picture. Clearing of the infiltrates is usually quite slow, taking up to six weeks.

Extrapulmonary manifestations of pericarditis, myocarditis, hepatitis, and even acute arthritis have been reported in severe cases.

Although the pulmonary findings are not specific or diagnostic and can simulate other conditions, such as Löffler's migratory pneumonia or granulomatous disease, the dense parenchymal involvement, the migratory bilateral nature of the infiltrates, and the clinical-roentgenologic disparity are often helpful diagnostic features.[14-17]

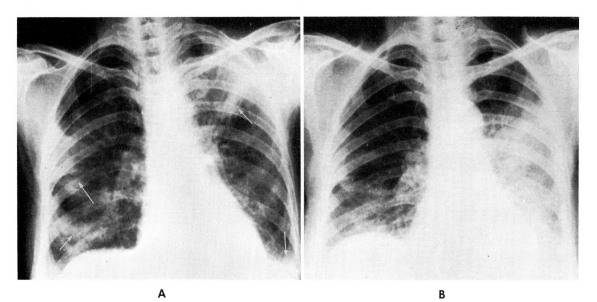

A B

Figure 6–12 **Psittacosis.**

A, Patchy pulmonary infiltrates involve the left upper lobe, the left and right bases, and the right midlung field *(arrows)*.

B, A few days later, pneumonia extends to the entire left lower lobe. There is partial clearing at the right base and complete clearing of the upper lobes. This patchy involvement is common in psittacosis.

Lymphogranuloma Venereum (Lymphopathia Venereum)

Roentgen findings are usually limited to the large bowel, especially the rectum.

In the acute proctocolitis stage, the rectum and entire colon may be affected, but unevenly. The involved areas appear rigid and show a loss of both haustrations and mucosal pattern. There may be severe spasm.

In the chronic stage, anorectal strictures are the commonest radiographic finding. They are most often tubular, continuous from the anal area, and of varying length. Although strictures are usually limited to the rectum, the sigmoid may also be involved; in rare cases the stricture may involve the entire left colon.

The lumen is smoothly reduced and may become string-like. The proximal colon may be dilated in severe cases. Thin horizontal barium projections are frequent and represent short blind sinus tracts; fistulous tracts and perianal abscess can often be demonstrated.

The more severe cases may resemble granulomatous colitis, but this disease rarely produces rectal strictures. The rectal involvement and barium spicules projecting from the lumen of many lymphopathia strictures will distinguish the latter from strictures due to irradiation and other causes.[18]

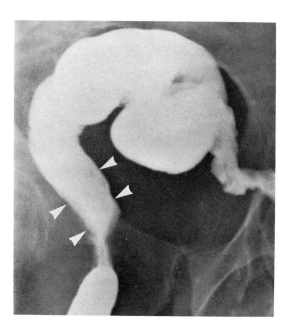

Figure 6–13 **Lymphogranuloma Venereum: Barium Enema.** There is a diffuse narrowing of the rectum *(arrowheads).* The rectum is smooth, and there is a gradual increase to normal caliber at the rectosigmoid junction. In most cases only the rectum is affected, but occasionally the sigmoid is involved.

Diffuse inflammatory stricture limited to the rectum is almost pathognomonic of chronic lymphogranuloma venereum.

Figure 6–14 **Lymphogranuloma (Lymphopathia) Venereum: Rectal Stricture.** The entire lower rectum is smoothly narrowed *(white arrows)* and widens gradually to the dilated sigmoid *(black arrow).* The barium projections *(arrowheads)* are short blind sinus tracts.

A rectal stricture with blind horizontal sinus tracts is highly suggestive of lymphogranuloma venereum.

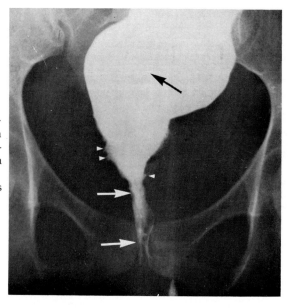

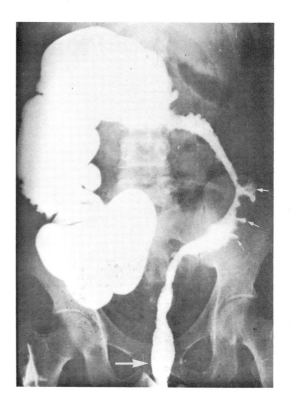

Figure 6–15 **Lymphogranuloma (Lymphopathia) Venereum: Extensive Involvement.** The entire left colon, including the rectum, is greatly narrowed up to the midtransverse area. The strictures are smooth, and the colon is shortened and shows no definite mucosal pattern.

The barium projections *(small arrows)* perpendicular to the lumen represent sinus tracts. The colon proximal to the strictures is considerably dilated.

This degree of involvement is unusual in this disease. A similar picture can be produced by a severe granulomatous colitis, but the narrowed rectum *(large arrow)* is extremely rare in the latter condition.

Measles (Rubeola)

Roentgenologic findings are limited to the chest. Hilar adenopathy is frequent, being found in almost two thirds of the patients examined roentgenologically. Interstitial infiltrates, most apparent at the bases, are seen in about one fifth of the patients. Bronchopneumonia and atelectasis are not uncommon.

Chest x-ray examinations are also useful in uncovering secondary infections, especially staphylococcal pneumonia.

A fulminating and often fatal interstitial pneumonia—giant-cell pneumonia—may occur in children afflicted with diseases of the reticuloendothelial system, especially leukemia. This pneumonia is caused by the measles virus, but skin lesions may be absent.[19, 20]

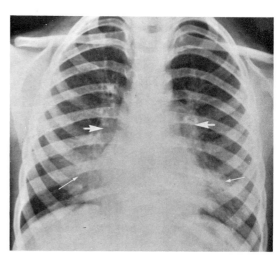

Figure 6–16 **Measles Pneumonia.** Infiltrate *(small arrows)* extends from both hila to the bases. Prominent hilar shadows are due to adenopathy *(large arrows).*

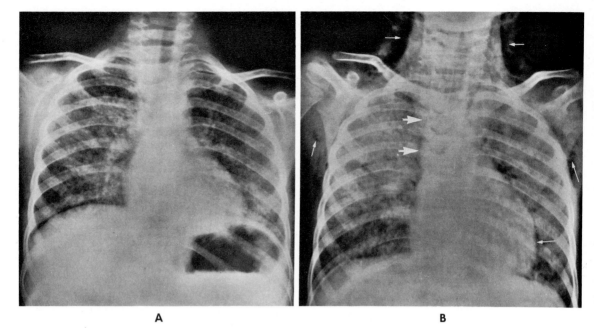

A   B

Figure 6–17 **Measles: Fatal Giant-Cell Pneumonia in Child with Leukemia.**

A, Widespread bilateral interstitial infiltration is more marked on the right side.

B, Several days later, pulmonary infiltrates have become confluent. Enlargement of the superior mediastinal nodes *(large arrows)* produces widening of the mediastinum. Air is present in the mediastinum, neck, and soft tissues of the thorax *(small arrows)*, from rupture of an area of interstitial emphysema into the mediastinum.

Rubella (Congenital Rubella)

There are no significant radiographic findings in childhood rubella. However, if rubella occurs during pregnancy, the fetus may be infected through transplacental passage of the virus. A variety of embryonal defects can develop; these defects include congenital heart disease, cataracts, and bone abnormalities. Hypoplasia of the pulmonary arteries occurs in more than half of these enfants.

At birth the metaphyses of the distal femora are chiefly affected; the proximal tibial and humeral metaphyses may also be involved. Radiographically, the metaphyseal margin is irregular and poorly defined, and the trabeculae are coarsened. Longitudinal areas of sclerosis and demineralization are interspersed. There is no periosteal reaction, and the rest of the bone shaft is normal.

Within about four weeks after birth, the metaphyseal appearance returns to normal; the zone of provisional calcification becomes sharp and regular, and the trabeculations become normal. The episode of rubella may be marked by a line of density that migrates, with bone growth, toward the shaft—the so-called line of temporary growth disturbance.

It is uncertain whether the bone changes represent a mild rubella osteomyelitis or whether they indicate merely a disturbance of endochondral bone formation caused by the virus. Similar bone changes have been reported in congenital cytomegalic inclusion disease.

Intracranial calcifications are often identified microscopically but are only rarely seen on the skull radiographs.[21-25]

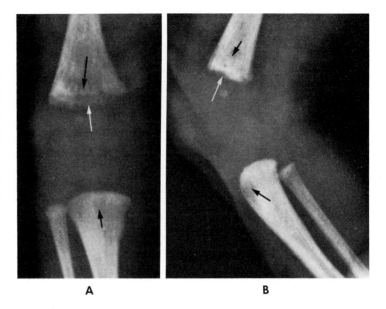

Figure 6–18 **Rubella Embryopathy: Metaphyseal Changes.** In these (*A*) anteroposterior and (*B*) lateral films of the knee of a newborn whose mother contracted rubella during pregnancy, an irregular metaphyseal line of the lower femur is visible (*white arrows*). Lucent areas are seen in the metaphysis of the femur and tibia (*black arrows*). There were similar changes in the other knee. The changes are characteristic.

A B

Smallpox (Variola)

There are usually no significant roentgenologic changes in uncomplicated smallpox, but rarely the virus may invade bone. Variola osteomyelitis occurs in less than one per cent of the cases, usually in children, within six weeks after onset of disease. The usual finding is a destructive lesion of the metaphysis associated with marked periosteal reaction. Frequently the adjacent joint and its bony surfaces are affected. Variola osteomyelitis often causes growth disturbances and marked late bone deformities, often with ankylosis of the involved joint.[26, 27]

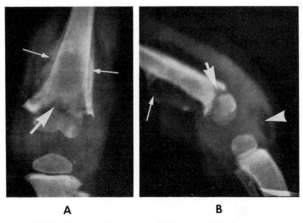

A B

Figure 6–19 **Variola Osteomyelitis: Lower Femur and Knee.** (*A*) Anteroposterior and (*B*) lateral films of the knee show destruction of the distal femoral metaphysis (*large arrows*), with marked periosteal proliferation (*small arrows*). The lytic process extends along the femoral shaft from the metaphysis. There is soft tissue swelling around the joint and displacement of the fat pad (*arrowhead*) due to intra-articular effusion. (Courtesy P. Cockshott and M. McGregor: J. Fac. Radiol., *10*:57, 1959.)

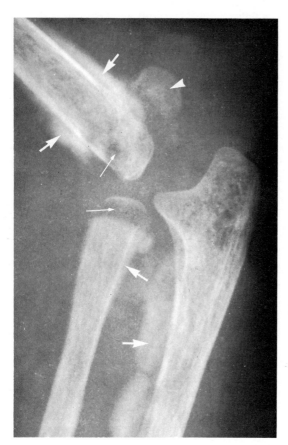

Figure 6–20 **Severe Variola Osteomyelitis and Arthritis of Elbow: Joint Disorganization.** There are large areas of lysis in the lower humerus and upper radius (*small arrows*); periosteal reaction is seen in all the bones (*large arrows*) and is most marked in the radius and humerus. The humeral epiphysis (*arrowhead*) is completely dislocated, and there is extensive soft tissue swelling surrounding the elbow joint.

Although the osteomyelitis in smallpox radiographically resembles bacterial osteomyelitis, the former is not due to secondary infection but is directly caused by the virus. Variola osteomyelitis characteristically settles in periarticular bones and invades the adjacent joint. (Courtesy P. Cockshott and M. McGregor: J. Fac. Radiol., *10*:57, 1959.)

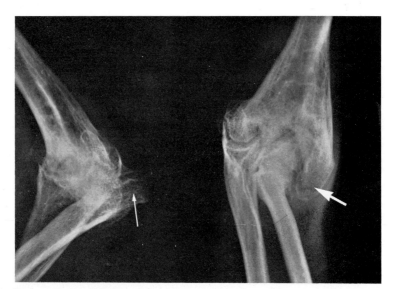

Figure 6–21 **Late Deformity and Subluxation Due to Smallpox Osteomyelitis.** The elbow is markedly deformed, with narrowed and irregular joint spaces especially between the ulna and humerus. The medial epicondyle of the humerus (*large arrow*) has overgrown, and a cubitus valgus deformity exists. The radius is subluxated (*small arrow*). Motion of the elbow was extremely limited. (Courtesy Dr. K. S. Bose, Calcutta.)

Poliomyelitis

There are no significant roentgenologic changes during the active phase of uncomplicated paralytic poliomyelitis. In the chronic stage the roentgenologic findings occur primarily as the result of muscular weakness and paralysis. There is evidence of bone atrophy (osteoporosis and decreased width of bone), scoliosis, and loss of muscular soft tissue volume; in severely paralyzed extremities, soft tissue calcification and ossification may develop, although this is relatively uncommon.

In almost half the patients with bulbar poliomyelitis who are under respirator care for chronic respiratory paralysis, one or more rib erosions develop, probably from constant scapular pressure. The superior surface of the third or fourth rib is the commonest site; occasionally the seventh, eighth, or ninth rib is eroded.

Nearly all the permanent respirator patients eventually develop calcification of the cartilaginous discs in the dorsolumbar spine, usually after five years or more in the respirator. Beginning in the margins, complete calcification of the discs eventually develops. Bony bridging and a "bamboo-spine" occur in some patients. Narrowing of the sacroiliac joint is frequent. The findings closely resemble ankylosing spondylitis.

Recurrent pneumonitis and focal atelectasis, due to retention of bronchial secretions, are frequent complications in bulbar respiratory patients.[28-31]

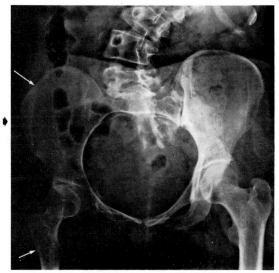

Figure 6-22 **Late Deformities Due to Poliomyelitis.** Atrophy of the entire right hemipelvis is associated with osteoporosis, thinning of the cortex *(arrows)*, and decreased size of the right pelvis and femur. The marked scoliosis of the lumbar spine is the result of spinal muscular weakness. Poliomyelitis was a fairly common cause of scoliosis before vaccine prophylaxis was introduced.

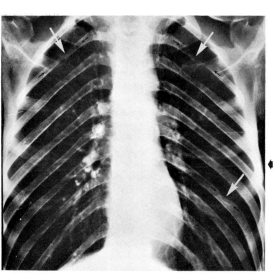

Figure 6-23 **Rib Erosions in Bulbar Poliomyelitis.** Erosion and thinning of the third posterior rib bilaterally *(arrows)* are due to scapular pressure on the paralyzed chest wall. There is also erosion of the superior border of the left eighth rib *(arrow)*. (Courtesy Dr. C. Burnstein, Buffalo, New York.)

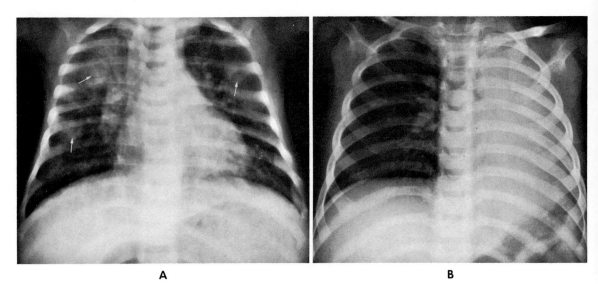

A   **B**

Figure 6–24 **Bulbar Poliomyelitis: Pneumonitis and Atelectasis in Two Patients.**

A, Bilateral patchy areas of pneumonitis *(arrows),* probably due to retained secretions, appear in infant with respiratory paralysis.

B, Complete atelectasis of the left lung is shown in a child with bulbar poliomyelitis and respiratory paralysis.

The atelectasis is the result of obstruction of the left main bronchus by thick mucus. There is a marked shift of the heart and mediastinum into the left chest, which has become dense as a result of collapse of the lung and superimposition of the cardiac mass. Surprisingly, the trachea remains in the midline. (Courtesy Dr. L. Pesce, Naples, Italy.)

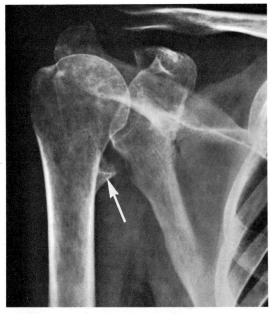

Figure 6–25 **Soft Tissue Ossification in Paralytic Poliomyelitis.** The triangular bone *(arrow)* arose as an amorphous calcification in the soft tissues. It probably is analogous to soft tissue calcium deposits found in longstanding paraplegia. Note the spotty areas of demineralization in the head of the humerus; these are caused by disuse. (Courtesy Dr. F. M. Hooper, Little Bay, Australia.)

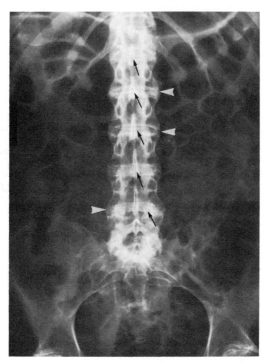

Figure 6–26 **Bulbar Poliomyelitis: Spine Changes After Prolonged Respirator Existence.** The intervertebral cartilages of the lumbar spine are extensively calcified *(arrows)*. The bamboo appearance of the spine is due to calcification of the lateral ligaments *(arrowheads)*. Both sacroiliac joints are obliterated. These changes are apparently the result of prolonged inactive recumbent existence in a respirator.

Extensive disc calcification is also seen in ochronosis. The other spine changes in the illustration are virtually indistinguishable from those of ankylosing spondylitis. (Courtesy Dr. D. Gilmartin, Melbourne, Australia.)

Coxsackie and Echoviral Disease

Infection by these viruses, particularly the Coxsackie, often involves pleural and pericardial surfaces. The myocardium may be affected by Coxsackie, sporadically in adults and epidemically in newborns and infants.

The x-ray spectrum includes simple pneumonitis, myocarditis, pleuritis with effusion, and pericarditis, occurring either individually or in combination. Both pericardial effusion and myocarditis produce enlargement of the cardiac silhouette, but generally the pulmonary vasculature is normal in the former and increased and prominent in the latter. A distinction, however, may be difficult radiographically. If indicated, angiocardiography, intravenous carbon dioxide study, or isotope scan can be employed for confirming pericardial effusion. Pericarditis occurs much more often than myocarditis.

The pneumonic infiltrates are nonspecific in appearance and location; but when infiltrates are accompanied by cardiac enlargement without pulmonary congestion, Coxsackie infection should be considered.

Resolution of the lesions, particularly the pericarditis, is slow.[2, 3, 7, 32]

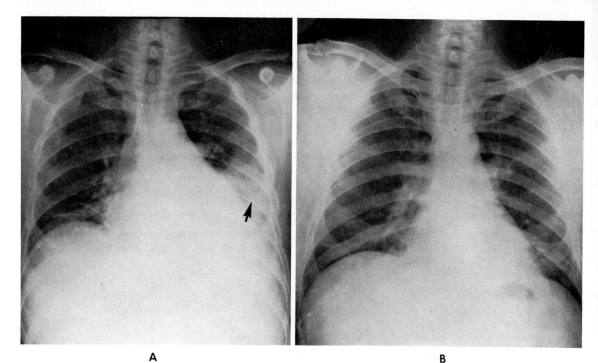

A   B

Figure 6–27 **Pericarditis and Pleuritis: Coxsackie Virus.**

A, The heart is enlarged due to pericardial effusion, and there is density at the left base due to pleural effusion and parenchymal infiltrate *(arrow)*.

B, Four weeks later, the heart is of normal size, and the lung fields and pleural cavities are normal.

The association of pleural and pericardial effusion in a febrile patient suggests a Coxsackie infection.

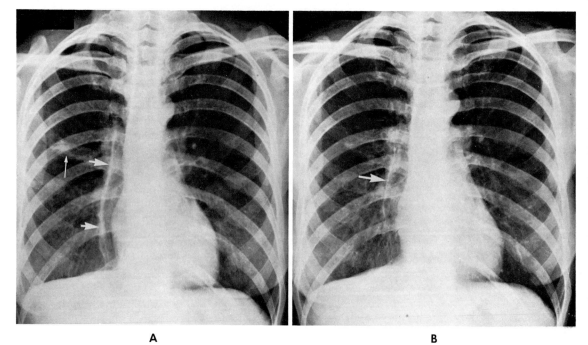

A   B

Figure 6–28 **Pneumonitis and Pleuritis: Echovirus.**

A, The long narrow area of density *(large arrows)* paralleling the heart shadow represents thickening of the posterior mediastinal pleura. A patch of pneumonitis *(small arrow)* is seen at the base of the right upper lobe.

B, One week later the pneumonia cleared, and the line of mediastinal pleurisy *(arrow)* decreased in length and width. It disappeared within the next 10 days.

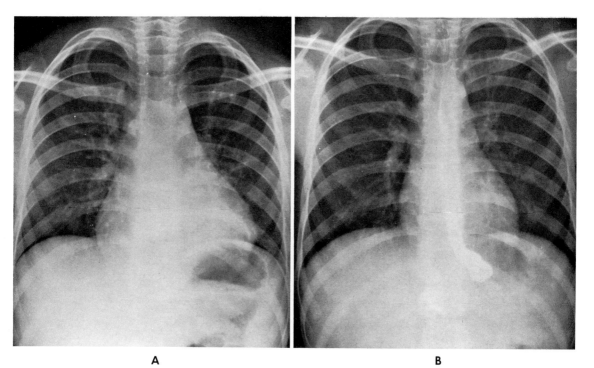

A   B

Figure 6–29 **Coxsackie Myocarditis in Adult.**

A, Cardiac enlargement, caused mainly by left ventricular dilatation and associated with increased vascular markings in the upper lobes, is indicative of mild congestion. No pneumonic involvement is seen.

B, One month later, after full recovery, the heart has returned to normal size and shape. The pulmonary vasculature is now completely normal.

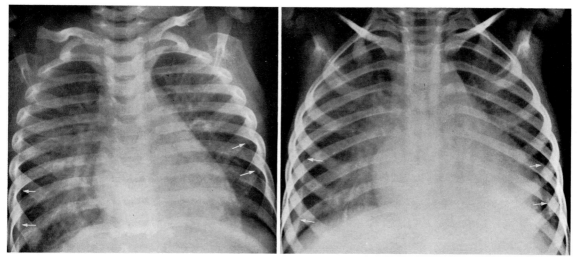

Figure 6–30 **Epidemic Coxsackie Myocarditis in Two Infants.** Generalized cardiomegaly with evidence of severe pulmonary congestion, and pleural stripes along the lateral chest walls *(arrows)* are present in both infants and are characteristic of myocardial weakness. The fatality rate is high. (Courtesy Dr. J. Munk, Haifa, Israel.)

Dengue and Related Diseases

Pulmonary involvement occurs in almost half the cases, but it is often unsuspected or masked by other symptoms of this prostrating disease.

Nonspecific bronchopneumonic infiltrates are the earliest and commonest roentgenologic findings. They may be bilateral, and large confluent areas can develop. Pleural effusion is uncommon except in severe cases.[33]

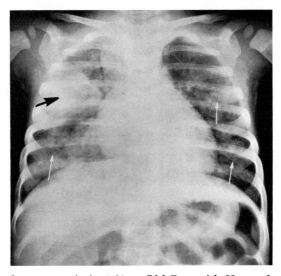

Figure 6–31 **Bilateral Bronchopneumonia in 3 Year Old Boy with Hemorrhagic Fever.** There is a large confluent pneumonic patch in the right upper lobe *(black arrow)*, patchy areas at the right base *(white arrow)*, and interstitial involvement throughout the left lung *(white arrows)*. (Courtesy Dr. E. Nelson, Bangkok, Thailand.)

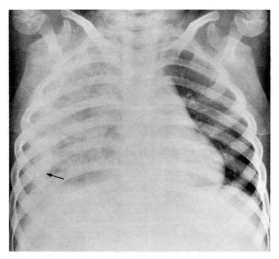

Figure 6–32 **Pneumonia and Pleural Effusion in Dengue.** A large pleural effusion from the right side opacifies the right chest and obscures the underlying bronchopneumonia. The fluid along the chest wall appears as a linear area of density *(arrow)*. The heart is shifted to the left. Pleural involvement is usually seen only in cases with extensive pulmonary disease. (Courtesy Dr. E. Nelson, Bangkok, Thailand.)

Varicella

Varicella pneumonia is extremely rare in children but occurs in about one sixth of adults afflicted with varicella. This complication usually arises three to six days after the appearance of the rash and is characterized by rapid development of acute respiratory symptoms, cyanosis, hemoptysis, and marked toxicity. In up to 20 per cent of cases the pneumonia proves fatal. The virus itself, not secondary invaders, is the etiologic agent.

Radiologically there are usually extensive bilateral fluffy miliary nodular infiltrates that are less marked at the apices. These alveolar densities vary from one millimeter to over a centimeter in diameter and in more severe cases tend to coalesce in the hilar or perihilar region. Hilar adenopathy is apparent in about 10 per cent of the patients.

In most cases the infiltrates clear completely within two to three weeks, but in some patients, small soft nodules persist for months or years. Many of the nodules may calcify two or more years after the acute episode, presenting a picture of scattered small calcifications that is quite similar to that of healed histoplasmosis. If there is a prior history of severe adult varicella, scattered calcifications are presumptive evidence of healed varicella pneumonia, particularly if the histoplasmin skin test is negative.[34-38]

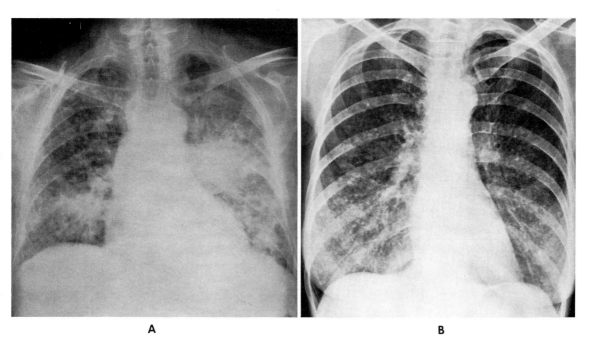

A   B

Figure 6–33 **Varicella Pneumonia in Two Adults.**

A, There are diffusely scattered, rather coarse, miliary infiltrates; large confluent areas of pneumonia are superimposed in the perihilar areas and at the bases. The patient was 67 years old.

B, Diffuse coarse miliary infiltrations in the lungs had persisted with relatively little change for over a month. The patient was a 55 year old woman with varicella pneumonia.

This miliary type of bronchopneumonia, in which confluence occurs, is seen in more severe cases of varicella pneumonia. (Courtesy Dr. R. Husebye, St. Paul, Minnesota.)

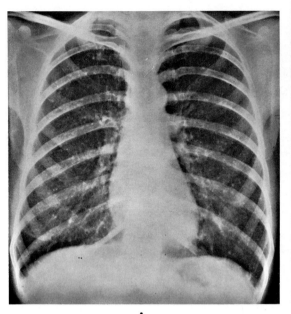 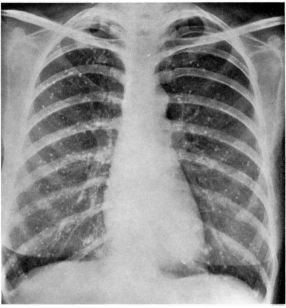

A   B

Figure 6-34 **Pulmonary Calcifications Following Varicella Pneumonia.**

A, Chest film made several years after severe adult varicella reveals widespread small nodular densities, many of which are calcified. The histoplasmin test was negative.

B, Six years later there has been a marked increase in the number and density of the calcifications.

In an individual with a history of adult varicella and a negative histoplasmin test, scattered pulmonary calcifications are most probably due to a previous varicella pneumonia in which the lesions had failed to resolve and had undergone fibrosis and eventual calcification. The frequency of late calcifications following varicella pneumonia is unknown. (Courtesy A. F. Knyvett, Brisbane, Australia.)

Cytomegalic Inclusion Disease

Generalized cytomegalovirus infection can occur in newborns and infants. The brain is involved in about 20 per cent of the cases, and in a large number of these, calcifications develop around the lateral ventricles. The demonstration of punctate bilateral calcifications so arranged as to suggest lateral ventricular dilatation is highly suggestive, although similar calcifications, often in the basal ganglia, are also seen in toxoplasmosis, primary hypoparathyroidism, pseudohypoparathyroidism, and tuberous sclerosis. Roentgenologic differentiation may be difficult or impossible.

Occasionally, nonspecific interstitial pneumonitis and diffuse bony sclerosis may be present at birth; in these cases the bones eventually undergo demineralization and atrophy. Metaphyseal changes similar to those seen in congenital rubella have been reported. Hepatomegaly and splenomegaly may be found in association with jaundice.

In patients under intensive immunosuppressive therapy, cytomegalovirus pneumonia may occur, characterized by nodular and alveolar infiltrates.[25, 39]

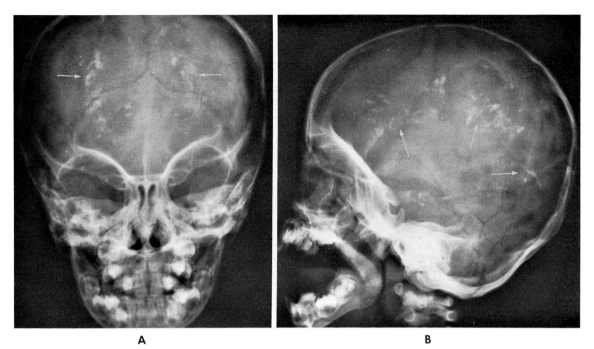

A   **B**

Figure 6–35 **Cytomegalic Inclusion Disease in Infant: The Skull.**   There are irregular small punctate calcifications *(arrows)*. Note that on both the posteroanterior *(A)* and lateral *(B)* films, the distribution of these calcifications produces a shape conforming to that of dilated lateral ventricles, which were present.

These findings are typical of cytomegalic inclusion disease. In the present case the calcifications were in the walls of the dilated lateral ventricles.

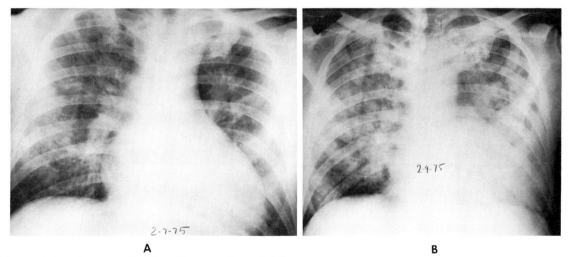

A   **B**

Figure 6–36 **Cytomegalic Virus Pneumonia in Adult.**

A, Scattered infiltrates appeared in a 44 year old man who was being treated with steroids for chronic myeloproliferation disease (myeloid metaplasia).

B, Two days later, there was a marked increase in the diffuse bilateral alveolar infiltrates, without pleural reaction. Death occurred 48 hours later.

Immunosuppressive therapy is almost always the predisposing factor of cytomegalic viral infections in adults. The rapid development of diffuse bilateral alveolar infiltrates, often with earlier nodular densities, is characteristic but not diagnostic.

RICKETTSIAL DISEASES

Q Fever

With the exception of Q fever, significant roentgenologic findings are unusual in rickettsial diseases. Occasionally, nonspecific pneumonias may develop, but these are frequently caused by secondary infection.

In about half the cases of Q fever, a pneumonia occurs that is radiographically unlike ordinary viral pneumonia. There is dense segmental or lobar consolidation, resembling pneumococcal pneumonia. Bilateral involvement occurs in about 25 per cent of patients. Hilar adenopathy is rare. A fibrinous pleural exudate, which clears within one to ten weeks, develops in about a third of the cases.[40, 41]

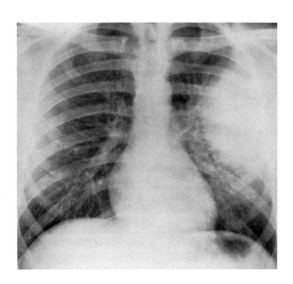

Figure 6–37 **Lobular Consolidation in Q Fever.** There is a dense homogeneous area of consolidation that is located in the left upper lateral aspect of the lung and extends to the pleural surface, resembling pneumococcal pneumonia. The findings are characteristic of Q fever. (Courtesy Dr. P. D. Hoeprich, Salt Lake City, Utah.)

Figure 6–38 **Bilateral Pneumonia and Pleurisy in Q Fever.** Mottled density over much of the right lower lobe is associated with a pleural reaction in the right costophrenic angle *(black arrow)*. A smaller homogeneous area of dense infiltration is seen in the left lower lobe *(white arrow)*. (Courtesy Dr. P. D. Hoeprich, Salt Lake City, Utah.)

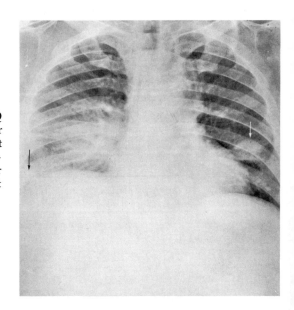

BACTERIAL DISEASES

Pneumococcal Pneumonia

Infection of the lung by aspiration of *Diplococcus pneumoniae* results in an alveolar or air-space pneumonia that affects especially the basilar segments and the posterior segments of the upper lobes. A homogeneous dense consolidation of a lobar segment or an entire lobe is usual. Invariably, the consolidation is limited by some portion of the visceral pleura, which usually also becomes involved. Often, air bronchograms are seen, diagnostic of fluid or exudates in the alveoli. Sometimes, a mild degree of atelectasis due to bronchial involvement and obstruction is noted. With resolution, the density becomes mottled, as scattered alveolar aeration occurs. Radiographic clearing of consolidation is generally complete in from one to three months. However, pleural densities, stranding, and some volume loss may persist for up to four months, especially in individuals who are over 50 years of age or have chronic airway disease. Slow resolution is not indicative of underlying malignancy.

The commonest complication is pleurisy with effusion, and this occurs in less than 10 per cent of cases. The effusions are small and sterile.

Empyema develops in less than 2 per cent of the patients and may be radiographically indistinguishable from a sterile effusion. The thicker empyema fluid has a tendency to loculate, forming circumscribed extrapulmonic densities with a convex central border. Resolution of empyema is slow, and late pleural calcification commonly develops.

Pulmonary abscess is now a rare complication. At first it appears as a circumscribed density in the resolving pneumonic area; with its extension into a bronchus, a pathognomonic air-fluid level arises.

The toxic ileus that frequently accompanies pneumonia is manifested by dilated loops of small and large bowel and a distended abdomen. Occasionally, abdominal symptoms mask the primary pulmonary disease, and the abdominal film findings should suggest a chest examination.

Rarer complications with radiographic findings include bronchopleural fistulas, pericarditis with effusion, mastoiditis, sinusitis, acute gastric dilatation, and congestive heart failure.[42, 43, 43a]

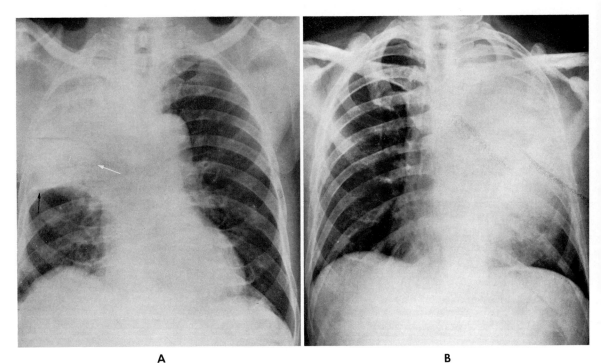

A   B

Figure 6–39   **Pneumococcal Pneumonia of Upper Lobes in Two Patients.**

A, There is complete consolidation of the right upper lobe. Note the homogeneous area of density abruptly terminating at the minor fissue *(black arrow).* There is no change in volume; the heart and mediastinum are not shifted. An air bronchogram *(white arrow)* confirms the presence of alveolar exudate.

B, The homogeneous area of consolidation in the entire left upper lobe extends to the diaphragm medially; the single fissure of the left lung extends obliquely as far as the diaphragm. The left heart border is obscured. Lateral film *(not shown)* defines the area of consolidation.

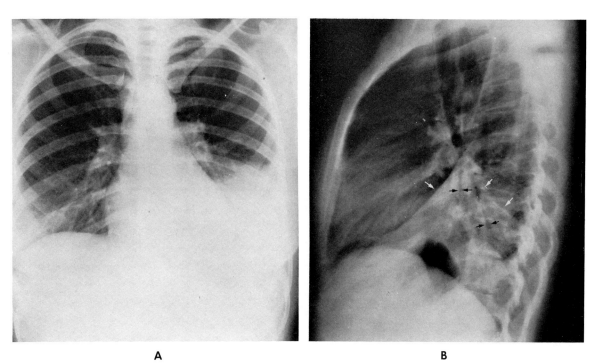

A   B

Figure 6–40 **Segmental Pneumococcal Pneumonia: Left Lower Lobe.** The homogeneous area of density at the left base involves the basal segments of the left lower lobe, obscuring the diaphragm. On the posteroanterior film *(A)* this appearance simulates an effusion. On lateral view *(B)* the segmental consolidation is evident *(white arrows)*, limited anteriorly by the long fissure. An air bronchogram is clearly seen *(black arrows)*.

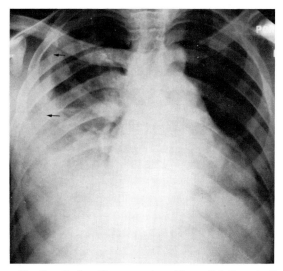

Figure 6–41 **Empyema Complicating Lobar Pneumonia.** Consolidation in the right lower lobe is somewhat obscured by extensive purulent effusion. The diaphragm is also obscured.

With loculation of empyema and clearing of pneumonia, a characteristic sharply demarcated area of density *(arrows)* is seen against the pleural surface.

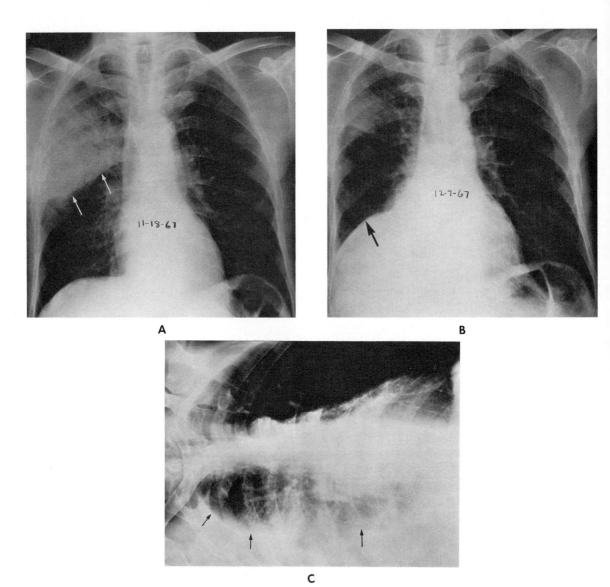

Figure 6–42 **Pneumococcal Pneumonia: Late Purulent Effusion.**

A, Chest film made on the fifth day of a lobar pneumonia reveals a fairly homogeneous consolidation of the anterior segment of the right upper lobe, sharply demarcated by the interlobal fissure *(arrows)*. There is no pleural reaction, and the right costophrenic angle is clear and sharp.

B, Reexamination, made three weeks later because of low grade fever, shows considerable resolution of the pneumonic density. The right diaphragm *(arrow)* appears elevated, suggesting some complication. The costophrenic angle remains sharp.

C, A right lateral decubitus film demonstrates a large amount of pleural fluid *(arrows);* what appeared to be an elevated right diaphragm was an infrapulmonic effusion (see p. 559 and Fig. 9–146). The fluid was purulent.

Empyema or lung abscess should be suspected if fever recurs during clinical resolution of a lobar pneumonia.

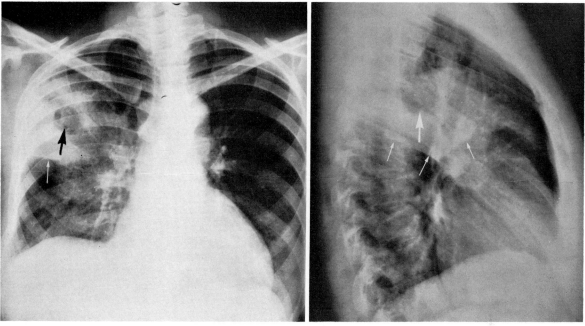

A   B

Figure 6–43 **Pneumococcal Pneumonia: Abscess Formation.**

A, There is consolidation of the anterior and posterior segments of the right upper lobe, and also a large abscess cavity *(black arrow)* containing a fluid level. The well-defined lower border of the consolidation *(white arrow)* represents the short horizontal fissure between the middle and upper lobes.

B, In the lateral film, consolidation extends to the fissure of the lower lobe *(two small arrows)* and to the short fissure anteriorly *(small arrow).* The abscess cavity and fluid level are in the middle of the area of consolidation *(large arrow).*

Abscess generally occurs during the later stages of pneumococcal pneumonia. It is an infrequent complication since the advent of antibiotics.

Klebsiella (Friedländer's) Pneumonia

This acute alveolar pneumonia usually appears as a dense confluent lobar or lobular consolidation, although it usually begins as a bronchopneumonia. There is marked predilection for an upper lobe, but lower lobe and bilateral involvement is not rare. The volume of the affected lobe or portion of lobe is often increased because of severe interstitial edema. This may cause bowing of an adjacent fissure away from the consolidation, a finding sometimes seen only on a lateral film. The advancing border of the consolidation is surprisingly sharp and distinct.

Cavitation occurs frequently and rapidly; an early appearing rarefaction in an acute lobar consolidation is suggestive of Klebsiella infection. The cavities may be multiple, may increase rapidly in size, and may become thin-walled. Occasionally, lung gangrene occurs. Pleural reaction and empyema are unusual but can occur. After resolution, the lobe may undergo fibrosis and volume loss.

An acute upper lobe consolidation with bulging fissures, a sharp advancing border, and early cavity formation strongly suggest the diagnosis of Klebsiella pneumonia. Of all the other lobar or pseudolobar pneumonic consolidations, only plague pneumonia frequently exhibits bulging fissures.

In chronic Klebsiella pneumonia, multiple areas of fibrosis and cavitation may simulate tuberculosis. The infection is usually superimposed on underlying chronic lung disease such as emphysema and bronchitis.[7, 43, 44]

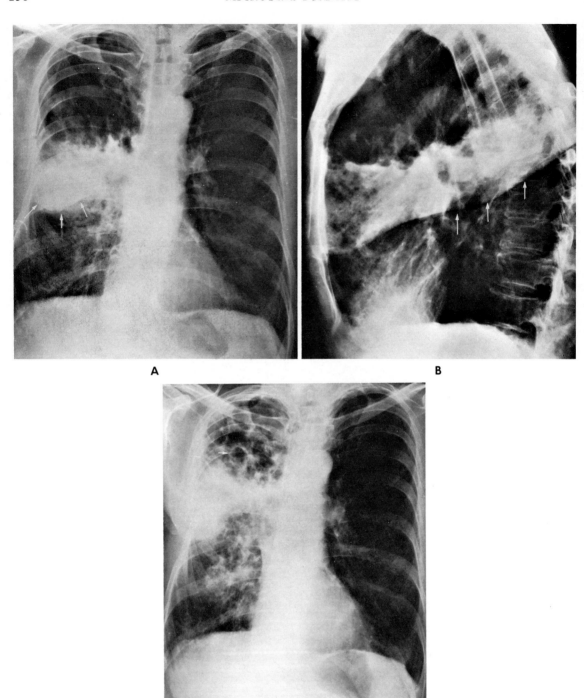

Figure 6–44 **Klebsiella Pneumonia.**

A, There is a dense area of consolidation above the fissure in the inferior portion of the right upper lobe. Bulging of the fissure *(arrows)* is due to the increased volume caused by consolidation. The patient also had longstanding pulmonary emphysema.

B, On the lateral film the relation of the area of consolidation to the long and short fissures is indicated. The entire length of the inferior portion of the upper lobe is involved. The fissure is again seen to bulge *(arrows).* Note the rather sharp margins of the upper border of the consolidation.

C, Numerous thin-walled cavities in the upper lung field *(arrow)* above the original consolidation are seen one week later. The disease has also extended to the right lower lung field.

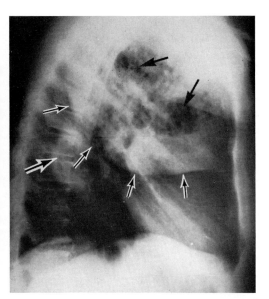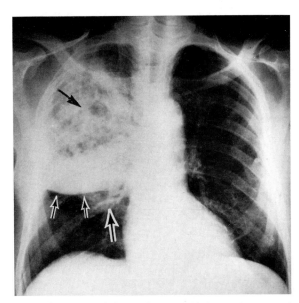

Figure 6–45 **Klebsiella Pneumonia.**   Chest films show involvement of the right upper lobe with marked downward bowing of its fissures *(small black-white arrows)* due to the swollen, engorged lobe; these phenomena are often seen in Klebsiella pneumonia.

The upper half of the lobe is undergoing necrosis and cavitation *(black arrows)*. The disease has spread to the lower lobe *(large black-white arrows)*.

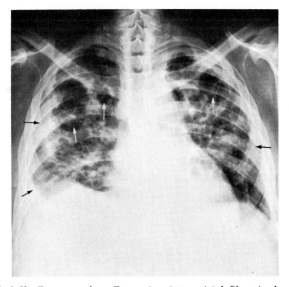

Figure 6–46 **Chronic Klebsiella Pneumonia.**   Extensive interstitial fibrotic densities are seen throughout both lung fields. There are numerous cavities of varying sizes, mainly in the upper lobes *(white arrows)*, and extensive pleural thickening along both lateral walls *(black arrows)*. In this case, Klebsiella pneumonia failed to undergo resolution. Both roentgenologically and clinically, the disorder frequently simulates advanced tuberculosis.

Pseudomonas and Other Gram-Negative Pneumonias

Serious pulmonary infections due to gram-negative bacteria in hospitalized patients are not uncommon. Pseudomonas is the most common offender; other infecting organisms include klebsiella, *Escherichia coli*, proteus, and bacteroides. Factors predisposing to the gram negative infections are prolonged antimicrobial therapy, immune deficiencies especially during extensive periods of immunosuppressive therapy, and general debilitation occurring in alcoholics, postoperative states, and so forth.

In pseudomonas pneumonia there is initially a *nodular* bronchopneumonia due to acinar consolidation. The irregular acinar nodules become confluent, and lucencies due to microabscesses become apparent. While there is a tendency to lower lobe involvement, bilateral diffuse patches of alveolar involvement are fairly frequent. Larger abscesses, up to 10 cm. in diameter, may rapidly develop. A moderate pleural reaction usually occurs, evidenced by localized pleural densities or a small effusion. The early nodularity, the lower lobe preponderance, the confluence with small lucencies, and the later abscess formation are suggestive findings.

Klebsiella pneumonia (see p. 157) characteristically causes massive alveolar and lobar consolidation with increase in lung volume resulting in bulging of the adjacent fissures. Cavitary abscess formation is frequent. Large effusions and empyemas are common.

The other gram-negative pneumonias are much less frequent, although no less life-threatening. The radiographic picture is usually similar to that of pseudomonas infection.[45, 46]

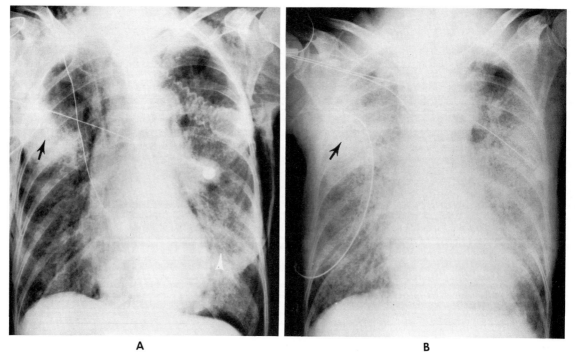

A   B

Figure 6–47 **Pseudomonas Pneumonia.**

A, Involvement is characteristically bilateral. A confluent pneumonia in the right upper lobe *(black arrow)* and early nodular infiltrates in the right lower lung are associated with diffuse alveolar and small nodular densities *(white arrowhead)* in the left lung. The small lucencies within the involved lung are probably uninvolved acini.

B, In two days, the alveolar involvement has become more diffuse and generalized. Note the characteristic small lucencies within the infiltrated areas. The confluent area *(arrow)* later became an abscess.

Pleural reaction is generally not marked in Pseudomonas pneumonia.

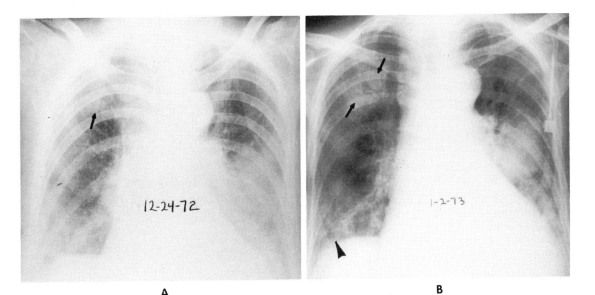

A   B

Figure 6–48 **Pseudomonas Pneumonia.**

A, There is an extensive alveolar infiltrate involving most of the left lower lobe, with obliteration of the left diaphragm. A small infiltrate is seen in the right upper lobe *(arrow)*.

B, Nine days later, there is some resolution of the left lower lobe density, with many lucent areas within the infiltrate. The right upper lobe infiltrate has increased *(arrows)*. Infiltrates are seen above the right diaphragm *(arrowhead)*.

C, Ten days later, there is a residual sharply bordered density in the left midlung field, which contains small lucencies *(white arrowhead)*. The right upper lobe lesion has undergone cavitation *(arrow)*. A mild pleural reaction is noted at the right base *(black arrowhead)*.

Multiple areas of alveolar consolidation, later cavitation, and rather modest pleural reaction are characteristics of the gram-negative pneumonias, particularly Pseudomonas pneumonia.

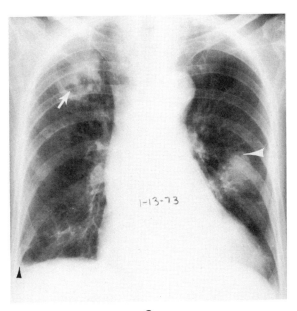

C

Streptococcal Diseases

No significant roentgenologic findings are produced by the majority of streptococcal infections, which involve mainly soft tissues and lymph nodes, and sometimes the bloodstream and meninges.

Pulmonary involvement, although now quite rare, can produce segmental consolidation and suppuration and give rise to a roentgenologic picture that is similar to the more common staphylococcal pneumonia, although pneumatoceles are less frequent. Dependent portions of the lung are most frequently involved, sometimes with volume loss. Empyema occurs in over half the cases of streptococcal pneumonia, especially if the pneumonia is untreated. Streptococcal pneumonia most often occurs as a secondary complication of other pulmonary disease, especially viral infections.[7, 47]

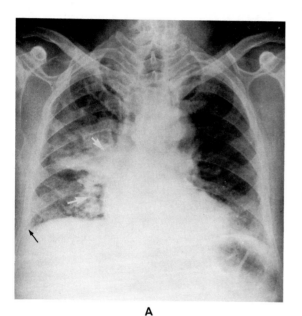

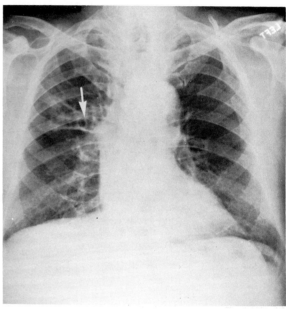

A

B

Figure 6–49 **Streptococcal Pneumonia with Abscess Formation.**

A, A confluent area of bronchopneumonia extends from the right hilum into the right upper *(white arrow)* and lower *(white arrow)* lung fields. Early pleural involvement causes the right costophrenic angle to appear hazy *(black arrow).* The left lung is not affected.

B, After vigorous antibiotic therapy, consolidation and pleural reaction have cleared. The anterior segment of the right upper lobe is honeycombed with multiple cavitations *(arrow).*

C, Chronic suppurated segment is clearly shown on lateral view. The lobe is shrunken and filled with numerous cavities *(arrow).*

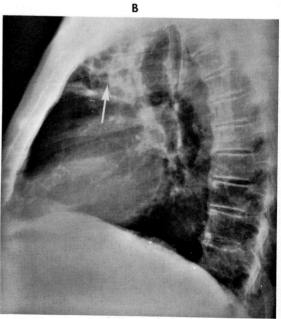

C

Rheumatic Fever

There are no roentgenographic findings in most cases of acute rheumatic fever.

If severe myocarditis develops, there will be generalized cardiac enlargement, often with fluctuation in heart size as the disease activity varies. Pulmonary vascular congestion and pulmonary edema may appear if congestive heart failure develops.

Rheumatic pneumonitis is a grave but fortunately rare complication. Most often it presents as bilateral confluent homogeneous alveolar densities. Since myocarditis and cardiac enlargement are often present in these severe cases, the pneumonitis may closely simulate pulmonary edema, and the distinction is often difficult. Some differentiating characteristics of the pneumonitis are the changing appearance of the persistent infiltrates, the clear or lucent zone that often separates the pneumonitis from the hilar area, and the occasional unilateral involvement. At times, unequivocal segmental and lobular pneumonic infiltrates occur. Pleural effusion rarely accompanies the pneumonitis.

The polyarticular symptoms are not attended by x-ray changes in the joints, other than periarticular swelling. Rarely, after resolution of the acute rheumatic fever and the active polyarthritis, deformities of the hands and/or feet, known as Jaccoud's arthritis, may develop. There will be ulnar deviation at the metacarpo- or metatarsophalangeal joints without osseous changes. This is probably a form of periarticular fibrosis.[48-50]

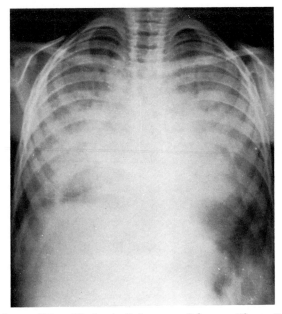

Figure 6–50 **Rheumatic Myocarditis with Acute Pulmonary Edema.** The patient is a 4 year old boy. Cardiomegaly is not evident because of pulmonary edema that has resulted in diffuse densities extending from both hila toward the periphery. The picture is characteristic of pulmonary edema. When pulmonary edema occurs in association with rheumatic fever, it may obscure the underlying pneumonitis.

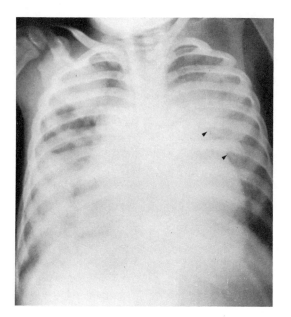

Figure 6–51 **Bilateral Rheumatic Pneumonitis and Myocarditis.** The heart is enlarged. There is extensive bilateral infiltration, which is segmental on the right. Although this picture is indistinguishable from that of pulmonary edema, the changing pattern of the densities over a two-month period confirms the pneumonic nature of the process. Also note the suggestion of a clear zone in the left hilar area *(arrowheads)*. The cardiomegaly was due to associated myocarditis.

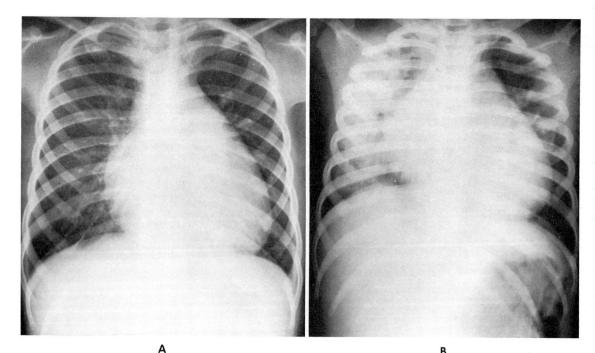

A   B

Figure 6–52 **Development of Rheumatic Pneumonitis: Myocarditis.**

A, On the patient's admission the lung fields were clear. Cardiac enlargement is apparent and due to the rheumatic myocarditis.

B, Twenty-two days later there is an extensive homogeneous area of infiltration on the right side. Note the clear zone between the consolidation and the heart; this is generally not seen in pulmonary edema. Cardiomegaly has increased. The child did not survive. (Courtesy D. Goldring: J. Pediatr., *53:*547, 1958.)

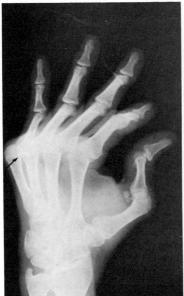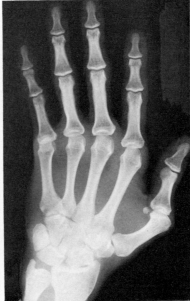

Figure 6–53 **Jaccoud's Arthritis Associated with Subsiding Rheumatic Fever.**   Ulnar deviation and a flexion deformity of the metacarpophalangeal joints, with subluxation of the fifth metacarpal *(arrow)*, are characteristic in this rare arthritis, which occurs with subsiding rheumatic fever. Note the periarticular swelling. (Courtesy H. L. Twigg and B. F. Smith: Radiology, *30*:417, 1963.)

Staphylococcal Diseases

Staphylococcal infection can involve any tissue or organ, but significant roentgenologic findings occur most often in the lungs, bones, joints, and occasionally the kidney.

Staphylococcal pneumonia is more frequent in children. It develops very rapidly, and extensive alveolar infiltrates may appear in a matter of hours. The smaller bronchi are filled with exudate, and air bronchograms are infrequent. Rapid involvement of multiple lung areas commonly occurs. Early involvement of the pleura leads to pleural effusion and often to empyema. Pneumatoceles are characteristic of the disease, particularly in children. These are thin-walled, single or multiple spherical air cysts within the parenchyma. They are of variable size and occasionally are large enough to fill a hemithorax. Fluid levels sometimes are seen. Rupture frequently occurs, producing a pneumothorax. Pneumatoceles may persist radiographically for months after recovery, although usually they disappear within six weeks. Lung abscesses develop more often in adults. A presumptive diagnosis of staphylococcus pneumonia can be made in a child with rapidly developing pneumonia in association with pleural effusion, pneumatoceles, or a pneumothorax. In adults, involvement is usually bilateral. Abscesses, often multiple, commonly develop, frequently with an irregular ill-defined inner wall. Pneumatoceles are less frequent in adults than in children.

Staphylococcal osteomyelitis is usually hematogenous. It may be present clinically up to 10 days before x-ray changes appear. Soft tissue swelling and obliteration of fat lines may be the earliest findings; later, areas of ill-defined bone lysis with or without periosteal reaction develop. Occasionally the periosteal reaction precedes the lytic lesions. As the disease becomes more chronic, periosteal new bone is laid down, and sclerosis develops around the lytic areas. Dead bone — sequestra — may appear as dense islands within the lytic areas. The periosteal new bone may form a surrounding shell, the involucrum.

Staphylococcal osteomyelitis resulting from extension of soft tissue infection causes early destruction of cortex and underlying bone.

Involvement of a joint causes effusion, which may at first actually widen the joint space, especially in children. Rapid cartilage destruction causes joint narrowing, and periarticular demineralization and bone destruction often occur. Bony ankylosis of the joint may develop.[7, 51–54]

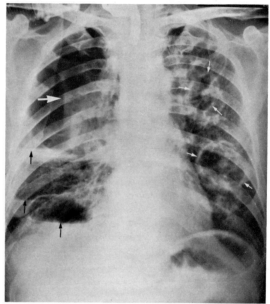

Figure 6–54   **Staphylococcal Pneumonia in Adult.**   There are several thin-walled pneumatoceles in the left lung *(small white arrows),* and there is also hydropneumothorax of the right chest *(large white arrow)* with several fluid levels *(black arrows)* at the right base. Infiltrates are scattered throughout both lungs. This combination of findings is highly suggestive of staphylococcal pneumonia.

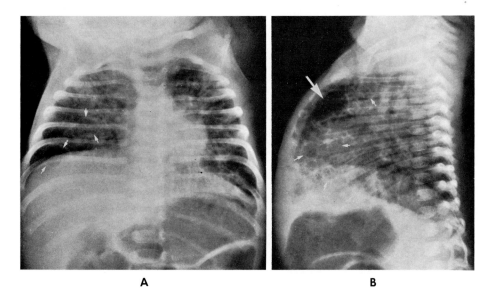

A   B

Figure 6–55 **Staphylococcal Pneumonia in Infant.**

A, Posteroanterior view shows diffuse infiltration of both lung fields with some areas of confluence in the upper lobes. Numerous thin-walled pneumatoceles are scattered throughout; the larger are more easily identified *(arrows).* Note areas of interstitial emphysema, best seen at the bases.

B, Lateral view clearly demonstrates the numerous pneumatoceles *(small arrows).* A large pneumatocele *(large arrow)* fills the anterior superior space.

Frequently a pneumatocele ruptures and produces pneumothorax. The presence of pulmonary infiltrates, pneumatoceles, and pneumothorax is characteristic of staphylococcal pneumonia in children.

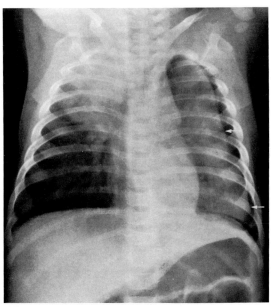

Figure 6–56 **Staphylococcal Pneumonia in Youth: Pneumothorax.** There is homogeneous consolidation of the right upper lobe and of the entire left lung, with pneumothorax of the entire left lung *(arrows).* The extensive consolidation prevents further collapse of the left lung. The pneumothorax is due to rupture of a pneumatocele, although pneumatoceles cannot be identified on this film.

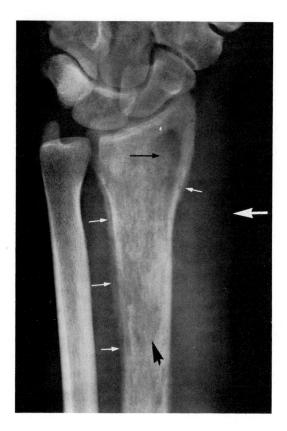

Figure 6–57 **Subacute Osteomyelitis of Lower Radius.** Soft tissue swelling *(large white arrow)* is usually the earliest finding. The lytic area is poorly defined *(small black arrow)*; other areas of lysis are seen in the medullary cavity *(large black arrow)*. Periosteal reaction is fairly extensive *(small white arrows)*. Note well-defined bone trabeculae in the adjacent uninvolved ulna.

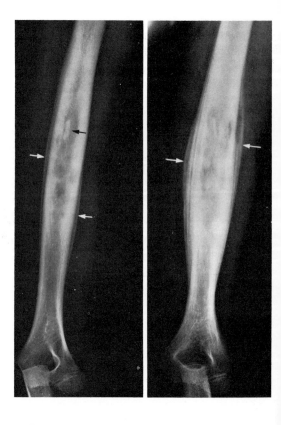

Figure 6–58 **Chronic Active Osteomyelitis of Humerus.** There is marked thickening of the cortex, and several layers of new periosteal bone formation are seen *(white arrows)*. Widening of the shaft is an indication of the chronicity of the process. The outline of the medullary shaft is irregular, and the lytic areas are partially obliterated by the new formation. An island of dead bone (sequestrum) is present *(black arrow)*. These findings indicate that the process is in an active phase.

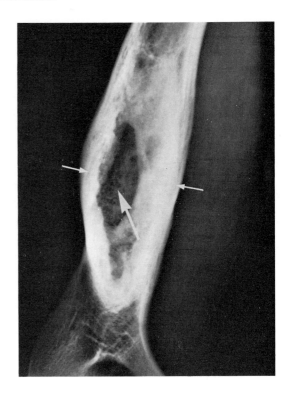

Figure 6–59 **Chronic Osteomyelitis of Humerus.** There is marked cortical thickening, reactive dense bone (*small arrows*), and a large defect (*large arrow*) in the widened shaft. Such cortical thickening associated with areas of lysis is characteristic of a chronic osteomyelitis of almost any infective origin. In many cases the extensive cortical thickening may completely obliterate the lytic areas; planigraphy or an overpenetrated film may aid in demonstrating these areas.

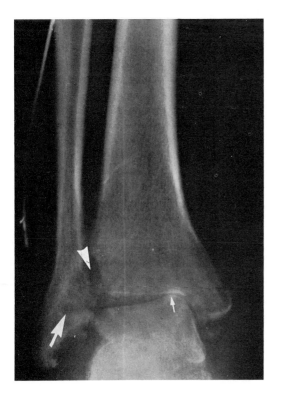

Figure 6–60 **Septic Arthritis and Osteomyelitis Secondary to Staphylococcal Septicemia in a Diabetic.** Rapid suppuration of the joint cartilage has led to pronounced narrowing of the joint space (*small arrow*) and lateral subluxation of the fibula (*arrowhead*). There are several areas of subarticular absorption. The lower fibula and tibia (*large arrow*) have been damaged, and there is demineralization of the joint. Soft tissue swelling, although marked, is not readily apparent in this film.

In pyogenic arthritis there is usually rapid destruction of the articular cartilage. The joint spaces become narrowed early in the course of disease, and there is extensive demineralization around the involved joint.

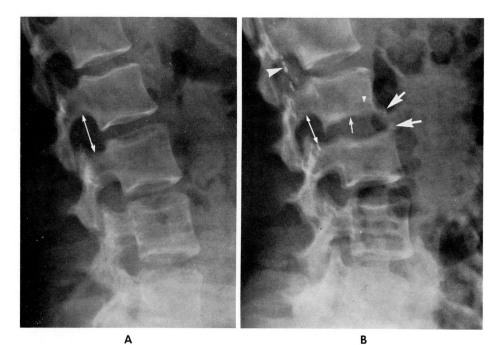

| A | B |

Figure 6–61 **Staphylococcal Osteomyelitis of Lumbar Spine (Infectious Spondylitis).**

A, No abnormalities are apparent in this lateral lumbar spinal film of a young woman who had severe lumbar pain and fever.

B, Five weeks later the film shows destruction of the inferior surface of L2 *(small arrow);* the lytic area extends into the body *(small arrowhead).* The superior surface of L3 is also involved and has become eroded, concave, and slightly sclerotic. Reactive exostoses have developed in the anterior aspects of the involved vertebrae *(large arrows).* Although the interspace appears widened, there has actually been destruction of the intervertebral cartilage, as evidenced by the decreased height of the neural foramen *(double-headed arrow)* (compare with *A).* Opaque oil in the spinal canal *(large arrowhead)* is a residue from a previous myelogram.

Radiographically, pyogenic osteomyelitis of the spine characteristically appears weeks after onset of clinical symptoms, progresses rapidly, and leads to bone destruction and irregularity of the intervertebral space and to reactive sclerosis and spurring. A paraspinal soft tissue mass is usually seen. With healing, there is usually decreased vertebral height and narrowing of the intervertebral space.

Staphylococcal osteomyelitis frequently cannot be distinguished from tuberculous spondylitis except by the rapid progression of the pyogenic process.

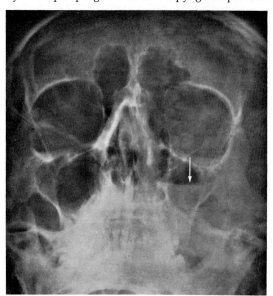

Figure 6–62 **Acute Staphylococcal Maxillary Sinusitis.** A suppurative exudate with an air-fluid level appears in the left maxillary antrum *(arrow).* When the patient is recumbent, a diffuse density in a paranasal sinus can be seen. The density can be caused by extensive mucosal thickening or edema, or by exudate. Therefore, erect views are necessary in order to distinguish fluid from extensive mucosal thickening.

Gonococcal Disease

Roentgenologic changes are not seen in acute gonococcal urethritis, but they may appear in the complications of arthritis or pelvic inflammatory disease.

Gonococcal arthritis may initially affect multiple joints, particularly the knees, wrists, elbows, and ankles. Usually, periarticular soft tissue swelling is the only roentgen finding. More severe and progressive involvement is usually limited to a single joint. The articular margins become hazy and indistinct; joint narrowing from cartilage destruction may appear. The articular margins may become eroded, and periarticular osteoporosis can occur. The roentgen appearance is identical to that of other types of suppurative arthritis; in the more protracted cases it resembles that of joint tuberculosis.

Pelvic inflammatory disease in the female can lead to tubo-ovarian abscesses. A mass may be seen that may produce pressure or narrowing of a bowel loop. Occlusion of the outer end of one or both fallopian tubes often develops and this can be demonstrated on hysterosalpingography.

Urethral stricture may be a sequela of the urethritis; it is readily demonstrated on urethrography.[55, 56]

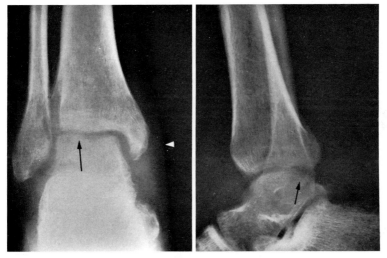

Figure 6–63 **Acute Gonococcal Arthritis.** The joint space is narrowed, and the cortex appears fuzzy *(arrow)*, but no bony destruction is seen. There is periarticular soft tissue swelling *(arrowhead)* and moderate periarticular osteoporosis. The appearance is similar to that of any other pyogenic arthritis.

Bacterial Meningitis

Although there are no specific radiologic findings in the various meningitides, the skull, sinuses, mastoid, and chest should be examined as possible sites of origin of the meningeal infection. If brain abscess is suspected, a CT scan is indicated.

Primary Haemophilus influenzae Pneumonia

Primary *Haemophilus influenzae* pneumonia can occur in adults but is more common in infants. Radiographically it is usually lobar or lobular, and unilateral, simulating pneumococcal pneumonia. However, unlike the latter, extensive pleural reaction is usual and is manifested as fibrinous exudate with a thick pleural line; free fluid is scanty. The pleural "thickening" may dominate the picture.

Unlike staphylococcal pneumonia in the young, the process is static rather than progressive, and pneumatoceles do not occur. Unless antibiotic therapy is promptly instituted, empyema may develop. Other complications include metastatic abscesses, septic arthritis, and meningitis.

In adults with pre-existing chronic pulmonary disease, such as asthma, chronic bronchitis in heavy smokers, or bronchiectasis, a complicating hemophilus pneumonia presents as a diffuse miliary bronchopneumonia, without lobar consolidation or significant pleural reaction.

A lobar pneumonia with prominent pleural reaction in an infant or young child should suggest the possibility of *Haemophilus influenzae* pneumonia.[7, 57, 58]

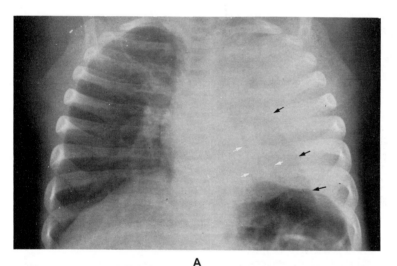

A

B

Figure 6–64 Haemophilus influenzae **Pneumonia and Pleurisy in Two Infants.**

A, There is complete lobar consolidation of the left lung, with an air bronchogram *(white arrows)* in the lower lobe. Extensive pleural reaction is seen along the entire lateral chest *(black arrows),* but only a small amount of fluid was obtained by thoracocentesis.

B, In the second infant, extensive pleural reaction is evidenced by the density *(arrows)* along the left lateral chest. No definite pneumonia ever developed in this patient. No pleural fluid could be obtained after numerous thoracocenteses. (Courtesy Dr. D. Altman: Radiology, *86*:701, 1966.)

Whooping Cough (Pertussis)

Roentgenologic findings are limited to the lung; there may be focal atelectasis, hyperinflation, segmental pneumonia, or interstitial infiltrates. In about one third of these cases dense pulmonary markings fan out from the hila and tend to obscure the sharp cardiac contour. These usually occur during the paroxysmal phase of the disease. A similar "shaggy heart" contour may occur in aspiration pneumonia, cystic fibrosis, measles pneumonia, and many interstitial pneumonias. Hilar adenopathy can be noted in about one third of cases.[37, 59]

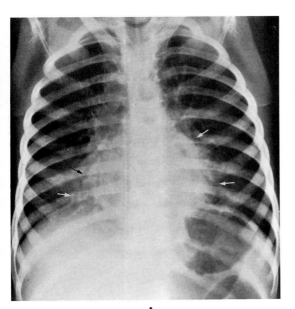 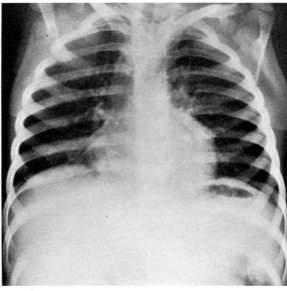

A   B

Figure 6–65 **Pertussis Pneumonia: "Shaggy Heart."**

A, Densities radiating from the hila obscure the cardiac border *(white arrows)* and produce the "shaggy heart" appearance. Air bronchogram at the right base *(black arrow)* confirms pneumonic consolidation of the alveoli. The lungs are hyperinflated; the diaphragm is low.

B, This film, made after clearing, demonstrates the normal sharply outlined cardiac border. The fan-shaped densities seen in this type of pneumonia are thought to be the result of bronchial plugging by mucus. (Courtesy Dr. M. Barnhard: Am. J. Roentgenol. Radium Ther. Nucl. Med., *84:*445, 1960.)

Donovanosis (Granuloma Inguinale)

This common venereal disease of women, caused by the microorganism *Donovania granulomatis*, is characterized by genital lesions; extragenital lesions are rare.

Very infrequently, systemic dissemination occurs and bone lesions may appear. The distal ends of long bones and almost any flat bone may be involved. Radiographically, these lesions are osteolytic, with some marginal sclerosis developing in longstanding cases. Periosteal reaction may be seen, usually during the healing phase.

These lesions can resemble fibrous defects, osteolytic metastases, myeloma, or coccidioidomycosis.[60]

Clostridial Infections

Gas gangrene is recognized radiographically by lucent streaks or bubbles of gas within swollen soft tissues. These lucencies cannot be radiographically distinguished from air introduced through a penetrating injury. However, in the latter instance the air is seen immediately after the injury and tends to decrease steadily. In gas gangrene the gas lucencies do not appear until a few days after onset of clinical symptoms, and they increase rather rapidly unless specific therapy is promptly instituted.

Enteritis necroticans is a severe jejunoileitis caused by *Clostridium perfringens*. It is characterized by edema, congestion, and ulceration of the bowel wall. In the acute stage abdominal films will show a paralytic ileus with dilated thickened loops. Extraluminal air-fluid collections can result from complicating bowel perforation. In the subacute or chronic stage, the jejunal loops may be rigid and "pipe-like," with thickened walls and obliterated mucosal pattern. Segmental stenoses and internal fistulas can occur. The late fibrotic stage can present a sprue-like small bowel pattern.

Infrequently, clostridial gas-forming infections can involve the uterus, urinary tract, gallbladder, and even the lungs. A necrotizing lobar pneumonia may occur. Empyema and pyopneumothorax (gas formed by bacilli) are sometimes seen. Gas formation can also occur with *Escherichia coli* infections.[61, 62]

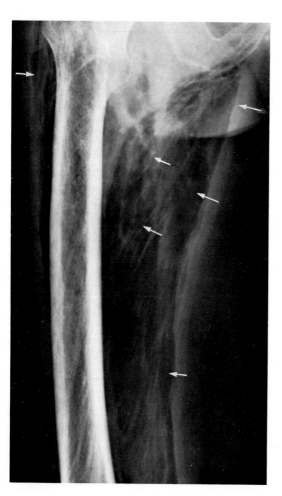

Figure 6–66 **Gas Gangrene of Thigh Caused by Contamination of Wound.** Streaky translucencies *(arrows)* in the muscle bundles and in the subcutaneous tissues of the thigh are accumulations of gas liberated in the tissues by clostridia.

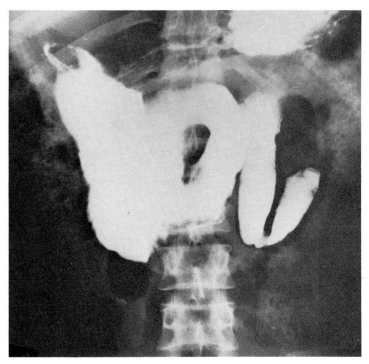

Figure 6–67 **Enteritis Necroticans: Late Jejunal Changes.** The jejunal loops are rigid and somewhat dilated, and show loss of mucosal pattern.

The conglomerate barium in the right upper quadrant was due to an internal fistula. (Courtesy D. J. Bassett, Gosford, Australia.)

Figure 6–68 **Postoperative Gas Gangrene of Abdominal Wall.** Recumbent view of abdomen reveals localized accumulations of gas *(arrows)* that were associated with infection of the anterior abdominal wall caused by *Clostridium welchii.*

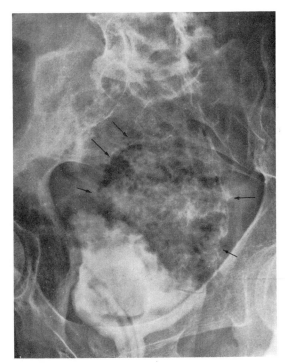

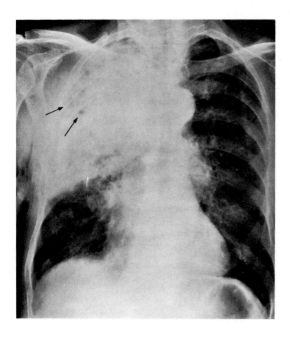

Figure 6–69 **Pneumonia: Lung Abscess Liberation of Gas Due to** *Clostridium perfringens.* There is massive consolidation of the right upper lobe with multiple small lucent areas *(arrows)* that were pockets of gas within a large noncommunicating abscess. Pulmonary involvement by clostridia is quite rare.

Tetanus

Some complications of tetanus may produce roentgenologic findings. Severe compression fractures of the mid-dorsal vertebrae are not uncommon following convulsions. A toxic myocarditis may lead to cardiomegaly and pulmonary edema. Pneumothorax, rib fractures, atelectasis, and pneumonitis are encountered. Most often these are complications of therapy.

Occasionally, focal calcification develops in a muscle—a myositis ossificans probably from muscle injury during tetanic convulsions.[63, 64]

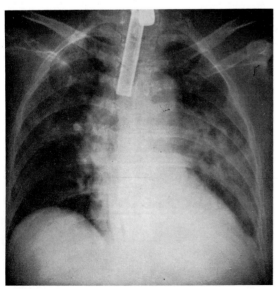

Figure 6–70 **Tetanus: Myocarditis and Pulmonary Edema.** There is cardiomegaly associated with patchy areas of increased density in the left lung; these areas are probably due to unilateral pulmonary edema. The chest picture returned to normal with clinical recovery. (Courtesy R. Walley and K. Evans: Br. J. Radiol. *40:*729, 1963.)

Typhoid Fever

Toxic ileus usually accompanies the acute stages of the disease. Abdominal films will usually show considerable small bowel distention with relatively little large bowel dilatation. Splenomegaly is often recognized.

Barium studies, although rarely done during the acute stages, may show dilated jejunal loops with edematous folds. The terminal ileum is most often abnormal and appears spastic, rigid, and segmented, with thickened folds and nodular defects due to hypertrophied Peyer's patches; ulcerations are sometimes demonstrated. Not infrequently, the cecum is spastic and deformed. The radiographic appearance alone can be similar to other inflammatory ileal conditions (regional enteritis, tuberculosis, amebiasis, and so forth), but the clinical and laboratory features of typhoid fever are quite specific. The ileum (and cecum) return to normal after successful therapy.

Perforation of the terminal ileum occurs in 5 per cent of cases, and evidence of free intraperitoneal air and intestinal ileus will then be seen on erect or decubitus abdominal films.

Pyelitis, cholecystitis, and osteomyelitis may complicate the disease, but their radiographic findings are not specific for typhoid fever.[65, 66]

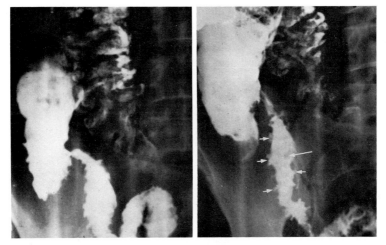

Figure 6–71 **Typhoid Fever: Hypertrophied Peyer's Patches and Ulcerations.** The hypertrophied lymphoid follicles appear as multiple lucent areas *(long arrow)* within the barium-filled terminal ileum. Many ulcers *(short arrows)* are seen projecting from both sides of the ileum, producing a ragged border. These ulcers can perforate. (Courtesy A. Laporte, Paris, France.)

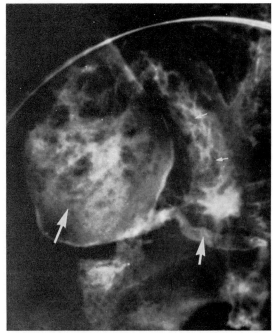

Figure 6–72 **Typhoid Fever: Hypertrophied Peyer's Patches.** Close-up view of ileocecal area reveals numerous lucent areas of varying size (*small arrows*) throughout the terminal ileum; these are swollen Peyer's patches (lymphoid follicles). The patches are always enlarged in active typhoid fever.

The cecum and appendix are indicated by large arrows. (Courtesy A. Laporte, Paris, France.)

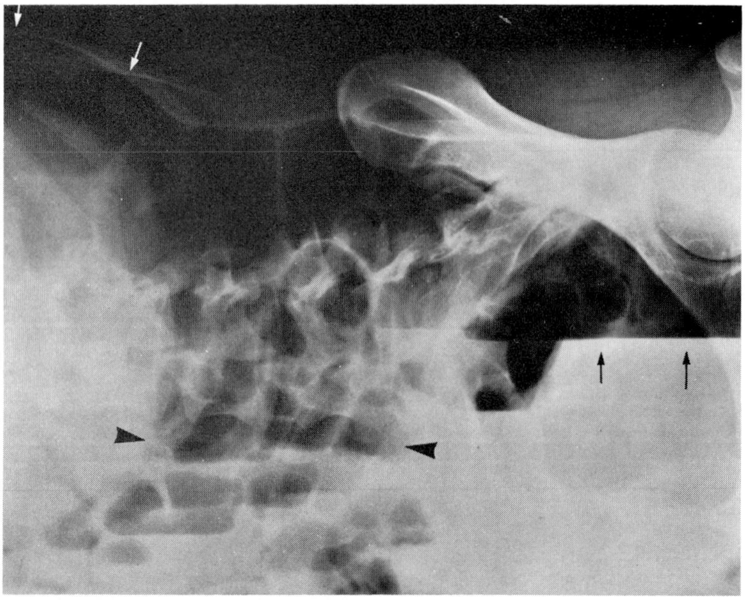

Figure 6–73 **Acute Perforation of Typhoid Ulcer.** A decubitus view of the abdomen reveals large amounts of free intraperitoneal air (*white arrows*) and air-fluid levels in the bowel (*black arrows*). Note pronounced distention of small bowel (*arrowheads*). These findings are nonspecific, indicative only of a perforated bowel.

Other Salmonella Infections

Infections can occur in almost any organ or tissue as a sequela of nontyphoid salmonella bacteremia.

Pulmonary involvement occurs most often in debilitated or immune deficient individuals. It is characterized by alveolar infiltrates that are often diffuse and bilateral. Lung abscess or empyema is a frequent complication (see *Pseudomonas and Other Gram-Negative Pneumonias,* p. 160).

Salmonella gastroenteritis develops after ingestion of heavily contaminated food and can involve the large bowel, or more frequently the small bowel. The small bowel appears atonic, edematous, and nodular; ulcerations may occur. A diffusely involved colon will be atonic and edematous. Focal areas of extremely edematous folds with superficial ulceration are often seen, sometimes in several separate segments. The radiographic appearance may initially suggest granulomatous or ulcerative colitis; an acute amebic or shigella colitis can produce a similar picture. With recovery, the colon gradually returns to normal, although changes may persist for a prolonged period. Stool cultures remain positive for a considerable interval after clinical recovery.

In the osteomyelitis which often complicates sickle cell disease, salmonella is frequently the offending organism.[67-69]

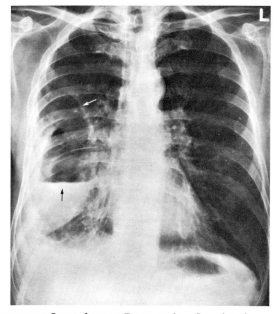

Figure 6–74 **Salmonella Empyema Secondary to Bacteremia.** Loculated empyema along the right lateral chest wall, with an air-fluid level *(black arrow),* which resulted from introduction of air during thoracocentesis. The medial wall of the empyema is indicated by a white arrow. Fluid at the right base obliterates the diaphragm and costophrenic angle.

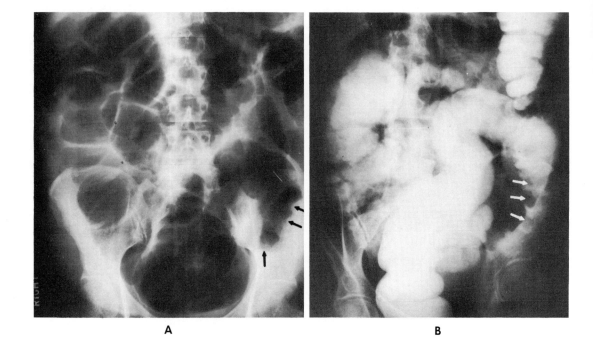

A   B

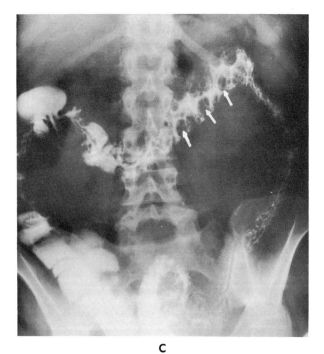

C

Figure 6–75 **Salmonella Enterocolitis.**

A 48 year old woman was admitted with a three-day history of high fever, diarrhea, distention and abdominal pain.

A, Abdominal film shows marked distention of the entire colon, including the rectum. Irregular indentations and narrowing of the sigmoid *(arrows)* are apparent.

B, Barium enema discloses an irritable, spastic sigmoid, with multiple large defects from mucosal edema *(arrows)*. These are similar to the thumbprinting of ischemic colitis. The mucosal condition of the rest of the distended colon could not be adequately evaluated. Salmonella was cultured from the stools.

C, After a few weeks of antibiotic therapy and clinical improvement, the colon still shows some residual areas of mucosal thickening in the transverse colon *(arrows)*, but the rest of the colon is practically normal.

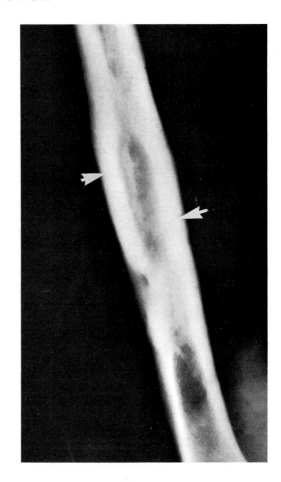

Figure 6–76 **Chronic Osteomyelitis Due to Salmonella: Sickle Cell Anemia.** The shaft of the humerus is widened, the cortices are considerably thickened (*arrows*), and the medullary cavity is narrowed and irregular. Note the diaphyseal location, unusual in ordinary pyogenic osteomyelitis but often seen in patients with sickle cell disease. The radiographic picture is that of nonspecific chronic osteomyelitis. The diaphyseal location and roentgen or clinical evidence of sickle cell disease suggests the probable etiologic agent.

Enteric Bacterial Infections

This group of gram-negative bacteria is the principal cause of urinary tract infections. Acute infections usually produce no roentgen changes. With recurrent infections, intravenous pyelography is needed to rule out predisposing obstructive or congenital lesions. (For the findings in pyelonephritis, see p. 810.)

Following prostatectomy or urinary tract infections, osteomyelitis of the spine may occur. Thinning of the intervertebral disc is the earliest finding, and is followed by decalcification and loss of definition of the adjacent vertebral margins. Paravertebral soft tissue mass (abscess) is small or absent. Differentiation from tuberculous spondylitis is often difficult. The disc and bone changes in enteric bacterial spondylitis develop more rapidly than do those of tuberculosis, and demineralization is minimal. Paravertebral masses are more prominent and frequent in tuberculosis.

Rarer enteric bacterial infections include pneumonitis and suppurative arthritis.[70, 71]

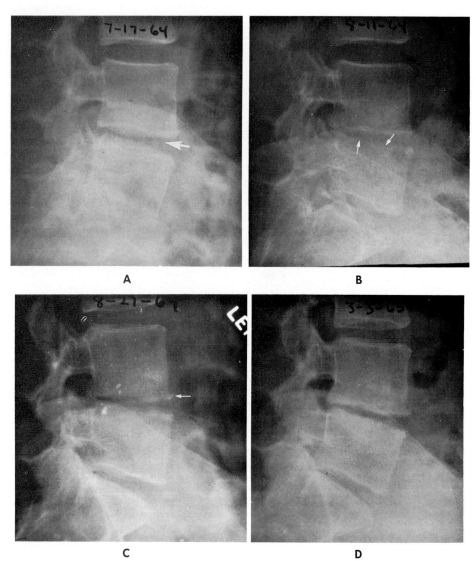

Figure 6–77   **Spondylitis Caused by** *Escherichia coli,* **Following Cystitis: Progressive Changes and Healing.**

A, Lateral view of the lumbar spine was made three weeks after onset of acute cystitis because of recurrent fever and back pain. The interspace between L4 and L5 *(arrow)* was somewhat narrowed, but the adjacent vertebrae appear intact.

B, Three weeks later, symptoms have not abated, and there is further narrowing of the interspace. Demineralization has begun in the margins of the adjacent vertebrae, and lysis has begun in L5 *(arrows).* The superior margin is somewhat irregular. Antibiotic therapy was instituted at this time.

C, Two weeks later there is greater irregularity of the vertebral margins adjacent to the interspace. The tiny spur at the inferior anterior margin of L4 *(arrow)* has increased in size. Progression of changes is more rapid and demineralization less marked than is the case in tuberculosis.

D, Five months later, long after full clinical recovery, the cortices of the vertebra are normal, but the intervertebral space is greatly narrowed and irregular. No further changes were noted on serial studies.

Without the use of antibiotics, there would have been progressive bone destruction, and the deformity would have been considerably more pronounced.

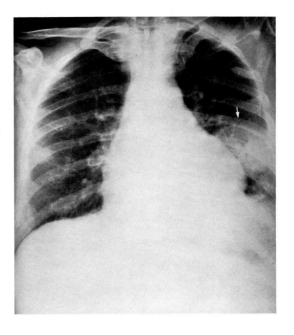

Figure 6–78 **Pneumonia Due to Proteus Bacillus.** There is a large area of consolidation in the left lower lung field *(arrow)* and another patch of consolidation just above the diaphragm. The patient had unrelated cardiomegaly.

The roentgenologic picture is nonspecific, and diagnosis can be made only by bacteriologic examination of the sputum.

Bacillary Dysentery (Shigellosis)

The radiographic findings depend upon the stage and severity of the colon involvement: In active, moderately severe disease there is edema of the entire mucosa, producing spasm and nodular irregularities of the bowel wall. In more severe forms, focal areas of ulcerations may be seen; they usually are not deep and only rarely extend into the muscularis. At this stage the shigellosis resembles acute widespread ulcerative or granulomatous colitis. Salmonella colitis has an identical appearance (see p. 180).

With recovery from the acute attack, the bowel returns to normal.[69, 72]

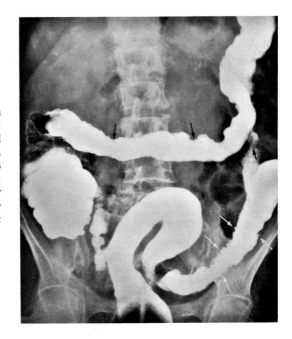

Figure 6–79 **Subacute Bacillary Dysentery: Barium Enema Study.** The transverse, descending, and proximal sigmoid colon are affected. Irregularity of the wall due to mucosal edema and spasm *(black arrows)* is seen throughout these segments. Superficial ulcerations *(white arrows)* are close together in the proximal sigmoid. The rectosigmoid and distal sigmoid appear uninvolved, in contrast to the findings in ulcerative colitis where the rectum and sigmoid are affected first and generally most severely.

Plague (Pasteurella pestis Infection)

Inhalation plague pneumonia produces dense alveolar consolidation frequently similar to pneumococcal pneumonia. More extensive and bilateral involvement may sometimes resemble pulmonary edema. Consolidation is usually massive and rapid, developing within a few hours of clinical onset, in contrast to inhalation tularemic pneumonia, in which at least 48 hours intervene. The lobe may be edematous, producing some bowing of the adjacent pleural fissures, similar to the findings in Klebsiella pneumonia. Pleural effusion may occur. If untreated, there may be rapid spread to other lobes and a fatal outcome.

In the more common ulceroglandular type, there may be mediastinal and hilar enlargement due to adenopathy without pulmonary involvement.[73-75]

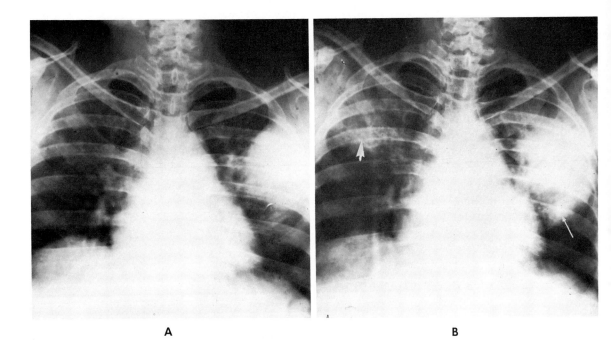

A   B

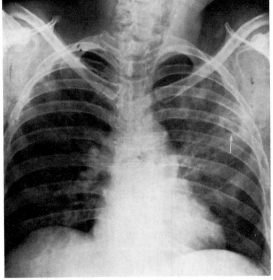

Figure 6–80 **Plague Pneumonia.**

A, Dense homogeneous area of consolidation in axillary portion of the left upper lobe is indistinguishable from that of typical pneumococcal pneumonia.

B, Six days later, despite the use of antibiotics, the lesion has spread, extending downward (*small arrow*) and upward. Medial extension has caused disappearance of the sharp medial border. A further area of consolidation is seen in the right upper lobe (*large arrow*).

C, A week later, with continued intensive antibiotic therapy, there has been marked clinical improvement, which is seen on radiography. The consolidation on the left has almost disappeared, but a cavity with an air-fluid level (*arrow*) has developed. The lesion on the right has virtually cleared. (Courtesy Dr. F. McCrumb, Jr., Baltimore, Maryland.)

C

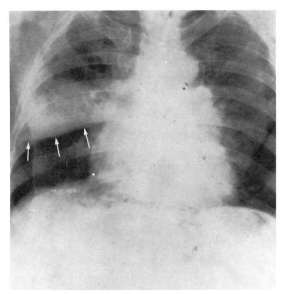

Figure 6–81 **Plague Pneumonia.**   There is homogeneous consolidation of the anterior segment of the right upper lobe. The sharp inferior border represents the limiting short fissure. The slight downward bowing of this fissure *(arrows)* is caused by volume increase of the involved segment owing to edema and hemorrhage, conditions that are characteristic of plague pneumonia.

The bilateral hilar enlargement is caused by inflammatory adenopathy, a frequent finding in plague pneumonia.

The finding of increased volume of a consolidated lobe or lobule is also commonly seen in Klebsiella pneumonia.

Tularemia (Pasteurella tularensis Infection)

The significant roentgenologic findings are limited to the chest. In the ulceroglandular type, lung changes, if present, usually appear after a week or 10 days of illness. The lesions may be single or multiple focal consolidations, often with pleural effusion and hilar adenopathy.

Primary inhalation tularemia may produce pulmonary changes as early as 48 hours after clinical onset. The pneumonic lesions are single or multiple, rounded or oval infiltrates with somewhat indistinct margins; they occur in any lobe. Unilateral hilar adenopathy is seen in up to half the cases and is almost always associated with either a pneumonia or a pleural effusion. The latter occurs in one third to one half of cases. Abscess formation is rare. In an endemic area, ovoid consolidations with unilateral adenopathy may suggest the diagnosis, but the roentgenologic findings are too variable and will simulate other inflammatory or neoplastic conditions.[76-79]

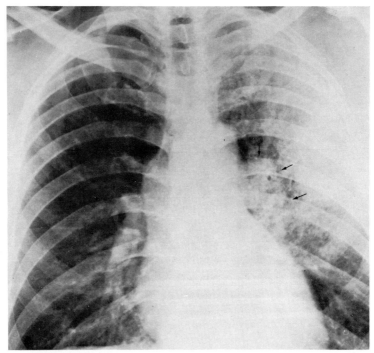

Figure 6–82 **Inhalation Tularemia.**   There is an extensive ill-defined area of consolidation at the periphery of the left upper lobe. The left hilar nodes *(arrows)* have become enlarged. The lesion assumed a typical oval shape a few days later. (Courtesy Lt. Col. E. Overholt, U.S. Army, Walter Reed Hospital.)

Melioidosis

In this relatively rare disease endemic in southeast Asia and caused by *Pseudomonas pseudomallei (Malleomyces pseudomallei)*, the radiologic findings are most often in the chest. Bone involvement is sometimes seen in the chronic form of melioidosis.

The acute form is commonly manifested by an acute toxic pneumonitis. Nodular poorly defined infiltrates appear and rapidly progress to a widespread bronchopneumonia with diffuse areas of fluffy consolidation. However, death from overwhelming septicemia may occur without pulmonary changes.

In the subacute form there are often persistent upper lobe infiltrates with occasional cavitation, mimicking tuberculosis. Sometimes widespread miliary lesions suggestive of miliary tuberculosis are seen. The clinical picture also mimics tuberculosis. Persistent nonspecific pulmonary infiltrates are common in chronic melioidosis.

Widespread extrapulmonary abscesses and infective granulomas may appear in the late subacute or chronic phase, often with hepatomegaly and splenomegaly.

Osteomyelitis is a frequent complication, presenting as an ill-defined lytic area with little or no bone reaction. Draining soft tissue sinuses may develop from the infected bone.

Paravertebral abscess and perinephric abscess are not rare and produce recognizable radiographic changes.

Often, the pulmonary lesions are found in a totally asymptomatic patient.

The nonspecific clinical and roentgenographic findings make diagnosis entirely dependent on organism identification or on a rising agglutination titer.[80-83]

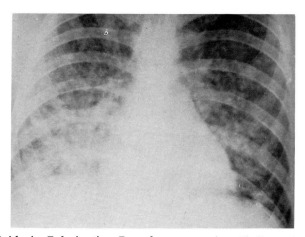

Figure 6–83 **Acute Melioidosis: Fulminating Bronchopneumonia.** Fluffy nodular densities are diffusely scattered throughout both lung fields, with confluent lesions at the right base. This young soldier died 12 hours later. A film made 36 hours previously showed only a few small nodular densities.

Diffuse bronchopneumonia is the most common radiographic finding in acute melioidosis. With rapid and overwhelming septicemia, death may occur before pulmonary lesions develop. (Courtesy Captain M. C. Patterson, M.D., U.S.A.F.)

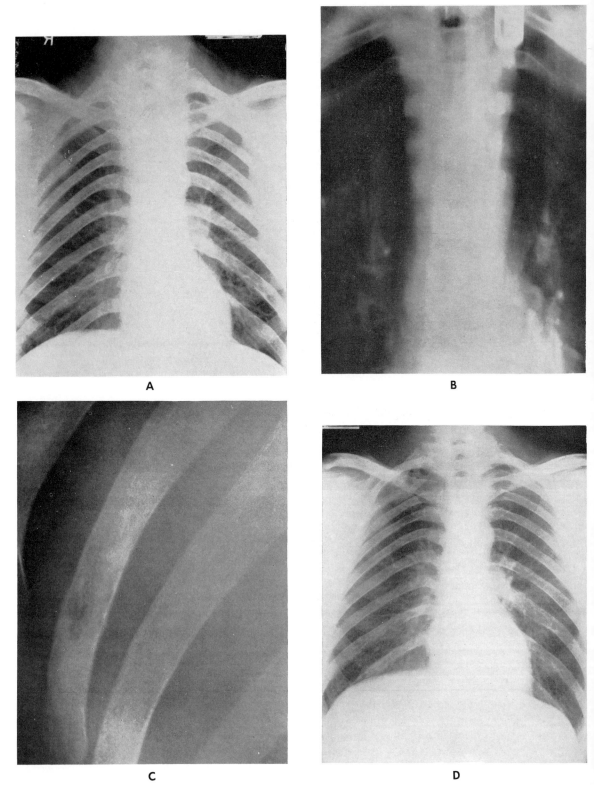

A

B

C

D

Figure 6–84   See legend on the opposite page.

Anthrax

Roentgenologic changes in the chest can occur after direct inhalation of spores (inhalation anthrax), which are rapidly carried to the mediastinal nodes. The earliest and most characteristic roentgen finding is mediastinal widening due to enlarged mediastinal nodes and a hemorrhagic mediastinitis. A diffuse bronchopneumonia, fibrinous pleurisy, and pleural effusion may subsequently develop.[84]

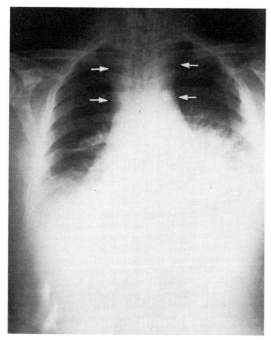

Figure 6–85 **Inhalation Anthrax.** There is diffuse widening of the mediastinum *(arrows)*, as well as bilateral basal bronchopneumonia with effusion on the left side; the parenchymal lesions developed after the mediastinitis. (Courtesy P. Brachman, Atlanta, Georgia.)

Figure 6–84 **Mediastinal and Bony Changes in Chronic Melioidosis.**

A, The superior mediastinum is widened, extending from the arch of the aorta into the base of the neck, due to abscess formation.

B, Tomogram reveals the paravertebral location of the abscess.

C, Enlarged view of the eighth rib shows an area of bony destruction with minimal sclerosis, which was due to melioidosis osteomyelitis.

D, Film after therapy shows restoration of normal mediastinal width. (Courtesy G. Jones and J. Ross: Br. J. Radiol., *36*:415, 1963.)

See illustration on the opposite page.

Listeriosis

Listeriosis may occur as an intrauterine infection, which causes a high neonatal mortality rate. This neonatal form is called *granulomatosis infantiseptica;* radiographically, it may present either as a bronchopneumonia or as a diffuse irregular miliary pulmonary nodulation with a tendency to coalescence. In adults, pneumonia, pleurisy, or pericarditis may occur.[85]

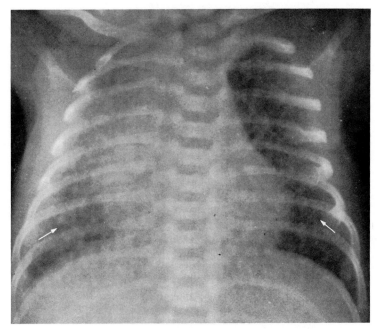

Figure 6–86 **Listeriosis in Newborn (Granulomatosis Infantiseptica).**　Coarse granular nodular infiltrates are seen throughout both lung fields *(arrows)* and are conglomerate in the right upper lobe. The mortality rate in this infection is high. (Courtesy B. Brogdon: Radiology, *79*:415, 1962.)

Brucellosis

Infection with brucella organisms can occur in any organ but the commonest roentgenologic findings are seen in the spine, especially the lower dorsal and upper lumbar bodies. Rarely, there is destructive arthritis, osteomyelitis of the long bones, renal involvement, and pneumonia with effusion.

Spine lesions may be marginal or central. Marginal lesions are poorly defined and may be found in the anterior and superior margins of the vertebral bodies. With progression, sclerosis and trabecular irregularities occur, and the disc spaces become narrowed. Proliferative bone changes usually develop rather rapidly. Paravertebral soft tissue swelling is not unusual. In the rare central type, there are areas of rarefaction in the midportion of the vertebra, leading to collapse of the vertebral body.

Roentgenologic differentiation of brucellosis from tuberculous spondylitis is extremely difficult.

Renal brucellosis simulates tuberculosis clinically and radiographically.[86-88]

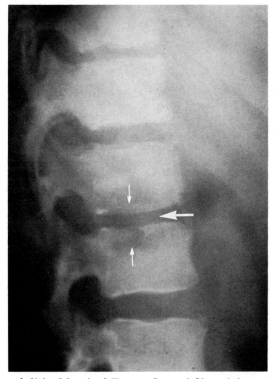

Figure 6–87 **Brucellosis Spondylitis: Marginal Type.** Lateral film of the spine reveals slight but definite narrowing of the intervertebral space *(large arrow)* between L2 and L3. There are ill-defined areas of rarefaction adjacent to this interspace on the inferior surface of L2 and the superior portion of L3 *(small arrows)*. The process occurred in a young man with brucellosis and progressed to further bone destruction. Roentgenologically, brucellosis spondylitis cannot be distinguished from tuberculosis. (Courtesy F. Zammit, Malta.)

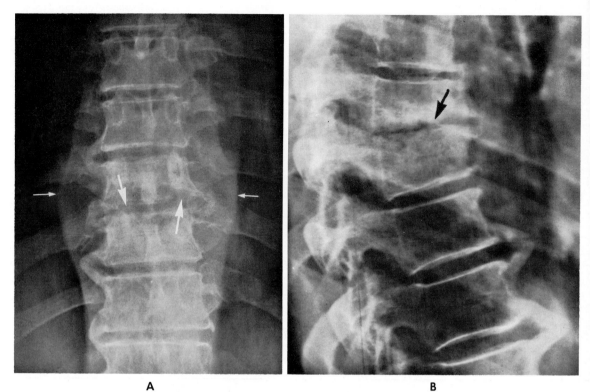

A   B

Figure 6–88 **Advanced Brucellosis Spondylitis.**

A, Localized fusiform widening of the paraspinal soft tissue shadows *(small arrows)* is due to a paravertebral abscess. The disc is narrowed between the ninth and tenth dorsal bodies. The margins of these vertebrae are irregular, and there has been destruction of bone *(large arrows)* and proliferation at the lateral margins of the interspace.

B, Lateral projection shows the narrowed irregular intervertebral space *(arrow)* and destruction of the inferior margin of T9, producing a decrease in its height. Note the mottled appearance of the two vertebrae and the hypertrophic reaction anteriorly. This lesion is indistinguishable from tuberculous spondylitis. (Courtesy F. Zammit, Malta.)

Primary Tuberculosis

The major manifestations of primary tuberculosis occur in the chest. They are found most commonly in children and young adults. The roentgenologic findings reflect fairly accurately the pathologic process. Usually a small focus of exudative tuberculous pneumonia appears, most often in the upper or midlung field. The lesion enlarges peripherally and appears as a homogeneous density with poorly defined margins. The draining lymph nodes at the hilum or mediastinum become enlarged and may often dominate the x-ray picture. This combination of a focal parenchymal lesion and unilateral enlarged hilar or mediastinal nodes is the *primary complex.* Occasionally, unilateral or bilateral nodal enlargement is seen without a parenchymal lesion; this is probably primary lymph node tuberculosis. Hematogenous spread, usually from a caseating node, can lead to tuberculosis of almost any organ or tissue.

Generally the primary lesions heal slowly, with calcification replacing both the parenchymal and nodal lesions. This combination of calcifications is the *healed primary complex.* Occasionally the primary lesions may progress to extensive tuberculous pneumonia and cavitation. The enlarged primary tuberculous nodes may also increase in size and produce bronchial compression or erosion and obstructive atelectasis.

The primary site may be the pleura, causing a persistent pleural effusion, often with no parenchymal disease. Without therapy, the effusion persists for months. Primary pleural tuberculosis is most frequent in adults.

In up to 10 per cent of adults, active primary tuberculosis progresses directly into reinfection apical or upper lobe tuberculosis.[89-93]

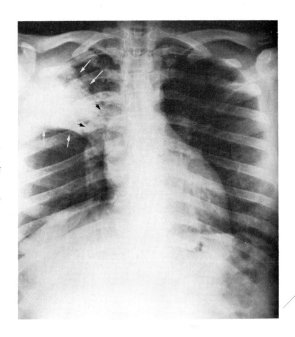

Figure 6–89 **Active Primary Tuberculous Complex in a Young Male.** There is a pneumonic consolidation in the right upper lobe *(white arrows)* and a large node in the right hilum *(black arrows)*. The upward bowing of the horizontal fissure *(lower white arrows)* indicates associated atelectasis from bronchial compression by the nodes.

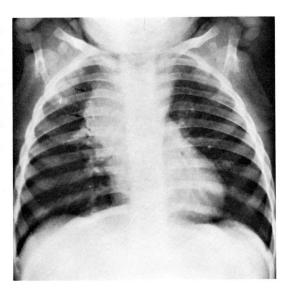

Figure 6–90 **Primary Tuberculous Complex with Massive Nodal Enlargement.** There is a small focus of pneumonia in the right upper lung field *(white arrow)*. The right upper mediastinum is greatly enlarged owing to adenopathy *(black arrows)*. Some enlarged nodes are also seen in the left upper mediastinum.

Tuberculoma of the Lung

The tuberculoma is a sharply circumscribed nodular density that may be due to either primary or reinfection tuberculosis. It may appear in any portion of the lung but is most frequent in upper lobes. It frequently contains mottled or concentric calcification. Tuberculomas are usually 1 to 3 cm. in diameter and may remain constant or may grow very slowly. They are rounded or lobulated, and small discrete satellite densities are often seen in the vicinity. Calcification helps distinguish a solitary tuberculoma from a neoplasm. However, planigraphy may be necessary to demonstrate faint calcification within a tuberculoma. If calcification cannot be shown, distinction from neoplasm may be impossible. (See also *Tuberculoma of the Brain,* p. 354.)[94]

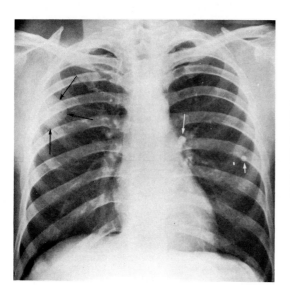

Figure 6–91 **Healed Primary Complex with Associated Tuberculoma.** The small round calcification in the left midlung field *(short white arrow)* in association with calcified nodes *(long white arrow)* in the left hilum represents the classic healed calcified primary complex. The sharply circumscribed homogeneous density in the right upper lobe *(black arrows)* is a noncalcified tuberculoma.

Reinfection Pulmonary Tuberculosis

The earliest roentgenologic finding is a hazy, poorly marginated alveolar infiltration that is most often localized to the apical or apical-posterior segment of an upper lobe. A solitary infiltrate in an anterior segment of an upper lobe rarely proves to be tuberculous, but any persistent apical lesion, however small, has a high probability of being tuberculous. Small lesions are easily overlooked or obscured by the clavicle or a rib. Often an apical lordotic view is helpful.

Lower lobe tuberculosis without involvement of an upper lobe (basal tuberculosis) is infrequent; the superior segment of a lower lobe is the usual site, but rarely, the lesions first appear in the extreme base, often producing infiltrates having a transverse linear pattern. The lesions usually appear as a nonspecific chronic pneumonitis. Cavitation can occur. Isolated lower lobe tuberculosis simulates many other conditions, and radiographic diagnosis may be difficult. However, cavitation in a superior segment of a lower lobe is quite suggestive of tuberculosis.

If a tuberculous infiltrate does not heal or become stabilized, the lesion may increase and undergo cavitation. Rapidly progressing upper lobe involvement can give rise to a diffuse pneumonic-like consolidation or even to a bloodstream invasion (miliary tuberculosis).

Tuberculous cavities initially appear as irregular lucencies within a homogeneous density. Subsequently the lucencies become round or oval with moderately thick walls. Fluid levels are usually absent. Single or multiple cavities become apparent. Planigraphy is often employed to determine the presence and extent of cavitation.

Following cavitation, there may be bronchogenic spread to other portions of either lung; this is the commonest pathway of pulmonary dissemination. The bronchial markings in the affected areas become thickened and fuzzy, and small patchy infiltrations appear. These may also undergo cavitation. Untreated tuberculosis generally progresses quite slowly, and radiographic changes may not be apparent for several months or longer. Lesions may become stabilized and remain unchanged for years.

Simple pleural effusion without parenchymal lesions may be the initial finding in either primary or reinfection tuberculosis. A persistent effusion in a young adult frequently proves to be tuberculous; parenchymal lesions may eventually develop. Pleural effusion can also occur secondary to parenchymal tuberculosis. These effusions may become secondarily infected with pyogenic organisms (thereby becoming a purulent empyema) and often loculate. Pleural calcification frequently develops.

Healing of the pulmonary lesions is manifested by diminished infiltration, decrease of cavity size, fibrosis, and often calcification. The fibrosis appears as a linear strand-like pattern, associated with volume loss. This causes the adjacent structures—usually the trachea, an interlobular fissure, or a hilum—to retract towards the fibrotic area. Compensatory emphysema may appear in other parts of the lung. In patients with fibrosis and calcification without overt cavitation (chronic fibroid tuberculosis), disease activity cannot be determined without serial films taken over extended periods of time.

Tuberculous bronchiectasis can develop from chronic parenchymal disease or from endobronchial spread. The multiple bronchiectatic lucencies are usually small, ovoid, thin-walled, and without much surrounding reaction, and they appear to follow the course of the bronchus. Distinction from a true tuberculous cavity may require bronchography.

Other pulmonary complications include pneumothorax, empyema, bronchopleural fistula, and bronchial stenosis.[95-98]

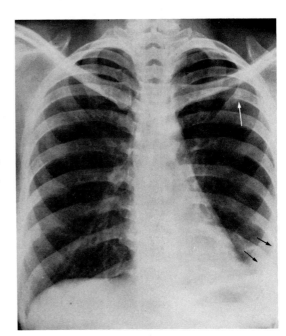

Figure 6–92 **Minimal Tuberculous Infiltration.** There is poorly defined infiltration in the left upper lobe posteriorly *(white arrow),* with associated pleural effusion at the left base *(black arrows).*

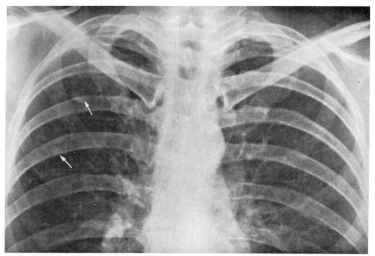

Figure 6–93 **Minimal Reinfection Tuberculosis.** Two small areas of infiltrative density *(arrows)* in the right upper lobe represent minimally active tuberculous lesions. Sputum examination was positive for tubercle bacilli, but the physical findings were negative.

Complete resolution occurred following two months of intensive antibiotic therapy.

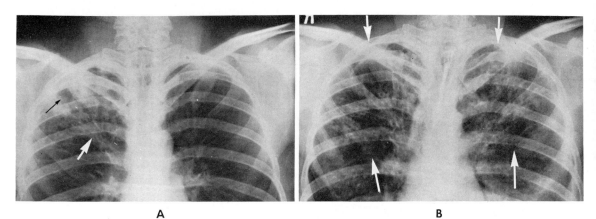

A

B

Figure 6–94 **Progressive Pulmonary Tuberculosis and Healing.**

A, Oval thick-walled cavity *(black arrow)* in the right upper lung field and an extensive surrounding parenchymal infiltration are shown. Note the linear strands of fibrosis *(white arrow)* extending from the elevated right hilum. This represents moderately advanced active tuberculosis. The left lung is clear.

B, Four months later the cavity and active exudate in the right upper lobe have been replaced by fibrosis *(lower left arrow)*. A pneumothorax has developed in the right upper chest *(upper left arrow)*, and there is extensive bronchogenic spread *(right arrows)*.

C, One year later, after antituberculous therapy, only minimal residual fibrosis remains in the right apex *(black arrow)*. Note elevation of right hilum *(white arrows)* due to volume shrinkage of the right upper lobe.

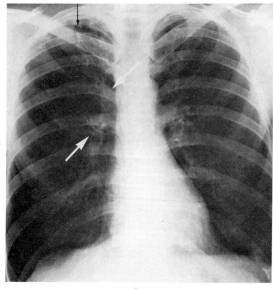

C

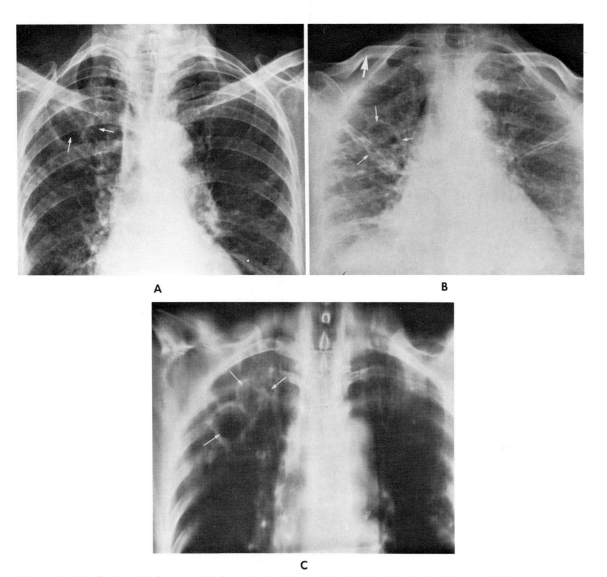

Figure 6–95 **Cavitary Pulmonary Tuberculosis: Conventional Film, Lordotic View, and Planigram.**

A, Conventional film: Moderate infiltration appears in the right upper lobe. The two large, ill-defined lucent areas *(arrows)* within this area suggest cavitation, but clear-cut cavity borders cannot be identified. The clavicle and ribs obscure much detail.

B, Lordotic view: The clavicle *(large arrow)* is seen well above the involved area. A cavity *(small arrows)* can now be identified, although its borders are still poorly defined. Another area of lucency above this cavity suggests cavitation.

The lordotic view projects the clavicle above the lung apices, allowing better delineation of apical disease. This view is frequently employed when apical infiltrates are suspected.

C, Planigram: Contiguous cavities *(arrows)* are clearly demonstrated.

Planigraphy is frequently essential for demonstrating cavitation within an area of tuberculous infiltration, especially in chronic fibroid tuberculosis.

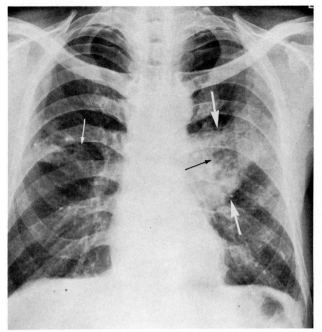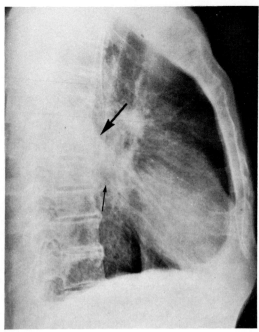

| A | B |

Figure 6–96 **Cavitary Lower Lobe Pulmonary Tuberculosis.**

A, The broad band of density in the left midlung field *(large white arrows)* contains an ill-defined lucency *(black arrow)* that was a cavity. Note area of infiltration in the right midlung field *(small white arrow)*; this is probably from bronchogenic spread. The apices of the upper lobes are clear.

B, On lateral view the large density and cavity within are seen to be in the superior segment of the left lower lobe *(arrows)*. The diagnosis of tuberculosis was made from sputum examination.

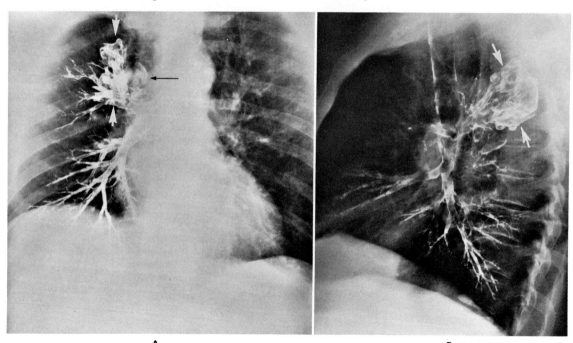

| A | B |

Figure 6–97 **Tuberculous Bronchiectasis and Cavitation: Bronchogram.**

A, The contrast-filled upper lobe bronchi are irregularly dilated and crowded *(white arrows)*, a characteristic picture in cylindrical bronchiectasis in a fibrotic shrunken area. Note absence of filling of the peripheral bronchioles and alveoli; this is frequently seen in tuberculous bronchiectasis, in contrast to ordinary bronchiectasis. A cavity filled with contrast material is somewhat obscured on the posteroanterior film; it is partially seen near the spine *(black arrow)*.

B, The large cavity is well demonstrated on lateral view *(arrows)*. Note characteristic posterior location.

198

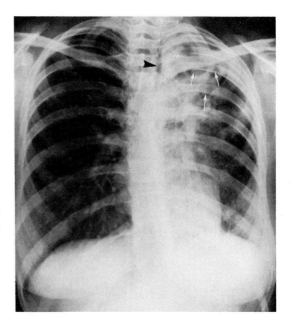

Figure 6–98 **Moderately Advanced Unilateral Tuberculosis with Cavitation.**   There is extensive infiltration of the left upper lobe, as well as numerous lucent areas of cavitation *(arrows).* Note evidence of volume shrinkage: the trachea is deviated to the left *(arrowhead),* the left hilar shadow is elevated, and the interspaces in the left upper chest are narrowed. Volume shrinkage is indicative of chronicity and fibrosis. There is also evidence of old pleural reaction with blunting of the left costophrenic angle. The right lung is clear but shows compensatory hyperinflation.

Unilateral upper lobe tuberculosis with volume shrinkage may simulate atelectasis due to bronchogenic carcinoma.

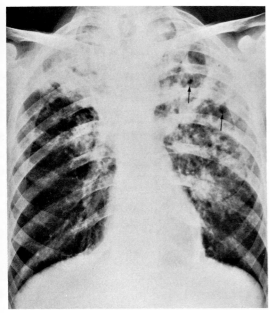

Figure 6–99 **Advanced Bilateral Pulmonary Tuberculosis.**   Extensive bilateral upper lobe infiltration is more marked on the left. There are numerous areas of lucency in both upper lobes, some due to bronchiectasis *(arrows)* and others due to cavitation. The full extent of cavitation in advanced cases is often appreciated only on planigraphy. There is bronchogenic spread to the left midlung field.

Both hilar shadows are elevated as a result of upper lobe fibrosis, and there is compensatory emphysema in the lower lobes. Frequently the heart shadow is small.

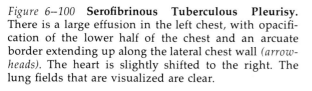

Figure 6–100 **Serofibrinous Tuberculous Pleurisy.**
There is a large effusion in the left chest, with opacification of the lower half of the chest and an arcuate border extending up along the lateral chest wall *(arrowheads).* The heart is slightly shifted to the right. The lung fields that are visualized are clear.

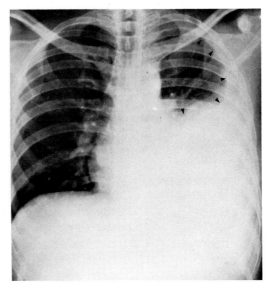

Tuberculosis of the Esophagus

This rare complication is usually the result of direct extension from adjacent mediastinal tuberculous lymph nodes. The roentgenologic findings are those of a chronic esophagitis and are nonspecific. Alteration and effacement of the mucosa, frequent ulceration, and eventual stricture may develop. Any portion of the esophagus may be involved.[99]

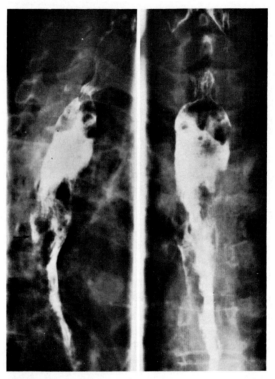

Figure 6–101 **Tuberculosis of Esophagus.** Ulceration *(black arrow)* and edematous, infiltrated, and thickened mucosal folds are seen. The more distal area exhibited spasm and irritability.

Tuberculosis of the Stomach

Roentgenologically this rare condition simulates carcinoma. There may be ulceration with indurated margins, or a diffuse involvement of the gastric wall, mimicking a linitis plastica or lymphoma. The pyloric region is the most common site of involvement, and pyloric stenosis may result. Active cavitary pulmonary tuberculosis is usually present.[100]

Intestinal Tuberculosis

Intestinal tuberculosis is most often secondary to active pulmonary tuberculosis. In 90 per cent of cases it localizes in the ileocecal area; the ascending colon is also frequently involved. Initially, irritability of the involved segments causes rapid emptying, and the area appears poorly filled. Later, the mucosal folds thicken and stiffen, and contour alterations may be seen. The cecum is shortened, and spiculations are noted. The terminal ileum is atonic and dilated. Ulcerations may appear in the affected segments. Fibrosis, shortening, and areas of narrowing and stenosis can develop in more advanced stages.

Differentiation from regional ileitis is difficult and often impossible. However, in the latter disease the ileum is often more uniformly narrowed and cecal involvement is frequently minimal or absent. The presence of pulmonary tuberculosis suggests the diagnosis of ileocecal tuberculosis. In blacks, ileocecal disease is more often tuberculous; regional enteritis is extremely rare in this ethnic group.

Infrequently, segmental colon involvement by hyperplastic tuberculosis occurs, usually without an associated ileocecal or pulmonary tuberculosis. Radiologically, the involvement presents as an area of narrowing or smooth-walled stricture, sometimes causing obstruction. The transverse colon and the sigmoid are the most common sites. Multiple areas of involvement can occur. In the absence of ileocecal and pulmonary lesions, radiographic distinction from a malignancy or other inflammatory strictures is often impossible.[101-103]

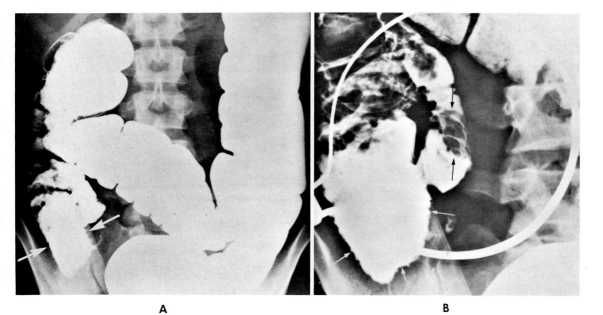

A   B

Figure 6–102 **Ileocecal Tuberculosis.**

A, The cecum is narrowed and contains spiculations *(arrows)* due to thickened folds. The ileocecal area was irritable; the ascending colon is almost empty because of marked irritability.

B, Enlarged view of the ileocecal area demonstrates the shortened cecum and the spiculated edges *(white arrows).* The terminal ileum has thick stiff folds *(black arrows).*

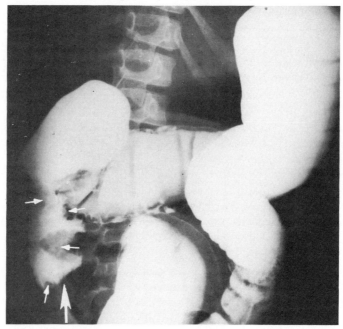

Figure 6--103 **Tuberculosis of Cecum and Ascending Colon in 8 Year Old Girl.** The cecal tip is quite irregular *(large arrow)* and irregular large defects are seen in the cecum and ascending colon *(small arrows).* These were masses of tuberculous tissue.

Tuberculosis of Lymph Nodes

Tuberculous lymph nodes enlarge during their active phase. With healing, the nodes become smaller and often undergo stippled or concentric calcification.

Radiographically, noncalcified nodes in the hilum or mediastinum may appear as lobulated or solitary masses, frequently unilateral and indistinguishable from masses or adenopathy of other causes. In the abdomen, large noncalcified retroperitoneal nodal masses may sometimes be mistaken for a neoplastic disease. Associated spine or sacroiliac inflammatory involvement may be a clue to the correct diagnosis of these masses, but lymphangiography will disclose the nodal nature of such lesions.

Calcified tuberculous nodes are very frequently encountered in the lung, hilum, mediastinum, mesentery, and cervical area. Their characteristic stippled and irregular calcification and location usually allows radiographic identification without difficulty.[104]

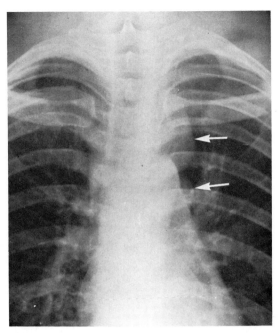

Figure 6–104 **Hilar and Mediastinal Adenopathy in Pulmonary Tuberculosis.** There is a large lobulated density in the left upper mediastinum *(upper arrow)*, and also a smaller enlarged left hilar node *(lower arrow)*. In contrast to lymphomatous adenopathy, tuberculous node enlargement is usually unilateral.

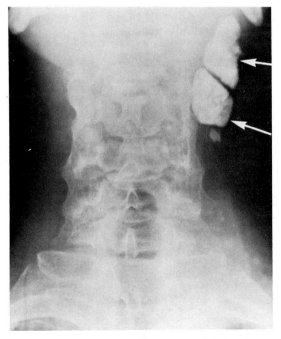

Figure 6–105 **Calcified Tuberculous Nodes in Neck.** On the left, there are densely calcified nodes *(arrows)*, which are characteristically mottled or stippled.

Tuberculous Pericarditis

This complication is most frequently caused by direct extension from pulmonary lesions, but it may be hematogenous. Radiographically the cardiac silhouette is enlarged, the lateral borders appear to bulge, and pulsations are decreased. The pulmonary vasculature generally remains normal. The base of the heart may remain normal in contour because of adhesions.

Routine chest films may fail to distinguish tuberculous pericarditis from an enlarged heart, and angiography or intravenous carbon dioxide study may be necessary. Direct pericardiocentesis and air instillation will demonstrate the true size of the heart and the thickness of the pericardium. Cardiac isotope scanning and ultrasound are now being used for diagnosis.

In late stages of the disease, the fluid may be absorbed, and an adhesive or a constrictive pericarditis can result. Pericardial calcification is not uncommon. (See also *Pericarditis with Effusion* and *Constrictive Pericarditis,* pp. 706 and 711.)

An enlarged cardiac silhouette in the presence of pulmonary tuberculosis suggests the possibility of tuberculous pericarditis.[105]

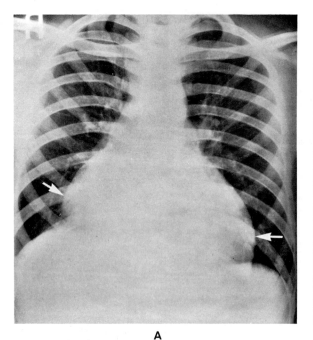 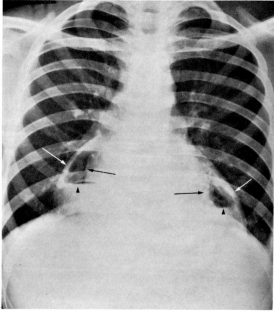

A   B

Figure 6–106 **Tuberculous Pericarditis.**

A, There is marked enlargement of the cardiac silhouette to the left and right *(arrows),* and bulging of the right side. The base of the heart remains normal. Note the normal pulmonary vasculature.

B, After pericardiocentesis and air instillation, the true cardiac size is apparent *(black arrows).* An air-fluid level on both sides *(arrowheads)* has replaced the fluid. Note the thickness of the pericardium *(white arrows).* In noninflammatory pericardial effusion the pericardium tends to remain thin, whereas inflammatory pericarditis usually causes considerable thickening.

Tuberculous Peritonitis

In the majority of cases no radiographic abnormalities are present in the abdomen or small bowel.

If ascitic fluid is present in considerable quantities (wet form), the abdomen may appear hazy. The bowel loops may be dilated and air-filled and appear centrally bunched in the supine position.

Mesenteric thickening and adhesions can occur in the dry form of the disease and lead to fixation of bowel loops, segmental dilatation and irregularity, an ill-defined mucosal pattern, and hypomotility. Segmentation of the barium column may occur. The space between some small bowel loops may be widened. Massive mesenteric thickening and bunching can simulate a neoplastic mass.

These roentgen changes are nonspecific and can occur in other types of chronic inflammatory or neoplastic peritonitis. Diagnosis requires peritoneal biopsy.[106]

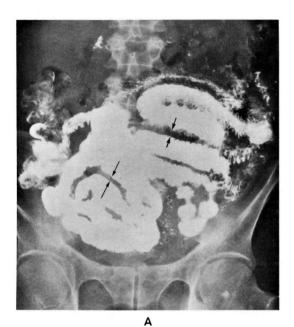 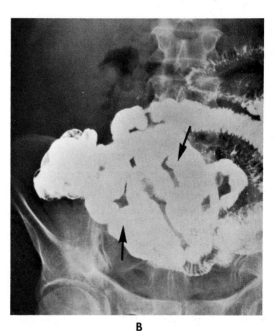

A　　　　　　　　　　B

Figure 6–107 **Tuberculous Peritonitis.** Films of a small bowel study made at five hours *(A)* and six hours *(B)* show evidence of hypomotility; the barium is almost entirely in the distal jejunum and ileum. Some distal ileal loops *(large arrows)* are clustered in the lower abdomen and are mildly dilated with a poorly defined mucosal pattern. The space between some loops *(small arrows)* is somewhat increased.

These are all nonspecific findings. In most cases of tuberculous peritonitis there are no small bowel changes.

Tuberculosis of the Urinary Tract

Renal tuberculosis is usually hematogenous and may be unilateral or bilateral. Roentgen evidence of pulmonary tuberculosis is seen only in about half the patients.

The early lesions involve the cortex or corticomedullary area and produce no radiographic changes. Papillary necrosis, a basic lesion of the disease, causes a fuzzy moth-eaten caliceal irregularity, often with a small pericaliceal cavity. These granulomatous cavities, even when large, may or may not communicate with the collecting system; nephrotomography may delineate them more clearly.

Later, infundibular strictures with dilated calices, cicatricial obliteration of one or more calices, and calcification of the intrarenal granulomatous masses appear, producing a more characteristic roentgen picture. Renal function may be greatly decreased in advanced cases.

Although other forms of unilateral or bilateral chronic pyelonephritis may produce a similar urographic picture, destructive caliceal changes, scattered parenchymal calcifications, and infundibular strictures are extremely suggestive findings.

Angiographically, the findings are nonspecific. Obliterated vessels are seen in focal areas without neovasculature or encasement, nonspecific evidence of parenchymal inflammation.

An affected ureter may show a beaded or nodular appearance with some dilatation or may present as a fibrotic stricture that may lead to an obstructive uropathy. Ureteral calcification can sometimes occur.

Bladder involvement may cause irregularities of the wall or may present merely as a smooth contracted bladder with a thickened wall.[107-110]

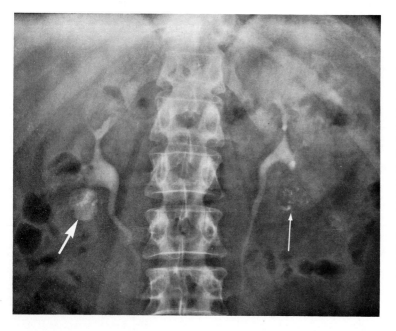

Figure 6–108 **Bilateral Renal Tuberculosis: Pyelogram.** There are mottled small irregular calcifications in the lower pole of the left kidney *(small arrow)*, and the lower infundibulum and calix are amputated. Similar calcifications on the right overlie a large irregular cavity filled with contrast material *(large arrow)*. The middle calix is irregular and poorly delineated. Both kidneys function normally.

Multiple caliceal irregularities and distortions associated with parenchymal renal calcifications strongly suggest renal tuberculosis.

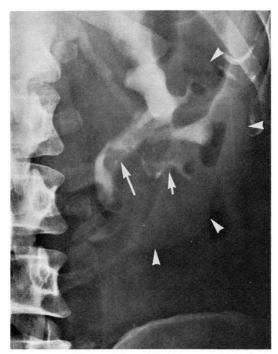

Figure 6–109 **Moderately Advanced Unilateral Renal Tuberculosis: Pyelogram.** The lower calix is amputated, and the middle infundibulum and calix are deformed and distorted beyond recognition *(short arrow)*. Caseous masses, invading the renal pelvis and infundibula to the upper calices *(long arrow)*, cause dilatation of the upper calices. Calcification has not yet developed, and the renal outline is still normal *(arrowheads)*.

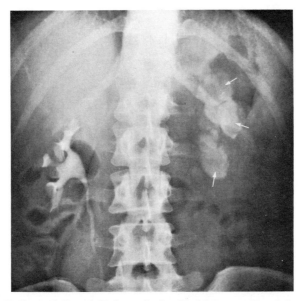

Figure 6–110 **Advanced Unilateral Renal Tuberculosis: Pyelogram.** The left kidney is nonfunctioning because of the presence of tuberculous pyelonephritis cavities *(arrows)* that have undergone calcification. The right kidney appears normal. Although pyelography demonstrates caliceal distortion, irregularity, and destruction, calcification seen on plain abdominal films may be the earliest indication of renal tuberculosis. However, calcifications are a relatively late finding.

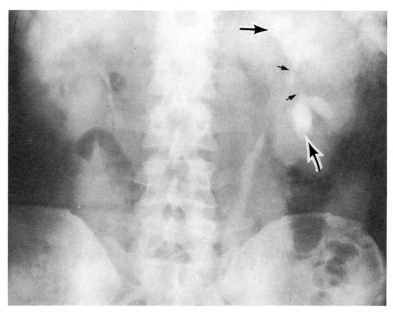

Figure 6–111 **Unilateral Tuberculous Pyelonephritis and Ureteritis.** Intravenous pyelogram reveals sac-like calices *(black-white arrows)* with strictures of the infundibula *(small arrows)*. The diseased left ureter is irregularly dilated. The right urinary tract was normal.

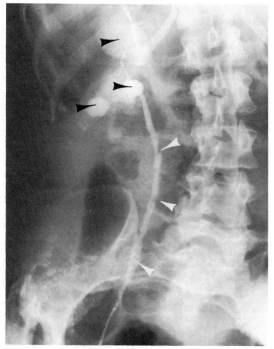

Figure 6–112 **Tuberculosis of Kidney and Ureter.** Hydronephrotic sacs of varying size *(black arrowheads)* are due to advanced renal tuberculous pyelonephritis, with narrowed infundibula. The ureter is irregular, with intraluminal defects *(white arrowheads)*. This appearance is characteristic of ureteral tuberculosis.

 The right hip changes were due to congenital dislocation.

Tuberculosis of the Musculoskeletal System

The earliest roentgenologic finding in tuberculous arthritis is periarticular swelling caused by joint effusion. Gradually osteoporosis develops around the joint, particularly at the articular margins. Destruction of cartilage and subarticular bone occurs late and generally begins in the non–weight-bearing portions of the joint. The bone lesions are usually lytic and sharply circumscribed; there is little sclerosis. The articular surfaces develop an indistinct, fuzzy appearance and joint destruction ensues.

Tuberculous osteomyelitis can occur in almost any bone. In adults, the lesions are generally lytic, destructive, and poorly marginated, usually with little or no periosteal reaction. Sometimes well-circumscribed cystic lesions are encountered, the so-called cystic tuberculosis. When the tuberculous lesions are associated with reactive sclerosis and periosteal proliferation, the appearance is identical with that of any chronic pyogenic osteomyelitis. In children, the reactive form is more common, and draining sinuses from the involved bones may occur. Involvement of the smaller bones of the hands or feet, especially in children, often produces a characteristic appearance. The shaft is expanded, showing both lytic and reactive areas, and there may be a fine layer of periosteal reaction. The epiphysis is spared. This expanded appearance has given rise to the term *spina ventosa*. In adults, tuberculous dactylitis also occurs but without significant bone expansion and with considerable periosteal reaction and endosteal sclerosis. Bone tuberculosis is associated with a high percentage of positive chest findings, particularly in children.

Tuberculosis of the spine most frequently begins in the vertebral body adjacent to the intervertebral disc, but it may begin in any portion of the vertebra. There is irregular destruction of the cancellous bone, ultimately involving the cortex. The vertebral body may eventually collapse. The destructive changes in the cartilaginous disc develop more slowly than in pyogenic osteomyelitis. However, narrowing of the intervertebral space eventually occurs, and a localized kyphosis—a gibbus—may follow vertebral collapse. Abscess of the paravertebral soft tissues is very frequent and is seen as fusiform soft tissue swelling around the vertebra. With healing, the abscess may calcify, and the involved bones will show some recalcification and mild sclerosis.

A lytic involvement of a vertebral body associated with a paraspinal soft tissue mass and a narrowed interspace is strongly suggestive of tuberculous spondylitis.[109–116]

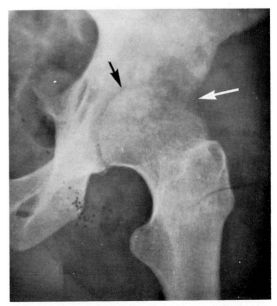

Figure 6–113 **Tuberculosis of Hip.** The articular bone margins are hazy and indistinct, and the joint space is narrowed *(black arrow)* owing to destruction of the articular cartilage. These are relatively late findings. Note the marked periarticular demineralization. The irregular areas of bone lysis *(white arrow)* in the lateral aspect of the femoral head and adjacent acetabulum are most pronounced in the nonweight-bearing portions of the joint. This is a characteristic feature of tuberculous arthritis.

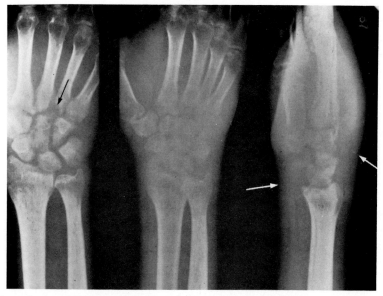

Figure 6–114 **Tuberculosis of Wrist.** Soft tissue and synovial swelling *(white arrows)* gives the wrist a fuzzy appearance. The edges of the carpal bones and the bases of the metacarpals are indistinct, owing to bone resorption. The articular cartilages and joint spaces are fairly well preserved. There is bony demineralization, and clearly defined areas of destruction are seen in the articular margins of the metacarpals *(black arrow).*

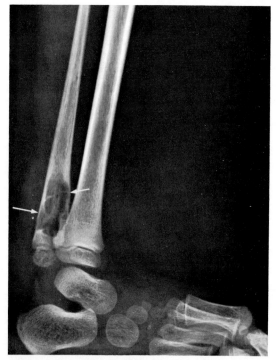

Figure 6–115 **Cystic Bone Tuberculosis.** Cyst-like bone destruction is seen in the lower diaphysis of the fibula, with expansion and thinning of the cortex *(arrows).* There is no sclerosis.

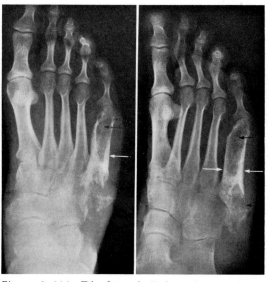

Figure 6–116 **Diaphyseal Tuberculosis (Spina Ventosa).** There is marked medullary destruction and expansion of the fifth metatarsal *(black arrows),* along with increased cortical density and thickness *(white arrows).* A more distinct area of bone destruction is seen at the base of this bone *(black arrowhead).*

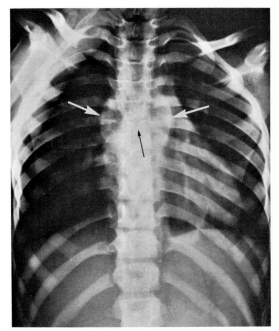

Figure 6–117 **Tuberculous Spondylitis: Cold Abscess.** Fusiform soft tissue swelling (*white arrows*) is caused by a paraspinal abscess associated with tuberculous spondylitis of T6 and T7. The disc space is narrowed (*black arrow*), and T7 is partially collapsed.

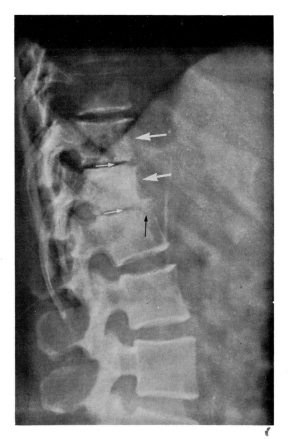

Figure 6–118 **Moderately Advanced Tuberculous Spondylitis.** The bodies of T11, T12, and L1 are destroyed, and the anterior margins of T11 and T12 show loss of substance and irregularity (*large white arrows*). The anterior superior border of L1 is also destroyed (*black arrow*).

The intervertebral spaces (*small white arrows*) are markedly narrowed, and there is kyphosis of this segment.

Characteristically there is progressive marginal bone destruction followed by cartilaginous destruction that leads to loss of the adjacent intervertebral space. In pyogenic spondylitis the changes are similar, but the disc space is narrowed and bone is destroyed within a few days or weeks.

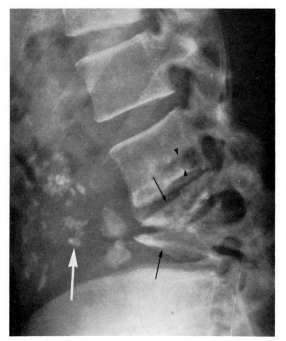

Figure 6–119 **Advanced Tuberculous Spondylitis.**
There is marked destruction of the bodies of L3 and L4
(black arrows), with complete destruction of the intervertebral disc and pronounced kyphosis (gibbus formation). There are also destructive lesions in the inferior portion of L2 *(arrowheads)* and narrowing of the
disc space between L2 and L3. Note the multiple irregular calcifications in the soft tissues anterior to the
spine *(white arrow);* these calcifications are within a
large cold abscess.

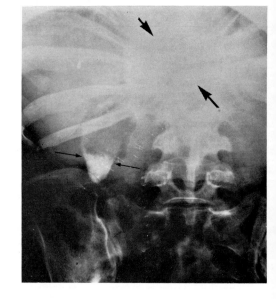

Figure 6–120 **Old Tuberculous Spondylitis with
Gibbus and Calcified Cold Abscess.** The ribs appear
to radiate from a central point, which is the site of a
marked gibbus *(large arrows).* The calcified cold abscess
(small arrows) simulates a dye-filled gallbladder. A
tuberculous abscess often extends along the psoas
muscles. The diaphragm serves as a barrier against
upward or downward spread of a tuberculous abscess.

Disseminated Hematogenous Tuberculosis: Miliary Tuberculosis

In the acute generalized form, pulmonary lesions may not appear until days or
weeks after clinical onset. There is uniform granularity or fine discrete nodulation
uniformly distributed throughout both lung fields. The early lesions may cause
some thickening of the interstitial markings and are often recognized only retro-
spectively. The lesions tend to become ovoid as they increase in size. Often those in
the upper lobe appear somewhat larger than those in the base. Pleural effusion is
not infrequent. In chronic miliary tuberculosis the nodules become larger, vary in
size, and tend to conglomerate. Hepatosplenomegaly may develop.

As a rule there are no other associated pulmonary tuberculous lesions, but occasionally hematogenous tuberculosis may develop in a patient with advanced pulmonary tuberculosis. The miliary lesions may then be obscured but should be recognized in previously unaffected lung areas.

With appropriate therapy, the miliary lesions will disappear in weeks or months.[95, 117, 118]

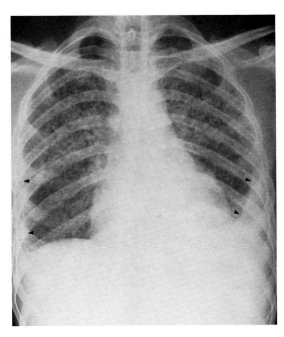

Figure 6–121 **Acute Miliary Tuberculosis.** Fine discrete nodules are uniformly distributed throughout both lung fields. Those in the upper lobe appear somewhat larger than those in the lower lobe. There is a pleural effusion at the left base and a pleural stripe on the right *(arrowheads)*. Enlargement of the cardiac silhouette is caused by tuberculous pericarditis.

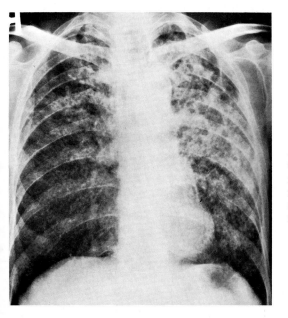

Figure 6–122 **Chronic Miliary Tuberculosis.** In addition to fibroulcerative lesions in the left upper lobe and right apex, there are fine nodules throughout the remaining lung fields due to hematogenous spread. The extreme bases are relatively clear, a frequent finding in miliary tuberculosis.

Diseases Due to Mycobacteria Other Than M. Tuberculosis and M. Leprae

Endemic pulmonary infections caused by anonymous mycobacteria are being increasingly recognized.

Although the epidemiology, development, and response to therapy of these infections are different from tuberculosis, radiographically the lesions of the two diseases are usually indistinguishable. Often, however, the cavities of the anonymous mycobacterial infections are thin-walled; pericavitary infiltrates are minimal, and pleural reaction is rare. Nodular lesions, calcified hilar nodes, and bronchogenic spread are frequently encountered. A paucity of symptoms in the presence of cavitary lesions is somewhat suggestive.

Bone and renal involvement is a rare occurrence. These lesions are radiographically identical to tuberculosis.[119-124]

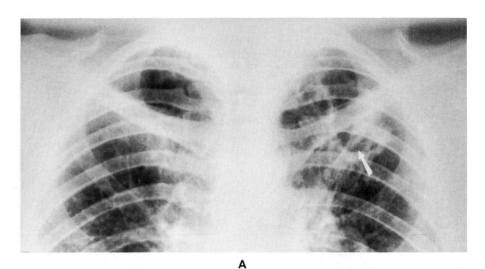

A

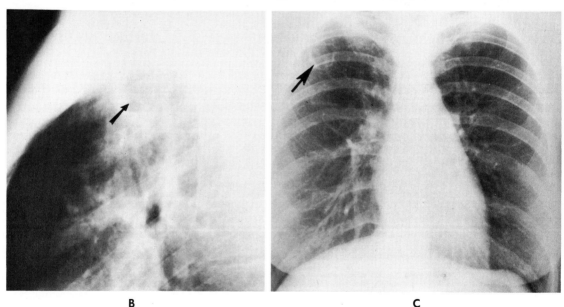

B   C

Figure 6–123 **Atypical Mycobacterial Pulmonary Lesions in Two Patients.**

A, The infiltration and cavity *(arrow)* in the left upper lobe are indistinguishable from tuberculosis. No other involvement was apparent.

B, The anterior location of the cavity *(arrow)* is unusual for classic tuberculosis. *Mycobacterium intracellulare* was the causative organism.

C, A solitary thin-walled cavity *(arrow)* was the only pulmonary lesion in another patient. The organism isolated was *M. kansasii.*

Although atypical mycobacterial lung lesions are usually indistinguishable from tuberculosis, there is a tendency to early cavitation, which is often thin walled. Bronchogenic spread is very uncommon in atypical mycobacterial infection. Response to antituberculous therapy is often slow or poor.

Leprosy (Hansen's Disease)

There are two distinct types of bone lesions: neurotrophic and osteomyelitic. In the more common neurotrophic form the changes primarily involve the distal bones of the hands and feet. There is progressive resorption of bone, beginning in the distal phalanges in the hands and in the metatarsophalangeal bones in the feet. The metatarsal shafts become thin and frequently develop pointed ends. Occasionally single Charcot joints are seen, most often in the tarsus.

In the infrequent (3 to 5 per cent) osteomyelitic type there is periosteal and endosteal reaction, most often in the ulna and fibula, producing increased thickness and density of the bone.

The neurotrophic resorptive bone changes are apparently due to nerve involvement, usually with superimposed trauma and secondary infection, whereas the periosteal and endosteal changes are due to direct bone involvement by the organism.[125-129]

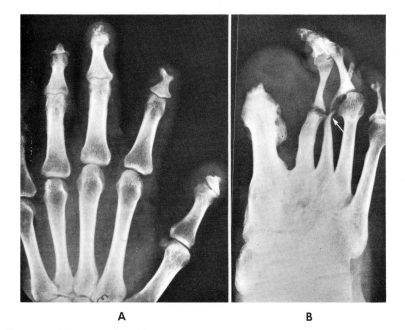

A   B

Figure 6–124   **Leprosy: Neurotrophic Changes.**

A, There is almost complete absorption of the distal phalanges and soft tissues. The remaining borders of the bones are smooth.

B, There is resorption of bones around the first, second, and third metatarsals, with thinning of the shafts and pointed bone remnants (*arrow*). The phalanges of the fourth and fifth toes are similarly involved. All the phalanges of the great toe have been absorbed. (Courtesy K. Bose, Calcutta.)

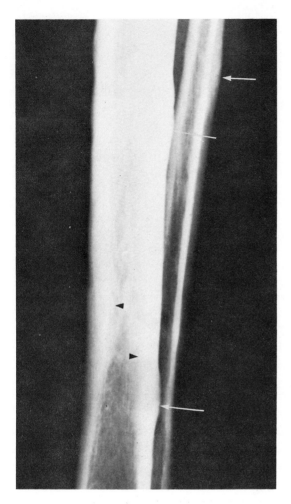

Figure 6–125 **Periosteal and Endosteal Reaction in Leprosy.**   There is marked endosteal thickening of the tibia *(arrowheads),* causing narrowing of the medullary cavity and a marked increase in bone density. Irregularity caused by local periosteal thickening is seen in the tibia and fibula *(arrows).*

These changes are nonspecific and can appear in any low grade chronic osteomyelitis. (Courtesy Prof. L. Oliva, Sienna, Italy.)

SPIROCHETAL INFECTIONS

Syphilis

The earliest and most frequent radiographic bone changes in congenital infantile syphilis occur in the metaphyses; a diaphyseal osteoperiostitis is somewhat less frequent. Symmetric bilateral involvement of several long bones of the extremities is usual.

Thickening of the epiphyseal plate may occur, frequently accompanied by a transverse zone of lucency across the entire metaphysis. This lucent line is not specific and is seen in other conditions of infancy that cause growth disturbances. However, the transverse lucent zone is a highly suggestive finding in an infant under six months. A destructive metaphysitis first manifests as a lytic area at the medial cortical corner of the metaphysis; this is a specific finding. The lysis may extend deep into the bone and may even produce fracture. Occasionally the epiphyseal plate becomes irregular and serrated, simulating early rickets. The epiphyseal ossification centers are generally spared.

The diaphyseal osteoperiostitis may produce small lytic cortical and subcortical defects, but periosteal reaction, lamellated or uniform, usually dominates the picture. In some infants, periosteal reaction is the sole finding. Infrequently, diaphyseal irregularity and medullary widening occur. The bone lesions of congenital syphilis generally heal during infancy, even without treatment.

In juvenile or latent congenital syphilis, diaphyseal sclerosis and cortical thickening are the most characteristic findings. Any of the long bones may be affected. Involvement of the upper tibia produces the classic saber-shin deformity.

In acquired syphilis the cardiovascular lesions include aortitis, aortic aneurysm, and aortic valvular disease. These are discussed in Section 10, *Diseases of the Cardiovascular System*. Bone lesions may occur in the skull, spine, or long bones. In general the roentgenologic appearance is similar to that of chronic osteomyelitis, with both bone proliferation and bone destruction, but without sequestra. Occasionally the picture is that of diffuse bone lytic lesions with periostitis. Gumma formation produces ill-defined lytic areas, usually within a dense thickened bone.

The skull lesions are predominantly lytic and involve both tables, especially the outer table. Spicules of undestroyed bone often persist within the lytic lesions.

Neurotrophic disintegration of a joint may develop in the tabetic phase of late syphilis. Although the knee and hip are most commonly involved, other areas, including the vertebral bodies, can be affected. The neurotrophic joint (Charcot joint) characteristically shows progressive disintegration, the articular ends becoming smoothly eroded and eburnated. There is soft tissue swelling and much calcific debris around the joint. Subluxation is common. The absence of pain allows movement of the joints, thereby preventing osteoporosis. The roentgenologic picture is quite characteristic.

Rarely, an atrophic type of Charcot joint may develop in which there is no periarticular calcification or debris. Only the smoothly and sharply eroded articular bone suggests the neurotrophic nature of the process.

Pulmonary gummatous involvement is quite rare. The lesions can simulate a chronic pneumonitis or a neoplastic mass.[130-133]

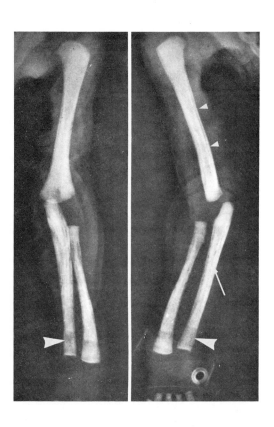

Figure 6–126 **Congenital Syphilis in Infant.** There are broad bands of decreased density proximal to the metaphyseal ends of both radii and ulnas *(large arrowheads)*. The long bones, especially in the forearms, demonstrate osteoperiostitis, with both sclerotic bone *(small arrow)* and scattered areas of lucency. The periosteal layering is best seen in the left humerus *(small arrowheads)*. Metaphyseal infractions have not developed; they usually are a later but more specific finding.

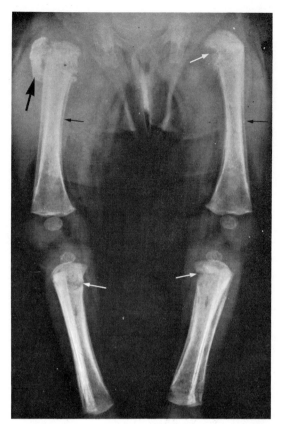

Figure 6–127 **Congenital Syphilis in Older Infant.**
All the bones manifest pronounced periosteal proliferative layering *(small black arrows)* as well as defects on
the medial side near the metaphyscal ends *(small white
arrows)*. The defect extends deep into the shaft in the
left upper femur; the metaphysis of the right upper
femur has been broken off and lies lateral to the shaft
(large black arrow). Small scattered lytic areas are seen
in both tibias.

Osteoperiostitis and submetaphyseal marginal
lytic areas are characteristic roentgenologic features of
congenital syphilis, but the juxtametaphyseal bands of
decreased density are less specific findings. Note the
bilateral symmetric involvement.

Figure 6–128 **Congenital Syphilis: Severe Metaphysitis.** Irregular lytic areas
are limited to the metaphyses of both
femurs and tibias. Characteristic medial
infractions are present *(arrows)*. There
was no periosteal reaction.

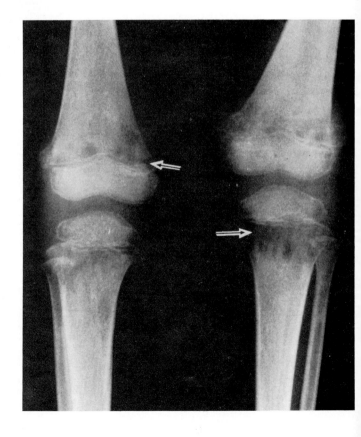

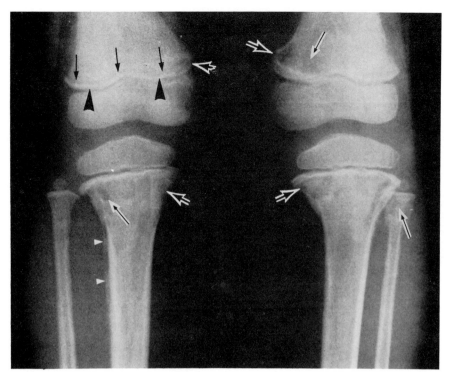

Figure 6–129 **Congenital Syphilis in Late Infancy.** The characteristic bone changes include: (1) increased density of the epiphyseal plate *(black arrowheads),* (2) transverse submetaphyseal band of decreased density *(black arrows),* (3) metaphyseal lytic areas *(smaller black-white arrows),* (4) medial cortical infractions *(larger black-white arrows),* and (5) periosteal reaction *(white arrowheads).*
  The epiphyseal centers are characteristically uninvolved.

Figure 6–130 **Bone Changes in Acquired Syphilis: Two patients.**

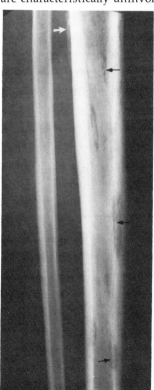

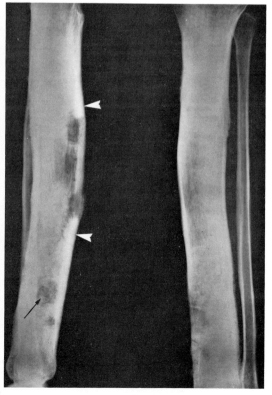

A, In a young male with skin lesions of secondary syphilis there are multiple scattered irregular lucent areas *(black arrows),* which are primarily in the cortex. Some periosteal reaction is apparent *(white arrow).* Similar changes were present in many other bones of the extremities. These are all areas of acute and subacute osteoperiostitis of the secondary stage of syphilis.

B, The tibia is dense, thickened, and irregular *(arrowheads),* the picture of chronic osteomyelitis of syphilis. The irregular lytic areas *(black arrow)* was an intraosseous gumma. The other lytic areas were also gummas, one of which had been biopsied (absent cortex).

Chronic osteomyelitis and gumma formation are seen in the tertiary stage of syphilis.

A   B

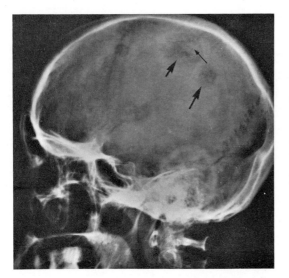

Figure 6–131 **Gummas of Skull: Late Syphilis.** Two circumscribed lytic lesions in the parietal bone *(large arrows)* are caused by gummas. Note uninvolved bone islands within the lytic area *(small arrow)*. Both tables are involved. The lytic areas result from gummatous destruction; the denser areas are reactive bone. The roentgenologic appearance of syphilis of bone is similar to that of other types of osteomyelitis, especially tuberculosis.

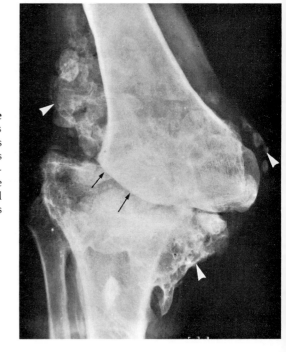

Figure 6–132 **Neurotrophic (Charcot) Knee in Late Syphilis.** The knee joint is deranged, and there is subluxation of the tibia laterally. The femur is smoothly eroded *(arrows)*. Soft tissue and ligamentous calcifications are seen throughout the joint area *(arrowheads)*. These changes are typical, demonstrating the characteristic bone erosion, joint disorganization, and soft tissue calcification. Note that mineral density is fairly well preserved.

Yaws

Bone changes are seen in about one sixth of the cases, exuberant periosteal new bone formation occurring most frequently and producing a picture indistinguishable from that of syphilis. However, in yaws, the bones of the forearms and hands are most often involved. Periosteal new bone formation may be associated with multiple lytic areas usually surrounded by endosteal areas of increased density. Granulomatous soft tissue nodules may occur late in the disease. Contiguous bone involvement can occur. Lesions of the axial skeleton are rare.[134–136]

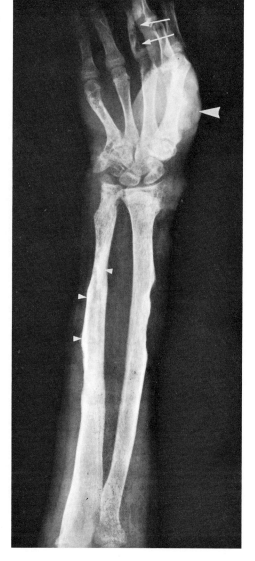

Figure 6–133 **Yaws.** A destructive lesion is seen in the proximal phalanx of the middle finger *(arrows),* without bone reaction. Multiple nodular soft tissue lesions can be seen adjacent to the subluxated thumb *(large arrowhead).* There is extensive periosteal new bone formation *(small arrowheads)* in the ulna, and scalloping of the midradius owing to extension of the soft tissue lesions to the bone.

Tropical Ulcer

This chronic ulcer, caused by the fusiform bacillus and Vincent's spirochete, usually occurs in the mid or lower leg. Children and young adults are most often affected.

In about 20 per cent of cases the underlying anterior tibia develops a fusiform periosteal reaction. This later blends with the cortex to produce cortical thickening, which may become extremely dense and thick—the classic ivory ulcer osteoma. The thickening may completely encircle the bone and even extend to involve the entire shaft. The "osteoma" may undergo saucerization and sequestration. In some cases, medullary bone may replace the thickened cortex, giving rise to the cancellous ulcer osteoma with bone deformity. Osteoporosis in the distal portion of the affected tibia is a frequent occurrence. The overlying soft tissues become thinned and atrophic.

Depending upon the stage of development, the radiographic appearance may simulate a nonspecific acute or chronic osteomyelitis, fibrous dysplasia, or Paget's disease. The total picture readily permits correct diagnosis.

In a longstanding, untreated tropical ulcer, malignant changes are not infrequent. These are characterized by the appearance of a soft tissue mass and a destructive erosion of the previously thickened underlying bone.[61, 137, 138]

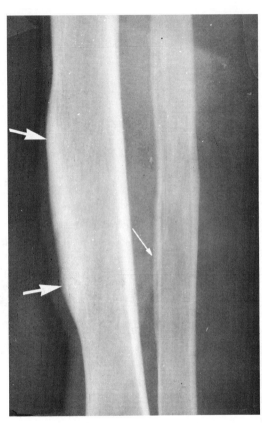

Figure 6–134 **Tropical Ulcer of Leg.** There is fusiform cortical thickening *(large arrows)* of the anterolateral aspect of the midtibia, and a periosteal layering of the fibula *(small arrow)*. A large ulcerating lesion of several years' duration was in the overlying soft tissues. (Courtesy M. Ghigo, Alassio, Italy.)

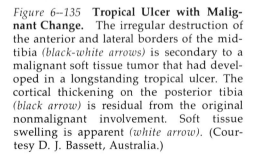

Figure 6–135 **Tropical Ulcer with Malignant Change.** The irregular destruction of the anterior and lateral borders of the midtibia *(black-white arrows)* is secondary to a malignant soft tissue tumor that had developed in a longstanding tropical ulcer. The cortical thickening on the posterior tibia *(black arrow)* is residual from the original nonmalignant involvement. Soft tissue swelling is apparent *(white arrow).* (Courtesy D. J. Bassett, Australia.)

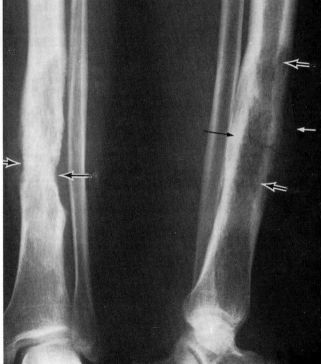

THE MYCOSES

Histoplasmosis

Pulmonary lesions occur in about 50 per cent of individuals clinically infected with *Histoplasma capsulatum*. In early milder cases a single ill-defined patch of peripheral infiltrate appears most often in the lower lung fields, usually in association with hilar adenopathy. These findings, called primary or benign histoplasmosis, are somewhat analogous to the primary complex of tuberculosis.

In the more acutely ill patients, small poorly defined densities may appear rapidly and diffusely in both lungs, often with bilateral hilar adenopathy. Such lesions are particularly common in younger patients.

The more chronic form of the disease strikingly resembles tuberculosis, although the histoplasmosis lesions develop at a faster rate. Areas of consolidation, sometimes quite large, usually develop in the apical and posterior apical segments. These often break down into thick-walled cavities. Focal areas of bronchiectatic lucencies may appear in late stages. Pleural reaction is common in adults, but it rarely occurs in children.

The pulmonary lesions heal with fibrosis, scarring, and, often, calcification. Stable coin lesions, similar to a tuberculoma, may remain unchanged indefinitely. The well-circumscribed histoplasmoma, usually less than 3 cm. in diameter, is most often in the lower lung fields. Central rounded calcification, if present, is virtually pathognomonic of a healed histoplasma lesion. Widespread disease may heal with characteristic small scattered calcifications in both lungs. Calcification of the hilar nodes is common, and these may infrequently cause bronchial obstruction.

Chronic mediastinitis can occur. Usually there are no roentgen findings until the superior vena cava is obstructed, and then the distended collateral veins (azygos group) may cause mediastinal widening (see *Acute Mediastinitis*, p. 571). Sometimes the granulomas are large enough to cause mediastinal widening, radiographically simulating Hodgkin's disease.

Hepatomegaly and splenomegaly are often seen in the disseminated form, especially in children. Small rounded calcifications in the spleen may appear.

Rarer complications include adrenal involvement (which may even produce an Addison syndrome), pericarditis, chronic osteomyelitis, and central nervous system infection. In some endemic areas, skeletal lesions have been reported in almost two thirds of the cases.[139-145]

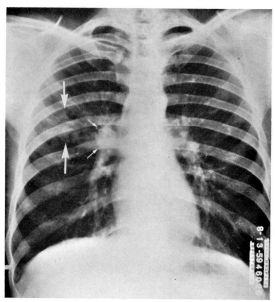

Figure 6–136 **Histoplasmosis: Primary Complex.**   There are irregular infiltrations in the right lung *(large arrows)* associated with enlargement of the right hilar nodes *(small arrows)*. The left hilum is also prominent. The picture is similar to the primary complex of tuberculosis.

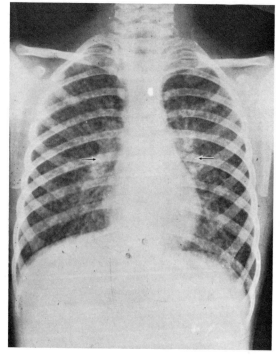

Figure 6–137 **Acute Disseminated Histoplasmosis.**   Chest film of an acutely ill young male reveals diffuse small nodular densities uniformly scattered throughout both lung fields. The hilar shadows were prominent, due to hilar adenopathy *(arrows)*. A mild splenomegaly was noted.

Complete resolution of the parenchymal lesions and a decrease in size of the hilar shadows occurred within three weeks.

The acute disseminated form of histoplasmosis can simulate sarcoidosis or miliary tuberculosis, although hilar adenopathy is rare in the latter. If some of the lesions undergo discrete calcification, a picture similar to that shown in Figure 6–139 can result.

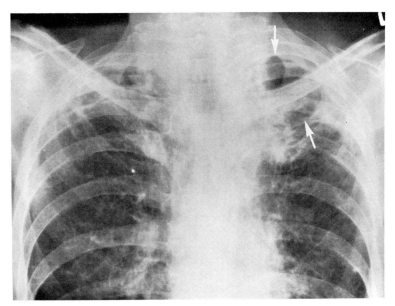

Figure 6–138 **Chronic Cavitary Histoplasmosis.** Fibroid infiltration appears in both upper lobes, but it is more pronounced on the left and is associated with cavitation *(arrows)*. Pleural thickening is noted at the left apex. This picture is indistinguishable from that of tuberculosis.

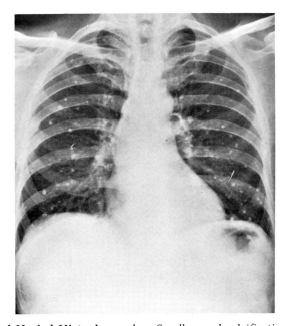

Figure 6–139 **Disseminated Healed Histoplasmosis.** Small round calcifications are scattered throughout both lung fields—the characteristic feature of healed histoplasmosis. Often there are similar calcifications in the spleen.

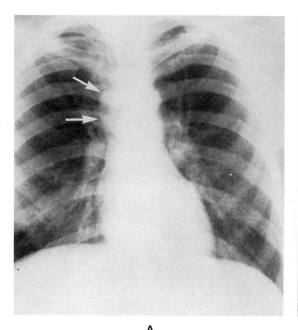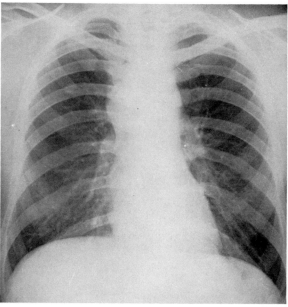

A

B

Figure 6–140   **Histoplasmosis: Progressive Chronic Mediastinitis.**

A, A chest film made one year prior to onset of clinical symptoms reveals a mass *(arrows)* that has caused widening of the right mediastinum.

B, This film, made a few months after the onset of malaise, enlargement of the cervical and axillary nodes, and symptoms of superior vena caval obstruction, shows further widening of the mediastinum, and the widening is now bilateral. The cause was a caseating granuloma containing *Histoplasma capsulatum.*

Mediastinitis due to histoplasmosis may develop without significant radiographic changes, or it may cause progressive widening of the mediastinum, simulating Hodgkin's disease. Histoplasmal mediastinitis cannot be distinguished radiographically from tuberculous mediastinitis. (Courtesy F. Lull: Radiology, 73:367, 1959.)

Coccidioidomycosis

Coccidioidomycosis is endemic in the western and southwestern United States. Primary pulmonary involvement occurs in a high percentage of cases, and the course of the pulmonary disease may vary from a few days to many months. The lung changes in primary coccidioidomycosis are highly variable.

A segmental, patchy, poorly marginated pneumonia occurs in about half the cases, frequently accompanied by ipsilateral hilar adenopathy and, less often, by pleural effusion. Pleural effusion occurs more often in children. Multiple transient cavities may appear, and usually disappear within a few months.

In the more disseminated form, widespread fluffy infiltrates are seen, sometimes with a nodular or miliary appearance.

The primary infiltrates may clear completely, but they often leave residual disease such as one or more cavities, nodular granulomas, or fibrotic scarring. About half of the residual cavities are thin-walled without surrounding reaction, and this is a characteristic and suggestive appearance. The thick-walled cavities are nonspecific, often suggesting tuberculosis. The nodular granulomas may or may not calcify and are of varying size.

Although residual thin-walled cavities are suggestive of coccidioidomycosis, the other pulmonary findings are nonspecific.

Bone involvement occurs in about 20 per cent of cases of the disseminated form of coccidioidomycosis. Any bone may be affected, but there is a tendency toward involvement of bony eminences and ends of long bones. The lesions are usually punched-out and sharply marginated lytic lesions that can resemble tuberculosis or even metastatic malignancy. Later, a sclerotic margin appears, but periosteal reaction is rare. Soft tissue swelling due to an overlying abscess or draining sinus is often seen. Extension to an adjacent joint can bring destruction to the fibrocartilage and articular bone margin, and this resembles arthritis of tuberculous origin.[146-150]

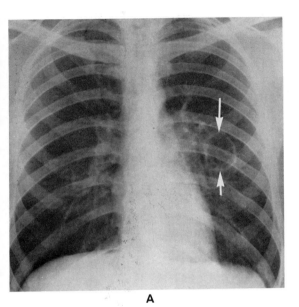
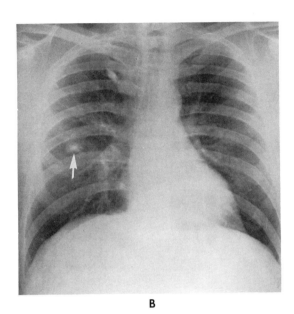

A   B

Figure 6–141 **Coccidioidomycosis: Residual Lung Changes in Two Patients.**

A, There is a characteristic thin-walled cavity in the left lung *(arrows),* without surrounding lung reaction or fluid level.

B, A circumscribed nodular density with central calcification *(arrow)* is shown. The right hilar nodes are enlarged.

While the nodular lesion with calcification cannot be distinguished from a tuberculoma, a thin-walled cavity without surrounding reaction is characteristic of coccidioidomycosis. (Courtesy Dr. M. Small, East Orange, New Jersey.)

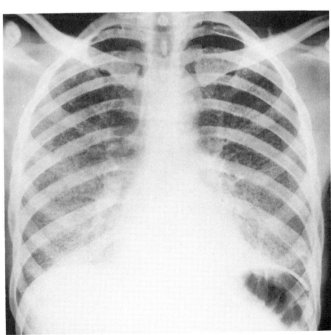

Figure 6–142 **Disseminated Hematogenous Coccidioidomycosis.** Diffuse miliary nodulation and fine linear densities are seen throughout both lung fields. The appearance is identical to that of miliary tuberculosis.

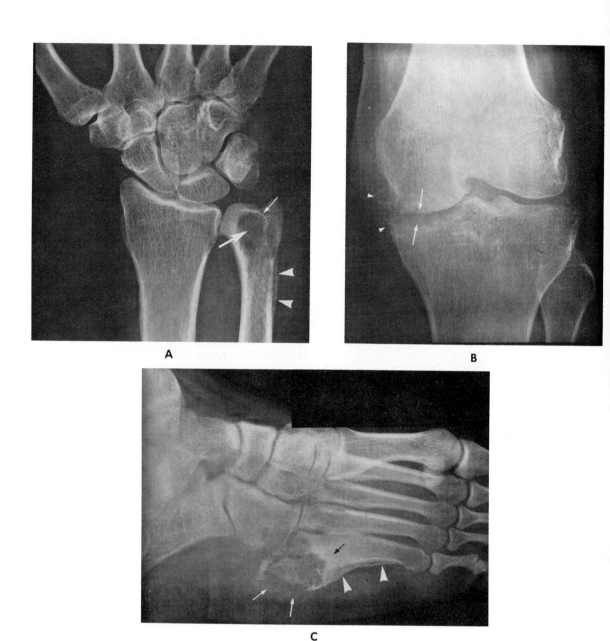

A

B

C

Figure 6–143 **Coccidioidomycosis of Bone and Joint: Two Patients.**

A, The sharply marginated lytic lesion in the distal ulna *(large arrow)* has a sclerotic upper margin *(small arrow)*. Periosteal reaction is evident *(arrowheads)*. The adjacent soft tissues were swollen and contained draining nodules.

B, In the swollen knee of the same patient, the cortical articular surfaces of the medial tibia and femur have been eroded and demineralized by an acute coccidioides arthritis *(arrows)* Some cartilage destruction has caused slight joint space narrowing. There is no reactive sclerosis. The osteophytes *(arrowheads)* are old and unrelated. Tuberculosis could produce an identical picture.

C, The base of the fifth metatarsal of another patient is expanded by a large irregular lytic lesion that has destroyed the cortex *(white arrows)*. The adjacent bone is sclerotic *(black arrow)*, and periosteal new bone is seen along the shaft *(arrowheads)*. The cuboid is also involved. The surrounding soft tissues were swollen and contained a painful plantar ulcer. (Courtesy Dr. George Jacobson, Los Angeles, California.)

Blastomycosis

The infection, which apparently originates in the lungs in 95 per cent of cases, is widely disseminated by hematogenous or lymphatic routes. The chief roentgenologic findings are in the chest and skeletal system. The skin, urinary tract, liver, spleen, and, least often, the central nervous system can be involved clinically.

The chest picture is quite variable and not characteristic. Circumscribed nodules or nondescript areas of bronchopneumonia may appear in any portion of the lungs. Cavitation may occur but is infrequent. Hilar adenopathy is common. Localized pleural thickening is often seen, but effusion is rare. Healing occurs with fibrosis, but calcification is uncommon. The nonspecific roentgen picture can sometimes simulate tuberculosis. Unilateral hilar adenopathy may suggest bronchogenic neoplasm.

In acute pulmonary blastomycosis, bronchopneumonic soft alveolar infiltrates may be seen. These may diminish or disapper within a few weeks.

Bone involvement occurs in almost half the cases. An intense periosteal reaction combined with bone destruction produces the picture of a nonspecific chronic osteomyelitis. Sclerosis and irregular widening of the bone may dominate the findings. Secondary soft tissue abscesses and draining sinuses are common, in contrast to tuberculous or luetic osteomyelitis. An adjacent joint may be secondarily involved. In the spine, a psoas abscess may appear before bone changes are apparent.[151-156]

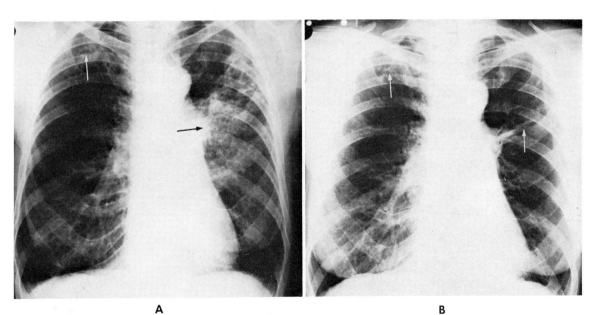

A   B

Figure 6–144 **Bilateral Pulmonary Blastomycosis.**

A, A large linear and alveolar infiltrate extends from the left hilum into the upper and middle lung fields. The left hilar nodes are enlarged *(black arrow)*. A smaller infiltrate at the right apex contains a small cavity *(white arrow)*. These lesions cannot be distinguished from tuberculosis.

B, One year after healing, there are fibrotic scars *(arrows)*. Note the absence of calcification. (Courtesy W. Boswell: Am. J. Roentgenol. Radium Ther. Nucl. Med., *81*:224, 1954.)

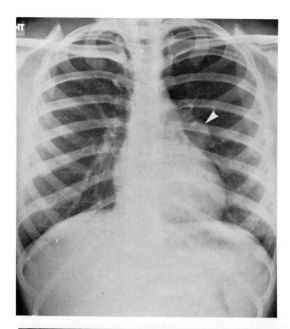

Figure 6–145 **Blastomycosis: Hilar Mass.** The sharply circumscribed nodular mass in the left hilum *(arrowhead)* was a collection of enlarged lymph nodes involved by blastomycosis. The picture simulates bronchogenic carcinoma.

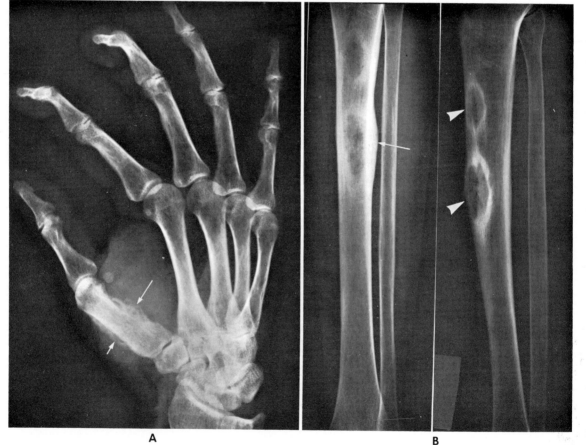

A   B

Figure 6–146 **Active Blastomycosis Osteomyelitis.**

A, There is destruction of the medial aspect of the first metacarpal *(long arrow),* associated wth irregular periosteal new bone formation. Similar periosteal reaction is seen on the lateral aspect of the shaft *(small arrow).*

B, In the tibia there are large circumscribed lytic areas *(arrowheads)* surrounded by sharply defined sclerotic zones. The full extent of periosteal thickening is visualized on anteroposterior view *(arrow)* (Courtesy W. Boswell: Am. J. Roentgenol. Radium Ther. Nucl. Med., *81*:224, 1954.)

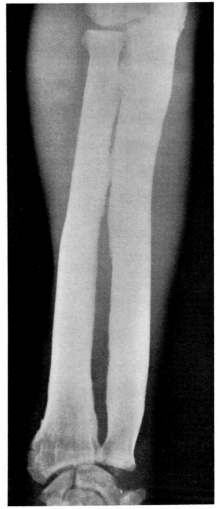

Figure 6–147 **Chronic Sclerosing Osteomyelitis: Blastomycosis.**
The radius and ulna are diffusely thickened and irregular due to
extensive periosteal new bone formation. No lytic areas are seen.
Blastomycotic bone involvement cannot be differentiated radiographically from many other forms of chronic osteomyelitis.

Cryptococcosis (Torulosis)

Cryptococcosis is world-wide in distribution. Cryptococcal infections may be
primary, but more often they are secondary invaders in patients with debilitating
diseases, particularly those with diffuse involvement of the reticuloendothelial system or those undergoing prolonged immunosuppressive therapy.

The commonest radiographic findings are in the lungs; bone involvement
occurs in 5 to 10 per cent of disseminated cases. Other tissues or organs, particularly the central nervous system, can be affected.

There is a wide spectrum of pulmonary manifestations, including solitary infiltrations, ill-defined masses, and multiple nodular densities. A solitary irregular
cryptococcal mass may simulate carcinoma. Cavitation occasionally occurs. Hilar
adenopathy is a late manifestation. Pleural reaction is sometimes found.

In the skeleton, any bone may be affected, but there is a predilection for bony
eminences. Radiographically, the lesions are usually well-circumscribed lytic
areas without much sclerosis or periosteal reaction. The lack of sclerosis and
periosteal reaction may help distinguish this disease from other fungal bone
infections. Occasionally the lesions can resemble metastatic malignancy.[153, 157–159]

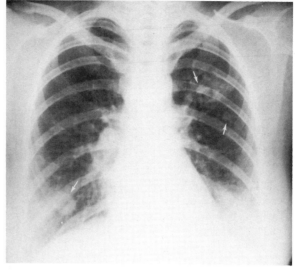

A

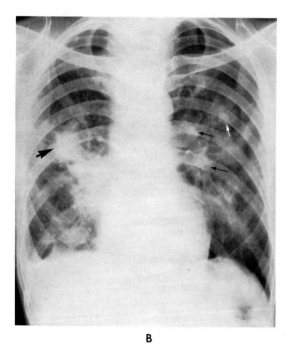

B

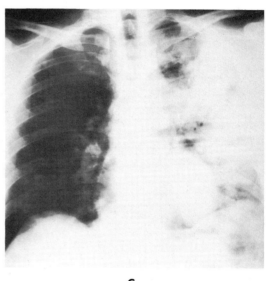

C

Figure 6–148 **Cryptococcosis of the Lung: Three Patients.**

The variability and nonspecificity of the radiographic lesions in pulmonary cryptococcosis are illustrated by these three moderately to severely advanced cases.

A, Nineteen year old female. An infiltrative lesion in the left upper lobe is associated with ill-defined nodular lesions in its vicinity and a large nodule in the right base *(white arrows).* Nodal enlargement is causing fullness in the right hilum. No pleural reaction is noted.

B, Twenty-five year old male. There are widespread ill-defined infiltrates in both lungs; some confluent *(large black arrow),* others nodular *(white arrow),* and many linear. The left hilar nodes are enlarged *(small black arrows).* The right hilum also is apparently enlarged. Mild pleural reaction is seen at the right base.

C, Fifty year old male. Extensive ill-defined confluent densities in the left lung and enlargement of the left hilum are apparent. The right lung is entirely clear. No pleural reaction is seen.

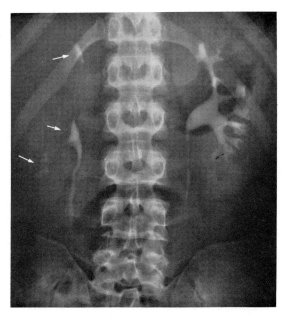

Figure 6–149 **Cryptococcal Granuloma of Kidney Complicating Sarcoidosis and Steroid Therapy: Pyelogram.** The patient received steroid therapy over a long period. The right kidney is enlarged, and a large mass has caused elongation and displacement of the upper and lower calices *(arrows)*. The middle calices are obliterated. Radiographic diagnosis was a large neoplasm, but the mass proved to be a huge cryptococcal granuloma.

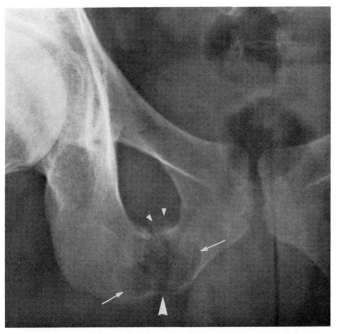

Figure 6–150 **Cryptococcosis of Bone.** The circumscribed, irregularly shaped lytic lesion *(small arrows)* at the ischiopubic junction shows no sclerosis. Some periosteal reaction *(small arrowheads)* appears in the upper margin. The inferior cortex is partially eroded *(large arrowhead)*.

The bone lesions of cryptococcosis are predominantly lytic and can simulate metastatic malignancy. (Courtesy Dr. George Jacobson, Los Angeles, California.)

Actinomycosis

Actinomycosis can produce radiographic changes in the lungs, pleura, bones, and gastrointestinal tract.

Pulmonary lesions are found in about 15 per cent of cases. These lesions are usually confluent areas of consolidation that are often near the hilum or superior segment of the lower lobe. Sometimes there is cavitation. Late fibrosis and lung contraction may develop. Extensive pleural involvement can occur, but large accumulations of fluid are uncommon. Extension to the chest wall occurs in more than half the cases, causing soft tissue swelling and often involving a rib with periosteal reaction or destructive changes. External sinus tracts often occur.

Bone involvement arises from invasion by an adjacent soft tissue lesion. The mandible is the most frequent site, since cervicofacial actinomycosis is the most common form of the disease. The radiographic appearance is that of nonspecific subacute or chronic osteomyelitis with areas of destruction and with sclerosis. Periosteal reaction and new bone formation may be extensive. Multiple bones are often involved, and draining sinuses are frequently seen.

Intestinal involvement generally occurs without associated lung or bone lesions. The cecum and appendix are the commonest sites. The radiographic appearance is one of nonspecific inflammation. Often there is an accompanying soft tissue mass and a sinus tract to the overlying skin.

Whereas the pulmonary, bone, or cecal lesions are radiographically nonspecific, their association with one or more sinus tracts should suggest the possibility of actinomycosis.[160-162]

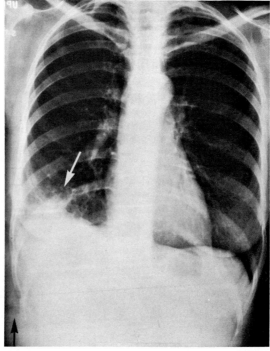

Figure 6-151 **Pulmonary Actinomycosis: Pleural and Soft Tissue Involvement.**   There is a small area of infiltration at the right base *(white arrow);* the costophrenic angle is obscured by pleural reaction. The infection has penetrated the chest wall with resultant soft tissue swelling. The air tract *(black arrow)* is in a sinus extending to the skin.

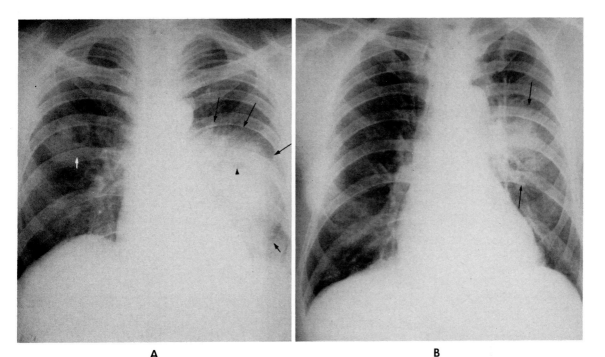

A   B

Figure 6–152 **Two Cases of Pulmonary Actinomycosis.**

    A, A large homogeneous density is seen in the left hilum and lower lung field *(black arrows),* with an air-fluid level in an abscess *(arrowhead).* There is pleural reaction at the left base. Another lesion in the right lung *(white arrow)* has hazy borders, and cavitation has probably begun.

    B, A large ill-defined density *(arrows)* is seen in the superior segment of the left lower lobe of the lung. No other areas were involved.

    In the absence of a draining sinus the pulmonary findings are nonspecific.

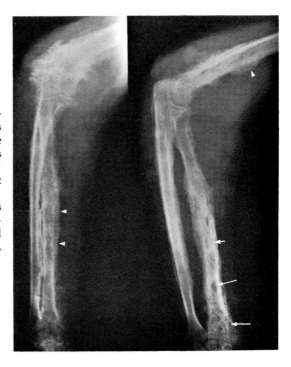

Figure 6–153 **Extensive Actinomycosis of Bone Secondary to Skin Lesions.** There are multiple lytic areas in the humerus, radius, and ulna *(short arrow).* The darker, sharply demarcated lytic areas are fistulous tracts *(long arrows)* to the skin; periosteal reaction is extensive and quite ragged *(arrowheads);* elbow joint spaces are fairly well preserved.

    The soft tissues are swollen; the actinomycosis originated here and extended into the bone. The picture is that of nonspecific, extensive, subacute and chronic osteomyelitis. (Courtesy M. Varadarajan, Madras, India.)

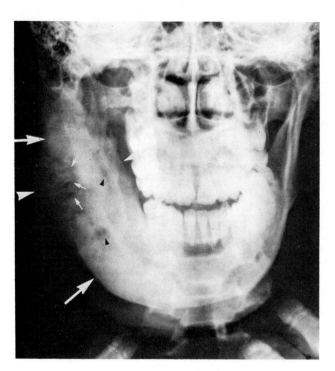

Figure 6–154 **Actinomycosis of Mandible.**
The right mandible of this 15 year old boy is
widened and sclerotic *(large white arrows)* and
contains numerous lytic areas *(black arrowheads)*. The bone thickening is due to subperiosteal new bone formation. Several horizontal lucent sinus tracts *(small white arrows)*
extend from the bone into the swollen soft tissues *(large white arrowhead)*. The cervicofacial
region is the most common site of actinomycosis.

Lytic-sclerotic changes and subperiosteal
new bone formation are nonspecific findings
of many forms of chronic osteomyelitis, but
multiple sinus tracts are more characteristic of
A. bovis infection. (Courtesy of Dr. George
Jacobson, Los Angeles, California.)

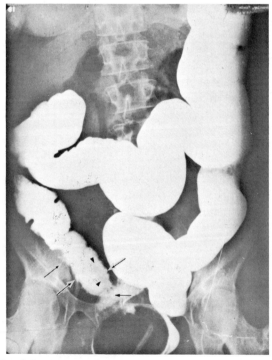

Figure 6–155 **Actinomycosis of the Cecum and Appendix.**   The cecum is narrowed, elongated, and fixed
in a low pelvic position. The walls are irregularly serrated *(black-white arrows)*, and the lumen contains a
faint lucent mass *(arrowheads)*, which is granulomatous tissue. The irregular barium collection at the cecal
tip *(black arrow)* is a fistula from the appendix to the abdominal wall.

The cecum and the appendix are the most common sites of gastrointestinal actinomycosis. Granulomatous colitis can produce a similar picture, including fistula formation.

Nocardiosis

This infection is found throughout the world. It is virtually always an opportunistic invader, most often in patients on steroid therapy or with an underlying malignancy. Initial involvement is always in the lungs, from which there may be hematogenous spread to the skin and other organs. Brain abscess is not uncommon. In over half the cases the disease is confined to the chest.

Radiographically there are single or multiple ill-defined chronic patches of nonspecific infiltration, often with cavitation. Pleural involvement is common and may lead to empyema. Occasionally there is extension to the ribs and chest wall. An indolent form, one having chronic nodular fibrosis and occasional cavitation, is also encountered. Pulmonary nocardiosis cannot be distinguished from tuberculosis or from many other fungal diseases without identification of the causative microorganism.

Prolonged immunosuppressive therapy may predispose to pulmonary nocardiosis.[163, 164]

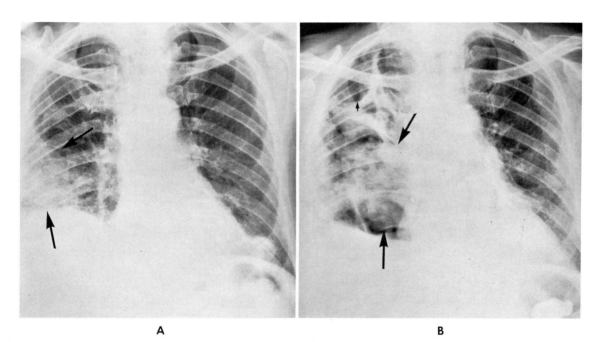

A   B

Figure 6–156 **Pulmonary Nocardiosis: Rapid Progression.**

  A, A homogeneous infiltrate is seen at the right base *(arrows)*, with pleural involvement.

  B, One month later there is extensive disease of the entire right lung, with confluent bronchopneumonia in the right lower lobe *(upper large arrow)* and multiple cavitations without fluid levels in the right upper lobe *(small arrow)*. Pleural reaction has increased. Excavation is developing in the lower lobe *(lower large arrow)*. This picture simulates advanced tuberculosis. (Courtesy Dr. J. Murray, Los Angeles, California.)

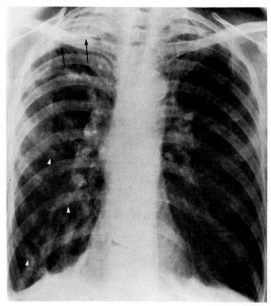

Figure 6–157 **Indolent Fibrosis in Nocardiosis.** There is bilateral apical pleural thickening *(black arrows)* with linear and nodular areas of fibrosis *(arrowheads).* Both hila are elevated, and basal emphysema is present. The findings cannot be distinguished from those of chronic tuberculosis. (Courtesy Dr. J. Murray, Los Angeles, California.)

Sporotrichosis (Sporothrixosis)

Sporotrichosis is usually limited to the skin. In disseminated sporotrichosis, pulmonary and osseous changes can occur but are exceedingly rare. The skeletal system may be involved by hematogenous spread or by direct extension from the skin lesions. With hematogenous spread to joints, there will be periarticular swelling and, later, loss of joint space with articular irregularities and erosions. The small joints of the hands and feet are most often affected. The radiologic appearance may simulate a destructive rheumatoid or an infectious arthritis such as tuberculosis.

The pulmonary lesions appear as areas of chronic pneumonitis, nodular masses, or fibrosis and cavitation; hilar adenopathy is frequent. None of the roentgenologic findings is specific.[165-168]

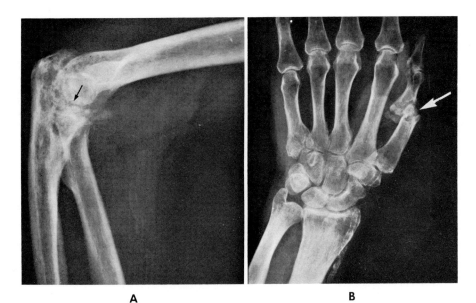

A   **B**

Figure 6–158 **Sporotrichosis: Bone and Joint Lesions: Hematogenous Spread.**

A, There is irregular narrowing of the joint space *(arrow).* Irregular destructive lesions are seen in the subarticular portion of the ulna.

B, Subluxation of the metacarpal phalangeal joint of the thumb *(arrow)* is present, with lytic changes in the juxta-articular bones. There is also calcification of the arterial vessels around the wrist, but this is not related to the sporotrichosis. (Courtesy Dr. P. Hodges, Ann Arbor, Michigan.)

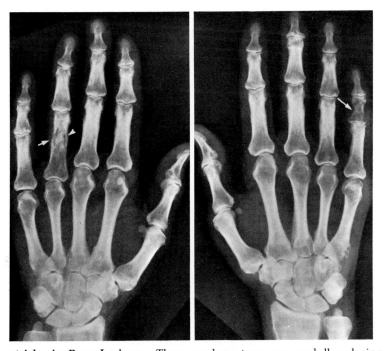

Figure 6–159 **Sporotrichosis: Bone Lesions.**   There are hematogenous medullary lesions in the proximal phalanx of the fourth finger of one hand and in the middle phalanx of the fifth finger of the other hand *(arrows).* The lesions are lytic, and there is no joint involvement.

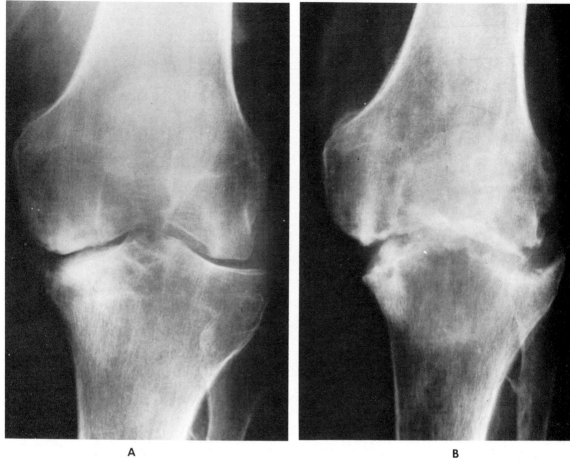

A   B

Figure 6–160 **Systemic Sporotrichosis: Progressive Destructive Arthritis.**

    A, Joint narrowing, eburnation, and erosions are seen on the medial half of the knee joint.

    B, Two years later, the entire joint space is irregularly narrowed, and large erosions are present.

    Tuberculous arthritis can present a similar appearance, but is usually accompanied by greater osteoporosis. (Courtesy K. D. Pearson, Santa Rosa, California.)

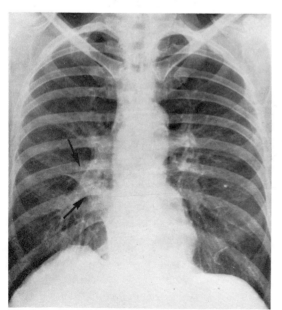

Figure 6–161 **Sporotrichosis of Lungs.** Enlargement of the right hilar nodes has caused atelectasis of the superior segment of the right lower lobe *(arrows).* Hilar adenopathy is thereby obscured. (Courtesy Dr. P. Hodges, Ann Arbor, Michigan.)

Mucormycosis

Mucormycosis is rare and occurs most often in patients with diabetes or leukemia. The paranasal sinuses are almost always infected, and clouding and thickening of the mucous membrane are evident.

In the chest there is a nonspecific lobular pneumonia that is usually associated with vasculitis. The latter may lead to pulmonary infarction and cavitation. The lesions are often multiple and may be a combination of large nodules, diffuse irregular alveolar infiltrates, and cavitating areas. Pleural effusion is uncommon. The findings are quite similar to those of candidiasis, aspergillosis, or cryptococcosis. Ulceration of the ileum and cerebrovascular thrombosis especially of the internal carotid artery can result from intestinal or cerebral vasculitis.

The combination of lung changes and sinusitis may simulate Wegener's granulomatosis.[169-171]

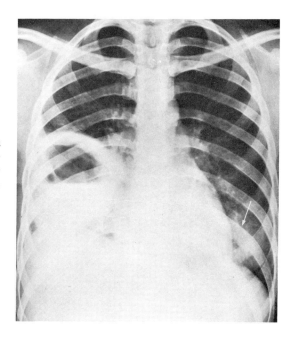

Figure 6–162 **Mucormycosis: Abscess Formation in a Diabetic.** In the right lower lung field there is a thick-walled abscess with an air-fluid level; a pulmonary infiltrate appears at the left base *(arrow).* (Courtesy H. Blakenberg, Huntersville, North Carolina.)

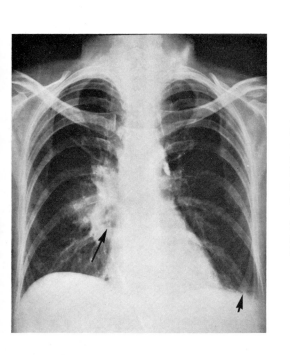

Figure 6–163 **Mucormycosis: Cavitation and Infiltraton in a Diabetic.** There is perihilar pneumonitis on the right, with involvement of the anterior segment of the upper lobe and the apical segment of the lower lobe. An abscess with an air-fluid level *(long arrow)* is seen in the apical lower lobe segment. There is minimal pneumonitis at the left base *(short arrow).*

Although the pulmonary findings are nonspecific, their appearance in a diabetic should suggest either mucormycosis or tuberculosis. (Courtesy Dr. D. Gabriele, New Haven, Connecticut.)

Maduromycosis (Madura Foot, Mycetoma)

Maduromycosis generally directly invades the feet and occurs most often in tropical regions. It is the most common fungal infection involving bony structures. Direct extension from a soft tissue mycetoma causes the bone involvement. The tarsometatarsal area of the foot is most commonly affected. In rarer cases the hand and subsequently the carpometacarpal bones are the affected sites.

In early bone lesions there are poorly defined osteolytic areas. Spread to adjacent bones and joints is common. Periosteal reaction is minimal. Eventually, diffuse lytic areas destroy and deform the bones and often produce bizarre filiform and undulating shapes. Many of the lucent areas are sinus tracts. Areas of endosteal sclerosis are common, giving a mixed lytic-sclerotic appearance, but sequestra almost never develop. Joint spaces usually become narrowed, indistinct, and eventually obliterated. Soft tissue swelling is a prominent feature of maduromycosis.[172, 173]

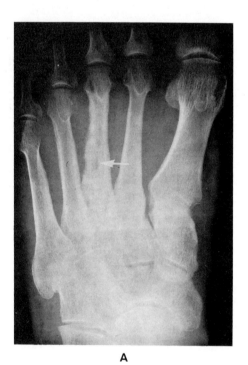 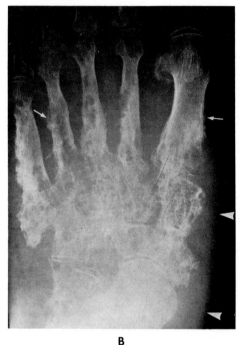

A   B

Figure 6–164 **Maduromycosis: Early and Late Changes in Foot.**

A, The sclerotic third metatarsal has an irregular proximal lateral border and contains some lytic areas *(arrow).* The appearance is consistent with many forms of chronic osteomyelitis.

B, Two years later the tarsal and metatarsal bones are diffusely involved by predominantly lytic (cortical and medullary) lesions. Some periosteal reaction is seen along the first and fourth metatarsals *(arrows).* The tarsometatarsal joints are virtually obliterated. There is marked soft tissue swelling *(arrowheads).*

Diffuse lytic involvement of the tarsometatarsal bones, with some periosteal reaction in a swollen foot, is characteristic of madura foot. (Courtesy Dr. George Jacobson, Los Angeles, California.)

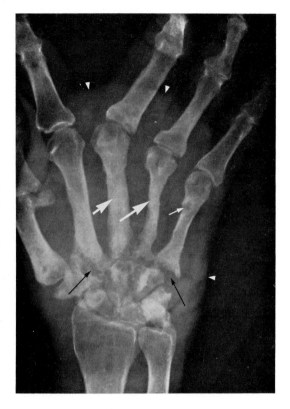

Figure 6–165 **Maduromycosis of the Hand (Myce-toma).** The distal carpal bones and proximal metacarpals, which have been directly invaded by the soft tissue granuloma, are chiefly affected. The articular surfaces are irregularly eroded *(black arrows),* and the medullary bone is dense and shows loss of normal trabeculation *(large white arrows).* Note the lucent trough *(small white arrow),* a sinus tract from the fifth metacarpal.

Cortical reaction is not conspicuous. The fourth metacarpal is irregularly eroded and thin, and there is extensive soft tissue swelling *(arrowheads).*

(Courtesy K. Varadarajan, Madras, India.)

Aspergillosis

Pulmonary infection by Aspergillus spores most often occurs as a complication of immunosuppressive therapy, in debilitated patients, or as a secondary invader of chronic lung disease. Infrequently, it is a primary pathogen causing extensive pneumonitis.

The radiographic findings are quite variable and nonspecific. Lobar consolidation, bronchopneumonic infiltrates, thick-walled cavities, and solitary mass-like lesions can occur, often in combination. Upper lobe lesions can simulate tuberculosis. The aspergilli may form a movable mass (aspergilloma or fungus ball) within a bronchiectatic or other cavitary lesion, producing a characteristic roentgen appearance of a cavity containing a rounded density surrounded by a crescent of air. The aspergilloma may move with positional changes of the patient. Pleural reaction is uncommon.

Allergic bronchopulmonary aspergillosis is a fairly distinct entity, occurring in adult asthmatics and associated with blood eosinophilia. It may be considered a form of pulmonary infiltrates with eosinophilia (P.I.E.) (see p. 121). Radiographically, the most common and characteristic findings are segmental linear areas of dilated bronchi, either air-containing with thickened walls (parallel line shadows) or opaque with secretions suggestive of cylindrical bronchiectasis. Hairline ring shadows (dilated bronchi on end) and honeycombed densities are also common. Occasionally massive consolidation is seen. Bronchography characteristically discloses dilated proximal bronchi with normal distal segments or complete distal occlusion. The infiltrates slowly resolve, and with clearing there may be a shrunken lobe or lobule.

Focal densities suggestive of dilated bronchi in an adult asthmatic with blood eosinophilia are highly suggestive of allergic bronchopulmonary aspergillosis. This can be confirmed by bronchial brushings or skin tests.[174-179]

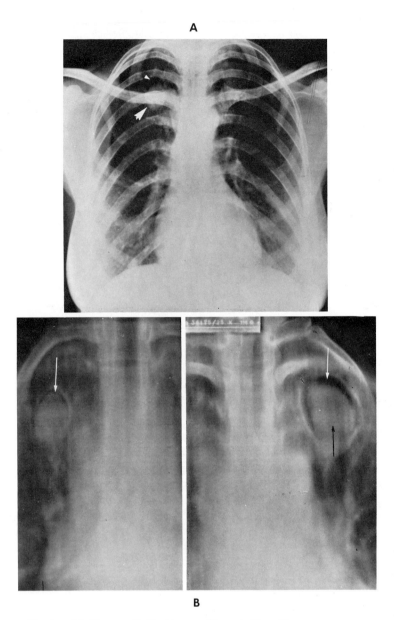

Figure 6--166 **Aspergillosis with Fungus Balls (Aspergilloma): Two Cases.**

A, Fungus ball appears as a round density *(large arrowhead)* in the right upper lobe, above which is the crescentic air shadow *(small arrowhead)*.

B, Laminography in another patient shows bilateral large fungus balls (aspergillomas) *(black arrow)* in the smooth-walled cavities; air crescents surround the densities *(white arrows)*.

Often the position of fungus balls is altered with changes in posture.

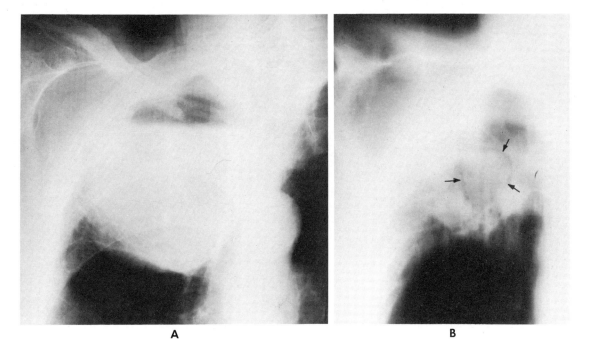

A   B

Figure 6–167 **Aspergilloma in Tuberculous Cavity.**

A, A large density with fluid level developed in a patient with prior cavitary tuberculosis in this area. This density proved to be a Klebsiella abscess superimposed upon the tuberculous cavity.

B, Tomogram made after two weeks of therapy reveals decrease of the lesion, but a rounded density surrounded by air *(arrows)* is now apparent. This was an aspergilloma (mycetoma). No Klebsiella organisms remained.

Mycetomas are usually aspergillomas and can occur in a variety of cavitary pulmonary diseases as a noninvasive opportunistic form of aspergillosis.

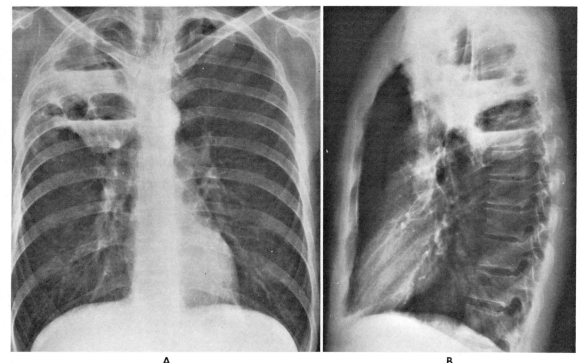

A   B

Figure 6–168 **Aspergillosis with Multiple Cavities.**

A, Multiple cavities with fluid levels are limited to the right upper lobe. At first the disorder resembled pneumonic consolidation.

B, Lateral view demonstrates consolidation of apical and posterior segments. The cavities are located posteriorly.

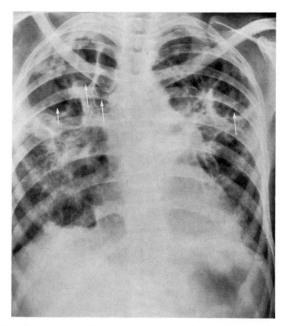

Figure 6–169 **Advanced Bilateral Aspergillosis.** There is diffuse bilateral involvement, with extensive fibrosis in the lower lobes and large contiguous cavities replacing both upper lobes *(arrows)*. Enlargement of the heart and dilation of the main pulmonary artery are attributable to secondary pulmonary hypertension. The picture is indistinguishable from that of advanced tuberculosis.

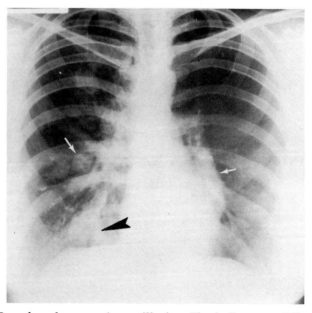

Figure 6–170 **Allergic Bronchopulmonary Aspergillosis.** The bulbous perihilar densities *(white arrows)* are mucus-filled dilated bronchi. The branching density at the right base *(arrowhead)* is characteristic. The patient was a chronic asthmatic. (Courtesy Dr. G. Simon, London, England.)

Candidiasis (Moniliasis)

Opportunistic infection by *Candida albicans* is most often seen in debilitated or terminal patients, or after prolonged steroid, immunosuppressive, or antibiotic therapy. The infection may be systemic and disseminated, or localized. Radiographic changes can occur in the lung, esophagus, or kidney.

In pulmonary candidiasis there may be peribronchial thickening, but more often there are ill-defined nodular bronchopneumonic densities of various sizes and shapes. The apices may be spared; two or more lobes are usually involved. Massive consolidation may occur. Hilar adenopathy is common. Serial films generally reveal changes in size and location of the pulmonary lesions.

Esophageal infection produces a characteristic picture, the upper third of the esophagus being most often involved. The earliest findings are irritability and spasm. Later the mucosa becomes granular and "bubbly"; and ultimately ulcerations develop, thereby giving an irregular jagged appearance to the lumen, with complete loss of normal mucosa. Nodular mucosal thickening produces a "cobblestone" appearance. Narrowing and rigidity may develop. In less advanced cases, the esophagus can be rapidly restored to normal with therapy.

Renal candidiasis is usually a hematogenous infection from systemic candidiasis, but it can also be primary, from ascending urinary tract infection. A radiographic appearance of papillary necrosis or chronic pyelonephritis usually develops. Fungus balls, presenting as shaggy lucencies, may appear in the opacified renal pelvis and calices.[176, 180–182]

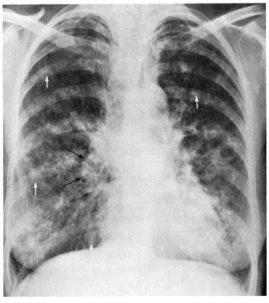

Figure 6–171 **Pulmonary Candidiasis: Extensive Pulmonary Involvement.** There are ill-defined nodular densities *(white arrows)* throughout both lung fields, as well as diffuse fibrosis. The apices are relatively spared. Hilar adenopathy is apparent on the right *(black arrows).*

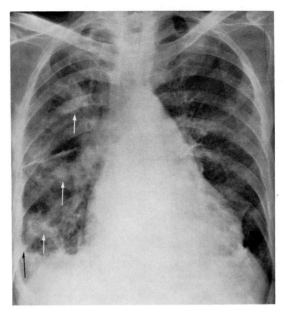

Figure 6–172 **Pulmonary Candidiasis.** There are large ill-defined patchy densities (*white arrows*) in the right lung field and pleural reaction (*black arrow*) at the right base. There is reticular interstitial fibrosis in the left lung. Cardiomegaly was caused by unrelated rheumatic heart disease. Changing patterns of infiltration frequently occur during serial studies.

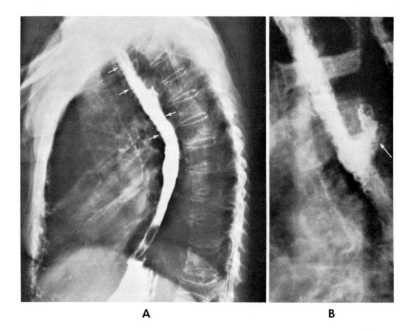

A   B

Figure 6–173 **Esophageal Moniliasis: Eighty Year Old Woman with Chronic Lymphatic Leukemia.**

A, The upper half of the barium-filled esophagus has a granular cobblestone appearance due to edematous mucosal folds. Multiple ulcerations give a serrated appearance to the walls of the esophagus (*arrows*).

B, Enlarged view shows multiple defects in the lumen, and ulcers in the wall. Note that the esophageal diverticulum (*arrow*) is also involved.

One week following nystatin therapy, the esophagus appeared normal on roentgenologic examination, and the dysphagia cleared. As a rule, esophageal moniliasis develops as a complication of a debilitating disease or of a disturbance of the immune mechanism.

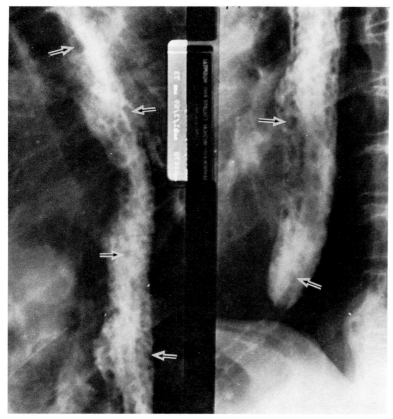

Figure 6–174 **Moniliasis of the Esophagus in Elderly Diabetic.** The esophagus is somewhat atonic and dilated. Diffuse inflammatory mucosal edema has produced round, oval, and linear lucencies throughout the entire esophagus *(arrows)*. No ulcerations are apparent. Esophageal varices may have a similar appearance; but they are rarely so extensive.

REFERENCES

1. Fine, N. L., Smith, L. R., and Sheedy, P. F.: Frequency of pleural effusions in mycoplasma and viral pneumonias. N. Engl. J. Med., *283*:790, 1970.
2. Conte, P., Heitsman, E. R., and Markarian, B.: Viral pneumonia. Radiology, *95*:267, 1970.
3. George, R. B., et al.: Roentgenographic appearance of viral and mycoplasmal pneumonias. Am. Rev. Resp. Dis., *96*:1144, 1967.
4. Rice, R. P., and Loda, F.: Roentgenographic analysis of respiratory syncytial virus pneumonia in infants. Radiology, *87*:1021, 1966.
5. Galloway, R. W., and Miller, R. S.: Lung changes in recent influenzal epidemic. Br. J. Radiol., *32*:28, 1959.
6. DelBuono, M. S.: Asiatic flu. Fortschr. Roentgenstr., *90*:171, 1959.
7. Scanlon, G. T., and Unger, J. D.: The radiology of bacterial and viral pneumonias. Radiol. Clin. North Am., *11*:317, 1973.
8. Wegmann, T., et al.: Pulmonary infiltrations in adenovirus infections. Schweiz. Med. Wochenschr., *89*:882, 1959.
9. Gold, R., et al.: Adenoviral pneumonia and its complications in infancy and childhood. J. Can. Assoc. Radiol., *20*:218, 1969.
10. Herbert, D. H.: The roentgen features of Eaton agent pneumonia. Am. J. Roentgenol. Radium Ther. Nucl. Med., *98*:300, 1966.
11. Stenström, R., Jansson, E., and von Essen, R.: Mycoplasma pneumonias. Acta Radiol., *12*:833, 1972.
12. Zwad, H. D.: Radiologic lung changes in mycoplasma pneumonia. Fortschr. Roentgenstr., *117*:413, 1972.

13. Thombs, D. D.: Cold agglutinin positive pneumonia. A review of thirty cases in children. Ohio Med. J., *63*:1171, 1967.
14. Bacon, A. P. C.: Psittacosis. Lancet, *2*:376, 1953.
15. Schaffner, W., et al.: The clinical spectrum of endemic psittacosis. Arch. Intern. Med., *119*:433, 1967.
16. Brezina, K.: Roentgen findings in pulmonary ornithosis. Fortschr. Geb. Roentgenstr. Nuklearmed., *107*:631, 1967.
17. Cornog, J. L., and Hanson, C. W.: Psitticosis as a cause of miliary infiltrates of the lung and hepatic granulomas. Am. Rev. Resp. Dis., *98*:1033, 1968.
18. Annamunthodo, H., and Marryatt, J.: Barium studies in intestinal lymphogranuloma venereum. Br. J. Radiol., *34*:53, 1961.
19. Fawcitt, J., and Parry, H. E.: Lung changes in pertussis and measles in childhood. Br. J. Radiol., *30*:76, 1957.
20. Sokolowski, J. W., et al.: Giant cell interstitial pneumonia. Am. Rev. Resp. Dis., *105*:417, 1972.
21. Rabinowitz, J. G., et al.: Osseous changes in rubella embryopathy. Radiology, *85*:494, 1965.
22. Grayston, B., et al.: Abnormalities following gestational rubella. J.A.M.A., *202*:95, 1967.
23. Tang, J. S., Kauffman, S. L., and Lynfield, J.: Hypoplasia of pulmonary arteries in infants with congenital rubella. Am. J. Cardiol., *27*:491, 1971.
24. Rowen, M., Singer, M. I., and Moran, E. T.: Intracranial calcification in the congenital rubella syndrome. Am. J. Roentgenol. Radium Ther. Nucl. Med., *115*:86, 1972.
25. Graham, C. B., Thal, A., and Wassum, C. S.: Rubella-like bone changes in congenital cytomegalic inclusion disease. Radiology, *94*:39, 1970.
26. Bose, K. S.: Osteo-articular lesions in smallpox. J. Indiana State Med. Assoc., *31*:151, 1958.
27. Cockshott, W. P., and McGregor, M.: The natural history of osteomyelitis variolosa. Clin. Radiol., *10*:57, 1959.
28. Pesce, L.: Radiological examination in acute anterior poliomyelitis. Ann. Radiol. Diagn., *36/2*:143, 1963.
29. Hooper, F. M.: A case of soft tissue ossification following poliomyelitis. J. Coll. Radiol. Aust., *7*:198, 1963.
30. Bernstein, C., Loesner, W., and Manning, L.: Erosive rib lesions in paralytic poliomyelitis. Radiology, *70*:368, 1958.
31. Gilmartin, D.: Cartilage calcification and rib erosion in chronic respiratory poliomyelitis. Clin. Radiol., *17*:115, 1966.
32. Munk, J., and Lederer, K. T.: Radiologic observations of 33 cases of primary interstitial myocarditis. Clin. Radiol., *9*:195, 1958.
33. Nelson, E. R.: Hemorrhagic fever. J. Pediatr., *56*:101, 1960.
34. Brailey, A. G., Jr., and Husebye, K. O.: Varicella pneumonia with prolonged roentgenographic change. Minnesota Med., *41*:695, 1958.
35. Esswein, J. G., and Domenico, V. P.: Hemorrhagic varicella pneumonia. Ann. Intern. Med., *53*:607, 1960.
36. Knyvett, A. F.: The pulmonary lesions of chickenpox. Q. J. Med., *35*:313, 1966.
37. Fraser, R. G., and Paré, J. A. P.: *Diagnosis of Diseases of the Chest.* Philadelphia, W. B. Saunders Co., 1970.
38. Glick, N., Levin, S., and Nelson, K.: Recurrent pulmonary infarction in adult chickenpox pneumonia. J.A.M.A., *222*:173, 1972.
39. Allen, J. H., and Riley, H. D.: Generalized cytomegalic inclusion disease, with emphasis on roentgen diagnosis. Radiology, *71*:257, 1958.
40. Jacobson, G., et al.: Roentgen manifestations of Q fever. Radiology, *53*:739, 1949.
41. Hoeprich, P. D., et al.: Q-fever. J.A.M.A., *170*:180, 1959.
42. Frazer, R. G., and Wortzman, G.: Acute pneumococcal lobar pneumonia: the significance of nonsegmental distribution. J. Can. Assoc. Radiol., *10*:37, 1959.
43. Danner, P. K., McFarland, D. R., and Felson, B.: Massive pulmonary gangrene. Am. J. Roentgenol. Radium Ther. Nucl. Med., *103*:548, 1968.
43a. Jay, S. J., Johanson, W. G., Jr., and Pierce, A. K.: The radiographic resolution of Streptococcus pneumoniae pneumonia. N. Engl. J. Med., *293*:798, 1975.
44. Felson, B., et al.: Roentgen findings in acute Friedländer's pneumonia (Klebsiella pneumonia). Radiology, *53*:559, 1949.
45. Renner, R. R., Cocarro, A. J., et al.: Pseudomonas pneumonia; a prototype of hospital based infection. Radiology, *105*:555, 1972.
46. Unger, J. D., Rose, H. D., and Unger, G. F.: Gram negative pneumonia. Radiology, *107*:283, 1973.
47. Meyers, H. I., and Jacobson, G.: Staphylococcal pneumonia in children and adults. Radiology, *74*:670, 1960.
48. Goldring, D., et al.: Rheumatic pneumonitis. J. Pediatr., *53*:547, 1958.
49. Twigg, H. L., and Smith, B. F.: Jaccoud's arthritis. Radiology, *80*:417, 1963.
50. Murphy, W. A., and Staple, T. W.: Jaccoud's arthropathy reviewed. Am. J. Roentgenol. Radium Ther. Nucl. Med., *118*:300, 1973.
51. Neligan, G. A., and Warrick, C. K.: Value of radiology in diagnosis and management of pyogenic osteitis in childhood. J. Fac. Radiologists, *5*:112, 1953.

52. Ceruti, E., Contreras, J., and Neira, M.: Staphylococcal pneumonia in childhood. Am. J. Dis. Child., *122*:386, 1971.
53. Kuperman, A. S., and Fernandez, R. B.: Subacute staphylococcal pneumonia. Am. Rev. Resp. Dis., *101*:95, 1970.
54. Gerardo, F., and Boisset, B.: Subpleural emphysema complicating staphylococcal and other pneumonias. J. Pediatr., *81*:259, 1972.
55. Keiser, H., et al.: Clinical forms of gonococcal arthritis. N. Engl. J. Med., *279*:234, 1968.
56. Forrester, D. M., and Nesson, J. W.: *The Radiology of Joint Disease*. Philadelphia, W. B. Saunders Co., 1973.
57. Vinik, M., et al.: Hemophilus influenza pneumonia. Radiology, *86*:701, 1966.
58. Tillotson, J. R., and Lerner, M. A.: Hemophilus influenzae bronchopneumonia in adults. Arch. Intern. Med., *121*:428, 1968.
59. Barnhard, H. G., and Kniker, W. T.: Roentgenologic findings in pertussis with particular emphasis on 'shaggy heart' sign. Am. J. Roentgenol. Radium Ther. Nucl. Med., *84*:445, 1960.
60. Kirkpatrick, D. J.: Donovanosis (granuloma inguinale): A rare cause of osteolytic bone lesions. Clin. Radiol., *21*:101, 1970.
61. Basset, D. J.: Radiology in the tropics. Aust. Radiol., *11*:16, 1967.
62. Matousek, V.: Radiographic manifestations of clostridial infection encountered at Queen Elizabeth Hospital, Adelaide. Aust. Radiol., *13*:387, 1969.
63. Evans, K. T., and Walley, R. V.: Radiological changes in tetanus. Br. J. Radiol., *36*:729, 1963.
64. Femi-Pearse, D., and Olowu, A. O.: Myositis ossificans: A complication of tetanus. Clin. Radiol., *22*:89, 1971.
65. Chérgié, E., et al.: [Radiologic aspects of small intestine in typhoid fever.] J. Radiol. Electrol. Med. Nucl., *34*:522, 1953.
66. Francis, R. S., and Berk, R. N.: Typhoid fever. Radiology, *112*:583, 1974.
67. Ingegno, A. P., et al.: Pneumonia associated with acute salmonellosis. Arch. Intern. Med., *81*:476, 1948.
68. Hook, E. W., et al.: Salmonella osteomyelitis in patients with sickle-cell anemia. N. Engl. J. Med., *257*:403, 1957.
69. Farman, J., Rabinowitz, J. G., and Meyers, M. M.: Roentgenology of infectious colitis. Am. J. Roentgenol. Radium Ther. Nucl. Med., *119*:375, 1973.
70. Richards, A. J.: Nontuberculous pyogenic osteomyelitis of the spine. J. Can. Assoc. Radiol., *11*:45, 1960.
71. Faegenburg, D., et al.: Colon Bacillus pneumonia. Am. J. Roentgenol. Radium Ther. Nucl. Med., *107*:300, 1969.
72. Rappaport, E. M., and Rappaport, E. O.: Typhoid enterocolitis simulating chronic bacillary dysentery. N. Engl. J. Med., *242*:698, 1950.
73. Douglas, R. F.: Chest roentgenology in early diagnosis of infectious disease (plague pneumonia). Milit. Med., *128*:104, 1963.
74. Mengis, C. L.: Plague. N. Engl. J. Med., *267*:543, 1962.
75. Sites, V. R., and Poland, J. D.: Mediastinal lymphadenopathy in bubonic plague. Am. J. Roentgenol. Radium Ther. Nucl. Med., *116*:567, 1972.
76. Overholt, E. L., and Tigertt, W. D.: Roentgen manifestations of pulmonary tularemia. Radiology, *74*:758, 1960.
77. Avery, F. W., and Barnett, T. B.: Pulmonary tularemia. Am. Rev. Resp. Dis., *95*:584, 1967.
78. Miller, R. P., and Bates, J. H.: Pleuropulmonary tularemia. Am. Rev. Resp. Dis., *99*:31, 1969.
79. Katunaric, D.: The radiologic picture of visceral tularemia. Ann. Radiol., *11*:43, 1968.
80. Jones, G. P., and Ross, J. A.: Melioidosis. Br. J. Radiol., *36/426*:415, 1963.
81. Patterson, M. C., et al.: Acute melioidosis in a soldier home from South Vietnam. J.A.M.A., *200*:117, 1967.
82. Spotnitz, M., et al.: Melioidosis pneumonitis. Analysis of nine cases of a benign form of melioidosis. J.AM.A., *202*:950, 1967.
83. Poe, R. H., Vassallo, C. L., and Domm, B. M.: Melioidosis: The remarkable imitator. Am. Rev. Resp. Dis., *104*:427, 1971.
84. Plotkin, S. A., and Brachman, P. S.: Inhalation anthrax. Am. J. Med., *29*:992, 1960.
85. Moore, P. H., and Brogdon, B. G.: Granulomatosis infantiseptica (listeriosis). Radiology, *79*:415, 1962.
86. Zammit, F.: Undulant fever spondylitis. Br. J. Radiol., *31*:683, 1958.
87. Weber, H. H.: Brucellosis and echinococcosis in Argentina. Schweiz. Med. Wochenschr., *88*:567, 1958.
88. Petereit, M. F.: Chronic renal brucellosis: A simulator of tuberculosis. Radiology, *96*:85, 1970.
89. Arany, L. S.: Ghon focus of tuberculous reinfection. Am. J. Roentgenol. Radium Ther. Nucl. Med., *81*:207, 1959.
90. Stead, W. W., et al.: The clinical spectrum of primary tuberculosis in adults. Ann. Intern. Med., *68*:731, 1968.
91. Rabinowitz, S. A.: Primary cavitating tuberculosis in children. Clin. Radiol., *23*:483, 1972.
92. Weber, A. L., Bird, K. T., and Janower, M. L.: Primary tuberculosis in childhood with particular emphasis on changes affecting the tracheobronchial tree. Am. J. Roentgenol. Radium Ther. Nucl. Med., *103*:123, 1968.

93. Giammona, S. T., et al.: Massive lymphadenopathy in primary pulmonary tuberculosis in children. Am. Rev. Resp. Dis., *100*:480, 1969.

94. Bleyer, J. M., and Marks, J. H.: Tuberculomas and hamartomas of the lung. Am. J. Roentgenol. Radium Ther. Nucl. Med., 77:1013, 1957.

95. Hardy, J. B.: Tuberculosis in white and Negro children. In *Roentgenologic Aspects of the Harriet Lane Study*. Cambridge, Mass., Harvard University Press, 1958, Vol. I.

96. Johnson, T. M., McCann, W., and Davey, W. N.: Tuberculous bronchopleural fistula. Am. Rev. Resp. Dis., *107*:30, 1973.

97. Freundlich, I. M., and Capp, M. P.: Granulomatous disease of the lungs. Radiol. Clin. North Am., *11*:295, 1973.

98. Berger, H. W., and Granada, M. G.: Lower lung field tuberculosis. Chest, *65*:522, 1974.

99. Rubinstein, B. M., et al.: Tuberculosis of esophagus. Radiology, 70:401, 1958.

100. Pinto, R. S., Zausner, J., and Beranbaum, E. R.: Gastric tuberculosis. Am. J. Roentgenol. Radium Ther. Nucl. Med., *110*:808, 1970.

101. Brombart, M., and Massion, J.: Diagnosis of ileo-cecal tuberculosis. Am. J. Dig. Dis., 6:589, 1961.

102. Chawla, S., Mukerjee, P., and Bery, K.: Segmental tuberculosis of the colon. Clin. Radiol., *22*:104, 1971.

103. Lewis, E. A., and Kolawole, T. M.: Tuberculous ileo-colitis in Ibidan. Gut, *13*:646, 1972.

104. Johnson, J. C., Dunbar, J. D., and Klainer, A. S.: The pseudotumor of retroperitoneal tuberculous lymphadenitis. Am. J. Roentgenol. Radium Ther. Nucl. Med., *111*:554, 1971.

105. Janovsky, R. C., et al.: Recurrent tuberculous pericarditis. Ann. Intern. Med., *37*:1268, 1952.

106. Brackin, J. F., Jr., et al.: Roentgen appearance of small intestine in tuberculous peritonitis before and after streptomycin therapy. Am. J. Roentgenol. Radium Ther. Nucl. Med., *68*:887, 1952.

107. Lattimer, J. K.: Roentgenographic classification of tuberculous lesions of the kidney. Am. Rev. Tuberc., *67*:604, 1953.

108. Becker, J. A., Weiss, R. M., and Latimer, J. K.: Renal tuberculosis: The role of nephrotomography and angiography. J. Urol., *100*:415, 1968.

109. Giustra, P. E., Watson, R. C., and Hulman, H. S.: Arteriographic findings in various stages of renal tuberculosis. Radiology, *100*:597, 1971.

110. Kollins, S. A., Hartman, G. W., et al.: Roentgenographic findings in urinary tuberculosis. Am. J. Roentgenol. Radium Ther. Nucl. Med., *121*:487, 1974.

111. Poppel, M. H., et al.: Skeletal tuberculosis: a roentgenographic survey. Am. J. Roentgenol. Radium Ther. Nucl. Med., *70*:936, 1953.

112. Kessler, A. D., et al.: Cystic tuberculosis of the bones in children. Am. J. Dis. Child., *88*:201, 1954.

113. Frecker, B. E.: Some radiologic aspects of tuberculous disease of bones and joints. Med. J. Aust., *1*:606, 1952.

114. Feldman, F., Auerbach, R., and Johnston, A.: Tuberculosis dactylitis in the adult. Am. J. Roentgenol. Radium Ther. Nucl. Med., *112*:460, 1971.

115. Godoy, I. A. A., and Sierra, C. M.: Osteoarticular tuberculosis in childhood. Rev. Mex. Radiol. (Span.), *26*:19, 1972.

116. Cremin, B. J., Fisher, R. M., and Levinsohn, M. W.: Multiple bone tuberculosis in the young. Br. J. Radiol., *43*:638, 1970.

117. Park, B. J.: Miliary tuberculosis. Am. Rev. Resp. Dis., 77:601, 1958.

118. Berger, H. W., and Samortin, T. G.: Miliary tuberculosis; diagnostic methods with emphasis on the chest roentgenogram. Chest, *58*:586, 1970.

119. Tsai, S. H., Yue, W. Y., and Duthoy, E. J.: Roentgen aspects of chronic pulmonary mycobacteriosis. Am. J. Roentgenol. Radium Ther. Nucl. Med., *90*:306, 1968.

120. Danigelis, J. A., and Long, R. E.: Anonymous mycobacterial osteomyelitis. Radiology, *93*:353, 1969.

121. Heitzman, E. R., Bornhurst, R. A., and Russell, J. P.: Disease due to anonymous mycobacteria; Potential for specific diagnosis. Am. J. Roentgenol. Radium Ther. Nucl. Med., *103*:533, 1968.

122. Seibert, C. E., and Tabrisky, J.: Radiological features of pulmonary atypical mycobacterial infections. Br. J. Radiol., *42*:140, 1969.

123. Wiot, J. F., and Spitz, H. B.: Atypical pulmonary tuberculosis. Radiol. Clin. North Am., *11*:191, 1973.

124. Rauscher, C. R., Kerby, G., and Ruth, W. E.: A ten year clinical experience with Mycobacterium kansasii. Chest, *66*:17, 1974.

125. Oliva, L., and Farris, G.: Endostoses and periostoses in leprosy. Radiol. Med. Milan, *43*:1174, 1957.

126. Basu, S. P., et al.: Angiography in leprosy. Ind. J. Radiol., *14*:180, 1960.

127. Faget, O. H., and Mayoral, A.: Bone changes in leprosy. Radiology, 42:1, 1944.

128. Enna, C. D., Jacobson, R. R., and Rausch, R. O.: Bone changes in leprosy. Radiology, *100*:295, 1971.

129. Newman, H., Casey, B., et al.: Roentgen features of leprosy in children. Am. J. Roentgenol. Radium Ther. Nucl. Med., *114*:402, 1972.

130. Mukherji, B. B.: Skeletal lesions of infantile syphilis. Ind. Med. J., *28*:355, 1957.

131. Dziadiw, R., et al.: Pulmonary gumma. Radiology, *103*:59, 1972.

132. Cremin, B. J., and Fisher, R. M.: The lesions of congenital syphilis. Br. J. Radiol., *43*:333, 1970.

133. Coblentz, D. R., et al.: Roentgenographic diagnosis of congenital syphilis in the newborn. J.A.M.A., *212*:1061, 1970.

134. Cockshott, W. P., and Davies, A. G. M.: Tumoural gummatous yaws. J. Bone Joint Surg., *42B*:785, 1960.

135. Riseborough, S. W., Joske, R. A., and Vaughn, B. F.: Hand deformities due to yaws. Clin. Radiol., *12*:109, 1961.

136. Jones, B. S.: Doigt en lorgnette and concentric bone atrophy associated with healed yaws osteitis. J. Bone Joint Surg., *54*:341, 1972.

137. Ghigo, M., and Magrini, M.: Osteo-periosteopathies secondary to tropical ulcers. Radiol. Med., *47*:719, 1961.

138. Kolawole, T. M., and Bohrer, S. P.: Ulcer osteoma — bone response to tropical ulcer. Am. J. Roentgenol. Radium Ther. Nucl. Med., *109*:611, 1970.

139. Schwarz, J., Baum, G. L., and Straub, M.: Cavitary histoplasmosis complicated by fungus ball. Am. J. Med., *31*:692, 1961.

140. Schwarz, E.: Regional roentgen manifestations of histoplasmosis. Am. J. Roentgenol. Radium Ther. Nucl. Med., *87*:865, 1962.

141. Lull, G. F., and Winn, D. R., Jr.: Histoplasmal mediastinitis. Radiology, *73*:367, 1959.

142. Goodwin, R. A., et al.: Early chronic pulmonary histoplasmosis. Am. Rev. Resp. Dis., *93*:47, 1966.

143. Fairbank, J. T., Tampas, J. P., and Longstreth, G.: Superior vena caval obstruction in histoplasmosis. Am. J. Roentgenol. Radium Ther. Nucl. Med., *115*:488, 1972.

144. Forrest, J. V.: Common fungal diseases of the lungs. II. Histoplasmosis. Radiol. Clin. North Am., *11*:163, 1973.

145. Christoforidis, A. J.: Radiologic manifestations of histoplasmosis. Am. J. Roentgenol. Radium Ther. Nucl. Med., *109*:478, 1970.

146. Small, M. J.: Late progression of pulmonary coccidioidomycosis. Arch. Intern. Med., *104*:730, 1959.

147. Rhangos, W. C., and Chick, E. W.: Mycotic infections of bone. South. Med. J., *57*:664, 1964.

148. Greendyke, W. H., Resnick, D. L., and Harvey, W. C.: Primary coccidioidomycosis. Am. J. Roentgenol. Radium Ther. Nucl. Med., *109*:491, 1970.

149. Dalinka, M. K., Dinninberg, S., et al.: Roentgenographic features of osseous coccidioidomycosis. J. Bone Joint Surg., *53*:1157, 1971.

150. Sagel, S. S.: Common fungal diseases of the lungs. I. Coccidioidomycosis. Radiol. Clin. North Am., *11*:153, 1973.

151. Boswell, W. L.: Roentgen aspects of blastomycosis. Am. J. Roentgenol. Radium Ther. Nucl. Med., *81*:224, 1959.

152. Abernathy, R. S.: Clinical manifestations of pulmonary blastomycosis. Ann. Intern. Med., *51*:707, 1959.

153. Rosen, R. S., and Jacobson, G.: Fungus disease of bone. Semin. Roentgenol., *1*:370, 1966.

154. Poe, R. H., Vassallo, C. L., et al.: Pulmonary blastomycosis versus carcinoma. Am. J. Med. Sci., *263*:145, 1972.

155. Sarosi, G. A., et al.: Clinical features of acute pulmonary blastomycosis. N. Engl. J. Med., *290*:540, 1974.

156. Armstrong, J. D.: Common fungal diseases of the lung. III. Blastomycosis. Radiol. Clin. North Am., *11*:169, 1973.

157. Hawkins, J. A.: Cavitary pulmonary cryptococcosis. Am. Rev. Resp. Dis., *84*:579, 1961.

158. Meighan, J. W.: Pulmonary cryptococcosis mimicking carcinoma of the lung. Radiology, *103*:61, 1972.

159. Gordonson, J., Birnbaum, W., et al.: Pulmonary cryptococcosis. Radiology, *112*:557, 1974.

160. Warthin, T. A., and Bushueff, B.: Pulmonary actinomycosis. Arch. Intern. Med., *101*:239, 1958.

161. Varadarajan, M. J.: Actinomycosis of bone. Punjab Med. J., *10*:321, 1961.

162. Flynn, M. W., and Felson, B.: The roentgen manifestations of thoracic actinomycosis. Am. J. Roentgenol. Radium Ther. Nucl. Med., *110*:707, 1970.

163. Murray, J. F., et al.: Changing spectrum of nocardiosis. Am. Rev. Resp. Dis., *83*:315, 1961.

164. Grossman, C. B., Bragg, D. G., and Armstrong, D.: Roentgen manifestations of pulmonary nocardiosis. Radiology, *96*:325, 1970.

165. Mikkelsen, W. M., et al.: Sporotrichosis. Ann. Intern. Med., *47*:435, 1957.

166. Ridgeway, N. A., et al.: Primary pulmonary sporotrichosis. Am. J. Med., *32*:153, 1962.

167. Winter, T. Q., and Pearson, K. V.: Systemic sporothrixosis. Radiology, *104*:579, 1972.

168. Mohr, J. A., et al.: Primary pulmonary sporotrichosis. Am. Rev. Resp. Dis., *106*:260, 1972.

169. Gabriele, O. F.: Mucormycosis. Am. J. Roentgenol. Radium Ther. Nucl. Med., *83*:227, 1960.

170. Blankenberg, J. W., and Verhoeff, D.: Mucormycosis of lung. Am. Rev. Tuberc., *79*:357, 1959.

171. Bartrum, R. J., Watneck, M., and Herman, P. G.: Roentgenographic findings in pulmonary mucormycosis. Am. J. Roentgenol. Radium Ther. Nucl. Med., *117*:810, 1973.

172. Davies, A. G. M.: Bone changes in Madura foot (mycetoma). Radiology, *70*:841, 1958.

173. Arredondo, H. G., and Ceballos, J. L.: Unusual location of mycetoma. Radiology, *78*:72, 1962.

174. Goldberg, B.: Radiologic appearances in pulmonary aspergillosis. Clin. Radiol., *13*:106, 1962.

175. Hertzog, P., et al.: Polymorphism of bronchopulmonary aspergillomas. J. Fr. Med. Chir. Thorac., *15*:557, 1961.

176. Vlakhov, K., and Karparov, M.: Bilateral pulmonary aspergilloma and pulmonary moniliasis. J. Radiol., *41*:34, 1960.
177. McCarthy, D. S., Simon, G., and Hargreave, F. E.: Radiological appearances in allergic broncho-pulmonary aspergillosis. Clin. Radiol., *21*:366, 1970.
178. Allergic pulmonary aspergillosis. (Editorial.) J.A.M.A., *216*:670, 1971.
179. Genoe, G. A., Morello, J. A., and Fennessy, J. J.: The diagnosis of pulmonary aspergillosis by the bronchial brushing technique. Radiology, *102*:51, 1972.
180. Kaufman, S. A., Scheff, S., and Levene, G.: Esophageal moniliasis. Radiology, *75*:726, 1960.
181. Clark, R. E., Minagi, H., and Palubinskas, A. J.: Renal candidiasis. Radiology, *101*:567, 1971.
182. Boldus, R. A., Brown, R. C., and Culp, D. A.: Fungus balls in the renal pelvis. Radiology, *102*:555, 1972.

SECTION
7

PROTOZOAN AND HELMINTHIC DISEASES

PROTOZOAN DISEASES

Amebiasis

The colon is the primary site of involvement. In the milder acute cases, the barium enema studies may be negative or may show merely localized spasticity and irritability. Any portion of the colon may be involved. The cecum and ascending colon are favored sites. Severe cases may show swollen edematous mucosa, thumbprinting, haustral irregularities, and sawtooth projections, but the actual ulcerations are rarely seen. The acute fulminating case may simulate ulcerative, ischemic, or granulomatous colitis both clinically and radiographically.

The more chronic form of amebiasis resembles a segmental, often multisegmental, granulomatous colitis. The lumen is narrowed, and the mucosa is granular or pseudopolypoid. Cecal localization is frequent. A narrowed and irritable cone-shaped cecum, a patulous ileocecal valve, and an uninvolved terminal ileum suggest amebiasis.

A chronic amebic granulomatous mass (ameboma) can occur in any part of the colon and closely simulates a neoplastic mass. A preserved mucosa may suggest a nonmalignant lesion, but distinction is usually not possible. If ameboma is considered, a dramatic response to emetine therapy (one to two weeks) will resolve the issue. Definitive diagnosis of amebiasis is made by demonstrating the parasite in the stool.

Hepatic abscess, especially of the right lobe, is the most frequent complication. The right diaphragm is elevated and fixed, usually anteriorly. The costophrenic angle remains sharp unless the diaphragm is infected. With secondary infection of the abscess, air-fluid levels may develop, thereby simplifying roentgenologic localization. Longstanding localized liver abscesses may undergo calcification.

Pleuropulmonary involvement, the second most common complication, is usually due to direct extension from the liver but, rarely, may be of hematogenous origin.

The pneumonic infiltrate resulting from direct extension is usually in the right base, just above the elevated right diaphragm. Pleural fluid frequently develops and may obscure the underlying lung involvement. If abscess occurs, it is usually in the anterior basal segment, an unusual location for ordinary bacterial abscess. Communication with a bronchus will lead to an air-fluid level. Bronchogenic spread to another segment of lung is possible. Abscess from hematogenous infection can occur in any lung segment.

Rarely, an acute purulent amebic pericarditis may appear.[1-6]

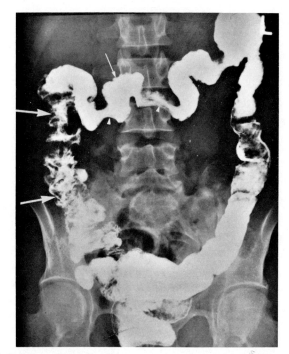

Figure 7-1 **Amebiasis of Colon.** Multiple spastic areas are seen in the transverse colon (*arrowheads*). Spicules (*small arrow*) are also evident; some of these may be ulcers. There is considerable mucosal edema and thickening in the ascending colon (*large arrows*). This picture is nonspecific and may be seen in any case of acute inflammatory colitis. In many cases of acute amebiasis, the colon appears normal.

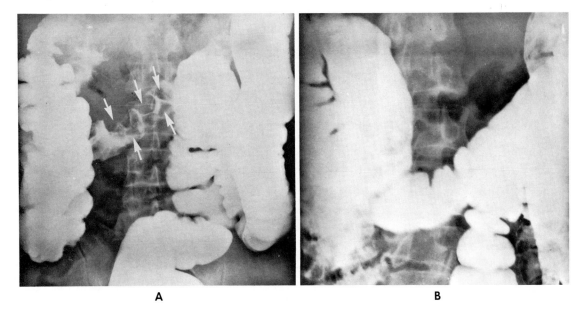

A   B

Figure 7–2 **Amebiasis: Barium Enema.**

A, The entire transverse colon is somewhat narrowed and filled with numerous contiguous filling defects *(arrows).* These were inflammatory masses (amebomas). The stools were loaded with endamebae.

B, Six weeks later, following intensive antiamebic therapy, the appearance of the transverse colon has returned to normal. The patient was clinically completely recovered.

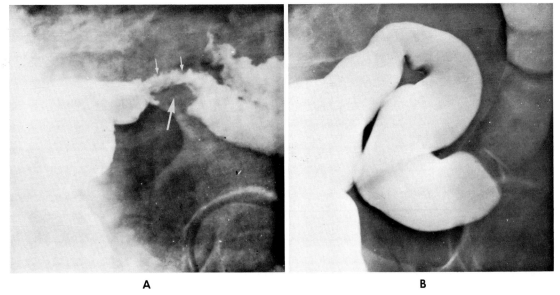

A   B

Figure 7–3 **Amebiasis with Ameboma of Sigmoid, Simulating Neoplasm and Diverticulitis.**

A, The large defect *(large arrow)* in the sigmoid is strongly suggestive of neoplasm. However, multiple serrations *(small arrows)* on the opposite wall raise the possibility of an inflammatory process. Amebiasis was confirmed by stool examination.

B, Several months later, after complete clinical recovery with antiamebic therapy, the sigmoid appears entirely normal. The ulcerations and ameboma have completely disappeared.

When ulcerations are not seen, a solitary ameboma may simulate a neoplasm. In uncertain cases antiamebic therapy (emetine) can be used as a therapeutic test and may obviate surgery.

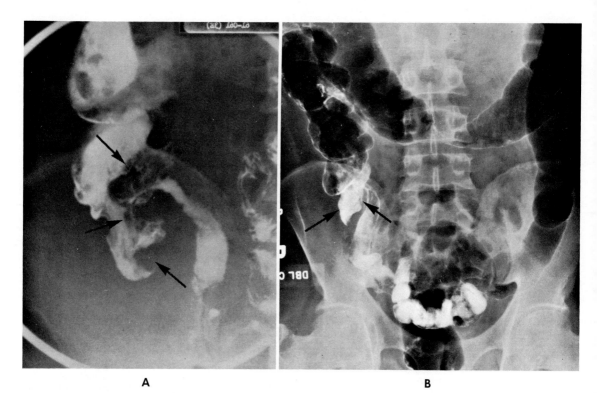

A   B

Figure 7–4   **Two Cases of Chronic Cecal Amebiasis.**

A, Granulomatous masses *(arrows)* produce filling defects in the ileocecal area and the cecal tip. In the granulomatous form the filling defects (amebomas) simulate neoplasms.

B, The cecum is narrowed, shortened, and conical; haustra are absent *(arrows)*. The ileocecal valve is patulous. This cone-shaped deformity of the cecum with an uninvolved terminal ileum is highly suggestive of chronic amebiasis, and is seen in about one third of the active cases.

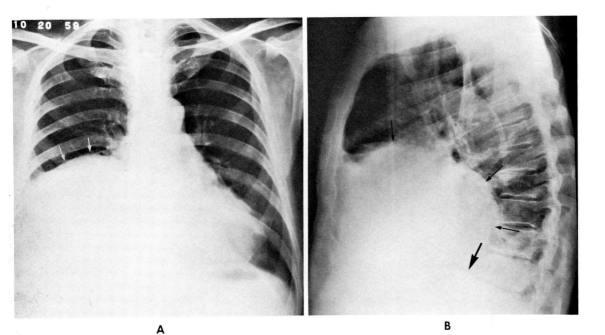

A B

Figure 7–5 **Amebic Abscess of Liver.**

A, There is marked elevation of the right hemidiaphragm *(arrows),* with minimal reaction in the costophrenic angle. Diaphragmatic motility is severely limited.

B, The extent of the right diaphragmatic elevation is seen in lateral film *(small arrows).* The contour is irregular, owing to pleural reaction. Diaphragmatic elevation is mainly anterior. The position of the left hemidiaphragm is indicated by the large arrow. Linear areas of plate-like atelectasis can be seen in the lung anteriorly.

Elevation of the right hemidiaphragm, especially anteriorly, is the most common roentgenologic finding in amebic liver abscess.

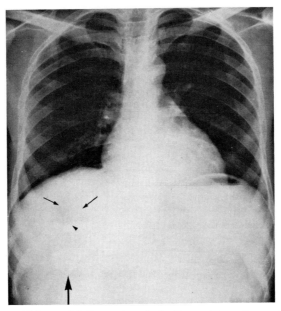

Figure 7–6 **Amebic Abscess of Liver with Secondary Infection.** There is an air-fluid level within the liver substance *(arrowhead).* The upper limit of the cavity is outlined by air *(small arrows).* The full extent of the cavity is shown by the opaque medium that has been instilled into the abscess *(large arrow).*

Air-fluid levels rarely occur in amebic liver abscess unless secondary infection supervenes. (Courtesy L. Whittaker, Nairobi, Kenya.)

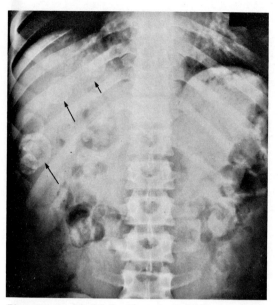

Figure 7–7 **Multiple Chronic Calcified Amebic Abscesses of Liver.** Multiple round calcifications *(arrows)* are seen in the liver parenchyma and result from longstanding inactive liver amebic abscesses. Such calcification is uncommon. (Courtesy L. Whittaker, Nairobi, Kenya.)

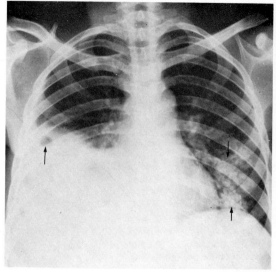

Figure 7–8 **Amebic Lung Abscess: Pneumonitis.** There is an air-fluid level in the right lower lung field *(arrow)* in an amebic abscess arising by direct extension from the liver; there is associated pleural reaction. Bronchogenic spread to the left lower lobe caused patchy pneumonitis *(arrows)*.

The lungs may be involved either by direct extension from the liver or by hematogenous spread; in the latter case the picture is indistinguishable from that of any lung abscess.

Giardiasis

Giardia lamblia is a common and usually harmless parasite in the duodenum and upper small bowel, but heavy infestations can produce inflammatory changes and cause diarrhea and even malabsorption. Agammaglobulinemia or dysgammaglobulinemia is often a predisposing factor. Small intestinal biopsy is usually necessary for diagnosis.

The roentgen alterations in symptomatic giardiasis are usually limited to the duodenum and jejunum. The valvulae conniventes become thickened and distorted, and the attendant irritability and spasm may narrow the lumen. Increased secretions cause blurring of the folds and some segmentation, sometimes producing a radiographic malabsorption pattern. The ileum usually remains normal. If an underlying immunoglobulin deficiency exists, small nodules may appear in the small bowel (see p. 28).

Similar changes in the duodenum and jejunum are seen in strongyloidiasis. However, in strongyloidiasis, ulcerations and permanent fibrotic narrowing can develop (see p. 282), whereas in giardiasis, the upper small bowel returns to normal after therapy.[7–9] Other conditions that can cause similar radiographic changes include Crohn's disease and eosinophilic gastroenteritis.

Figure 7–9 **Giardiasis of Small Bowel: Two Patients.** The jejunum reveals spasm, irritability, and thickened distorted folds (valvulae) in both patients. Thickened folds in the more distal small bowel are also present, but less marked. (From Marshak, R. H., and Lindner, A. E.: *Radiology of the Small Intestine.* 2nd ed. Philadelphia, W. B. Saunders Co., 1976.)

Malaria

Radiographic evidence of splenomegaly is found in most cases of subacute or chronic malaria. In pernicious falciparum malaria, chest films may disclose the pattern of diffuse alveolar pulmonary edema without cardiac enlargement. This is usually a late complication in fatal cases, and closely resembles shock lung clinically, radiologically, and histologically. Rarely, an acute interstitial edema, without alveolar consolidation, develops; this may respond to treatment.[10-13]

Chagas' Disease

Cardiomyopathy is found in about one half of the chronic cases. It is manifested roentgenologically by cardiac enlargement, usually of all chambers but predominantly of the left side. The aortic knob is relatively small, and the pulmonary vasculature is normal until the changes of congestive failure supervene. Pulmonary embolization is a frequent complication.

Megacolon is the next most common roentgenologic finding. Usually the colon is dilated down to the sigmoid. Irritability and spasm of the distal sigmoid, which occur in the early stage, are reversible, but later there is permanent narrowing. The proximal colon becomes dilated, so that radiographically the disorder is practically indistinguishable from Hirschsprung's megacolon. Fecal impaction and sigmoid volvulus are frequent complications.

Esophageal spasm is seen early in the disease, progressing to permanent narrowing in the lower segment with marked dilatation of the proximal esophagus. At this stage the roentgenologic picture is quite similar to that of achalasia of the esophagus.

The similarities between Chagas' disease and Hirschsprung's megacolon and achalasia of the esophagus probably result from the involvement of the submucosal nerve plexuses. This involvement is similar in all three disorders and causes loss of contractility and relative narrowing of the involved segment. In Chagas' disease, the nerve plexuses in the aperistaltic segments are invaded by parasites; in Hirschsprung's disease and in achalasia, the segments are aganglionic because of faulty development.

Occasionally atonicity and dilatation of the biliary ducts have been observed in Chagas' disease.[14-18]

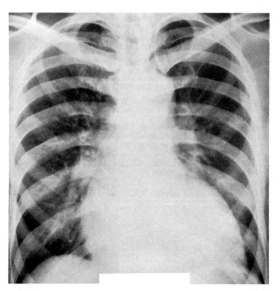

Figure 7–10 **Chronic Chagas' Disease: Cardiopathy.** Enlargement of the heart is due mainly to left ventricular dilatation. The aortic knob is small in proportion to the left ventricle. There is no vascular congestion. This picture is indistinguishable from that of many other cardiopathies. (Courtesy B. Fishbein, Port-of-Spain, Trinidad, West Indies.)

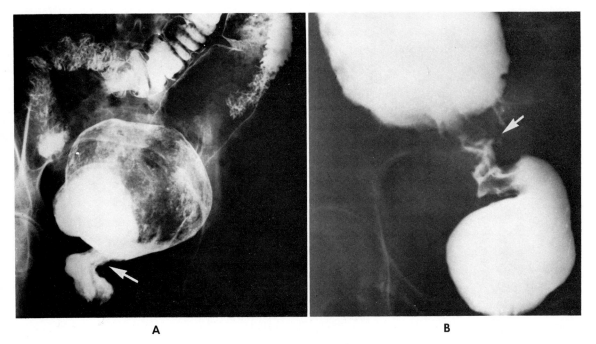

A   B

Figure 7–11 **Chronic Chagas' Disease: Two Cases of Megacolon.**

A, There is marked dilatation of the sigmoid colon above the narrowed segment of the rectosigmoid *(arrow).*

B, The area of apparent narrowing is in the upper rectum and sigmoid *(arrow).* These segments are of nearly normal caliber but appear narrow by contrast with the dilated bowel proximally. Hirschsprung's disease manifests a similar roentgenologic picture. (Courtesy M. Riera, Santiago, Chile.)

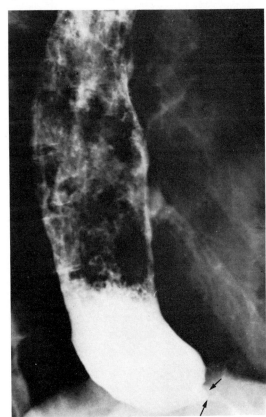

Figure 7–12 **Chronic Chagas' Disease: Megaesophagus.** The esophagus is greatly dilated, containing air and nonopaque residue, and tapers to a narrow segment *(arrows).* The picture is identical to that of achalasia of the esophagus.

Toxoplasmosis

In congenital toxoplasmosis, multiple discrete intracranial calcifications are frequently found in association with hydrocephalus or microcephaly. Although the calcifications may be diffusely scattered in the brain substance, they tend to be most numerous in the posterior portion of the brain. Curvilinear streaks of calcium in the basal ganglia and thalamus are the most characteristic but the least frequent form. Calcification of the lining of the ventricles can also occur. The calcifications may be indistinguishable from those of cytomegalic inclusion disease. Hydrocephalus develops in about 80 per cent of the cases.

Acquired toxoplasmosis of adults is widely disseminated and can affect almost any organ. There may be nonspecific pulmonary infiltrates with hilar adenopathy.[19-23]

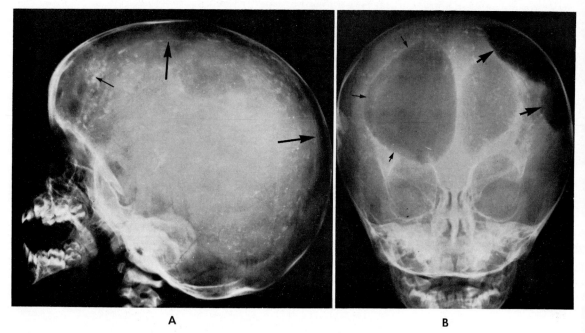

A   B

Figure 7–13 **Congenital Toxoplasmosis: Intracranial Calcifications and Hydrocephalus.**

A, Lateral film shows widely disseminated small calcifications *(small arrow)* and linear plaque-like calcifications *(large arrows)* lining the ventricular system. Such calcifications are the most characteristic finding.

B, Air studies reveal a huge hydrocephalus with greatly dilated ventricles *(small arrows)* and air in the dilated subarachnoid spaces *(large arrows).* Note the grouping of the calcifications around the ventricles. (Courtesy S. Fisher: Am. J. Roentgenol. Radium Ther. Nucl. Med., 59:816, 1948.)

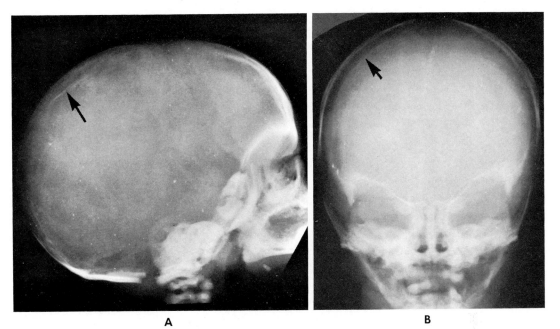

A B

Figure 7–14 **Congenital Toxoplasmosis of Skull: Curvilinear Calcifications and Hydrocephalus.** A curvilinear calcific plaque is seen in the right posterior parietal area close to the vault *(arrow).* It lines the greatly dilated lateral ventricle. Cytomegalic inclusion disease may produce similar roentgenologic findings. (Courtesy A. Tucker: Am. J. Roentgenol. Radium Ther. Nucl. Med., *86*:458, 1961.)

Pneumocystis carinii Pneumonia
(Plasma Cell Pneumonia)

This uncommon protozoan pneumonia most often appears as a complication in premature or debilitated infants, sometimes in epidemic form. Older children or adults with an impaired immunity response (those taking immunosuppressive drugs or those with immune globulin disorders) are also susceptible.

In infants there is progressive bilateral perihilar infiltration, usually alveolar and associated with hyperinflation. Air bronchograms virtually always develop and may be the earliest sign of alveolar consolidation. More extensive consolidation may give a "ground-glass" density with lucent interspersed areas of interstitial and peripheral emphysema. Hilar adenopathy and pleural fluid are generally absent. In severe cases, increasing dyspnea and cyanosis, with little or no fever, are associated with total lung opacification. Pneumothorax or pneumomediastinum is an ominous development.

In older children and adults, the infection starts as perihilar bilateral infiltrates. A diffuse acinar distribution may simulate interstitial disease, but later coalescence and air bronchograms reveal the alveolar nature of the process. Rarely, the disease begins as a unilateral lobar consolidation. Pleural reaction is rare.

The radiographic appearance may resemble cytomegalovirus or opportunistic bacterial pneumonia, pulmonary edema or hemorrhage, aspiration pneumonia, and alveolar proteinosis. Definitive diagnosis generally requires lung biopsy.[24-29]

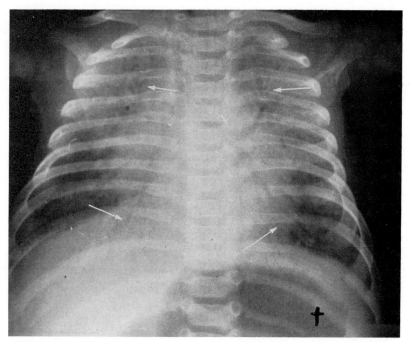

Figure 7–15 **Pneumocystis carinii Pneumonia: 2 Month Old Infant.** A somewhat granular perihilar consolidation is giving an airless appearance to the upper lung fields. The bases are hyperaerated. Air bronchograms throughout the lungs *(arrows)* indicate the extensive alveolar consolidation. This film was made after a week of illness; the infant died.

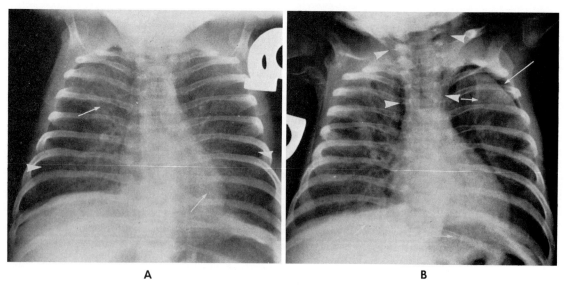

A   **B**

Figure 7–16 **Pneumocystis carinii Pneumonia: 3 Month Old Infant.**

A, Fine granular densities extend from the perihilar area; the right upper lobe is more involved than the others. Air bronchograms are faintly visible in this lobe and at the left base *(arrows)*. The lung peripheries *(arrowheads)* are hyperlucent, and the diaphragms are low, findings indicative of hyperaeration and emphysema. Note the absence of hilar node enlargement.

B, Three days later a left pneumothorax *(long arrow)* and a pneumomediastinum *(arrowheads)* have developed from the rupture of an emphysematous bleb. Air bronchograms are clearly visible in the consolidated outer portion of the left upper lobe and at the bases *(short arrows)*. This infant died.

Progressive perihilar granular infiltrates with air bronchograms are the most common and most characteristic findings in this condition. (Courtesy Dr. H. Gregoire, Montreal, Canada.)

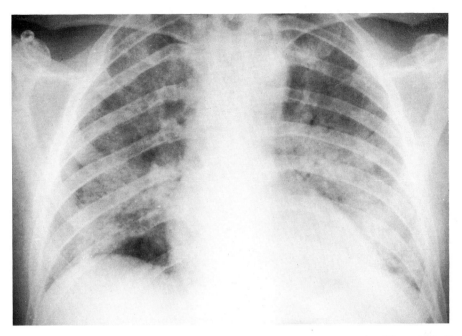

Figure 7–17 **Pneumocystis carinii Pneumonia Complicating Immunosuppressive Therapy in Renal Transplant Patient.**   Bilateral acinar and alveolar densities extend from both hila into the lung peripheries. More diffuse consolidation, associated with many air bronchograms, occurred in a few days. Lung biopsy confirmed the radiologic impression of *Pneumocystis carinii* pneumonia.

Pneumocystis carinii pneumonia is among the many opportunistic organism infections that can complicate prolonged immunosuppressive therapy.

HELMINTHIC DISEASES

Paragonimiasis

The pulmonary changes are usually unilateral and slowly progressive. An entire lung or only an upper lobe may be affected. Bilateral involvement occurs in about 20 per cent of patients.

In over half the cases, one or more lesions may show a characteristic ring shadow, formed by an eccentric cystic cavity within a nodular lesion. There is little or no perifocal reaction. Tracts or burrows often arise from these cavities and are best demonstrated on laminograms. Irregular nodular densities up to 3 to 4 cm. in diameter with poorly defined borders are common. Linear infiltrates occur most often adjacent to the typical ring shadows, especially in the lower lung fields. Pleural thickening is seen in about 20 per cent of cases.

Radiographic distinction from cavitating tuberculosis is difficult, but the absence of perifocal reaction adjacent to the ring shadows is suggestive of paragonimiasis. In cases without a characteristic ring shadow, the roentgen findings can imitate a wide variety of infectious diseases.

Cerebral paragonimiasis is found in only one per cent of patients with pulmonary involvement. The cerebral abscesses are calcified in about half of these cases. The calcification is usually unilateral and may present as amorphous punctate densities, as ill-defined, small, nodular collections, or as congregated, multiple, larger oval lesions with a denser periphery, resembling soap bubbles. The latter is the most common and characteristic manifestation.

Evidence of increased intracranial pressure is seen in about one third of patients with cerebral involvement. In over half the cases, unilateral cortical and subcortical atrophy and arachnoiditis can be demonstrated by air studies. Angiography may show a vascular shift, due to atrophy, away from the side of the lesion, but no abnormal vessels are seen.

The cerebral punctate and nodular calcifications can resemble those of toxoplasmosis, cysticercosis, tumor, or tuberous sclerosis. However, in an endemic area, unilateral cortical atrophy and calcifications are highly suggestive of cerebral paragonimiasis.[30-37]

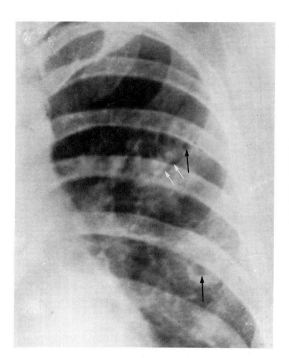

Figure 7–18 **Pulmonary Paragonimiasis with Ring Shadows.** There are ill-defined nodular densities *(black arrows)* containing eccentric lucent areas that form the characteristic ring shadows. A linear lucency *(white arrows)* extending into the upper ring shadow represents a burrow tract. (Courtesy R. Suwanik, Dhonburi, Thailand.)

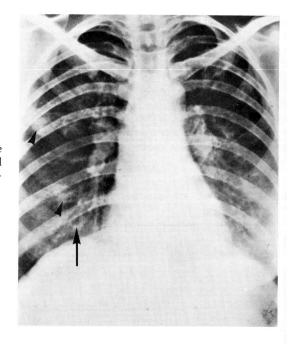

Figure 7–19 **Pulmonary Paragonimiasis.** There are ill-defined nodular densities *(arrowheads)* and a typical ring shadow with an eccentric lucency *(arrow).* (Courtesy R. Suwanik, Dhonburi, Thailand.)

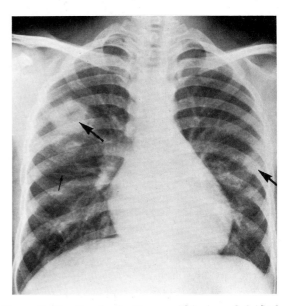

Figure 7--20 **Pulmonary Paragonimiasis.** There is a nodular density on the left pleura *(large arrow)* and a ring shadow in the right upper lobe *(large arrow).* Many linear infiltrates are seen in the lower lobes *(small arrow).*

The findings in these three cases (Figs. 7--18, 7–19, 7–20) are characteristic of chronic paragonimiasis. (Courtesy S. Yang, Taipei, Taiwan.)

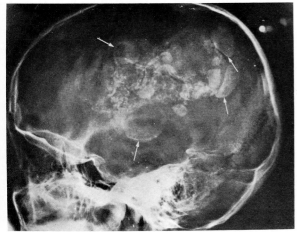

Figure 7--21 **Cerebral Paragonimiasis: Calcified Abscesses.** There is an extensive collection of rounded and oval calcifications of various sizes, producing a "soap-bubble" appearance. Increased peripheral density is seen in many of the lesions *(arrows).* All these calcifications were in the right parietal lobe.

This type of calcification is the most characteristic—and is almost pathognomonic—of paragonimiasis. When punctate and small nodular calcifications occur alone, differentiation from tuberous sclerosis, toxoplasmosis, cysticercosis, and glioma must be considered. (Courtesy Dr. Shin J. Oh, Minneapolis, Minnesota.)

Figure 7--22 **Cerebral Paragonimiasis: Calcified Abscesses and Cerebral Atrophy.** A pneumoencephalogram demonstrates enlargement of the right lateral ventricle *(large arrow),* with some shift of the ventricles to the right; these findings are characteristic of right-sided cortical atrophy. The collection of rounded calcifications *(small arrows)* in the right parietal area consists of calcified paragonimiasis abscesses. This combination of unilateral calcified masses and cerebral atrophy is highly suggestive of cerebral paragonimiasis.

Pneumoencephalography is frequently unsuccessful because acute cerebral paragonimiasis may produce a basal arachnoiditis. (Courtesy Dr. Shin J. Oh, Minneapolis, Minnesota.)

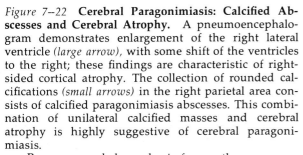

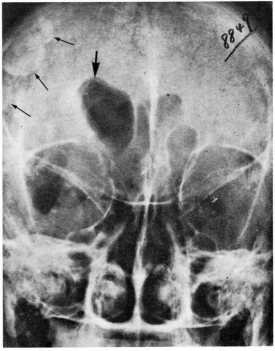

Schistosomiasis (Bilharziasis)

The ova of the Schistosoma worm lodge in the terminal arterioles and venules and produce a perivascular edema, an obliterative vasculitis, and, eventually, tissue fibrosis and granulomas.

Significant radiographic findings may occur in the lungs, in the gastrointestinal tract, and in the urinary tract. Massive splenomegaly is usually seen in advanced cases.

Schistosoma mansoni is the usual cause of pulmonary involvement. In the lungs the perivascular inflammatory reaction is manifested by small nodular and linear densities that are often diffusely distributed in a miliary pattern. Later, an obliterative arteritis develops, leading to pulmonary hypertension and cor pulmonale. This is characterized by a progressive dilatation of the main and major pulmonary arteries and a disparate decrease of peripheral arterial size and number. Right ventricular enlargement develops. These progressive changes can lead to enormous or even aneurysmal dilatation of the main pulmonary artery. Often the nodular and linear parenchymal densities diminish or disappear in this stage.

Colon involvement and, particularly, rectal involvement are frequent; *S. mansoni* or *S. japonicum* are causative worms. Segmental colonic lesions, more often in the right colon, are characterized by spasm, mucosal edema, and, sometimes, atonia. Ulceration may occur, and fibrotic strictures are not unusual. Granulomatous polyps are often detected, especially in the rectum. The roentgen appearance of colonic schistosomiasis is quite similar to that of granulomatous colitis.

In urinary tract involvement by *S. haematobium*, the wall of the bladder and the lower ureters are favorite sites. Initial bladder wall edema is followed by fibrosis and calcification. The bladder may become small or deformed by segmental fibrosis, and granulomatous polyps projecting into the lumen may simulate carcinoma on contrast films. The wall calcification is characteristically thin and linear and may give a laminated appearance to the emptied bladder. Ureterovesical reflux is common.

Linear calcification of the lower ureters is a later finding. The calcification may outline a dilated or narrowed ureter on a plain film.

Hydroureter and hydronephrosis often develop either from ureterovesical reflux or from lower ureteral stricture.

Liver infestation may produce periportal cirrhosis with the roentgen manifestations of portal hypertension. (See *Cirrhosis of the Liver*, p. 1109.) However, bilharzial liver fibrosis is mesenchymal in contrast to the parenchymal fibrosis of Laennec's cirrhosis, and there are somewhat different findings on angiography and splenoportography. The intrahepatic arterial supply is deficient in the bilharzial liver.[38-46]

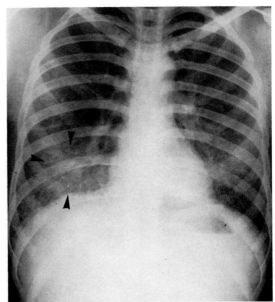

Figure 7–23 **Schistosomiasis (Bilharziasis): Active Pulmonary Infestation.** Fine miliary nodules are scattered throughout both lung fields, and conglomerate densities are seen at the bases *(arrowheads)*. This is the acute phase of pulmonary reaction to the ova in the arterioles and capillaries; cor pulmonale has not yet developed.

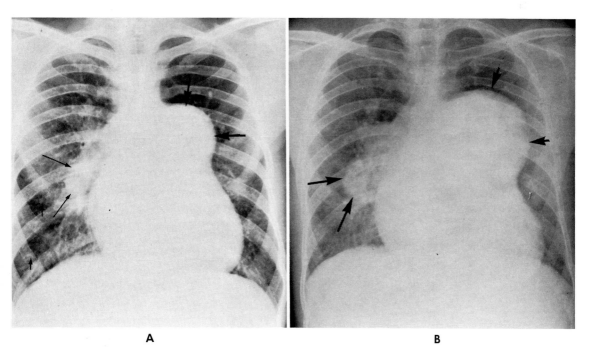

A   B

Figure 7–24 **Pulmonary Schistosomiasis: Progressive Pulmonary Hypertension and Cor Pulmonale.**

    A, There are numerous linear and nodular densities in the pulmonary parenchyma *(short arrows)*. The heart is enlarged, and the right pulmonary arteries are greatly enlarged *(medium arrows)*. The main pulmonary artery *(large arrows)* is almost aneurysmal.

    B, Three years later the main pulmonary artery *(short arrows)* has increased further in size; the heart and right pulmonary artery *(large arrows)* are also much larger. The more distal pulmonary markings are considerably decreased, owing to obliteration of the peripheral pulmonary vasculature. The extreme dilatation of the pulmonary artery caused by pulmonary schistosomiasis is unequaled by almost any other disease. (Courtesy Z. Farid, Cairo, Egypt.)

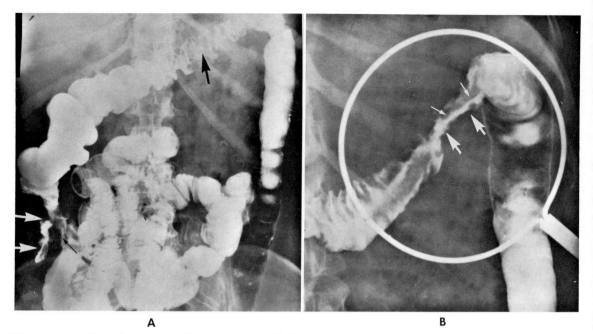

A   **B**

Figure 7–25 **Granulomatous Schistosomiasis of Colon.**

A, This is a barium enema study of a 40 year old woman with a five-year history of abdominal pain, nausea, and vomiting. The cecum and ascending colon are contracted and irregular *(large white arrows),* and heavy folds are evident. The terminal ileum *(small black arrow)* appears to be normal. The distal transverse colon *(large black arrow)* was also irritable and showed heavy folds.

Ova of *Schistosoma mansoni* were found in the stool. At surgery, a large schistosoma granuloma of the cecum, ascending colon, and appendix was resected.

B, A barium enema study was done two years later, because the patient experienced recurrent pain and diarrhea. The previously irritable segment of the transverse colon has become narrowed *(large arrows)* and irregular and contains a greatly thickened fold *(small arrows).* Another schistosoma granuloma was found within and around the involved segment. (Courtesy W. Seaman: Radiology, *85*:682, 1965.)

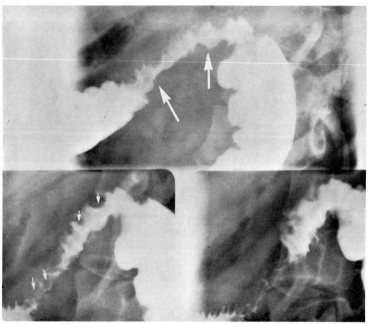

Figure 7–26 **Schistosomiasis of Colon.** A barium enema study (three spot films) of a 29 year old female who had acute bloody diarrhea, nausea, and vomiting shows segmental narrowing of the distal transverse colon *(large arrows),* which was irritable and contained many thickened folds *(small arrows).* This irritability causes the changing configuration seen in the films.

The resected segment was congested, edematous, and covered with fibrin. *Schistosoma mansoni* ova were found.

The colonic lesions of schistosomiasis radiologically resemble those of tuberculosis, amebiasis, and granulomatous colitis, and sometimes those of malignancy. Colonic polyps are often found in patients infested with *Schistosoma mansoni.* (Courtesy W. Seaman: Radiology, *85*:682, 1965.)

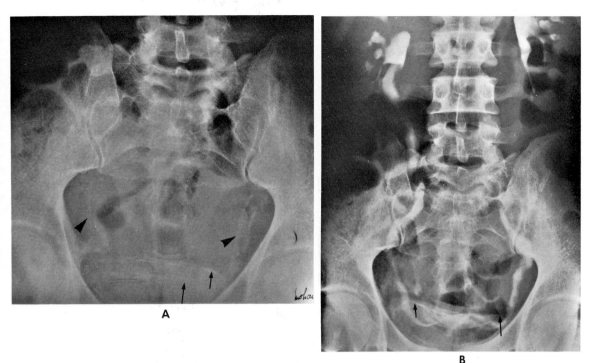

Figure 7–27 **Urinary Schistosomiasis.**

A, There are characteristic transverse laminated linear calcifications of the bladder wall *(small arrows),* and longitudinal calcifications outline the lower ureters *(arrowheads).*

B, Intravenous urography demonstrates dilatation of both ureters, especially the right one. The right kidney is moderately hydronephrotic. The bladder calcification *(arrows)* is somewhat obscured by the contrast material in the bladder. (Courtesy P. Mallaret, Cannes, France.)

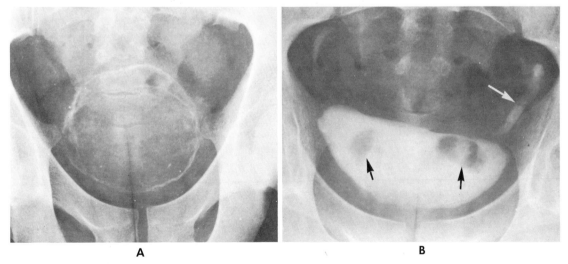

Figure 7–28 **Schistosomiasis of Urinary Bladder: Two Patients.**

A, Plain film of the pelvis reveals a ring calcification, which represents calcification of the wall of a distended bladder. When the bladder is emptied, the calcifications appear as transverse horizontal lines.

B, In another patient there are two large filling defects *(black arrows)* in the opacified bladder. These were bilharzial polyps. The left lower ureter *(white arrow)* was persistently dilated owing to fibrosis of its lower end. No bladder or ureteral calcification is present. (Courtesy M. M. Al-Ghorab, Alexandria, Egypt.)

Echinococcosis (Hydatid Disease)

Cystic hydatid disease due to the tapeworm *Echinococcus granulosus* is the commonest form, and dogs, wolves, and cattle are the intermediate hosts. Involvement and radiographic findings are most common in the liver and the lungs. Kidney and bone lesions are considerably less frequent.

With liver involvement, a nonspecific hepatomegaly and elevation of the right diaphragm may be the only findings. Later, calcification develops in the outer and inner wall of the cyst, giving a characteristic radiographic appearance. Usually there is a sharp thin rim calcification of the outer cyst wall and a somewhat irregular calcified inner wall. Smaller internal calcifications are not unusual. The liver cysts may reach 10 cm. in diameter; multiple cysts may be present. Liver involvement occurs in up to 80 per cent of cases.

Celiac angiography is extremely helpful in localizing and identifying uncalcified or partly calcified hydatid cysts in the liver and spleen. Vessels are stretched around the avascular cyst, and an opacified halo is usually seen outlining the rim of the lesion. This rim probably represents the ectocyst and is virtually diagnostic of the disease.

Pulmonary lesions are found in about one fourth of cases. These are round or oval densities that may attain huge proportions. Sometimes, if air seeps in between the double wall of pericyst and endocyst, a characteristic halo or crescent of air surrounds the fluid-filled endocyst. If bronchial communication occurs, there will be air-fluid levels and, sometimes, rapid emptying of the cyst. Peripherally located cysts may rupture into the pleural cavity and produce a hydrothorax or a hydropneumothorax. Unlike hydatid cysts in other tissues, the pulmonary lesions rarely calcify.

The kidney lesions are found in less than 5 per cent of cases. When unruptured and uncalcified, the renal cyst may produce the pressure distortion, displacement, and elongation of calices that is indistinguishable from that due to an ordinary cyst. Incomplete calcification of the wall may suggest renal malignancy. When ruptured, the cyst can communicate with adjacent calices, and great distortion of the parenchyma and collecting system may suggest renal tuberculosis or even a diffuse neoplastic infiltration. Most renal hydatid cysts expand anteriorly.

Bone lesions occur least frequently (less than 2 per cent of cases) and may involve flat or long bones. Usually there are rather extensive multilocular, irregular lytic areas without reactive or sclerotic bone changes. However, secondary infection may eventually lead to sclerosis. Except in endemic areas, the true nature of the lesion is rarely suspected.

Alveolar hydatid disease is due to *Echinococcus multilocularis;* cats, mice, and small rodents are the intermediate hosts. The parasite usually remains localized in the liver. Hepatomegaly without calcifications occurs in about 20 per cent of the cases. Liver calcifications, which occur in about 70 per cent of cases, are quite different from cystic hydatid calcification. Typically, there are small radiolucencies outlined by calcific densities 2 to 4 mm. in diameter. These lucencies are scattered in larger areas of amorphous calcification. These findings are considered diagnostic of alveolar hydatid disease but must be distinguished from other hepatic calcifications, particularly those from primary and metastatic hepatic tumors (see hepatic calcifications in *Basic Roentgen Signs*).

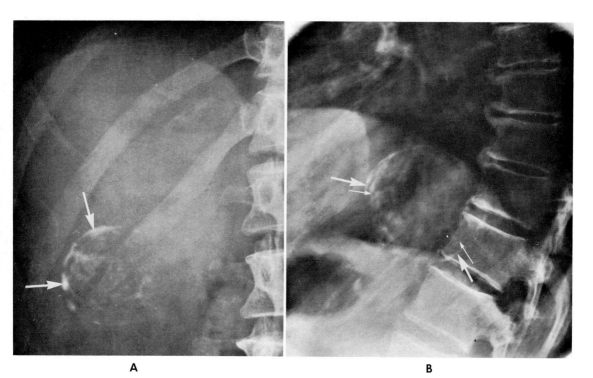

A	**B**

Figure 7–29 **Two Cases of Calcified Echinococcus Cyst of Liver.**

   A, There is a single, round, partially calcified echinococcus cyst in the inferior portion of the right lobe. The peripheral calcification *(arrows)* is somewhat thick, and there is very little calcification of the inner wall. Complete calcification occurs only with total quiescence.

   B, A cyst of similar size in the posterior liver substance just below the diaphragm shows the thicker outer wall calcification *(large arrows)* and the less well-defined inner wall calcification *(small arrows)*.

   Note in both cases the sharp outer margins and the irregular inner margins of the ring calcifications; this is a characteristic finding.

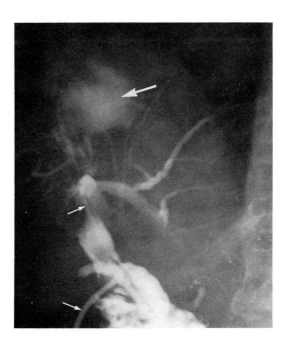

Figure 7–30 **Echinococcus Disease of Liver and Communication with Biliary Tree: T-Tube Cholangiogram.** This is a T-tube *(small arrows)* cholangiogram in a patient who had echinococcus disease of the liver and biliary tract symptoms. The echinococcus cystic cavity *(large arrow)* penetrated the biliary tree. Note the irregular outline of the cavity, caused by proliferating scolices and daughter cysts within the wall of the main lesion.

   Surgical intervention is necessary in such a case to prevent widespread seeding of the disease in the liver, biliary tree, and bloodstream (via the intestines).

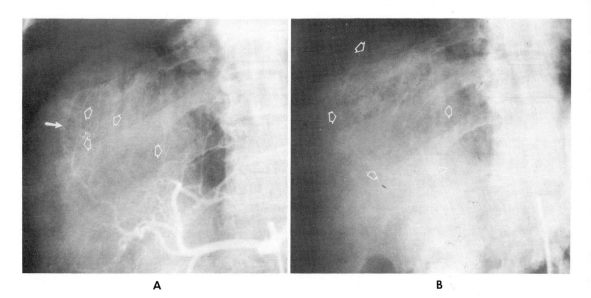

A   **B**

Figure 7–31   **Uncalcified Echinococcus Cyst of Liver: Angiogram.**

A, Celiac arteriogram shows stretched hepatic vessels *(open arrows)* in the upper liver. Opacification of the wall of the cyst *(solid arrow)* has started to appear.

B, Venous phase of the study discloses the opacified circular wall of the cyst *(arrows)* with relative lucency (absence of hepatographic density) within the cyst. (Courtesy Drs. I. Garti and V. Deutch, Tel-Aviv, Israel.)

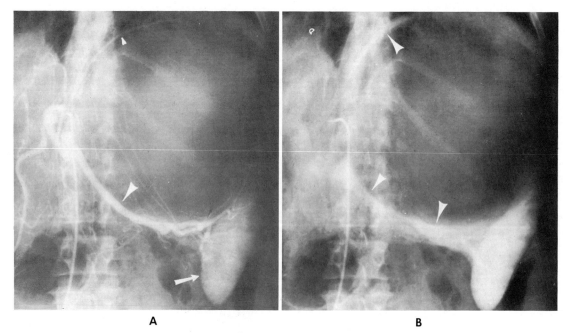

A   **B**

Figure 7–32   **Noncalcified Hydatid Cyst in Spleen: Celiac Arteriography.**

A, During the arterial phase, the splenic *(large arrowhead)* and left gastric arteries *(small arrowhead)* are draped around a huge circular avascular mass. The uninvolved portion of the spleen *(arrow)* is opacified.

B, During the capillary phase, a thin circular rim of density *(arrowheads)* is apparent. This rim of density (pericyst) is characteristic of an echinococcus cyst. (Courtesy G. K. Rizk, Beirut, Lebanon.)

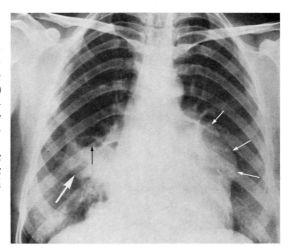

Figure 7–33 **Echinococcus of Lung, Heart, and Peri-cardium.** The round, sharply demarcated homogeneous density (*large arrow*) in the parenchyma of the right lower lung contains an air-fluid level (*black arrow*) that has resulted from rupture into a bronchus. Calcification does not develop. This type of lesion may completely empty through the bronchus and disappear.

The lobulated bulge on the left side of the cardiac shadow (*small white arrows*) was attributable to cardiac and pericardial involvement; the lesions increased in size. (Courtesy P. Pitzorno, Sassari, Italy.)

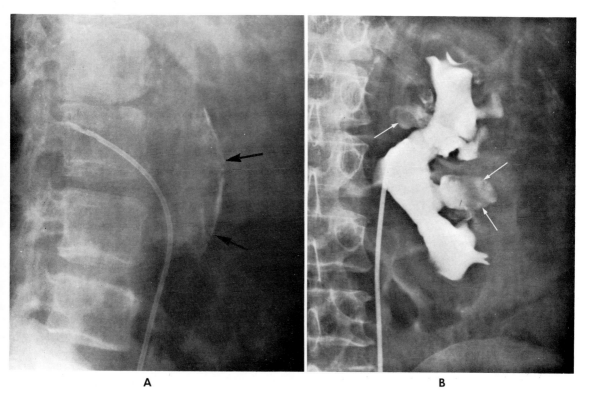

A   B

Figure 7–34 **Extensive Echinococcus Disease of Kidney.**

    A, There is a large, crescentic area of calcification (*arrows*) in the anterior portion of the kidney; this represents the anterior wall of a large cyst that has penetrated the renal parenchyma.

    B, Retrograde pyelogram reveals extensive destruction of the calices, with irregular dilation (*arrows*) resulting from parenchymal infiltration and seeding of the scolices.

    Although the caliceal changes closely resemble the destructive changes of tuberculosis, the calcified anterior wall is more compatible with an echinococcus cyst.

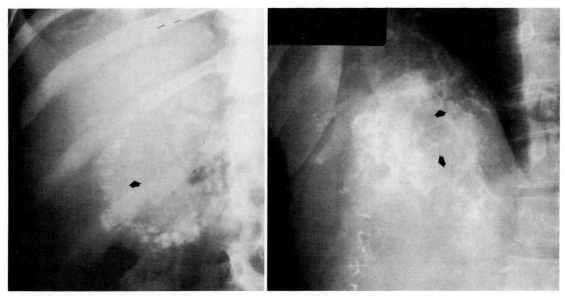

Figure 7–35 **Alveolar Hydatid Disease of Liver: Two Patients.**   In both patients there are scattered irregular calcifications in the liver. Although these appear amorphous, closer inspection reveals a lucent center in many of the calcifications *(arrows)*. This lucent center is highly characteristic of alveolar hydatid lesions and represents intact cysts with calcified walls. The amorphous deposits are within degenerating cysts.

Only the liver is involved in alveolar hydatid disease. (Courtesy Dr. W. M. Thompson, Durham, North Carolina.)

Cysticercosis

Cysticercosis is caused by infestation with the cyst stage of the tapeworm, usually *Taenia solium.* Significant radiographic findings can often be seen in the muscles and intracranially.

After months or years the cysts in the muscle may die and calcify, producing characteristically dense elliptical shadows with a noncalcific central area. Tiny round calcifications and elliptical lesions up to 2.5 cm. may be found. The long axis of the cyst follows the muscle planes, producing a distinctive pattern that may help to distinguish these from the somewhat similar soft tissue calcifications of Ehlers-Danlos syndrome. (See p. 1419.)

Central nervous system involvement occurs in practically every active case. In about half of these, intracranial calcifications develop in the brain tissue. Typically there is a slightly eccentric spherical calcification, 1 to 2 mm. in diameter (the calcified scolex), surrounded by a calcified sphere 7 to 12 mm. in diameter. This appearance may help differentiation from the calcifications of tuberous sclerosis and toxoplasmosis.

Spherical soft tissue masses representing live cysts can be demonstrated in the cisternas and ventricles during pneumoencephalography in about 10 per cent of cases. These lesions almost never calcify; the individual mass may become large enough to produce obstructive hydrocephalus.

The demonstration of dead calcified cysts does not rule out persisting active infestation.[56-58]

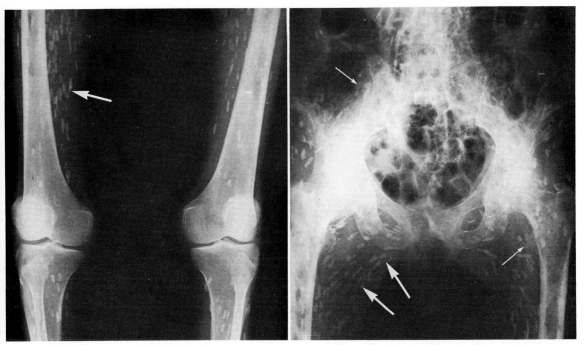

Figure 7–36 **Extensive Calcified Cysticercosis of Muscle.** Numerous elliptical calcifications, evident in the muscle planes of the pelvis and lower extremities, are variable in size and are aligned along the long axis parallel to the muscle planes *(large arrows)*. Nearly all the calcifications have a small lucent center *(small arrows)*. When viewed on edge, they resemble a ring calcification. Note the associated soft tissue atrophy. (Courtesy T. Keats: Missouri Med., *58:*457, 1961.)

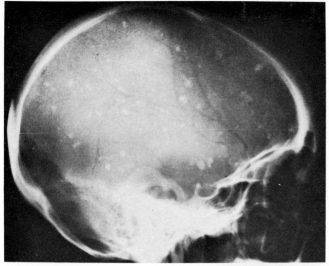

A

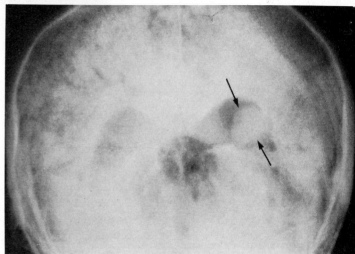

B

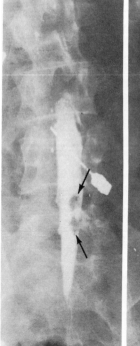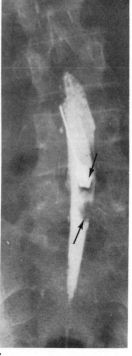

C

Figure 7–37 **Cysticercosis of Central Nervous System in Three Patients.**

A, Somewhat irregular nodular calcifications are scattered throughout the brain substance. These represent dead cysts.

Scattered brain calcifications are also seen in tuberous sclerosis, toxoplasmosis, cerebral paragonimiasis, cytomegalic inclusion disease, and primary hypoparathyroidism (see illustrations under these headings). The location and shape of the individual calcifications may sometimes help to distinguish these conditions.

B, Pneumoencephalogram of another patient reveals a rounded mass *(arrows)* within the posterior portion of the slightly dilated left ventricle. Lateral view *(not shown)* disclosed similar but smaller masses in the left temporal horn and in the cisterna chiasmatis. These were all active cysticercus vesicles.

Distinction of a cysticercus vesicle from an intraventricular or intracisternal tumor can be made only on clinical grounds, particularly if only a single lesion is uncovered.

C, Myelogram of a third patient reveals two well-circumscribed defects *(arrows)* in the cauda equina area; these were cysticercus vesicles in the subarachnoid space. (Courtesy Dr. Guillermo Santin, Mexico City, Mexico.)

Trichuriasis (Whipworm Infection, Trichocephaliasis)

The adult worm, 30 to 50 mm. in length, lives primarily in the cecum but readily invades the appendix and terminal ileum. In heavy infestations the entire colon and rectum may be invaded. Eosinophilia and microcytic anemia are common findings.

The coiled male or whip-shaped female worms, surrounded by excess mucus, produce linear translucencies and a pinwheel appearance, giving a characteristic granularity on the air-contrast barium enema study. The mucus may cause flocculation of the barium in the fully filled colon.[59-61]

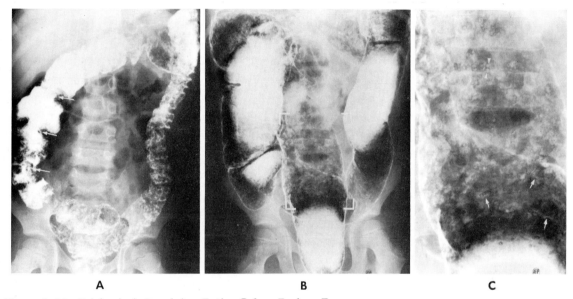

A   B   C

Figure 7–38   **Trichuriasis Involving Entire Colon: Barium Enema.**

This 7 year old boy had profuse rectal bleeding caused by a heavy infestation of whipworms (*Trichuris trichiura*).

A, There is a diffuse granular mucosal pattern, caused by innumerable worms adherent to the mucosa. The flocculation of the barium (*arrows*) is caused by the mucus admixed with the worms.

B, In the air-contrast film, innumerable tiny densities are seen in the air column; these represent the worms diffusely scattered throughout the colon.

C, Magnification of the marked-off rectosigmoid area in *B* demonstrates the individual coiled male worms imbedded in mucus (*arrowheads*), and the whip-shaped female parasites (*arrows*). The worms are attached to the mucosa.

The roentgen pattern resembles lymphoid hyperplasia of the colon. (Courtesy Dr. M. M. Reeder, Washington, D.C., and Dr. J. Astacio, San Salvador; Armed Forces Institute of Pathology Photograph.)

Strongyloidiasis

This tropical parasitic infection is also endemic in the southern United States.

The duodenum and jejunum are the favorite sites of involvement. In the acute phase duodenal dilatation is common and is associated with edematous folds in the duodenum and jejunum. Segmental irritability and spasm occur. In more severe and chronic disease, nodular defects, areas of mucosal atrophy and ulcerations, rigidity, and tubular narrowing are evident in the duodenum and jejunum. The ileum is usually not affected. The radiographic appearance may simulate granulomatous enteritis, lymphosarcoma, or eosinophilic gastroenteritis. However, the long segments of tubular narrowing are fairly characteristic of strongyloidiasis.

Generally, the small bowel returns to normal following successful antihelminthic therapy.

Occasionally gastric involvement occurs, usually in the antrum, producing edema, rigidity, and nodularity. Infrequently, a patchy Löffler's pneumonia occurs.[62-64]

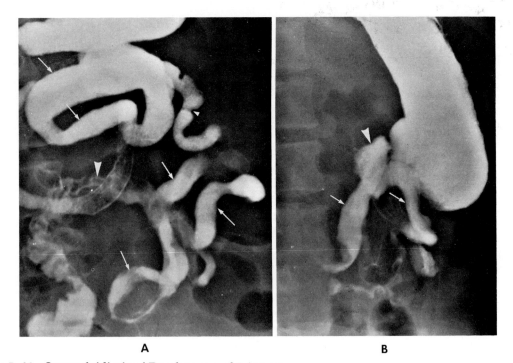

A   B

Figure 7–39 **Strongyloidiasis of Duodenum and Jejunum.**

A, The upper small bowel appears smooth and tubular without a mucosal pattern *(arrows).* The narrowing is almost uniform, but areas of local constriction can be seen *(small arrowhead).* Note the abrupt transition to a more nearly normal mucosal pattern *(large arrowhead).*

B, The duodenal loop *(arrows)* has a similar appearance. The duodenal bulb is also deformed *(arrowhead)* by a duodenal ulcer. (Courtesy G. Arantes Pereira, Rio de Janeiro, Brazil.)

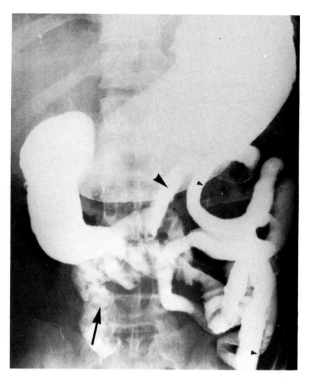

Figure 7–40 **Strongyloidiasis of Proximal Small Bowel.** The fourth portion of the duodenum *(large arrowhead)* and the proximal jejunum are uniformly narrowed and tubular, strikingly similar to the preceding illustration. There are small marginal serrations *(small arrowheads)* throughout the involved loops, apparently due to submucosal ulcerations. Greatly thickened walls cause the separation of these loops. The involvement is continuous, and the distal small bowel is normal *(arrow)*. The stomach and duodenum proximal to the narrowed bowel are dilated.

Ascariasis

The worm can often be demonstrated in the small bowel by contrast studies and occasionally by plain films of the abdomen. The body of the parasite is clearly outlined as a cylindrical defect in the barium column. If the worm ingests the barium, a string-like density of barium will outline the parasite's intestinal canal. This is usually seen on delayed films made up to 24 hours after an oral barium meal.

Plain films of the abdomen occasionally show the worms as linear streaks of density contrasting with the air in the bowel. The worms may be single or multiple and have a predilection for the distal jejunum. There are usually no alterations in mucosal pattern or motility, although in rare cases a deficiency pattern may be seen. Large masses of worms can produce a mechanical intestinal obstruction.

Respiratory symptoms and pulmonary changes may occur 10 to 14 days after massive ingestion of ova. Localized or diffuse alveolar infiltrates may appear, often in the lung periphery. Occasionally hilar enlargement also occurs. The infiltrates are quite variable, changeable, and migratory. Asthma-like symptoms and blood eosinophilia are usual. This clinical-radiologic complex is considered a form of Löffler's pneumopathy (see p. 000). Prompt and dramatic response to steroids usually occurs.[65-67]

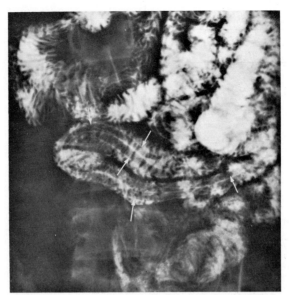

Figure 7–41 **Ascariasis of Small Intestine.** Multiple worms are outlined (*arrows*) in the middle segment; they contrast clearly with the barium. No abnormalities of the mucosal pattern are apparent.

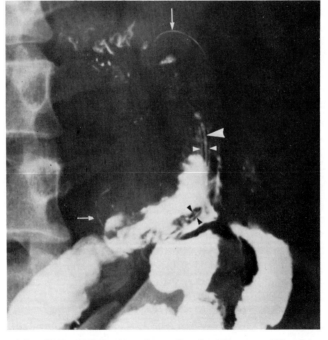

Figure 7–42 **Ascariasis of Small Bowel: Barium Ingestion by Worms.** The thin curved lines of density (*small arrows* and *large arrowhead*) are barium within the digestive tubes of several worms. The bodies of the worms, when surrounded by intestinal barium, can be identified as lucent areas (*small arrowheads*).

    This film was part of a small bowel series (five-hour film). A 24-hour film, free from small bowel barium, may often demonstrate the worm-ingested barium more dramatically. The worm is more likely to ingest barium if the patient has been fasting prior to the study.

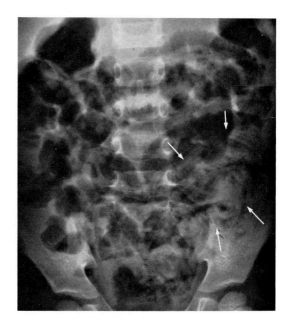

Figure 7–43 **Ascarid Masses in Small Bowel Producing Partial Obstruction.** Note the spiral densities in the area bounded by the arrows. These were masses of worms that had accumulated in the dilated loops. Although obstruction was incomplete, surgical intervention was necessary. (Courtesy W. Wong, Hongkong.)

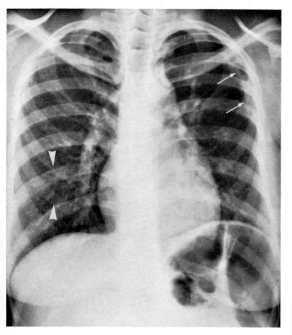

Figure 7–44 **Ascariasis of Lung.** There is a patch of pneumonitis and pleural reaction in the left upper chest *(arrows)*. Larvae of Ascaris were found on biopsy. Markings at the bases *(arrowheads)* are unusually prominent.

The pulmonary findings in Ascaris infestation are nonspecific and may alter rapidly.

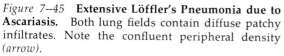

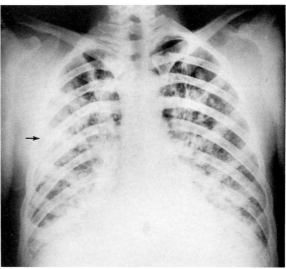

Figure 7–45 **Extensive Löffler's Pneumonia due to Ascariasis.** Both lung fields contain diffuse patchy infiltrates. Note the confluent peripheral density *(arrow)*.

This young man had fever, cough, cyanosis, and eosinophilia. Ascaris larvae were found in his sputum. (Courtesy Drs. A. J. Harrold and J. A. Phills, Montreal, Canada.)

Loiasis

Larvae of the filaria *Loa loa* (African eye worm) are introduced into man through the bite of the mango fly. The mature worm has a predilection for the eye and subcutaneous tissues. The dead worms usually undergo calcification and may be seen radiographically as fine coiled or filamentous calcification several milli-meters in length. These fine calcifications are difficult to visualize, except in thinner portions of the hand or foot.[58, 68]

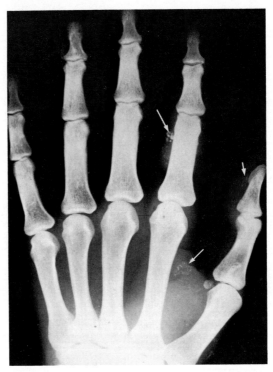

Figure 7–46 **Calcified Filaria in Soft Tissues of Hand (Loiasis).**   There are coils of calcified worms (*arrows* in the soft tissues. Calcification occurs only after the parasite dies. (Courtesy I. Williams, London, England.

Filariasis

Filaria bancrofti or *F. malayi* can infect the lymphatics or connective tissues in man.

Characteristic roentgen findings on lymphography are seen in chyluria and in filarial edema, the two most common complications of chronic filariasis.

In chyluria, a fistulous connection develops between the renal lymphatics and the renal pelvis. On lymphography there is retrograde filling of dilated tortuous renal lymphatics via the para-aortic nodes. These channels empty very slowly, mainly into the renal pelvis.

The para-aortic nodes and channels are also usually abnormal. At first they are prominent, with mottled defects in the enlarged nodes. Later the nodes become granular and small. Eventually there is nonfilling of these nodes.

The thoracic duct is usually beaded and tortuous with segmental areas of dilatation and narrowing. It is rarely completely occluded.

In lymphedema and elephantiasis, the inguinal and pelvic nodes are occluded by the filaria. During the earlier lymphedema stage, the nodes are enlarged with dilated sinusoids. The superficial lymphatics are dilated and varicose. In the later elephantiasis stage, the normal channels are obliterated and there is an extensive network of filled superficial channels (dermal backflow). (See Fig. 10–206.)[69–71]

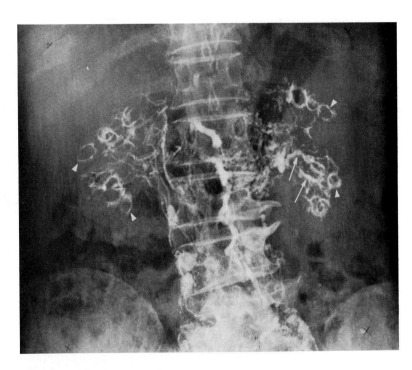

Figure 7–47 **Chronic Filariasis: Chyluria.** Lymphangiogram discloses bilateral filling of the renal lymphatics by reflux from the abnormal para-aortic nodes and channels. Many of the renal lymphatic channels are dilated and tortuous *(arrows)*. The circular densities *(arrowheads)* are caused by contrast material surrounding the base of the renal papillae. The opaque material in the abnormal renal lymphatics will slowly enter the renal collecting system via lymphatorenal fistulas. (Courtesy Dr. M. Akisada, Tokyo, Japan.)

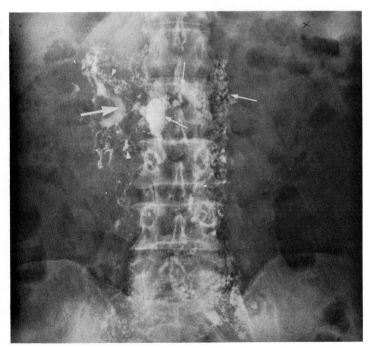

Figure 7–48 **Chronic Filariasis: Chyluria: Lymphangiogram.** The opacified right renal lymphatics are tortuous and dilated *(arrowheads)*. The renal pelvis is opacified *(large arrow)* because of lymphatopelvic fistulization, which is the lesion responsible for chyluria.

The para-aortic nodes and channels *(small arrows)* are decidedly abnormal. With progression of lymphatic involvement, opacification of the para-aortic nodes and channels decreases or becomes entirely absent. (Courtesy Dr. M. Akisada, Tokyo, Japan.)

Dirofilariasis

The common heartworm of dogs, *Dirofilaria immitis,* can occasionally be transmitted to man, probably through a mosquito vector. The worm may be trapped and die in a small arteriole of the lung, leading to a focal pulmonary infarct that presents as a radiologic coin lesion. The nodule is usually solitary and remains uncalcified. Differentiation from other coin lesions can be made only by recognition of the worm from a biopsy specimen.[72, 73]

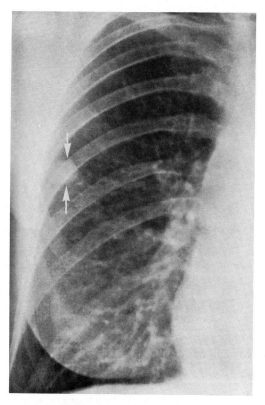

Figure 7–49 **Pulmonary Dirofilariasis: Coin Lesion.**   The sharply marginated noncalcified rounded density *(arrows)* in the right upper lobe abuts the pleural surface. It is indistinguishable from primary or metastatic malignancy or from any noncalcified granuloma. A resection biopsy disclosed necrosis and granulomatous change. Emboli containing Dirofilaria were identified in degenerated blood vessels.
     Similar solitary coin lesions have been the characteristic finding in recent reports of pulmonary dirofilariasis. (Courtesy Dr. C. A. Beskin, Baton Rouge, Louisiana.)

Dracontiasis (Dracunculiasis)

The *Dracunculus medinensis* (guinea worm) lives in the soft tissues. After its death it undergoes calcification. The region of the groin and the joints of the lower extremities are the favorite sites, but any soft tissue may be involved. The calcified worms appear as linear, oval, and serpiginous opacities, often coiled, and range from 1 to 40 mm. in length. In longstanding cases, ankylosis may develop in the joints adjacent to the calcifications.[58, 74]

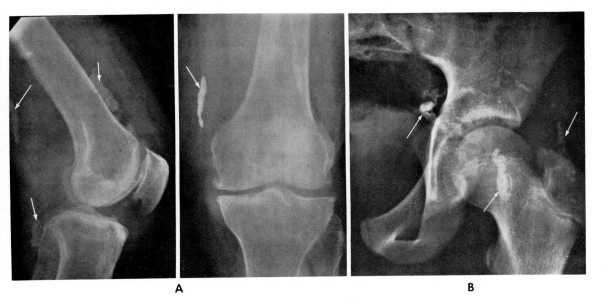

Figure 7–50 **Calcified Dracunculiasis.**

A, There are numerous irregular calcifications *(arrows)* in the soft tissues around the knee joint.

B, Similar calcifications are in the soft tissues in the groin *(arrows).*

In areas in which dracunculiasis is endemic, such roentgenologic findings strongly suggest the diagnosis. (Courtesy A. Fardour, Shiraz, Iran.)

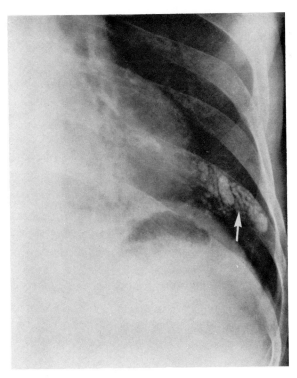

Figure 7–51 **Calcified Dracunculi in Chest Wall.** The coiled calcification *(arrow)* is the dead parasite in the soft tissues of the chest wall. This type of coiled calcification is characteristic of dracunculiasis.

Hookworm Disease (Ancylostomiasis, Miner's Anemia)

A deficiency or malabsorption pattern of the small bowel in hookworm infestation has been sporadically reported. A recent large series, however, fails to confirm this association.[75]

Cutaneous Larva Migrans (Creeping Eruption)

In this condition the cat and dog hookworm *(Ancylostoma braziliense)* enters and migrates along the skin. In about one third of cases, the larvae may reach the lung and cause transient patchy pulmonary infiltrates with a blood eosinophilia (Löffler's pneumopathy; see p. 121). The pulmonary lesions begin about one week after the skin eruption and may continue for up to four weeks.[76]

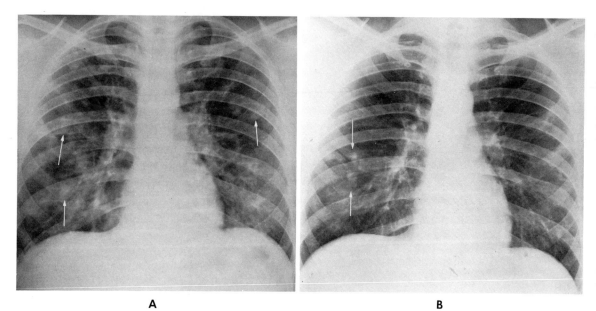

| A | B |

Figure 7–52 **Creeping Eruption: Löffler's Pneumonia.**

A, There are diffuse patches of nodular and linear infiltrates *(arrows)* scattered throughout both lung fields. Eosinophilia was noted.

B, Two weeks later the type and location of the infiltrates have changed *(arrows).*

Löffler's pneumonia is nonspecific and is seen in a number of allergic and parasitic disorders. (Courtesy E. Kalmon, Oklahoma City, Oklahoma.)

Löffler's Tropical Eosinophilia

See *Eosinophilic Pneumonia,* page 121.

REFERENCES

1. Whittaker, L. R.: Intrathoracic complications of ruptured amebic liver abscess. East Afr. Med. J., *40*:95, 1963.
2. Rodrigues, D., et al.: Chronic cardiopathy in Chagas' disease. Bol. Soc. Brasil Radiol., *2*:41, 1958.
3. Wilson, E. S.: Pleuropulmonary amebiasis. Am. J. Roentgenol. Radium Ther. Nucl. Med., *111*:518, 1971.
4. Tchang, S.: Amebiasis in northern Saskatchewan: Radiological aspects. Can. Med. Assoc. J., *99*:688, 1968.
5. Hardy, R., and Scullin, D. R.: Thumbprinting in a case of amebiasis. Radiology, *98*:147, 1971.
6. Balikian, J. P., Uthman, S. M., and Khour, N. F.: Intestinal amebiasis. Am. J. Roentgenol. Radium Ther. Nucl. Med., *122*:245, 1974.
7. Marshak, R. H., Ruoff, M., and Lindner, A. E.: Roentgen manifestations of giardiasis. Am. J. Roentgenol. Radium Ther. Nucl. Med., *104*:557, 1968.
8. Babb, R. R., Peck, O. C., and Vescia, F. G.: Giardiasis. J.A.M.A., *217*:1359, 1971.
9. Ochs, H. D., Ament, M. E., and Davis, S. D.: Giardiasis with malabsorption in X-linked agammaglobulinemia. N. Engl. J. Med., *287*:341, 1972.
10. Spitz, S.: Pathology of acute falciparum malaria. Mil. Surgeon, *99*:555, 1946.
11. MacCallum, D. K.: Pulmonary changes resulting from experimental malaria infection in hamsters. Arch. Pathol., *86*:681, 1968.
12. Heineman, H. S.: The clinical syndrome of malaria in the United States. Arch. Intern. Med., *129*:607, 1972.
13. Godard, J. E., and Hansen, R. A.: Interstitial pulmonary edema in acute malaria. Radiology, *101*:523, 1971.
14. Fishbein, B., and Sutton, R. N. P.: Chagas' disease in West Indies. Lancet, *1*:330, 1963.
15. Fonseca, L. C.: Digestive tract in Chagas' disease. Rev. Brasil Radiol., *3*:1, 1960; *2*:132, 1959.
16. Aticas, A., et al.: Megaesophagus, megacolon and Chagas' disease in Chile. Gastroenterology, *44*:433, 1963.
17. Reeder, M. M., and Simão, C.: Chagas' myocardiopathy. Semin. Roentgenol., *4*:374, 1969.
18. Todd, I. P., et al.: Chagas' disease of the colon and rectum. Gut, *10*:1009, 1969.
19. Remington, J. S., et al.: Toxoplasmosis in the adult. N. Engl. J. Med., *262*:180, 1960; *262*:237, 1960.
20. Tucker, A. S.: Intracranial calcification in infants (toxoplasmosis and cytomegalic inclusion disease). Am. J. Roentgenol. Radium Ther. Nucl. Med., *86*:458, 1961.
21. Fisher, S. H.: Toxoplasmosis with cerebral calcification. Am. J. Roentgenol. Radium Ther. Nucl. Med., *59*:816, 1948.
22. Gombert, H. J.: Clinical and roentgenologic picture of chronic pulmonary toxoplasmosis. Fortschr. Roentgenstr., *78*:728, 1953.
23. Müssbichler, H.: Radiologic study of intracranial calcifications in congenital toxoplasmosis. Acta Radiol., *7*:369, 1968.
24. Robillard, G., Bertrand, R., et al.: Plasma cell pneumonia in infants; review of 51 cases. J. Can. Radiol., *16*:161, 1965.
25. Feinberg, S. B., Lester, R. G., and Burke, B. A.: The roentgen findings in Pneumocystis carinii pneumonia. Radiology, *76*:594, 1961.
26. Jacobs, J. B., et al.: Needle biopsy in Pneumocystis carinii pneumonia. Radiology, *93*:525, 1969.
27. Forrest, J. U.: Radiographic findings in Pneumocystis carinii pneumonia. Radiology, *103*:539, 1972.
28. Fischer, R., et al.: Pneumocystis carinii pneumonia in adults. Dtsch. Med. Wochenschr., *94*:2135, 1969.
29. Doppman, J. L., Geelhoed, G. W., and Devita, V. T.: Atypical radiographic features in Pneumocystis carinii pneumonia. Radiology, *114*:39, 1975.
30. Yang, S. P., et al.: The clinical and roentgenologic courses of pulmonary paragonimiasis. Dis. Chest, *36*:494, 1959.
31. Kim, S. K.: Cerebral paragonimiasis; a report of 47 cases. Arch. Neurol., *1*:30, 1959.
32. Suwanik, R., and Harinsuta, C.: Pulmonary paragonimiasis. Am. J. Roentgenol. Radium Ther. Nucl. Med., *81*:236, 1959.
33. Oh, S. J.: Roentgen findings in cerebral paragonimiasis. Radiology, *90*:292, 1968.
34. Béland, J. E., et al.: Paragonimiasis (the lung fluke): Report of four cases. Am. Rev. Resp. Dis., *99*:261, 1969.
35. Gutman, R. A., et al.: Radiographic changes accompanying successful therapy of pulmonary paragonimiasis. Am. Rev. Resp. Dis., *99*:255, 1969.
36. Higashi, K., et al.: Cerebral paragonimiasis. J. Neurosurg., *34*:515, 1971.
37. Ogakwu, M., and Nwokolo, C.: Radiologic findings in pulmonary paragonimiasis as seen in Nigeria. Br. J. Radiol., *46*:699, 1973.
38. Chait, A.: Schistosomiasis mansoni; roentgenologic observations in a nonendemic area. Am. J. Roentgenol. Radium Ther. Nucl. Med., *90*:688, 1963.
39. Farid, Z., et al.: Chronic pulmonary schistosomiasis. Am. Rev. Tuberc., *79*:119, 1959.
40. Medina, J. T., et al.: Schistosomiasis mansoni involving the colon. Radiology, *85*:682, 1965.
41. Tarabulcy, E. Z.: Radiographic aspect of urogenital schistosomiasis (bilharziasis). J. Urol., *90*:470, 1963.

42. James, W. B.: Urological manifestations in Schistosoma haematobium infestation. Br. J. Radiol., *36*:40, 1963.
43. Al-Ghorab, M. M.: Radiologic manifestations of genito-urinary bilharziasis. Clin. Radiol., *19*:100, 1968.
44. Hidayat, M. A., and Wahid, H. A.: A study of the vascular changes in bilharzic hepatic fibrosis and their significance. Surg. Gynecol. Obstet., *132*:997, 1971.
45. Viana, R. L., and Martins, J.: Hepatic arteriography, the hemodynamic factor in the study of hepatosplenic bilharzia fibrosis. S. Afr. Med. J., *46*:96, 1972.
46. Phillips, J. F., Cockrill, H., et al.: Radiographic evaluation of patients with schistosomiasis. Radiology, *114*:31, 1975.
47. Pitzorno, P., and Perassi, F.: [On a case of hydatidosis of the heart.] Radiol. Med., *47*:438, 1961.
48. Pitzorno, P.: Presumptive and definite radiologic findings in hydatid disease of the liver. Stud. Sassar., *41*:1, 1963.
49. Teplick, J. G., et al.: Echinococcus of the kidney. J. Urol., *78*:323, 1957.
50. Latham, W. J.: Hydatid disease. Part I and II. J. Fac. Radiol., *5*:65, 1953; *5*:83, 1953.
51. Delahaye, R. P., Laurent, H., and Massoubre, A.: Radiologic aspects of osseous hydatidosis. J. Radiol. Electrol. Med. Nucl., *48*:269, 1967.
52. Thompson, W. M., Chisholm, D. P., and Tank, R.: Alveolar hydatid disease. Am. J. Roentgenol. Radium Ther. Nucl. Med., *116*:345, 1972.
53. Garti, I., and Deutsch, V.: The angiographic diagnosis of echinococcus of the liver and spleen. Clin. Radiol., *22*:465, 1971.
54. Rizk, G. K., Tayyarah, K. A., and Ghandur-Mnaymneh, L.: Angiographic changes in hydatid cysts of liver and spleen. Radiology, *99*:303, 1971.
55. Balikian, J. B., and Mudarris, F. F.: Hydatid disease of the lungs. Am. J. Roentgenol. Radium Ther. Nucl. Med., *122*:692, 1975.
56. Keats, T. E.: Cysticercosis: roentgen manifestations. Missouri Med., *58*:457, 1961.
57. Santin, G., and Vargas, J.: Roentgen studies of cysticercosis of central nervous system. Radiology, *86*:520, 1966.
58. Reeder, M. M.: Tropical diseases of the soft tissues. Semin. Roentgenol., *8*:47, 1973.
59. Reeder, M. M., and Hamilton, L. C.: Tropical diseases of the colon. Semin. Roentgenol., *3*:62, 1968.
60. Boon, W. H., and Hoh, T. K.: Severe whipworm infestation in children. Singapore Med. J., *2*:34, 1961.
61. Fisher, R. M., and Cremin, B. J.: Rectal bleeding due to Trichuris trichiura. Br. J. Radiol., *43*:214, 1970.
62. Pereira, G. A., et al.: Intestinal strongyloidiasis. Rev. Brasil Radiol., *3*:127, 1960.
63. Berkman, Y. M., and Rabinowitz, J.: Gastrointestinal manifestations of strongyloidiasis. Am. J. Roentgenol. Radium Ther. Nucl. Med., *115*:306, 1972.
64. Louisy, C. L., and Barton, C. J.: The radiological diagnosis of Strongyloides stercoralis enteritis. Radiology, *98*:535, 1971.
65. Wong, W. T.: Intestinal obstruction in children due to Ascaris simulating intussusception. Br. J. Surg., *49*:300, 1961.
66. Osborne, D. P., et al.: Solitary pulmonary nodule due to Ascaris lumbricoides. Dis. Chest, *40*:308, 1961.
67. Phills, J. A., Harrold, A. J., et al.: Pulmonary infiltrates, asthma and eosinophilia due to Ascaris suum infestation in man. N. Engl. J. Med., *286*:965, 1972.
68. Williams, I.: Calcification in loiasis. J. Fac. Radiol., *6*:142, 1954.
69. Akisada, M.: Filarial chyluria in Japan. Radiology, *90*:311, 1968.
70. Kanetkar, A. V., et al.: Lymphangiographic patterns of filarial edema. Clin. Radiol., *17*:258, 1966.
71. Rajaram, P. C.: Lymphatic dynamics in filarial chyluria and prechyluric state—lymphographic analysis of 52 cases. Lymphology, *3*:114, 1970.
72. Beskin, C. A., Colvin, S. H., Jr., and Beaver, P. C.: Pulmonary dirofilariasis. J.A.M.A., *198*:665, 1966.
73. Feld, H.: Dirofilaria immitis (dog heartworm) as a cause of a pulmonary lesion in man. Radiology, *108*:311, 1973.
74. Ghigo, M., and Magrini, M.: Roentgen findings in calcified dracunculosis. Radiol. Med., *45*:953, 1959.
75. Chuttani, H. K., Puri, S. K., and Misra, R. C.: Small intestine in hookworm disease. Gastroenterology, *53*:381, 1967.
76. Kalman, E. H.: Creeping eruption with transient pulmonary infiltration. Radiology, *62*:222, 1954.

8

DISORDERS OF THE NERVOUS SYSTEM

NEW TECHNOLOGIES IN NEURORADIOLOGY

Computerized Tomography

Computerized tomography (CT) is a major technologic breakthrough and has proved to be a superb noninvasive tool for neuroradiology. Stated simply, detectors (not x-ray film) pick up the radiation emanating from two "slices" of tissue per revolution, each "slice" being up to 1.3 cm. thick. A total of 28,000 or more readings are obtained for each slice. These readings are carried to the computer programmed to solve a huge number of simultaneous equations.

The underlying principle is the slight difference in radiation absorption coefficient between tissues of various densities, thus permitting differentiation of cerebrospinal fluid, gray matter, and white matter, as well as abnormal densities. The computer provides a digital printout of calculated density values for each "slice." This is also transformed to a density image to be viewed on a brightness modulated cathode ray tube (CRT). Operator controls allow the viewer to selectively display various density levels and alter contrast to enhance visualization. Photographs are made with a Polaroid camera for permanent record.

Distinguishable densities are obtained for bone, heavy metal, cerebrospinal fluid, blood, gray matter, white matter, and air. The density of many tumors and areas of hemorrhage is often greatly increased following intravenous injection of water-soluble contrast media.

In a high percentage of cases, the CT will provide exact location and size of tumors, hemorrhage, infarction, and larger malformations. Hydrocephalus, cerebral atrophy, porencephaly, and ventricular displacement are readily demonstrated. Calcifications, even when not apparent on conventional skull films, are strikingly visible on a scan.

Serial scans permit evaluation of change in size of tumors, hemorrhage or abscess, effects of surgery or irradiation, and tumor recurrence. Localization of a shunt and post-shunt effects on ventricular size are readily obtained. The progress of atrophy and some degenerative conditions can be followed.

The CT scan appears to be the simplest and most accurate screening procedure for suspected intracranial pathologic conditions.

The CT scan findings will usually obviate the need for air studies and will often decrease the necessity for diagnostic cerebral arteriographic studies. However, the vascular studies will delineate the vascular supply of a lesion whose extirpation is contemplated; this information is not found on the scan. It can be confidently expected that further technical advances will increase the diagnostic information on the CT scan.

The anatomic details of a normal scan are illustrated below. The findings in various intracranial conditions are discussed and illustrated under specific headings.[1-7]

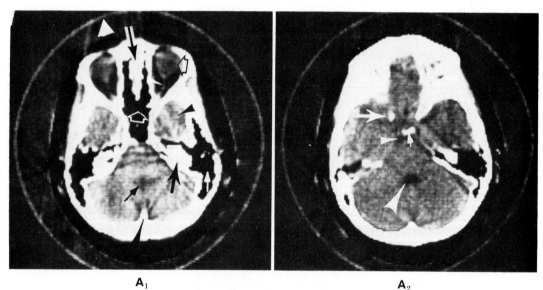

A_1 A_2

Figure 8–1 **Normal CT Scans: Anatomic Landmarks.**
Note: on a CT scan, the right side of the patient is the right side of the scan, as viewed.

A_1, Section at base of brain, through orbits. Orbital globe *(open black arrow)*; optic nerve *(small white arrowhead)*; bony nasal septum *(black-white arrow)*; air in posterior nasal cavity and nasopharynx *(open white arrow)*; air in mastoid cells *(white arrow)*; middle fossa *(small black arrowhead)*; petrous bone *(large black arrow)*; fourth ventricle *(small black arrow)*; internal occipital protuberance *(large black arrowhead)*.

A_2, Basal section, 1.3 cm. more rostral. Dorsum sella *(small white arrow)*; basal cisterns *(small white arrowhead)*; anterior clinoid *(large white arrow)*; fourth ventricle *(large white arrowhead)*.

Illustration continued on the opposite page.

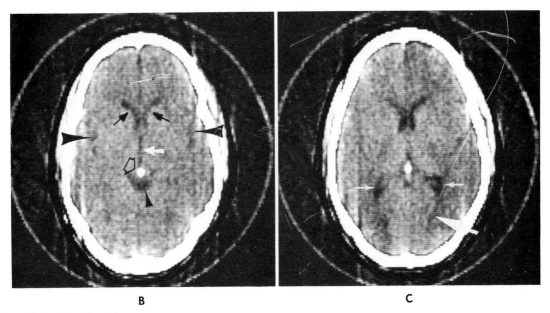

B   C

Figure 8–1   (Continued)

B, Section through anterior horns of lateral ventricles. Anterior horns *(small black arrows)*; third ventricle *(white arrow)*; pineal *(open black arrow)*; cisterna of corpora quadrigemini *(small arrowhead)*; sylvian cisterns *(large black arrowheads)*.

C, Section through anterior and posterior lateral ventricles. Posterior portion of ventricular bodies *(small arrows)*; posterior horn *(large arrow)*.

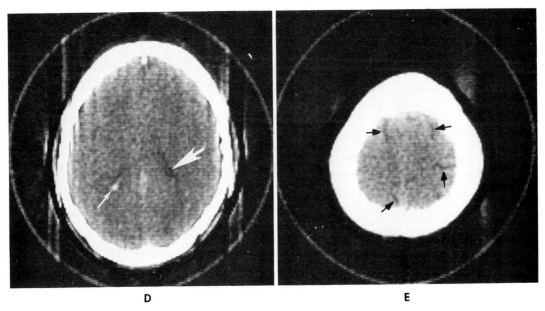

D   E

Figure 8–1   (Continued)

D, A more rostral section through posterior ventricular bodies. Left choroid plexus calcification *(small white arrow)*; right posterior portion of lateral ventricle *(large white arrow)*.

The pineal and choroid plexuses usually contain sufficient calcification to be clearly evident on a CT scan even though not apparent on regular skull films.

E, Section near top of brain. Normal sulci *(black arrows)*.

EXTRAPYRAMIDAL DISORDERS

Parkinson's Disease

No consistent or specific neuroradiologic changes are reported. However, if the symptoms of Parkinson's disease are associated with dementia, a communicating normal pressure hydrocephalus may be present, and all symptoms may be greatly alleviated by a shunt.

Chronic constipation and loss of intestinal tone are common complications of Parkinson's disease. In some patients, abdominal distention and obstipation develop as a result of massive dilatation of the colon. Abdominal films of these patients will show a marked gaseous distention of the entire colon; this picture often simulates a low mechanical obstruction, but barium enema will disclose a redundant, atonic, dilated, but nonobstructed colon. Not infrequently a volvulus of the redundant sigmoid occurs and produces a true mechanical obstruction. In these cases the barium enema will demonstrate the obstruction by revealing the beak-shaped tapering of the lumen that is characteristic of volvulus.

An unexplained megacolon in an adult may be clarified if a history of parkinsonism is obtained.

No specific changes on the CT scan have been reported. However, the scan will often show nonspecific brain atrophy, particularly when the disease develops in older individuals.[8, 9]

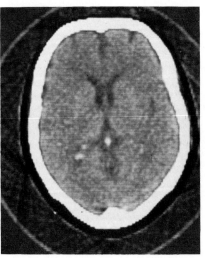   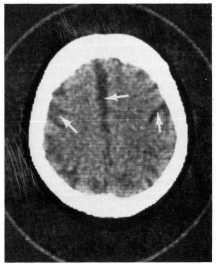

A   B

Figure 8–2 **Parkinson's Disease: CT Scan.** In this 75 year old man with a three year history of Parkinson's disease, the scan reveals a completely normal ventricular system *(A),* but marked enlargement of the sulci and commissure *(arrows, B)* is indicative of cortical atrophy.

Findings of atrophy are common in older patients with Parkinson's disease, but such findings are entirely nonspecific.

CEREBROVASCULAR DISEASES

Cerebral Ischemia and Infarction

Cerebrovascular occlusion is most often due to cerebral atherosclerosis and thrombosis. Emboli are much less common. The middle cerebral artery and the internal carotid artery are the commonest sites. Stenotic occlusion without thrombosis is quite common. More than one artery is often found to be occluded.

Definitive radiographic diagnosis is made by cerebral angiography. The study will show an abrupt cutoff of the occluded vessel; late collateral filling of the ischemic area is a confirmatory finding. Frequently there is a prolonged arterial and arteriovenous phase, and stasis in the opacified collateral vessels. Diagnostic error or uncertainty may result from local arterial spasm or from anatomic variations especially of the anterior cerebral arteries, both of which often arise from the same side. Conclusive evidence of occlusion of the smaller vessels is frequently difficult or impossible, and persistent avascular brain areas may be the only radiologic clue.

Embolic occlusion is relatively uncommon. It is generally associated with the mural thrombi of heart disease but can also result from a detached peripheral thrombus passing through a patent foramen ovale. Angiographic distinction of thrombosis from embolism is usually not possible.

On the CT scan, brain infarcts can often be identified as focal areas of diminished scan density. The earliest finding may be a mottling of the area as a result of edema; this gradually becomes a more homogeneous low density lesion. The nonhomogeneous early appearance corresponding to the distribution of a larger vessel may help to differentiate infarct from other mass lesions. CT scans are positive in up to 80 per cent of cases of cerebral ischemia and infarction. Absence of ventricular shift may help distinguish a large infarct from a low density tumor. The infarcted area does not usually increase in density following injection of contrast material.[10-14]

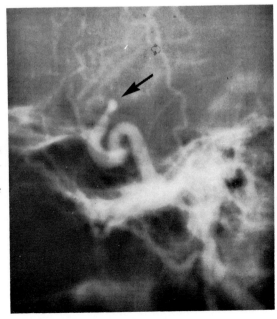

Figure 8–3 **Occlusion of Internal Carotid Artery: Arteriogram.** The opacified carotid siphon is entirely occluded just beyond its cavernous portion (*arrow*). This is the most frequent intracranial site of internal carotid occlusion. The carotid tapers proximal to the occlusion. The smaller branches near the occluded vessel arise in the external carotid system.

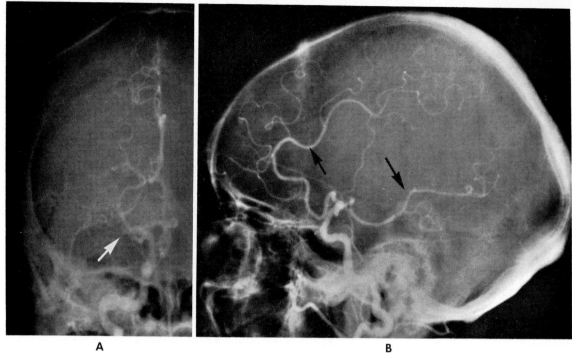

A

B

Figure 8–4 **Occlusion of Middle Cerebral Artery: Arteriogram.**

A, In anteroposterior view the right middle cerebral artery is occluded just distal to the bifurcation *(arrow)* of the internal carotid artery.

B, In lateral view the anterior and posterior cerebral arteries *(arrows)* are clearly demonstrated; the middle cerebral artery is absent. Later films demonstrate extensive collateral vessels arising in the anterior and posterior cerebral arteries.

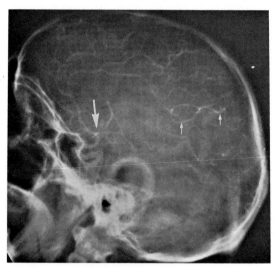

Figure 8–5 **Occlusion of Middle Cerebral Artery with Filling of Anastomoses: Arteriogram.** Lateral view of the late arterial phase of a cerebral arteriogram indicates occlusion of the proximal segment of the middle cerebral artery *(large arrow)*. There is late opacification of peripheral branches of the middle cerebral artery *(small arrows)*; these branches are filled by collaterals from the anterior cerebral artery. Collateral filling is confirmatory of proximal occlusion of a major cerebral artery.

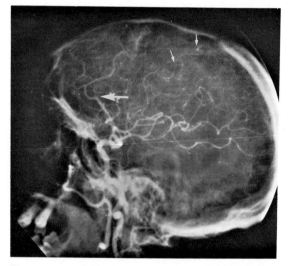

Figure 8–6 **Occlusion of Pericallosal Branch of Anterior Cerebral Artery: Arteriogram.** Lateral projection of late arterial phase of cerebral arteriogram discloses that there is nonfilling of the occluded pericallosal artery *(large arrow)* at the bifurcation of the anterior cerebral artery. Small distal branches of the occluded pericallosal artery *(small arrows)* are opacified via collaterals from the middle cerebral artery. Collateral filling is confirmatory evidence of vascular occlusion; when a larger artery is not opacified on a unilateral arteriogram because of anatomic variation, collateral filling does not occur.

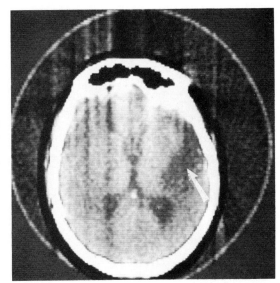

Figure 8–7 **Brain Infarct: CT Scan.** There is a large irregular area of decreased density in the right temporal area *(arrow)* without shift of midline structures.

Injection of contrast material does not significantly increase the density of infarcted areas.

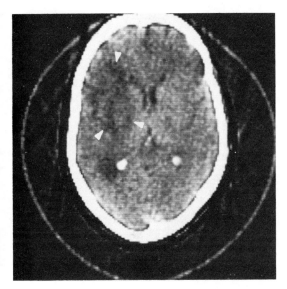

Figure 8–8 **Temporal Lobe Infarct: CT Scan.** An irregular, poorly marginated, mottled area of diminished density *(arrowheads)* in the left temporal lobe is not associated with any ventricular displacement. The finding is characteristic of an infarct. There was a history of a prior cerebrovascular accident.

Cerebral Transient Ischemia Attacks (TIA)

These attacks may demonstrate a variety of neurologic symptoms and are due to extracranial arterial disease. Internal carotid atherosclerotic plaques and narrowing are the most common causes; similar involvement of the vertebral and/or basilar artery is also a major causative factor. Occasionally, an atherosclerotic plaque may become ulcerated and detached, leading to a more permanent neurologic deficit.

Radiographic investigation of TIA requires opacification of the aortic arch and its branches in the mediastinum and neck. Atherosclerotic plaques and/or narrowing of the internal carotid, vertebral, or basilar artery are readily demonstrated. Carotid involvement is seen most often at the bifurcation of the common carotid. Vertebral artery compromise usually occurs close to its origin from the arch. A correlation of the signs and symptoms that are evident during an attack, together with the radiographic arterial abnormalities, will generally indicate whether the radiographic findings are clinically significant.

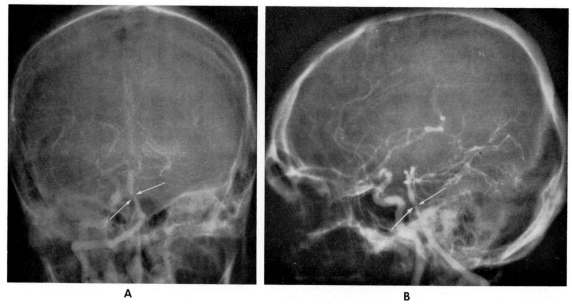

A   B

Figure 8–9 **Arteriosclerotic Stenosis of Basilar Artery: Brachial Arteriogram.**

A, Anteroposterior view demonstrates filling of the basilar artery, which is considerably narrowed in its midportion (*arrows*), and shows slight poststenotic dilatation. Despite the basilar stenosis, the posterior cerebral arteries are well filled.

B, Lateral view again indicates localized stenosis and arteriosclerotic elongation of the basilar artery (*arrows*). Severe stenosis, as well as occlusion, can cause symptoms of cerebral ischemia.

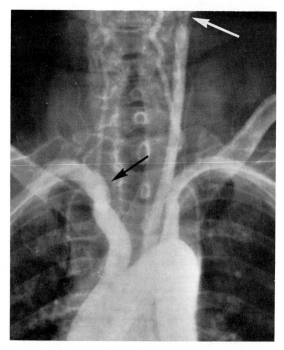

Figure 8–10 **Occlusion of Common Carotid Artery: Brachiocephalic Arteriogram.**   The right common carotid artery (*black arrow*) remains unfilled due to thrombotic occlusion just above its origin from the innominate artery. The left internal carotid artery at the bifurcation (*white arrow*) is also narrowed.

The most common extracranial site of carotid occlusion is the bifurcation high in the neck. Extracranial stenosis or occlusion of the carotid artery can cause stroke or attacks of transient ischemia.

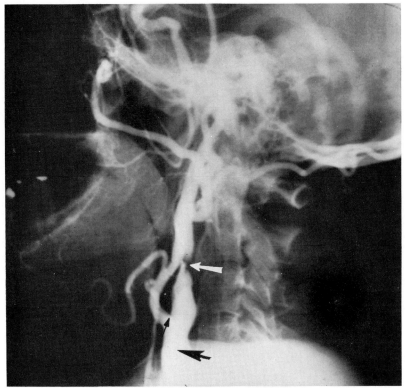

Figure 8–11 **Transient Ischemic Attacks: Internal Carotid Artery Disease.** Selective opacification of the common carotid artery *(large black arrow)* via a catheter in the aortic arch discloses an area of localized stenosis and irregularity *(white arrow)* in the internal carotid artery a short distance above the bifurcation *(small black arrow)* of the common carotid. This was an irregular atherosclerotic plaque.

Irregular atherosclerotic plaques in the internal carotid artery, usually close to the bifurcation, are frequent causes of recurrent attacks of cerebral ischemia.

Subclavian Steal Syndrome

The syndrome results from occlusion of the left subclavian artery proximal to the origin of the left vertebral artery.

Arteriography of the aortic arch will demonstrate opacification of the right common carotid, right vertebral, and left common carotid arteries. The left subclavian and left vertebral arteries are not filled during injection. A short time later, however, the contrast material from the right vertebral and basilar arteries fills the left vertebral artery from above downward, producing a late opacification of this vertebral artery and of the subclavian artery segment distal to the obstruction. This hemodynamic change can lead to a relative basilar artery insufficiency.

Although a similar situation can occur on the right side, it is extremely rare because of the anatomic arrangement of the right common carotid and vertebral arteries.[15-17]

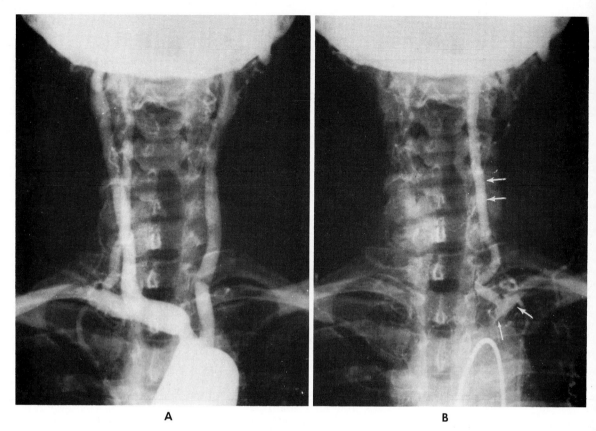

A   B

Figure 8–12 **Subclavian Steal Syndrome: Arch Aortogram.**

A, The arch study opacifies all the brachiocephalic branches except the left subclavian and its left vertebral artery branch.

B, Later, after contrast material has left the aorta and brachiocephalic branches, the left vertebral artery *(two arrows)* fills from above into the left subclavian *(large arrow).* The left subclavian artery is occluded and tapered at its origin *(small arrow).*

Intracranial Hemorrhage

Intracranial hemorrhage occurs considerably less frequently than cerebral thrombosis. Arteriosclerosis and hypertension are the principal causes of bleeding, but aneurysm, arteriovenous malformation, or tumor is sometimes the etiologic factor; their radiographic features are discussed elsewhere.

Although diagnosis and localization is generally a clinical problem, cerebral angiography is often necessary for uncovering the cause and location of the hemorrhage.

Intracranial or intraventricular hemorrhage is easily recognized on the CT scan as an area of increased density. The margins are irregular and indistinct, and a zone of decreased density, probably due to edema, surrounds the hemorrhage.

On the angiogram, the bleeding vessel may be spastic or nonfilled. There is often prolonged arterial opacification if the brain has become edematous. The intracerebral hematoma, if sizable, will produce displacement of adjacent vessels in a manner similar to that of any intracerebral mass lesion, but without tumor stain or abnormal vessels. Pneumoencephalogram may show ventricular displacement by a nonspecific mass lesion.[5, 6, 18–20]

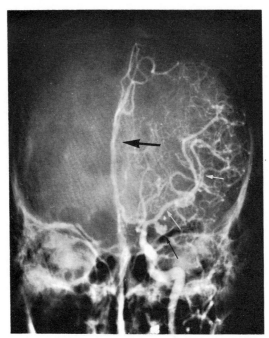

Figure 8–13 **Hematoma of Temporal Lobe: Arteriogram.**   Anteroposterior projection indicates displacement of the anterior cerebral artery *(large black arrow)* to the right and elevation of the proximal segments of the middle cerebral artery *(white arrows)* caused by a temporal lobe hematoma. An aneurysm in the first portion of the middle cerebral artery *(small black arrow)* was the source of bleeding. Temporal lobe tumor produces identical vascular displacement.

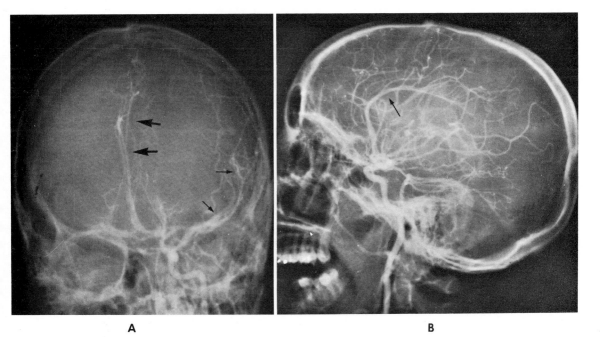

A   B

Figure 8–14 **Hematoma of Thalamus: Arteriogram.**

A, In anteroposterior projection there is a shift of the anterior cerebral artery *(large arrows)* to the right, and lateral displacement of the middle cerebral artery and its branches *(small arrows).* A deep hemispheric mass was present and proved to be a thalamic hematoma.

B, Lateral view indicates loss of normal undulations of the anterior cerebral artery *(arrows),* probably from dilatation of the ventricular system.

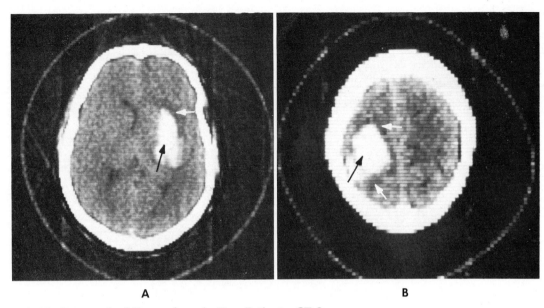

A   B

Figure 8–15   **Intracerebral Hemorrhage in Two Patients: CT Scan.**

A, A circumscribed area of increased density *(black arrow)* represents clotted blood deep in the right temporal lobe. A zone of edema around the blood is seen as decreased density *(white arrow)*. The ventricles are slightly shifted to the left.

B, A similar finding in another patient with hemorrhage into the left parietal lobe.

Cerebral Aneurysms

The commonest intracranial aneurysm is the saccular or berry type, which is usually of congenital origin. Multiple berry aneurysms are not rare. The much less frequent fusiform aneurysm may be of luetic, arteriosclerotic, traumatic, or mycotic origin. Calcification occasionally occurs in the wall of an arteriosclerotic aneurysm, but skull films are generally unrevealing.

The location, size, and shape of an aneurysm is readily disclosed by cerebral angiography. Bilateral studies are indicated if bleeding occurs without localizing signs. Berry aneurysms are usually small and round, and often a small stalk is present. These aneurysms can attain a diameter of up to 3 cm. but rarely exceed 1 cm. They occur most frequently at vascular bifurcations; the anterior and posterior communicating arteries and the origin of the ophthalmic artery are favorite sites. In children, the most common site of congenital aneurysm is the bifurcation of the internal carotid artery.

Active bleeding is suggested by arterial spasm in the vicinity of an opacified aneurysm and by irregularities of the aneurysmal contour. Bleeding most often extends into the subarachnoid space, but intracerebral hematomas can also occur, evidenced by vascular displacement.

Larger aneurysms may occasionally be visualized as a circumscribed area of increased density on the CT scan. However, a small aneurysm will be undetected on the scan.[5, 18, 21-23]

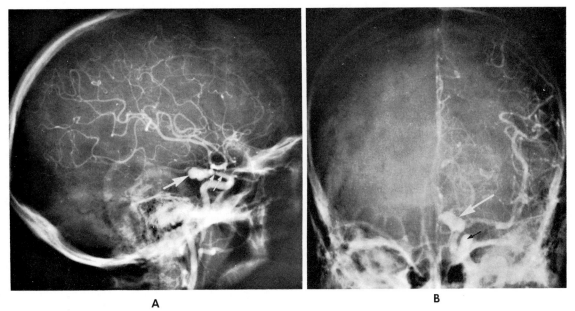

| A | B |

Figure 8–16 **Saccular (Berry) Aneurysm of Posterior Communicating Artery: Carotid Arteriogram.**

A, Lateral view during opacified arterial phase demonstrates the contrast-filled aneurysm *(large arrow)* in the posterior communicating artery—one of the most frequent sites of aneurysm. The vessel proximal to the aneurysm posteriorly is the posterior cerebral artery. Spasm and narrowing of the carotid artery *(small arrows)* adjacent to the aneurysm indicate active bleeding, as does the irregular contour of the aneurysm.

B, The location and size of the aneurysm *(white arrow)* are clearly shown in anteroposterior view. In certain projections a sharply angulated vessel *(black arrow)* can simulate an aneurysm; anteroposterior, lateral, and, occasionally, oblique projections help prevent such errors, especially in respect to smaller sharply curved vessels.

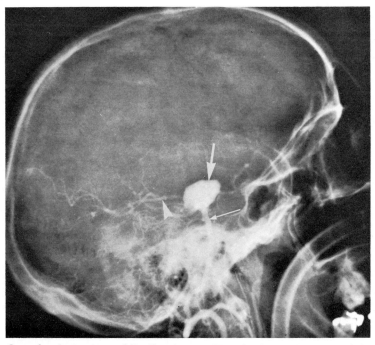

Figure 8–17 **Large Saccular Aneurysm of Basilar Artery: Vertebral Arteriogram.** A large aneurysm *(large arrow)* is present at the junction of the basilar artery *(small arrow)* and the posterior cerebral artery *(arrowhead).*

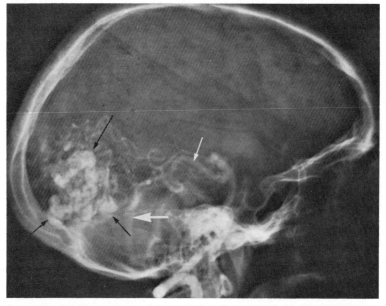

Arteriovenous Malformation of the Brain

Arteriovenous angioma is the most common form of the several types of arteriovenous malformations. It consists of a coiled mass of arteries and veins and is generally cortical or subcortical in location. It is usually supplied by a major artery and is thus well visualized during angiography.

Occasionally, calcification of the angiomatous vessels can be seen on plain films. Superficial lesions may cause thinning of the overlying bone and large irregular vascular grooves in the bone. In most cases, however, the skull findings are negative, and diagnosis can be made only by cerebral angiography.

There is rapid filling of a dilated main feeding artery and of the convoluted vascular collection. The draining veins are filled almost simultaneously, a striking and characteristic finding. Occasionally spontaneous regression or thrombosis of a malformation occurs.

On the CT scan, larger malformations may appear as circinate areas of increased density, but often only after intravenous injection of contrast material. Ordinarily, however, the smaller malformations will not be apparent on the scan, unless there is some calcification within the vessels of the lesion.

Cavernous angiomas are small masses of blood-filled spaces without a separate arterial or venous supply; often opacification does not occur during angiography. Capillary angiomas are rare but are a constant feature of Sturge-Weber disease. (See p. 394.)

In so-called Japanese cerebrovascular disease, angiomatous formations are associated with narrowing of the intracranial portion of the internal carotid artery.[5, 6, 18, 24-26]

Figure 8–18 **Arteriovenous Malformation of Occipital Lobe: Vertebral Angiogram.** The posterior cerebral artery *(small white arrow)* and its branches are filled, and there is a collection of tortuous dilated vessels *(black arrows)* in the region of the occipital lobe. This is the typical angiographic picture of arteriovenous malformation. There is early venous drainage *(large white arrow)*.

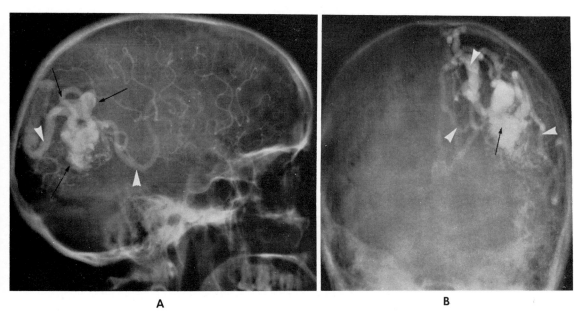

Figure 8–19 **Arteriovenous Malformation of Parieto-Occipital Area: Carotid Angiogram.**

A, Lateral view in late arterial phase demonstrates the tortuous vessels of the malformation *(arrows),* with multiple draining veins *(arrowheads).* This venous filling in the arterial phase is due to arteriovenous shunting.

B, Anteroposterior view demonstrates the large number of draining venous channels *(arrowheads)* and tortuous vessels of the malformation *(arrow).*

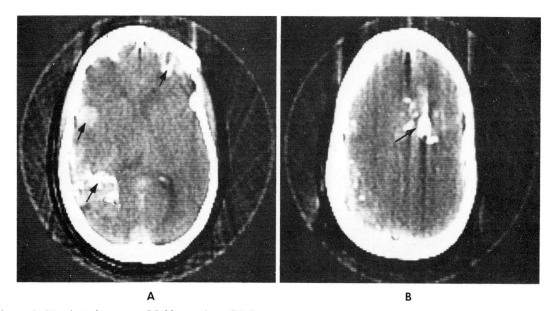

Figure 8–20 **Arteriovenous Malformation: CT Scan.**

A, Scan made after intravenous administration of contrast material shows scattered high density ovoid and serpiginous collections *(black arrows)* on both sides.

B, At a more cephalad level, another collection of angiomatous vessels *(black arrow)* is seen, probably in the basal ganglia.

Before injection of contrast material, these arteriovenous malformations were considerably less dense.

Pseudobulbar Palsy

A fairly constant roentgenologic finding in pseudobulbar palsy is the prolonged filling of the valleculae and pyriform sinuses when barium swallow is attempted. This is due to weakness of the hypopharyngeal and pharyngeal muscles that initiate deglutition. Although this is usually demonstrable with fluoroscopy and spot roentgenologic films, it is perhaps better evaluated with cineradiographic or videotape studies of the esophagus.[18]

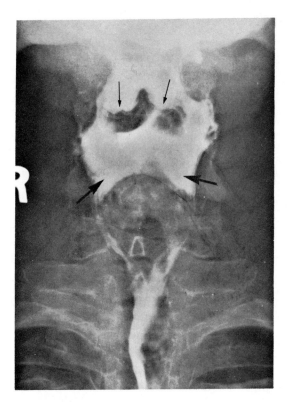

Figure 8–21 **Pseudobulbar Palsy: Deglutition.** In a film of the neck made several minutes after barium swallow, contrast material is shown filling the pyriform sinuses *(large arrows)* and the valleculae *(small arrows)*. Normally, these structures are rapidly emptied following barium swallow, but in pseudobulbar palsy and in other conditions where there is weakness of the muscles of deglutition, the pyriform sinuses and valleculae empty very slowly.

Cerebral Venous Lesions

Thrombosis of the cerebral veins or dural sinuses cannot always be conclusively demonstrated by routine cerebral angiography, since opacification of the sinuses is inconstant in these studies. Only the deep cerebral veins are constantly opacified.

Dural sinus thrombosis is usually due to adjacent inflammatory disease or to a neoplasm but can also occur after trauma. Oral contraceptive medication has been implicated in some cases. Venous sinusography, in which a catheter is placed in the anterior portion of the superior sagittal sinus, opacifies the superior sagittal and transverse sinuses and can demonstrate occlusion of these channels.

Thrombosis of the dural sinuses can cause increased intracranial pressure and papilledema and is one of the causes of "pseudotumor," in which the angiographic or pneumographic studies appear normal in spite of increased intracranial pressure. (See *Benign Intracranial Hypertension*, p. 354.)[13, 27-29]

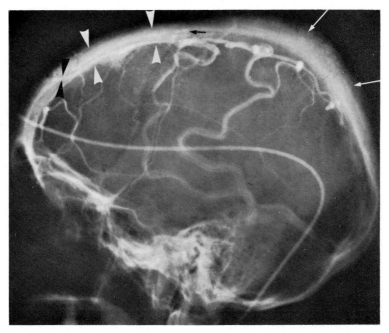

Figure 8–22 **Occlusion of Superior Sagittal Sinus by Meningioma: Sinusography.** A catheter is inserted into the anterior portion of the superior sagittal sinus *(black arrowheads)*, and the injected contrast material fills the anterior half of this sinus *(white arrowheads)*. Abrupt blockage of the sinus *(black arrow)* is attributable to a meningioma that has caused severe cortical reaction and thickening *(white arrows)* in the posterior parietal area. The area is drained by collateral superficial veins, which are normally not opacified during sinusography.

Subdural Hematoma

Although trauma is the usual cause, in many cases skull fracture may not occur. Bleeding arises from ruptured veins between the dura and the leptomeninges. The parietal and temporal regions are the commonest sites; occipital hematomas can also occur. Although most often unilateral, bilateral subdural hematomas are not rare.

Plain skull films may demonstrate a lateral shift of the calcified pineal, but this may be absent if there are bilateral hematomas.

On the CT scan, identification of a subdural hematoma depends on the age and stage of the lesion. In early lesions the increased density of clotted blood is seen along the convexity of the brain. With absorption of fluid into the clot, the density may become identical to that of brain tissue, and a shift of the ventricular system may be the only clue to the presence of a hematoma. Later lesions become less dense and can be readily identified by their location and contour. Serial scans are useful for assessing the progress of a subdural hematoma or disclosing recurrence after surgery.

Acute subdural hematoma is also readily diagnosed by cerebral anteriography. The frontal or anteroposterior projection provides the most information. As a rule, one observes a characteristic picture: The smaller peripheral arterial branches are displaced from the inner table at the site of the hematoma. Associated with this is contralateral displacement of the opacified anterior cerebral artery. This shift may be absent if the hematoma is small, but absence of shift is often due to another hematoma on the other side, and contralateral arteriography is then indicated. (See Fig. 8–111.)

CT scans as well as angiography may also disclose evidence of an associated intracerebral hematoma, which the neurosurgeon can evacuate during surgery.

In acute or subacute subdural hematoma, the defect between the skull and the displaced peripheral arteries is crescent-shaped; in chronic lesions it usually becomes more fusiform. (See Figs. 8–112 and 8–113.) However, exceptions to this time factor versus hematoma shape do occur.

Pneumoencephalography is less informative, usually demonstrating compression and displacement of the homolateral ventricle, which is indicative of a nonspecific intracranial mass lesion. Subarachnoid filling on the affected side may be poor or absent.[5, 6, 30-34]

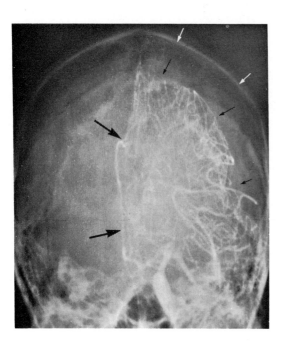

Figure 8–23 **Large Left Parietal Subdural Hematoma: Arteriogram.** Frontal arteriogram demonstrates marked displacement of the peripheral vessels *(small black arrows)* from the inner tables *(white arrows)*. The crescentic space between the tables and vessels is filled by a large subdural hematoma. There is considerable shift of the anterior cerebral artery to the right *(large black arrows)*.

Evidence of a subdural hematoma of moderate size, without a shift of the anterior cerebral vessel, should suggest another hematoma on the opposite side.

Figure 8–24 **Right Subdural Hematoma: Pneumonencephalogram.** The right lateral ventricle *(black arrow)* is depressed and shifted to the left. The interventricular septum is also shifted *(arrowhead)*. There is no air in the subarachnoid space on the right.

The roof of the right lateral ventricle is depressed and flattened as a result of being compressed under the falx cerebri.

The skull has been fractured in the right temporoparietal area *(white arrow)*.

This pneumoencephalographic picture is nonspecific for any mass lesion: arteriography or CT scan can make a more definitive diagnosis of subdural hematoma.

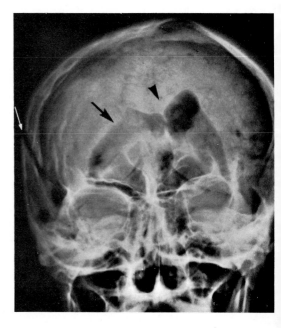

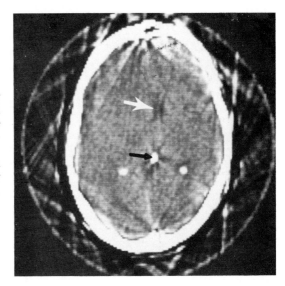

Figure 8–25 **Subdural Hematoma without Density Change: CT Scan.** The ventricles *(white arrow)* and pineal *(black arrow)* are considerably shifted to the right, but no area of abnormal density is clearly apparent on the left.

Not infrequently, the subdural hematoma has a scan density identical to that of brain tissue and may not "light up" after administration of contrast material. Ventricular shift will be the only clue to its presence.

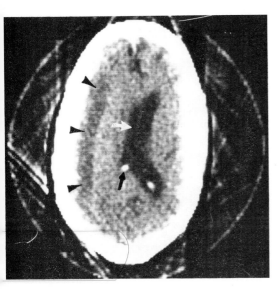

Figure 8–26 **Subdural Hematoma: CT Scan.** The long, smoothly marginated area of decreased density *(arrowheads)* paralleling the left skull border represents a subdural hematoma. The left ventricle is compressed *(white arrow)* and shifted to the right. The calcified left choroid *(black arrow)* has been displaced forward and to the right.

Figure 8–27 **Bilateral Frontal Subdural Hematomas in a Child: CT Scan.** Bilateral symmetric bands of decreased density *(arrows)* are seen adjacent to the frontal bones on both sides. These were subdural hematomas. The lateral and third ventricles were mildly dilated.

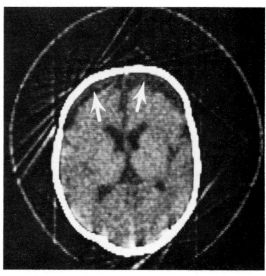

INFECTIONS AND INFLAMMATORY DISEASES OF THE CENTRAL NERVOUS SYSTEM AND ITS COVERINGS

Meningitis

In acute bacterial and tuberculous meningitis, there are no significant findings on plain skull films.

Narrowing and occlusion of arteries at the base and periphery of the brain are often seen angiographically in both purulent and tuberculous meningitis. Abnormalities in circulatory dynamics, including prolonged arteriovenous circulation, premature venous filling, and hypervascular patterns, are frequently encountered.

In chronic meningitis, especially of tuberculous origin, adhesions and thickening of the arachnoid at the base of the brain may occlude the foramina of Luschka and Magendie or even the cisterna magna, leading to obstructive hydrocephalus, which includes dilatation of the fourth ventricle. Occlusion of the subarachnoid pathways around the brain stem can also produce ventricular dilatation (communicating hydrocephalus).

Angiographic findings of increased sweep of the anterior cerebral artery (dilated ventricles) plus narrowed or occluded arteries at the base of the brain and in focal peripheral areas are highly suggestive of chronic tuberculous meningitis.[35, 36]

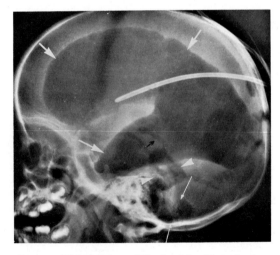

Figure 8–28 **Hydrocephalus Due to Tuberculous Meningitis: Ventriculogram.** Lateral projection in a child demonstrates greatly dilated lateral ventricles *(large white arrows)* and a moderately dilated fourth ventricle *(arrowheads)*. The aqueduct is very faint *(black arrow)*. The small amount of air in the cisterna magna *(thin white arrows)* indicates that obstruction is not complete. Tuberculous meningitis is a fairly frequent cause of obstruction of the foramina of Luschka and Magendie.

Intracranial Abscess

Focal suppuration of the brain substance can occur by direct extension from adjacent structures. Thus, infections of the ear and petrous pyramids can lead to cerebellar or temporal lobe abscess, and suppurative frontal sinusitis may cause frontal lobe abscess. Additionally, suppurative thrombophlebitis extending into the brain, penetrating wounds, and hematogenous septic emboli are possible causes of intracranial abscess. A septic embolus is most commonly a consequence of pulmonary suppuration and may cause single or multiple scattered brain abscesses.

Plain radiographs may be helpful in disclosing evidence of mastoiditis, sinusitis, or osteomyelitis. Rarely, gas-forming organisms produce air-fluid levels in the brain abscess. Calcification of a chronic abscess is an infrequent occurrence.

The CT scan will disclose an area of decreased density, often mottled and irregular. Distinction from a tumor is difficult and depends mostly on the clinical picture. There is generally no significant density increase following intravenous injection of contrast material.

Changes in the abscess due to antibiotic therapy can be readily followed by serial scans.

Arteriography and pneumoencephalography demonstrate a space-occupying lesion that usually is indistinguishable from a neoplasm or other mass. Vascular blush or stain within the lesion is a frequent finding in neoplastic lesions but rarely occurs in an abscess. However, a hypervascular capsular blush is seen in about 50 per cent of brain abscesses.[5, 6, 18, 36-38]

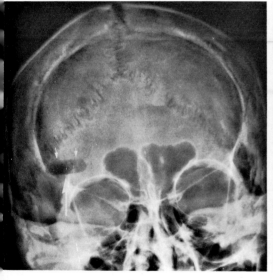 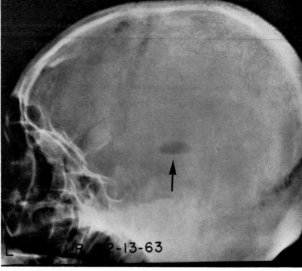

A   B

Figure 8–29 **Temporal Lobe Abscess Secondary to Mastoiditis: Air-Fluid Levels.** In posteroanterior *(A)* and lateral *(B)* erect views, there is a small accumulation of air *(small white arrows)* in the region of the right temporal lobe, with a fluid level *(large arrows)*. Note the diastasis of the sagittal suture from increased intracranial pressure.

Free air is a rare finding in brain abscess. In plain skull films it is more likely to result from a penetrating brain wound in which air is introduced from outside. (Courtesy Dr. J. Zizmor, New York, New York.)

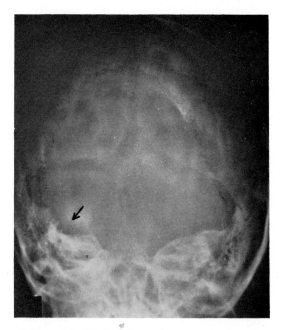

Figure 8–30 **Brain Abscess and Petrositis: Plain Film.** Anteroposterior view reveals a large concave area of bone destruction *(arrow)* involving the superior margin of the right petrous pyramid. This was due to a suppurative process that originated in the mastoid cells. The area of bone destruction cannot be distinguished from tumor erosion.

Brain abscess due to petrositis has become very rare since the advent of antibiotics. (Courtesy Drs. J. H. Allen and C. A. Cobb, Jr., Nashville, Tennessee.)

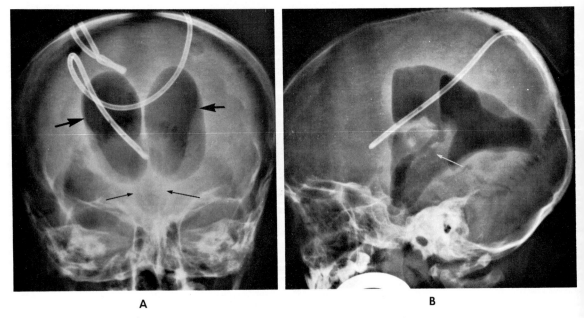

A   B

Figure 8–31 **Obstructive Hydrocephalus Due to Brain Abscess: Ventriculogram.**

A, In anteroposterior view there is dilatation of the lateral ventricles *(large arrows)* and of the third ventricle *(small arrows).* There is no air below this point.

B, Lateral view with patient prone demonstrates obstruction of the aqueduct of Sylvius *(arrow)* just below the dilated third ventricle. An abscess of the posterior fossa was found at surgery.

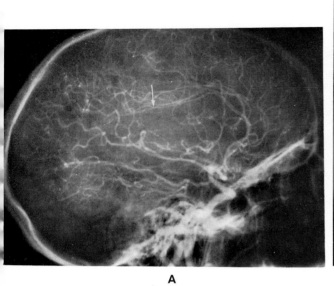

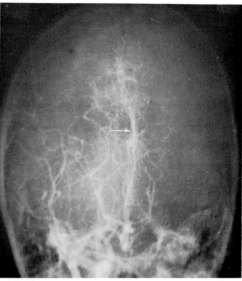

A   B

Figure 8–32 **Right Parietal Parasagittal Brain Abscess: Brachial Arteriogram.**

A, Lateral view of a right brachial arteriogram discloses downward displacement of the callosal branches *(arrow)* of the right anterior cerebral artery.

B, Frontal view discloses left lateral displacement *(arrow)* of these callosal branches. These findings suggest a mass lesion in the parasagittal region of the right parietal area. The lesion proved to be an abscess that probably was secondary to chronic pulmonary suppuration. The patient was a 54 year old man.

In the absence of a vascular tumor stain, angiographic differentiation between tumor and any other type of mass lesion is generally impossible.

igure 8–33 **Brain Abscess: CT Scan.**   The irregular area f diminished density *(large arrow)* on the right was due o an abscess. There is some shift of the anterior horns *small arrow).* The pineal is not displaced.

Progress of an abscess can be readily determined by erial scans.

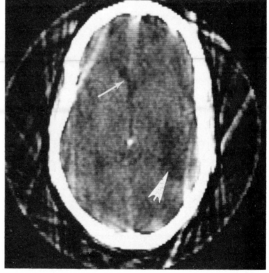

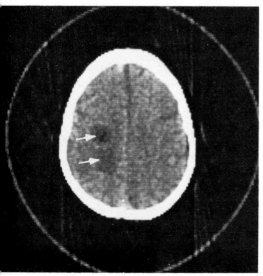

Figure 8–34 **Brain Abscess: CT Scan.** The rounded lucent area and the less regular lucency behind it *(arrows)* are portions of an abscess in the left parietal lobe. The patient has congenital heart disease and endocarditis.

Spinal Epidural Abscess

On plain films there is often evidence of pyogenic or tuberculous spondylitis (see pp. 211 and 212), since most epidural abscesses develop by direct extension from an infectious spondylitis. The bone and intervertebral changes are seen more frequently in tuberculous cases, since the bone changes in acute pyogenic spondylitis may not appear before the epidural abscess is well established.

On myelography, narrowing and partial obstruction of the oil column is evident at the level of the bone changes. The column may appear as thin linear densities on each side of the cord. Complete block can occur.

The myelographic appearance is similar to that of other extradural lesions (see p. 415), but the associated bone changes of infection will suggest an inflammatory mass.[18, 39-41]

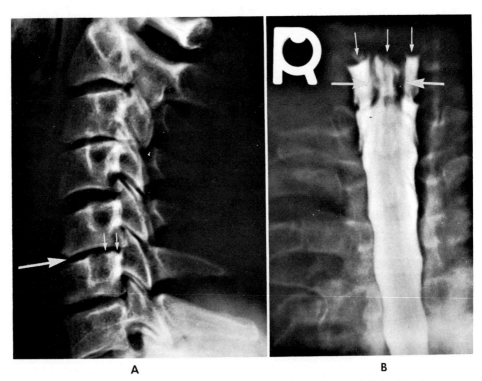

A   **B**

Figure 8–35 **Epidural Abscess of Spinal Canal Due to Tuberculous Spondylitis: Myelogram.**

A, In lateral view of the cervical spine there is narrowing of the vertebral interspace at C5–C6 *(large arrow),* and destructive erosion of the posterior aspect of the superior articular plate of C6 *(small arrows).* The combination of bone erosion and destruction with a narrowed intervertebral space suggests an infectious process.

B, Myelogram indicates complete block of the subarachnoid space at the level of C5 *(small arrows).* The cord *(large arrows)* is displaced slightly to the left.

The findings are those of extradural block. Occasionally they can be mimicked by an intradural extramedullary lesion. In the present case an epidural abscess due to tuberculous spondylitis was found at surgery.

Myelitis

Inflammation limited to the substance of the spinal cord usually produces no characteristic myelographic changes. However, if the cord becomes greatly swollen the myelographic appearance may simulate that of an intramedullary neoplasm. This pseudotumor-appearance has been described in acute infectious myelitis, acute necrotic myelopathy, and in neuromyelitis optica. In the latter condition, distinction from an intramedullary tumor will depend upon clinical evidence of associated cerebral or ocular dysfunction.

When cord inflammation is secondary to pyogenic or tuberculous spondylitis, there may be bone changes and myelographic evidence of spinal canal involvement. However, actual cord involvement cannot be demonstrated.[42]

Chronic Adhesive Spinal Arachnoiditis

Chronic adhesive spinal arachnoiditis may result from prior inflammation, but often the origin is obscure. Neurologic symptoms can occur from intraspinal root compression or from local myelitis with degeneration of spinal tracts. On myelography there may be partial or complete obstruction of the contrast column. The subarachnoid space is usually partially obliterated by the adhesions, and the contrast material may assume a stringy, globular, or "broken" appearance. Droplets are often trapped and fixed for long periods. Complete obstruction without these characteristic myelographic irregularities will simulate the appearance of an extradural tumor, but without cord displacement.[43-45]

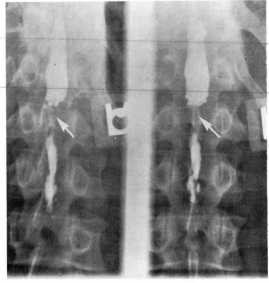

Figure 8–36 **Chronic Adhesive Arachnoiditis: Myelogram.** There is abrupt narrowing of the oil column in the region of L1 (*arrows*); and the subarachnoid space remains narrowed, somewhat irregular, and poorly filled. Thick arachnoid adhesions surround this area.

In arachnoiditis the opaque medium is often scattered and fragmented, similar to the "broken" oil column sometimes seen in normal persons; however, in arachnoiditis the findings remain constant on repeat myelogram.

Neurosyphilis

There are no characteristic roentgenologic changes in the central nervous system. The rare intracranial gumma produces the angiographic and pneumoencephalographic changes of a nonspecific mass lesion. Luetic basal arachnoiditis can lead to an obstructive hydrocephalus. In general paresis there may be evidence of brain atrophy. In longstanding tabes dorsalis, Charcot's disease is fairly common in one or more peripheral joints, especially the knees and hips. (See pp. 85 to 87 and pp. 406 and 407.)

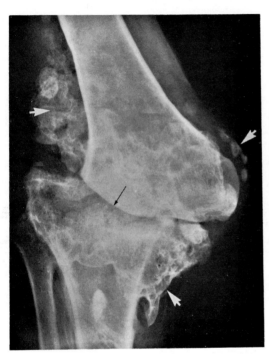

Figure 8–37 **Neurosyphilis: Tabes Dorsalis: Charcot Joint.** Derangement of the knee joint is characterized by destruction and fragmentation of the cartilage, lateral subluxation of the tibia, and smooth erosion of the lateral tibial condyle and the articular surfaces *(black arrow)*. Numerous calcifications in the soft tissues around the joint are present *(white arrows)*. Despite this disorganization, the bones are not demineralized.

The picture is characteristic of the hypertrophic form of Charcot's joint; in the atrophic form, calcified debris around the joint is lacking.

INTRACRANIAL TUMORS AND STATES CAUSING INCREASED INTRACRANIAL PRESSURE

Intracranial Tumors

In addition to routine skull radiographs, CT brain scans, cerebral angiography, and/or cerebral pneumography are necessary for the diagnosis and localization of the majority of intracranial mass lesions. Supplementary techniques, including laminography, positive-contrast ventriculography, isotope scan, and cisternography, are also sometimes employed.

Positive findings on routine skull films occur in approximately 50 per cent of cases. Increased intracranial pressure frequently causes demineralization and erosion of the dorsum sellae and, sometimes, enlargement of the sella; the latter finding may simulate an intrasellar tumor. These sellar changes are apparently due to pulsatile pressure from the adjacent anteroinferior recess of a dilated third ventricle. In children, widening of the sutures and increased convolutional markings are common findings of increased intracranial pressure. Occasionally, diffuse or localized osteoporosis of the vault is seen.

Roentgen findings that aid in tumor localization include tumor calcification, erosive or proliferative changes in adjacent bones, and displacement of a calcified pineal gland. Tumor calcification occurs in about 15 per cent of all brain tumors. Calcification is seen in about 50 per cent of oligodendrogliomas and in about 20 per cent of astrocytomas, but in less than 10 per cent of all gliomas. A high proportion of craniopharyngiomas exhibit suprasellar calcification. Often a specific type of tumor is suggested by the roentgen pattern of calcification.

Erosive or proliferative changes in adjacent bones frequently occur and accurately localize a tumor. Illustrative examples are widening and erosion of an internal auditory canal caused by an acoustic neurinoma, enlargement and erosions of the sella caused by intrasellar or suprasellar tumors, and thinning of a skull table caused by glioma or meningioma. Proliferative bone sclerosis, often with spiculed new bone formation, is seen in 25 to 40 per cent of meningiomas.

Lateral displacement of a calcified pineal gland will indicate the side occupied by a tumor; a lateral shift of 3 mm. or more is significant. Anteroposterior displacement can be determined by special measurements, but this displacement rarely occurs without a significant lateral shift. Infrequently, a calcified choroid plexus in the floor of one lateral ventricle is visibly displaced by a supratentorial tumor.

On the CT scans, a tumor appears as an area of decreased or increased density, or a combination of densities. On a μ density scale that runs from -500 to $+500$, the tumor density varies from 10 to 40. Tumors with increased density often show a surrounding halo of decreased density, presumably from adjacent edema. In a high percentage of cases, the tumor density will be increased, sometimes dramatically, following intravenous injection of contrast media. This density increase does not occur in cysts and is seen infrequently in infarcts or abscesses.

The CT scan will clearly indicate encroachment or displacement of the ventricular system.

Rarely, a tumor may have a density identical to that of surrounding brain and may not be apparent on the scan. Usually injection of contrast material will enhance the density of the tumor, making it clearly visible.

Cerebral arteriography is a useful procedure for determining the location, size, and vascular pattern of an intracranial mass. Carotid arteriography is employed for suspected supratentorial lesions, and vertebral arteriography is used to detect infratentorial tumors. Brachiocephalic or transfemoral arch arteriography will allow simultaneous opacification of the carotid and vertebral artery trees. The more common angiographic findings in brain tumor are the following:

1. Contralateral displacement of the midline anterior cerebral arteries and the internal cerebral vein.

2. Abnormal vessels, and tumor stain within the tumor area.

3. Displacement, uncoiling, and stretching of vessels in the vicinity of a tumor.

4. Early venous filling in the tumor area due to arteriovenous shunting of abnormal tumor vessels.

One or more of these findings are usually present, but the changes may be quite subtle, and recognition often requires considerable specialized skill and experience.

Air-contrast studies of the ventricular and subarachnoid spaces (pneumoen-cephalography and ventriculography) are less frequently employed since the advent of CT scans. However, air studies are useful in detecting posterior fossa lesions and for probing the cause of hydrocephalus. They can demonstrate aqueduct obstruction from infratentorial lesions, and growths within the ventricular system or the midbrain. Lesions in the vicinity of the optic chiasm, the sella, and the pons may obliterate or compress the basal cisterns, and hemispheric tumors will cause contralateral displacement of the lateral ventricles. Often, accurate localization of lesions by air studies may be difficult or impossible; moreover, air studies cannot distinguish a tumor from a nontumorous mass.[5, 6, 18, 46-54]

Orbital Tumors

Primary tumors of the orbit are uncommon and are often difficult to demonstrate radiographically. Most primary lesions arise in the posterior orbit behind the globe. A wide variety of pathologic entities, including glioma of the optic nerve, vascular malformations, meningioma, and dermoids, have been reported.

Invasion of the orbit by tumor extension from the sinuses or nasopharynx occurs more frequently than do primary lesions.

Plain films may show orbital bone erosions or calcifications in less than half the cases. Orbital venography may help demonstrate an expansive process behind the globe.

The CT scan is the most accurate radiographic modality for demonstration of orbital masses. On a CT scan through a normal orbit, the globe and optic nerve are easily identified, since they are surrounded by low density retro-orbital fat (see Fig. 8–1 A). A mass behind the globe will appear as an abnormal density. Comparison with the other orbit is useful and should always be made in questionable cases. Proptosis of the globe, even when not marked, is easily recognized on the scan.[54a]

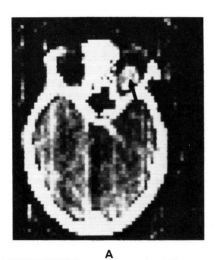 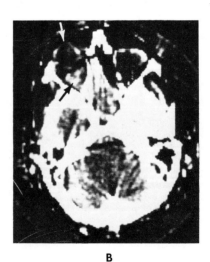

A   B

Figure 8–38   **Retro-Orbital Tumors in Four Patients: CT Scans.**

A, The dense circumscribed area in the right orbit *(arrow)* was a neurolemmoma. Only the anterior aspect of the globe is visualized, and it is definitely displaced anteriorly (proptosis) compared to the left globe.

B, The irregular density in the posterior left orbit *(black arrow)* was a sarcoma, which had arisen in the ethmoid sinus. The left globe *(white arrow)* is clearly proptosed.

Illustration continued on the opposite page.

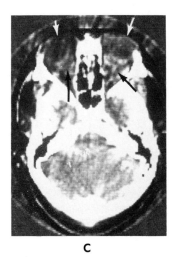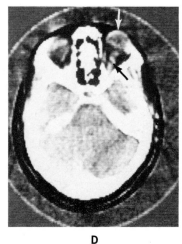

C   D

Figure 8–38   (Continued)

C, Irregular mottled densities *(black arrows)* fill the retro-orbital spaces of both orbits. Both globes *(white arrows)* are pushed forward. The lesions were lymphosarcomatous masses, and these were the first positive findings of the disease in this patient.

D, The serpiginous density along the right optic nerve *(black arrow)* was an optic nerve meningioma. The globe is displaced considerably forward *(white arrow)*.

The normally lucent (fat density) retro-orbital space (see Fig. 8–1 A_1) provides ready contrast to any tumor that arises in this area, and the CT scan is extremely accurate in detecting such lesions.

Meningiomas

Although meningiomas can occur in almost any intracranial area, their frequency and often distinctive radiographic characteristics warrant separate discussion.

This tumor is relatively benign and sharply circumscribed and almost always has a dural attachment or origin. A meningioma may grow as large as 20 cm. in diameter. It represents 15 per cent of all intracranial tumors and occurs most often in adults but occasionally does appear in children. Rarely, multiple intracranial meningiomas occur and are sometimes associated with neurofibromatosis.

The most frequent sites are the parasagittal area, the petrous ridge, the sphenoidal ridge, and over the anterior cerebral convexity. Intraventricular, posterior fossa, and retro-orbital meningiomas are infrequent; intraventricular lesions are less rare in children.

Plain film findings occur in 75 per cent of cases. Adjacent bony hyperostosis is seen in about 33 per cent of cases, and local bone erosions accompanied by increased vascular channels in the calvarium appear in about 25 per cent. Tumor calcification is found in 10 per cent of cases, appearing as granular psammomatous concretions or areas of dense calcification, or both. Larger lesions can cause pineal shift or increased intracranial pressure with atrophy of the dorsum sella.

The CT scan findings are often characteristic. There is a circumscribed mass with increased density throughout the lesion or around its periphery. Thickening of adjacent bone is frequently apparent.

Pneumography may disclose a nonspecific mass lesion, but angiography may show characteristic hypervascularity of the circumference or entire tumor and in 50 per cent of cases an intense homogeneous circumscribed tumor stain or blush.[5-7, 55-57]

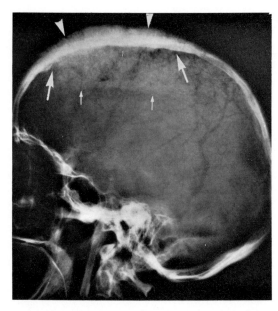

Figure 8–39   **Frontal Lobe Meningioma: Changes in Bone.**   There is irregular thickening of the inner table *(large arrows)*, perpendicular bone spiculation, and thickening of the outer table *(arrowheads)*. The diploic area has been obliterated by the reactive bone. The lesion extends laterally from the midline *(small arrows)*. These changes are typical of meningioma and often indicate the full extent of the lesion. Extension into the brain can be determined by angiographic studies.

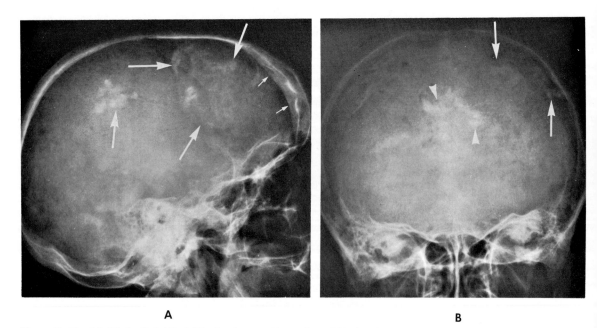

A   B

Figure 8–40   **Multiple Calcified Meningiomas: Frontal and Parietal Lobes.**

A, In lateral view a large and irregularly calcified mass is demonstrated in the frontal lobe *(large arrows)* and in the left parietal area *(large arrows)*. There is thickening of the internal skull table anteriorly *(small arrows)*.

B, Calcification of the frontal lobe meningioma *(arrowheads)* extends across the midline in anteroposterior view. Calcification in the parietal lobe appears to occur as two distinct lesions *(large arrows)*.

This case illustrates some of the less common features of meningioma: multiple lesions, calcification (seen in only 10 or 15 per cent of meningiomas), and minimal reaction of the adjacent bone.

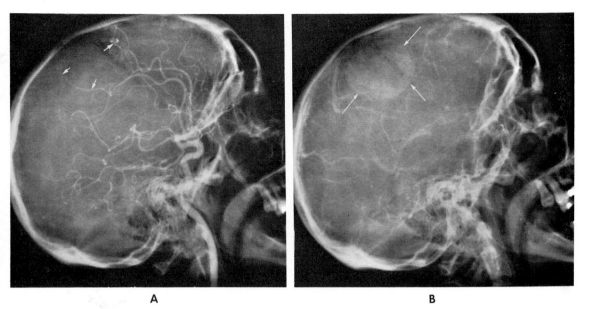

A   B

Figure 8–41 **Parietal (Parasagittal) Meningioma: Arteriogram.**

A, Curving and stretching of peripheral arteries *(arrows)* suggest presence of a mass lesion in the parietal area.

B, Tumor stain *(arrows)* persisting beyond the venous phase outlines the rounded tumor, which proved to be a meningioma. This diagnosis was suggested by the homogeneous opacification and by the configuration of the tumor stain, which are typical of meningioma.

There is neither sclerosis nor production of bone, since the tumor is not adjacent to a bone. Anteroposterior arteriogram *(not shown)* disclosed parasagittal location of the tumor.

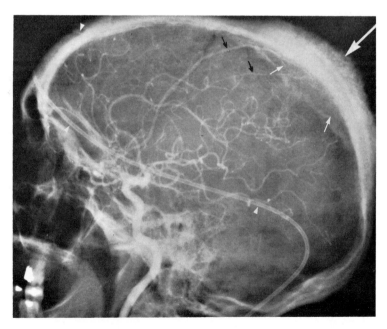

Figure 8–42 **Posterior Parietal Meningioma: Arteriogram.** Hyperostosis and spiculation *(large white arrow)* are characteristic of meningioma. The configuration of the tumor is clearly indicated by the triangular circumscribed tumor stain *(small white arrows)*. The lesion is fed by external carotid branches *(black arrows)*, which can be seen crossing the skull, and by the middle cerebral branches.

An external catheter had been inserted into the anterior superior sagittal sinus *(arrowheads)* in preparation for a sinogram.

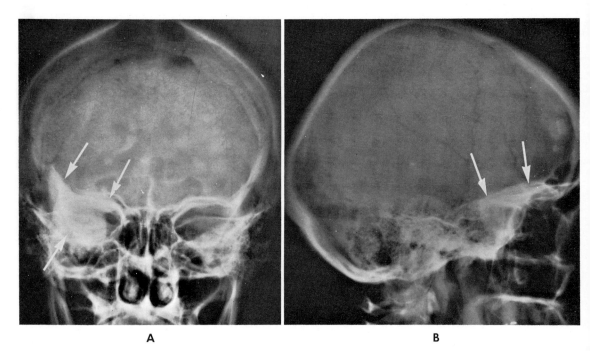

A   B

Figure 8–43 **Sphenoidal Ridge Meningioma.**

A, Considerable sclerosis involving the greater wing of the right sphenoid bone and the supraorbital plate *(arrows)* is evident in posteroanterior view.

B, In lateral film, the supraorbital sclerosis extends to the right anterior clinoid *(arrows).*

These bone changes are typical of sphenoidal ridge meningioma. In the present case, the tumor also involved the anterior tip and undersurface of the temporal lobe.

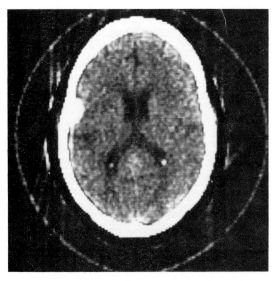

Figure 8–44 **Meningioma of Left Parietal Area: CT Scan.** The sharply marginated, dense area extending from the inner table was a highly vascular meningioma containing calcium.

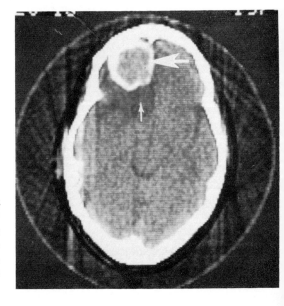

Figure 8–45 **Meningioma of Left Frontal Area: CT Scan.** The large sharply marginated mass with a high density periphery *(large arrow)* was a meningioma arising from the inner table of the frontal bone. A large, low density zone of edema *(small arrow)* is adjacent to the tumor.

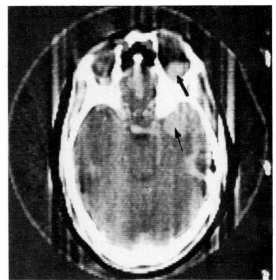

Figure 8–46 **Meningioma of Orbit and Sphenoid: CT Scan.** The rounded density in the posterior right orbit *(thick arrow)* and the larger density behind the thickened right sphenoid bone *(thin arrow)* were both due to a sphenoid meningioma. The density of these tumors was greatly enhanced by intravenous injection of contrast material.

Another collimated "cut" showed definite proptosis of the right orbital globe.

CT scans are the most reliable radiographic study for demonstrating an orbital tumor. (See Fig. 8–38*A*.)

Meningiomas adjacent to a bone are identified on the CT scan by their contiguity with the bone, the thickening of the bone, and the circumscribed density of the tumor.

Cerebellar Tumors

The rapidly growing cerebellar medulloblastoma is the most common malignant brain tumor of childhood. It is also the only common tumor that occasionally metastasizes, and distant lesions may appear in the spinal canal or in the bones; the latter are often osteoblastic. Astrocytoma is the next most common cerebellar tumor. Less frequent are ependymoma and hemangioblastoma. The latter is invariably present in the von Hippel–Lindau syndrome. Benign tumors of the cerebellum are very rare.

Cerebellar tumors often occlude the aqueduct or, less commonly, the fourth ventricle, causing obstructive hydrocephalus. In children, the skull films may show widened sutures, increased convolutional markings, and often erosions of the dorsum, all due to increased intracranial pressure. Tumor calcification is very uncommon. Plain skull films in the adult are generally unrevealing.

On the CT scan, a cerebellar tumor appears as an area of decreased or increased density in the posterior fossa. Intravenously injected contrast material frequently increases the scan density, a finding that helps distinguish a neoplastic lesion from a cyst, infarct, or abscess. Midline or larger hemispheric tumors will displace, distort, or obliterate the fourth ventricle. Obstructive hydrocephalus often occurs and is readily recognized on the scan.

Pneumography may show the tumor border impinging on the fourth ventricle, which may be dilated or displaced. If obstructive hydrocephalus is present, ventriculography must be employed to demonstrate the dilated lateral and third ventricles, and it may show obstruction of the aqueduct or fourth ventricle.

Vertebral angiography may identify cerebellar tumors by displacement of the posterior inferior cerebellar arteries or veins. A vascular blush or tumor stain may occur in highly vascular tumors. If obstructive hydrocephalus is present, the anterior cerebral artery will be stretched and the angle of the thalamostriate vein will be increased.[5, 6, 18, 58-61]

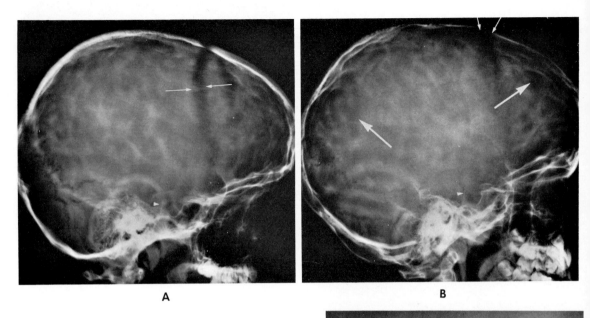

A
B

Figure 8–47 **Cerebellar Tumor with Increased Intra-cranial Pressure: Three Cases.**

A, The coronal suture is considerably widened *(arrows),* and the convolutional markings are increased because of an astrocytoma. The dorsum sellae *(arrowhead)* is intact.

B, In a child with an ependymoma of the fourth ventricle and cerebellum, the sutures are widened *(small arrows);* the convolutional markings are increased *(large arrows);* and there is erosion of the upper portion of the dorsum sellae *(arrowhead),* which tapers to a point. The increase in the convolutional markings is due to pressure on the skull from swollen cerebral convolutions.

C, A young adult had longstanding increased intra-cranial pressure from a hemangioblastoma of the cerebellum. There is enlargement of the sella with marked erosion and virtual disappearance of the dorsum *(long arrow).* The sella has a double bony floor *(short arrows).* These changes are due to pulsating pressure from

C

the anterior portion of the dilated third ventricle, and at times they may simulate a tumor in or around the sella. In adults, increased intracranial pressure can cause sellar changes but not suture diastasis or an increase in convolutional markings; the last two are found exclusively in children.

Tumor calcification is extremely rare in cerebellar or other infratentorial lesions.

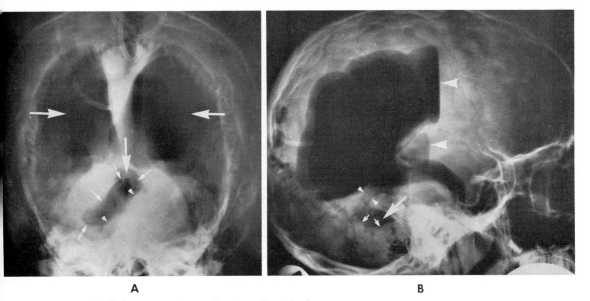

A B

Figure 8–48 **Medulloblastoma of Cerebellum: Ventriculogram.**

A, Frontal view discloses greatly dilated lateral and third ventricles *(large arrows).* The fourth ventricle is also greatly dilated *(small arrows),* and the lobulated borders of the tumor *(arrowheads)* project into the ventricle.

B, Lateral view taken with the patient's brow down discloses air-fluid levels in the dilated lateral and third ventricles *(large arrowheads).* The fourth ventricle *(large arrow)* is displaced forward, and the curved tumor border *(small arrows)* is visible in the inferior aspect of this ventricle. The aqueduct *(small arrowheads)* is dilated.

Separation of the sutures, prominent convolutional markings, and a thinned dorsum sellae are also present.

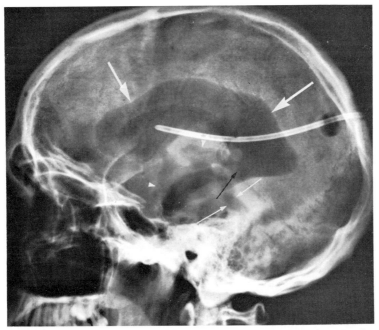

Figure 8–49 **Medulloblastoma of Cerebellum: Ventriculogram.** The lateral ventricles *(large arrows)* and the third ventricle *(arrowheads)* are greatly dilated. The air also outlines the aqueduct *(black arrow)* and fourth ventricle *(small white arrows);* both are displaced forward, and the aqueduct is kinked by the large cerebellar tumor.

There is a close relationship between the anteroinferior recess *(lower arrowhead)* of the dilated third ventricle and the dorsum sellae, which has been thinned by pulsations from this dilated recess.

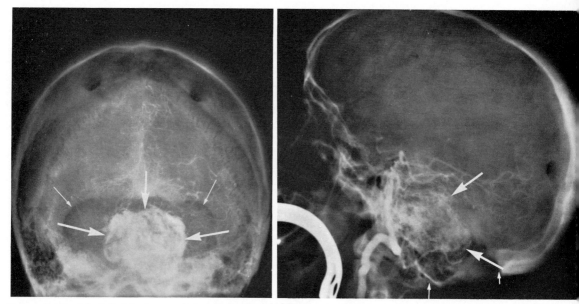

Figure 8–50 **Medulloblastoma of Cerebellum: Vertebral Arteriogram.**   The highly vascular tumor is completely opacified (tumor stain) *(large arrows)* during arteriography and is more clearly depicted in anteroposterior view. The tumor is in the midline; lateral view indicates its location anteriorly in the vermis of the cerebellum, from which it arose. The occipital bone defects *(small arrows)* are from previous exploratory surgery.

   The density of the tumor stain is dependent upon the vascularity of the tumor; stains of this intensity are unusual.

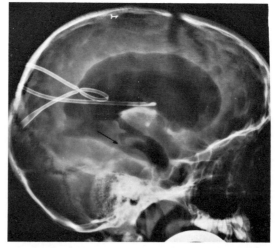

Figure 8–51 **Astrocytoma of Cerebellum in a Child: Ventriculogram.**   The lateral and third ventricles are dilated, and the upper portion of the aqueduct of Sylvius *(arrow)* is kinked and obstructed.

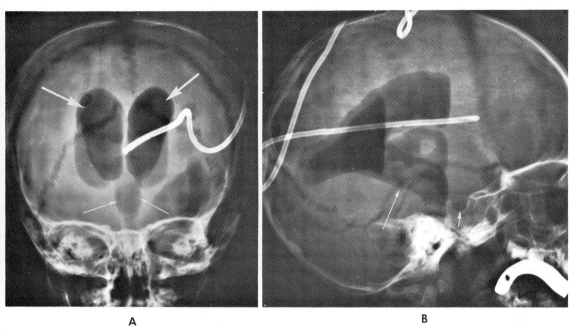

Figure 8–52   **Ependymoma of Fourth Ventricle in a Child: Ventriculogram.**

A, Anteroposterior view indicates symmetric dilatation of the lateral ventricles *(large arrows)* and dilatation of the third ventricle *(small arrows).*

B, In lateral view, the occluded aqueduct terminates abruptly *(long arrow).* There is diastasis of all the sutures. The normal cortical density—the "white line"—in the floor of the sella is lacking *(short arrow);* this is an indication of increased intracranial pressure.

Ependymomas frequently arise in the ventricles and can become quite large; in time, most of those arising in the fourth ventricle cause obstruction of the aqueduct and, subsequently, internal hydrocephalus.

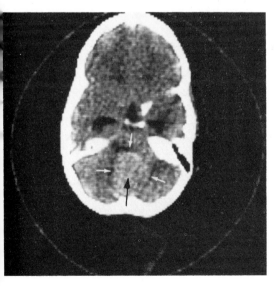

Figure 8–53   **Cerebellar Medulloblastoma: CT Scan.**

There is a rounded area of moderately increased density *(black arrow)* in the posterior fossa, surrounded by a zone of diminished density *(white arrows)* due to edema.

Moderate dilatation of the lateral and third ventricles was observed on more cephalad scan sections.

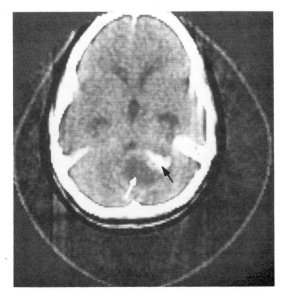

Figure 8–54   **Cerebellar Tumor: CT Scan.**

A rounded, lucent area *(white arrow)* in the posterior fossa and a triangular zone of increased density *(black arrow)* were both part of a glioma. The latter zone was considerably less dense before contrast material was administered intravenously.

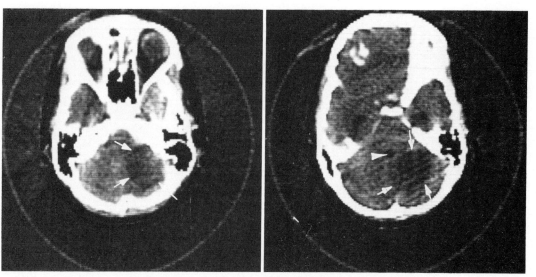

Figure 8–55 **Von Hippel-Lindau Disease: Hemangioblastoma of Cerebellum: CT Scan.** Ataxia, vertigo, and headaches developed in a 21 year old girl with Von Hippel-Lindau disease.

CT scans through the base disclose a rounded, somewhat circumscribed area of decreased density (*white arrows*) on both "slices" in the right cerebellar area. This proved to be an hemangioblastoma. The fourth ventricle (*arrowhead*) is normal in size and position. The other ventricles were also normal, since the right-sided tumor had not compromised or obstructed the aqueduct.

Frontal Lobe Tumors

Meningiomas and gliomas, especially astrocytomas, are the most frequent frontal lobe tumors. There may be localized calcification in both types. With meningiomas there may be characteristic changes—usually sclerosis and reactive bone formation—in the adjacent bony tables of the skull.

The CT scan will uncover most tumors of the frontal lobes as areas of altered density. The pressure effects on an adjacent frontal horn of the lateral ventricles are readily recognized.

Pneumography demonstrates a shift of the lateral ventricles away from the side of the lesions; often local compression of the anterior horn of one lateral ventricle aids in exact localization of the tumor.

Arteriography may disclose arterial displacement, abnormal vasculature, and tumor stain.[5, 6, 18, 34, 46, 48, 49, 51, 62]

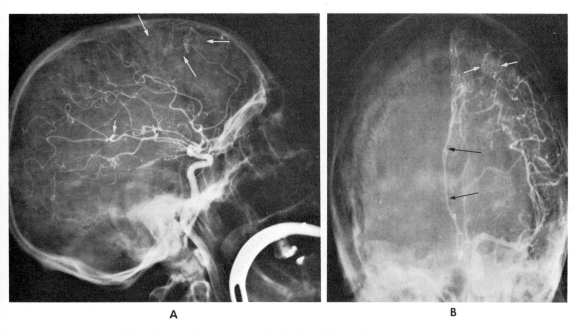

Figure 8–56 **Frontal Lobe Tumor: Astrocytoma: Left Carotid Arteriogram.**

A, In lateral view there are abnormal vessels *(long arrows)* supplying a small astrocytoma in the posterior frontal area. There is downward displacement of the anterior cerebral artery and the sylvian vessels *(small arrows)* of the middle cerebral artery.

B, Anteroposterior projection demonstrates lateral bowing (shift) of the anterior cerebral artery *(black arrows)*. A faint tumor stain outlines part of the lesion *(white arrows)*.

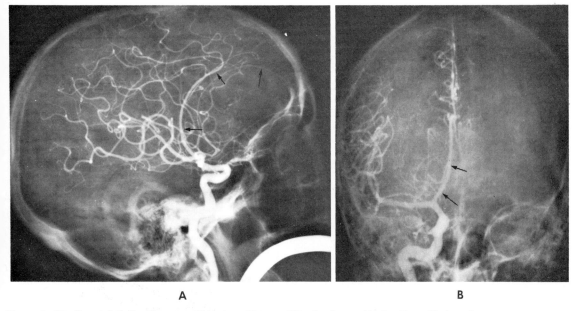

Figure 8–57 **Frontal Lobe Tumor: Olfactory Groove Meningioma: Right Carotid Arteriogram.**

A, Posterior bowing and stretching of the anterior cerebral artery *(arrows)* caused by the frontal midline tumor are seen on the lateral view.

B, The proximal portion of the right anterior cerebral artery *(arrows)* is elevated in anteroposterior view. The corresponding left anterior cerebral artery was similarly elevated, indicating the midline location of the tumor.

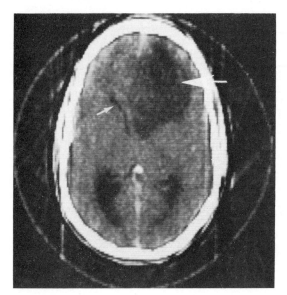

Figure 8–58 **Large Frontal Lobe Tumor: CT Scan.** A huge area of decreased density (even after intravenous injection of contrast material) *(large arrow)* is filling the entire right frontal lobe. The frontal horns of the lateral ventricles *(small arrow)* are markedly displaced to the left. The linear density below the pineal is a normal venous channel opacified by the intravenously administered contrast material.

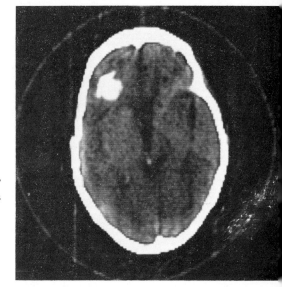

Figure 8–59 **Frontal Lobe Oligodendroglioma: CT Scan.** The very dense mass in the left frontal lobe was a partially calcified oligodendroglioma.

Parietal Lobe Tumors

The parietal lobe, including the frontoparietal area, is the most frequent site for brain tumors of all types, including the gliomas and the meningiomas.

The CT scan will usually demonstrate the lesion as an area of altered density (see under *Intracranial Tumors,* p. 318). Contralateral ventricular or pineal displacement is readily apparent.

Angiographically, tumors in the suprasylvian area are localized by their effect on the middle cerebral artery and its branches. Characteristically, on lateral view the tumors depress the looped vessels of the sylvian triangle, and on anteroposterior view they cause depression of the sylvian point. (The sylvian point, an angiographic landmark, is the most medial point of the highest loop of the sylvian vessel arcade.) The normal undulations of the middle cerebral artery or its branches may be lacking in the tumor vicinity. Generally, there is contralateral displacement of the anterior cerebral arteries, often with a "squared-off" shift, in contrast to the bowed shift that results from a frontal lobe tumor. Abnormal vasculature and tumor stain are frequent localizing findings.

Pneumography usually demonstrates a contralateral displacement of the lateral and third ventricles, often with compression of the body of the ipsilateral ventricle.[5, 6, 18, 46, 48, 49, 51, 62, 63]

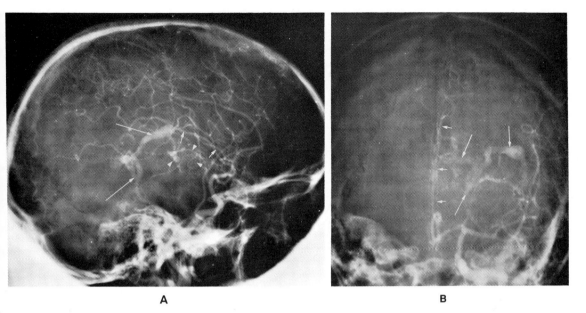

| A | B |

Figure 8–60 **Deeply Situated Parietal Lobe Astrocytoma: Arteriogram.**

A, Lateral view during arterial phase discloses a collection of abnormal vessels *(arrowheads)* and early, abnormal venous drainage *(long arrows).* The middle cerebral artery *(small arrows)* does not undulate and appears stretched and elevated, probably because of some trapping of the temporal horn with resultant dilatation or because of involvement of the temporal lobe by tumor.

B, In anteroposterior view a collection of early-filled abnormal veins *(long arrows)* occupies the deep parietal area on the left. The anterior cerebral artery *(small arrows)* is slightly displaced.

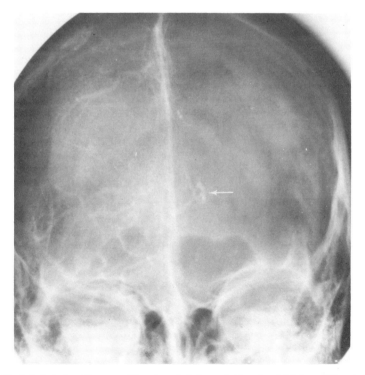

Figure 8–61 **Deep Parietal Lobe Tumor: Displaced Internal Cerebral Vein.** The internal cerebral vein *(arrow),* opacified during the venous phase of a right carotid arteriogram, is displaced markedly to the left of its normal midline location.

In more posteriorly located cerebral mass lesions, the internal cerebral vein displacement may sometimes be seen without similar shift of the anterior cerebral artery.

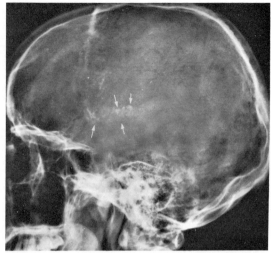

Figure 8–62 **Deep Frontoparietal Oligodendroglioma with Calcification.** In lateral view irregular plaques of calcification *(arrows)* are demonstrated in a large oligodendroglioma deep in the parietal lobe. Anteroposterior view *(not shown)* revealed the calcification to be on the right side rather than in the suprasellar area.

Of all the gliomas, oligodendrogliomas show the greatest frequency of calcification—about 50 per cent. Astrocytomas—next in frequency—exhibit calcifications in about 20 per cent.

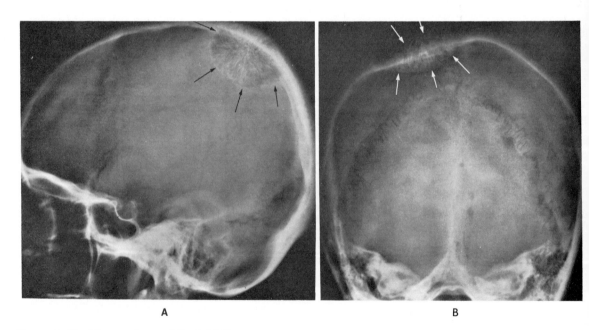

A   B

Figure 8–63 **Hemangioma of Parietal Bone.**

A, Area of translucency *(arrows)* has a sclerotic border and bony striae that represent bony laminae between the vessels of the hemangioma. The appearance is strongly suggestive of hemangioma.

B, In tangential anteroposterior view spiculation perpendicular to the tables *(arrows)* is characteristic. The involved area of bone is flattened. There were neurologic signs and symptoms, due to pressure on the posterior parietal lobe.

Most hemangiomas of bone arise in the skull and vertebrae.

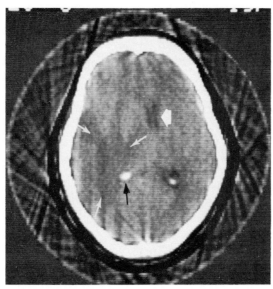

Figure 8–64 **Parietal Lobe Tumor: CT Scan.** A large irregular area of decreased density in the left parietal area *(small white arrows)* is a glioma that has displaced the ventricles *(broad white arrow)* and the left calcified choroid plexus *(black arrow)* to the right.

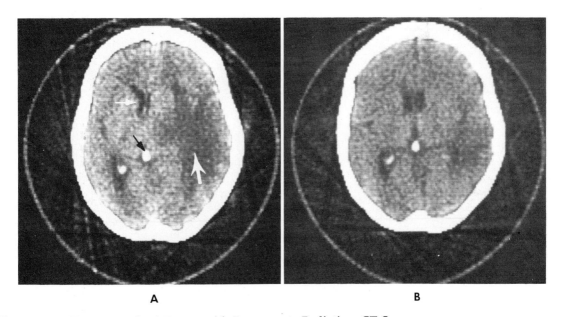

A   B

Figure 8–65 **Temporoparietal Tumor with Response to Radiation: CT Scan.**

A, The huge irregular area of diminished density *(large white arrow)* in the right temporoparietal area has markedly displaced the ventricles *(small white arrow)* and calcified pineal *(black arrow)* to the left.

B, After a few weeks of radiation therapy, the tumor is greatly decreased in size, and the shift of the ventricles and pineal is considerably reduced.

Temporal Lobe Tumors

The temporal lobe is a common site of a variety of gliomas and other tumors. Meningiomas arising from the sphenoidal ridge or from the greater wing of the sphenoid bone usually compress and elevate the lobe.

Skull films often reveal a pineal shift and, occasionally, gliomatous calcification. Sclerosis or erosive changes of the sphenoidal wing or ridge will localize a meningioma.

The CT scan will usually disclose an area of altered density in the temporal lobe, frequently associated with contralateral shift of the ventricular system and/or pineal (see under *Intracranial Tumors*, p. 318).

Angiography frequently reveals a contralateral shift of the midline anterior cerebral arteries, a common finding in the majority of hemispheric mass lesions. The localizing angiographic findings are the characteristic elevation and medial displacement of the middle cerebral artery and, often, an elongation and straightening of the carotid siphon. Localizing tumor stain or abnormal vessels are frequent. A meningioma can often be identified by its circumscribed, uniform, dense stain.

Pneumographic localization can be made if there is deformity, displacement, or abrupt cutoff of the temporal horn. Sphenoidal meningiomas generally elevate the anterior portion of the temporal horn. However, a less specific contralateral shift of the lateral ventricles is often the only finding.[5, 6, 18, 46, 48, 49, 51, 62]

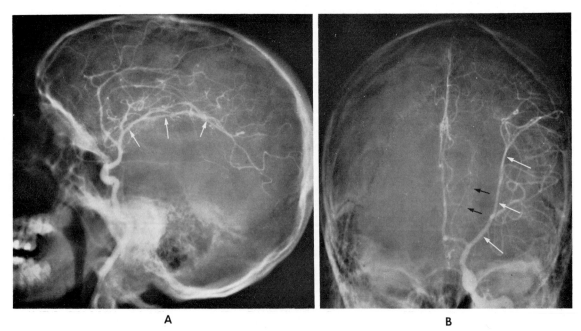

A   **B**

Figure 8–66 **Temporal Lobe Astrocytoma: Angiogram.**

A, There is marked elevation and stretching of the middle cerebral artery in lateral view *(arrows).*

B, In anteroposterior view, there is marked medial displacement and elevation of the middle cerebral artery *(white arrows).* The anterior choroidal artery *(black arrows)* is clearly visualized and displaced medially by the temporal lobe mass.

No tumor stain was apparent. Although tumor stain aids in identifying mass lesions, many tumors do not have abundant vasculature and do not produce a stain during angiography. Absence of tumor stain does not rule out a brain tumor.

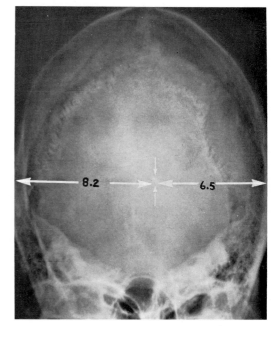

Figure 8–67 **Temporal Lobe Tumor: Pineal Shift.** The pineal gland is calcified and readily identified *(small arrows)* in anteroposterior view. It is shifted from its normal midline position to the left side. The distance from the center of the pineal calcification to the internal table is 8.2 cm. on the right and 6.5 cm. on the left. The difference, 1.7 cm., indicates a shift to the left of 8.5 mm. (half the difference). A large tumor was found in the right temporal lobe.

The occurrence of pineal calcification increases with age, and it may provide a highly significant clue to mass lesions of the hemispheres. A shift of over 2 mm. (4 mm. difference between the left and right measurements) in an otherwise symmetrical skull is highly suspicious. Anteroposterior shifts of the calcified pineal, as seen on lateral skull films, are more difficult to assess.

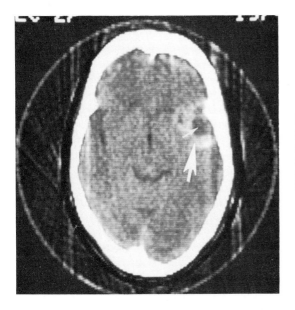

Figure 8–68 **Temporal Lobe Tumor: CT Scan.** A fairly circumscribed mass with a circumferential increased density (after intravenous injection of contrast material) *(large arrow)* and a lucent center *(small arrow)* proved to be a cystic glioma in the right temporal lobe. Only a slight ventricular shift is apparent.

Occipital Lobe Tumors

Only 2 per cent of primary brain tumors arise in the occipital lobe, although many posterior parietal lesions may encroach upon this area.

The occipital tumors are readily localized on the CT scan as an area of altered density (see under *Intracranial Tumors,* p. 318). Forward displacement of a choroid plexus may sometimes be seen. A posterior horn may be displaced or distorted.

Deformity and displacement of one or both posterior horns of the lateral ventricles may also be seen on air studies. Failure of a posterior horn to fill is often attributable to a normal anatomic variant, and undue significance should not be attached to this finding alone.

On arteriography, occipital lobe tumors may cause deformity or stretching of the posterior branches of the posterior cerebral artery. There may be tumor stain and early venous drainage locally.[5, 6, 18, 46, 48, 49, 51, 62]

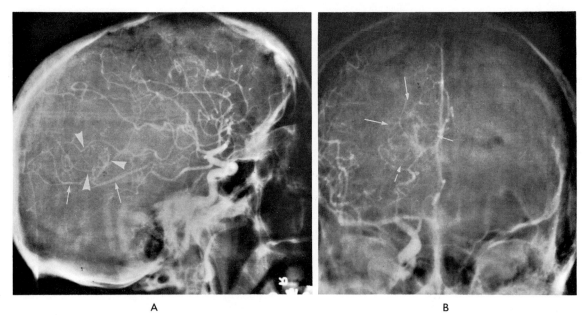

A   B

Figure 8–69 **Occipital Lobe Glioma: Arteriogram.**

A, Lateral view shows stretching and loss of normal undulations of the posterior cerebral artery *(arrows)*. A collection of abnormal vessels producing a faint tumor stain *(arrowheads)* marks the tumor area.

B, In anteroposterior view a faint tumor stain and abnormal vessels *(arrows)* are demonstrated in the vicinity of the right posterior cerebral artery.

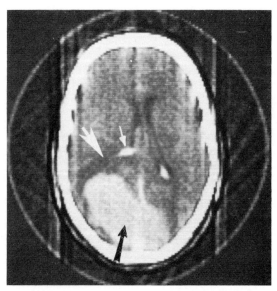

Figure 8–70 **Huge Occipitoparietal Tumor: CT Scan.**   The tumor density is very high *(black arrow)* and is surrounded by a zone of decreased density *(large white arrow)*, probably edema. The left choroid plexus is displaced forward *(small white arrow)*, and the lateral ventricles are displaced to the right. The linear density extending from the tumor in the midline is probably a large draining vein.

Tumors in or Around the Third and Lateral Ventricles

Tumors of the third ventricle often produce obstruction of the aqueduct or foramen of Monro, and subsequent internal hydrocephalus. When neighboring structures are involved, many neurologic symptoms arise. Among the more frequent of these relatively rare tumors are colloid cyst and pinealoma. Chiasmatic and thalamic gliomas and craniopharyngiomas also may extend into the area of the third ventricle.

On the CT scan, a mass lesion can often be identified within a lateral or third ventricle, especially if the abnormal density is seen surrounded by the low density of cerebrospinal fluid. The third ventricular tumors may appear as a midline density obliterating the normal linear outline of the third ventricle. Obstructive dilatation of the lateral ventricles is usually associated. Lateral ventricular tumors may be difficult to distinguish from callosal or basal ganglia lesions that are encroaching upon and distorting an adjacent ventricle.

The pneumogram may demonstrate a defect within or encroaching upon the third ventricle. Dilatation of the third and lateral ventricles is common, and severe hydrocephalus may develop. Angiography may be useful as an auxiliary procedure in order to disclose a tumor stain and the full extent of the lesion. The venous phase is often most revealing.

Most tumors in the lateral ventricles are gliomas, very often an ependymoma. The much rarer benign lesions include papilloma of the choroid plexus, epidermoid tumor, and meningioma. Tumors of the corpus callosum often extend into the ventricles and cause widening of the septum pellucidum. The lateral ventricle is dilated, and the large filling defect may extend across the midline into the adjacent ventricle, especially if the lesion originated in the septum pellucidum. Angiography, less revealing than the CT scan or the pneumogram, usually demonstrates only stretching of the anterior cerebral artery over the dilated ventricles. (See *Pinealoma,* p. 1259.)[18, 53, 64]

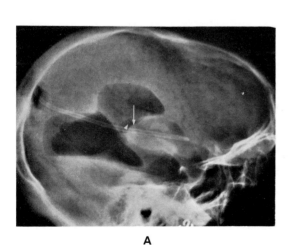
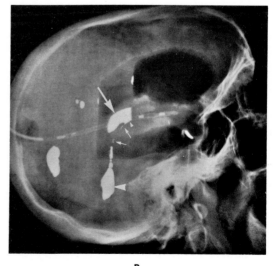

A   B

Figure 8–71 **Tumor of Third Ventricle: Air and Iodized Oil (Lipiodol) Ventriculogram.**

A, There is considerable dilatation of the lateral ventricles. A mass *(arrow)* fills the posterior portion of the third ventricle. A small pocket of air in the superior recess of this ventricle is demonstrated *(arrowhead).*

B, Lipiodol introduced into the ventricular system outlines the fourth ventricle *(large arrowhead),* the narrowed and posteriorly bowed upper aqueduct *(small arrows),* and the dilated suprapineal recess *(large arrow).* A smooth defect marks the roof of the tumor *(small arrowhead).* The nonopacified portion of the third ventricle is filled with tumor.

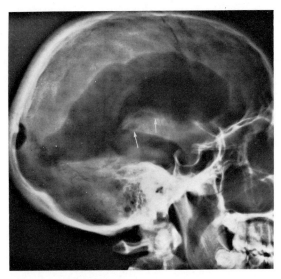

Figure 8–72   **Colloid Cyst of Third Ventricle: Ventriculogram.**   The lateral ventricles are greatly dilated. A mass (*arrows*) nearly fills the third ventricle, and there is little air in the superoposterior portion above the mass. The aqueduct is obstructed and is not visualized.

Tumors of the third ventricle almost always eventually occlude the aqueduct or the foramen of Monro and cause hydrocephalus; failure of the third ventricle to fill completely is a significant finding for tumor localization.

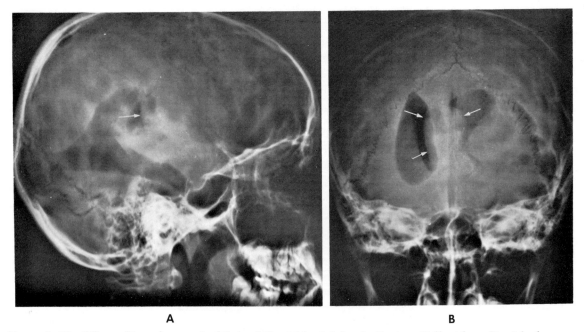

A   B

Figure 8–73   **Glioma (Ependymoma) of Lateral Ventricle Arising in Septum Pellucidum: Ventriculogram.**

A, In right lateral decubitus view there is a rounded filling defect (*arrow*) in the anterior portion of the lateral ventricle. Only the posterior portion and the temporal horn of the dilated ventricle are filled with air.

B, Anteroposterior view discloses a mass (*arrows*) projecting medially into the dilated right ventricle and the left ventricle (*arrows*). The space between the ventricles is considerably increased, a frequent finding in intraventricular gliomas; they tend to grow across the septum pellucidum into the opposite ventricle. This is particularly common when the lesion arises from the septum pellucidum or from the medial border of a ventricle. The foramen of Monro is frequently occluded, and this leads to ventricular dilatation.

In many cases of choroid plexus tumors of the lateral ventricle, there is hypersecretion of spinal fluid with resultant internal and external hydrocephalus—a unique finding in brain lesions.

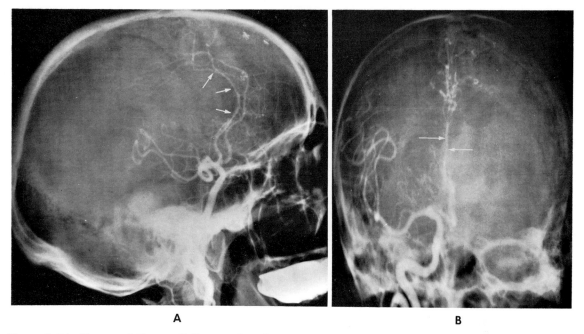

Figure 8–74 **Tumor of Corpus Callosum: Arteriogram.**

A, Lateral view reveals upward displacement and stretching of the anterior cerebral artery *(arrows).*

B, Anteroposterior view demonstrates unusual straight course of the anterior cerebral artery *(arrows),* which is stretched owing to dilatation of the lateral ventricles.

In the present case the findings were due to an anterior callosal tumor, but a very similar picture is seen in ventricular dilatation from any cause. Pneumographic studies are more informative in callosal tumors and may depict a mass encroaching upon the air-filled ventricles, or a widened septum pellucidum.

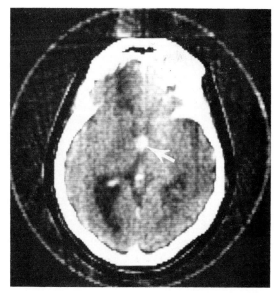

Figure 8–75 **Colloid Cyst of Third Ventricle: CT Scan.** The sharply marginated, rounded density *(white arrow)* is in midline, occupying the region of the third ventricle. This cyst caused surprisingly little obstruction, and the lateral ventricles are only minimally dilated.

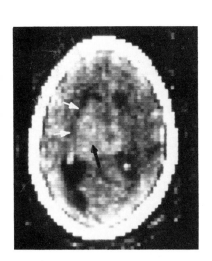

Figure 8–76 **Intraventricular Glioblastoma: CT Scan.** A large mass of somewhat increased density *(black arrow)* is arising from the body of the left lateral ventricle and displacing it to the left *(white arrows).* The tumor is extending across the midline and somewhat displacing the right lateral ventricle.

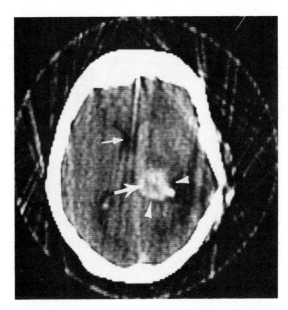

Figure 8–77 **Thalamic Glioma: CT Scan.** The dense mass (after intravenous administration of contrast material) is lying in the region of the right posterior thalamus *(large arrow)*. Lucent zones of edema *(arrowheads)* surround the tumor.
The ventricular system *(small arrow)* is displaced to the left.

Glioma of the Pons

This tumor most frequently occurs in young persons. Radiographic diagnosis is best made by pneumoencephalography, which demonstrates the anterior border of the tumor encroaching upon the pontine and interpeduncular cistern; posteriorly, the tumor causes elongation and displacement of the aqueduct. The fourth ventricle marks the posterior wall of the pons, and measurement of the distance between this wall and the posterior edge of the pontine cistern indicates the size of the pons. Normally it measures up to 4 cm., but it is usually significantly greater in the presence of a pontine tumor.

On the CT scan, a pontine tumor may cause a swelling or enlargement of the brain stem. The latter structure is often identified as a rounded midline area surrounded by a thin layer of lucent cerebrospinal fluid.

Vertebral arteriography may show anterior displacement of the basilar artery. Often the various branches of the several cerebellar arteries are displaced. Extrapontine tumors in this area are not in the midline, and they would displace the basilar artery laterally.[2, 5, 6, 18, 34, 46, 65]

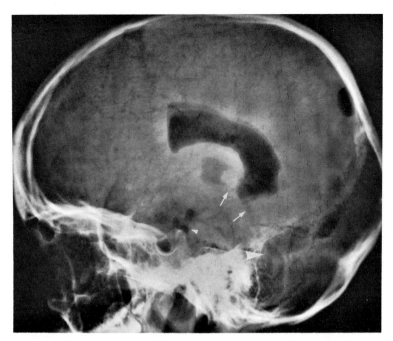

Figure 8–78 **Glioma of Pons in a Child: Pneumoencephalogram.** The aqueduct of Sylvius *(arrows)* is bowed and displaced posteriorly by the enlarged pons, but it is not occluded. The anterior border of the enlarged pons *(small arrowhead)* projects into the pontine cistern, which is thus virtually obliterated. The distance between this point and the anterior wall of the fourth ventricle *(large arrowhead)* marks the anteroposterior dimension of the pons, which in this case is considerably greater than normal. Pontine tumors generally are inoperable.

Acoustic Neuroma

Acoustic neuroma constitutes approximately 8 per cent of all intracranial tumors and about 80 per cent of cerebellopontine-angle lesions. Radiographic detection is extremely important, particularly for early lesions when clinical diagnosis is often equivocal.

On regular skull films the significant early findings include widening of the internal acoustic canal and slight erosion and demineralization of its bony walls. A difference in width between the left and right canal of more than 2 mm. is usually significant. Later, the more evident destruction of the surrounding bone simplifies the diagnosis. However, less than 50 per cent of these tumors can be definitely identified on plain skull films. Planigraphy of the petrous pyramids is far more effective in demonstrating early significant changes in the acoustic canal, and allows correct diagnosis of about 80 per cent of these tumors.

The CT scan will generally show a sharply circumscribed area of increased density adjacent to and usually behind the petrous bone. The density is usually increased following intravenous injection of contrast material. Some erosion of the petrous tip may be apparent.

Large lesions may extend into the cerebellopontine angle, encroach upon the aqueduct and fourth ventricle, and produce internal hydrocephalus. The pneumogram or CT scan will show ventricular dilatation and lateral pressure upon the fourth ventricle. In uncertain cases, opacification of the cerebellopontine cistern with positive contrast material will usually confirm the diagnosis. Since the contrast material normally fills the internal auditory canal, even smaller tumors limited to the canal can be demonstrated. Larger tumors that extend outside the canal will displace the opacified cistern medially.

Brachial arteriography may show elevation of the posterior cerebral and superior cerebellar arteries. The basilar artery may be displaced away from the lesion. Not infrequently, tumor stain or abnormal vessels will localize the tumor.

Plain films and planigraphy, supplemented by CT scans enhanced by contrast material and cerebellopontine cisternography, are the most reliable studies for diagnosis.[5, 18, 66-71]

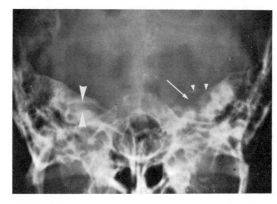

Figure 8–79 **Acoustic Neuroma.** There is bony erosion of the left acoustic meatus and canal (*arrow*) near the petrous tip. A thin shell of bone (*small arrowheads*) marks the roof of the widened and eroded canal. The acoustic canal on the right (*large arrowheads*) is normal.

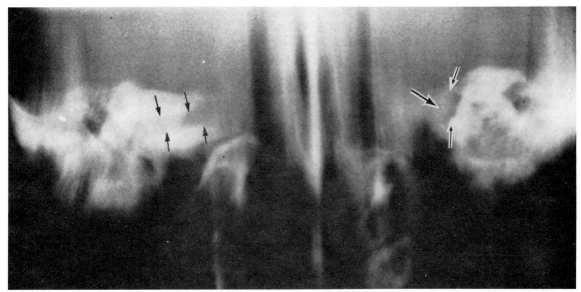

Figure 8–80 **Acoustic Neuroma: Tomograms of Temporal Bone.** On the normal right side, the auditory (acoustic) canal (*black arrows*) and the normal surrounding bone are clearly outlined on the tomogram.

On the left side an acoustic neuroma has destroyed the bony walls of the medial portion of the canal (*large black-white arrow*) and has widened the lumen of the midcanal (*small black-white arrows*).

A small acoustic tumor may produce subtle but significant alterations of the acoustic canal; they can be seen on tomograms but not on conventional films.

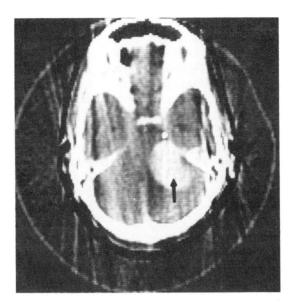

Figure 8–81 **Acoustic Neuroma: CT Scan.**   The large, sharply circumscribed density *(black arrow)* adjacent and posterior to the right petrous tip was a large acoustic neuroma.

Its location, sharp borders, and increased density after intravenous injection of contrast material are characteristic.

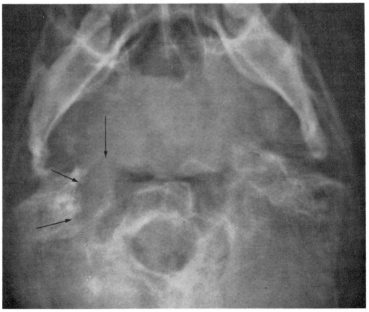

Figure 8–82 **Destruction of Petrous Pyramid by Glomus Jugulare.**   Occipitomental film demonstrates a rather large erosive defect in the apex of the right petrous pyramid *(arrows)*. This was caused by the rare glomus jugulare, but a similar roentgenologic appearance might be produced by a large acoustic neuroma.

Pituitary Adenomas

The chromophobe and the eosinophilic adenoma are the commonest pituitary intrasellar tumors giving rise to observable changes on skull films. Basophilic adenomas are rare and usually too small to cause alterations of the sella. Pituitary adenomas occasionally develop following bilateral adrenalectomy for Cushing's disease.

The sella turcica becomes enlarged and ballooned, and a double floor often develops as a result of unequal downward growth and erosion, which is usually best seen in anteroposterior view. The normal white line of cortical bone of the floor of the sella may be diminished or lost. Further enlargement may result in erosion of the dorsum sellae. The anterior clinoids, being lateral to the sella, are involved late in the course of disease, if at all. A very large adenoma can extend up into the anterior portion of the third ventricle and, in rare cases, cause obstructive hydrocephalus. Adenomas rarely calcify. (See also Fig. 16–4.) If the adenoma has extended beyond the sella, angiography may reveal "opening" of the carotid siphon curve and elevation of the proximal cerebral arteries.

An adenoma can sometimes be demonstrated on a CT scan made through the base of the brain. The tumor appears as a sharply circumscribed density, but may be difficult to appreciate because of the bone densities in this area. The scan may be informative if the adenoma is large enough and has extended beyond the sellar confines.[18, 53, 54, 72]

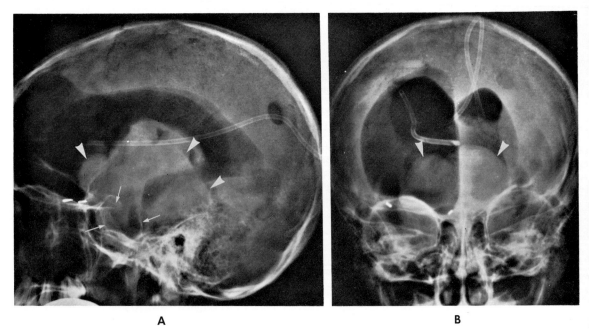

A   **B**

Figure 8–83   **Huge Chromophobe Adenoma with Suprasellar Extension: Ventriculogram.**

A, Lateral view discloses the characteristic sellar changes *(arrows)* produced by an intrasellar tumor. The sella is enlarged, the dorsum is thinned *(right small arrow)*, and the posterior clinoids are eroded. The floor of the sella is depressed into the sphenoid sinus.

The huge tumor *(arrowheads)* extends into the third ventricle area and has caused obstructive dilatation of the lateral ventricles. .

B, Anteroposterior view defines the border of the huge lobulated adenoma *(arrowheads),* and the dilated lateral ventricles. (The metallic clips around the right supraorbital area were placed there earlier during an unsuccessful attempt to remove the tumor.)

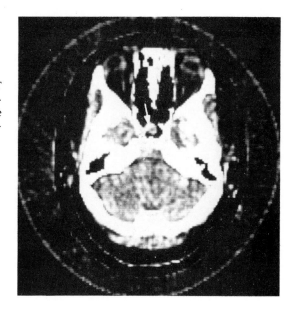

Figure 8–84 **Intrasellar Pituitary Adenoma: CT Scan.** A small rounded density (after intravenous injection of contrast material) is seen *(arrow)* within the bony confines of the sella turcica. This was an adenoma.

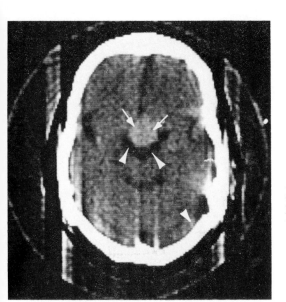

Figure 8–85 **Large Pituitary Adenoma: CT Scan.** The circumscribed area of increased density *(arrows)* is a large adenoma that has extended up into the lucent basal cisterns *(arrowheads)*.

Suprasellar and Parasellar Tumors

Although an intrasellar pituitary adenoma can extend into the suprasellar area, the commonest primary tumor of this region is the craniopharyngioma. This tumor is nearly always found in younger persons and does not produce hypopituitarism until very late in its course. About 70 per cent of craniopharyngiomas become calcified. The calcifications may assume various shapes and patterns but are usually stippled. The appearance and location of these calcifications are highly suggestive of the diagnosis. In rapidly growing craniopharyngioma, the calcifications are fine and flaky. Dense calcifications suggest a slower growing lesion. The tumor may directly erode the sella and simulate the findings of an intrasellar lesion.

Other tumors around the sella include gliomas of the optic chiasm, and meningiomas. The latter are usually lateral to the sella and may cause increased density in the sphenoid bone. Usually, pneumoencephalography demonstrates the suprasellar mass outlined by air in the basal cisterns. On the CT scan only fairly large lesions of the suprasellar area can be clearly recognized. Calcifications within a craniopharyngioma, however, are clearly demonstrated, even when such calcifications are too faint to be seen on conventional films. Angiography may show elevation of the carotid siphon and the anterior cerebral artery; this is most clearly seen on the anteroposterior arteriogram.[18, 53, 54, 72, 73]

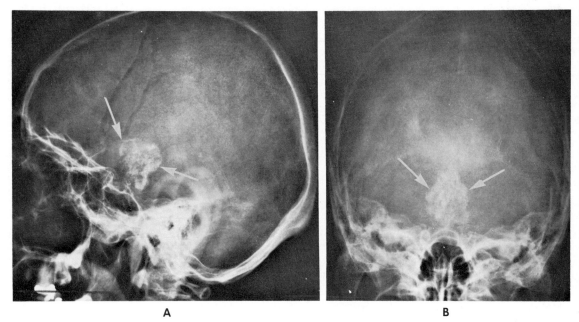

A   **B**

Figure 8–86   **Extensively Calcified Craniopharyngioma in a Child with Hypopituitarism.**

   A, The entire tumor is outlined as a calcified mass *(arrows)* in the suprasellar region. There is little or no change in the sella.
   B, In anteroposterior view the calcified tumor *(arrows)* is demonstrated in the midline just above the dorsum sellae.

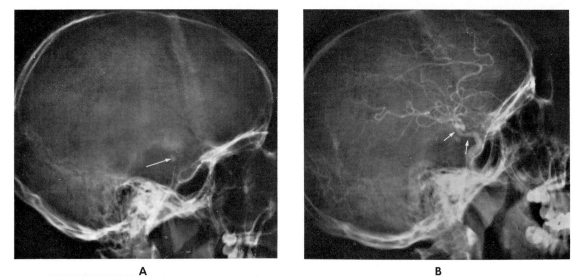

A   **B**

Figure 8–87   **Craniopharyngioma: Skull Film and Arteriogram.**

   A, Lateral view defines the small area of granular calcification *(arrow)* above the posterior clinoids of the dorsum sellae. No sellar changes are noted.
   B, Lateral arteriogram indicates elevation of the distal portion of the internal carotid *(arrows)* by the tumor mass. Calcification directly above the sella strongly suggests a craniopharyngioma, particularly in younger persons. Other suprasellar tumors often can be differentiated by the absence of calcification and by the off-center location on frontal projections.

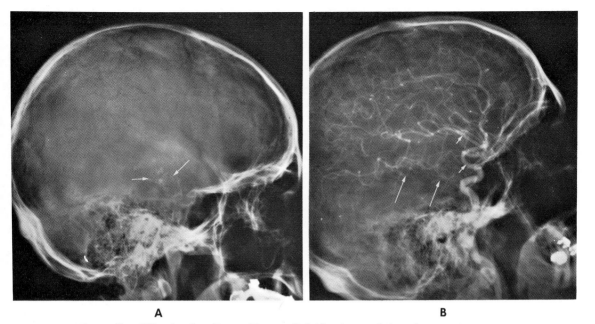

A   B

Figure 8–88 **Parasellar Oligodendroglioma: Tumor Calcification and Arteriogram.**

A, In lateral view there is irregular calcification above the sella *(arrows)*, simulating a suprasellar craniopharyngioma.

B, Left lateral arteriogram shows a collection of abnormal vessels in the tumor area *(long arrows)*. The distal portion of the internal carotid artery is elevated *(lower small arrow)*, and the proximal portion of the left middle cerebral artery is elevated and straightened *(upper small arrow)*.

The changes in the distal portion of the left carotid artery and in the proximal portion of the left middle cerebral artery allow localization of the lesion in the left parasellar area; tumor localization is more clearly demonstrated in anteroposterior arteriogram *(not shown)*.

Calcification occurs in oligodendrogliomas, but this tumor usually does not arise in the midline, in contrast to craniopharyngioma.

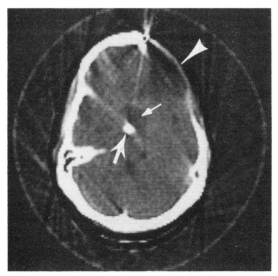

Figure 8–89 **Suprasellar Tumor: Recurrent Craniopharyngioma: CT Scan.**   The dense rounded area *(large white arrow)* in the suprasellar area is calcification in a recurrent craniopharyngioma. The anterior lucency *(small white arrow)* is uncalcified tumor and edema.

The extremely thin right frontal bone *(arrowhead)* is a surgical defect.

Metastatic Tumors of the Brain

Although metastases are reported to constitute about 20 per cent of all brain tumors, the true percentage is probably considerably higher. A primary neoplasm from almost any locus occasionally metastasizes to the brain, but the most common primary lesion is the bronchogenic carcinoma. For this reason a chest radiogram is essential whenever brain tumor is suspected.

Metastatic tumors may be single or multiple. Any area of the brain may be involved, and metastasis to the brain may or may not be associated with metastatic lesions of the skull.

On the CT scan a solitary metastatic lesion cannot be distinguished from a primary tumor. The metastatic lesion is most often an area of increased density generally surrounded by a large zone of low density (edema). This zone is usually greater than that seen in a primary tumor. However, like primary tumors, the density of a metastatic lesion may be low or identical to that of surrounding brain tissue, and not visible. Nearly all metastatic tumors show increased density after intravenous injection of contrast material. This should be done whenever a primary or metastatic lesion is strongly suspected and the regular scan appears negative.

In the proper clinical setting, multiple circumscribed areas of abnormal density are virtually diagnostic of metastatic brain disease.

The CT scan appears to be the most reliable method of detecting metastatic brain disease. It should probably be performed before definitive surgery for primary bronchogenic carcinoma is undertaken.

On angiographic or air studies a solitary metastasis is indistinguishable from a primary tumor in a given area. Multiple lesions suggest metastatic disease; these lesions are usually highly vascular and give rise to multiple tumor stains. Calcification of a brain metastasis is very rare; evidence of metastases to the bones of the skull suggests that the intracranial mass is of metastatic origin.

Metastatic neuroblastoma gives rise to an unusual and characteristic picture. Plaques of tumor tissue develop along the brain surface, so that the sutures widen and granular osteoporosis develops. There may be fine linear projections from both tables, and these are reactive changes attributable to diploetic and dural metastases. In infants and children these findings are virtually diagnostic, although leukemia occasionally causes similar changes.[18, 49, 62, 74, 75]

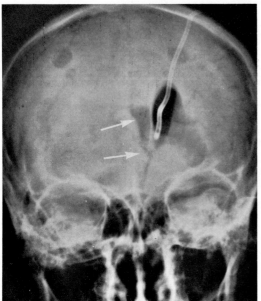

Figure 8–90 **Metastatic Tumor of Right Frontoparietal Area: Ventriculogram.** There is marked displacement of the lateral ventricles to the left, with evidence of a mass pressing against the right ventricle *(arrows)*. The roof of the right ventricle is depressed, chiefly because this ventricle has been forced under the falx, which remains in the midline.

This pneumographic appearance is nonspecific and is found in most sizable frontoparietal mass lesions.

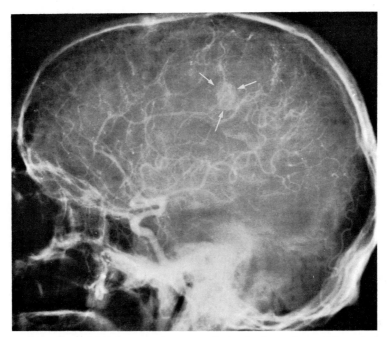

Figure 8–91 **Metastasis to Parietal Lobe: Angiogram.** Lateral view of carotid angiogram discloses a circumscribed tumor stain *(arrows)* that was due to a metastatic nodule from a carcinoma of the lung.

A sizable percentage of metastases to the brain yield a tumor stain on angiography; if single, the metastasis cannot be differentiated from a primary tumor or mass. Multiple mass lesions, with or without tumor stain, usually signify metastatic disease of the brain.

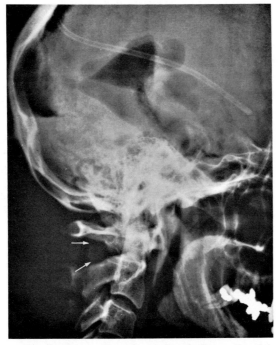

Figure 8–92 **Metastatic Breast Carcinoma to Vermis of Cerebellum: Combined Ventriculogram and Air Myelogram.** The patient had increased intracranial pressure with brain stem and cerebellar symptoms. The ventriculogram disclosed internal hydrocephalus due to a cerebellar mass that was occluding the aqueduct.

Air myelogram indicated that a rounded mass *(arrows)* was projecting into the spinal subarachnoid space. This mass was from herniation of a cerebellar tonsil.

Metastatic lesions to the cerebellum are infrequent.

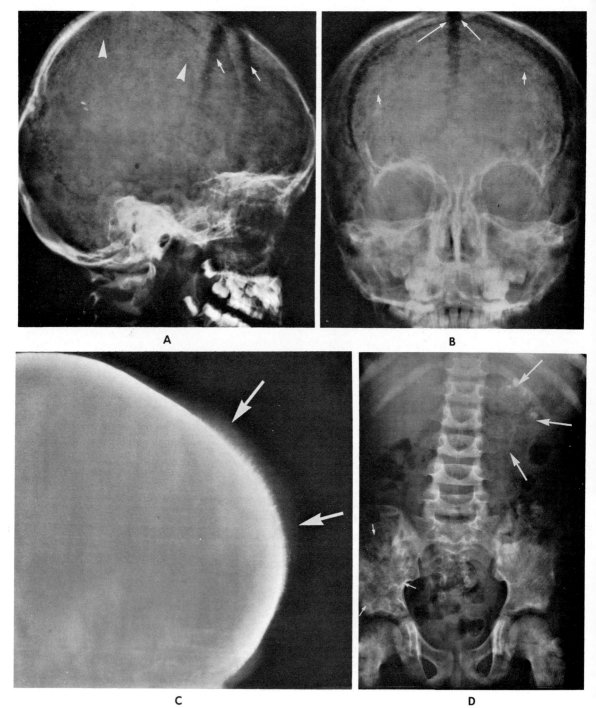

A

B

C

D

Figure 8–93 **Neuroblastoma with Cerebral Metastasis.**

    A, Lateral view shows separation of the coronal suture *(arrows).* Diffuse mottled osteoporosis has produced a granular appearance *(arrowheads).*

    B, Posteroanterior view discloses separation of the sagittal suture *(long arrows)* and diffuse spotty osteoporosis *(small arrows).*

    C, In soft tissue film, vertical bony striations *(arrows)* are demonstrated; these are often seen in metastatic neuroblastoma. They are reactive bone changes brought about by the underlying malignant plaques. Metastatic plaques along the brain surface are unique to metastatic neuroblastoma. In a child the findings are virtually pathognomonic of the disease, although, occasionally, similar findings are seen in leukemia.

    D, View of the abdomen shows irregular calcification in a neuroblastoma *(large arrows)* that obscures the left kidney shadow. The pelvic bones are diffusely involved by metastatic disease, so that there is considerable irregularity in the bone density *(small arrows).*

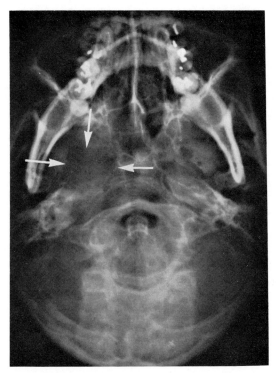

Figure 8–94 **Bone Metastasis of the Base of the Brain.** The patient had a reticulum cell sarcoma, and developed neurologic symptoms referable to the base of the brain.

A submentovertex view demonstrates a large irregular area of bone destruction in the right sphenoid bone *(arrows)* in the region of the foramen ovale. The lesion had invaded the base of the brain.

Metastases to the bones of the skull may occur in the absence of brain involvement, but when metastases to the bones are associated with neurologic symptoms, metastases to the brain substance or direct extension of the bone lesions to the brain has probably occurred.

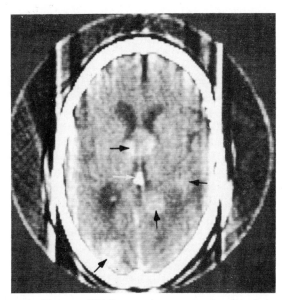

Figure 8–95 **Multiple Metastatic Tumor: CT Scan.** A scan made after intravenous injection of contrast material distinctly shows four masses of increased density *(black arrows)*. Zones of lucency due to edema surround most of the metastases. The pineal *(white arrow)* and the ventricles are not displaced. The centrally located anterior metastasis lies just behind the anterior horns, which are somewhat dilated.

Most metastatic lesions to the brain are highly vascular, and may show up only after contrast material is administered intravenously. Large zones of surrounding edema are characteristically encountered. Multiplicity of such lesions on the scan is virtually diagnostic of metastatic disease.

Tuberculoma of the Brain

This rare lesion is always secondary to an active or inactive tuberculosis focus elsewhere. It is generally solitary and is found most often in the cerebellum or cerebral cortex. It may range from a few millimeters to 3 to 4 cm. in diameter.

Calcification occurs in only 3 to 8 per cent of cerebral tuberculomas and is generally a sign of a healed or quiescent lesion. The calcification may appear as an incomplete or broken ring, or as a group of calcaneous deposits that simulate tumor calcification.

Angiographic or pneumographic studies will disclose a nonspecific mass lesion. Increased intracranial pressure and ventricular dilatation may result from a tuberculoma in the posterior fossa.

In an adult with evidence of tuberculosis, a brain tuberculoma should be included in the differential diagnosis of an intracranial mass lesion with or without calcification.[76-78]

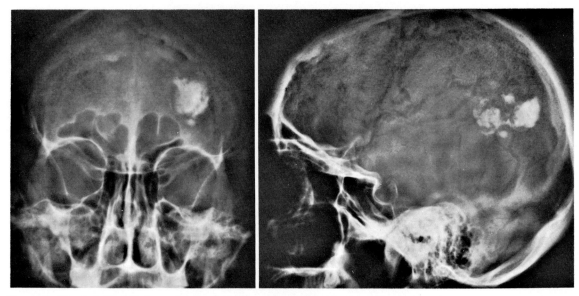

Figure 8–96 **Calcified Tuberculoma of Brain, Simulating Tumor.** In posteroanterior and lateral views numerous dense mottled calcifications are found in the left posterior parietal area. The patient was a 29 year old woman who had chronic pulmonary tuberculous lesions.

Although the brain lesion was not verified surgically, the history and roentgenologic findings strongly suggested a calcified tuberculoma.

Benign Intracranial Hypertension (Pseudotumor Cerebri)

This uncommon condition is characterized by increased intracranial pressure without dilatation of the ventricular system and is probably due to cerebral edema. It is apparently more common in females, and the majority of patients are under age 40. The cause is unknown in most cases; in a small percentage, the condition is apparently caused by dural sinus thrombosis. Pseudotumor from sinus thromboses has been a complication in several cases of Hughes-Stovin syndrome (see p. 548).

On regular skull films the increased intracranial pressure may cause demineralization of the dorsum sellae in adults, and increased digital markings or widened sutures in children.

On the CT scan, the ventricular system will appear small or normal, a significant finding in the presence of increased intracranial pressure.

Pneumography and angiography will also be normal, and can be obviated if a CT study is available.

If pseudotumor is due to dural sinus thrombosis, direct opacification, via catheter, of the superior sagittal sinus (sinusography) will demonstrate an occluded sagittal or transverse sinus and filling of numerous venous collaterals.

In the presence of increased intracranial pressure, a small or normal ventricular system with no midline shift is characteristic of pseudotumor cerebri.

Decrease of the intracranial pressure and long-term improvement have been achieved by lumbar subarachnoid–peritoneal shunt.[5, 6, 79-83]

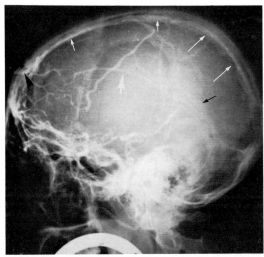

Figure 8–97 **Pseudotumor Cerebri Due to Thrombosis of Superior Sagittal Sinus: Sinusogram.** This study was made after a pneumoencephalogram had proved negative in a 28 year old man with chronic increased intracranial pressure.

The contrast material was injected via a catheter in the superior sagittal sinus *(black arrowhead)*. The anterior portion of the sinus *(short white arrows)* is filled, but the posterior portion is not opacified *(long white arrows)*.

Collateral venous channels are filled anteriorly, and the inferior sagittal sinus *(large white arrow)* and the straight sinus *(black arrow)* are also filled, thus corroborating a block in the superior sagittal sinus.

In a normal study there is filling only of the superior sagittal and transverse sinuses, which empty into the jugulars. No other venous channels are opacified. (Courtesy Drs. B. S. Ray and N. S. Dunbar, New York Hospital, New York, New York.)

Arnold-Chiari Malformation

Downward displacement of the cerebellar tonsils and brain stem through the foramen magnum may occlude the foramina of Luschka and Magendie and produce an obstructive hydrocephalus. Other congenital malformations of the upper cervical spine and foramen magnum are often associated: particularly, absence of the arches of the atlas and some degree of fusion with the occiput, and an enlarged foramen magnum. Other frequent associations include basilar impression, a shallow posterior fossa, and lumbar meningocele.

Pneumoencephalography or myelography is necessary to demonstrate the displaced brain stem and medulla. The cerebellar tonsils may be seen projecting into and narrowing the spinal subarachnoid space just below the foramen magnum. If obstruction is incomplete, pneumoencephalography will outline somewhat dilated ventricles. With complete obstruction, ventriculography is required to confirm ventricular dilatation and to demonstrate obstruction below the fourth ventricle, which is deformed, elongated, and displaced downward. The ventricular changes are readily seen on CT scan.

A high proportion of patients show a deformity of the anterior wall of the third ventricle and an unusually large massa intermedia. These findings are helpful when the fourth ventricle cannot be adequately visualized.[18, 84-87]

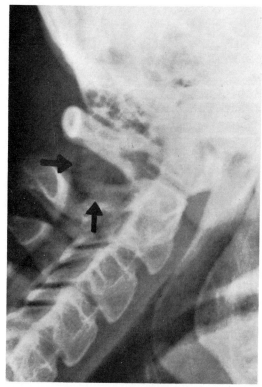

Figure 8–98 **Arnold-Chiari Malformation: Pneumoencephalogram.** Lateral film of the upper cervical area demonstrates narrowing of the subarachnoid air column just below the foramen magnum. A soft tissue mass *(arrows)* represents the herniated cerebellar tonsils protruding through the foramen.

Internal Hydrocephalus (Obstructive or Hypertensive Hydrocephalus)

In infants and children, internal hydrocephalus can occur from obstruction of the foramen of Monroe, the third ventricle, the sylvian aqueduct, the fourth ventricle, or the foramina of Luschka and Magendie (Dandy-Walker syndrome). The obstruction may be caused by congenital malformation, inflammatory process, neoplasm, previous surgery, or trauma.

Skull films will reveal an enlarged cranium with thinned tables and separated sutures. Increased convolutional markings appear if the ventricular dilatation continues or occurs after closure of the sutures. When the obstruction is proximal to the fourth ventricle, the posterior fossa is small and shallow. In the Dandy-Walker syndrome, in which there is a greatly dilated fourth ventricle, the posterior fossa is enlarged.

The CT scan is admirably suited for noninvasive investigation of hydrocephalus. Dilatation of the ventricles is readily apparent. In aqueductal stenosis, only the fourth ventricle remains small. In the Dandy-Walker syndrome, a huge fourth ventricle is apparent. The peripheral sulci and cisterns are generally small.

Distinction of obstructive hydrocephalus from the ventricular dilatation associated with brain atrophy can usually be made by the absence of dilated sulci and cisterns in obstructive hydrocephalus. Clinical distinction is also rarely difficult.

After shunting, serial scan will reveal the location of the shunt and its effect on ventricular size. Complicating postshunt collections of subdural blood or fluid resulting from too rapid brain shrinkage can easily be identified.

Ventriculography discloses the degree of ventricular dilatation and can usually demonstrate the site and sometimes the cause of the obstruction. Midline tomography may be especially helpful. Stenosis of the aqueduct of Sylvius is the commonest cause of obstruction and produces great dilatation of the third and lateral ventricles. In the Dandy-Walker syndrome the foramina of Luschka and Magendie are occluded and the entire ventricular system, particularly the fourth ventricle, is dilated. Small amounts of air introduced into the spinal subarachnoid space may delineate the lower end of the obstruction and will differentiate this entity from communicating hydrocephalus.

Cerebral angiography usually demonstrates the changes of nonspecific ventricular dilatation, which include stretching of the anterior cerebral arteries over the dilated lateral ventricles. On frontal view there is increased space between the anterior and middle cerebral arteries, inferolateral displacement and curvature of the lenticulostriate arteries, and increased curvature of the thalamostriate veins.

When ventriculoatrial tube shunt is used for decompression of internal hydrocephalus, both immediate and late complications may be demonstrable radiographically. Acute complications resulting from too rapid decompression include overlap of cranial bones, subdural hematoma, and acute pulmonary edema. Late complications include premature closure of the sutures, thickening of the calvaria, and thrombosis of the right atrium and superior vena cava. The latter can cause recurrent pulmonary emboli with eventual pulmonary hypertension and cor pulmonale.[1-6, 18, 88-91]

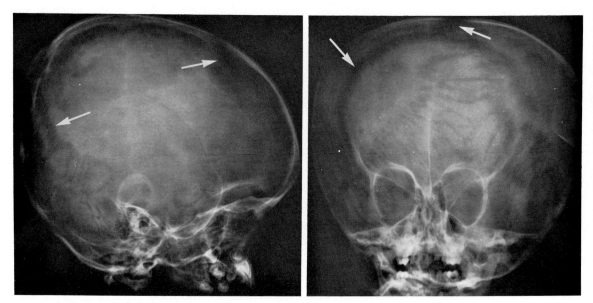

Figure 8–99 **Internal Hydrocephalus: Congenital.**

    A, Lateral view discloses characteristic disproportion between the enlarged cranium and the facial bones. The sutures are separated *(arrows).* The posterior fossa is disproportionately small.

    B, Posteroanterior projection demonstrates very clearly the disproportion between the cranium and the facial bones. Separation of the sutures is clearly demonstrated *(arrows).*

    The small size of the posterior fossa suggests that the obstruction lies above the fourth ventricle, probably in the aqueduct.

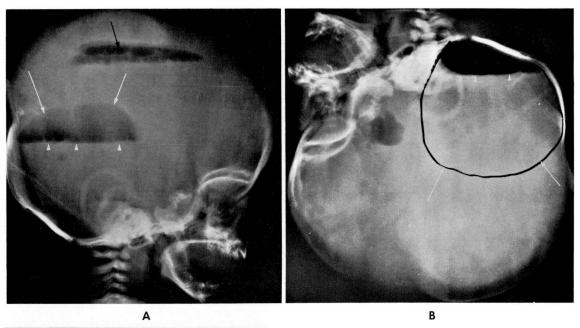

A

B

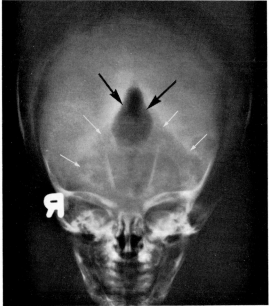

C

Figure 8–100 **Internal Hydrocephalus: Dandy-Walker Syndrome: Ventriculogram.** Lateral views of ventriculogram with the patient in the erect (*A*) and upside down (*B*) positions demonstrate enormous fourth ventricle (*white arrows*) with air-fluid levels (*arrowheads*). The black circle on the upside down film indicates the probable size of the ventricle. The dilated lateral ventricle is shown in *A* (*small black arrow*).

C, Occipital projection discloses enlargement of the third ventricle (*black arrows*) and massive enlargement of the fourth ventricle (*white arrows*).

The obstruction was in the foramina of Luschka and Magendie. On plain films the posterior fossa appears enlarged because of the greatly dilated fourth ventricle.

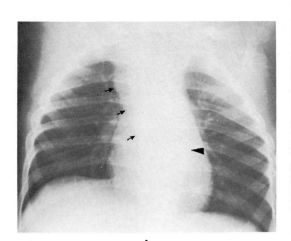
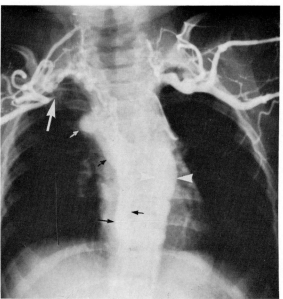

A   **B**

Figure 8–101   **Ventriculoatrial Shunt for Hydrocephalus: Thrombosis of Superior Vena Cava and of Right Subclavian and Innominate Veins.**

A, Plain chest film, made two and a half years after ventriculoatrial tube shunt, reveals soft tissue densities of a dilated azygos vein *(small arrows)* and of the hemiazygos vein *(arrowhead).*

B, Bilateral antecubital vein injection reveals nonfilling of the superior vena cava, obstruction of the right subclavian and innominate veins *(large white arrow),* and opacification of the dilated azygos *(small black and white arrows)* and hemiazygos *(arrowheads)* veins. (Courtesy Dr. G. J. Kurlander, Indianapolis, Indiana.)

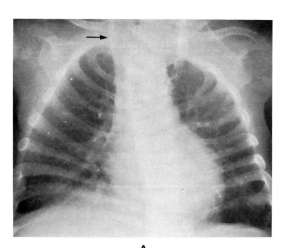
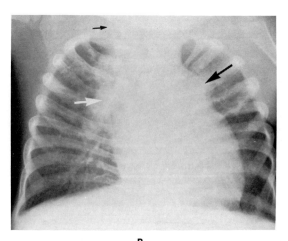

A   **B**

Figure 8–102   **Pulmonary Hypertension Following Ventriculoatrial Shunt.**

A, Five months after insertion of ventriculoatrial tube *(arrow),* the heart and lung fields appear normal.

B, Two and a half years later the heart is greatly enlarged, and the main pulmonary artery segment *(large black arrow)* is dilated. Both main pulmonary arteries are greatly enlarged *(white arrow),* although the left artery is obscured by the cardiac silhouette. The peripheral vessels attenuate rapidly. The picture, characteristic of severe pulmonary hypertension, results from small emboli repeatedly arising from the right atrium or superior vena cava. The ventriculoatrial tube is faintly visualized *(small arrow).*

It has been recommended that every patient with a ventriculoatrial shunt have periodic chest examinations for this serious complication; the shunt should be removed before cor pulmonale develops. (Courtesy Dr. G. J. Kurlander, Indianapolis, Indiana.)

Figure 8–103 **Moderate Obstructive Hydrocephalus: CT Scan.** There is symmetric moderate dilatation of the lateral ventricles and dilatation of the third ventricle *(arrow)*.

The fourth ventricle was small (seen on more caudal section). Aqueductal stenosis was the cause.

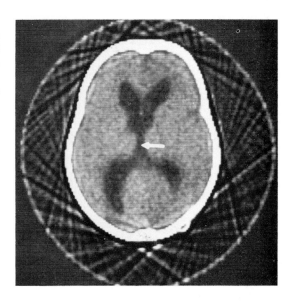

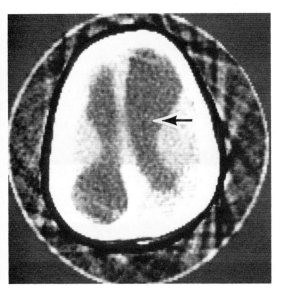

Figure 8–104 **Marked Obstructive Hydrocephalus: CT Scan.** The bodies of both lateral ventricles *(arrow)* are enormously dilated. Other sections showed dilatation of the third but not of the fourth ventricle. Aqueductal stenosis was the cause.

Figure 8–105 **Obstructive Hydrocephalus with Shunt Localization: CT Scan.** The lateral ventricles are enormously dilated and the shunt is easily identified *(arrows)*, extending from the anterior right lateral ventricle to the subarachnoid space.

The entire shunt is rarely seen in a single scan "cut," but its complete course can be readily visualized from several cuts at different levels.

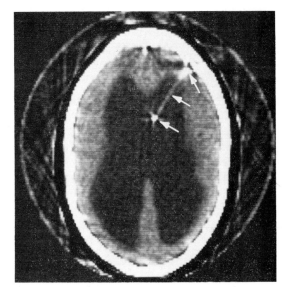

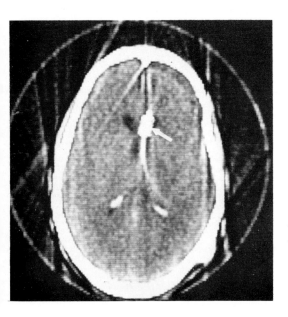

Figure 8–106 **Obstructive Hydrocephalus; Return of Ventricles to Near Normal After Shunt: CT Scan.** The ventricles, which had been moderately dilated, are now almost of normal size. The large shunt tip is clearly seen *(arrow)* in the right anterior horn.

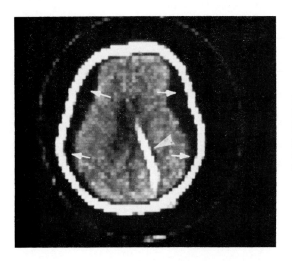

Figure 8–107 **Obstructive Hydrocephalus: Rapid Shunting Causing Brain Shrinkage: CT Scan.** The large shunt (*arrowhead*) has drained the previously greatly dilated ventricles so rapidly that the brain has shrunk and "hygromatous" fluid (*arrows*) surrounds the hemispheres.

INJURIES OF THE HEAD AND SPINE

Skull Fracture, Subdural and Epidural Hematoma, Traumatic Arteriovenous Fistula, and Intracerebral Hemorrhage

Multiple skull projections, including tangential views, may be needed to demonstrate some fractures. Trauma to the brain cannot be correlated with the size and extent of the fracture; conversely, severe cerebral damage frequently occurs in the absence of fracture.

The CT scan and arteriography are used to investigate possible complications, including subdural and epidural hematoma, focal intracranial hemorrhage or edema, and post-traumatic arteriovenous fistula.

The CT findings in subdural hematoma and intracranial hemorrhage are discussed elsewhere (see pp. 302 and 309). Focal edema will appear as an irregular area of diminished density. An arteriovenous fistula may be difficult to recognize but may become apparent after intravenous injection of contrast material.

In an acute unilateral subdural hematoma, arteriography discloses a crescentic avascular defect between the inner table and the inwardly displaced peripheral cortical vessels. The anterior cerebral artery is shifted to the opposite side unless the hematoma is quite small. Absence of a shift in the presence of a sizable hematoma should suggest a second hematoma on the opposite side. These findings are seen on the frontal projection. Lateral films are generally unrevealing unless the middle cerebral arterial group is elevated by a subtemporal hematoma.

As the hematoma becomes chronic, it usually becomes fusiform, and the avascular defect may assume a lenticular shape with a convex medial border, in contrast to the usual concave medial border of an acute lesion.

An epidural hematoma, which forms rapidly after head injury, usually occurs over the parietotemporal convexity and is generally associated with a fracture extending across the middle meningeal artery area. The angiographic picture resembles a subdural hematoma, but the avascular zone is usually of a lenticular shape, thereby simulating a chronic subdural hematoma radiographically but not clinically. Other angiographic findings, although seen in a minority of cases, include displacement or spasm of the middle meningeal artery, extravasation of contrast material, and displacement of a major venous sinus away from the skull table.

Intracranial hemorrhage and edema, if bilateral and diffuse, are difficult to recognize angiographically unless serial examinations are performed. Unilateral or focal hemorrhage or edema produces changes resembling an intracranial mass lesion. Occasionally a contused area of brain tissue will show contrast staining and early venous filling.

Traumatic arteriovenous fistulas generally arise between the internal carotid artery and the cavernous sinus. On a carotid arteriogram there is opacification of the cavernous sinus during the arterial filling phase. The ophthalmic vein is greatly dilated and rapidly opacified by reverse flow from the cavernous sinus. If the fistula is large, the rapid shunting prevents satisfactory opacification of the intracranial arterial tree.[30, 31, 92-94]

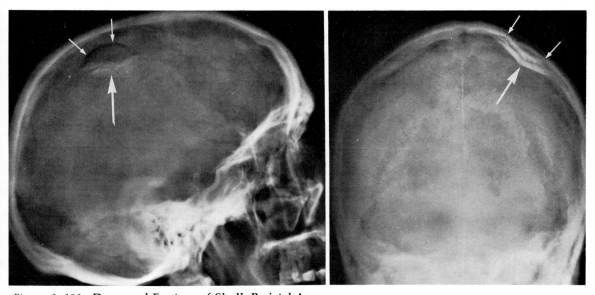

Figure 8-108 **Depressed Fracture of Skull: Parietal Area.**

A, In lateral view a curved radiolucent line *(small arrows)* is shown in the parietal region, with an adjacent area of increased density *(large arrow)* crossed by several small fracture lines. The area of increased density signifies overlapping of bones due to depression of the tables.

B, Anteroposterior projection (tangential to the lesion) clearly shows the fractures and the depression *(arrows)* of the inner and outer tables. The depression sometimes involves the inner table only if the outer table recoils into normal position.

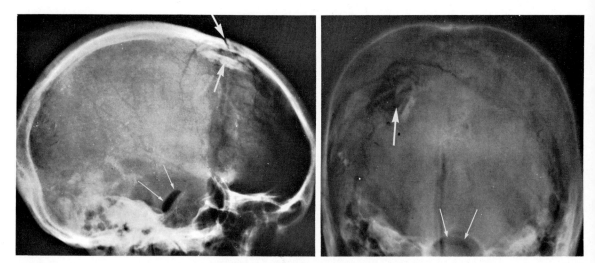

Figure 8–109 **Depressed Fracture of Skull, with Air at Base and in Subarachnoid Space.** This typical, depressed skull fracture shows the lucent area and double density due to the depressed fragment *(large arrows)*. The fracture has extended to the subarachnoid space, so that air has entered the space and filled the cisterns *(small arrows)*.

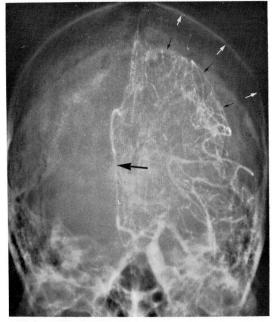

Figure 8–110 **Acute Subdural Hematoma: Arteriogram.** The terminal vessels *(small black arrows)* have separated from the inner tables *(small white arrows),* and a large clot fills the space. The anterior cerebral arteries *(large black arrow)* are shifted to the other side. These findings are characteristic of a unilateral subdural hematoma. If there is little or no displacement of the anterior cerebral artery, a second hematoma on the opposite side should be considered.

In the acute hematoma the shape is crescentic, since the hematoma follows the contour of the brain.

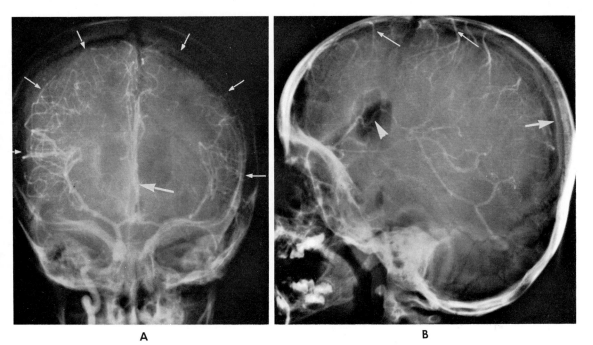

A   B

Figure 8–111   **Bilateral Acute Subdural Hematoma: Arteriogram.**

A, This is an anteroposterior projection of a bilateral angiogram. The peripheral cerebral vessels have been shifted away from the inner table on both sides *(small arrows)* by extensive bilateral subdural hematomas. The anterior cerebral arteries have not been shifted *(large arrow)*.

B, In a lateral view, venous phase, the sagittal sinus *(large arrow)* and its tributary veins *(small arrows)* are *not* separated from the inner table. This finding confirms that the hematoma is subdural and not epidural. (Compare with Fig. 8–114.)

Air in the ventricles *(arrowhead)* is from an earlier pneumoencephalogram.

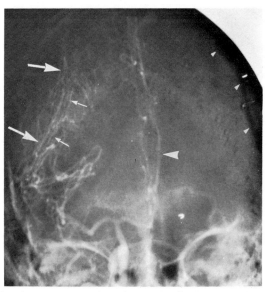

Figure 8–112   **Chronic Subdural Hematoma: Arteriogram.**   Anteroposterior view of right carotid arteriogram discloses that the peripheral vessels *(large arrows)* are shifted away from the inner table in the parietal area, and the inward bowing indicates the presence of a fusiform subdural mass. The displaced vessels are crowded together *(small arrows)*. The anterior cerebral artery is bowed to the left *(large arrowhead)*. At surgery a fusiform chronic subdural hematoma was found in the right parietal area.

Fracture and metal suture on the left side *(small arrowheads)* are from the original injury and subsequent removal of an acute subdural hematoma on the left side.

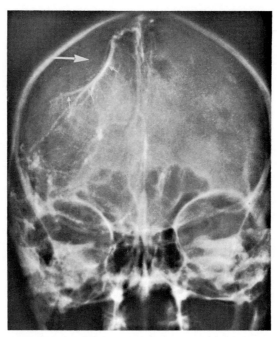

Figure 8–113 **Chronic Subdural Hematoma.** In venous phase of angiogram the brain tissue and vessels are displaced medially by a large subdural mass in the avascular area *(large arrow)*. The bowed vein of Trolard *(small arrows)* defines the convex medial border of the hematoma. The venous structures are in their normal midline position, suggesting the possibility of another subdural hematoma on the opposite side.

A subdural hematoma that becomes chronic often liquefies, absorbs fluid, and becomes fusiform, characteristically producing medial bowing of the displaced vessels. In acute subdural hematoma the displaced brain and vessels retain their normal convexity, paralleling the skull tables. However, differentiation cannot always be made.

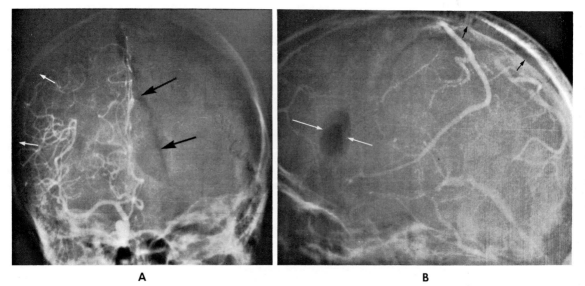

A   B

Figure 8–114 **Epidural Hematoma: Arteriogram.**

A, Anteroposterior projection shows a crescentic space between the terminal vessels of the hemisphere and the inner table in the right parietal region *(white arrows)*. There is no evidence of midline shift. This picture is indistinguishable from that of subdural hematoma. Linear skull fracture *(black arrows)* is evident.

B, In lateral projections the sagittal sinus and adjacent veins are displaced *(black arrows)* from the inner table, indicating the epidural location of the hematoma. Air *(white arrows)* in the anterior horn is from an earlier pneumoencephalogram.

Epidural hematomas are less common than subdural hematomas and are almost always associated with skull fracture.

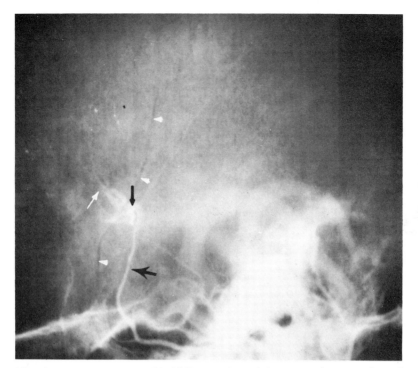

Figure 8–115 **Epidural Hematoma: Tear of Middle Meningeal Artery.** There is a long fracture line *(white arrowheads)* in the temporoparietal area. The opacified anterior branch of the middle meningeal artery *(large black arrow)* shows abrupt narrowing and extravasation of contrast material *(small black arrow)* where it crosses the fracture line. A later film showed contrast material within the fracture line.

Opacified veins *(white arrow)* indicate a partial arteriovenous post-traumatic fistula.

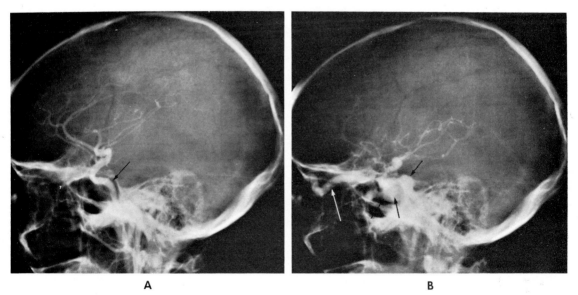

A   B

Figure 8–116 **Traumatic Cavernous Sinus Fistula: Arteriogram.**

A, Early phase of a cerebral arteriogram shows normal filling of the carotid siphon *(arrow)* and its branches.

B, In later arterial phase, there is complete opacification of the cavernous sinus *(black arrows)* and the ophthalmic vein *(white arrow).* Limited opacification of the intracranial arteries is caused by shunting of contrast material into the veins.

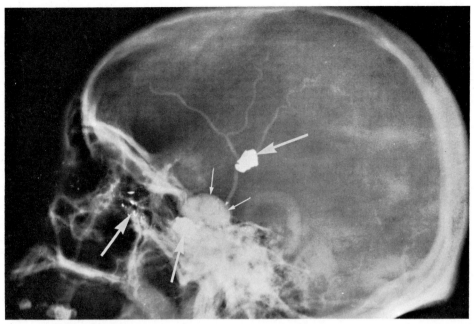

Figure 8–117 **Traumatic Cavernous Sinus Fistula: Arteriogram.** There are two large metallic bodies and several smaller ones in the cranium *(large arrows)*; most of them are impacted in the base of the skull. Opacification of the cavernous sinus *(small arrows)* has occurred during the arterial phase; there was no filling of the intracranial branches owing to the shunt. Traumatic arteriovenous fistula can follow nonpenetrating trauma to the skull or, most frequently, direct injury to the carotid artery and cavernous sinus by a projectile.

Hematomyelia

Hemorrhage into the spinal cord tissue usually follows trauma and always produces acute neurologic symptoms. Plain radiographs are useful in determining whether or not there is associated fracture or dislocation. With myelography, the spinal cord may appear dilated, causing narrowing of the subarachnoid space and, sometimes, complete obstruction of the oil column. The history aids in differentiation from other intramedullary lesions.[95]

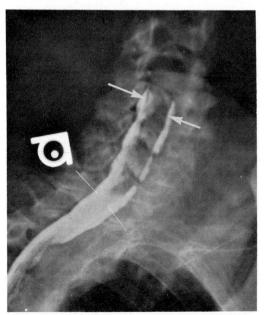

Figure 8–118 **Hematomyelia of Cervical Cord Following Trauma: Myelogram.** Paraplegia developed in an adult following trauma to his neck. The myelogram indicates considerable widening of the cord in the midcervical area *(large arrows)*; compare with appearance of cord in lower cervical area, between small arrows. There was a complete block at the level of the cord swelling, and oil could not flow beyond this point. Swelling of the cord demonstrated on a myelogram suggests hematomyelia if there has been recent trauma.

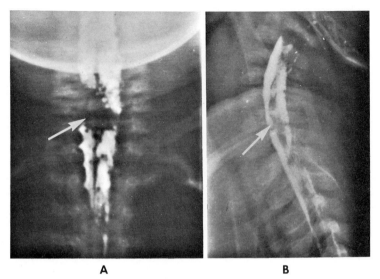

A   B

Figure 8–119   **Traumatic Hematomyelia in a Child: Myelogram.**

   A, In anteroposterior view there is a localized central defect *(arrow)* in the contrast column at the lower cervical level; it was caused by expansion of the spinal cord by intramedullary hemorrhage.

   B, Lateral view shows the fusiform defect *(arrow)* in the expanded cord.

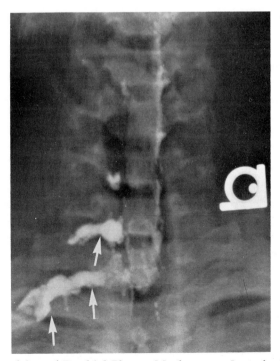

Figure 8–120   **Traumatic Avulsion of Brachial Plexus: Myelogram.**   Irregular long outpouchings *(arrows)* of contrast medium extend along the right side at the interspaces of C5–C6 and C6–C7. These outpouchings of contrast medium are re-formations of the subarachnoid lining through the dural tears at the site of nerve root avulsion. The findings are characteristic.

Paraplegia

Roentgenograms often disclose the nature and location of the underlying lesion in cases of paraplegia. Plain films and myelograms may disclose evidence of trauma, primary or metastatic tumors, or infectious processes. (See pp. 406 to 423.)

Many complications of paraplegia may be disclosed radiographically. Cord bladder, a very common complication, appears in the early stage as a large distended bladder; there is pronounced retention of urine. When the condition becomes chronic, the bladder is shrunken, contracted, and highly trabeculated, and there is often evidence of secondary infection or calculus formation. Frequently, vesicoureteral reflux, hydronephrosis, and pyelonephritis develop and can be recognized radiographically.

In longstanding paraplegia, soft tissue calcification around the paralyzed joints, particularly the hips, often develops. Subluxation of joints is a common occurrence.[96-98]

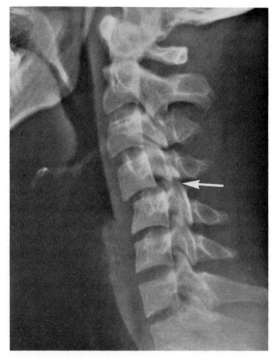

Figure 8–121 **Dislocation of Cervical Spine: Paraplegia.** Following trauma, lateral view shows anterior shifting of vertebral body C4 over C5 and sharp angulation of the canal, owing to dislocation of the corresponding intervertebral joints *(arrow)*. The lordotic curve is reversed at the level of the dislocation. Such an injury can cause compression of the cord and is a frequent cause of paraplegia.

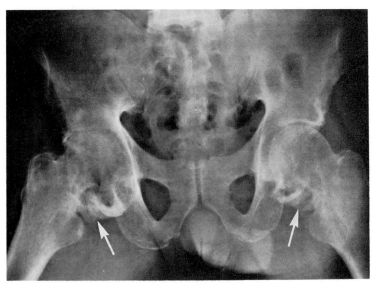

Figure 8–122 **Paraplegia: Soft Tissue Calcification.** Extensive soft tissue calcium deposition has occurred *(arrows)* about the lower aspect of both hip joints. Periarticular calcification in the chronic paraplegic can be minimal or extensive.

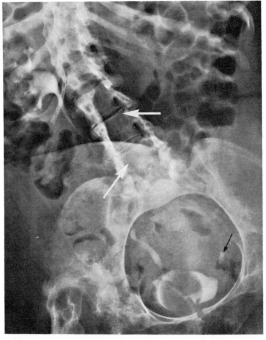

Figure 8–123 **Cord Bladder: Intravenous Pyelogram.** The contrast-filled bladder is small and contracted, and did not enlarge during urography. The rounded defect is a Foley catheter. A calculus *(black arrow)* in the left ureter has caused obstruction, with nonfilling of the left urinary collecting system. The paraplegia was due to a meningocele associated with severe spina bifida *(white arrows)*. There was associated congenital dislocation of the hips.

The early cord bladder is large and does not contract; later it becomes small and irregular. Stasis and reflux are common and encourage infection and the formation of calculi.

DEGENERATIVE AND HEREDOFAMILIAL DISEASES OF THE CENTRAL NERVOUS SYSTEM

Spina Bifida, Spinal Meningocele, and Meningomyelocele

An asymptomatic and innocuous cleft defect of the neural arch in the region of the spinous process (spina bifida occulta) occurs in approximately 20 per cent of individuals. These defects are most common in the lower lumbar and upper sacral spine, with the cervical spine next in order of frequency.

Larger lumbar and cervical defects are often accompanied by herniation of the meninges (meningocele) or of the meninges and a portion of the cord or nerve roots (meningomyelocele). In these cases the large bony defect also includes absence of the laminae and, often, an increased interpediculate distance, which is due to a widened spinal canal. The sac containing the meningocele or meningomyelocele can often be detected as a soft tissue mass posterior to the spine, usually in the lumbosacral area.

Occult meningoceles of the sacral area may produce no bony changes until late in infancy; detection requires myelography. Large meningomyeloceles in infants can cause herniation of the cerebellar tonsils. Other congenital defects frequently accompany meningocele, including dislocated hips, various osseous spinal defects, and visceral malposition.

Cord or root compression in a meningomyelocele can produce a neurogenic bladder and neurologic changes in the lower extremities. If surgery is contemplated, gas myelography with tomography is useful for identifying nerve roots and other neurogenic tissue within the sac.[18, 99-104]

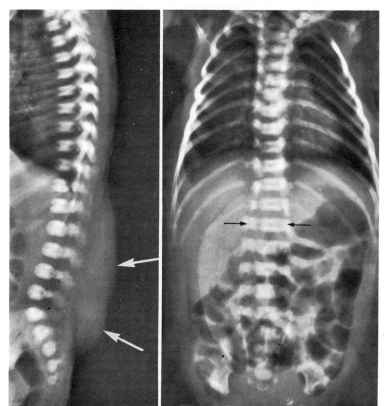

Figure 8–124 **Lumbar Meningocele: Spina Bifida.**

A, In an infant, lateral view discloses in the lumbosacral region a large soft tissue mass *(arrows)* due to a meningocele. Often these masses are smaller and less obvious, both clinically and radiographically.

B, Anteroposterior view demonstrates widening of the lumbar interpediculate space *(arrows)* due to a widened canal and unfused neural arches. The defects in the laminae cannot be detected before ossification.

A   B

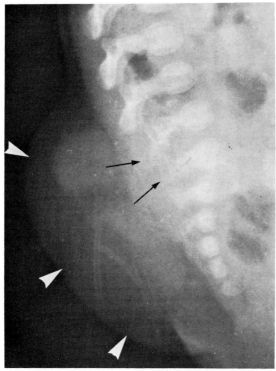

Figure 8–125 **Meningomyelocele.** In a newborn, lateral view discloses an abnormal soft tissue mass in the lumbosacral region *(arrowheads)*. There are somewhat ill-developed spinous processes in this area *(arrows)*. In this case the cauda equina was also herniated.

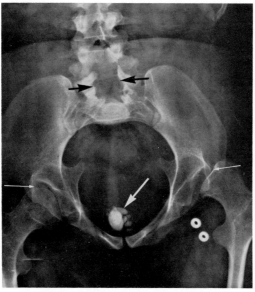

Figure 8–126 **Lumbosacral Spina Bifida: Neurogenic Bladder.** There is a large defect of the lamina and spinous processes of L5 and the upper sacral segments *(black arrows)*. Calculi of the bladder *(large white arrow)* are a consequence of stasis infection in this neurogenic bladder. There is dysplasia of the right hip and congenital dislocation of the left hip *(small white arrows);* various skeletal anomalies may coexist with spina bifida.

Diastematomyelia

Diastematomyelia is a congenital splitting of the cord by a bony or fibrocartilaginous septum. It occurs most frequently in the lower thoracic and upper lumbar segments and is often associated with skeletal and central nervous system anomalies.

Typically, there is fusiform widening of the canal, with an increase in interpediculate distance that extends over several segments. Associated anomalies are always encountered, including hemivertebrae, spina bifida, block vertebrae, and malformed neural arches. Scoliosis is very frequent. The presence of anomalies and the absence of pedicle erosion will help in differentiation from an expanding intraspinal mass lesion. In about half the cases the septum is ossified and can be seen on regular spine films or laminograms. This is a characteristic and specific finding.

Myelography is definitive, demonstrating the splitting of the contrast material around a round or oval midline defect.[18, 105-108]

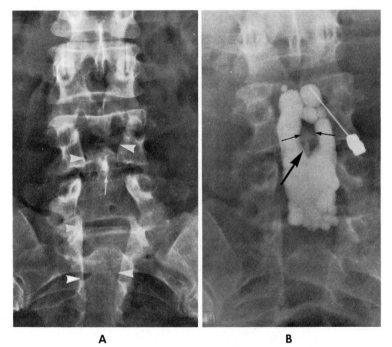

A   B

Figure 8–127 **Diastematomyelia.**

A, In an adult, frontal view indicates widening of the spinal canal in the lumbar region, with an extensive spina bifida *(arrowheads).* The ossified spicule at the level of L3 *(arrow)* represents the calcified portion of the septum.

B, Myelogram demonstrates characteristic division of the subarachnoid space *(small arrows).* The central defect represents the septum, only part of which is calcified *(large arrow).*

Syringomyelia

Syringomyelia refers to an enlargement of the central canal of the spinal cord, beginning most frequently in the cervical region and usually extending downward. As the disease progresses, widening of the spinal cord may ensue. The findings are dependent on the degree of enlargement of the cord. On myelography there may be fusiform dilatation of the cervical cord, with marked cord widening. The surrounding subarachnoid spaces may be narrowed or obliterated.

Further enlargement of the cord causes widening of the entire spinal canal, and erosion and flattening of the pedicles is demonstrable on plain films. The anteroposterior diameter of the spinal canal may be increased; usually, however, little or no change is seen on spinal radiographs. In early or minimal cases the myelogram also is normal. Frequently, the dilated cord collapses when the head is raised; this is best demonstrated by air myelography. This is virtually a pathognomonic finding and will exclude other intramedullary lesions.

Syringomyelia may be associated with other central nervous system deformities such as Arnold-Chiari syndrome, craniosynostosis, or basal arachnoiditis. In well-advanced cases, there may be radiographic evidence of a neuropathic (Charcot) joint, usually in an upper extremity. (See Fig. 4–48.)[109-113]

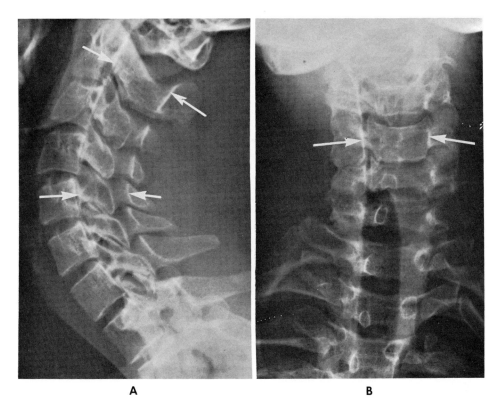

A   **B**

Figure 8–128   **Syringomyelia: Cervical Spine.**

A, Anteroposterior diameter of the spinal canal (*arrows*) is increased in the upper cervical area. There is increased lordosis.

B, Anteroposterior projection indicates an increase in the interpediculate distance (*arrows*) of the vertebral bodies of C2, C3, C4, and C5 due to widening of the spinal canal. The medial aspect of the pedicles is flattened.

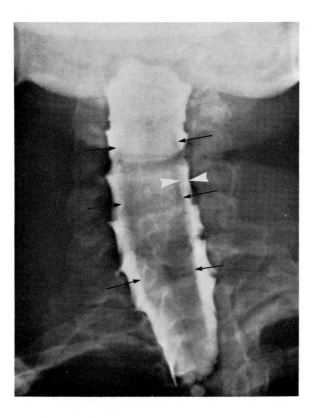

Figure 8–129 **Syringomyelia: Cervical Myelogram.** There is fusiform widening of the cord *(black arrows)* and narrowing of the subarachnoid space *(arrowheads)*, findings which may be indistinguishable from an intramedullary tumor.

Acrania

Prenatal roentgenologic diagnosis can be made on abdominal films. The fetal skeleton is intact, but the bones of the cranial vault are absent. Only the petrous pyramids are seen, and these are unusually prominent because of the absence of the other cranial bones. Frequently, the maternal uterine shadow is larger than expected owing to the polyhydramnios, which is often associated with severe fetal abnormalities.[18, 114]

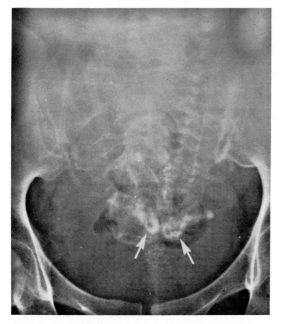

Figure 8–130 **Acrania.** Film of the maternal abdomen made during the last trimester discloses a well-developed fetal skeleton, except for the absence of the cranial vault. The petrous pyramids *(arrows)* are clearly seen. The absence of the bones of the vault lends prominence to the fetal petrous pyramids, producing a characteristic appearance.

Brain Atrophy

Brain atrophy occurs in a wide variety of conditions, prominent among which are senile dementia, and the presenile dementias such as Pick's disease and Alzheimer's disease. It is also found in Huntington's chorea, Parkinson's disease, general paresis, and many other diseases characterized by organic mental syndromes. The degree of atrophy disclosed on the CT scan or pneumoencephalogram often does not correlate with the severity of the mental dysfunction.

CT scans or pneumoencephalography will show some degree of dilatation of the ventricular system, especially of the lateral ventricles. The cerebral convolutional sulci become enlarged, and in severe cases the basal cisterns are also enlarged. If the atrophy is more marked unilaterally, the ipsilateral ventricle and sulci will be more dilated than the contralateral structures. In unilateral atrophy the dilated homolateral ventricle may be considerably displaced toward the atrophic side, causing bowing of the septum pellucidum. The actual size of the sulci may be better appreciated on the CT scan than on the pneumoencephalogram. (See *Cerebral Hypoplasia, Microcephaly, and Cerebral Hemiatrophy*, p. 379.) Distinction of cerebral atrophy from communicating low pressure hydrocephalus (see p. 390), in which ventricular dilatation is also present, can often be made by pneumography but may require a RISA cisternogram.

In diffuse atrophy, angiography will usually disclose only the nonspecific changes due to ventricular dilatation: namely, a straightening and widened curve of the pericallosal artery and a displacement of the septal and striothalamic veins. In severe cortical atrophy the peripheral vessels may not extend to the inner table, thereby simulating a subdural hematoma.[5, 6, 115-118]

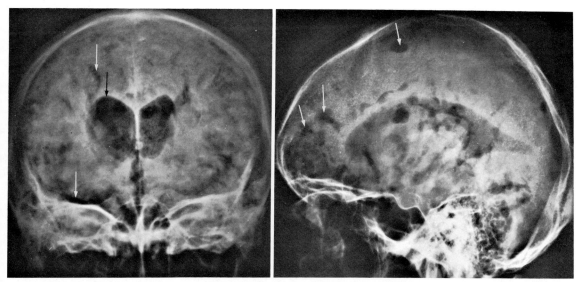

Figure 8–131 **Brain Atrophy in an Adult: Pneumoencephalogram.** The lateral ventricles are dilated, and their normally pointed superior margins are rounded. The right ventricle is more dilated and rounded *(black arrow)* than the left. The dilatation of the convolutional sulci *(white arrows)* is also greater on the right side, which was more atrophic than the left.

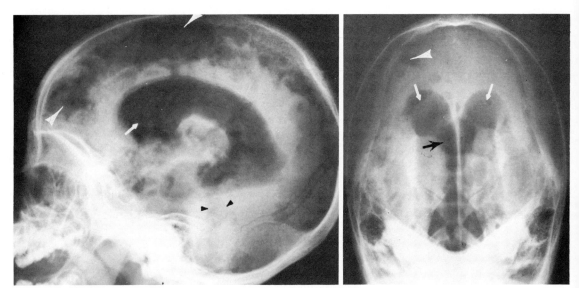

Figure 8–132   **Advanced Symmetric Cerebral Atrophy in a Child.**   There is dilatation of the third (*black arrow*) and fourth (*black arrowheads*) ventricles and symmetric dilatation of the lateral ventricles (*white arrows*). The subarachnoid sulci and pathways are greatly enlarged *(white arrowheads)*.
These are the classic pneumoencephalographic findings of severe cerebral atrophy.

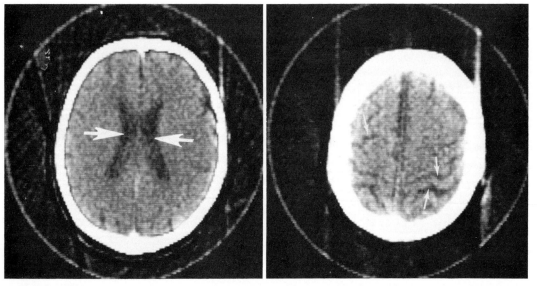

Figure 8–133   **Mild Generalized Brain Atrophy: CT Scan.**   The lateral ventricles are moderately dilated *(large arrows)* and the cortical sulci seen on a more rostral cut are enlarged *(small arrows)*.

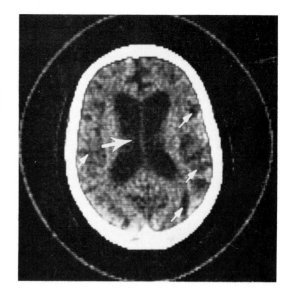

Figure 8–134 **Advanced Brain Atrophy: CT Scan.** The ventricles are greatly dilated *(large arrow),* and a large number of greatly dilated sulci *(small arrows)* are seen.

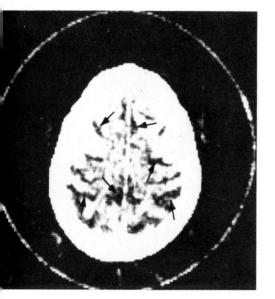

Figure 8–135 **Cortical Atrophy: CT Scan.** The greatly dilated sulci seen in a rostral "slice" appear as broad serpiginous lines of low density *(arrows).* The ventricles, seen on more caudad sections, were *not* dilated.

Enlarged sulci without ventricular dilatation presumably indicate cortical atrophy rather than generalized brain atrophy.

Cerebral Hypoplasia, Microcephaly, and Cerebral Hemiatrophy

Cerebral underdevelopment may be due to congenital hypoplasia or to atrophy secondary to brain injury or infection during gestation or early infancy. The hypoplasia or atrophy can be generalized, unilateral, or focal.

Generalized cerebral underdevelopment is accompanied by a small cranium, or *microcephaly.* The cranial bones become thickened. The sutures tend to close prematurely and become difficult to identify. The facial bones, although of normal size, appear disproportionately large. On CT scans and pneumoencephalographic studies, the ventricles and cisternal cavities are enlarged. In microcrania due to craniosynostosis, the skull is small and deformed and the sutures are obliterated; however, the bony thickening of the vault and the ventricular dilatation are much less marked than in congenital microcephaly.

In *cerebral hemiatrophy* or *unilateral hypoplasia* there are osseous changes on the affected side that compensate for the decreased brain volume. The petrous pyramid and orbit on the atrophic side are elevated, and the vault is flattened and thickened. There may also be homolateral decreased convolutional markings, deviation of the crista galli to the affected side, and unilateral enlargement of the air cells of the frontal sinus and mastoid.

On the CT scan, cerebral hypoplasia appears as a diffuse or localized thinning of peripheral brain tissue. The ventricular system is enlarged in proportion to the degree of hypoplasia. Diffuse hypoplasia with dilated ventricles may simulate hydrocephalus on the scan, but clinical distinction is quite evident.

In hemiatrophy, unilateral enlargement of the sulci and homolateral ventricle are associated with shift of the ventricular system to the affected side.

On pneumoencephalogram, the lateral ventricle on the affected side is dilated, its upper border is rounded, and it is shifted laterally. These changes are proportionate to the degree of underlying atrophy. The associated osseous changes will avoid confusing the picture with that of a mass lesion displacement of the ventricles.[30, 99, 119-122]

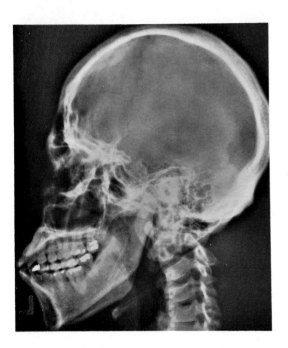

Figure 8–136 **Microcephaly in an Adult.** Lateral view demonstrates disproportion in size between the facial bones and the skull. The thickness of the cranial bones is increased. The suture lines are inconspicuous.

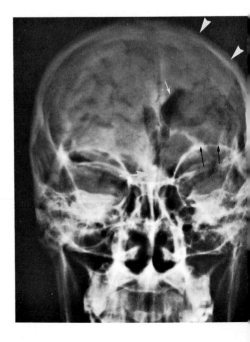

Figure 8–137 **Hemiatrophy: Pneumoencephalogram.** Anteroposterior projection demonstrates an enlarged left lateral ventricle that has blunted corners *(small white arrow)* and is shifted to the left side. The extreme slope of the cranial vault and the increased thickness of the bone in the left temporoparietal region are evident *(arrowheads)*. The left orbit is elevated *(black arrows),* and the crista galli is deviated to the affected side *(large white arrow).* The left frontal sinus is abnormally large. Petrous elevation is not apparent in this case. Ventricular shift due to a mass lesion is not associated with bony changes that would indicate volume loss.

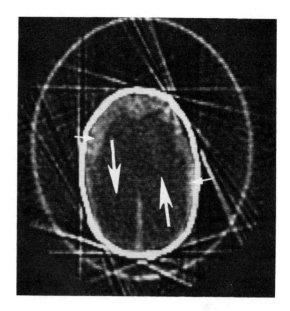

Figure 8–138 **Cerebral Hypoplasia in a Child: CT Scan.** The ventricles are huge *(large arrows),* and only a thin rim of brain tissue *(small arrows)* is present at the periphery.

Although obstructive hydrocephalus may give a similar scan appearance, the clinical setting permits ready differentiation.

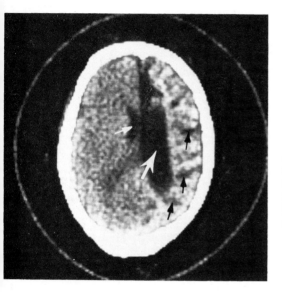

Figure 8–139 **Hemiatrophy: CT Scan.** There is enormous dilatation of the right lateral ventricle *(large white arrow)* associated with dilated sulci *(black arrows)* on the right. The entire supratentorial ventricular system is shifted to the right. The left lateral ventricle *(small white arrow)* and left brain tissue appear normal.

Agenesis of Corpus Callosum

Agenesis of the corpus callosum may be either an isolated brain defect or associated with agenesis of other areas. It may be complete or partial. Plain skull films are unrevealing, but pneumoencephalographic findings are characteristic and diagnostic. On frontal view, the lateral ventricles are widely separated, and the third ventricle is elevated and somewhat dilated. The superior aspect of the lateral ventricles often assumes a characteristic horn-like configuration with a concave roof, often termed a "batwing" appearance.

On angiograms the principal finding is the abnormal course of the anterior cerebral and pericallosal arteries. Instead of their normal smooth sweep around the genu of the corpus callosum, these vessels course directly upward and backward.

Not infrequently, a lipoma of the corpus callosum is associated with agenesis of this structure. The lipoma is often recognized on plain skull films by a midline area of lucency and by calcific streaks in its wall. Concomitant characteristic findings of agenesis of the corpus callosum are seen on pneumoencephalography.

When agenesis is associated with lipoma, the CT scans will show a circumscribed lucent area of very low (fatty) density in the vicinity of the anterior horns. If the lipomatous wall is partially calcified, there will be a highly dense linear zone in the rim of the lipoma.[123-128]

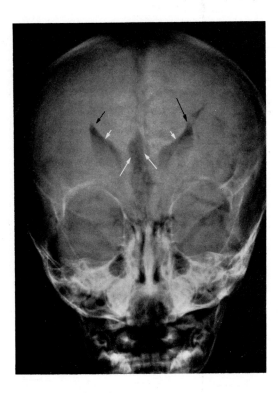

Figure 8–140 **Agenesis of Corpus Callosum: Pneu** **moencephalogram.** Frontal projection indicates ele vated location of the dilated third ventricle *(long white arrows)* and the increase in distance between the latera ventricles. The lateral ventricles have a concave media roof *(short white arrows)* with pointed superior horn: *(black arrows)*. These ventricular alterations are charac teristic.

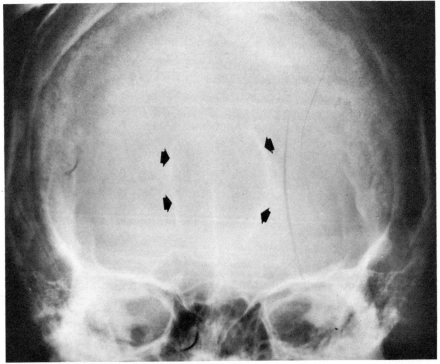

Figure 8–141 **Lipoma Associated with Agenesis of Corpus Callosum.** There is a symmetric midlin lucency with linear calcification of its lateral borders *(arrows),* findings virtually pathognomonic of lipoma c the corpus callosum. It is always associated with agenesis. (Courtesy Dr. S.M. Wolpert, Boston, Mas sachusetts.)

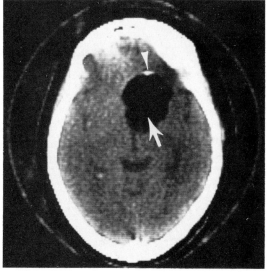

Figure 8–142 **Lipoma With Agenesis of the Corpus Callosum: CT Scan.** The large, sharply demarcated lucent area *(large arrow)* is extending to the right from midline. The μ density of the lesion was of fatty tissue (less than cerebrospinal fluid). A calcified density *(arrowhead)* is in the anterior wall of the lipoma.

Porencephaly

Porencephaly is a cyst-like cerebral defect that may communicate with the ventricular system.

The CT scan will show a lucent cavity of cerebrospinal fluid density. Other evidences of brain atrophy may be associated. In contrast to pneumoencephalography, even a noncommunicating porencephaly can be demonstrated.

Pneumoencephalography can also show the cavities if they communicate with the ventricular or subarachnoid system and fill with air.

Porencephaly may be a developmental anomaly or may occur following trauma or focal vascular brain destruction. The cavities are often associated with generalized brain atrophy or underdevelopment.[5, 6, 129]

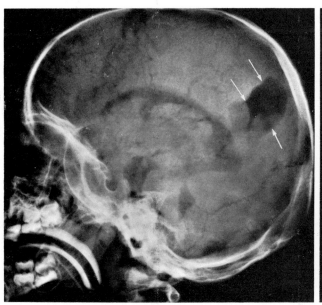
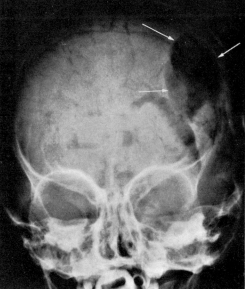

Figure 8–143 **Porencephaly in an Infant: Pneumoencephalogram.** A well-defined air-filled cavity *(arrows)* communicates with the ventricular system. The cyst is situated in the posterior parietal area, a frequent site.

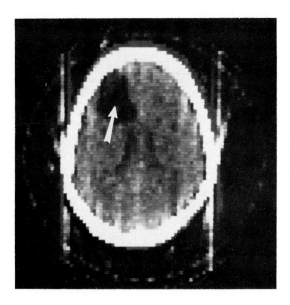

Figure 8–144 **Porencephalic Cyst: CT Scan.** The large circumscribed area of lucency *(arrow)* in the left frontal lobe has a μ density of cerebrospinal fluid. There was no ventricular shift.

Agenesis, Hypoplasia, and Atrophy of the Cerebellum

With decreased size or total absence of the cerebellum, enlargement of the cisterna magna and dilatation of the fourth ventricle are often seen on CT scan or pneumoencephalogram.

On the CT scan, the fourth ventricle is usually enlarged, often with prominent cerebellar sulci and an enlarged superior cerebellar cisterna. In severe degeneration, a cystic area of cerebrospinal fluid density may replace the cerebellar tissue.

On pneumography the atrophic cerebellum may be silhouetted by the surrounding air in the cerebellar sulci and cisternas, which are enlarged. Special manipulations of the head are usually necessary to fill these sulci with air. The third and lateral ventricles remain normal unless the cerebellar atrophy is associated with generalized cerebral atrophy. An atrophic cerebellum is sometimes found in Friedreich's ataxia.

Less severe degrees of hypoplasia of the cerebellum are difficult to detect because of the wide variability in the size of the cerebellar sulci and the cisterna magna in normal subjects.[5, 6, 130-133]

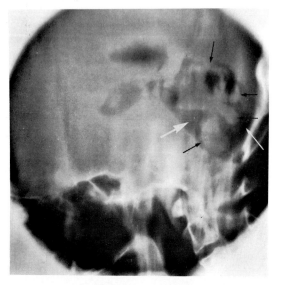

Figure 8–145 **Hypoplasia of Cerebellum: Pneumoencephalogram.** Midline tomography in lateral projection demonstrates the folia *(black arrows)* of the hypoplastic cerebellum, which are outlined by air in the widened subarachnoid spaces *(small white arrow).* The cisterna magna and the fourth ventricle *(large white arrow)* are also enlarged, and the fourth ventricle is posterior to its usual position. The third and lateral ventricles are normal, since the remainder of the brain was not atrophied.

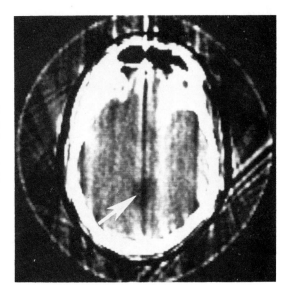

Figure 8–146 **Hypoplasia of the Cerebellum with Cerebellar Cyst: CT Scan.** A large, low density area *(arrow)* in the midline of the posterior fossa is a cerebellar cyst, associated with cerebellar hypoplasia.

A hypoplastic or atrophic cerebellum can sometimes be outlined on the CT scan by the prominent dilated cerebellar sulci that are often associated.

Premature Closure of Cranial Sutures (Craniosynostosis) and Craniostenosis

Closure of the sutures (craniosynostosis) before the age of 3 years restricts growth of the brain and skull (craniostenosis), and mental deficiency may follow.

Skull radiography can demonstrate the site and extent of premature closure and will rule out other causes of skull deformity. In early craniosynostosis the involved suture becomes narrowed and indistinct. Later, the suture margins become sclerotic, and this is followed by bony bridging and obliteration. The prematurely fused suture limits skull growth at right angles to the suture. The abnormal shape of the skull depends upon the location, degree, and age of onset of the premature closure. Rapid growth of a restricted brain may cause increased convolutional markings in the skull.

Various combinations and degrees of premature closure are possible, and a large variety of bizarre skull shapes can develop. The main groups are as follows:

1. Dolichocephaly, or scaphocephaly, is due to early fusion of the sagittal suture. The skull is narrowed and elongated.

2. Brachycephaly is caused by premature closure of the coronal suture. The skull is short, high, and wide.

3. Oxycephaly is due to early closure of both the sagittal and coronal sutures and is manifested by a short, narrow, but high skull.

4. Microcrania results from premature fusion of all the major sutures. It may be difficult to differentiate microcrania from microcephaly, in which early suture closure also occurs. Increased convolutional markings in microcrania, and radiolucencies between the interdigitating sutures in microcephaly may help to distinguish one entity from the other. In addition, the ventricles are usually normal in microcrania but are dilated in microcephaly.

5. Plagiocephaly refers to an asymmetric, flattened skull that results from fusion of the coronal and squamous sutures on only one side.[18, 119, 131, 134-136]

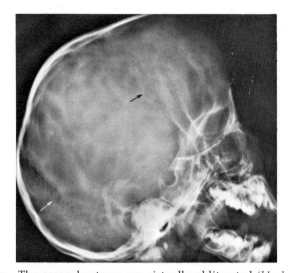

Figure 8–147 **Oxycephaly.** The coronal sutures are virtually obliterated *(black arrow),* but the lambdoidal suture *(white arrow)* is still patent. Anteroposterior growth is limited so that the head is turret-shaped.

Scrutiny of the films in infants and young children may disclose sutural abnormalities before deformity of the skull is evident.

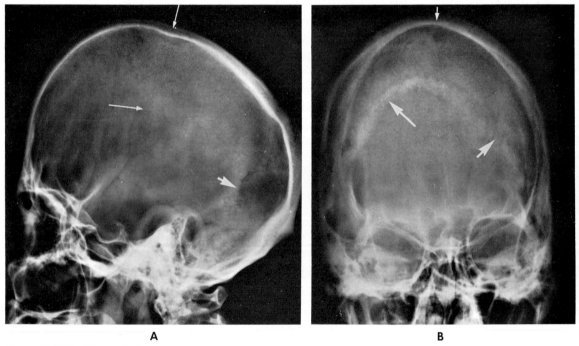

A   B

Figure 8–148 **Oxycephaly.**

A, Lateral view discloses a high vertex and a small anteroposterior diameter; the normal frontal prominence is lacking, as is the coronal suture *(small arrows indicate usual location).* The lambdoidal suture *(large arrow)* is partially open and sclerotic.

B, Anteroposterior view demonstrates the high vertex. The lambdoidal suture is closed and sclerotic on the right *(long large arrow)* and partially open on the left *(short large arrow).* The coronal suture is lacking.

The sagittal suture is also closed *(small arrow),* but sagittal closure apparently occurred later than did coronal closure, since the head is turret-shaped.

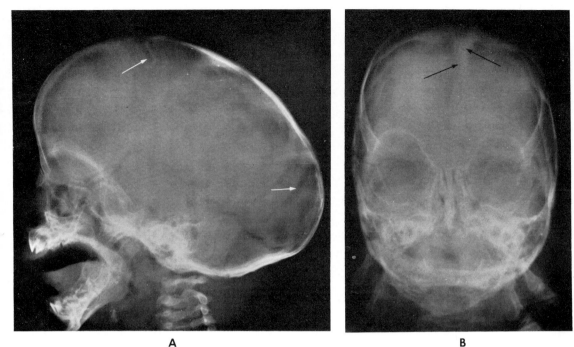

A   B

Figure 8–149 **Dolichocephaly.**

A, In lateral view the anteroposterior diameter is long as a result of premature closure of the sagittal suture. The coronal and lambdoid sutures are still open *(arrows).*

B, In anteroposterior view the closed sagittal suture is replaced by a line of increased density *(arrows).*

Meningoencephalocele and Cranial Meningocele

Prolapse of the meninges (meningocele) or of the meninges and the brain (meningoencephalocele) usually occurs through a defect of the bony calvarium or a fontanelle. Pneumoencephalography or CT scan is required to determine the degree of brain prolapse, while angiography may disclose which anatomic portion of the brain is in the hernia. The skull defect is usually in the midline and is associated with a soft tissue density in the parietal, occipital, frontal, or nasal region.

Plain films demonstrate defects in the bones, local demineralization or disappearance of normal anatomic outlines. Characteristically, the defect in the inner table is larger than that of the outer, indicating that pressure is exerted from inside the brain.[18, 120, 137]

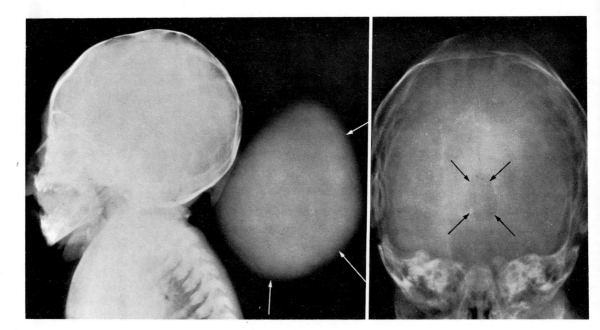

Figure 8–150 **Meningoencephalocele.**

A, The large soft tissue mass (*arrows*) in the posterior aspect of the skull represents a large meningoen-cephalocele.

B, Anteroposterior view reveals a defect (*arrows*) of the occipital bone, which is the site of herniation of the brain and meninges. A pneumoencephalogram would be necessary to determine the presence and amount of brain tissue in the herniation.

Mongolism (Down's Syndrome)

This condition, due to trisomy 21, is the most common of the trisomy syndromes.

The most consistent skeletal abnormality in early infancy is in the pelvis: the acetabular angle is decreased, the ilia are widened and flared, and the ischia are slender and elongated. These findings are particularly important for early diagnosis in doubtful cases. Other frequent skeletal abnormalities are as follows:

1. Hypoplasia of the middle phalanx of the fifth finger gives rise to the characteristic curved fifth finger.

2. There is squaring of the vertebral bodies: the superoinferior length becomes equal to or greater than the anteroposterior measurement.

3. Only 11 ribs are present instead of the normal 12.

4. Multiple ossification centers occur in the manubrium.

5. In the skull there may be microcrania, thinning of the bones of the vault, a shortened skull base and nasal bones, and delayed closure of the sutures and fontanelles. Hypoplasia of facial bones and poor pneumatization of the paranasal sinuses are often seen. Orbital hypertelorism is common.

Congenital heart disease, especially septal defects and atrioventricularis communis, is frequently associated. A higher than normal incidence of megacolon and of duodenal atresia is encountered in mongoloids.

There is a significantly increased incidence of acute leukemia.[138–142]

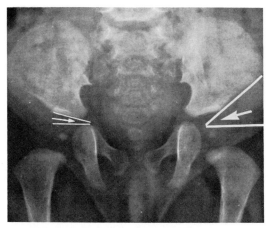

Figure 8–151 **Mongolism: The Pelvis.** Increased width and lateral flaring of both ilia give rise to the characteristic "elephant ear" appearance. The angle made by the slope of the lateral pelvis *(large arrow)* is much more acute than that in the normal pelvis. The acetabular slope *(small arrow)* is considerably decreased. The picture is virtually pathognomonic of mongolism and is helpful in diagnosing questionable cases.

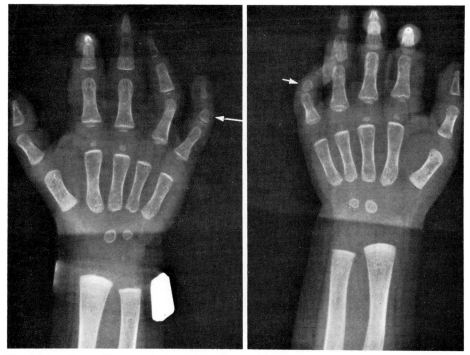

Figure 8–152 **Mongolism: Characteristic Deformity of the Fingers.** The middle phalanges *(arrows)* of both fifth digits are small and hypoplastic, thus producing characteristic bowing.

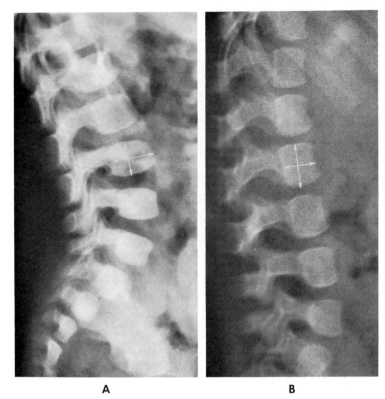

A   **B**

Figure 8–153 **Lumbar Vertebrae: Normal and Mongoloid.**

A, In a normal 1½ year old infant, lateral view demonstrates the normal dimensions and configuration of the lumbar vertebrae. The anteroposterior diameter of the bodies is longer than the vertical diameter (*double-headed arrows*).

B, In a 1½ year old mongoloid, the vertical diameter is longer than the anteroposterior diameter (*double-headed arrows*).

This alteration in the lumbar bodies is found in a high percentage of mongoloids, particularly those under 2 years of age. Although the finding is not pathognomonic, it may be helpful in uncertain cases. (Courtesy Drs. J. F. Rabinowitz and J. E. Moseley, New York, New York.)

Communicating Hydrocephalus (Occult Hydrocephalus; Normal Pressure Hydrocephalus)

In this condition of diverse origin, there is apparently diminished resorption of cerebrospinal fluid, leading to dilatation of the ventricular system, but with unimpeded communication between the ventricles and spinal fluid.

In most patients, the obliteration of the basal cisterns or subarachnoid pathways is secondary to previous subarachnoid hemorrhage or severe head trauma. Many of the afflicted older patients, however, have had no apparent antecedent episodes or injury. Clinically, there is a rather rapid progressive deterioration and dementia. In older individuals the clinical picture may be indistinguishable from that of ordinary senile dementia, although symptom progression in the latter is much more gradual.

The pneumoencephalogram in communicating hydrocephalus shows generalized ventricular dilatation. Air is absent over the hemispheric sulci, and there may be wide basilar cisternas. The presence of large amounts of subarachnoid gas over the cortex virtually excludes a diagnosis of communicating hydrocephalus; its absence, however, is not conclusive evidence of communicating hydrocephalus. An interesting and confirmatory finding is progressive enlargement of the lateral ventricles for several days following the air study. This does not occur in cerebral atrophy.

The pneumoencephalographic distinction between communicating hydrocephalus and cerebral atrophy is extremely important, since shunting procedures often arrest or diminish the deteriorating dementia symptoms of communicating hydrocephalus. Scanning the brain after the introduction of a radionucleotide (RISA) into the spinal fluid will usually help distinguish brain atrophy from communicating hydrocephalus.

The CT scan will reveal dilatation of the entire ventricular system, including the fourth ventricle. There may or may not be a paucity of visible cortical sulci. Absence of sulci should suggest the possibility of this entity. However, the scan findings are often indistinguishable from those of generalized atrophy. A RISA cysternogram is usually a necessary adjunct for making the diagnosis of communicating hydrocephalus.[143, 144]

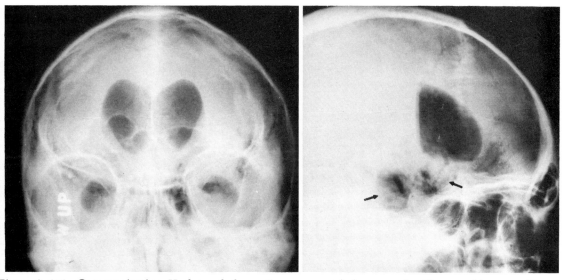

Figure 8–154 **Communicating Hydrocephalus.** Pneumoencephalogram of a 59 year old man with progressive mental deterioration shows well-marked dilatation of the entire ventricular system, well-filled basal cisterns (*arrows*), but absence of air in the subarachnoid pathways.

Marked improvement occurred after insertion of a shunt.

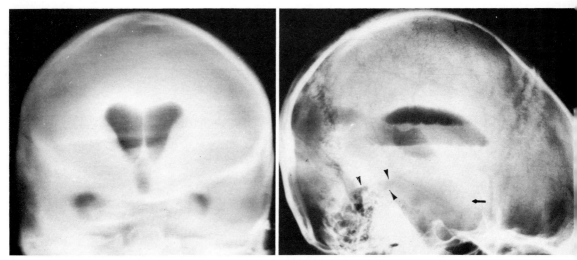

Figure 8–155 **Communicating Hydrocephalus: Post-Traumatic Obliteration of Basal Cisterns.** After severe head injury six months previously, this 18 year old girl began exhibiting symptoms of mental deteri oration.

Pneumoencephalogram shows moderate dilatation of all the ventricles but complete absence of air i the basal cisterns *(arrow)* and subarachnoid sulci.

Hemorrhage and scarring had obstructed the flow of fluid from the fourth ventricle *(arrowheads)* to th basal cisterns and subarachnoid sulci.

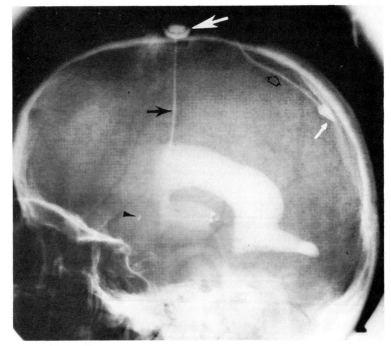

ⁱigure 8–156 **Communicating Hydrocephalus: Ventriculosinus Shunt Insertion.** The metal-tipped *(blac arrowhead)* shunt tube *(black arrow)* extends from a lateral ventricle into a check valve *(large white arrow* that is beneath the skin and from there *(open black arrow)* into the posterior portion of the sagittal sinu *(small white arrow).* Intraventricular contrast material is opacifying the entire shunt and valve, indicating potency and proper insertion.

The Spinocerebellar Ataxias

This group of diseases includes both heredofamilial and acquired conditions in which the cerebellum and/or the spinocerebellar tracts are hypoplastic or degenerated and atrophic. Friedreich's ataxia is the best known of this group.

Radiographically, cerebellar atrophy or hypoplasia is demonstrated by CT scans or posterior fossa pneumography, in which the enlarged cisterns and sulci outline a cerebellum of decreased size (see *Agenesis, Hypoplasia, and Atrophy of the Cerebellum*, p. 384).

In *Friedreich's ataxia,* in which the spinocerebellar tracts are primarily involved, a small cerebellum may or may not be seen. Scoliosis or kyphoscoliosis of the upper dorsal spine is commonly noted. Other findings include deformity of the feet (pes planus), cardiac enlargement due to cardiomyopathy, and a small cervical cord on myelography.

In *hereditary spastic ataxia,* there are disorders of esophageal motility owing to difficulty in initiating swallowing.[130-133, 145]

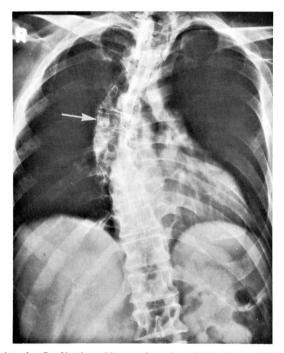

Figure 8–157 **Friedreich's Ataxia: Scoliosis.** Upper dorsal scoliosis *(arrow)* is often present in Friedreich's ataxia. Kyphosis may also occur. Both are due to a progressive muscular weakness.
Moderate cardiomegaly is due to myocardiopathy, which may be associated.

Sturge-Weber Disease (Encephalotrigeminal Angiomatosis)

The classic Sturge-Weber syndrome consists of a unilateral venous angioma of the leptomeninges associated with a nevus in the trigeminal area of the face. The characteristic roentgenologic findings are undulating plaques of calcification in the brain cortex, most often in the parietal area. This calcification is within the brain cortex and not in the angioma, and the plaques appear to follow the convolutions. This unusual appearing calcification is found in about two thirds of cases.

Cerebral arteriography reveals abnormalities in almost half the cases; in the older patient in whom calcification is advanced, angiographic abnormalities are less frequent. A capillary-venous angioma is the commonest finding, and arteriovenous malformations, arterial thrombosis, and circulatory anomalies are sometimes seen, especially in patients in whom calcification is absent. In most patients, there is an abnormal pattern of venous drainage from the involved area. The superficial cortical veins are decreased, and the deep veins are enlarged and tortuous, often with bizarre coursing collaterals. Pneumoencephalography often reveals cortical atrophy of the brain on the involved side.

The CT scan will reveal the focal calcifications and areas of focal atrophy, if present.[146, 147]

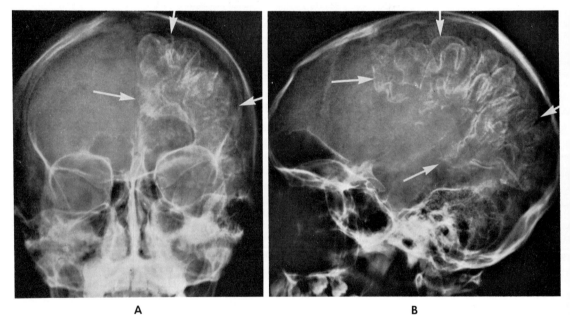

| A | B |

Figure 8–158   **Sturge-Weber Disease: Capillary Angioma, Atrophy of Brain, and Convolutional Calcification.**

A, In posteroanterior view extensive characteristic convolutional calcification in the left parietal and occipital area *(arrows)* terminates abruptly in the midline. The calcification is in the cortex and not in the angioma. The large space between the tables and the calcification is due to convolutional atrophy, another feature of this disease. Decreased volume on the left side is associated with elevation of the left orbit, left petrous pyramid, and left sphenoid ridge, as well as with enlargement of the left frontal sinus.

B, Lateral view clearly demonstrates the convolutional shape of the calcifications *(arrows)* and the characteristic parallel calcification in the cortex. As a rule the parietal and occipital areas are the most frequent sites of involvement.

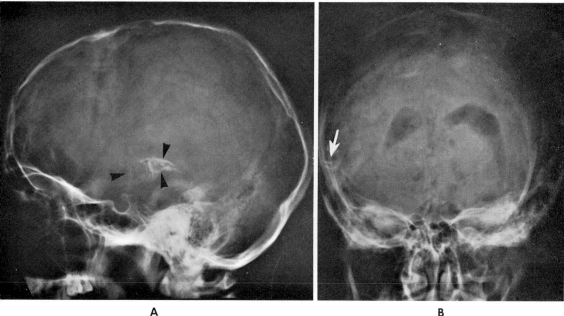

| A | B |

Figure 8–159 Sturge-Weber Disease.

A, The calcification in the inferior aspect of the right parietal lobe *(arrowheads)* shows the characteristic multiple curvilinear plaques in the convolutions of the cerebral cortex.

B, Anteroposterior film following pneumoencephalography shows the calcification near the edge of the skull *(arrow),* but the full extent of calcification cannot be appreciated on this view. The air-filled ventricular system is normal, and cortical atrophy has not yet developed.

Tuberous Sclerosis

This rare condition is characterized by multiple hamartomatous lesions in various organ systems. Fairly characteristic roentgen findings occur in the brain, skeletal system, kidneys, and lungs. The radiologic findings are important for diagnosis of tuberous sclerosis in those patients without the cardinal clinical triad of adenoma sebaceum, epilepsy, and mental deficiency.

In about one half the cases there are characteristic clusters of intracranial calcifications of varying size and shape. These are calcified sclerotic plaques, usually located in the basal ganglia and in the walls of the lateral ventricles. Less frequently the calcifications may appear in the cerebellum, sometimes assuming a vascular configuration and thereby simulating Sturge-Weber calcifications. Pneumoencephalography may reveal nodular protrusions into the gas-filled and often enlarged ventricles.

The principal skeletal changes consist of circumscribed areas of increased bone density, most characteristically in the posterior portion of the vertebral bodies and in the pedicles. In the hands and feet, involved in almost two thirds of cases, the roentgen findings are virtually pathognomonic. Cyst-like changes occur in the phalanges and are accompanied by irregular periosteal new bone formation in the metacarpals and metatarsals, a distinctive combination of findings.

Hamartomatous tumors, which are usually angiomyolipomas, of one or both kidneys occur in about half the cases. Conversely, about one half of all angiomyolipomas occur in patients with tuberous sclerosis. Radiographically, one or both kidneys are enlarged, and ill-defined lucent areas of tumor fatty tissue can sometimes be detected. On pyelography, the caliceal system is distorted and displaced; multiple bilateral tumors can produce a picture that mimics polycystic disease, and renal angiography may be necessary to distinguish between these two entities. The angiomyolipomas are very vascular, and the vessels are tortuous and irregular, often strongly resembling neoplastic vessels.

The lungs are least frequently involved. A diffuse, nonspecific interstitial fibrosis of varying severity is the usual finding. In advanced cases the lung has a honeycomb appearance, with interstitial emphysema and small bullae.

If calcific plaques are present in the ventricular walls or basal ganglia, these will be readily seen on the CT scan, even if their radiographic density is too low for detection on routine skull films. If there are large nodular protrusions into the lateral ventricles, these may be apparent on the scan.[148-155]

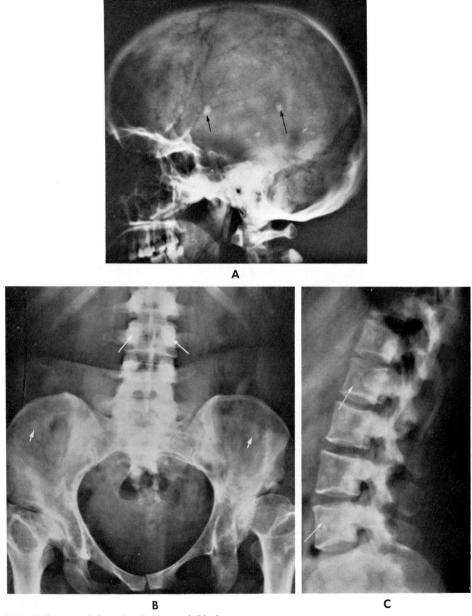

A

B   C

Figure 8–160 **Tuberous Sclerosis: Brain and Skeleton.**

A, Lateral view demonstrates numerous rounded calcifications *(arrows)* in the region of the lateral ventricles.

B, In anteroposterior view the pedicles and lumbar vertebrae are dense *(long arrows).* Rounded areas of increased density are seen scattered throughout the pelvic bones *(short arrows).*

C, Lateral view discloses osteosclerotic areas in the vertebral bodies *(arrows).* The increased bone density is most pronounced in the posterior portion of the bodies and in the pedicles; this is a characteristic finding. (Courtesy Dr. J. Hasegawa, Chicago, Illinois.)

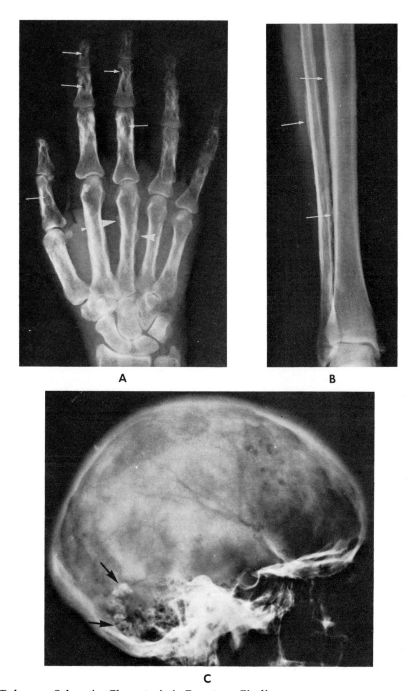

A

B

C

Figure 8–161 **Tuberous Sclerosis: Characteristic Roentgen Findings.**

This woman did not have mental deficiency, epilepsy, or skin lesions.

A and *B,* There are numerous cyst-like lucencies *(arrows)* in all of the phalanges. The borders of these "cysts" are localized areas of trabecular thickening. Irregular periosteal overgrowth *(large arrowheads)* and cortical thickening *(small arrowhead)* are seen in the metacarpals and along the shafts of the tibia and fibula *(arrows in B).*

C, There is wavy calcification *(arrows)* in the right cerebellar hemisphere; no supratentorial calcifications are seen.

Illustration continued on the following page.

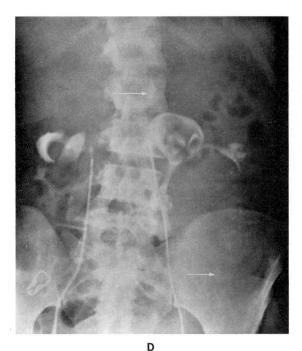

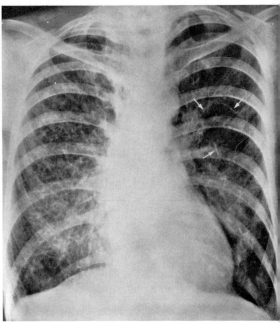

D   E

Figure 8–161 (Continued)

D, Bilateral retrograde studies disclose greatly enlarged kidneys with tremendous distortion of the pelvis and the calices, features that suggest extensive neoplastic involvement bilaterally. These lesions were angiolipomas.

Note the sclerosis of the pedicles and the sclerotic nodules in the pelvic bones *(arrows)*. These bone changes are better illustrated in Figure 8–160.

E, There is diffuse interstitial fibrosis in both lungs, with interstitial emphysema and a localized emphysematous "cyst" *(arrows)*.

This patient illustrates a representative spectrum of the radiologic changes in tuberous sclerosis: cyst-like changes in the phalanges, and irregular periosteal reaction in the more proximal long bones; sclerosis of posterior elements of the spine; angiolipomas of the kidney; brain calcification, although cerebellar calcification is much less common than cerebral calcification; and diffuse interstitial pulmonary fibrosis and emphysema.

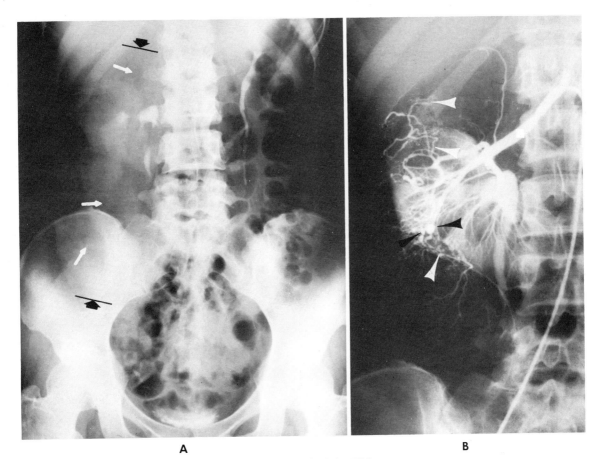

A   B

Figure 8–162 **Tuberous Sclerosis: Angiomyolipoma of Right Kidney.**

   A, The excretory urogram shows marked enlargement of the right kidney *(black lines and arrows)* with lucent areas *(white arrows)* above and below the compressed calices.

   B, Right renal arteriogram demonstrates neovasculature *(white arrowheads)* and vascular "lakes" *(black arrowheads)* in the mass.

   An angiomyolipoma, although benign, is hypervascular and has angiographic findings simulating a malignant tumor.

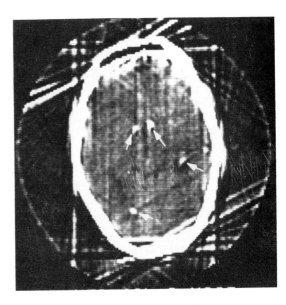

Figure 8–163 **Tuberous Sclerosis: CT Scan.** Scattered small densities *(arrows)* of calcific μ *(μ=density number on scanner scale)* are apparent. No significant ventricular changes were present.

   Calcifications in the brain are seen in several conditions, including tuberous sclerosis, and are not specific. Many calcifications that are readily apparent on a scan may not be visible on conventional skull films.

   The oblique dark and white streaks are artefacts due to motion of patient's head during the scan.

Huntington's Chorea

In this hereditary degenerative disease of the central nervous system, the pathologic findings are mainly confined to the basal ganglia and cerebral cortex.

The pneumographic studies and CT scan will demonstrate the changes of atrophy. In many cases, the atrophy of the caudate nucleus will lead to an absence of the caudate bulge in the lateral walls of the lateral ventricles. This is a suggestive but by no means specific finding.[155a]

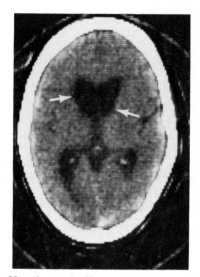
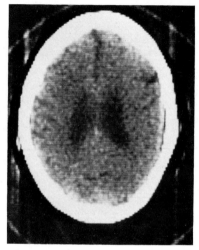

Figure 8–164 **Huntington's Chorea: CT Scan.** The lateral and third ventricles are dilated and the sulci are enlarged, characteristic findings of brain atrophy.

The posterior portions of the anterior horns *(arrows)* are unusually wide compared to the rest of the dilated ventricles. The lateral bulging *(arrows)* is apparently due to absence of the normal caudate convex impression because of atrophy of the caudate nucleus. Similar findings have been reported on pneumoencephalograms of patients with Huntington's chorea.

Klippel-Feil Syndrome

This syndrome is characterized by congenital fusion of two or more cervical vertebrae and is manifested by shortening and limited mobility of the neck. Deafness occurs in about one third of patients. The cervical fusion always involves the vertebral bodies and often the neural arches. The intervertebral space is absent or rudimentary.

The anteroposterior diameter of the vertebral bodies is characteristically narrower at the level of fusion, but the height of the fused vertebral bodies is generally equal to the combined height of these vertebrae and their interspace in the normal spine. Other developmental anomalies are often associated, particularly Sprengel's deformity, spina bifida, cervical ribs, and malformations of the occipital bone. Diverse anomalies of the central nervous system, cardiovascular system, and genitourinary tract are also frequent.[18, 156-159]

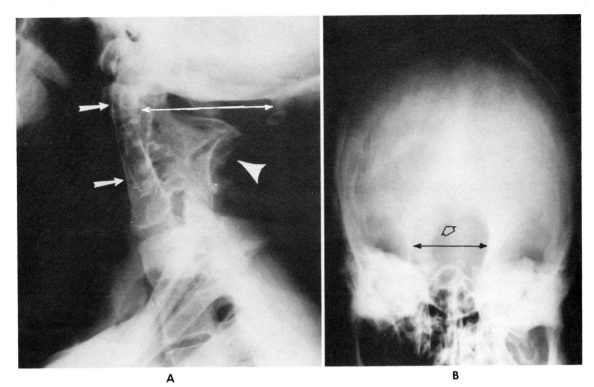

| A | B |

Figure 8–165 **Klippel-Feil Syndrome with Skull Anomalies.**

A young woman with short, poorly mobile neck and evidence of progressive mental alterations.

A, Lateral film of the cervical spine shows anteroposterior narrowing and fusion of C1 to C4 *(arrows)* and narrowed AP diameters. The posterior elements are deficient *(large arrowhead).* The occipital bone is deficient, and the foramen magnum *(double-arrowed line)* is greatly enlarged.

B, Skull film shows the large foramen magnum *(double-arrowed line)* and an additional bone defect *(open arrow).*

Mild hemiatrophy (hypoplasia?) was found on pneumoencephalography.

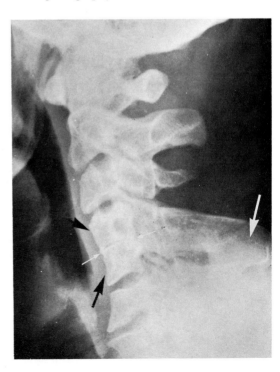

Figure 8–166 **Klippel-Feil Syndrome.** Lateral projection indicates fusion *(black arrows)* of the fourth and fifth vertebral bodies, their laminae, and their spinous processes. The anteroposterior diameter of the bodies at the level of the rudimentary disc *(small white arrows)* is decreased. The cervical spine appears somewhat short, but the shortness is accentuated clinically because the fused spinous process *(large white arrow)* is greatly enlarged and very long. The patient also had Sprengel's deformity.

Sprengel's Deformity

This congenital deformity may be seen alone or associated with the Klippel-Feil syndrome. The scapula is elevated and anteriorly rotated. The scapula is often fixed in this abnormal position by an accessory omovertebral bone that extends from the vertebra to the medial margin of the scapula, with which it articulates.[160, 161]

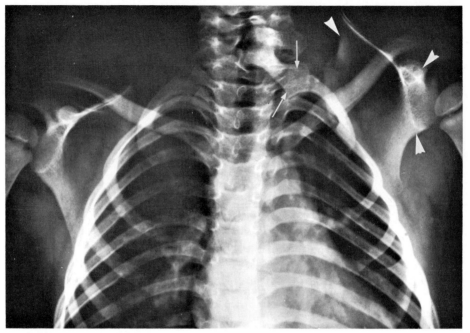

Figure 8–167 **Sprengel's Deformity.** The left scapula is elevated *(arrowheads)* and fixed because of articulation with an omovertebral bone *(arrows)*.

Basilar Impression and Platybasia

Basilar impression refers to invagination of the foramen magnum into the cranial cavity and is virtually always associated with platybasia, which is a flattening and abnormal angulation of the base of the skull. The primary form is congenital and is often associated with cervical anomalies such as Klippel-Feil syndrome and occipitalization of the atlas. It may also be associated with Arnold-Chiari malformation, aqueductal stenosis, and syringomyelia.

The acquired form results from conditions causing bone softening, such as Paget's disease, osteomalacia, or severe osteoporosis. The neurologic signs, which are due to cord or medullary pressure, may simulate multiple sclerosis, syringomyelia, or even a posterior fossa tumor.

Skull radiographs may show an elevated odontoid process and clivus, a deformity of the foramen magnum, and a flattened skull base. Numerous radiologic measurements have been devised for confirming significant elevation of the odontoid process. Frequently utilized is McGregor's line, which is drawn from the posterior margin of the hard palate to the posterior edge of the foramen magnum (occipital bone). If the odontoid process extends more than 4.5 mm. above this line, basilar impression should be suspected.

Angiography usually discloses vascular dysplasias and anomalies of origin and course of the vertebral artery. Posterior curving and some stretching are often seen.[162-164]

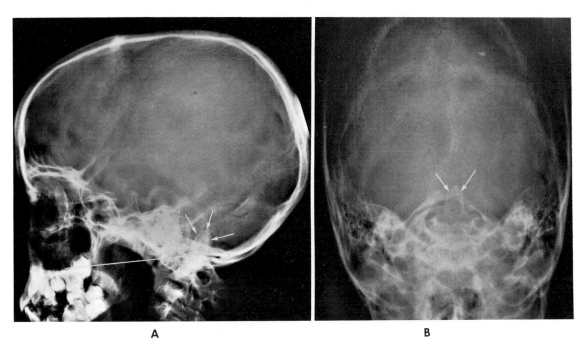

A   B

Figure 8–168 **Platybasia: Congenital.**

A, In a child, lateral projection demonstrates flattening of the base of the skull. Most of the odontoid processes *(arrows)* appears above Chamberlain's line *(horizontal white line)* (basilar impression). The posterior fossa is unusually small.

B, Occipital projection demonstrates abnormal high position of the odontoid process *(arrows).*

Figure 8–169 **Platybasia: Secondary to Paget's Disease.** The base of the skull is flattened; the odontoid process *(arrows)* projects well above Chamberlain's line *(horizontal white line).* The skull is flattened as a result of softening of the bones due to Paget's disease. The thickened sclerotic skull tables are characteristic of Paget's disease.

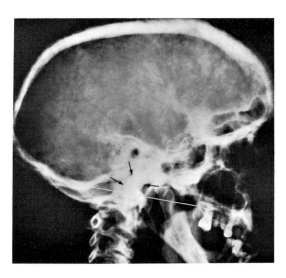

The "Empty" Sella Turcica

In this condition the sella turcica is globular and may be symmetrically enlarged, leading to an erroneous diagnosis of pituitary tumor.

The "empty" sella results from a congenital absent or incomplete diaphragma sellae (the dural covering of the sella turcica). The pulsations of the cerebrospinal fluid cause a globular remodeling and often an enlargement of the sella and a flattening of the pituitary gland. These changes can occur in patients with a normal cerebrospinal fluid pressure, but they are more marked in the presence of benign intracranial hypertension, as in cases of pseudotumor cerebri and the Pickwickian syndrome.

Radiographically, the "empty" sella turcica is globular in shape and may be normal in size or symmetrically expanded with a thinned, slightly bowed dorsum, suggestive of an intrasellar tumor. However, the cortical "white" line of the sella floor remains intact in the "empty" sella. Pneumoencephalography will reveal air in the upper portion of the sella, and the flattened pituitary can be identified between the air and the sellar floor. Special head maneuvers may be necessary to permit entrance of air into the sella. In any patient with a symmetrically enlarged sella, pneumoencephalography should be done before surgery or radiation is instituted.

On a CT scan, a rounded low density area may be seen occupying the rostral portion of the sellar area.

Although usually no symptoms are associated with the "empty" sella, in some patients optic nerve symptoms occur, apparently owing to prolapse of the chiasm into the empty sella. Cerebrospinal fluid rhinorrhea has also been reported. The "empty" sella appears to occur predominantly in obese adult women with diabetes and hypertension.[165-167]

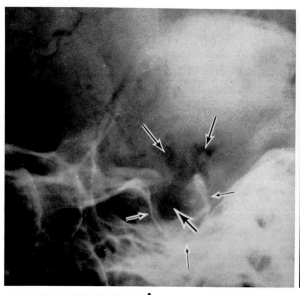 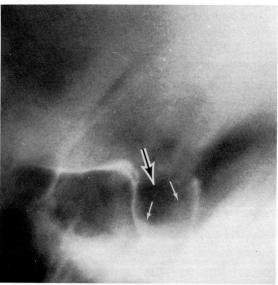

| A | B |

Figure 8–170 **"Empty" Sella with Chiasmal Symptoms.**

A, Pneumoencephalogram of a young woman with visual disturbance shows an enlarged globular sella turcica *(small arrows).* On prior plain skull films this was thought to be due to a pituitary tumor.

The air in the chiasmatic and interpeduncular cisterns *(middle-sized arrows)* appears to extend deep into the sella *(large arrow).*

B, Midline tomogram of the air study shows that the sella is actually over half filled with air. The compressed pituitary gland *(small arrows)* is outlined by the air.

A subsequent episode suggesting acute chiasmal pressure necessitated surgery. The optic chiasm was found stretched and prolapsed into the "empty" sella *(large arrow)*; the diaphragma sellae was completely absent.

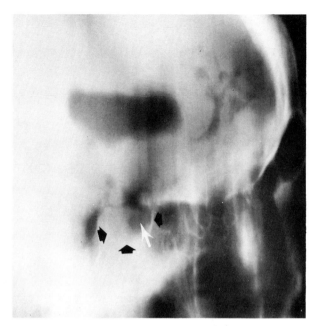

Figure 8–171 **"Empty" Sella in 72 Year Old Female.** Lateral midline tomogram of encephalogram shows a somewhat globular sella *(black arrows)* with air extending into the anterior superior portion of the sella *(white arrow).*

There were no apparent symptoms due to this abnormality.

Idiopathic Focal Epilepsy

In patients with a focus of epileptiform activity on the electroencephalogram (EEG), all relevant radiographic studies are usually negative. However, in about 25 per cent of cases, detailed carotid arteriograms will reveal some abnormality of the central sulcus branches of the middle cerebral artery. One or more of these branches may be absent, narrowed, or abruptly cut off. The latter suggests vascular occlusion, while the narrowed or absent sulcus vessels are probably a congenital defect. However, clinical correlation with these angiographic changes is often lacking.

The CT scan generally reveals no specific changes in *idiopathic* epilepsy. However, in a sizable number of cases clinically labeled as idiopathic epilepsy, the CT scan will disclose an underlying pathologic condition, such as tumor, atrophy, porencephalic cyst, and so forth.

Myelopathy from Cervical Canal Narrowing

A developmentally narrowed sagittal diameter of the cervical bony canal may be asymptomatic or may cause mild symptoms of myelopathy due to constricting effects on the cord. However, if posterior cervical osteophytes develop in later life, signs and symptoms of focal cervical cord compression will occur more often and more severely in the patient with a narrowed canal.

There is considerable variation in the sagittal diameter of the normal canal, depending upon the size and habitus of the individual. In normal persons the average diameter is about 16 mm., while in patients with cervical myelopathy the canal is less than 13 mm. The narrowed canal is best studied with the neck in full extension. A simple radiographic estimate can be made by comparing the sagittal diameters of the canal and adjacent cervical bodies; normally these are almost equal.

Myelography demonstrates the size of the cord in relationship to the canal and, if osteophytes are present, discloses the degree of local compression of the subarachnoid space or cord.[169, 170]

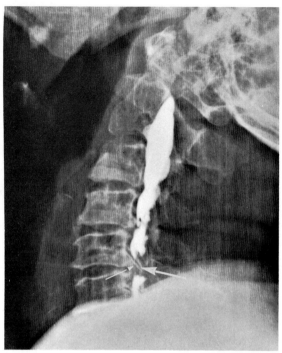

Figure 8–172 **Narrowed Cervical Canal: Myelogram.** The cervical canal is quite narrowed, as seen by the myelogram and the measurements of the cervical bony canal. The spur *(small arrow)* is almost completely obstructing the already narrowed oil column *(large arrow)*.

MECHANICAL LESIONS OF THE NERVE ROOTS AND SPINAL CORD

Spinal Canal Lesions

Conventional films and myelography are necessary for localization and diagnosis. Planigraphy and sometimes stereoscopic views are also employed.

The findings on regular spine films will depend upon the site of origin and the size of the lesion. Lesions can arise within the cord (intramedullary) or within the dural space (intradural extramedullary), or can affect the spinal canal from outside the dura (extradural lesions).

Smaller intramedullary lesions produce no changes, but larger lesions can sufficiently expand the cord to widen the canal. This is evidenced by an increased interpediculate distance and medial flattening of the pedicles.

Extramedullary but intradural lesions may occasionally cause erosion of an adjacent pedicle or of the posterior vertebral margin. Large intradural extramedullary lesions may also widen the interpediculate space. A neurofibroma may widen or erode a neural foramen as the tumor grows out of the canal.

With the exception of a herniated disc, extradural lesions are most often malignant tumors that produce destructive bone changes. Inflammatory extradural masses are usually associated with the findings of the underlying spondylitis.

Positive-contrast myelography is essential for localizing intraspinal lesions and may help to distinguish between intramedullary, intradural, and extradural lesions. Intramedullary masses cause widening of the cord, with symmetric narrowing of the surrounding subarachnoid space. Extramedullary intradural masses displace the cord, narrowing the contralateral subarachnoid space while widening the space in the vicinity of the tumor. Extradural lesions will compress the subarachnoid space and may displace the entire canal contralaterally. A herniated disc will compress the subarachnoid space anteriorly or laterally, at an interspace level, and will often elevate or obliterate a homolateral nerve sleeve. Complete subarachnoid block can occur with any of these types of lesions but is most frequent in the larger extradural lesions. With total block, it may be impossible to determine myelographically whether the lesion is intramedullary, intradural, or extradural.[171-174]

Intramedullary Lesions

Plain film changes are observed only if the lesion is large enough to cause widening of the spinal canal. An increase in the interpedicular distance and flattening of the medial borders of the pedicles may then be apparent.

The myelogram demonstrates expansion of the cord, which encroaches symmetrically upon the subarachnoid space in all directions. With complete obstruction there is an abrupt concave termination of the column of contrast medium. The appearance should be similar in all radiographic projections. The cord may be involved at any level, and expansion may encompass several segments.

Gliomas are most frequent, and among these, ependymoma and astrocytoma are the commonest. Ependymomas may arise at any level but most frequently originate in the lower cord or the cauda equina. Dermoids, neurolemmomas, vascular tumors, inflammatory masses, and syringomyelia or hydromyelia are also seen. Intramedullary lesions, however, are the least frequent of the spinal lesions.[18, 171, 172, 174-177]

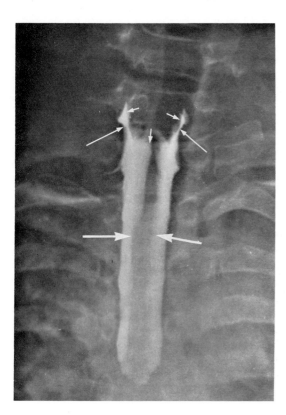

Figure 8–173 **Intramedullary Glioma of Cervical Cord: Myelogram.** Symmetric expansion of the cord *(short arrows)* has caused uniform displacement and narrowing of the subarachnoid thin space *(long thin arrows).* The normal cord and subarachnoid space are shown below *(large arrows).* The spinal cord is not displaced.

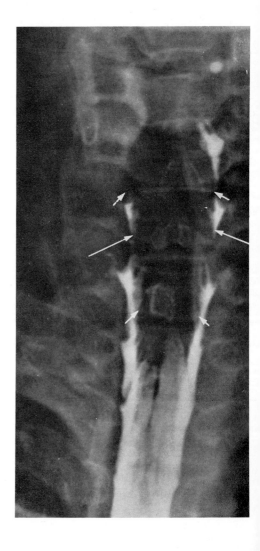

Figure 8–174 **Intramedullary Glioma of Cervical Cord: Myelogram.** The expanded cord *(short arrows)* has caused symmetric narrowing of the subarachnoid space *(long arrows)* of both sides, which is characteristic of an intramedullary lesion. The lesion extends over more than three vertebral segments.

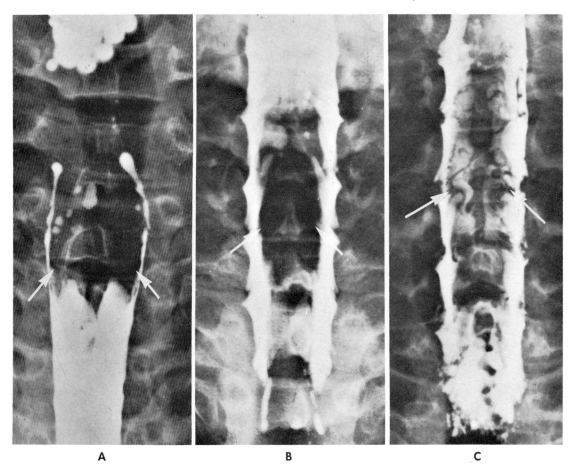

A   B   C

Figure 8–175 **Three Cases of Intramedullary Tumor:** *A,* **Dermoid;** *B,* **Ependymoma;** *C,* **Astrocytoma: Myelogram.** The three lesions are different, yet they produce similar myelographic changes. In each case there is widening of the cord *(arrows)* with symmetric narrowing of the surrounding contrast-filled subarachnoid space. (Courtesy Drs. G. Lombardi and A. Passerini, Milan, Italy.)

Extramedullary Intradural Lesions

The majority of these lesions are benign. Neuroma, neurofibroma, and meningioma are the most frequently found tumors; less common are hemangioma, dermoid, varicosities, lipoma, subarachnoid cyst, and metastatic glioma. Meningioma is the most common lesion in the thoracic canal; it occurs most often in middle-aged females.

On myelography, the cord is usually displaced and compressed against the wall opposite the lesion, causing narrowing of the contralateral subarachnoid space and widening of the space on the ipsilateral side. The spinal canal may bulge toward the pedicle on the side opposite the lesion. A filling defect or complete block of the subarachnoid space may develop.

Bone changes are infrequent. However, neuroma and neurofibroma frequently cause pressure erosion of adjacent vertebrae (see p. 419), since these lesions often extend extradurally.

Varicosities and hemangiomas appear as tortuous and serpiginous filling defects on the myelogram.[18, 171, 172, 174, 175, 177, 178]

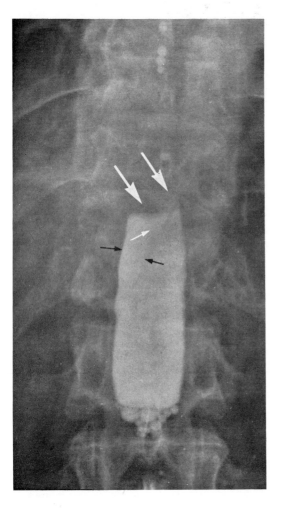

Figure 8–176 **Intradural Extramedullary Meningioma with Complete Block: Myelogram.** The spinal canal *(large arrows)* is completely obstructed, and the cord is displaced to the left *(small white arrow)*. The subarachnoid space is wider on the right *(black arrows)*. These changes are characteristic of an intradural extramedullary lesion.

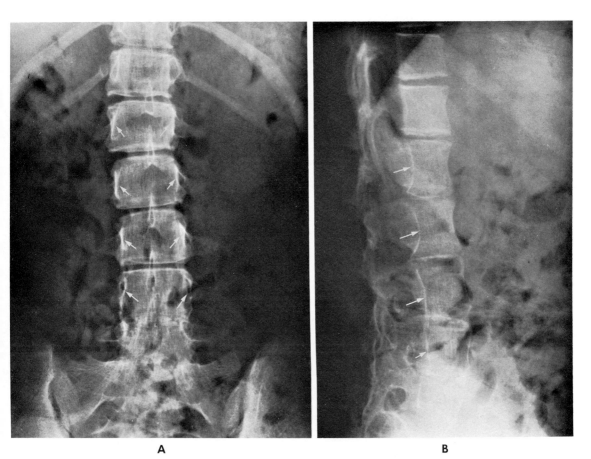

A   B

Figure 8-177 **Extensive Dermoid of Lumbar Canal: The Spine.**

A, A long intradural tumor has caused widening of the spinal canal, flattening of the pedicles *(arrows),* and an increase in the interpedicular distance.

B, Lateral view shows marked erosion of the posterior aspect of the lumbar vertebral bodies *(arrows)* by the enlarged, widened canal, a change that is characteristic of a large intradural lesion. Dermoids, however, are usually much smaller and generally produce fewer and less marked erosive changes.

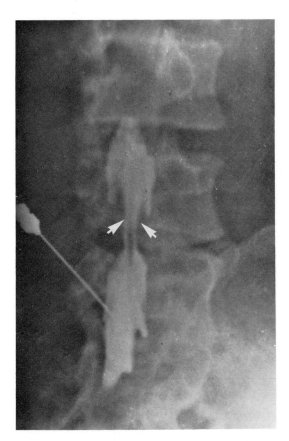

Figure 8–178 **Metastatic Medulloblastoma in Spinal Canal: Myelogram.** A bilateral defect has caused compression and narrowing of the subarachnoid space at L4 *(arrows)*. The defect closely resembles an extradural lesion. A metastatic medulloblastoma surrounds the canal in the region of the cauda equina. Other primary brain tumors can metastasize to the spinal canal, but the medulloblastoma is the most frequent one to do so. These lesions are usually multiple and nodular. Rarely, a metastasis may be intramedullary.

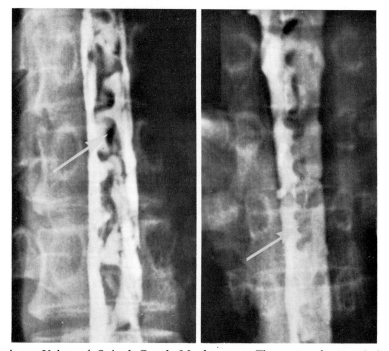

Figure 8–179 **Varicose Veins of Spinal Canal: Myelogram.** There are characteristic circinate defects *(arrows)* in the dorsal canal that are due to the presence of numerous intradural varicosities. Vascular anomalies of the cord and canal rarely produce bony changes.

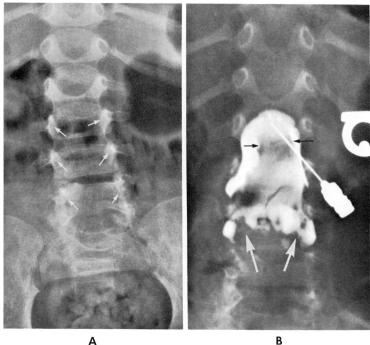

A   B

Figure 8–180 **Intradural Lipoma in an Infant.**

A, The pedicles of the lower three lumbar vertebral bodies are flattened, and the interpedicular distance is increased *(arrows)* by an expanding intradural lipoma.

B, Anteroposterior myelogram demonstrates the displaced cord *(black arrows)* and the widened subarachnoid space on the right, both of which are characteristic of an intradural extramedullary lesion. There is a filling defect and complete block of the column *(white arrows).*

C, Lateral view discloses anterior displacement of the subarachnoid space *(arrows).*

Lipomas develop more frequently in children than in adults. Often they are quite long and may cause the type of change illustrated here.

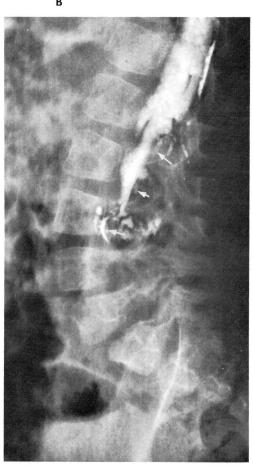

C

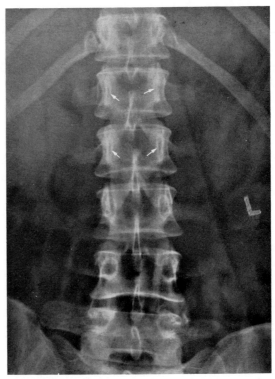

Figure 8–181 **Intradural Arachnoid Cyst: Changes in Bone.**   In anteroposterior view of the lumbar spine, flattening of the pedicles of L1 and L2 *(arrows)* is demonstrated; increased interpedicular distance is also shown. A large arachnoid cyst caused these bone changes, and complete block was seen on myelography.

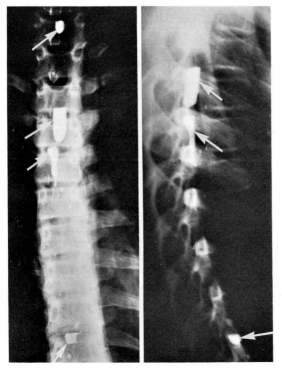

Figure 8–182 **Multiple Small Intradural Arachnoid Cysts: Myelogram.**   Anteroposterior and lateral erect views demonstrate four contrast-filled areas in the dorsal canal *(arrows)*. They are within arachnoid cysts and are situated posteriorly.

Subarachnoid cysts may be single or multiple. The larger ones can cause complete obstruction to the oil column and erosion of the pedicles, as illustrated in Figure 8–181. The smaller cysts usually fill with the opaque material, but erect views generally are needed to demonstrate residual filling after the remainder of the oil descends to the lumbar canal. Primary arachnoid cysts develop most commonly in the dorsal canal. Irregular cyst-like accumulations of opaque material may also be seen in severe arachnoiditis. (Courtesy Drs. G. Perrett, I. Green, and J. Keller, Iowa City, Iowa.)

Extradural Lesions

Extradural lesions arise in the spine and compress or invade the spinal canal and its contents. Disc herniation is the commonest extradural lesion but is considered separately as a clinical and roentgenologic entity (see p. 423).

The majority of spinal tumors that encroach upon the canal are malignant. Metastatic tumors are most frequent; lymphoma, myeloma, or, occasionally, a primary sarcoma also occurs. Infrequently, extradural pressure is caused by a bone hemangioma or chordoma. The nerve root tumors, which usually originate intradurally, can extend through the neural foramina and cause extradural encroachment. Other causes of extradural encroachment include tuberculous or pyogenic spondylitis and epidural abscess or hematoma.

Regular spine films usually reveal bone changes from the spinal tumor or infection; bone destruction from metastatic malignancy is the commonest finding. Often a paravertebral soft tissue mass is seen.

Myelography may show focal indentation and narrowing of the oil column. The subarachnoid space may be displaced. Partial or complete block is quite frequently encountered. In the absence of bone changes or paravertebral mass, myelographic distinction between an extradural and an intradural extramedullary lesion may be impossible.[174, 177, 179, 180]

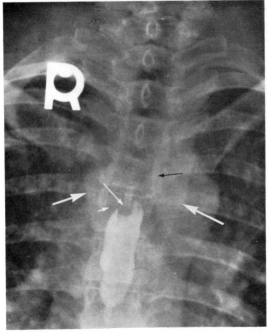

Figure 8–183 **Metastatic Tumor with Compression of Cord: Myelogram.**   There is a soft tissue mass *(large white arrows)* at T4 and T5. The pedicle of T4 has been destroyed *(black arrow)*. Complete block of the subarachnoid space has occurred. Asymmetry is caused by displacement of the cord and canal to the left *(long thin white arrow)*. The distance between the right pedicle of T5 and the subarachnoid space *(short white arrow)* is increased. Coupled with myelographic evidence of complete or partial block, the finding of associated vertebral destruction or paravertebral masses provides strong evidence for an extradural lesion.

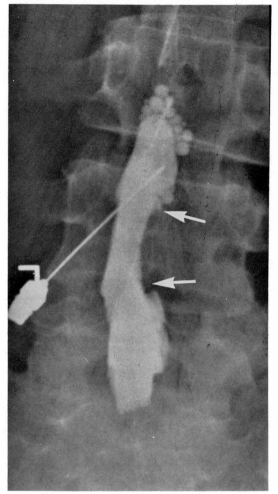

Figure 8–184 **Extradural Tumor: Hodgkin's Disease: Myelogram.**   There is a filling defect in the lateral aspect of the subarachnoid space in the lumbar area. The subarachnoid space has been pushed away from the pedicle *(arrows)*. Definite radiographic evidence of involvement of the vertebra by Hodgkin's disease is lacking. Compression and displacement of the subarachnoid space is characteristic of an extradural lesion.

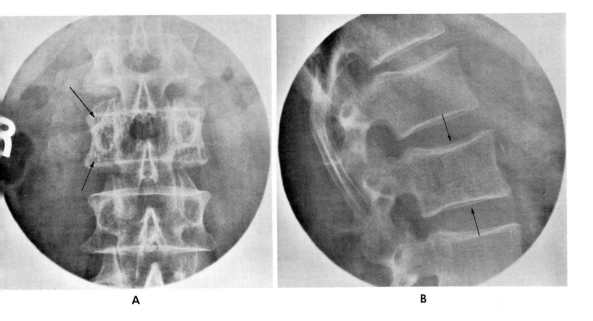

A   B

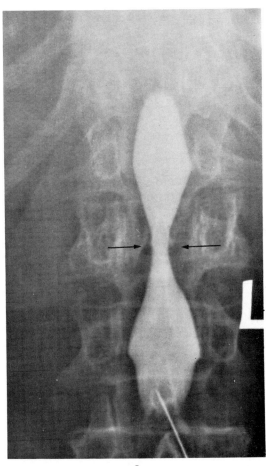

C

Figure 8–185 **Hemangioma of Vertebra: Myelogram.**

A and *B,* The typical coarse vertical trabeculation of the vertebral body *(arrows)* and pedicles is disclosed in spot films of L3.

C, The subarachnoid space at L2 is narrowed bilaterally, and the distance between the pedicles and lateral wall of the subarachnoid space *(arrows)* is increased. This picture results from extension of the vertebral hemangioma to the dura. Generally, hemangiomas remain limited to the vertebral body and produce no neurologic symptoms.

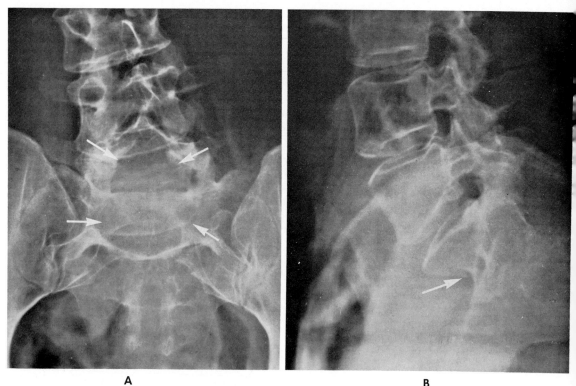

A

B

Figure 8–186   **Chordoma of the Sacrum.**

A, In anteroposterior view a large destructive lesion has involved the upper segments of the sacrum *(arrows).*

B, In a lateral film the anterior aspect of the affected segment is indicated by the arrow. Characteristically, chordoma arises anterior to the vertebral body. The most frequent sites of this low grade malignant lesion are the base of the skull and the sacrococcygeal area.

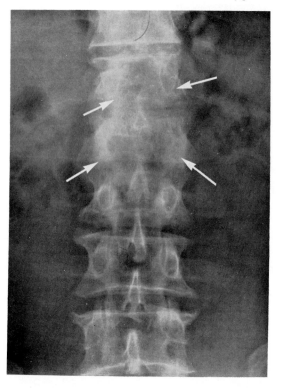

Figure 8–187   **Extradural Lesion: Osteogenic Sarcoma.** A mixed osteolytic and osteoblastic lesion of the bodies and pedicles of T12 and L1 produces a mottled appearance *(arrows).* The lesion is an osteogenic sarcoma originating in L1. It caused neurologic symptoms by invasion of the spinal canal.

Neuroma (Neurinoma and Neurofibroma)

Spinal neuromas or neurinomas may arise at any level. They may cause osseous changes as they enlarge; however, enlargement of the neural foramina may be due to an associated congenital abnormality. Enlargement of the intervertebral foramina provides an important diagnostic clue. The tumor is usually dumbbell-shaped and compresses the subarachnoid space, the nerve roots, and the cord.

Multiple neurinomas or neurofibromas occur in von Recklinghausen's disease. These tumors arise from both spinal nerve roots and peripheral nerves; bone involvement is most commonly encountered in the spine or ribs. On myelography there is a defect in the column of contrast medium at the level of the nerve roots. Features of an intradural or extradural lesion, or both, are seen. Multiple small bilateral defects of the subarachnoid space at the levels of the nerve roots are characteristic of widespread neurofibromatosis. (See also pp. 1411 to 1413.)[18, 171, 172, 174, 175, 181, 182]

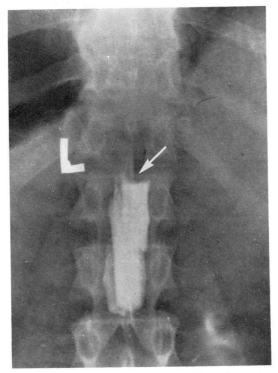

Figure 8-188 **Neurinoma of the Spinal Canal: Myelogram.** There is a complete block at L1, with asymmetry of the subarachnoid space *(arrow);* the cause is an intradural lesion. The defect is sharply marginated, abrupt, and eccentric. Neurinomas and meningiomas are the most common intradural extramedullary lesions.

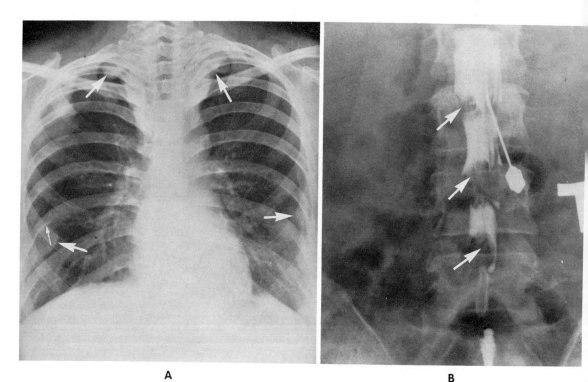

A   **B**

Figure 8–189   **Neurofibromatosis: Chest Film and Myelogram.**

A, There are many neurofibromatous masses *(upper large arrows)* in the area of the upper thoracic spine, and there are also soft tissue tumors of the chest wall *(lower large arrows)*. The inferior surface of a rib on the right *(small arrow)* has been eroded by a neurofibroma of the intercostal nerve. In many cases paraspinal masses seen in neurofibromatosis are associated thoracic meningoceles of varying sizes.

B, Myelogram shows multiple bilateral intradural filling defects at the level of the nerve roots in the lumbar region *(arrows)*. Multiple defects are characteristic of neurofibromatosis.

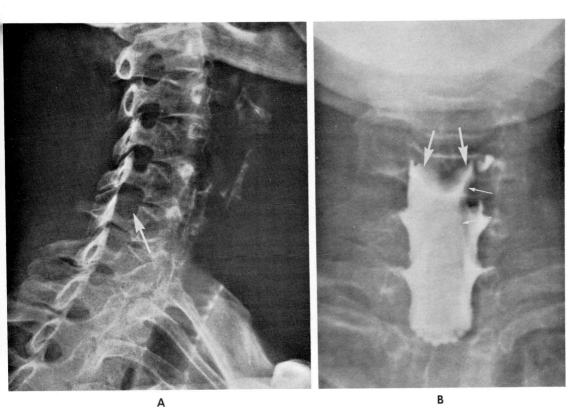

A **B**

Figure 8–190 **Neurofibroma of Cervical Spine: Myelogram.**

A, In plain film, oblique view, there is marked enlargement of the intervertebral foramen between C5 and C6 on the right *(arrow)* due to a neurofibroma.

B, The subarachnoid space *(large arrows)* is completely blocked by a large intradural portion of the tumor. The extradural portion of the tumor *(small arrows)* caused displacement of the column of contrast material. A neurofibroma may be intradural, extradural, or both; when it is mixed, it is dumbbell-shaped.

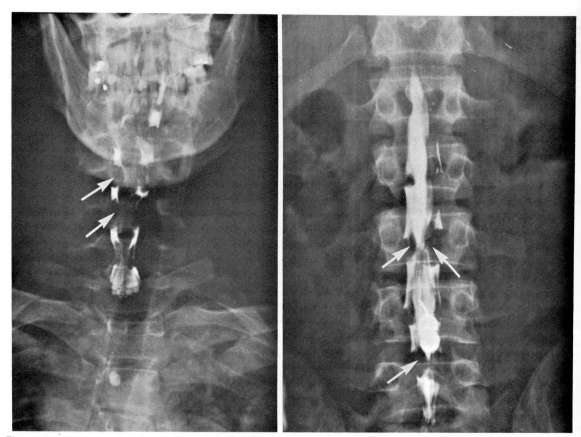

Figure 8–191 **Extensive Neurofibromatosis: Myelogram.** The cervical and lumbar myelograms indicate bilateral defects of the subarachnoid space at almost every intervertebral level *(arrows)*. This appearance is characteristic.

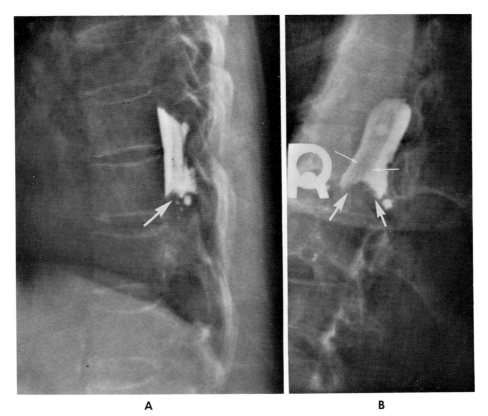

A   **B**

Figure 8-192   **Intradural Perineural Fibroblastoma: Myelogram.**

A, Lateral view discloses complete block *(arrow)* in the lower thoracic region.

B, In addition to complete obstruction *(large arrows),* an oblique view shows the cord displaced anteriorly *(between small arrows);* the subarachnoid space is asymmetrical.

These features are characteristic of extramedullary intradural lesions; in this case, the lesion is a perineural fibroblastoma. (Courtesy of Dr. Arlyne Shockman, Veterans Administration Hospital, Philadelphia, Pennsylvania.)

Degenerative Disc Disease and Disc Herniation

The radiographic changes of degenerative disc disease may or may not be associated with posterior or lateral herniation of an intervertebral disc. Definitive radiographic diagnosis of herniation can be made only by myelography.

Both degenerative disc disease and disc herniation occur most frequently in the lower lumbar spine (L4-L5, L5-S1) and the lower cervical area (C5-C6, C6-C7). Disc herniation in the dorsal spine is infrequent; occasionally it is a complication of Scheuermann's disease (juvenile vertebral epiphysitis). Multiple disc herniations are not rare.

A degenerated disc, with or without herniation, leads to narrowing of the involved interspace and, usually, to marginal spurs and sclerosis. In the cervical spine, posterior spurs can encroach upon the neural foramina (see *Cervical Spondylosis,* p. 426). Acute herniation of a disc will not produce radiographic changes, and a completely normal spine study does not exclude the diagnosis.

On myelography, disc herniation produces a smooth extradural defect at a lateral or anterior portion of the oil column at the level of the intervertebral space.

Frequently a nerve sleeve is elevated or totally "amputated" (not opacified) on the side of the herniation. If the myelographic changes are not at the interspace level, the lesion is more likely to be a tumor than a herniated disc. Complete subarachnoid block from disc herniation is extremely rare.

Central disc herniation can often be appreciated or distinguished from intramedullary tumor only on the cross table lateral myelogram films; these should be made in all myelographic examinations. Rarely, a herniated disc, especially at L5-S1, may not produce recognizable myelographic changes. In capacious wide canals in the lower lumbosacral area, decubitus films may be needed to demonstrate the defect.[183-187]

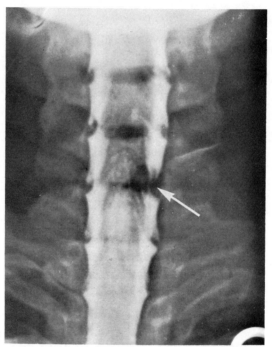

Figure 8–193 **Herniated Cervical Disc: Myelogram.** Anteroposterior projection discloses the characteristic localized extradural defect at the level of C5–C6 on the left *(arrow),* the most frequent site of herniated cervical disc. The defect is at the exact level of the root sleeve, which is characteristic of disc lesions. A similar defect not at the intervertebral level would most likely be neoplastic.

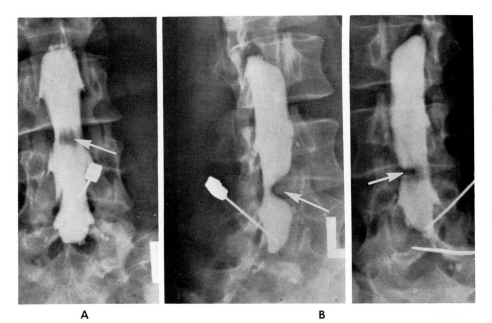

A   B

Figure 8–194 **Lumbar Herniated Disc: Myelogram.**

A, In anteroposterior view a central filling defect *(arrow)* is demonstrated in the contrast column at the level of L4–L5.

B, Oblique projections disclose a circumscribed extradural defect that is typical of a central disc *(arrows).*

In a case like the one illustrated, oblique or lateral views are necessary to confirm the location of the lesion. A sharp defect localized to the intervertebral level is characteristic of a herniated disc.

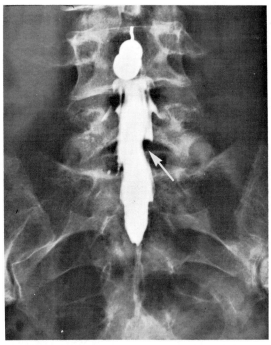

Figure 8–195 **Lumbar Herniated Disc: Myelogram.**   There is a lateral defect in the contrast column *(arrow)* at the level of L5–S1, with amputation of the root sleeve at this level. The picture is characteristic of a lateral disc. In smaller herniations the persistent unilateral upward displacement of a nerve sleeve may be the only abnormality.

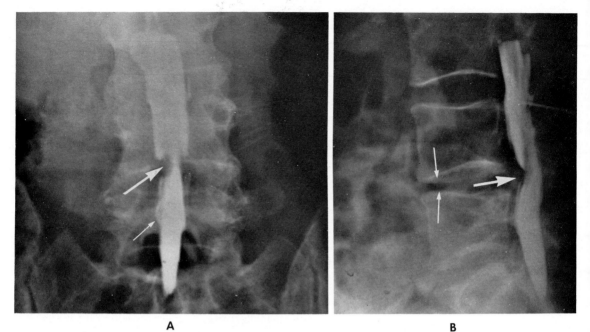

A   B

Figure 8–196 **Lumbar Herniated Disc: Myelogram.**

A, In anteroposterior view, bilateral defects of the contrast column at the level of L4–L5 signify a large disc herniation *(large arrow)*. On the right side at L5–S1 the nerve sleeve *(small arrow)* is elevated, suggesting a second smaller disc protrusion at this level.

B, Lateral projection demonstrates extradural location of this defect at L4–L5. The outline is sharp, and there is gradual displacement of the subarachnoid space from the wall of the spinal canal *(large arrow)*. A narrowed disc space is present *(small arrows)*. As a rule, in longstanding herniation, a narrow intervertebral space and local spurring are associated findings.

Cervical Spondylosis (Degenerative Disc Disease)

The lower cervical spine is the commonest site of degenerative disc disease, and radiographic changes are found in the majority of older people. The C5-C6 and C6-C7 interspaces are most frequently affected. Radicular pain characteristic of the "cervical syndrome" is often associated, but marked radiographic changes without symptoms are common.

Radiographically there is narrowing of the affected interspace or interspaces due to thinning of the cartilaginous disc. Reactive bone spurs appear at the anterior and posterior vertebral margins, often with surrounding sclerosis. Oblique views will demonstrate the extent of encroachment of the bony spurs into the neural foramina; these posterior spurs may cause radiculitis due to nerve root pressure. Chronic disc herniation is often associated with the radiographic changes of spondylosis.

Myelography is usually negative in spondylosis without disc herniation, although unusually large posterior spurs can produce a pressure defect on the anterior oil column. This may cause neurologic symptoms, particularly if the anteroposterior diameter of the bony spinal canal is less than 1.3 cm. (See also *Spondylosis*, p. 98, and *Myelopathy from Narrowed Cervical Canal*, p. 405.)[170, 188–191]

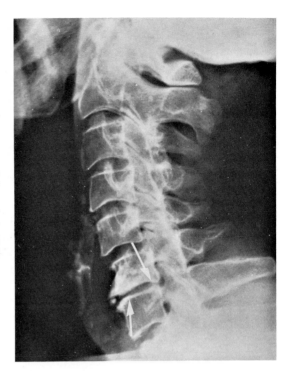

Figure 8–197 **Cervical Spondylosis.** Lateral view demonstrates narrowing of the intervertebral space between C6 and C7 (arrows). There are prominent hypertrophic spurs anteriorly, and some sclerosis of the anterior vertebral margins. No significant spurring is seen posteriorly, although oblique views are necessary to demonstrate small posterior spurs.

The regular curve normally made by the posterior margins of the cervical vertebrae is abruptly broken by the apparent posterior position of C6 and C7, which is a result of fixation and limitation of motion of these two cervical bodies by the disease process.

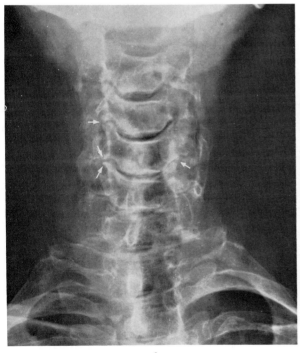

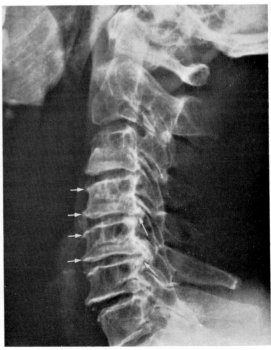

A   B

Figure 8–198 **Severe Cervical Spondylosis.**

A 61 year old man experienced neck and arm pain and neurologic disturbances in the lower extremities.

A, Anteroposterior view: There is spurring (arrows) along the margins of the small joints (joints of Luschka).

B, Lateral view: Several intervertebral spaces are narrowed, especially C3–C4, C4–C5, and C5–C6. There is reactive sclerosis and anterior and posterior bony spurs (arrows).

Legend continued on the following page.

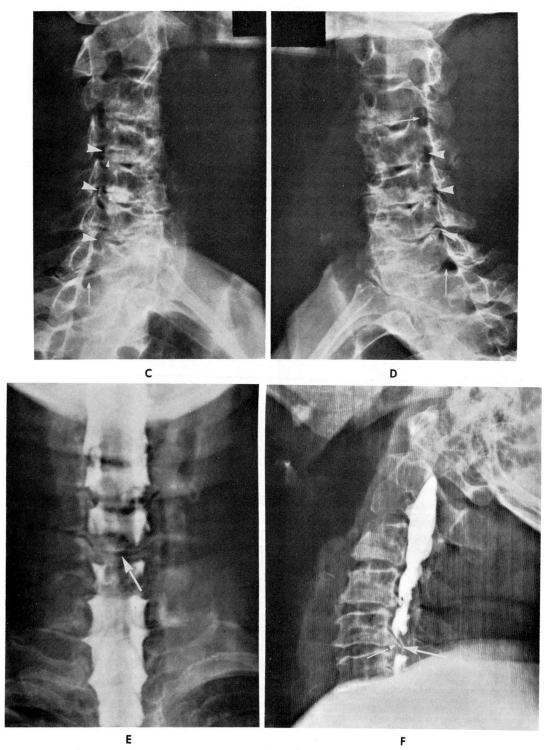

Figure 8–198 Continued.

C and D, Oblique views: Extensive posterior spurs encroach upon the neural foramina on both sides (*large arrowheads*) and virtually obliterate them. The uninvolved neural foramina appear normal (*small arrows*). The small joints of Luschka (*small arrowheads*) have become narrowed in many vertebral segments.

E, Anteroposterior myelogram: There is a large filling defect (*arrow*) at the C5–C6 level. If only this projection is viewed, the defect might be considered to be due to either an intradural or an extradural lesion.

F, Lateral myelogram: The filling defect (*large arrow*) is due to pressure from a large posterior spur (*small arrow*) arising from C5.

Marked improvement in neurologic symptoms occurred after surgical removal of the spur. There was no associated disc herniation.

Causalgia: Sudeck's Atrophy

Sudeck's atrophy is a rather rare complication of a fracture, especially of the wrist, elbow, or ankle, and involves the bone and soft tissues distal to the fracture site. The clinical onset of pain, swelling, and dysfunction occurs approximately one to two months following the fracture and precedes the bone changes by four to eight weeks.

Osteoporosis of the distal bones is usually most pronounced around the joints and is characterized by overall demineralization and scattered small areas of cystic rarefaction. The cortex is relatively intact, and the joint spaces are unaltered. In severe cases, however, the borders of the smaller bones appear to vanish, giving a pseudofusion appearance.

The radiographic appearance is indistinguishable from that of severe osteoporosis of disuse, but the late onset of the clinical and roentgen changes (after a fracture) is highly characteristic.

In the more acute forms, the bones usually revert to normal, but in the chronic forms, permanent changes (thick scanty trabeculae) may result.[192-194]

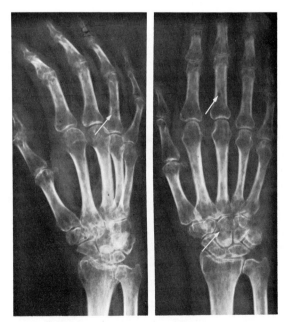

Figure 8–199 **Causalgia: Sudeck's Atrophy.** There is considerable osteoporosis of the bones of the hand, with cyst-like areas of spotty demineralization *(arrows)*. The cortical outlines and the articular surfaces and spaces are preserved. Osteoporosis is most evident in the periarticular portions. These changes developed several weeks after a fracture of the elbow. The picture is nonspecific and is indistinguishable from other forms of severe osteoporosis.

Thoracic Outlet Syndrome (Scalenus Syndrome) and Cervical Rib

The thoracic outlet syndrome is due to compression of the subclavian vessels and/or the brachial plexus by the scalenus anticus muscle or a cervical rib. The latter may be the only radiographic finding, but it occurs in less than one third of patients with the syndrome and is also frequently present in asymptomatic individuals. Occasionally a bifid rib or fusion of the first and second ribs is seen.

If the symptoms are vascular rather than neurologic, subclavian or brachial arteriography will confirm the diagnosis.

There usually will be localized compression or torsion of the artery in the scalenus tunnel; the narrowing is frequently associated with poststenotic dilatation. Most significant, however, is demonstration of further marked narrowing or even complete obliteration of the artery during hyperabduction maneuvers that cause contraction of the scalenus and pectoralis minor muscles (Adson or modified Adson maneuvers). Rarely, a chronically compressed subclavian artery will become thrombosed, and there will be complete occlusion on arteriography.

Arteriography can also demonstrate other causes of subclavian ischemia such as atherosclerosis or extrinsic involvement from neighboring disease.[195-199]

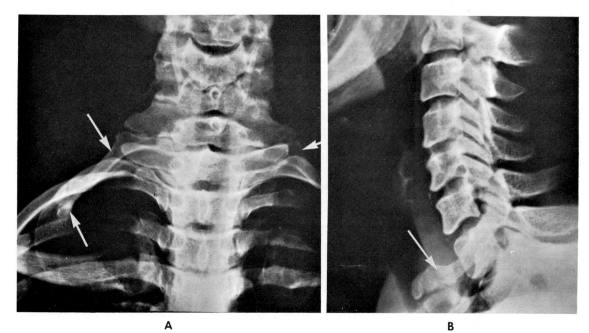

A   **B**

Figure 8–200 **Cervical Rib.** Anteroposterior *(A)* and lateral *(B)* views show bilateral cervical ribs *(arrows),* which are longer on the right. Most cervical ribs are asymptomatic; however, they may be a factor in the thoracic outlet syndrome.

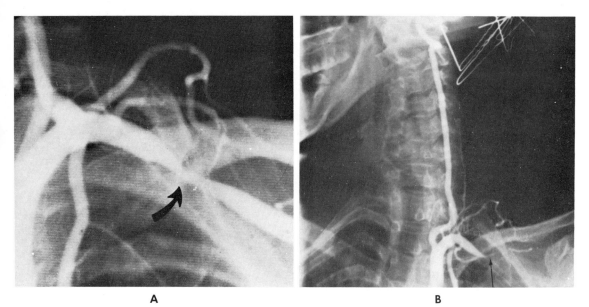

A   B

Figure 8–201   **Scalenus Syndrome: Arteriogram.**

A, The course of the subclavian artery is narrowed as it passes through the scalenus tunnel *(arrow)*.

B, Pinpoint block *(arrow)* occurs during a modified Adson maneuver. The vertebral artery overfills because of the subclavian block. (Courtesy Dr. E. K. Lang, Indianapolis, Indiana.)

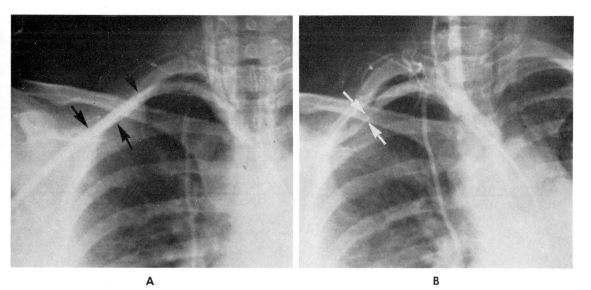

A   B

Figure 8–202   **Thoracic Outlet Syndrome: Thoracic Aortogram.**

The patient, a woman, complained of a numbness and tingling in the right arm, which occurred most often during sleep.

A, With the patient relaxed and the chin in midline, the aortic arch and right subclavian artery are opacified; the right subclavian artery is normal *(arrows)*.

B, With the chin extended and the head turned sharply to the right, opacification demonstrates marked compression and narrowing of the right subclavian artery *(arrows)* in the region corresponding to the scalenus tunnel. This position of the chin and head caused discomfort in the right arm during the study.

Hypertrophic Interstitial Neuritis

There are no significant neuroradiologic findings in the usual case of peripheral neuritis. However, in the rare hypertrophic interstitial neuritis, due to hyperplasia of the sheath of Schwann, involvement of the spinal nerve roots may produce a characteristic myelographic picture. In the oil column there will be parallel linear radiolucencies due to thickened nerve roots. At the interspaces, transverse bar defects are almost constantly seen. These findings are virtually pathognomonic and may be the only clue to diagnosis in cases without peripheral nerve enlargement.[200, 201]

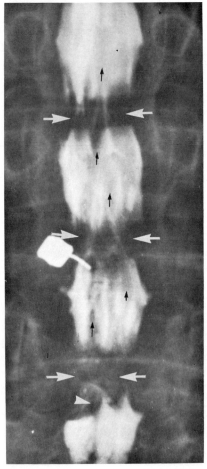

Figure 8–203 **Hypertrophic Interstitial Neuritis in 63 Year Old Male: Myelogram.**   The needle is at L3–L4. At all the lumbar interspaces there are transverse bar defects *(white arrows)* due to protrusions of the annulus fibrosus. The parallel linear translucencies in the oil column *(black arrows)* are the thickened nerve roots of the cauda equina. The defect at L4–L5 *(white arrowhead)* is an expanded spinal nerve root projecting into the canal.

This myelographic picture is virtually diagnostic of hypertrophic interstitial neuritis. (Courtesy Dr. V. C. Hinck, Portland, Oregon.)

Narrow Lumbar Canal Syndrome

Acute and chronic low back pain and sciatica may be due to congenital narrowing of the lower lumbar spinal canal.

On conventional films, if the lower three lumbar bodies reveal an anteroposterior diameter of the canal less than 15 mm. and an interpediculate measurement less than 25 mm., narrowing of the canal should be considered.

The myelographic findings most often show prominent posterior disc protrusive defects at the lumbar interspaces, often with lateral defects. Occasionally partial or complete obstruction is encountered.

Laminectomy and diskectomy generally will relieve these symptoms.[202]

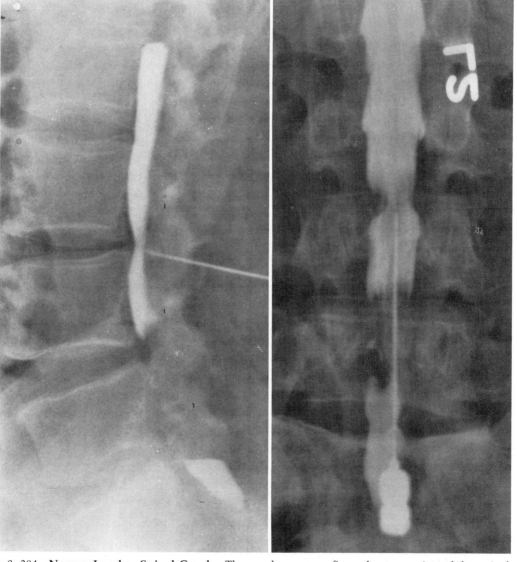

Figure 8–204 **Narrow Lumbar Spinal Canal.** The myelogram confirms the narrowing of the spinal canal in both diameters. This was apparent on regular spine films by decreased interpediculate distance, narrowed anteroposterior diameter of canal, and small intervertebral foramina.

The narrowed oil column is obstructed at L4–L5. Laminectomy and diskectomy produced clinical recovery. (Courtesy Dr. G. H. Roberson, Boston, Massachusetts.)

DISEASES OF MUSCLE AND NEUROMUSCULAR JUNCTION

Progressive Muscular Dystrophies

Duchenne's pseudohypertrophic muscular dystrophy is the most common of the many forms of progressive muscular dystrophy *without myotonia.*

The most specific roentgen finding in all types is the replacement of muscle fiber by orderly lines of radiolucent fat, giving the affected muscle a striated appearance. The total muscle mass is not decreased. In the pseudohypertrophic form, the muscle volume is actually increased even before fat replacement occurs. By contrast, in wasting diseases, the muscle volume decreases and fat replacement is more haphazard and irregular.

A characteristic increased anteroposterior diameter of the fibula is commonly seen in the Duchenne pseudohypertrophic form. In the advanced stages of most forms, there is an increased incidence of scoliosis, contractures, and bone atrophy, with cortical thinning, narrowing, and osteoporosis.

A myocardiopathy leading to congestive failure can occur in the Duchenne form.[98, 203-207]

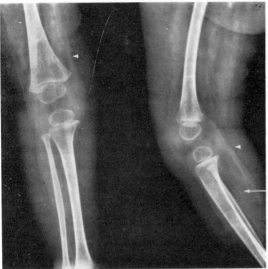
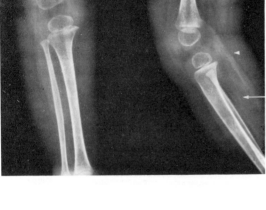

Figure 8–205 **Progressive Muscular Dystrophy.** Lucent stripes within the muscle represent bands of fatty infiltration *(arrow).* The soft tissue volume is unaffected. The muscle fascial planes are clearly demarcated from the subcutaneous tissue *(arrowheads).* Osteoporosis is minimal. The findings are typical of muscular dystrophy.

Figure 8–206 **Progressive Muscular Dystrophy: Scoliosis.** The apex of the curve is in the dorsal region. The scoliosis is secondary to muscular weakness and is seen fairly often in muscular dystrophy.

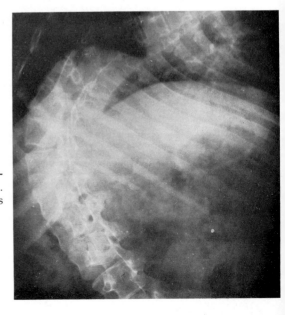

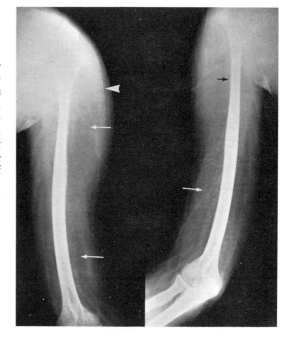

Figure 8–207 **Pseudohypertrophic Muscular Dystrophy.** There is an increase in the total muscle volume associated with lucent streaks of fatty infiltration *(white arrows)*. The subcutaneous fat is not increased. There is a sharp demarcation between the muscle fascial planes and the subcutaneous tissue *(arrowhead)*. Atrophic narrowing of the upper humeral shaft *(black arrow)* is secondary to muscular weakness. An increase in the muscle mass with fatty infiltration is typical of the pseudohypertrophic form of muscular dystrophy. (Courtesy Dr. A. Lewitan, Jewish Chronic Disease Hospital, Brooklyn, New York.)

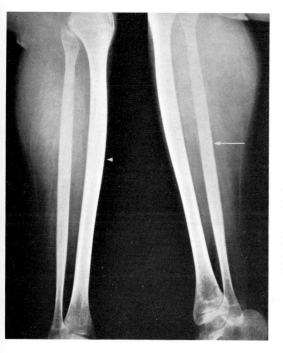

Figure 8–208 **Pseudohypertrophic Muscular Dystrophy.** The anteroposterior diameter of the fibula is widened *(arrow)*. The fibula is almost as wide as the tibia, which is thinned and atrophic *(arrowhead)*. Fatty infiltration of the muscles is not as marked as in the preceding case. There has been no increase in the amount of subcutaneous fat. Serial soft tissue films depicting progressive fatty infiltration indicate the progress of the disease. (Courtesy Dr. A. Lewitan, Jewish Chronic Disease Hospital, Brooklyn, New York.)

Myotonia Atrophica (Myotonic Dystrophy)

Striated muscle is predominantly involved, although smooth muscle is also affected. Cataracts, baldness, and testicular atrophy occur in most afflicted males. Radiographic changes can appear in the musculoskeletal system, in the gastrointestinal tract, and in the chest.

Skeletal muscle atrophy is manifested radiographically by loss of muscle mass and corresponding increase in fat between the bundles, similar to the appearance in other dystrophies.

Thickening of the calvarium, a small sella turcica, large frontal sinuses, and an elongated mandible are seen in the majority of patients. The thickening of the calvarium may be diffuse or localized to the frontal bone. The localized form appears similar to hyperostosis frontalis interna, which is a normal finding especially in females.

In the gastrointestinal tract, disturbances in deglutition are manifested by prolonged filling of the pyriform sinuses with contrast material, a finding also seen in bulbar or pseudobulbar disease. Often the entire esophagus is dilated, with poor peristaltic activity. Persistent dilatation of atonic small bowel loops and sometimes of the colon is not unusual. For some unknown reason, the sigmoid remains undilated or even appears narrowed.

Acute and chronic pulmonary changes can develop from aspiration pneumonia secondary to the dysphagia. Impaired diaphragmatic excursions can cause hypoinflation, basal atelectatic plaques, and recurrent pneumonias. Pulmonary hypertension and right heart failure are late complications. Rarely, myocardial involvement can cause cardiomegaly and congestive failure. Pectus excavatum is also a frequent finding.[207-212]

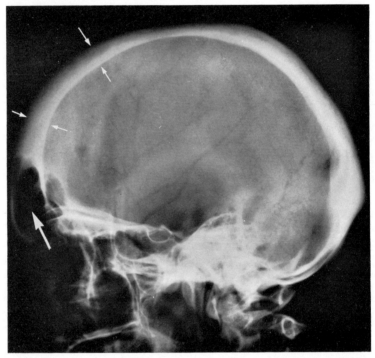

Figure 8–209 **Myotonia Atrophica: Cranial Hyperostosis.** There is diffuse thickening of the calvarium (*small arrows*) and moderate enlargement of the frontal sinuses (*large arrow*). The sella is somewhat small. Although hyperostosis is frequently found in myotonia atrophica, it is not diagnostic, since it often occurs in normal persons, especially in females. (Courtesy Dr. G. DiChiro, National Institutes of Health, Bethesda, Maryland.)

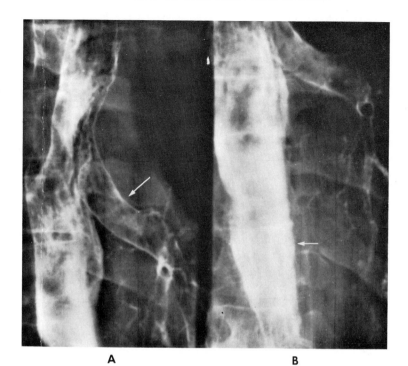

A **B**

Figure 8–210 **Myotonia Atrophica: Dysphagia and Aspiration.**

A, Some of the ingested barium has been aspirated into a bronchus *(arrow),* owing to lack of normal esophageal motility.

B, The esophagus is atonic and somewhat dilated *(arrow);* prolonged retention of the barium in the esophagus is due to lack of peristalsis.

Decreased esophageal peristalsis is the commonest gastrointestinal manifestation of myotonia atrophica.

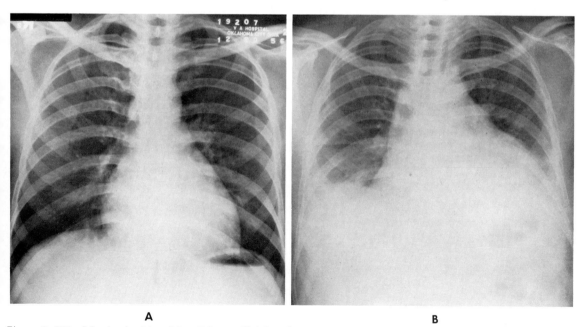

A **B**

Figure 8–211 **Myotonia Atrophica: Myocardial Involvement.**

A, Initially, the heart and lungs are essentially normal.

B, There is progressive cardiomegaly and congestive heart failure. Fluid at the bases has obliterated the costophrenic angles. The picture is nonspecific and indistinguishable from that of the cardiopathy associated with many other conditions.

Myasthenia Gravis

Thymic enlargement from thymic hypertrophy or thymoma may be detected radiographically in approximately 10 per cent of patients with myasthenia gravis. Up to 75 per cent of all thymomas are associated with myasthenia gravis. They are found in the anterior mediastinum at the junction of the heart and great vessels, arising as sharply demarcated oval densities, 25 per cent of which may contain linear or amorphous calcium deposits. A minority of thymomas have ill-defined borders, and these usually prove to be malignant and invasive.

Thymic enlargement may be difficult to see in routine projections, and multiple projections and tomographic films may be needed.

Dysphagia, a common symptom of myasthenia, is due to weakness of the muscles of deglutition. Barium studies may show prolonged retention of the barium in the valleculae and pyriform sinuses. Cineradiography or videotape studies more accurately demonstrate this disturbance. Improvement after drug therapy is also well demonstrated by cineradiography.[213-216]

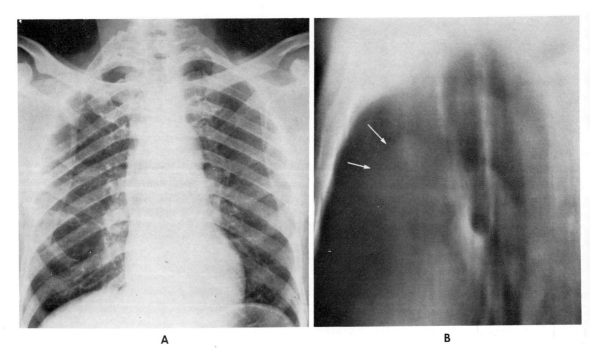

A   **B**

Figure 8–212   **Myasthenia Gravis: Thymoma.**

A, Posteroanterior view indicates possible mediastinal widening, although no definite lesion can be demonstrated.

B, Lateral planigram clearly demonstrates an anterior lobulated mediastinal mass *(arrows),* which was a thymoma.

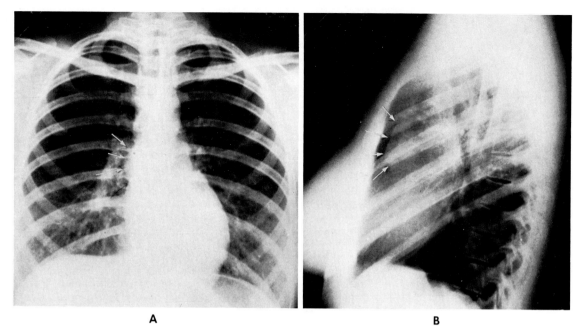

A   B

Figure 8–213 **Myasthenia Gravis: Thymoma.**

  A, Posteroanterior projection discloses a small mediastinal density just to the right of the midline *(arrows).*

  B, Lateral projection shows the anterior mediastinal density *(arrows),* which proved to be a thymoma at surgery.

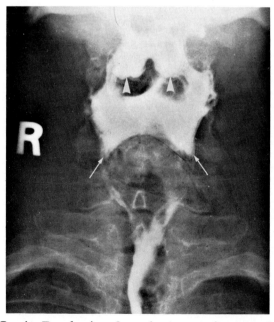

Figure 8–214 **Myasthenia Gravis: Dysphagia.**   Several minutes after ingestion of barium there is retention in the valleculae *(arrowheads),* the pyriform sinuses *(arrows),* and the upper esophagus. Following therapy with neostigmine, deglutition returned to normal. Prolonged retention of barium in the pyriform sinus occurs in any condition in which weakness of the muscles of deglutition is a factor.

Calcinosis

Multiple diffuse calcium deposits in the subcutaneous fat, muscles, and tendons can produce muscular and mobility disturbances. These calcium deposits are usually merely one manifestation of a generalized disease, such as scleroderma, dermatomyositis (polymyositis), Raynaud's disorder, chronic renal failure, and other metabolic disturbances (gout, hyperparathyroidism, hypervitaminosis D, and so forth). In scleroderma, especially when accompanied by Raynaud's phenomenon, the calcium deposits are usually in the hands or feet (see p. 40, Figs. 3–8 and 3–9). The diffuse deposits in dermatomyositis (see p. 41) involve subcutaneous tissue, fascia, and muscles.

Progressive widespread calcifications of fat and, later, of muscles are features of a rare idiopathic disease called *calcinosis universalis.* In an even rarer condition, *calcinosis circumscripta,* idiopathic small calcium collections appear in scattered subcutaneous areas.

Massive deposits of calcium in periarticular soft tissues occur in a rare condition of unknown origin called *tumoral calcinosis.* The calcific tumors occur in healthy individuals, usually in the first two decades of life. There is evidence that these tumors arise from bursa or tendon sheaths. The hips, elbows, shoulders, ankles, wrists, and feet are frequent sites. Involvement is usually multiple and somewhat symmetric. The lesions are very dense, with a round or oval contour and irregular rough borders. Distinction from calcinosis circumscripta is made by the periarticular location of tumoral calcinosis. In vitamin D intoxication, milk-alkali syndrome, and secondary hyperparthyroidism, somewhat similar calcific deposits can appear. However, no underlying disease is associated with tumoral calcinosis.[217-219]

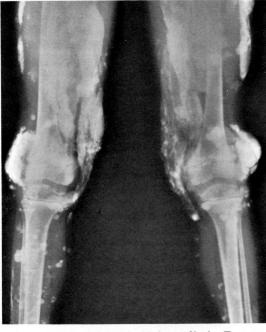

Figure 8–215 **Calcinosis Universalis in Dermatomyositis.** There are dense plaques of calcium in the skin, subcutaneous tissues, tendons, and muscles. The patient was a child who had dermatomyositis.

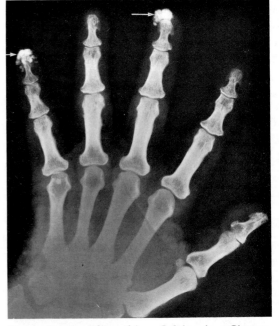

Figure 8–216 **Idiopathic Calcinosis Circumscripta.** There are dense granular collections of calcium in the region of the distal phalanges of the second and fifth fingers *(arrows).* There is no radiographic evidence of soft tissue atrophy on the films and no clinical evidence of Raynaud's phenomenon. No cause for the calcinosis was found.

Myositis Ossificans

In post-traumatic myositis ossificans, calcification occurs within the injured muscle, which sometimes may be in close apposition to bone. A lucent zone separates the calcification from the bone, which will distinguish it from a juxtacortical osteosarcoma. Calcification and ossification also occur in the soft tissues of paraplegics.

Progressive myositis ossificans (Münchmeyer's disease), which begins at birth or in early life, is characterized by plates or columns of ossification in tendons, fascia, ligaments, and muscle sheaths. Band-like densities of bone first appear in the neck and back. Commonly, digital anomalies such as great toe microdactyly and synostoses are seen.[220-222]

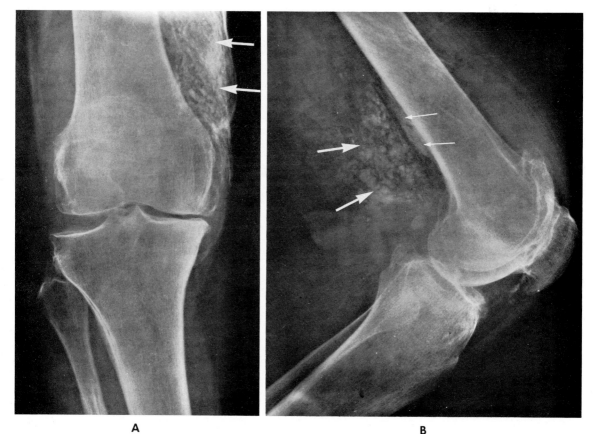

A   **B**

Figure 8–217 **Traumatic Myositis Ossificans.**

A, Posteroanterior view demonstrates laminated calcifications, which are in the muscle bundles medial to the femur and which are secondary to intramuscular hemorrhage *(large arrows).*

B, In lateral view, the soft tissue calcifications *(large arrows)* are posterior to the knee, and a post-traumatic calcified subperiosteal hemorrhage is seen *(small arrows).*

REFERENCES

1. Ambrose, J.: Computerized transverse axial scanning. Br. J. Radiol., *46*:1023, 1973.
2. New, P. F. J., Scott, W. R., et al.: Computerized axial tomography with the EMI scanner. Radiology, *110*:109, 1974.
3. Scott, W. R., New, P. F. J., et al.: Computerized axial tomography of intracerebral and intraventricular hemorrhage. Radiology, *112*:73, 1974.
4. Momose, K. J., New, P. F. J., et al.: Use of computed tomography in ophthalmology. Radiology, *115*:361, 1975.
5. Hahnemann Medical College and Hospital—Department of Diagnostic Radiology: First Conference on Computerized Axial Tomography. November, 1974.
6. Hahnemann Medical College and Hospital—Department of Diagnostic Radiology: Second Conference on Computerized Axial Tomography—Role of CTT scan in pediatrics. April, 1975.
7. Hahnemann Medical College and Hospital—Department of Diagnostic Radiology: Third Conference on Computerized Axial Tomography—Role of CTT scan in ophthalmology. June, 1975.
8. Caplan, L. H., et al.: Megacolon and volvulus in Parkinson's disease. Radiology, *85*:73, 1965.
9. Sypert, G. W., Leffman, H., and Ojemann, G. A.: Occult normal pressure hydrocephalus manifested by parkinsonism-dementia complex. Neurology, *23*:234, 1973.
10. Ring, B. A.: Angiographic recognition of occlusions of isolated branches of the middle cerebral artery. Am. J. Roentgenol. Radium Ther. Nucl. Med., *89*:391, 1963.
11. Schechter, M. M., and Zingesser, L. H.: The radiology of basilar thrombosis. Radiology, *85*:23, 1965.
12. Mount, L. A., and Taveras, J. M.: Angiographic demonstration of collateral circulation of cerebral hemispheres. Arch. Neurol., *78*:235, 1957.
13. Martin, J. P.: Venous thrombosis in the central nervous system. Proc. R. Soc. Med., *37*:383, 1944.
14. Alter, M., et al.: Cerebral infarction: Clinical and angiographic correlations. Neurology, *22*:590, 1972.
15. Ashby, R. N., Karas, B. G., and Cannon, A. H.: Clinical and roentgenographic aspects of the subclavian steal syndrome. Am. J. Roentgenol. Radium Ther. Nucl. Med., *90*:535, 1963.
16. Weber, M. B.: Pitfalls in the diagnosis of subclavian steal syndrome. Illinois Med. J., *136*:237, 1969.
17. Gammill, S. L., Font, R. G., and Nice, C. M.: The subclavian steal syndrome, or, curse you, Red Baron. Semin. Roentgenol., *6*:127, 1971.
18. Taveras, J. M., and Wood, E. H.: *Diagnostic Neuroradiology.* 2nd ed. Baltimore, Williams & Wilkins Co., 1974.
19. McKissock, W.: Some aspects of subarachnoid hemorrhage—a symposium. Br. J. Radiol., *32*:79, 1959.
20. Chase, N. E.: Radiology of cerebral vascular disease. Bull. N. Y. Acad. Med., *39*:790, 1963.
21. Bull, J. V. D.: Contribution of radiology to the study of intracranial aneurysms. Br. Med. J., *2*:1701, 1962.
22. Thompson, J. R., Harwood-Nash, D. C., and Fitz, C. R.: Cerebral aneurysms in children. Am. J. Roentgenol. Radium Ther. Nucl. Med., *118*:163, 1973.
23. Benoit, B. G., and Wortsman, G.: Traumatic cerebral aneurysms. J. Neurol. Neurosurg. Psychiatry, *36*:127, 1973.
24. Anderson, F. M., and Kerbin, M. A.: Arteriovenous anomalies of the brain: revue and presentation of 37 cases. Neurology, *8*:89, 1958.
25. Iraci, G., Marin, G., et al.: Further observations on the so-called Japanese cerebrovascular disease. Am. J. Roentgenol. Radium Ther. Nucl. Med., *115*:35, 1972.
26. Eisenman, J. I., Alekoubides, A., and Pribram, H.: Spontaneous thrombosis of vascular malformation of the brain. Acta Radiol., *13*:77, 1972.
27. Isfort, A., and Engelmeier, M. P.: Zur Diagnostik akuter intrakranieller Phlebothrombosen unter besonderer Berucksichtigung angiographischer Verfahren. (The diagnosis of acute intracranial phlebothrombosis with special reference to angiographic methods.) Fortschr. Neurol. Psychiat., *31*:121, 1963.
28. Buchanan, D. S., and Brazinsky, J. H.: Dural sinus and cerebral venous thrombosis: Incidence in young women receiving oral contraceptives. Arch. Neurol., *22*:440, 1970.
29. Vines, F. S., and Davis, D. O.: Clinical-radiological correlation in cerebral venous occlusive disease. Radiology, *98*:9, 1971.
30. Norman, O.: Angiographic differentiation between acute and chronic subdural and extradural hematoma. Acta Radiol., *46*:371, 1956.
31. Gannon, W. E.: Roentgen signs of resolving subdural hematoma. Radiology, *79*:420, 1962.
32. Radcliffe, W. B., Guinto, F. C., et al.: Subdural hematoma shape. Am. J. Roentgenol. Radium Ther. Nucl. Med., *115*:72, 1972.
33. Radcliffe, W. B., Guinto, F. C., and Scatliff, J. H.: Cerebral and extracerebral hematomas. Semin. Roentgenol., *6*:103, 1971.
34. Davis, D. O., and Pressman, B. D.: Computerized tomography of the brain. Radiol. Clin. North Am., *12*:297, 1974.
35. Lehrer, H., et al.: Angiographic triad in tuberculous meningitis. Radiology, *87*:829, 1966.

36. Leeds, N. E., and Goldberg, H. L.: Angiographic manifestations in cerebral inflammatory disease. Radiology, 98:595, 1971.
37. Tutton, K., and Ollenshaw, R.: Radiological investigation of intracranial abscesses. Br. J. Radiol., 31:571, 1958.
38. Carey, M. E., Chou, S. N., and French, L. S.: Experience with brain abscesses. J. Neurosurg., 36:1, 1972.
39. Decker, H. G., Shapiro, S. W., and Porter, H. R.: Epidural tuberculous abscess simulating herniate lumbar intervertebral disc. Ann. Surg., 149:249, 1959.
40. Bell, D., and Cockshott, W. P.: Tuberculosis of vertebral pedicles. Radiology, 99:43, 1971.
41. Fraser, R. A. R., et al.: Spinal subdural empyema. Arch. Neurol., 28:235, 1973.
42. Parker, J. J., and Anderson, W. B.: Myelitis simulating spinal cord tumor. Am. J. Roentgenol. Radium Ther. Nucl. Med., 95:942, 1965.
43. Lombardi, G., Passerini, A., and Migliavacca, F.: Spinal arachnoiditis. Br. J. Radiol., 35:314, 1962.
44. Mulvey, R. B.: Unusual myelographic pattern of arachnoiditis. Radiology, 75:778, 1960.
45. Smith, R. W., and Loeser, J. D.: A myelographic variant in lumbar arachnoiditis. J. Neurosurg., 36:441, 1972.
46. Neuroradiologić Symposia:
   No. 2, Acta Radiol., 34, 1950.
   No. 3, Acta Radiol., 40, 1953.
   No. 4, Acta Radiol., 46, 1956.
   No. 5, Acta Radiol., 50, 1958.
   No. 6, Acta Radiol., (Diagnosis) 1, 1963.
47. Leeds, N. E.: Supratentorial space-occupying lesions. Semin. Roentgenol., 5:138, 1970.
48. Zimmerman, H. M.: The ten most common types of brain tumors. Semin. Roentgenol., 6:48, 1971.
49. Scatliff, J. H., Giunto, F. C., and Radcliffe, W. B.: Vascular patterns in cerebral neoplasms and their differential diagnosis. Semin. Roentgenol., 6:59, 1971.
50. Pribram, H. F. W.: Intratentorial masses. Semin. Roentgenol., 5:152, 1970.
51. Blatt, E. S., and Hehman, K. N.: Angiography of supratentorial mass lesions. Semin. Roentgenol., 6:70, 1971.
52. Davis, D. O., and Roberson, G. H.: Angiographic diagnosis of posterior fossa lesions. Semin. Roentgenol., 6:89, 1971.
53. Mishkin, M. M.: Juxtasellar mass lesions. Semin. Roentgenol., 5:165, 1970.
54. Hilal, S. K.: Angiography of juxtasellar masses. Semin. Roentgenol., 6:75, 1971.
54a. Hanafee, W. N., and Dayton, G. O.: The roentgen diagnosis of orbital tumors. Radiol. Clin. North Am., 8:403, 1970.
55. Appel, L.: Plain film aspects of intracranial meningiomas. J. Belg. Radiol., 53:108, 1970.
56. Gold, L. H. A., et al.: Intracranial meningiomas. Neurology, 19:873, 1969.
57. Lloyd, G. A. S.: Radiology of primary orbital meningioma. Br. J. Radiol., 44:405, 1971.
58. Gol, A., and McKissock, W.: Cerebellar astrocytoma. J. Neurosurg., 16:287, 1959.
59. McRae, D. L., and Elliott, A.: Radiological aspects of cerebellar astrocytomas and medulloblastomas. Acta Radiol., 50:52, 1958.
60. Kruyff, E., and Munn, J. D.: Posterior fossa tumors in infants and children. Am. J. Roentgenol. Radium Ther. Nucl. Med., 89:951, 1963.
61. Melmon, F. L., and Rosen, S. W.: Lindau's disease. Am. J. Med., 36:595, 1964.
62. Abrams, H. L. (ed.): Angiography. Boston, Little, Brown and Company, 1961.
63. Kricheff, I. I., and Taveras, J. M.: Angiographic localization of suprasylvian space-occupying lesions. Radiology, 82:602, 1964.
64. Boudreau, R. P.: Primary intraventricular tumors. Radiology, 75:867, 1960.
65. Hueng, Y. P., and Wolf, B. S.: Angiographic features of brain tumors and differential diagnosis from fourth ventricle tumors. Am. J. Roentgenol. Radium Ther. Nucl. Med., 110:1, 1970.
66. Schmitz, A. L., and Haveson, S. B.: Roentgen diagnosis of eighth nerve tumors. Radiology, 75:531, 1960.
67. Valvassori, G. E.: The radiologic diagnosis of acoustic neuroma. Arch. Otolaryng., 83:582, 1966.
68. Valvassori, G. E.: The abnormal internal auditory canal; the diagnosis of acoustic neuroma. Radiology, 92:449, 1969.
69. Goree, J. A., et al.: Percutaneous retrograde brachial angiography in the diagnosis of acoustic neurinoma. Am. J. Roentgenol. Radium Ther. Nucl. Med., 92:829, 1964.
70. Valvassori, G. E.: The diagnosis of acoustic neuromas. Semin. Roentgenol., 4:171, 1969.
71. Lapayowker, M. S., and Cliff, M. M.: Bone changes in acoustic neurinomas. Am. J. Roentgenol. Radium Ther. Nucl. Med., 107:652, 1969.
72. New, P. F. J.: The sella turcica as a mirror of disease. Radiol. Clin. North Am., 4:75, 1966.
73. Bartlett, J. R.: Craniopharyngiomas: An analysis of some aspects of symptomatology, radiology and histology. Brain, 94:725, 1971.
74. Dilenga, D., Simon, J., and David, M.: Angiographical appearances of cerebral metastases as observed with rapid automatic seriograph. Ann. Radiol., 8:163, 1965.
75. Sellwood, R. B.: The radiological approach to metastatic cancer of the brain and spine. Br. J. Radiol., 45:647, 1972.
76. Ramamurthi, B., and Varadarajan, M. G.: Diagnosis of tuberculoma of the brain. Clinical and radiological correlation. J. Neurosurg., 18:1, 1961.
77. Berlin, L.: Tuberculoma of the brain. Am. J. Roentgenol. Radium Ther. Nucl. Med., 90:1185, 1963.

78. Sinh, G., Pandya, S. K., and Dastur, D. K.: Pathogenesis of unusual intracranial tuberculoma and tuberculous space-occupying lesions. J. Neurosurg., *29*:149, 1968.
79. Ray, B. S., and Dunbar, H. S.: Thrombosis of the dural venous sinuses as a cause of "pseudo-tumor cerebri." Ann. Surg., *134*:376, 1951.
80. Jacobson, H. G., and Shapiro, J. H.: Pseudotumor cerebri. Am. J. Roentgenol. Radium Ther. Nucl. Med., *82*:202, 1964.
81. Vander Ark, G. D., Kempe, L. G., and Smith, D. R.: Pseudotumor cerebri treated with lumbar-peritoneal shunt. J.A.M.A., *217*:1832, 1971.
82. Newborg, B.: Pseudotumor cerebri treated by rice reduction diet. Arch. Intern. Med., *133*:802, 1974.
83. Boddie, H. G., et al.: Benign intracranial hypertension. Brain, *97*:313, 1974.
84. Shapiro, R., and Robinson, F.: The roentgenographic diagnosis of the Arnold-Chiari malformation. Am. J. Roentgenol. Radium Ther.Nucl. Med., *73*:390, 1955.
85. Gooding, C. A., Carter, A., and Hoare, R. D.: New ventriculographic aspects of the Arnold-Chiari malformation. Radiology, *89*:626, 1967.
86. Yu, H. C., and Deck, M. D. F.: The clivus deformity of the Arnold-Chiari malformation. Radiology, *101*:613, 1971.
87. Liliequist, B.: Encephalography in the Arnold-Chiari malformation. Acta Radiol., *53*:17, 1960.
88. Juhl, J. H., and Wesenberg, R.: Radiological findings in congenital and acquired occlusions of the foramina of Magendie and Luschka. Radiology, *86*:801, 1966.
89. Kurlander, G. J., and Chua, G. T.: Roentgenology of ventriculo-atrial shunts for the treatment of hydrocephalus. Am. J. Roentgenol. Radium Ther. Nucl. Med., *101*:157, 1967.
90. Swischuk, L. E., Meyer, G. A., and Bryan, N.: Infantile hydrocephalus and cerebral angiography. Am. J. Roentgenol. Radium Ther. Nucl. Med., *115*:50, 1972.
91. Sartor, K., Hill, B. J., and Jerva, M. J.: Hydrocephalus: The radiology of ventriculo-atrial shunts and their complications. Fortschr. Roentgenstr., *117*:381, 1972.
92. Leslie, E. V., Smith, B. H., and Zoll, J. G.: Value of angiography in head trauma. Radiology, *78*:930, 1962.
93. Schechter, M. M., and Zingesser, L. H.: Special procedures in the management of traumatic lesions of the head and neck. Radiol. Clin. North Am., *4*:53, 1966.
94. Kindt, G. W., and Gabrielsen, T. O.: Angiographic demonstration of brain injury without significant mass lesion. J. Neurosurg., *35*:296, 1971.
95. Mayer, J. A.: Extradural spinal hemorrhage. Can. Med. Assoc. J., *89*:1034, 1963.
96. Damanski, M.: Heterotopic ossification in paraplegia. A clinical study. J. Bone Joint Surg., *43B*:286, 1961.
97. Kaufer, C.: Osseous neoformations in paraplegics. (Ueber knoecherne Neubildungen bei Paraplegikern.) Deutsch Med. Wochenschr., *90/38*:1674, 1965.
98. Wright, V., Catterall, R. D., and Cook, J. B.: Bone and joint changes in paraplegic men. Ann. Rheum. Dis., *24/5*:419, 1965.
99. Caffey, J.: *Pediatric X-ray Diagnosis.* 6th ed., Chicago, Year Book Medical Publishers, Inc., 1972.
100. Schaffer, A. J.: *Diseases of the Newborn.* 3rd ed., Philadelphia, W. B. Saunders Company, 1971.
101. Schwidde, J. T.: Spina bifida; survey of two hundred and twenty-five encephaloceles, meningoceles, and myelomeningoceles. Am. J. Dis. Child., *84*:35, 1952.
102. de Anquin, C. E.: Spina bifida occulta with engagement of the fifth spinous process. J.Bone Joint Surg., *41B*:486, 1959.
103. Hertzog, E., Bamberger-Bozo, C., and Rougerie, J.: Myelomeningoceles. Neuroradiologic study of 115 cases. Ann. Radiol., *14*:827, 1971.
104. Vasant, C., and Darab, K. D.: Meningoceles and meningomyeloceles (ectopic spinal canal). J. Neurol. Neurosurg. Psychiatry, *33*:251, 1970.
105. Seaman, W. B., and Schwartz, H. G.: Diastematomyelia in adults. Radiology, *70*:692, 1958.
106. Hilal, S. K., Marton, D., and Pollack, E.: Diastematomyelia in children. Radiology, *112*:609, 1974.
107. Allan, J. D.: Diastematomyelia. Arch. Neurol., *20*:309, 1969.
108. Kein, H. A., and Greene, A. F.: Diastematomyelia and scoliosis. J. Bone Joint Surg., *55*:1425, 1973.
109. Wells, C. E. C., Spillane, J. D., and Bligh, A. S.: The cervical spinal canal in syringomyelia. Brain, *82*:23, 1959.
110. Marks, J. H., and Livingston, K. E.: The cervical subarachnoid space, with particular reference to syringomyelia and the Arnold-Chiari deformity. Radiology, *52*:63, 1949.
111. Klefenberg, G., and Saltzman, G. F.: Gas myelographic studies in syringomyelia. Acta Radiol., *52*:129, 1959.
112. Logue, V.: Syringomyelia: A radiodiagnostic and radiotherapeutic saga. Clin. Radiol., *22*:2, 1971.
113. Conway, L. W.: Hydrodynamic studies in syringomyelia. J. Neurosurg., *27*:501, 1967.
114. Boffano, M., and Cignetti, P.: Diagnosi radiologica prenatale di anenecefalia. Contributo casuistico. (Prenatal roentgenologic diagnosis of anencephaly. Case report.) Minerva Ginec., *15*:606, 1963.
115. Bergström, K., and Lodin, H.: Angiography in senile cerebral atrophy. Acta Radiol., *4/2*:187, 1966.
116. Schnitker, M. T., and Ulrich, R. P.: Observations on the 24 hrs. pneumoencephalogram with special reference to the diagnosis of cortical atrophy. Radiology, *70*:15, 1958.

117. Jackson, H., and Law, C. W.: The size of the lateral ventricles of the brain, with particular reference to the evaluation of cerebral atrophy. Med. Proc. Johannesburg, 7:129, 1961.
118. Schechter, M. M.: Pneumography in brain atrophy. Semin. Roentgenol., 5:196, 1970.
119. Dyke, C. G., Davidoff, L. M., and Masson, C. B.: Cerebral hemiatrophy and homolateral hypertrophy of the skull and sinuses. Surg. Gynecol. Obstet., 57:588, 1933.
120. Campbell, J. A.: Roentgen aspects of cranial configurations. Radiol. Clin. North Am., 4/1:11, 1966.
121. Schwartz, A., and Levy, S.: Evolution of roentgenologic skull changes with unilateral loss of brain substance in children. Neurology, 12:133, 1962.
122. Smith, D. R., Kempe, L. G., and Dunaway, R.: Prenatal agenesis of one cerebral hemisphere. J. Neurosurg., 30:80, 1969.
123. Wollschlaeger, G., et al.: Lipoma of the corpus callosum. Am. J. Roentgenol. Radium Ther. Nucl. Med., 86:142, 1961.
124. Holman, C. B., and MacCarty, C. S.: Cerebral angiography in agenesis of corpus callosum. Radiology, 72:317, 1959.
125. Hankinson, J., and Amador, L. V.: Agenesis of the corpus callosum diagnosed by pneumoencephalography. Br. J. Radiol., 30:200, 1957.
126. Davidoff, L. M., and Dyke, C. G.: Agenesis of the corpus callosum, its diagnosis by encephalography; report of 3 cases. Am. J. Roentgenol. Radium Ther. Nucl. Med., 32:1, 1934.
127. Wolpert, S. M., Carter, B. L., and Ferris, E. J.: Lipomas of the corpus callosum. Am. J. Roentgenol. Radium Ther. Nucl. Med., 115:92, 1972.
128. Loeser, J. D., and Alvord, E. C.: Agenesis of the corpus callosum. Brain, 91:553, 1968.
129. Naef, R. W.: Clinical features of porencephaly: a review of thirty-two cases. Arch. Neurol. Psychiat., 80:133, 1958.
130. Greenfield, J. H.: *The Spino-cerebellar Degenerations.* Oxford, Blackwell Scientific Publications, 1954.
131. LeMay, M., and Abramowicz, A.: The pneumoencephalographic findings in various forms of cerebellar degeneration. Radiology, 85:284, 1965.
132. Strecker, E. P., Hodges, F. J., and James, A. E.: Cerebellar atrophy: Cisternographic and pneumoencephalographic analysis. Am. J. Roentgenol. Radium Ther. Nucl. Med., 115:760, 1972.
133. Joubert, M., et al.: Familial agenesis of the cerebellar vermis. Neurology, 19:813, 1969.
134. Moss, M. L.: The pathogenesis of premature synostosis in man. Acta Anat., 37:351, 1959.
135. Tod, P. O., and Yelland, J. D. N.: Craniostenosis. Clin. Radiol., 22:472, 1971.
136. Simmons, D. R., and Peyton, W. T.: Premature closure of the cranial sutures. J. Pediatr., 31:528, 1947.
137. Gilmor, R. L., et al.: Angiographic assessment of occipital encephaloceles. Radiology, 103:127, 1972.
138. Caffey, J., and Ross, S.: Roentgen features of pelvic bones in mongolism. Am. J. Roentgenol. Radium Ther. Nucl. Med., 80:458, 1958.
139. Rabinowitz, J. G., and Moseley, J. E.: Lateral lumbar spine in Down's syndrome. Radiology, 83:74, 1964.
140. Roche, A. F., Seward, F. S., and Sunderland, S.: Nonmetrical observations on cranial roentgenograms in mongolism. Am. J. Roentgenol. Radium Ther. Nucl. Med., 85:659, 1961.
141. James, A. E., Merz, T., et al.: Radiological features of the most common autosomal disorders. Clin. Radiol., 22:417, 1971.
142. Mortensson, W., and Hall, B.: Abnormal pelvis in newborn infants with Down's syndrome. Acta Radiol., 12:847, 1972.
143. Benson, D. F., LeMay, M., et al.: Diagnosis of normal pressure hydrocephalus. N. Engl. J. Med., 283:609, 1970.
144. Rovit, R. L., Schechter, M. M., et al.: Progressive ventricular dilatation following pneumoencephalography: A radiologic sign of occult hydrocephalus. J. Neurosurg. 36:50, 1972.
145. Walker, J., Singer, K., and Baker, P.: Disorders of esophageal motility in a family with hereditary spastic ataxia. Neurology, 19:1212, 1969.
146. Peterman, A. F., et al.: Encephalotrigeminal angiomatosis (Sturge-Weber). J.A.M.A., 167:2169, 1958.
147. Bentson, J. R., Wilson, G. H., and Newton, T. H.: Cerebral venous drainage pattern of the Sturge-Weber syndrome. Radiology, 101:111, 1971.
148. Brown, B. St. J.: Tuberous sclerosis. J. Can. Assoc. Radiol., 12:1, 1961.
149. Holt, J. F., and Dickerson, W. W.: Osseous lesions of tuberous sclerosis. Radiology, 58:1, 1952.
150. Crosett, A. D.: Roentgenologic findings in the renal lesion of tuberous sclerosis. Am. J. Roentgenol. Radium Ther. Nucl. Med., 98:739, 1966.
151. Viamonte, M., Jr., et al.: Angiographic findings in a patient with tuberous sclerosis. Am. J. Roentgenol. Radium Ther. Nucl. Med., 98:723, 1966.
152. Milledge, R. D., Gerald, B. E., and Carter, W. J.: Pulmonary manifestations of tuberous sclerosis. Am. J. Roentgenol. Radium Ther. Nucl. Med., 98:734, 1966.
153. Teplick, J. G.: Tuberous sclerosis. Radiology, 93:53, 1969.
154. Green, G. J.: The radiology of tuberous sclerosis. Clin. Radiol., 19:135, 1968.
155. Bigot, R., et al.: Isolated angiomyolipomas of the kidney. Arteriographic study of 10 cases. J. Radiol. Electrol. Med. Nucl., 52:789, 1971.

155a. Gath, I., and Vinje, B.: Pneumoencephalographic findings in Huntington's chorea. Neurology, *18*:991, 1968.

156. McRae, D. L.: Significance of abnormalities of the cervical spine. Am. J. Roentgenol. Radium Ther. Nucl. Med., *84*:3, 1960.

157. Shoul, M. I., and Ritvo, M.: Clinical and roentgenological manifestations of Klippel-Feil syndrome (congenital fusion of cervical vertebrae, brevicollis); report on 8 additional cases and review of literature. Am. J. Roentgenol. Radium Ther. Nucl. Med., *68*:369, 1952.

158. Ramsay, J., and Bliznak, J.: Klippel-Feil syndrome with renal agenesis and other anomalies. Am. J. Roentgenol. Radium Ther. Nucl. Med., *104*:476, 1972.

159. Palant, D. I., and Carter, B. L.: Klippel-Feil syndrome and deafness. Am. J. Dis. Child., *123*:218, 1972.

160. Köhler, A., and Zimmer, E. A.: *Borderlands of the Normal and Early Pathologic in Skeletal Roentgenology.* 10th ed., Grune & Stratton, Inc., 1956, p. 423.

161. Blair, J. D., and Wells, P. O.: Bilateral undescended scapula associated with omovertebral bone. J. Bone Joint Surg., *39A*:201, 1957.

162. Bull, J. W. D., Nixon, W. L. B., Pratt, R. T. C., and Robinson, P. K.: Platybasia. Brain, *82*:10, 1959.

163. Paradis, R. W., and Sax, D. S.: Familial basilar impression. Neurology, *22*:554, 1972.

164. Klaus, E., and Urbanek, K.: Angiographic findings of the vertebral artery in basilar impression. Fortschr. Roentgenstr., *116*:378, 1972.

165. Kaufman, B.: The "empty" sella turcica — a manifestation of the intrasellar subarachnoid space. Radiology, *90*:931, 1968.

166. Bernasconi, V., Giovanelli, M. A., and Papo, I.: Primary empty sella. J. Neurosurg., *36*:157, 1972.

167. Shore, I. N., DeCherney, A. H., et al.: The empty sella syndrome. J.A.M.A., *227*:69, 1974.

168. Vermess, M., Stein, S. C., et al.: Angiography in idiopathic focal epilepsy. Am. J. Roentgenol. Radium Ther. Nucl. Med., *115*:120, 1972.

169. Bechar, M., Front, D., et al.: Cervical myelopathy caused by the narrowing of the cervical spinal canal. Clin. Radiol., *22*:63, 1971.

170. Nurick, S.: The pathogenesis of the spinal cord disorder associated with cervical spondylosis. Brain, *95*:87, 1972.

171. Lombardi, G., and Passerini, A.: Spinal cord tumors. Radiology, *76*:381, 1961.

172. Epstein, B. S.: Spinal canal mass lesions. Radiol. Clin. North Am., *5*:185, 1966.

173. Banna, M., and Gryspeerdt, G. L.: Intraspinal tumors in children. Clin. Radiol., *22*:17, 1971.

174. Traub, S. P.: Mass lesions in the spinal canal. Semin. Roentgenol., *7*:240, 1972.

175. Shapiro, I. H., and Jacobson, H. G.: Differential diagnosis of intradural and extradural spinal canal tumors. Radiology, *76*:718, 1961.

176. Barone, B. M., and Elvidge, A. R.: Ependymomas: A clinical survey. J. Neurosurg., *33*:428, 1970.

177. Tenner, M. S.: Myelography of non-mass lesions in the spinal canal. Semin. Roentgenol., *7*:277, 1972.

178. Murray, R. O.: Intradural arachnoid cyst. Br. J. Radiol., *32*:689, 1959.

179. Epstein, B. S.: Spinal canal mass lesions. Radiol. Clin. North Am., *4*:185, 1966.

180. Pear, B. L.: Spinal epidural hematoma. Am. J. Roentgenol. Radium Ther. Nucl. Med., *111*:155, 1972.

181. Gautier-Smith, P. C.: Clinical aspect of spinal neurofibromas. Brain, *90*:359, 1967.

182. Curtis, B. H., et al.: Neurofibromatosis with paraplegia: Report of 8 cases. J. Bone Joint Surg., *51A*:843, 1969.

183. Lindblom, K.: Intervertebral disc degeneration considered as a pressure atrophy. J. Bone Joint Surg., *39A*:933, 1957.

184. Teng, P., and Papatheodorou, C.: Myelographic findings in spondylosis of the lumbar spine. Br. J. Radiol., *36*:122, 1963.

185. Holman, C. B.: The roentgenologic diagnosis of herniated intervertebral disc. Radiol. Clin. North Am., *4/1*:171, 1966.

186. Mixter, W. J., and Barr, J. S.: Rupture of the intervertebral disc with involvement of the spinal canal. N. Engl. J. Med., *211*:210, 1934.

187. Peterson, H. O., and Kieffer, S. A.: Radiology of intervertebral disk disease. Semin. Roentgenol., *7*:260, 1972.

188. Payne, E. E., and Spillane, J. D.: The cervical spine; anatomico-pathological study of 70 specimens using a special technique with particular reference to the problem of cervical spondylosis. Brain, *80*:571, 1957.

189. Pallis, C., and Jones, A. M.: Cervical spondylosis: incidence and implications. Brain, *77*:274, 1954.

190. Wilson, G., Weidner, W., and Hanafee, W.: Comparison of gas and positive contrast in evaluation of cervical spondylosis. Am. J. Roentgenol. Radium Ther. Nucl. Med., *97*:648, 1966.

191. Levine, R. A., et al.: Cervical spondylosis and dyskinesias. Neurology, *20*:1194, 1970.

192. Plewes, L. W.: Sudeck's atrophy in hand. J. Bone Joint Surg., *38B*:195, 1956.

193. Bierling, G., and Reisch, D.: Sudeck's atrophy after fractures. Fortschr. Röntgenstr., *82*:1, 1955.

194. Conti, R.: Post-traumatic osteoporosis and post-traumatic Sudeck's syndrome. Minerva Radiol., *15*:111, 1970.

195. Lang, E. K.: Arteriographic diagnosis of the thoracic outlet syndrome. Radiology, *84*:296, 1965.

196. Rosenberg, J. C.: Arteriographic demonstration of compression syndromes of the thoracic outlet. South. Med. J., *59*:400, 1966.
197. Haimovici, H., and Caplan, L. H.: Arterial thrombosis complicating the thoracic outlet syndrome; arteriographic considerations. Radiology, *87*:457, 1966.
198. Lang, Erich K.: Neuromuscular compression syndromes. Dis. Chest, *50*:572, 1966.
199. Urschel, H. C., Jr., and Razzuk, M. A.: Management of the thoracic outlet syndrome. N. Engl. J. Med., *286*:1140, 1972.
200. Hinck, V. C., and Sachdev, N. S.: Myelographic findings in hypertrophic interstitial neuritis. Am. J. Roentgenol. Radium Ther. Nucl. Med., *95*:947, 1965.
201. Rao, C. V. G. K., et al.: Dejerine-Sottas syndrome in children (hypertrophic interstitial poly-neuritis). Am. J. Roentgenol. Radium Ther. Nucl. Med., *122*:70, 1974.
202. Robinson, G. H., Llewellyn, H. J., and Taveras, J. M.: The narrow lumbar spinal canal syndrome. Radiology, *107*:89, 1973.
203. Lewitan, A., and Nathanson, L.: Roentgen features of muscular dystrophy. Am. J. Roentgenol. Radium Ther. Nucl. Med., *73*:226, 1955.
204. Kaufman, H.: A new roentgen finding in pseudohypertrophic muscular dystrophy. Am. J. Roentgenol. Radium Ther. Nucl. Med., *89*:970, 1963.
205. Welsh, J. D., Haase, G. R., and Bynum, T. E.: Myotonic muscular dystrophy. Arch. Intern. Med., *144*:669, 1964.
206. Wahi, P. L., et al.: Cardiopathy in muscular dystrophy. Dis. Chest, *53*:79, 1968.
207. Gay, B. B., and Weems, H. S.: Roentgenologic evaluation of disorders of muscle. Semin. Roentgenol., *8*:25, 1973.
208. DiChiro, G., and Caughey, J. E.: Skull changes in eighteen cases of dystrophia myotonia, Acta Radiol., *54*:22, 1960.
209. Kohn, N. N., Faires, J. S., and Rodman, T.: Unusual manifestations due to involvement of invol-untary muscles in dystrophia myotonica. N. Engl. J. Med., *271*:1179, 1964.
210. Pruzanski, W., and Profis, A.: Pulmonary disease in myotonic dystrophy. Am. Rev. Resp. Dis., *91*:874, 1965.
211. Krain, S., and Rabinowitz, J.: The radiologic features of myotonic dystrophy. Clin. Radiol., *22*:462, 1971.
212. Lee, K. F., Lin, S. R., and Hodes, P. J.: New roentgenologic findings in myotonic dystrophy. Am. J. Roentgenol. Radium Ther. Nucl. Med., *115*:179, 1972.
213. Murray, J. P.: Deglutition in myasthenia gravis. Br. J. Radiol., *35*:43, 1962.
214. Hillenius, L., and Mosetitsch, W.: The diagnosis of thymomas. Fortschr. Roentgenstr., *99*:28, 1963.
215. Wilkins, E. W., Jr., et al.: Cases of thymoma at Massachusetts General Hospital. J. Thorac. Cardiovasc. Surg., *52*:322, 1966.
216. Leigh, T. F., and Weens, H. S.: Roentgen aspects of mediastinal lesions. Semin. Roentgenol., *4*:59, 1969.
217. Shanks, S. C., and Kerley, P.: *Textbook of X-ray Diagnosis.* 4th ed. Philadelphia, W. B. Saunders Company, 1969, Vol. 4.
218. Yaghmai, I., and Mirbod, P.: Tumoral calcinosis. Am. J. Roentgenol. Radium Ther. Nucl. Med., *111*:573, 1971.
219. Naidich, T. P., and Siegelman, S. S.: Paraarticular soft tissue changes in systemic diseases. Semin. Roentgenol., *8*:101, 1973.
220. Varadarajan, M. G., and Daniel, A.: Myositis ossificans progressiva. Current Med. Pract., *4*:520, 1960.
221. Norman, A., and Dorfman, H. D.: Juxtacortical circumscribed myositis ossificans: Evolution and radiologic features. Radiology, *96*:301, 1970.
222. Illingworth, R. S.: Myositis ossificans progressiva. Arch. Dis. Child., *46*:264, 1971.

SECTION

9

RESPIRATORY DISEASE

THE DIAPHRAGM

Paralysis of the Diaphragm

Paralysis of a hemidiaphragm may be partial or complete. The involved diaphragm is elevated, and its respiratory excursions are limited or paradoxical during fluoroscopy. Movement in a direction opposite to that of the normal diaphragm (paradoxical motion) is indicative of complete paralysis. The heart and mediastinum are usually displaced to some degree to the opposite side.

Tumor or inflammatory disease in or adjacent to the mediastinum with involvement of the phrenic nerve is a common cause of diaphragmatic paralysis. Careful roentgen study of the mediastinum may uncover the cause in unexplained cases. Diseases of the cord or central nervous system can also lead to diaphragmatic paralysis.

Subphrenic inflammatory conditions can cause elevation and complete immobilization of a hemidiaphragm.[1]

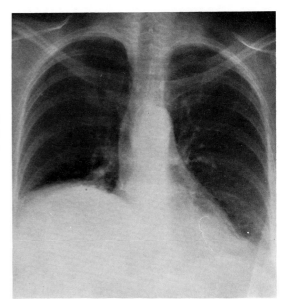

Figure 9–1 **Paralysis of the Diaphragm.** The right hemidiaphragm is uniformly elevated. Fluoroscopy demonstrated paradoxical upward movement on inspiration.

Traumatic Diaphragmatic Hernia

Herniation of abdominal organs into the thorax following rupture of the diaphragm is usually the result of a nonpenetrating injury. In over 95 per cent of cases, herniation is into the left thorax. Usually the stomach or colon, or both, enter the thorax. With small diaphragmatic tears, the viscera in the thorax and the diaphragm shadow can be readily distinguished. In large hernias the diaphragmatic shadow is often obscured by the visceral densities in the chest, and the superior border of the herniated viscera may resemble a diaphragm. On posteroanterior films the picture may simulate eventration of the diaphragm. However, in traumatic hernia a complete diaphragm is not seen on lateral views, whereas in eventration the entire elevated diaphragm can easily be identified. In questionable cases the diagnosis can be made by diagnostic pneumoperitoneum, which will identify the location of the diaphragm and will produce a pneumothorax in the presence of a diaphragmatic rupture.

Barium studies may be necessary to identify the chest densities as gastrointestinal structures.[2-4]

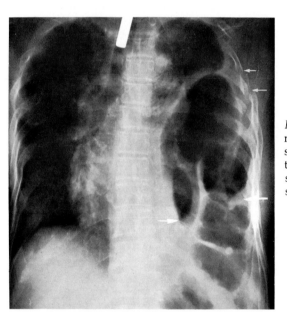

Figure 9–2 **Large Traumatic Hernia.** A large segment of splenic flexure (*large arrows*) and stomach is seen within the left chest. Note the multiple rib fractures (*small arrows*). The heart and mediastinum are shifted to the right. The diaphragmatic outline is obscured by the colon and cannot be identified.

Figure 9–3 **Small Traumatic Hernia.** The stomach is displaced to the left and elevated above the diaphragm into the chest (*arrows*) following traumatic rupture of the anterior portion of the left diaphragm. In ordinary hiatal hernia the stomach is always close to the midline. In questionable cases in which the diaphragm cannot be outlined, a pneumoperitoneum usually helps clarify the diagnosis.

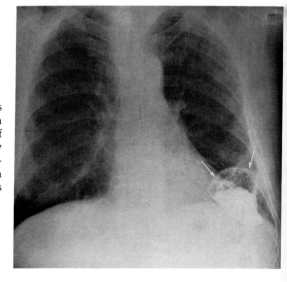

Congenital Hernia (Bochdalek's Hernia)

Most significant congenital hernias of abdominal viscera into the chest occur through the foramen of Bochdalek (pleuroperitoneal foramen) in the posterolateral muscular segment of the diaphragm. In the newborn, serious cardiorespiratory embarrassment or bowel strangulation can occur.

Most frequently the herniation is found in the left hemithorax and may consist of stomach, small and large bowel, and spleen. The diaphragm may not be identified or may be only partially visualized. Multiple gas shadows in the involved hemithorax, with fluid levels when the patient is in the erect position, are characteristic findings. The heart and mediastinum are displaced to the opposite side. There will be diminished or absent gas shadows in a scaphoid abdomen.

In the rare case in which plain film findings are inconclusive, a barium meal study will identify the stomach or small bowel in the chest.

Small hernias may be asymptomatic and may be discovered by accident on a chest film in adults. Occasionally, the kidney alone may herniate and simulate an intrathoracic mass.[5, 6]

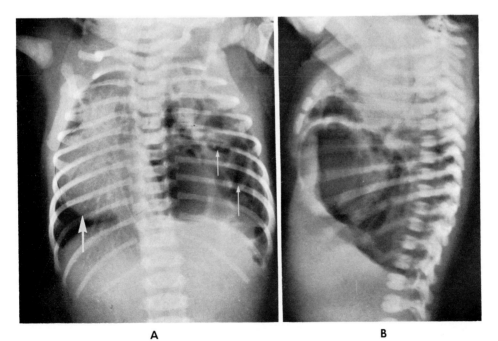

A   **B**

Figure 9-4 **Congenital Diaphragmatic Hernia: Pleural Peritoneal Hernia in Newborn (Bochdalek's Hernia).**

A, In posteroanterior view, multiple round gas shadows within the chest represent bowel (*small arrows*). The diaphragm is not seen. The heart is shifted markedly to the right (*large arrow*). There is no gas in the abdomen.

B, Lateral view also demonstrates the presence of bowel gas shadows in the chest.

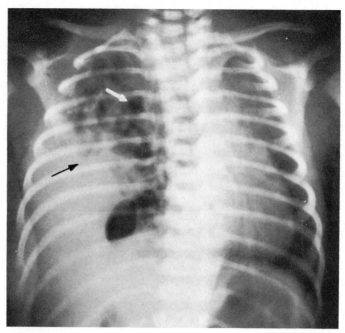

Figure 9–5 **Congenital Hernia Through Foramen of Bochdalek.** Chest film of this newborn shows mottled and irregular lucencies in the right chest (*white arrow*), which represent herniated bowel. The homogeneous density *(black arrow)* was herniated liver. The heart is displaced to the left. Note absence of the right diaphragmatic shadow.

Hernia Through Foramen of Morgagni

Herniation of an abdominal structure through the foramen of Morgagni produces a radiographic density of varying size in the anterior cardiophrenic angle of the chest, most often on the right side. The hernial sac usually contains only omentum and presents as a rounded, sharply marginated, homogeneous density that often cannot be distinguished from pericardial cyst or tumor. Planigraphy may disclose fat lucency within the mass, a suggestive finding.

Traction from the herniated omentum often causes elevation and a peaked upward angulation of the transverse colon, which is best demonstrated on an erect film of a barium enema study. There may also be upward displacement of the distal stomach and duodenum. In doubtful cases, an erect film made after diagnostic pneumoperitoneum may show air in the thoracic density, confirming the diagnosis. Occasionally there is bowel in the hernia, which can be identified by the bowel markings or gas, or by contrast gastrointestinal examination.

A mass density in the anterior costophrenic angle and an elevated, peaked transverse colon are findings highly suggestive of omental herniation through the foramen of Morgagni. These findings will differentiate this disorder from a pericardial cyst.[7]

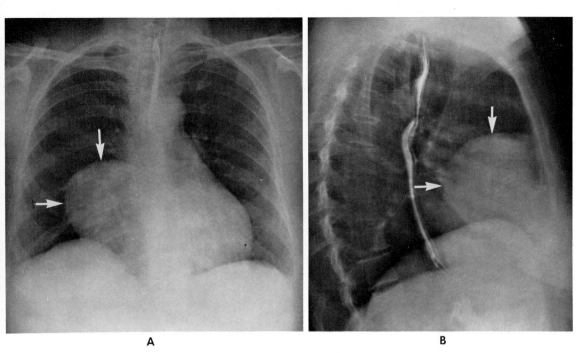

A   B

Figure 9-6 **Morgagni Hernia.**

  A, In posteroanterior view, a round density *(arrows)* obliterates the right cardiac border.
  B, Lateral view indicates the anterior location of the mass *(arrows)*.
  This hernia cannot be distinguished from a solid mass. In some cases, fat or bowel markings are seen within the density. A pneumoperitoneum may be diagnostic. A hernia must be considered in the differential diagnosis of a lesion in this location.

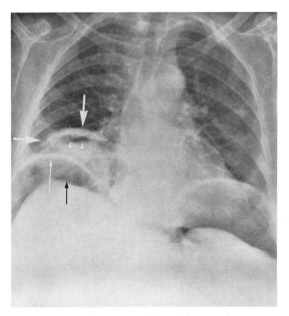

Figure 9-7 **Morgagni Hernia: Pneumoperitoneum: Erect View.** There is a smooth round right anterior cardiophrenic mass *(large white arrows)*. Following a diagnostic pneumoperitoneum, air has entered the mass *(arrowheads)*, confirming the diagnosis of a Morgagni hernia. There is air beneath both diaphragms, clearly delineating the right hemidiaphragm *(small white arrow)* and the upper border of the liver *(black arrow)*.

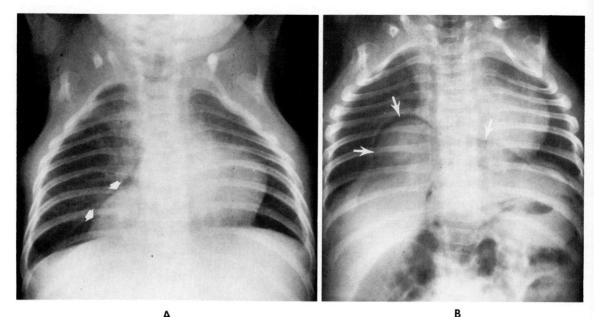

A B

Figure 9–8 **Morgagni Hernia: Pneumoperitoneum.**

A, The density in the right cardiophrenic area (*arrows*) was anterior (on lateral film), suggesting a hernia through the foramen of Morgagni.

B, After a diagnostic pneumoperitoneum, intraperitoneal air can be seen outlining the hernia *(arrows).*

Eventration

Eventration results from muscular atrophy and atonicity of a hemidiaphragm, usually the left. The involved diaphragm becomes thinned and markedly elevated, and some limitation of respiratory excursions can be seen fluoroscopically. In severe cases, movement may be entirely absent or even paradoxical, simulating a totally paralyzed diaphragm. The heart is generally displaced to the opposite side.

In the more common left hemidiaphragmatic eventration, the stomach and splenic flexure are elevated and may mimic a traumatic or congenital hernia through the diaphragm. A distinction can usually be made by identification of the eventrated diaphragm above these viscera; this is usually best appreciated on lateral view. In doubtful cases oral barium studies may make the distinction: in eventration the gastric contours are normal, whereas in traumatic hernia the fundus and antrum are crowded together in the hernial sac.[8]

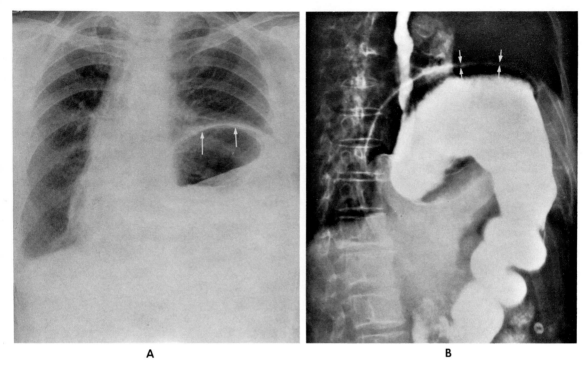

A B

Figure 9–9 **Eventration.**

A, On posteroanterior view there is marked elevation of the left hemidiaphragm (*arrows*), with a shift of the heart to the right. Paradoxical motion of the diaphragm was noted on fluoroscopy.

B, The barium-filled stomach lies beneath the elevated left diaphragm. Notice the decreased thickness of the diaphragm (*arrows*). Radiographically, a traumatic hernia may simulate an eventration.

Hepatodiaphragmatic Interposition

In interposition, the large bowel (or, rarely, the small bowel) is between the liver and the diaphragm. It may be due to congenital abnormal mobility of the colon, megacolon, relaxation of suspensory ligaments in a ptotic liver, or elevation of the diaphragm from eventration or paralysis. The bowel gas beneath the right diaphragm may simulate a pneumoperitoneum or subdiaphragmatic abscess. Barium enema studies readily confirm the diagnosis and, in questionable cases, distinguish interposition of the large bowel from that of the small bowel.

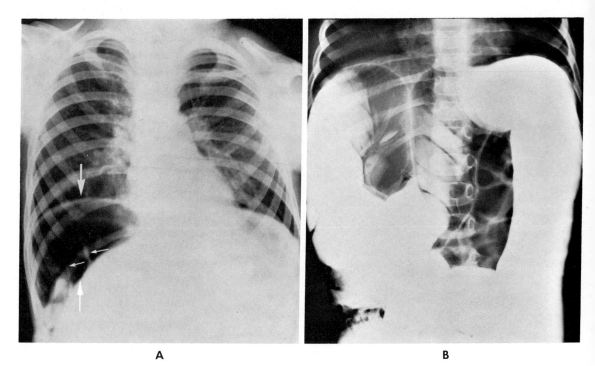

A   B

Figure 9–10 **Hepatodiaphragmatic Interposition.**

   A, There is air beneath the right diaphragm (*large arrows*). While this picture superficially resembles a pneumoperitoneum, the bowel markings (*small arrows*) suggest that this is colonic gas between the liver and diaphragm.
   B, Barium enema confirms the presence of large bowel between the liver and diaphragm.

AIRWAY OBSTRUCTION

Acute Bronchitis

   There are no significant x-ray abnormalities associated with acute bronchitis in adults. In infants and children, acute bronchitis is often associated with hyperinflation. Roentgenograms are useful only to exclude associated or complicating pneumonitis.

Chronic Bronchitis

   The radiographic changes in chronic bronchitis are minimal and usually nonspecific; occasionally there is an increase in peribronchial markings, especially at the bases, owing to thickened bronchial and peribronchial tissue.
   In a significant number of patients with chronic bronchitis there are roentgenologic findings of emphysema: low, flat diaphragms, a narrow vertical heart with a prominent pulmonary artery segment, and a decreased number of peripheral vascular shadows.

About three fourths of patients have abnormalities evident on bronchography, including (1) dilatation of the bronchial glands, leading to irregular outpouchings from the bronchial lumens, most often seen in the walls of major bronchi; (2) bronchiolar diverticulosis (bronchiolectasis), appearing as small cyst-like outpouchings from the terminal bronchi; (3) irregularity and distortion of the bronchial wall that frequently produce a beaded appearance; (4) atrophy of the bronchial mucosa causing annular areas of slight widening and narrowing of the lumen; and (5) abrupt termination of smaller branch bronchi with square or truncated endings. Other findings include bronchiolar spasm, increased secretions, and scattered areas of contrast-filled alveoli.[9]

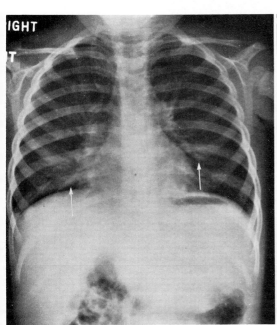

A

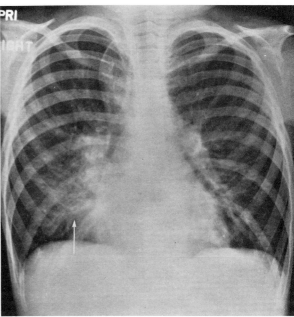

B

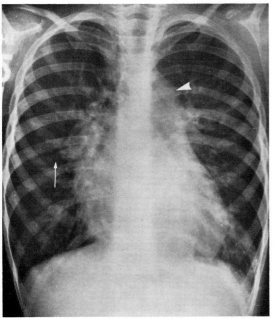

C

Figure 9–11 **Chronic Bronchitis: Mucoviscidosis with Progressive Changes.**

A, There is hyperinflation of the lungs and pneumonitis at the right base. The bronchial walls appear thickened *(arrows).*

B, In later film, areas of pneumonitis are more pronounced, and there are small areas of atelectasis—a combination brought about by the thick secretions and superimposed infection. Note thickening and increased density of the bronchial walls *(arrow),* which are indicative of chronic bronchitis.

C, Two years later, emphysema is evident. Cor pulmonale and pulmonary hypertension are suggested by the prominent main pulmonary artery *(arrowhead)* and abrupt narrowing of the peripheral vasculature *(arrow).*

Recurrent bouts of infection usually occur in mucoviscidosis and characteristically lead to peribronchial fibrosis, hyperinflation, and focal areas of atelectasis.

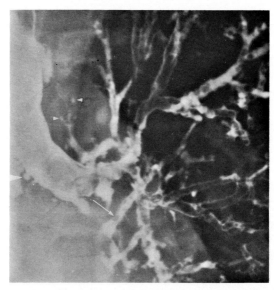

Figure 9–12 **Chronic Bronchitis: Bronchogram.**   The subject was a 42 year old chronic smoker with a persistent cough. Bronchographic changes characteristic of chronic bronchitis include dilatation of the bronchial glands in the main bronchus *(large arrowhead)*, dilatation of the terminal bronchioles *(small arrowheads)*,
and areas of spasm with bronchial wall irregularity *(arrow)*.

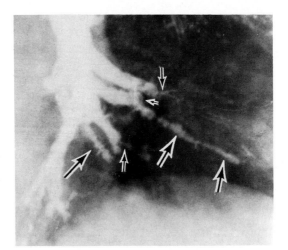

Figure 9–13 **Chronic Bronchitis: Bronchogram.**   Close-up film of the left lower lobe bronchial tree shows
marked irregular narrowing and beading of several of the larger bronchi *(large arrows)*, abrupt narrowings
at bifurcations *(middle-sized arrows)* and total cutoff of some branches by secretory plugs *(small arrow)*. Many
of the smaller bronchi could not be opacified because of thick secretory contents. These are the bronchographic changes of severe chronic bronchitis.

Pulmonary Emphysema

Pulmonary emphysema results from irreparable destruction of pulmonary septal tissue distal to the terminal nonrespiratory bronchiole. Abnormally enlarged
air spaces replace the normal lung tissue. Roentgen changes may directly or indirectly reflect these anatomic changes. The roentgen findings include hyperinflation, abnormalities of the pulmonary vasculature, cardiac changes, and local
bullae and blebs.

The exact cause of pulmonary emphysema is unknown. Like chronic bronchitis, it occurs more commonly in heavy cigarette smokers. Homozygous alpha-1-antitrypsin deficiency is a rare familial condition that is associated with severe emphysema that frequently occurs before age 40; this is usually a predominantly lower lobe emphysema. Heterozygous alpha-1-antitrypsin deficiency is much more common and is associated with emphysema only in cigarette smokers. There also appears to be an unusually high incidence of liver cirrhosis and primary hepatoma in adults with emphysema due to alpha-1-antitrypsin deficiency.

The commonest roentgen signs of emphysema result from hyperinflation, which is almost always present in moderate to severe cases but which may be absent in milder cases. However, hyperinflation can occur without airway obstruction, especially in the aged. The roentgen changes due to hyperinflation include an increased anteroposterior chest diameter, an increase in both the size and lucency of the retrosternal space, low flat diaphragms, hyperlucent lung fields, and a small cardiac shadow. Limitation of diaphragmatic excursion to less than 2 cm. is a fairly reliable finding. Serrated diaphragms are seen in moderate to severe cases and are probably the result of muscle fiber hypertrophy from the increased respiratory effort. Localized bullae are frequent but can occur without generalized emphysema. Air trapping, or increased residual volume, can often be demonstrated by combined inspiratory and expiratory film.

The degree of hyperinflation recognized on the chest roentgenograms may not correlate accurately with the extent of emphysema. Therefore, roentgen findings other than hyperinflation are important for more accurate diagnosis. A significant finding is reduction in the number and caliber of the peripheral arteries, especially in those beyond the third or fourth branching. This peripheral vascular attenuation can be detected on regular chest films in about one half of the cases, but it is demonstrated more clearly and frequently by planigraphy. Wedge arteriography of the lung most accurately reveals these vascular changes. Often, small associated arteriovenous shunts are seen. Redistribution of the vascularity to the less damaged lung areas frequently occurs. In panacinar emphysema the basal segments are frequently severely involved, and the increased vascularity to the upper lobes may simulate the vascular appearance of mitral stenosis or early congestive failure. Prominence of the main and central pulmonary arteries is observed in almost every case of severe emphysema and is a manifestation of pulmonary hypertension. The heart in emphysema appears elongated and has a narrow transverse diameter. If cor pulmonale develops, there may be evidence of right ventricular enlargement.

Another less easily recognized form is "increased markings" emphysema, which is usually associated with chronic bronchitis. The bronchovascular markings are increased, and evidence of hyperinflation is minimal or moderate. However, cardiac enlargement is common, and enlarged central pulmonary arteries are virtually always seen.

Not infrequently, however, the chest roentgenogram findings may appear completely normal in some patients with proven mild to moderate emphysema.

Cardiac failure with interstitial pulmonary edema can obscure the radiolucent appearance of emphysema. The lung compliance is diminished by interstitial fluid, and the diaphragms appear less depressed than before failure. Alveolar and interstitial pulmonary edema may be patchy in an emphysematous lung, with no involvement of the hypovascular segments. In centrilobular emphysema, which predominantly involves the upper lobes, there is no vascular redistribution to the upper lobes in congestive failure.[1, 10-19]

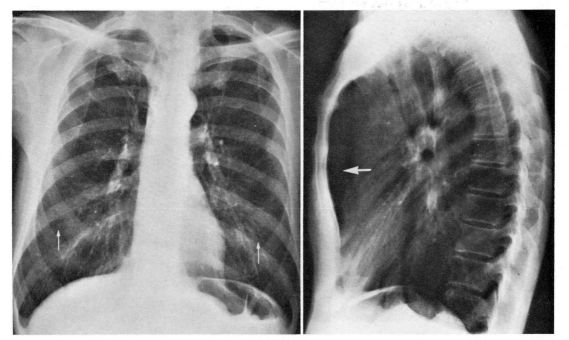

Figure 9–14 **Emphysema.** The diaphragms appear low but are normally arched, even though there is severe emphysema. The heart is vertical. The anteroposterior diameter of the chest is increased. The retrosternal space is more translucent *(large arrow).* There is a marked decrease in the size and number of the peripheral arteries *(small arrows)* This is the most characteristic finding in emphysema.

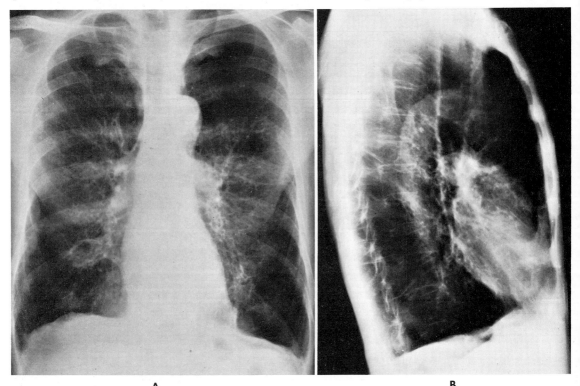

A   **B**

Figure 9–15 **Advanced Pulmonary Emphysema.**

A, In posteroanterior view, the size and number of peripheral pulmonary vessels are reduced. There is unequal distribution of the vessels, and marked translucency of the lung bases and the left upper lung field. The diaphragms are flattened and low.

B, Lateral chest film reveals marked retrosternal lucency and only slightly increased anteroposterior diameter of the chest. The flattening of the diaphragms is clearly demonstrated.

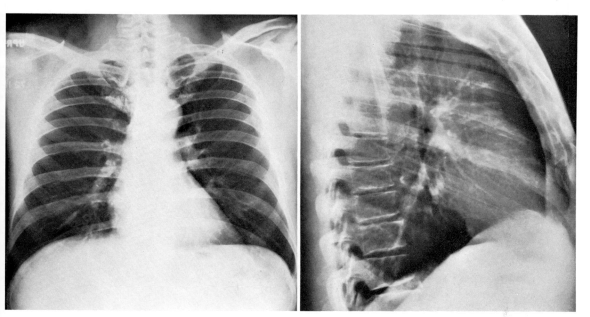

Figure 9–16 **Emphysema: Normal Chest Appearance.** Emphysema proved by pulmonary function studies is present, with normal findings on x-ray examination. In the centrilobular type of emphysema, without associated hyperinflation, there are frequently no roentgenographic abnormalities.

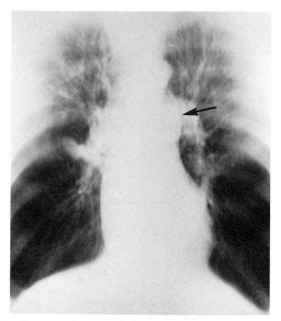

Figure 9–17 **Pulmonary Emphysema in Association with Alpha-1-Antitrypsin Deficiency in a Young Man.** A midthorax planigram reveals striking hyperlucency of both lower lung fields, associated with flattened and depressed diaphragms. The lower lobe vascular shadows are sparse, small, and virtually absent in the periphery. By contrast, the upper lobe vessels are larger and abundant, and they clearly extend to the lung periphery. The main pulmonary artery is quite enlarged (*arrow*), which is indicative of cor pulmonale.

Diffuse emphysema limited to the lower lobes in a younger individual is suggestive of alpha-1-antitrypsin deficiency. Lower lobe involvement is most suggestive of panacinar emphysema. (Courtesy Dr. N. G. Hepper, Mayo Clinic, Rochester, Minnesota.)

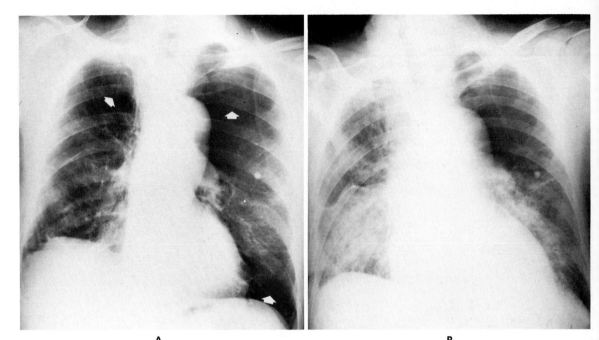

A	B

Figure 9–18 **Cardiac Failure in Emphysematous Patient: Absence of Pulmonary Edema in Areas of Hypoperfusion.**

A, Prior to cardiac failure, severely emphysematous segments are apparent in both upper lobes and at the extreme left base (*arrows*).

B, With onset of cardiac failure, alveolar densities of pulmonary edema are present only in more vascular portions. The emphysematous areas are not involved, producing an atypical distribution of the pulmonary edema.

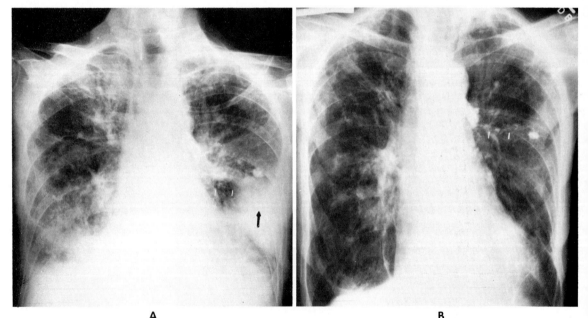

A	B

Figure 9–19 **Emphysema and Congestive Failure: Compliance Loss and Patchy Infiltrates.**

A, The chest film of an 80 year old man in congestive failure shows extensive basal infiltrates obscuring the diaphragms, which do not appear depressed. The upper lobes are relatively spared. The homogeneous density on the left (*arrow*) was fluid in the long fissure.

B, Two days later, after clinical recovery, the chest shows severe emphysema with depressed diaphragms. Fibrotic changes are seen, but the congestive changes have disappeared.

Loss of lung compliance during congestive failure will elevate the diaphragms and obscure the emphysema. Absence of congestive changes in the more severe emphysematous areas will produce patchy densities easily confused with pneumonitis.

462

Bronchiectasis

The diffuse form of bronchiectasis usually occurs in both lower lobes and often in the lingula and right middle lobe. The cause is uncertain in most cases. Recurrent lung infection is a predisposing factor, and a high incidence of bronchiectasis occurs in the dysgammaglobulinemias and in mucoviscidosis. Bronchiectasis in association with congenital dextrocardia and sinusitis is a recognized clinical triad called Kartagener's syndrome.

Focal bronchiectasis is usually a complication of a localized pulmonary infection such as pneumonia, abscess, tuberculosis, or atelectasis.

Chest films may disclose nothing more than heavy basal markings. In more advanced cases there may be honeycomb lucencies within areas of interstitial densities, occasionally with a fluid level. Pneumonic densities may result from secretions within the small cavities or from superimposed pneumonia, a frequent complication. Chronic fibrosis and volume loss causing crowding together of the markings may be late findings. Bronchiectasis should be considered in a patient with chronic productive cough and persistent basal densities.

Definitive diagnosis and extent of involvement are determined by bronchographic studies. Cylindrical bronchiectasis is characterized by dilatation of the peripheral bronchi and bronchioles. If this form occurs during an acute pneumonia or a temporary atelectasis, it may be reversible. Bulbous dilatation of terminal small bronchi may occur in more severe forms (varicose bronchiectasis). Saccular bronchiectasis, in which the terminal bronchioles and alveoli are dilated, is irreversible. The two types may coexist. Bronchographic opacification of all the bronchiectatic areas is sometimes prevented by thick secretions.

A bronchopleural fistula is an occasional complication of bronchiectasis and may lead to recurrent or chronic hydropneumothorax or empyema. Bronchography will demonstrate the fistulous connection.[20-24]

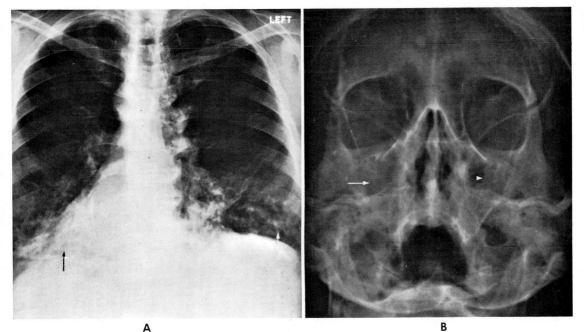

A   B

Figure 9–20 **Kartagener's Syndrome (Bronchiectasis, Dextrocardia, and Sinusitis).**

A, Posteroanterior film discloses dextrocardia, cyst-like lucencies, and increased markings at both bases (*arrows*) due to bronchiectasis.

B, Water's view of the sinuses reveals the right maxillary sinus (*arrow*) obscured by haziness, and the mucosa in the left maxillary sinus (*arrowhead*) thickened by chronic sinusitis.

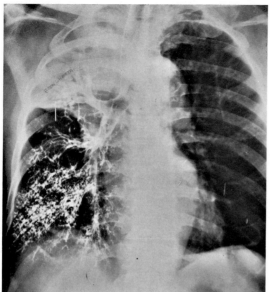

Figure 9–21 **Reversible Cylindrical Bronchiectasis in Upper Lobe Pneumonia.** There is an extensive area of cylindrical bronchiectasis (*arrow*) in the consolidated right upper lobe.

Two months after resolution of the pneumonia the bronchogram was normal. Cylindrical bronchiectasis may occasionally be completely reversible when the primary process clears, but saccular bronchiectasis is not reversible.

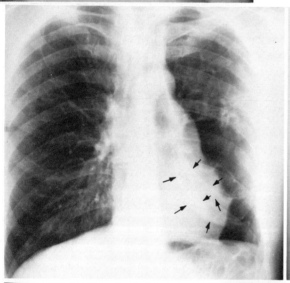

A

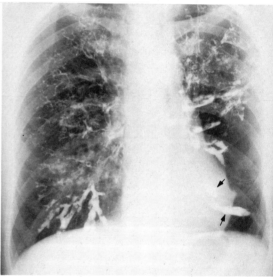

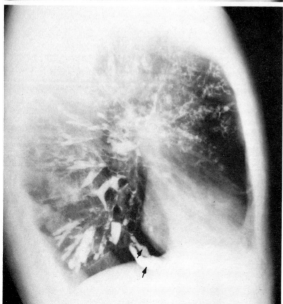

C

Figure 9–22 **Advanced Cylindrical Bronchiectasis**

A, On the regular chest film of a 45 year o man with chronic cough, the only clue to advance bronchial disease is branched sausage-like densiti (*arrows*) behind the heart. These resemble the dens ties of mucoid impaction or bronchopulmona aspergillosis (see Figs. 9–42, 9–43, and 6–170).

B and *C,* Bronchograms disclose that these de sities are dilated bronchi (*arrows*) and are part of widespread cylindrical bronchiectasis of both low lobes. There were no saccular or cystic bronchiect tic areas.

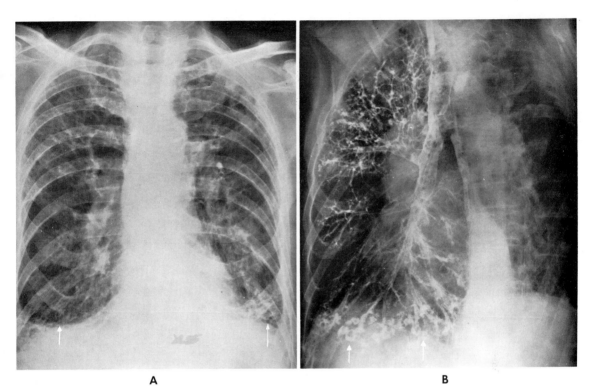

A   **B**

Figure 9–23   **Saccular Bronchiectasis: Plain Film and Bronchogram.**

A, Numerous small saccular lucencies and heavy markings at the bases (*arrows*) are caused by bilateral bronchiectasis.

B, Bronchography in the left oblique position demonstrates cystic dilatation of the distal bronchi (*arrows*), characteristic of saccular bronchiectasis.

Although plain films of the chest may suggest the diagnosis, as in this case, not uncommonly the films are negative even in the presence of extensive bronchiectasis. Bronchography is necessary in all cases for definitive diagnosis and determination of the extent of the disease process.

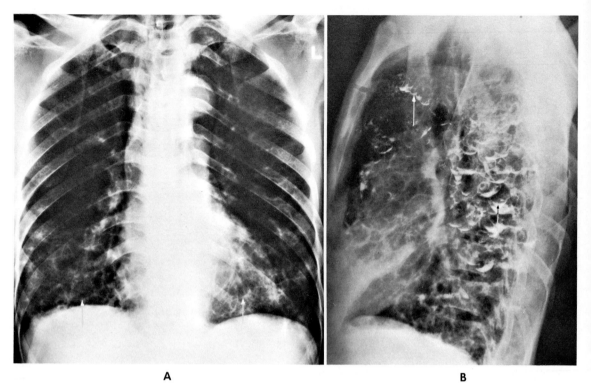

A   **B**

Figure 9–24 **Cystic Bronchiectasis.**

A, Plain film reveals large cystic lucencies, heaviest at the bases (*arrows*), and thickened bronchial walls. The upper lobes appear relatively normal.

B, Bronchography demonstrates extensive areas of cystic dilatation (*arrows*) with fluid levels throughout the entire lung. Cystic bronchiectasis may be generalized and is then considered to be a form of congenital cystic dysplasia of the lung.

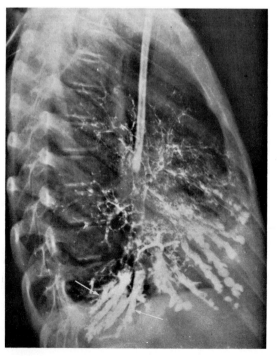

Figure 9–25 **Cylindrical and Saccular Bronchiectasis.** Bronchogram demonstrates the wide distal bronchi (*arrows*) that are characteristic of cylindrical bronchiectasis. Although occasionally reversible, cylindrical bronchiectasis more often is permanent and associated with saccular lesions. Notice the saccular dilatation of the more anterior lower lobe bronchi.

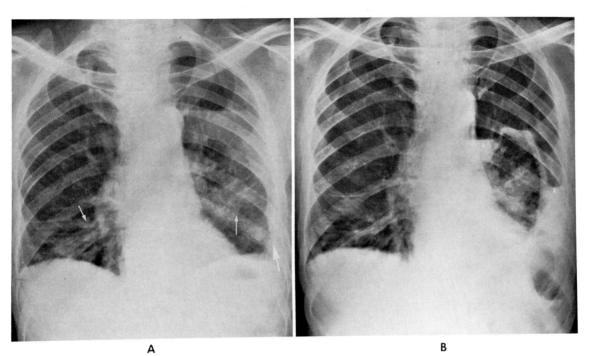

A   B

Figure 9–26 **Cylindrical Bronchiectasis in Agammaglobulinemia: Complicating Bronchopleural Fistula.**

A, Bilateral thickened basilar markings (*small arrows*) are due to bronchiectasis. There is fluid in the left chest (*large arrow*).

B, A bronchopleural fistula, an unusual complication, has developed with formation of empyema. Notice the fluid level (*arrowheads*).

Bronchial Asthma

During an acute asthmatic attack there is usually pulmonary hyperinflation. This leads to flattening of the diaphragms, increased lucency of the lung fields, and an increase in the retrosternal space. The pulmonary vasculature remains normal. Inspissated mucous plugs may cause transient parenchymal densities due to segmental atelectasis. When pneumonitis is superimposed, one may see a miliary pattern, although as a rule pneumonia associated with asthma is indistinguishable from ordinary bronchopneumonia. Between attacks the findings are normal. Rarely, severe coughing and straining in an acute attack can lead to pneumothorax, interstitial emphysema, and/or pneumomediastinum. (See also *Aspergillosis,* p. 243, and *Mucoid Impaction of the Bronchi,* p. 477.)[25, 26]

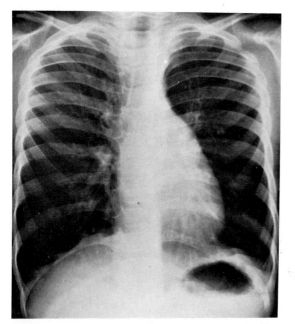

Figure 9–27 **Acute Bronchial Asthma in a Child.** There is evidence of hyperinflation of the lungs; the diaphragms are flattened and extend down to the eleventh posterior rib. The normal pulmonary vasculature differentiates hyperinflation from emphysema. This film was made during an acute asthmatic attack. The hyperinflation cleared with clinical recovery.

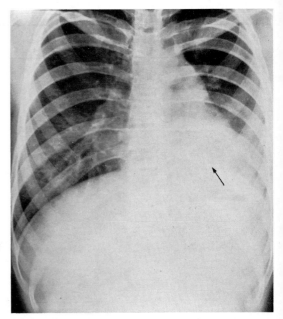

Figure 9–28 **Bronchial Asthma: Complicating Atelectasis.** In posteroanterior view, an area of increased density posterior to the heart obliterates the diaphragm (*black arrow*). The heart is shifted to the left. The interspaces on the left appear to be narrowed.

The density represents an area of atelectasis secondary to a mucous plug from an acute asthmatic attack. This is not uncommon, especially in younger children.

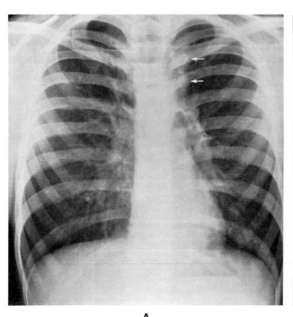

A   B

Figure 9–29 **Bronchial Asthma: Segmental Atelectasis.**

A, The atelectatic segment of the left apex presents as a density against the left upper mediastinum (*arrows*). This developed during an acute attack, and probably was the result of a mucous plug.

B, One week later the atelectatic segment has reexpanded, with subsequent disappearance of the mediastinal density.

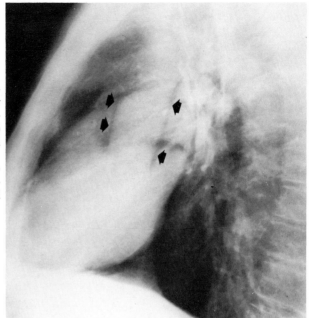

Figure 9–30 **Pneumomediastinum in an Asthmatic Patient.** Lateral chest film made shortly after a severe asthmatic attack discloses mediastinal air outlining the ascending aorta (*arrows*) and pulmonary artery. The air disappeared within a few days.

This form of "spontaneous" pneumomediastinum is probably due to a rupture of a mediastinal air bleb caused by severe straining and coughing.

Atelectasis

Complete occlusion of a bronchus is most often due to neoplasm, foreign body, or inspissated mucous plugs. Distal to the occlusion, the alveolar and bronchiolar air is absorbed, and the affected segment collapses, producing a homogeneous ground-glass opacity. Occasionally the air in the smaller bronchi is not absorbed, so that air bronchograms may be observed. The lower lobes collapse medially and posteriorly. An atelectatic left upper lobe appears as an anterior and medial density. The smaller right upper lobe collapses against the midmediastinum, and may simulate a mediastinal mass. If contiguous with the heart or diaphragm, an atelectatic segment will obliterate the borders of these structures (silhouette sign).

Usually there is a compensatory shift of adjacent structures toward the atelectatic area; the extent of shift depends on the size and location of the atelectasis. Generally, the homolateral diaphragm is elevated, the heart and mediastinum are shifted toward the lesion, and the hilum is displaced toward the atelectatic segment. If an entire lobe or a major portion of a lobe is collapsed, its pleural borders may shift and become bowed, conforming to the contour of the collapsed segment. The interspaces are narrowed on the affected side, and there is usually compensatory hyperinflation of the uninvolved aerated segments.

Nonobstructive atelectasis is generally due to lung compression by an adjacent lesion such as tumor, pleural fluid, large bullae, or pneumothorax. These lesions will dominate the roentgen picture. Nonobstructive atelectasis may also take place when there is impairment of the pulmonary surfactant system. This occurs after cardiopulmonary bypass, and after a number of insults, such as shock or trauma, which produce the adult respiratory distress syndrome. Cicatricial fibrosis may also lead to volume loss.[27, 28]

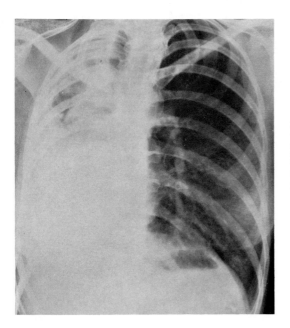

Figure 9–31 **Massive Atelectasis: Postoperative Mucous Plug.** The atelectatic right lung is homogeneously dense and obscures the heart shadow. The heart, mediastinum, and trachea are shifted into the right chest. The right interspaces are narrowed, and the right diaphragm is elevated.

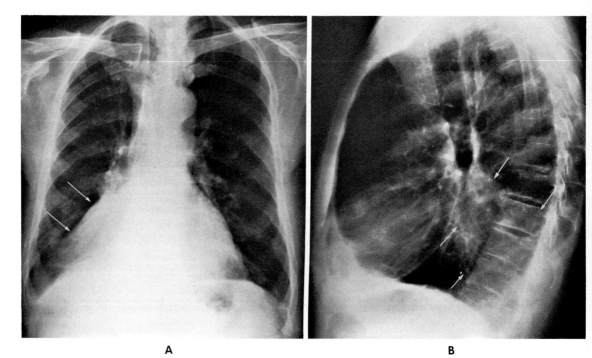

A   B

Figure 9–32 **Right Lower Lobe Atelectasis: Bronchogenic Carcinoma.**

A, There is an obstructive bronchogenic carcinoma of the right lower lobe bronchus with collapse of several basal segments, obliterating the medial portion of the diaphragm. The collapsed segment (*arrows*) lies close to the heart, which is shifted somewhat to the right. The interspaces on the involved side are narrowed.

B, Lateral view reveals the wedge of collapsed tissue in the lower posterior lung field (*arrows*). The major fissure (adjacent to the most anterior arrow) is bowed posteriorly, indicating partial collapse of the lower lobe.

In older individuals the heart and mediastinum tend to become fixed, and less dramatic shifts occur with atelectasis.

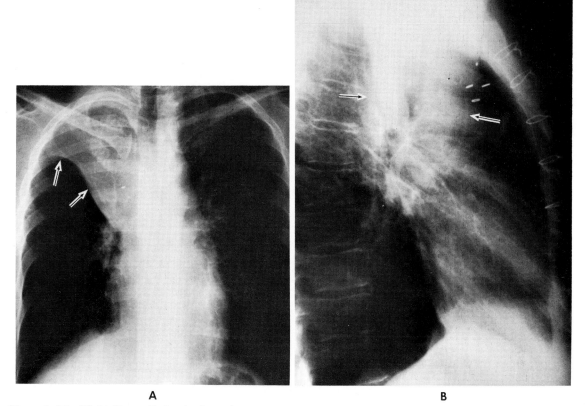

A   B

Figure 9–33 **Right Upper Lobe Atelectasis.**

A, There is homogeneous opacification of the right upper lobe. Its inferior border is elevated and bowed upward (*arrows*), a virtual pathognomonic sign of right upper lobe atelectasis.

B, The collapsed lobe (*arrows*) overlies the middle mediastinum on lateral view. The metallic clips and sternal sutures are from prior surgery.

A right upper lobe collapses against the midmediastinum, while a left upper lobe collapses against the anterior chest.

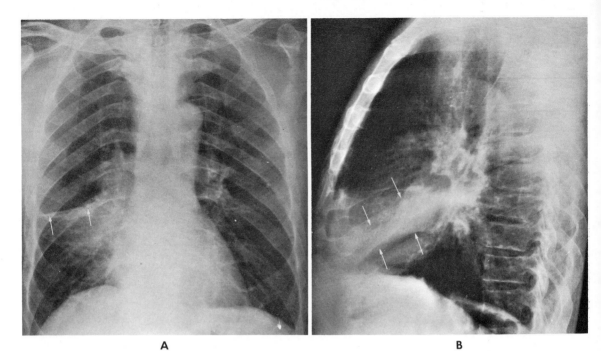

A	B

Figure 9–34 **Middle Lobe Atelectasis: Bronchogenic Carcinoma.**

A, Posteroanterior view demonstrates a density due to a contracted right middle lobe. The short fissure is depressed (*arrows*), and the volume of the consolidated middle lobe is considerably diminished. Characteristically, the right heart border is obscured.

B, On lateral view the contracted middle lobe appears as a band-like density (*arrows*).

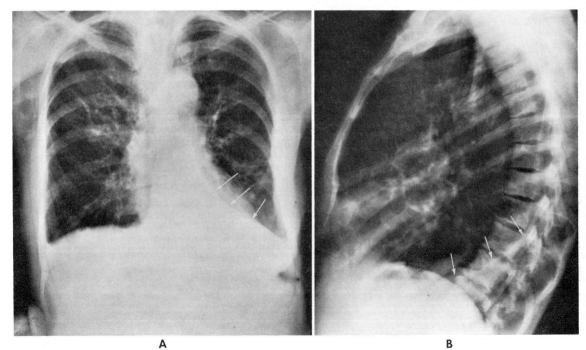

A	B

Figure 9–35 **Left Lower Lobe Obstructive Atelectasis.**

A, Complete collapse of the left lower lobe is demonstrated by the concave line behind the heart, which represents the border of the collapsed lobe (*arrows*). The left diaphragm is obscured by the collapsed lobe, a characteristic finding in left lower lobe atelectasis. Note compensatory hyperinflation of the left upper lobe.

B, Lateral view demonstrates collapsed lobe against the posterior chest wall (*arrows*). Only the right diaphragmatic shadow can be seen. While this loss of the diaphragmatic shadow is characteristic of lower lobe atelectasis, not infrequently one can see the diaphragm when the atelectatic lung shrinks toward the hilum.

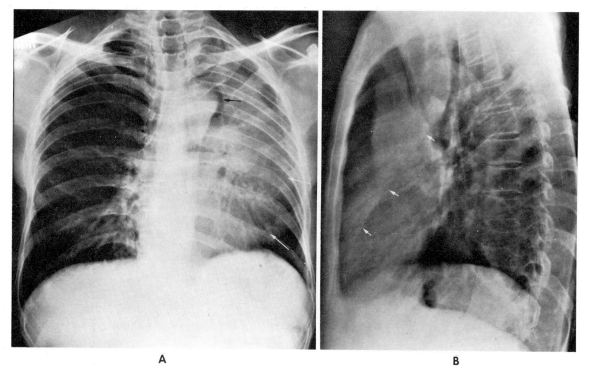

A **B**

Figure 9–36 **Left Upper Lobe Atelectasis.**

A, Posteroanterior view reveals a homogeneous ground-glass density in the left upper lobe (*white arrows*) with elevation of the left hemidiaphragm. The hilar shadow is also elevated. The heart and mediastinum are shifted to the left. The left cardiac borders are obscured by the adjacent collapsed lung. Note the narrowing of the interspaces on the left. The right upper lobe, enlarged by compensatory emphysema, has herniated across the mediastinum (*black arrow*).

B, On lateral view, the atelectatic upper lobe and the forward-displaced fissure (*arrows*) are clearly demonstrated.

Middle Lobe Syndrome

The right middle lobe syndrome refers to chronic atelectasis of this lobe from non-neoplastic causes. Generally, there is an inflammatory stenosis or bronchial obstruction from a node, usually tuberculous. However, the cause is often obscure. The middle lobe eventually becomes the seat of irreversible fibrosis and, often, bronchiectasis.

Radiographically, the collapsed middle lobe appears as a density at the right lower lung field, obliterating a portion of the right cardiac border. Often an apical lordotic view is necessary for demonstrating the atelectatic density. On lateral view, the right middle lobe area is occupied by a narrow band of density. Bronchography may show complete obstruction of the right middle lobe bronchus, or a fibrotic bronchiectatic lobe may be opacified through a narrowed bronchus.

A collapsed right middle lobe from neoplastic bronchial obstruction may produce an identical roentgen appearance.[29-31]

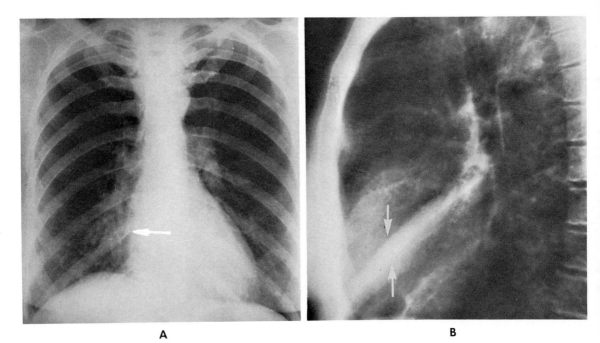

A   B

Figure 9–37 **Middle Lobe Syndrome.**

A, Posteroanterior view of the chest demonstrates obliteration of the normally sharp right heart border by a right middle lobe density (*arrow*).

B, Lateral view demonstrates more clearly a sharp dense band (*arrows*) representing an atelectatic right middle lobe segment. The upper border of the density is the displaced minor fissure.

Bronchography revealed severe stenosis of the right middle lobe bronchus, which prevented passage of opaque material. In less severe cases the lobar bronchus may be patent, but it is usually narrowed by inflammatory stricture or extensive pressure. During surgery, an extrinsic tuberculous node that compressed the right middle lobe bronchus was found.

Acute Epiglottitis

In children with croup due to acute upper airway obstruction, a lateral film of the soft tissues of the neck may reveal a swollen, thickened epiglottis and absence of air in the valleculae, findings indicative of acute epiglottitis. However, most cases of childhood croup are due to subglottic spasm, and the lateral neck film will reveal a normal epiglottic shadow but some subglottic narrowing of the trachea.

In adults, acute epiglottitis can occasionally occur with similar findings on the lateral neck film.[32-34]

Figure 9-38 **Acute Epiglottitis in 2 Year Old.** Lateral neck film of child with acute croup reveals a swollen rounded epiglottis (*small arrow*), which is partially obstructing the airway.

The hypopharyngeal air space above the epiglottis is unusually prominent (*large arrow*). The normal epiglottis is thin and has a tapering, pointed tip.

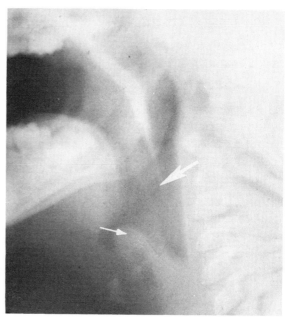

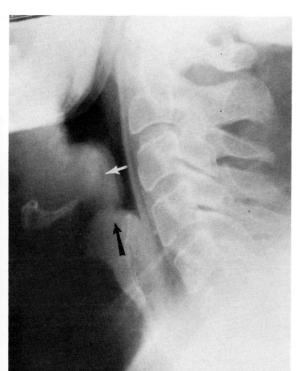

Figure 9-39 **Acute Epiglottitis in Adult.** Lateral neck film made after onset of respiratory difficulty reveals enormous swelling of the epiglottis (*white arrow*), which is almost completely blocking the airway. The subglottic air seen beneath a normal thin epiglottis is entirely obliterated. Subglottic edema (*black arrow*) is also present.

Clinically, this epiglottic swelling was thought to be a combination of infection and superimposed allergy.

Intrabronchial Foreign Bodies

Radiopaque foreign bodies are usually easily recognized on chest films. However, a high proportion of aspirated foreign bodies are nonopaque, and roentgen diagnosis must be made from the secondary changes in the lung parenchyma. Bronchography may be needed for confirmation and localization.

If the foreign body has caused complete bronchial obstruction, the trapped air is rapidly resorbed, the alveoli collapse, and the involved segment or lobe becomes atelectatic.

The roentgen picture will depend upon the location of the obstruction. With extensive volume shrinkage, the heart and mediastinum are shifted toward the affected side, the homolateral diaphragm becomes elevated, and the interspaces become narrowed on the affected side. The atelectatic segment presents as a parenchymal density (see *Atelectasis*, p. 469). With massive collapse, the normal lung may herniate across the mediastinum. Secondary pneumonia and lung abscess may develop in the affected lung segment.

Partial bronchial obstruction can cause obstructive emphysema, in which air enters during inspiration but remains partially trapped as the bronchus contracts during expiration. The affected lobe becomes hyperaerated, causing a shift of the heart and mediastinum toward the normal side. This shift becomes quite pronounced during forced expiration, when the emphysematous segment does not contract. A mediastinal shift toward one side during deep expiration, and a return toward the midline on deep inspiration is strong evidence of obstructive emphysema on the contralateral side. These findings are best demonstrated in children, in whom mediastinal mobility is quite marked. Decubitus films on the side of the obstructive emphysema show failure to decrease lung volume.

Obstructive emphysema may progress to atelectasis if the partial obstruction becomes complete.

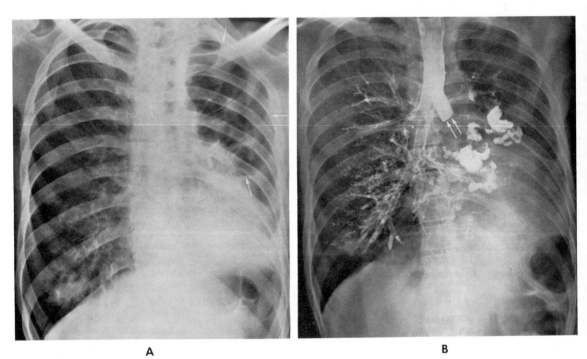

A   B

Figure 9–40 **Chronic Atelectasis: Foreign Body.**

A, This film shows atelectasis of the left lung attributable to aspiration of a nonopaque foreign body 14 years previously. The foreign body had not been removed. There is marked shift of the heart to the left. The borders of the heart are obscured by the atelectatic left lung. The left interspaces are narrowed. The aerated lung tissue seen in the left upper thorax is actually a herniated segment of the right lung *(arrows).*

B, The bronchogram demonstrates abrupt and complete occlusion of the left main bronchus *(arrows).* Note that all the opacified bronchi on both sides arise from the right bronchial tree. The irregular collections of contrast material on the left represent bronchiectatic cavities.

The roentgenologic findings in this case are almost indistinguishable from those seen in agenesis of the left lung. (Courtesy Dr. William Weiss, Philadelphia, Pennsylvania.)

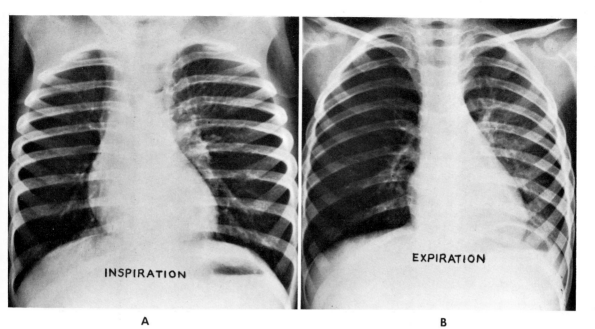

INSPIRATION

EXPIRATION

A   B

Figure 9–41 **Obstructive Emphysema: Foreign Body.**

A, Inspiratory film in a young child thought to have inhaled a foreign body reveals decreased pulmonary markings in the right lung, a finding suggestive of obstructive emphysema.

B, Expiratory film reveals normal decrease in lucency and volume of the left lung, but persistent hyperinflation of the right lung, with widened interspaces. There is a distinct shift of the heart and mediastinum to the left. These findings confirm the diagnosis of obstructive emphysema of the right lung. A nonopaque plastic bead was found in the right main bronchus.

In obstructive emphysema, inadequate expiration leads to persistent inflation of the affected lung, so that the heart and mediastinum are shifted to the opposite side when the normal lung deflates during expiration. If only a smaller segmental bronchus is involved, local persisting hyperinflation may be observed; mediastinal shift during inspiration and expiration may be minimal or absent. These shifts are more pronounced in younger individuals, since the mediastinum generally becomes less mobile in older people.

Obstructive emphysema caused by a foreign body, if untreated, generally progresses to atelectasis.

Mucoid Impaction of the Bronchi

Multiple and diffuse inspissated mucous plugs in segmental or small bronchi occur most often in asthmatics but can also develop in patients with cystic fibrosis and allergic pulmonary aspergillosis (see p. 243).

Large mucous plugs, usually in one segment, are often of unknown origin. Nonspecific symptoms of cough, wheezing, hemoptysis, or chest pain may be present.

The roentgenographic appearance is usually quite characteristic. One or more oval or oblong densities in the parenchyma are noted, frequently in an upper lobe. When multiple adjacent bronchi are involved, V- or Y-shaped densities or a grapelike cluster is apparent. Tomography may clarify the characteristic shapes of overlapping lesions. Distal atelectasis is uncommon in spite of the plugging. Bronchography will demonstrate focal obstruction, and if the plugs have been expelled, dilatation of the previously impacted bronchial segment will be seen.[35]

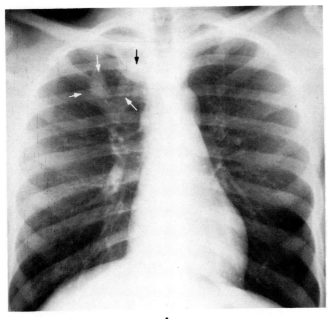

A

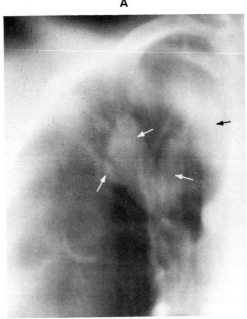

B

Figure 9–42 **Mucoid Impaction of Bronchus.** Chest film (*A*) and tomogram (*B*) reveal branching, ovoid densities (*white arrows*) in the right upper lobe. There is also a density in the right upper mediastinum (*black arrows*). The patient was a healthy 17 year old boy with a history of a single episode of mild hemoptysis.

The ovoid V-shaped densities are characteristic of mucoid impactions of a branching major bronchus. The mediastinal density proved at surgery to be an area of focal atelectasis.

This patient had no obvious cause for the mucoid impaction, such as asthma, cystic fibrosis, or allergic aspergillosis.

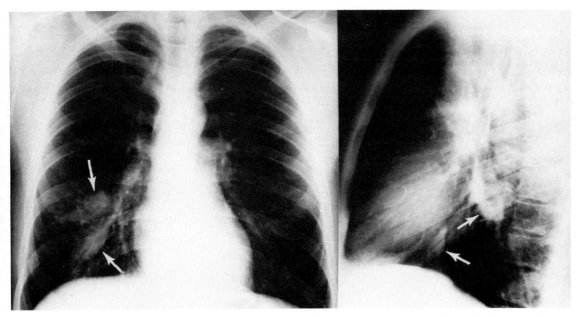

Figure 9–43 **Mucoid Impactions of Bronchi.** The ovoid densities in the right lower lung field *(arrows)* of an asymptomatic young man are fairly characteristic of mucoid impactions of bronchi. Similar findings occur in allergic bronchopulmonary aspergillosis (see Fig. 6–170). (Courtesy Drs. M. E. Whitcomb and S. S. Braman, Washington, D.C.)

Tracheobronchomegaly (Mounier-Kuhn Syndrome)

Tracheobronchomegaly is characterized by a dilatation of the trachea and major bronchi and is due to congenital (?) weakness of the elastic and muscular fibers in the tracheobronchial tree. Tracheal and bronchial collapse tends to occur with forced expiration or cough. The condition may be familial. Most of the patients are over 30 years of age. Disturbances of pulmonary ventilation with resultant chronic inflammatory bronchopulmonary disease are often associated.

The increased diameter of the trachea is recognized radiographically on adequately penetrated routine chest films, and it is most readily seen on lateral projection. High kilovoltage films or planigraphy will best show the dilated major bronchi. The pulmonary parenchyma may reveal areas of chronic inflammatory change. Bronchography discloses deep cartilaginous corrugations of the dilated trachea and bronchi, an appearance that is somewhat similar to a dilated intestinal loop. Diverticula-like outpouchings of the redundant tracheal mucosa may be opacified. Areas of saccular bronchiectasis and changes of chronic bronchitis are often found.[36-38]

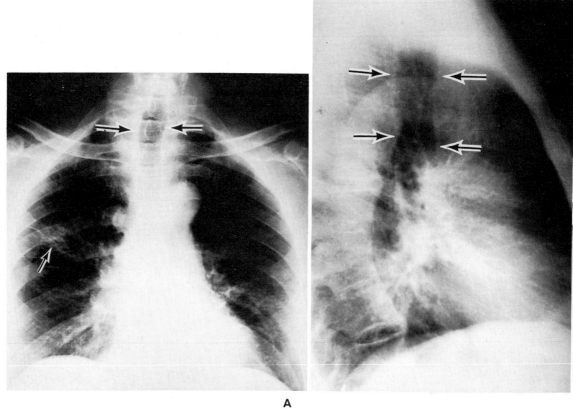

A

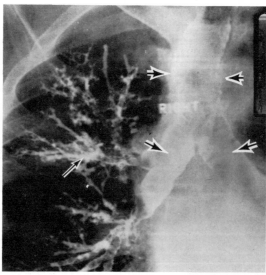

B

Figure 9–44 **Tracheobronchomegaly (Mounier-Kuhn Syndrome).**

A, Chest films show a widened tracheal airspace, most marked on the lateral film (*large arrows*). There is a chronic inflammatory infiltrate (*small arrow*) in the anterior segment of the right upper lobe.

B, Bronchogram shows the dilated trachea and major bronchi (*large arrows*). In the area of chronic inflammation, the bronchi are crowded together and dilated (*small arrow*).

ABNORMAL AIR SPACES

Giant Bullous Emphysema

In this condition the emphysematous bullae are bilateral, often begin in the apices of the upper lobes, and, with enlargement, compress the adjacent lung tissues. This disorder occurs almost exclusively in males and is also known as *vanishing lung*.

Radiographically the bullae appear as peripheral radiolucent air-filled sacs, varying from small cyst-like areas to giant lucencies replacing an entire upper lobe. The absence of lung markings is striking; fine fibrous septa are often seen within the larger bullae. In advanced cases, both upper lobes may be entirely replaced by giant bullae, and the lower lobes may be severely compressed. Not infrequently, spontaneous pneumothorax occurs from rupture of a bulla.

This disorder virtually never occurs in a nonsmoker, is often associated with generalized emphysema, and is probably a special form of "smoker's emphysema."

Surgical removal of a bullous upper lobe may relieve lower lobe compression and produce clinical improvement.[39-43]

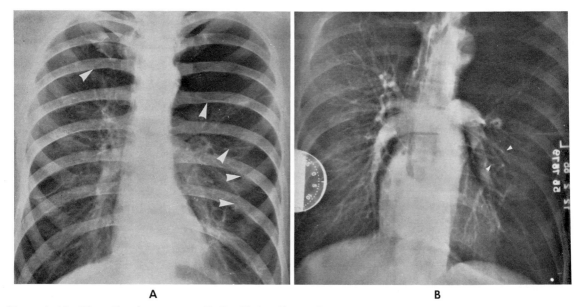

A   B

Figure 9–45 **Giant Emphysematous Bulla: Plain Film and Angiogram.**

A, Posteroanterior view demonstrates multiple giant emphysematous bullae in upper lobes (*arrowheads*). The left upper lobe is considerably more involved than the right, and the bullae are compressing the adjacent lung tissue. Fine fibrous septa and the absence of lung markings are characteristic.

B, Angiogram demonstrates striking diminution of the vessels in the involved lung segments and crowding of the vessels in the compressed adjacent lung segments (*arrowheads*).

Lung Cysts

The lung cyst is usually a solitary lesion in an otherwise normal lung; occasionally, several cysts are present. The size of a cyst is extremely variable. These lesions are congenital or developmental in origin, and a hereditary tendency has been observed. However, the cause of many lung cysts is unknown.

A solitary cyst presents as a rounded area of lucency with no internal markings, bounded by a thin wall, and surrounded by normal lung tissue. The cyst is filled with air although its communicating channel may be extremely small. Fluid may accumulate within the cyst, simulating a thin-walled abscess with an air-fluid level. If entirely fluid-filled, the rounded density will resemble a sharply marginated solid mass. An infected cyst may show a thickening of its wall in addition to fluid. Large cysts may compress the adjacent lung.

A cyst that develops during the course of a pulmonary infection is more aptly termed a pneumatocele, but it may be radiographically indistinguishable from a congenital cyst (see *Postpneumonic Pneumatocele*, p. 488). An emphysematous bulla is less regular in shape and does not have a uniform, circular wall. Multiple congenital cysts in an otherwise normal lung can readily be distinguished from polycystic lung, a diffuse disease (see p. 484).[39]

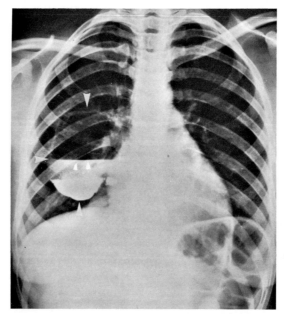

Figure 9–46 **Lung Cyst: Bronchogram.**   This is a chest film of a 21 year old female who had minimal chest symptoms. A large thin-walled cystic lesion is seen in the right lung field *(large arrowheads)*. Note the air-fluid level *(small arrowheads)* resulting from introduction of contrast material through the bronchi. A lung cyst may contain air or fluid, or both.

Bronchogenic Cysts

Bronchogenic cyst is congenital and results from abnormal budding or branching of the tracheobronchial tree. Symptoms, if present, are generally from pressure on the adjacent bronchus or esophagus.

The cyst is usually found in the mediastinum, but it may arise in the lung parenchyma. The mediastinal cyst is most frequently located beneath the carina and produces a smooth, homogeneous, circumscribed, spherical or oval density. The major portion of the cyst is usually posterior to the bifurcation of the trachea and may occasionally displace the trachea or bronchus. Cysts attached to the tracheobronchial tree may move with swallowing. There may be an indentation on the lateral esophagogram.

Less frequently, the cyst may arise within the lung parenchyma, in which case it appears as a nonspecific round mass. Only rarely is there bronchial communication, so that fluid levels are generally not seen. The lesion is usually indistinguishable from a simple lung cyst or a circumscribed mass lesion (see also Fig. 9–184).[39, 44–46]

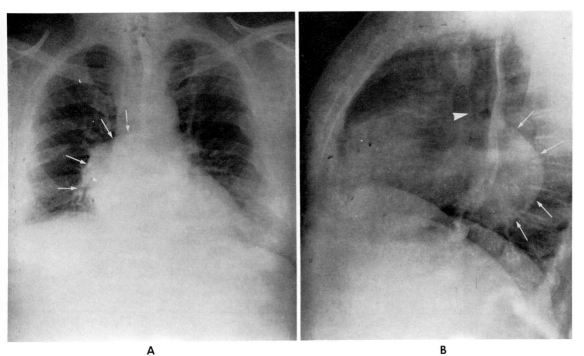

A   B

Figure 9–47 **Bronchogenic Cyst: Mediastinum.**

A, There is an abnormal density (*arrows*) just inferior and to the right of the carina. The right heart border (*arrowhead*) can be seen through the density.

B, Lateral films reveal the mass in the posterior half of the chest (*arrows*) behind the carina (*arrowhead*). A sharply bordered round mass in this location is characteristic of a bronchogenic cyst.

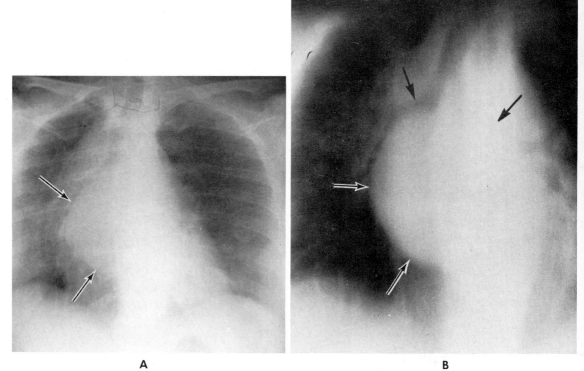

A   B

Figure 9–48 **Large Bronchogenic Cyst: Characteristic Location.**

A, On regular chest film, a large density (*arrows*) is seen within the right side of an enlarged cardiac shadow.

B, Planigram demonstrates a sharply marginated density (*black-white arrows*) that is elevating both main stem bronchi (*black arrows*) and markedly compressing the left main bronchus.

This was a large bronchogenic cyst beneath the carina. Its location, the sharp rounded contour, and the pressure effects on the stem bronchi are characteristic features of a bronchogenic cyst. However, a greatly enlarged left atrium can produce similar radiographic findings.

Polycystic Lung

This is a diffuse pulmonary disease also known as muscular hyperplasia, muscular cirrhosis, bronchiolar dilatation, and bronchiolar emphysema.

The lung fields may show a diffuse nodularity, and close inspection may reveal some of the nodules to be small cysts. Generally, the cysts remain small. Areas of interstitial densities and more conglomerate shadows are present. The radiographic involvement is not uniform; one lung or area may be predominantly affected, but the disease is bilateral. Progression is slow, and the lungs remain unchanged over extended periods. Diffuse involvement will simulate the appearance of the honeycomb lung (see p. 487). It has been postulated that the polycystic lung is an advanced stage of interstitial fibrosis rather than a distinct entity.

Bronchography reveals numerous areas of cystic bronchiolar dilatation that produce a distorted, irregular, pleated appearance. Normal arborization of the bronchial tree is obscured and distorted, and the entire tree appears foreshortened. Alveolar filling, which often occurs in the bronchograms of emphysema and chronic bronchitis, is rarely seen in polycystic lung. The bronchographic findings are highly characteristic.

The diffuse interstitial or conglomerate densities and the relatively small size of the cysts will distinguish polycystic lung from the lung with multiple congenital cysts.[47-49]

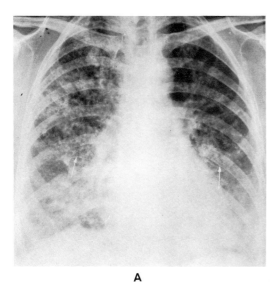
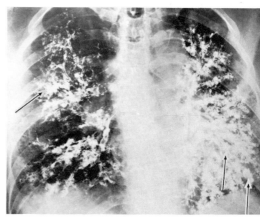

A   B

Figure 9–49   **Polycystic Lung: Diffuse Bronchiolar Dilatation: Two Patients.**

A, Chest film of a 57 year old woman with a 10-year history of chronic productive cough reveals diffuse fibrotic appearing densities and conglomerate areas in the right lung. Numerous small cystic collections (*arrows*) can be seen on careful inspection. Tuberculosis was suspected, but biopsy was characteristic of polycystic lung.

B, Bronchogram of another patient shows numerous collections of contrast material throughout the bronchial tree *(arrows),* which are indicative of a diffuse bronchiolectasis. The bronchial tree appears foreshortened. This bronchographic appearance is characteristic. (Courtesy Dr. A. J. Christoforidos, Columbus, Ohio.)

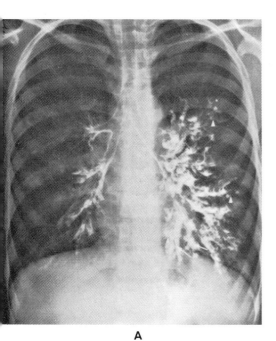
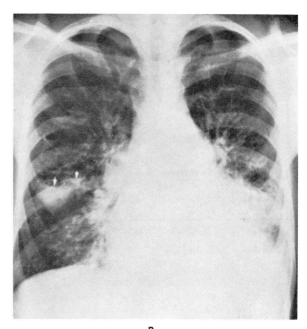

A   B

Figure 9–50   **Polycystic Lung.**

A, The aborization of the left bronchial tree is distorted by collections of contrast material in dilated bronchioles. Some unfilled cysts (*arrowheads*) are seen in the periphery. Some cylindrical bronchiectasis (*arrows*) is apparent at the left base. The right bronchial tree is foreshortened and poorly filled.

B, Four years later there are diffuse interstitial densities, especially at the bases. Larger cysts with fluid levels (*arrows*) are in the right lung. The heart and main pulmonary artery are enlarged—a cor pulmonale from diffuse interstitial disease.

Congenital Lobar Emphysema

This disease of the newborn may involve either an upper lobe or the right middle lobe. If it is not treated, severe dyspnea and death may result.

The emphysema results in an expanded hyperlucent lobe in which vascular markings are present but decreased. The markings help to differentiate congenital lobar emphysema from pneumothorax and congenital cysts, in which there are no vascular markings. The diaphragm on the affected side is depressed, the interspaces are widened, and the heart and mediastinum are displaced toward the normal side. The adjacent lung tissue becomes compressed. The overdistended lung segments may herniate across to the normal side. Congenital lobar emphysema is usually a surgical emergency, although some cases of spontaneous resolution have been reported.[44, 50, 51]

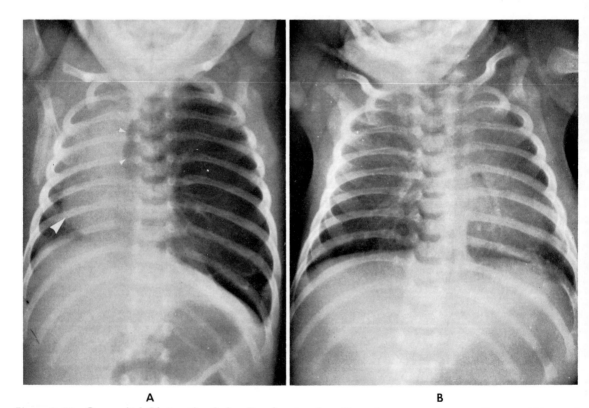

A   **B**

Figure 9–51 **Congenital Obstructive Lobar Emphysema in a Neonate.**

A, In an infant with severe dyspnea, a hyperlucent left hemithorax displaces the heart (*large arrowhead*) and mediastinum entirely into the right chest. The left intercostal spaces are widened, and the left diaphragm is depressed. The hyperinflated left upper lobe has herniated into the right thorax (*small arrowheads*). Note the sparse bronchovascular markings in the lucent left lung field. The picture is typical of congenital lobar emphysema.

B, The left upper lobe was removed. The heart and mediastinum are now in their normal positions, and the lung fields appear normal.

Honeycomb Lung

The "honeycomb" lung has a reticulated appearance that is due to a uniformly distributed diffuse interstitial fibrosis interspersed with cyst-like areas of dilated alveoli and bronchioles. Superimposed on the fibrosis may be a poorly defined nodularity. Complicating pneumothorax may result from a ruptured subpleural emphysematous bleb. This picture is most characteristically associated with histiocytosis X (eosinophilic granuloma), in which there is neither hilar adenopathy nor pleural involvement. Other causes of honeycomb lung are the collagen diseases, especially scleroderma, sarcoidosis, tuberous sclerosis, polycystic lung, and other conditions manifesting diffuse interstitial fibrosis. Frequently, however, no cause can be determined.[52, 53]

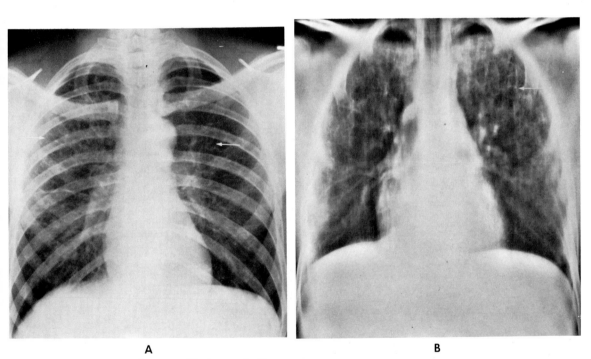

A   B

Figure 9–52 **Honeycomb Lung: Histiocytosis X.**

A, There are extensive bilateral, diffuse, interstitial fibrotic strands, nodularity, and multiple small cystlike areas *(arrows)*. The picture is typical.

B, Planigram more clearly demonstrates the honeycomb appearance. The cystic areas *(arrow)*, which are sometimes difficult to appreciate on routine films, are clearly defined on planigram (Courtesy Dr. William Weiss, Philadelphia, Pennsylvania.)

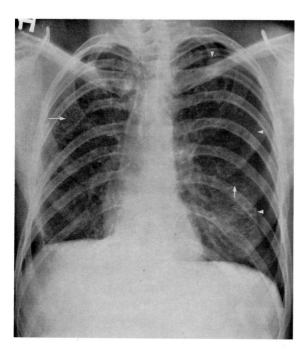

Figure 9–53 **Honeycomb Lung: Histiocytosis X with Pneumothorax.** Pneumothorax is present in the left lung (*arrowheads*). It is a common complication of honeycomb lung and is caused by rupture of a cyst into the pleura. There are multiple small cystic areas (*arrows*) and fine reticular and nodular densities, all of which are characteristic of honeycomb lung.

Postpneumonic Pneumatocele

Areas of local obstructive emphysema secondary to inflammatory narrowing of the smaller bronchi are best demonstrated during resolution of a pneumonic process. They appear as thin-walled, cyst-like structures. Spontaneous clearing may occur with resolution of the primary pneumonic process, or the pneumatocele may persist for months. They may arise in any septic pneumonia but are most common in staphylococcal pneumonia. Rupture of a pneumatocele may produce a pneumothorax (see p. 565).[54]

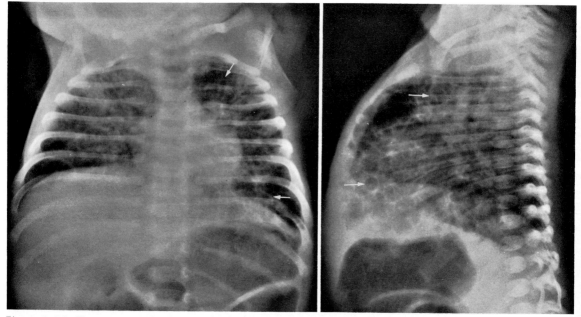

Figure 9–54 **Postpneumonic Pneumatocele: Staphylococcal Pneumonia.** Posteroanterior and lateral chest views reveal multiple thin-walled cyst-like pneumatoceles (*arrows*) of various size throughout both lung fields. These usually disappear spontaneously following recovery, but they may persist for months.

Lung Abscess

Abscesses occur most often as a complication of aspiration of food, vomitus, or foreign body; of bacterial pneumonia; or of bronchial obstruction. Abscesses may also be secondary to septicemia, and they occasionally develop in an infected pulmonary infarct. The abscess resulting from aspiration most frequently occurs in the dependent segments of the lung—the posterior segments of the upper lobe and the superior segments of the lower lobe.

The abscess first appears as a round but poorly defined area of segmental consolidation usually near the periphery of the lung. No fluid level is seen until bronchial communication is established. Planigraphy may be needed to demonstrate the lucent center of the abscess and to help differentiate an abscess from a cavitary bronchogenic carcinoma (see p. 491).[55]

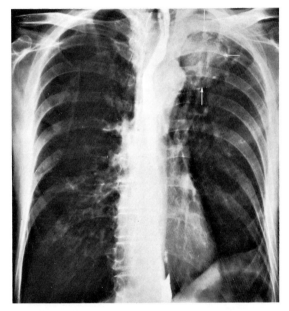

Figure 9–55 **Lung Abscess.** A round density with ill-defined borders in the left apical area (*arrows*) was a lung abscess that did not communicate with the bronchus. Radiographically the lesion simulates a bronchogenic carcinoma.

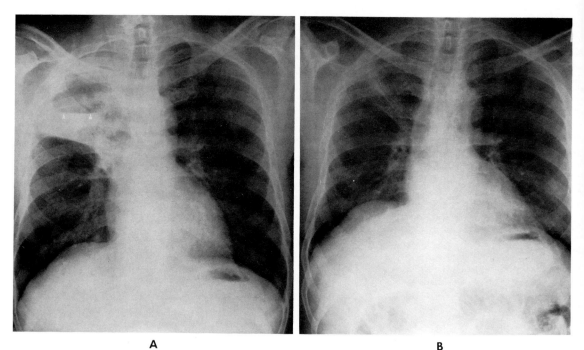

A   B

Figure 9–56 **Lung Abscess: Resolution.**

A, There is a large air-fluid level (*arrowheads*) within an otherwise homogeneous consolidation of the right upper lobe; this is typical of lung abscess.

B, Six weeks later the abscess has cleared, and only some residual fibrotic strands remain. Lung abscess can usually be cured by antibiotic therapy.

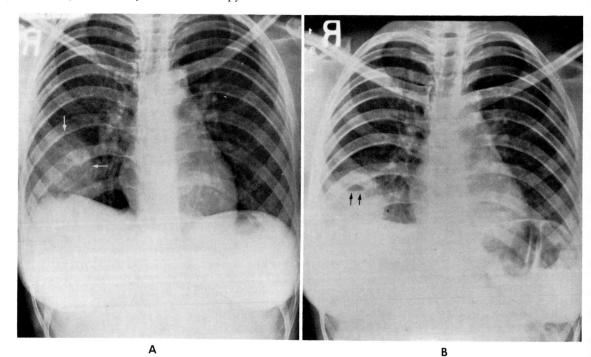

A   B

Figure 9–57 **Right Lower Lobe Abscess.**

A, A round ill-defined density in the right lower lobe extends to the pleural surface (*arrows*). The pleural reaction results in elevation and fixation of the right diaphragm.

B, Several weeks later an air-fluid level (*arrows*) characteristic of a lung abscess has developed.

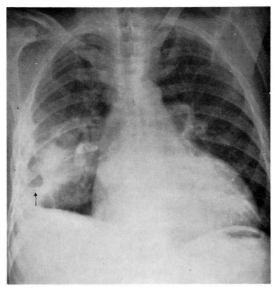

Figure 9–58 **Abscess and Pulmonary Infarct.** An air-fluid level (*arrow*) indicative of cavitation has developed in a pulmonary infarction. Cavitation occurs in only about 5 per cent of pulmonary infarcts. (Courtesy Dr. Arlyne Shockman, Veterans Administration Hospital, Philadelphia, Pennsylvania.)

Cavitary Bronchogenic Carcinoma

Bronchogenic carcinoma must be considered in any cavitary lesion that arises in a patient over 40 years of age. In cavitary carcinoma the walls are usually thick and irregular. A solid mass or nodule may be attached to the wall of the cavity or situated within it, and this appearance is suggestive of carcinoma. Distinction must be made between breakdown and excavation of the tumor itself and bacterial inflammatory abscesses that are secondary to bronchial obstruction and occur distal to the tumor.

Often one cannot distinguish cavitary carcinoma from a benign abscess. However, selective bronchography, bronchial washings, and transbronchial biopsy are often helpful in making a definitive diagnosis.[56]

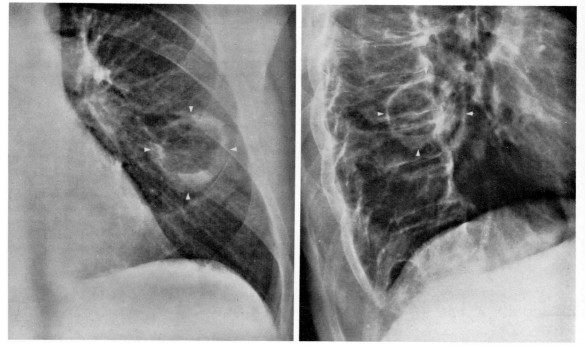

Figure 9–59 **Cavitary Bronchogenic Carcinoma.** Posteroanterior and lateral views disclose an area of increased density in the left lower lobe (*arrowheads*). The central lucency is due to tumor necrosis. The irregularity and nodularity of the wall of the cavity and its eccentric location are characteristic of tumor necrosis.

Unilateral Hyperlucent Lung (Swyer-James Syndrome)

This entity is thought to be secondary to repeated unilateral infections resulting in obliterative bronchitis and bronchiolitis. Collateral ventilation develops distal to the obliteration, causing secondary overdistention and emphysema.

Unilateral hyperlucent lung is characterized by abnormal radiolucency of a lung or lung segment. Superficially it resembles obstructive emphysema, but there is no volume increase, and no obstruction can be demonstrated in the major bronchus.

Radiologically there is abnormal radiolucency of one lobe or lung, and the lung may be of normal or decreased size. Volume during the various phases of respiration remains unchanged in the affected lung, so that the hyperlucency persists on forced expiration. This causes the mediastinum to shift toward the affected side on inspiration and to shift away from it on expiration. There is also decreased motion of the associated hemidiaphragm and rib cage. The hilar shadow is small, and the pulmonary vascular markings are decreased in size and number.

On bronchography there is a decrease of bronchial subdivisions, a lack of tapering, and evidence of bronchitis with or without proximal bronchiectasis. There is no evidence of obstruction of the larger bronchi.[57-59]

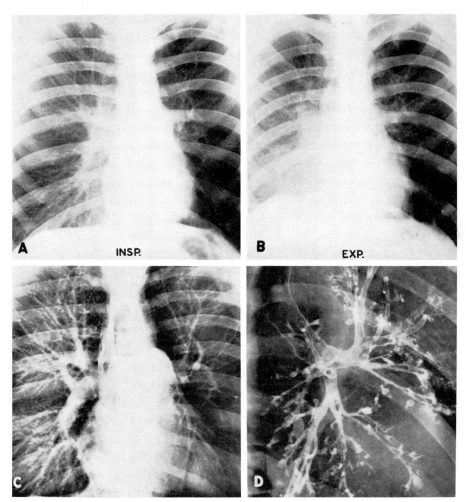

Figure 9–60 **Idiopathic Unilateral Hyperlucent Lung.**

A, The left lung is hyperlucent, the left hilum is small, and pulmonary vasculature is sparse.

B, Expiratory view discloses shift of the mediastinum toward the right. Notice absence of change in left hemidiaphragm and rib cage.

C, In pulmonary angiogram the left main pulmonary artery is small, and the peripheral pulmonary vessels are smaller and fewer than normal; there is a compensatory increase in caliber in vessels supplying the right lung.

D, Bronchogram demonstrates the unusual diffuse form of bronchiectasis and the absence of alveolar filling, which are characteristic. (Courtesy Dr. H. N. Margolin, Cincinnati, Ohio.)

Traumatic Lung Changes: Contusion and Cyst

Pulmonary contusion is the most common lesion resulting from closed chest trauma. Radiographically there may be patchy ill-defined parenchymal densities, peribronchial infiltrates, and, rarely, frank consolidation. Only the history of trauma, the occasional presence of rib fractures or pneumothorax, and the clinical course distinguish pulmonary contusion from pneumonitis. Clearing begins within two to three days and is usually complete within two weeks.

Pulmonary laceration may result from a penetrating wound or from severe blunt trauma. A round or spindle-shaped homogeneous mass density may develop, representing a clot within the lacerated area. Later, with retraction and liquefaction of the clot, a crescent of air may be seen at the periphery of the hematoma, or a fluid level may appear, resulting in a cystic appearance. A progressive decrease in size of the hematoma occurs if infection does not develop.[60-62]

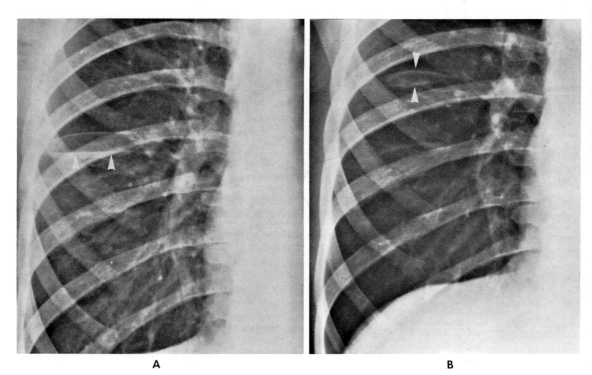

A   **B**

Figure 9–61 **Traumatic Lung Cyst.**

A, Posteroanterior view demonstrates cyst-like area with a fluid level (*arrowheads*) that developed following closed trauma to the chest.

B, Two weeks later the cyst was much smaller (*arrowheads*), and within six weeks the chest was normal. This is the typical sequence.

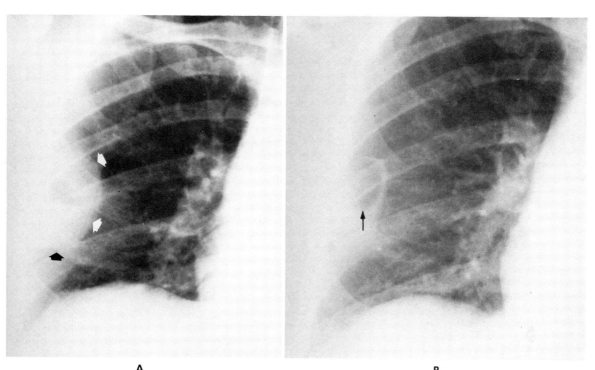

A **B**

Figure 9–62 **Pulmonary Hematoma and Cystic Liquefaction.**

A, Two days following chest trauma a rounded homogeneous density (*arrows*) has appeared in the lung periphery, lying against the pleural surface. This was a hematoma in the lung.

B, One week later, the density has decreased in size, and the lesion now contains an air-fluid level (*arrow*). The lesion was completely resolved eight days later.

Cystic change often develops during resolution of a pulmonary hematoma.

DIFFUSE LUNG DISEASE

Diffuse pulmonary involvement is found in a large number of diseases. Since there are only a limited number of possible radiographic patterns, definitive diagnosis usually cannot be made from the radiographic appearance alone. Some conditions are characterized by alveolar densities, others by interstitial infiltrates, and still others by diffuse nodulation. Recognition of these patterns and their possible causes will narrow the diagnostic possibilities. However, in many diseases mixed patterns may be found.

Consolidation of individual alveolar acinar units appears as small densities with ill-defined, somewhat irregular edges. Larger confluent alveolar densities exhibit fluffy "cotton candy" margins. The presence of an air bronchogram is pathognomonic of alveolar consolidation surrounding unoccluded small bronchi. Confluent alveolar densities often have a butterfly distribution, with broad lateral margins. Complete opacification of a pulmonary segment or lobe usually indicates alveolar consolidation.

Among the principal disseminated alveolar diseases are pulmonary edema of various causes, many bacterial and certain viral and rickettsial pneumonias, *Pneumocystis carinii* pneumonia, hyaline membrane disease, acute pulmonary hemorrhage, hemosiderosis, alveolar proteinosis, lymphoma, alveolar cell carcinoma, an exudative form of sarcoidosis, bronchogenic spread of tuberculous or fungal disease, alveolar microlithiasis, and periarteritis nodosa. Some of these are discussed in the following pages, and most of the remaining conditions will be found elsewhere in these volumes.

Interstitial infiltrates usually appear as linear densities that seem to accompany the bronchovascular markings. Fine interstitial infiltrates may be overlooked or mistaken for accentuated normal markings. Unlike bronchovascular markings, however, the interstitial infiltrates do not branch and do not necessarily diminish in diameter as they extend peripherally. Fine nodulation may sometimes accompany the linear densities. If a disease process extensively involves the interlobular septa, there will be a reticular lung pattern and Kerley-like transverse lines in the periphery. Small lucencies of emphysema can develop between chronic interstitial infiltrates, producing the honeycomb lung appearance.

Conditions with predominantly interstitial infiltrates include the various pulmonary fibroses—the pneumoconioses, collagen diseases, rheumatoid lung, histiocytosis X, and a form of pulmonary sarcoidosis—certain viral infections, and lymphangitic metastases. Honeycomb lung is characteristically seen in histiocytosis X, but can develop in other diseases with chronic diffuse fibrosis.

Diffuse nodular disease without other interstitial infiltrates produces another radiographic pattern. The nodules may be quite small (miliary) or up to several millimeters in diameter. The individual nodule is round or slightly oval, and its borders are clearly defined, in contrast to small alveolar (acinar) densities, which are irregular with ill-defined margins. Diffuse miliary nodules are seen in miliary tuberculosis, in the miliary form of sarcoidosis, in the early stage of diffuse metastatic disease, and in less common conditions like certain viral infections, the reticuloendothelioses, and showers of emboli.

Diffuse Interstitial Lung Disease

Diffuse interstitial lung disease can cause alterations of alveolar gas exchange by obliteration of capillaries and alveoli and by thickening of some alveolar walls. The resultant physiologic deficit has been termed *alveolar capillary block*. Alveolar capillary block is most often due to disseminated fibrotic disease of the lung. When alveolar capillary block is clinically present, the x-ray film may occasionally help in determining the cause.

The roentgenologic findings vary from normal appearing lung fields to advanced pulmonary disease, and frequently pulmonary biopsy may be needed. Generally, both lungs are diffusely involved, having a fine reticular pattern due to perialveolar thickening and fibrosis; in time, more pronounced fibrosis and small fibrotic nodules may develop.

Diseases that may lead to alveolar capillary block are (1) idiopathic disorders such as the Hamman-Rich syndrome; (2) granulomatoses such as sarcoidosis, fungus diseases, and tuberculosis; (3) histiocytosis X; (4) occupational disorders such as silo filler's disease, farmer's lung, silicosis, and asbestosis; (5) systemic diseases, especially the collagen group, such as scleroderma and rheumatoid arthritis; and (6) neoplastic diseases, including pulmonary lymphangitic metastases and alveolar cell carcinoma.

Although quite often the interstitial lung picture is nonspecific, certain basic interstitial patterns may be recognized and will help narrow the huge number of diagnostic possibilities. The *connective tissue pattern* consists of linear, nodular, lineonodular, and reticular changes with small rounded lucencies 2 to 10 mm. in diameter—the classic honeycomb lung pattern. This may appear in well-advanced sarcoidosis, collagen disease, histiocytosis X, pneumoconiosis, tuberous sclerosis, and so on. The *lymphatic pattern,* seen in lymphangitic metastases, pulmonary congestion, or interstitial pulmonary edema, consists of septal lines, linear lines, and lineonodular changes. The *bronchial pattern* of parallel lines, ring shadows, and coarse reticulation is seen in bronchiectasis, cystic fibrosis, and chronic bronchitis. The *miliary pattern,* with miliary mottling, small nodulation, and linear changes, is seen in occupational diseases, miliary tuberculosis, pulmonary hemosiderosis, and hematogenous spread of disease. Diffuse arterial or venous engorgement can usually be distinguished from linear interstitial infiltrates on a high quality film.[63-65]

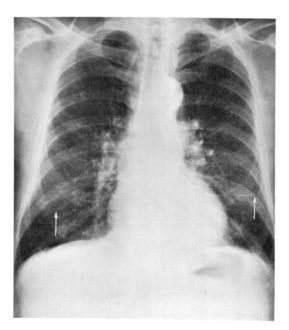

Figure 9-63 **Hamman-Rich Syndrome.** Linear fibrosis is most pronounced at the bases *(arrows),* and hilar and pleural changes are absent. In the Hamman-Rich syndrome, the initial fine reticular fibrosis usually progresses to confluent basilar shadows, and there is only moderate nodulation.

Although the alterations seen on the roentgenogram are not striking, the patient had a severe alveolar capillary block and dyspnea.

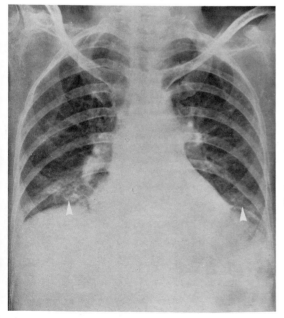

Figure 9-64 **Scleroderma.** Insterstitial fibrosis with multiple small cystic areas *(arrowheads)* is seen predominantly in the bases. These are characteristic but not diagnostic of scleroderma. The upper lung field is involved but to a smaller degree. The x-ray findings do not reflect the severity of alveolar capillary block. The diaphragms are high owing to loss of compliance, a common sequela of scleroderma.

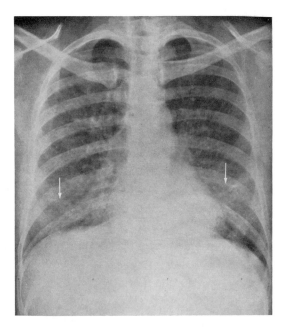

Figure 9-65 **Idiopathic Fibrosis.** There is fine inter-stitial fibrosis at both bases (*arrows*) in this patient who had a marked alveolar capillary block clinically. Following pulmonary biopsy, diagnosis of idiopathic pulmonary fibrosis was made.

Fibrotic strands may be differentiated from vascular markings by their failure to branch and their irregular thickness and density. Frequently, a fine nodularity is superimposed on the strands. The cause of pulmonary fibrosis generally cannot be determined by films alone, and pulmonary biopsy is needed.

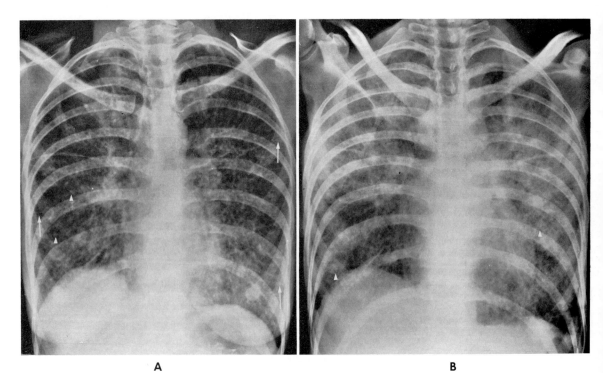

A   B

Figure 9-66 **Lymphangitic Metastases.**

A, A diffuse reticulated pattern radiates from both hila, with superimposed nodularity (*arrowheads*). The hilar shadows are enlarged, and the peripheral lymphatic vessels are prominent and engorged (*arrows*).

B, Three months later there is definite progression of the metastatic process, and the nodularity is more pronounced (*arrowheads*). The patient experienced severe dyspnea although physical findings in the chest were minimal.

Diffuse Alveolar Diseases of the Lung

Pulmonary Edema

Pulmonary edema is the commonest noninfectious cause of diffuse alveolar densities. Cardiac failure, both acute and chronic, is the cause of the vast majority of cases, but many other conditions can result in pulmonary edema.

The classic radiographic appearance consists of bilateral confluent alveolar perihilar densities, with a batwing or butterfly configuration. Often, however, the confluent densities are randomly scattered through the lung fields. In a small percentage of cases, the lesions are unilateral. Pathologic states that are associated with perfusion deficits, such as emphysema, lead to unusual distribution patterns of the pulmonary edema. Pleural effusion is usually absent or minimal. The densities are often indistinguishable from confluent areas of pneumonia, but the rapid development, the frequently changing appearance, and the sometimes sudden dramatic resolution of the pulmonary edema lesions may be differentiating features.

Cardiomegaly is present when pulmonary edema develops in the chronic cardiac or renal patient. Pulmonary edema can occur in a large group of noncardiac conditions, and in many of these the mechanism is not well understood. Alteration of the permeability of pulmonary capillaries or of alveolar membranes, or both, has been postulated. The radiographic appearance of the lungs is identical to that of pulmonary edema of cardiac origin, but cardiomegaly is usually absent. The long list of noncardiac causes of pulmonary edema or of the pulmonary edema-like picture includes acute glomerulonephritis, uremia, intracranial conditions (subarachnoid hemorrhage, epileptic seizures, head trauma), periarteritis nodosa, fluid or blood overload, drug hypersensitivity or overdose (nitrofurantoin, salicylates, heroin, and so forth), high altitude exposure, aspiration, diffuse pulmonary hemorrhage, noxious chemical inhalation (such as silo filler's disease and smoke inhalation), bacteremic shock, oxygen toxicity, adult respiratory distress syndrome, and near drowning. (See Figs. 1–6, 9–101, 9–103, 10–3, 11–12, and 11–28. See also *Basic Roentgen Signs*.)[66-71]

Pulmonary Alveolar Proteinosis

This rare condition of uncertain origin can occur in children and adults.

The progressive accumulation of a proteinaceous material within the pulmonary alveoli is responsible for the fine granular or feathery alveolar densities usually accompanied by an air bronchogram. The densities are most often bilateral and symmetric, and appear to be extending from the hila often into the lower third of the lung fields. Such an appearance can resemble pulmonary edema, but the normal heart size, the normal vasculature, and the chronicity of the alveolar densities in proteinosis should allow distinction. Hilar adenopathy does not occur.

The pulmonary picture may remain unchanged over several years, it may slowly progress, or it may regress. Regression usually begins in the periphery. There is a wide disparity between the extensive roentgen changes and the mild symptomatology. Pulmonary function studies, however, invariably show impairment of diffusion.

Diminution or disappearance of the alveolar densities can occur after intensive bronchial lavage with heparin and sodium chloride or repeated intermittent positive pressure breathing.[72-74]

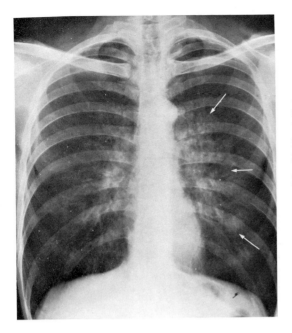

Figure 9–67 **Pulmonary Alveolar Proteinosis.** Chest film in a 33 year old man whose symptoms were weakness and cough shows bilateral perihilar feathery alveolar consolidation (*arrows*) more marked on the left. Vascular congestion and cardiomegaly are absent. The granular appearance is typical of an alveolar process. Lung biopsy confirmed the diagnosis.

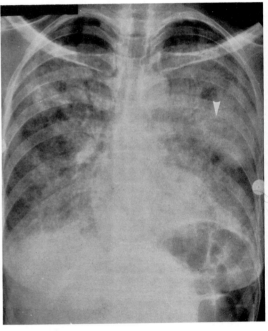

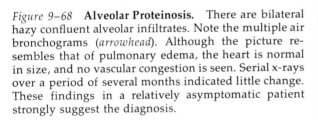

Figure 9–68 **Alveolar Proteinosis.** There are bilateral hazy confluent alveolar infiltrates. Note the multiple air bronchograms (*arrowhead*). Although the picture resembles that of pulmonary edema, the heart is normal in size, and no vascular congestion is seen. Serial x-rays over a period of several months indicated little change. These findings in a relatively asymptomatic patient strongly suggest the diagnosis.

Aspiration Pneumonia

Pulmonary changes due to aspiration of lipids, foreign bodies, or hydrocarbons are considered elsewhere. (See *Chronic Lipoid Pneumonia*, p. 519; *Intrabronchial Foreign Bodies*, p. 475; and *Smoke, Hydrocarbon, and Zinc Chloride Inhalation*, p. 526.)

Aspiration of esophageal contents, causing pulmonary infiltrates, can occur in patients with dysphagia due to any obstructive or neuromuscular esophageal disturbance. It occurs in 10 per cent of patients with achalasia. There is usually a lobular or lobar alveolar infiltrate, sometimes multiple. The midlung fields are most commonly involved. Resolution is rather slow, and the patient is rarely acutely ill, unless secondary infection supervenes. Abscess formation can occur. In chronic aspiration, variable areas of fibrosis may develop.

Aspiration of gastric contents is the most common cause of pulmonary infiltrates in certain clinical settings; e.g., in general anesthesia, tracheostomy, trauma, coma, and debilitation. The acute clinical picture of dyspnea and cyanosis may occur immediately or a few hours after the aspiration, depending on the quantity and acidity of the aspirate. This serious complication is known as

Mendelsohn's syndrome and is particularly serious if the pH of the aspirated gastric contents is less than 2.5. Within a short time after aspiration, multiple alveolar densities develop, appearing as a widespread bronchopneumonia or pulmonary edema, depending on the amount and acidity of the material. These areas of chemical alveolar pneumonia resolve very slowly. Massive aspiration pneumonia has a mortality rate of 70 per cent.

In the newborn, massive aspiration of meconium will cause focal patches or generalized areas of diminished aeration within an hour after delivery. A pleural reaction sometimes occurs. In most cases the lungs become clear within 24 hours. In a few infants, emphysema, pneumothorax, or pneumomediastinum may later develop, which is serious and sometimes fatal.[75-77]

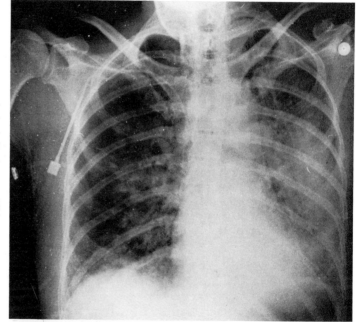

Figure 9–69 **Aspiration Pneumonia During Drug Coma.** After becoming comatose from an overdose of phenobarbital, the patient aspirated vomitus. The chest film shows patchy alveolar infiltrates in the right lung and a diffuse hazy infiltrate with air bronchograms in the left.

This aspiration pneumonia resolved slowly over a period of two and one half weeks.

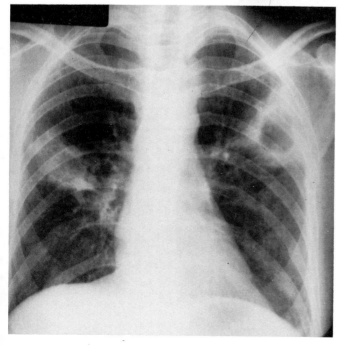

Figure 9–70 **Aspiration Pneumonitis and Abscess.** The alveolar infiltrate on the right and the cavitary abscess in the left upper lobe were due to aspiration of vomitus a few days previously.

Aspiration leads to alveolar densities, which may be focal or diffuse and bilateral, sometimes resembling pulmonary edema. Secondary infection can lead to breakdown and abscess formation.

Desquamative Interstitial Pneumonia (DIP)

In this primary interstitial pneumonia of unknown origin, there is proliferation and desquamation of alveolar lining cells and macrophages, causing alveolar densities radiographically. Later, a progressive interstitial fibrosis often develops.

The radiographic findings are variable, but most often there are bilateral, triangular ground-glass opacifications of the posteroinferior portions of the lower lobes. On posteroanterior films the right-sided involvement is usually more apparent. Less characteristically there may be small, scattered, and predominantly basal alveolar densities; accentuated basal markings may sometimes be the only finding. Progressive involvement in some cases causes volume loss, diffuse fibrosis, emphysema, and a honeycomb lung. In chronic cases, subpleural reticular densities are common and suggestive findings. Pleural reaction is uncommon.

The persistent cough and dyspnea often seem disproportionate to the radiographic findings, but pulmonary function studies disclose a significant diffusion defect. Definitive diagnosis can be made only by lung biopsy; the radiographic findings alone will simulate other chronic interstitial or alveolar-interstitial diseases.[78-81]

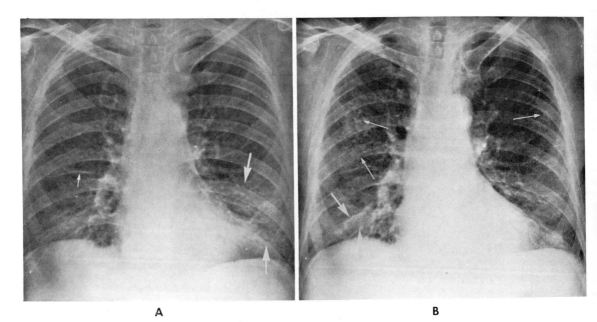

A   B

Figure 9–71 **Desquamative Pneumonia.**

A, Chest film in a 48 year old man whose symptoms included increasing shortness of breath, cough, and hoarseness reveals reticular densities in the left base (*large arrows*); the densities obscure the left lower cardiac border. Markings at the right base are prominent. The remaining lung fields appear normal. There is some thickening of the short interlobar fissure (*small arrow*). The disparity between the radiographic findings and the clinical symptoms was resolved by a lung biopsy, upon which the diagnosis of desquamative interstitial pneumonia was made.

B, Six years later there is little change in the left base. The triangular ground-glass density at the right base (*large arrows*) is characteristic. There are also fine reticular densities in both upper lung fields (*small arrows*). (Courtesy Dr. Peter Theodos, Philadelphia, Pennsylvania.)

Shock Lung (Adult Respiratory Distress Syndrome)

Shortly or immediately after septic or nonseptic shock, acute progressive and often fatal pulmonary insufficiency may supervene clinically. Some degree of circulatory collapse is often present.

At first, scattered ill-defined areas of alveolar consolidation appear. These densities resolve rapidly in patients who recover, but when clinical deterioration continues, the pulmonary lesions become more extensive and coalescent, with a radiographic appearance that is sometimes suggestive of patchy pulmonary edema. Cardiac size remains relatively unchanged. Histologically, the pulmonary lesions are combinations of alveolar collapse, alveolar congestion, edema, hemorrhage, and hyaline membrane formation. The exact cause of these changes is not entirely clear.

In this clinical setting it may be difficult to distinguish the radiologic appearance of shock lung from a diffuse bronchopneumonia, from alveolopathy due to high concentration oxygen therapy, or from fluid overload, cardiac pulmonary edema, or overwhelming aspiration pneumonia. Indeed, in some cases one or more of these factors may be partly responsible for the radiographic changes.[82, 83]

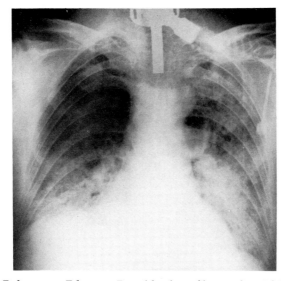

Figure 9–72 **Shock Lung: Pulmonary Edema.** Portable chest film made within six hours after the patient went into shock and hypoxia from a ruptured cecum reveals confluent perihilar alveolar densities most marked at the bases. The heart was not enlarged. The lung fields had been clear eight hours previously.

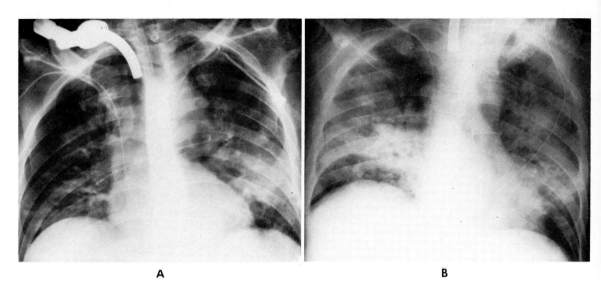

A   **B**

Figure 9–73   **Shock Lung in Acute Pancreatitis.**

   A, Chest film made on admission reveals a large area of patchy alveolar consolidation at the left base and some smaller infiltrates in the right lung. There is no cardiomegaly. The patient was in clinical shock.
   B, Less than 24 hours later, there are diffuse confluent alveolar densities in both lungs.
   The diffuse alveolar densities of the adult respiratory distress syndrome (shock lung) in the postoperative patient must be distinguished clinically from cardiac edema, pneumonia, aspiration, fluid overload, and oxygen toxicity.

Congenital Pulmonary Lymphangiectasis

   This rare disorder is characterized by markedly dilated lymphatic channels in the lung. The resultant bulky interstitial tissues mechanically interfere with alveolar expansion and cause neonatal cyanosis and respiratory distress.
   Radiographically there are bilateral, diffuse, arborizing, linear and reticular densities distributed along the vascular markings. Sometimes the appearance is more granular and nodular. Hyperinflation and areas of emphysema are constant features. The heart is not enlarged.
   The linear densities may resemble the venous engorgement of total anomalous venous return, but there is no cardiomegaly. The granular and miliary nodular form of the disease may be mistaken for hyaline membrane disease or neonatal atelectasis, both of which are far more common conditions.[84, 85]

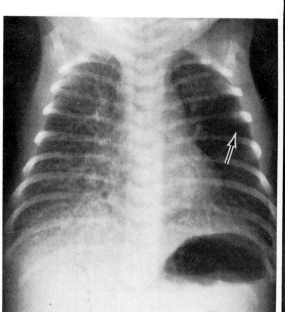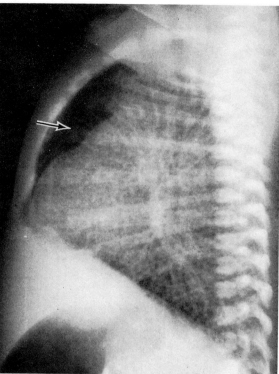

Figure 9–74 **Congenital Pulmonary Lymphangiectasis in Newborn.** Diffuse interstitial densities are radiating from both hila, but the heart is not enlarged. The lungs are hyperinflated, and the depressed, flattened diaphragms are best seen on the lateral view. A large emphysematous area is seen in the left chest (*arrows*), and scattered lucencies of interstitial emphysema are present in the right lung.

Idiopathic Pulmonary Hemosiderosis

This idiopathic disease, characterized by hemoptysis, iron deficiency anemia, cough, and dyspnea, usually follows a course of repeated exacerbations and remissions.

One third of patients have no abnormal chest findings during the initial attack. More often, bilateral small nodular alveolar infiltrates produce a fine ground-glass appearance in the lung fields, and they increase in size and number during an acute attack. Between attacks there is partial or complete clearing, since the hemosiderin is removed by macrophages.

After repeated attacks, reticular striations resembling interstitial fibrosis may develop. Hilar adenopathy has been reported but is rare. An identical lung picture is found in Goodpasture's syndrome (see p. 804), in which the pulmonary hemorrhages are associated with glomerulitis.[86, 87]

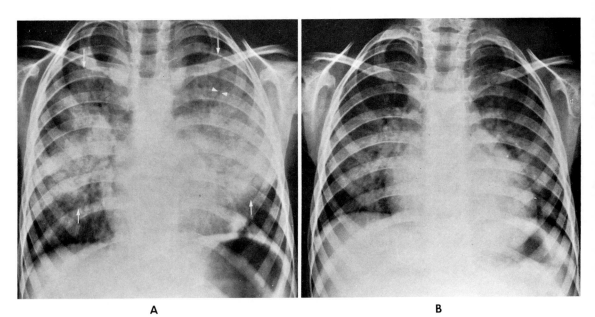

A   B

Figure 9–75   **Pulmonary Hemosiderosis: Acute Attack with Hemoptysis.**

A, Multiple fine granular nodules in the mid and lower lung fields have become confluent (*arrows*), thus accounting for the homogeneous ground-glass appearance. Note the air bronchogram (*arrowheads*), which indicates alveolar opacification.

B, Following clinical recovery there is considerable clearing of the granular densities.

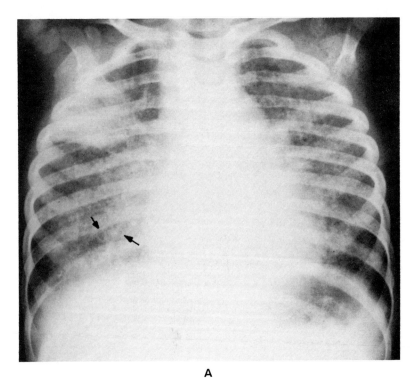

A

Figure 9–76   **Idiopathic Pulmonary Hemosiderosis: Recurrent Episodes in a 3 Year Old Girl.**

A, Shortly after the patient experienced an episode of fever and hemoptysis, a chest film reveals diffuse interstitial and alveolar acinar nodular densities, predominantly perihilar. A conglomerate density is in the right upper lobe. A few air bronchograms are apparent (*arrows*). The irregular small nodular densities are blood-filled acini. Hemosiderin in phagocytes and in the interstitial tissues is producing the interstitial densities.

Legend continued on the following page.

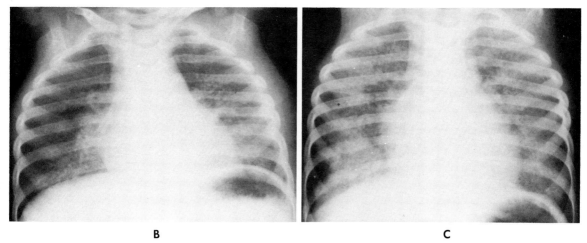

B   C

Figure 9–76 Continued

   B, Two months later, the lungs have cleared considerably.

   C, In a film taken six weeks after that shown in *B,* following another clinical flareup, diffuse alveolar densities are again seen.

Pulmonary Alveolar Microlithiasis

   In this rare familial disorder, minute calculi are deposited progressively and diffusely within the alveoli of both lungs. Although generally asymptomatic, the condition can sometimes produce respiratory symptoms in childhood, or may lead to late symptoms from complicating fibrosis, emphysema, or right heart failure.

   Radiographically there are diffuse miliary nodulations with an alveolar configuration. The lesions are always more dense than the miliary lesions of other diffuse lung diseases such as sarcoid, miliary tuberculosis, and pneumoconiosis. Neither pleural reaction nor hilar adenopathy has been reported.

   With progression of disease the nodules become more numerous and dense, especially in the lower lobes. Calculi may completely fill all the alveoli of individual acini, producing nodules up to 5 mm. diameter. The lung bases may become stony dense and completely obliterate the heart, diaphragms, and lower lung detail, even on overpenetrated films. Frequently the calcific nodules are concentrated just beneath a pleural surface, producing a dense linear lung border; the adjacent pleura may appear as a *lucent* band. Even the heart shadow may present as a relatively lucent area between the dense medial lung borders.

   The diffuse alveolar nodulation of calcific density is highly characteristic. When associated with obliterative density in the bases, the picture is virtually pathognomonic of microlithiasis.[88-90]

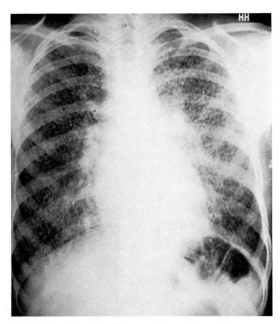

Figure 9–77 **Alveolar Microlithiasis: Early Findings.**   Chest film of an asymptomatic 17 year old boy was made because his brother had advanced microlithiasis with mild dyspnea and chronic cough.

There are diffuse and rather dense nodulations throughout both lung fields. The nodules have the hazy borders characteristic of alveolar consolidation. Involvement is not sufficient to obscure the cardiac or diaphragmatic borders. The density of the nodules is the only clue to the correct diagnosis radiographically; many other conditions have a similar roentgen appearance, but with less density to the individual nodule. (Courtesy J. P. Balikan, Beirut, Lebanon.)

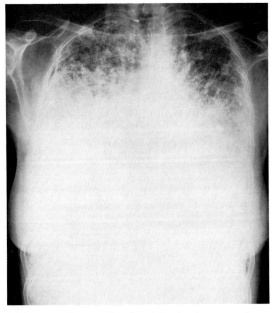

Figure 9-78 **Alveolar Microlithiasis.**   The streaky densities in the upper lung fields represent tiny discrete areas of calcification that are easily recognizable in the original film. Overlapping of these areas in the voluminous lower lobes obscures the entire lower chest. An overpenetrated film would demonstrate discrete calcifications in the lower lobes.

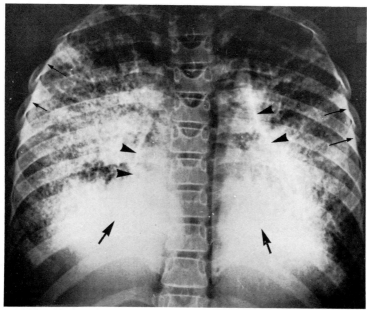

Figure 9–79 **Alveolar Microlithiasis: Advanced Disease.** Overpenetrated film of an 11 year old girl without respiratory symptoms reveals diffuse, dense, alveolar-shaped nodules. The apices are least involved, but the density at the bases (*large arrows*) is practically impenetrable due to enormous concentration of individual microliths. The cardiac and diaphragmatic borders are completely obscured by the basal densities.

The outer borders of the lungs are dense stripes (*small arrows*) in several areas, due to subpleural concentration of microliths. The medial lung borders can also be identified as dense lines (*arrowheads*), and the cardiac mass is the relatively lucent area between the dense medial lung borders. (Courtesy J. P. Balikan, Beirut, Lebanon.)

PHYSICAL AND CHEMICAL IRRITANTS

Pneumoconiosis

The clinical and radiographic picture in the various pneumoconioses depends upon the chemical composition, particle size, and concentration of the inhaled dust, and upon the length of exposure and individual susceptibility.

Of the inorganic dusts, silica (SiO_2), the causative agent of silicosis and anthracosilicosis, is the most potent producer of parenchymal lung fibrosis. Compound silicates, such as asbestos or talc, cause a lesser degree of fibrosis. Extensive fibrosis in a pneumoconiosis is eventually complicated by debilitating pulmonary emphysema. Pleural fibrosis is rare except in asbestosis.

Irritating inorganic dusts like beryllium or manganese can produce an initial chemical pneumonitis, which is often followed by late fibrosis and emphysema. Relatively inactive but radiopaque inorganic dusts like iron oxide or calcium compounds will produce no fibrosis but may cause pulmonary opacities that are due mainly to nodular deposits of the inhaled opaque dust.

Certain organic particles like sugar cane dust (bagassosis) or wheat dust (farmer's lung) may cause an acute allergic reaction with miliary infiltrates. These may clear completely, or they may lead to a coarse fibrosis. Linen dust (byssinosis) produces little or no infiltrates or fibrosis, but chronic bronchitis and emphysema may occur.

The linear and nodular fibrosis of most of the pneumoconioses is nonspecific; a history of occupational exposure to a specific dust is necessary for diagnosis.

Silicosis and Anthracosilicosis

In silicosis, a fine interstitial fibrosis that accentuates the bronchovascular markings may be the first detectable alteration. However, sometimes the early changes tend to obscure the orderly branching of the pulmonary vessels, and this is a subtle but significant finding. As fibrosis progresses, nodules 1 to 2 mm. in diameter develop throughout the lung fields. Small transverse lines of density in the region of the costophrenic angles — Kerley B lines — are fairly frequent.

The apices and bases are relatively spared and may appear hyperlucent from associated emphysema. The fibrotic nodules may increase in size but only rarely produce conglomerate shadows.

The hilar nodes may be enlarged early in the course of disease and later may develop a thin layer of calcification — the "egg shell" appearance — which is characteristic. Eventually, fibrous shrinkage may decrease hilar size.

In anthracosilicosis, in addition to nodular fibrosis, conglomerate fibrotic densities, often massive, may develop. These are usually bilateral, more or less symmetric, and surrounded by a layer of aerated lung tissue. They become closer to the hilum as peripheral emphysema increases. They usually develop in the posterior portion of the lungs. Rarely cavitation occurs. The massive areas of fibrosis generally have flattened lateral borders that tend to parallel the rib cage. If unilateral, the silicotic mass may simulate a carcinoma. Severe generalized emphysema and localized blebs are seen frequently.

Tuberculosis and pyogenic infections often complicate silicosis and modify the roentgenologic picture. If massive fibrosis is observed in the presence of nodular silicosis, if cavitation occurs, if the disease is markedly asymmetric, or if extensive pleural disease is found, superimposed tuberculous or pyogenic infection should be suspected.[91, 92]

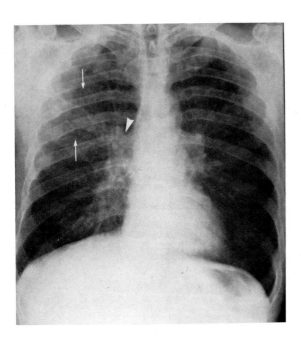

Figure 9–80 **Early Silicosis.** Interstitial nodulation (*arrows*) is seen; the apices and bases are spared. The hyperlucency of both bases is caused by emphysema, which is commonly associated with silicosis. The prominent hilar shadows (*arrowhead*) are due to adenopathy.

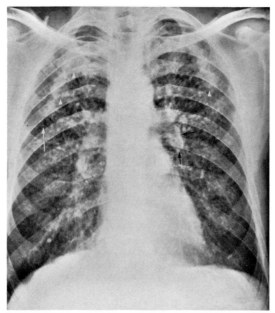

Figure 9–81 **Moderately Advanced Silicosis.**   Larger nodules (*white arrow*) are scattered throughout both lung fields. Note the basal emphysema and the "egg shell" calcification in the left hilum (*black arrows*). Symmetric groups of conglomerate nodules are present in both upper lung fields *(arrowheads)*.

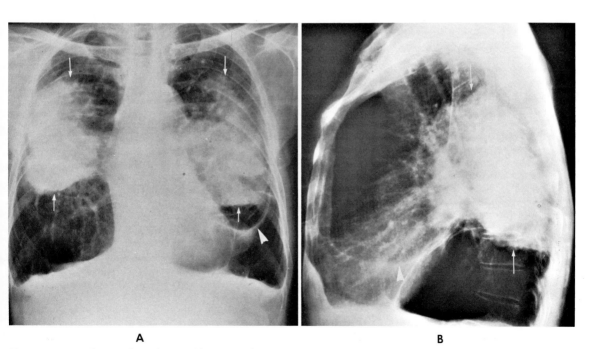

A   **B**

Figure 9–82 **Advanced Anthracosilicosis with Conglomerate Densities.**

A, Posteroanterior view reveals massive bilateral conglomerate fibrotic shadows (*arrows*). These are sharply demarcated, fairly symmetric, and surrounded by aerated lung tissue. Emphysema is pronounced, and blebs are present at the bases (*arrowhead*).

B, Lateral view indicates posterior location of the conglomerate densities (*arrows*). The wall of the large anterior emphysematous bleb overlies the left conglomerate density (*arrowhead*). Severe generalized emphysema as well as localized blebs is common in anthracosilicosis.

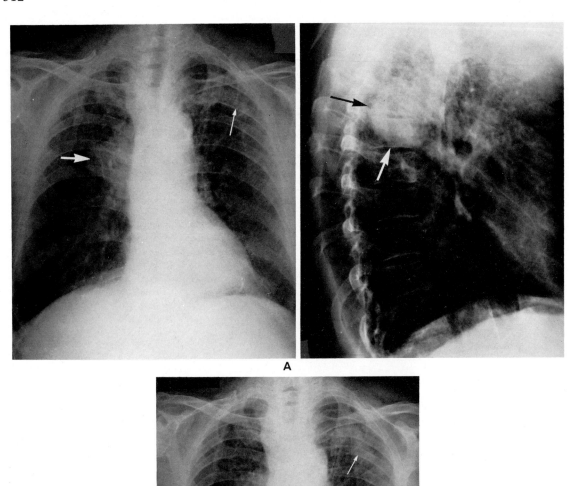

A

B

Figure 9–83   **Anthracosilicotic Density Simulating Lung Tumor.**

A, Chest films of a 63 year old man who complained of dyspnea disclose a somewhat circumscribed density (*large arrows*) in the right hilar region, but located considerably more posteriorly. The fibrotic appearing densities in the left upper lobe (*small arrow*) were considered to be old fibrotic granulomatous disease. The remaining lung fields were emphysematous. No significant fibrosis is evident.

The tentative diagnosis of pulmonary neoplasm was revised only after obtaining a history of 10 years' work as a coal miner 35 years previously.

B, Seven years later, the right hilar mass appears smaller. However this was due not to a real decrease in size but to an increasing peripheral pulmonary emphysema that crowded the mass more medially behind the heart. Its actual size on lateral view was unchanged.

The left upper lobe densities have now assumed the more usual appearance of an anthracosilicotic conglomerate mass (*small arrow*). Note that it, too, has "migrated" a bit more medially.

The above films illustrate the tumor-like appearance of a solitary anthracosilicotic mass, especially if the remaining lung fields show no fibrosis. The case also illustrates the apparent medial migration of anthracosilicotic masses as peripheral emphysema increases.

Asbestosis

The most frequent radiographic finding is pleural thickening, either alone or associated with diffuse basal parenchymal fibrosis. Parenchymal lesions without pleural changes are relatively uncommon.

The pulmonary fibrosis is usually most marked in the lower lung fields, and occasionally it is unilateral. The fibrotic densities tend to obscure the lower cardiac borders, producing the "shaggy heart" appearance in about 20 per cent of patients. As fibrosis progresses, interstitial emphysema develops, and the fibrotic densities may assume a more nodular appearance.

Hilar enlargement occurs in over half the cases and is usually vascular.

The pleural thickening is characteristically in the form of linear plaques, most often along the thoracic wall with sparing of the apices and costophrenic angles. The small plaques are often the earliest findings, but these may readily be overlooked or mistaken for companion rib shadows. Calcification of the plaques is a very late finding, rarely seen less than 20 years after initial exposure. It is a significant diagnostic finding but occurs in only about 20 per cent of patients.

The individual roentgen findings are nonspecific, and a history of occupational exposure is necessary for diagnosis. However, pleural thickening and calcification, basal fibrosis with a shaggy heart appearance, and hilar enlargement are highly suggestive findings.

There is an unusually high incidence of malignant mesotheliomas in patients with asbestosis. About 10 per cent of all malignant mesotheliomas occur with asbestosis. Bronchogenic carcinoma also is about three times more common in patients with asbestosis than in the rest of the population.[93-95]

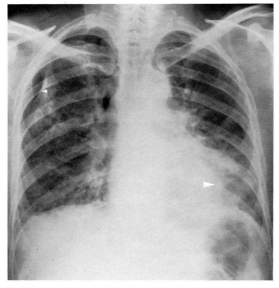

Figure 9–84 **Asbestosis with Chronic Pneumothorax.** There is diffuse nodular interstitial fibrosis and interstitial emphysema. Note the "shaggy heart" (*large arrowhead*). A chronic pneumothorax (*small arrowhead*) in the right upper lobe has developed as a result of rupture of an emphysematous bleb.

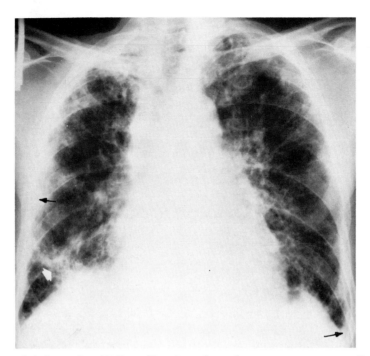

Figure 9–85 **Advanced Asbestosis.** Diffuse fibrosis and emphysema are apparent. Conglomerate areas of fibrosis are seen at the bases (*white arrow*). The borders of the enlarged heart are blurred because of adjacent fibrosis ("shaggy heart"). Pleural thickening is minimal (*black arrows*), and there is no pleural calcification. Twenty years previously, this man had worked for three years in asbestos dust.

In the absence of extensive pleural thickening or calcification, this picture of diffuse fibrosis is nonspecific. Diagnosis requires a history of exposure.

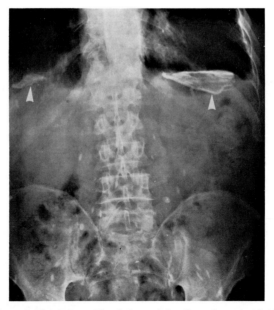

Figure 9–86 **Asbestosis: Pleural Changes.** The bilateral basilar pleural calcifications (*arrowheads*) in a 60 year old asbestos worker suggest asbestosis.

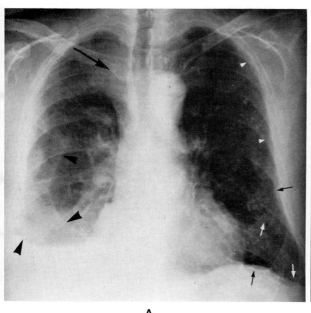 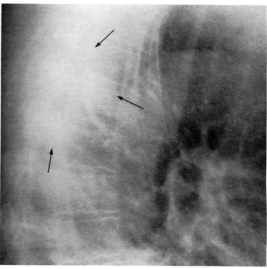

A   **B**

Figure 9–87 **Asbestosis with Malignant Pleural Mesothelioma.**

A, In the left chest there are fibrotic densities at the base (*small white arrows*), pleural thickening along the chest wall (*white arrowheads*), and calcified plaques on the pleural surfaces (*small black arrows*). These changes are fairly characteristic of asbestosis.

In the right chest there is extensive pleural reaction at the base (*black arrowheads*) and a superior mediastinal density *(large black arrow).* The latter is a tumor mass on the posterior pleural surface.

B, On lateral film, the pleural mass (*black arrows*) is vaguely outlined behind the aorta.

In persons exposed to asbestos, there is a disproportionate incidence of mesothelioma, even without clinical asbestosis.

Talcosis

The fine basilar nodular fibrosis that is seen is similar to the lesions of asbestosis, but significant pleural changes do not occur. Diagnosis hinges upon a history of exposure.[96]

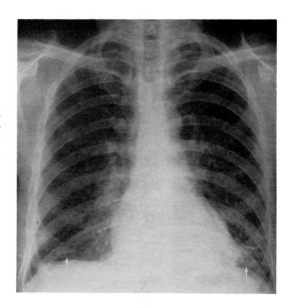

Figure 9–88 **Talcosis.** The fine nodular bilateral basal fibrosis (*arrows*) seen in this talc worker are somewhat similar to the appearance of asbestosis. Notice the shaggy heart; there were no pleural changes. (Courtesy Dr. William Weiss, Philadelphia, Pennsylvania.)

Berylliosis

Acute beryllium inhalation can produce a chemical pneumonitis with alveolar edema and hemorrhage, characterized by haziness or ground-glass alveolar densities of pulmonary edema, and followed by areas of "soft" infiltration. Enlarged hilar nodes are frequent. Spontaneous clearing usually occurs, but chronic lung changes may appear after a latent period of months or years.

The chronic changes are usually due to prolonged exposure, and they consist of diffuse granular and nodular densities; either type may occur alone, but more often both kinds of densities are found. Individual nodules are usually small but can reach a size of 1 cm. Interstitial emphysema may develop. The hilar nodes are usually enlarged. A chronic nodular lung picture with enlarged hilar shadows is the most characteristic appearance of chronic berylliosis, but is readily mistaken for nodular sarcoid.

In advanced cases linear fibrotic densities appear, and there may be segmental areas of contraction, most often in the upper lobes.

However, the roentgenologic findings alone are nonspecific, and a history of exposure to beryllium is necessary for diagnosis.[97-99]

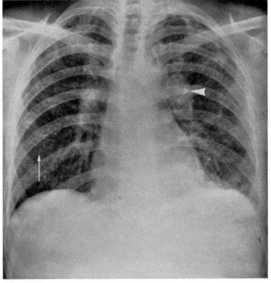

Figure 9–89 **Berylliosis.** In a 20 year old male who was exposed to beryllium fumes, there is extensive nodularity (*arrow*) of both lung fields due to disseminated beryllium granulomas. The hilar shadows are prominent (*arrowhead*) due to enlarged nodes. These findings are fairly characteristic of chronic berylliosis. (Courtesy Dr. Arlyne Shockman, Veterans Administration Hospital, Philadelphia, Pennsylvania.)

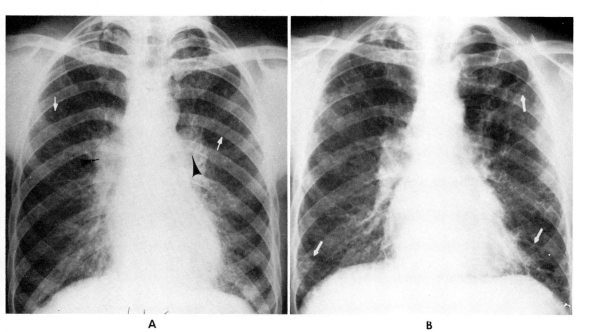

A   B

Figure 9-90 **Progressive Pulmonary Berylliosis.**

A, The lung fields of this 37 year old male working in a beryllium atmosphere for three years show a nodular pattern (*arrows*) with increased reticular markings. The hilar shadows (*arrowheads*) are enlarged.

B, Seven years later, the lungs show a marked coarse fibrosis (*arrows*), with many lucent areas of interstitial emphysema. Severe respiratory symptoms were present.

The radiographic appearance is nonspecific and can occur in progressive fibrosis resulting from many causes.

Siderosis

Prolonged inhalation of iron dust most often occurs in arc-welders. It causes sharply defined nodular densities that are uniformly distributed throughout both lung fields. These are due to localized accumulations of inhaled iron particles within the lung lymphatics rather than to fibrosis. The iron deposits are asymptomatic, since neither pulmonary fibrosis nor interstitial emphysema result from iron deposition. No hilar enlargement is seen. There may be slow regression of the lesions after cessation of exposure.[97]

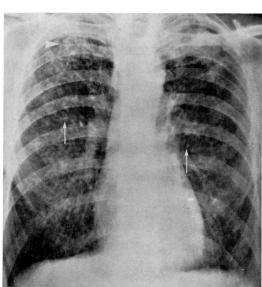

Figure 9-91 **Siderosis.** There are multiple, uniformly distributed, sharply demarcated nodules throughout both lung fields (*arrows*). The hila are normal in size, and there is no fibrosis or emphysema. The tuberculous lesions in both apices (*arrowhead*) are unrelated.

Farmer's Lung and Other Organic Dust Inhalation Diseases (Allergic Alveolitis)

A hypersensitive pulmonary reaction may occur after inhalation of various organic dusts. In this category are farmer's lung (hay and grain dust), bagassosis (dried sugar cane dust), maple bark-stripper's disease, pituitary snuff-taker's disease, and pigeon breeder's disease. The basic lesion is an allergic alveolitis.

The roentgen picture in all of these conditions is quite similar, although non-specific. If the exposure is massive and intermittent, there will be diffuse, ill-defined nodular shadows of alveolar consolidation, many of which become larger and more confluent. Occasionally, the picture resembles pulmonary edema. A peripheral subpleural location of many of these confluent densities is fairly common.

In many cases the condition clears gradually and completely, although there may be residual emphysema and fibrosis.

With more continuous low dosage exposure, diffuse or focal fibrosis occurs, with coarsened bronchovascular markings, fine nodulations, and a reticular pattern. Sometimes a fibrotic shrinkage of one or more lobes occurs. In advanced chronic cases, pulmonary hypertension and cor pulmonale can appear.[100-103]

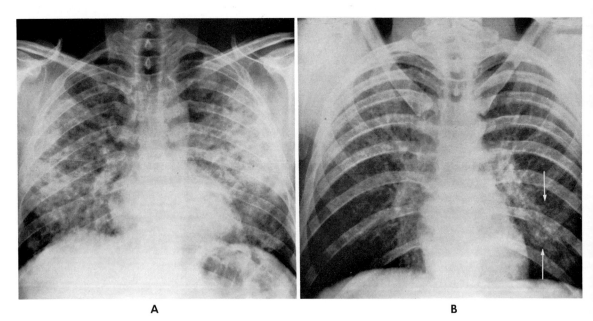

A   B

Figure 9–92   **Acute Farmer's Lung (Inhalation of Hay and Grain Dust).**

A, There are multiple confluent and nodular densities throughout both lung fields with sparing of the apices.

B, Months after the acute process had subsided, there is a characteristic picture of multiple fine nodules (*arrows*) superimposed on a reticular background. Note basilar emphysema. These changes are typical of the chronic form. (Courtesy Dr. Ralph C. Frank, Sacred Heart Hospital, Eau Claire, Wisconsin.)

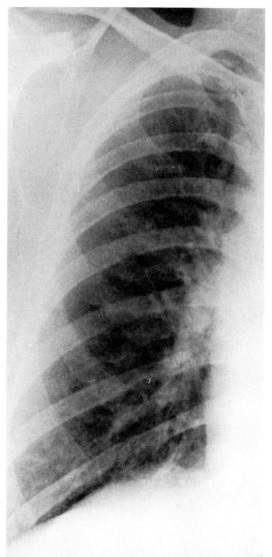

Figure 9–93 **Pigeon Breeder's Disease: Subacute Stage.** There is a diffuse, fine reticular pattern throughout the lung, with some accentuation of the bronchovascular markings. The picture is entirely non-specific, and diagnosis depends upon the history of exposure. (Courtesy J. D. Unger, Milwaukee, Wisconsin.)

Chronic Lipoid Pneumonia

Chronic aspiration of lipoid medicaments most often occurs in infants and debilitated elderly persons.

The early chest findings are small scattered alveolar densities, usually in the perihilar and lower lobe areas. The chronic pulmonary changes from continued aspiration depend upon the chemical nature of the lipid.

One form of the chronic lipoid lung is characterized by fine reticular linear densities with a spun-glass appearance. These are most marked at the lung bases and are accompanied by Kerley B lines. The latter are due to lipoid distention of the lymphatic vessels in the interlobular septa. Another type of lung reaction is the formation of one or more granulomatous-lipoid masses, which vary from a small nodule to a large sharply demarcated mass, and which are usually located in a lower lobe. The density may readily be mistaken for a pulmonary neoplasm. The lipoid granuloma and the reticulolymphatic forms may coexist.

Lipoid pneumonia should be considered in an individual with a history of oil ingestion if there is either a perihilar mass or bilateral lower lobe infiltrates with Kerley B lines.[104, 105]

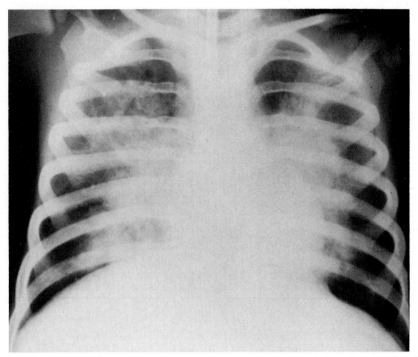

Figure 9–94 **Lipoid Pneumonia.** There are extensive diffuse fluffy perihilar alveolar densities. The lung periperhy is clear.
This child had been aspirating mineral oil for a few weeks.

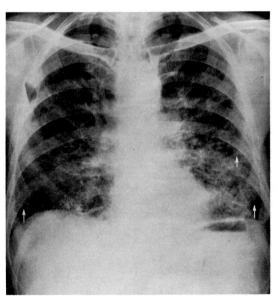

Figure 9–95 **Lipoid Pneumonia: Chronic Interstitial.** Chest film in a 53 year old male who ingested mineral oil over a long period of time demonstrates extensive bilateral reticular densities most marked at the bases. The short transverse lines of increased density in the lower lung periphery *(arrows)* (Kerley B lines) are the interlobular septa that have been thickened by distention of lymphatics with oil.

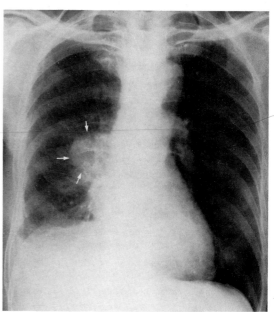

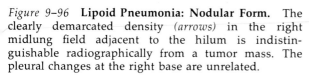

Figure 9–96 **Lipoid Pneumonia: Nodular Form.** The clearly demarcated density *(arrows)* in the right midlung field adjacent to the hilum is indistinguishable radiographically from a tumor mass. The pleural changes at the right base are unrelated.

Chemical Pneumonitis

The inhalation of any irritant material in sufficient quantities can produce chemical bronchiolitis and pneumonitis. If the agent is absorbed from the lungs, there may also be evidence of systemic disease.

Among the host of noxious vapors that may be accidentally inhaled, some are more commonly encountered, and their pulmonary effects have been described. These include nitrous oxide (silo filler's disease), mercury vapor, ordinary smoke, the hydrocarbons (especially kerosene), and zinc oxide.

The pathologic and radiographic changes vary with the irritative quality of the agent and the intensity and duration of exposure. Although some agents may produce a distinct radiographic picture, generally the findings of chemical pneumonitis are so similar that diagnosis cannot be made without obtaining the specific history.

Intense acute exposure usually causes perihilar infiltrates and pulmonary edema. Occasionally, complicating emphysema and pneumothorax may occur. With less exposure or less irritating chemicals, patchy pneumonitis may develop, often with interstitial emphysema. Chronic exposure causes nodular fibrosis and interstitial emphysema.

If a fatal outcome is averted following acute exposure, clearing of the pulmonary densities usually parallels clinical recovery. With more protracted exposure or severe respiratory damage, clearing may be slow; occasionally interstitial fibrosis and emphysema are permanent. Sometimes bacterial pneumonia may complicate the radiographic picture.

Nitrogen Oxide Inhalation: Silo Filler's Disease

Gaseous oxides of nitrogen may arise from decomposition of fresh silage (silo filler's disease) or may be present in smoke from combustion of organic nitrogenous matter. Inhalation of these gases produces an unusual sequence of events: an acute respiratory episode, recovery, and commonly a clinical relapse in three to six weeks. The mechanism of the late relapse is unknown.

Acute exposure will cause a chemical pneumonitis manifested radiographically as confluent perihilar pulmonary edema or, in milder cases, as patches of bronchopneumonic alveolar infiltrates. These develop 3 to 30 hours after exposure. Radiographic clearing parallels clinical recovery. However, within three to six weeks, another distressing respiratory episode may occur, associated with recurrence of patchy infiltrates or, more commonly, with miliary nodulation. The latter may simulate miliary tuberculosis. The densities are due to an obliterating bronchiolitis and may persist for prolonged periods. In some cases there may be progressive fibrosis and emphysema, with permanent respiratory impairment. Delayed involvement can occur even if the acute episode was quite mild and transient. Prolonged steroid therapy will favorably affect the clinical course and the radiographic picture. Definitive diagnosis requires a history of exposure to the gas.[106, 107]

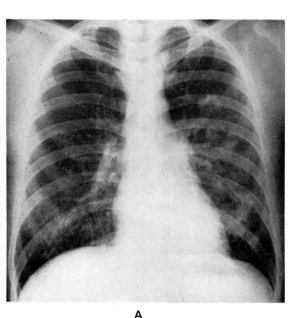

A

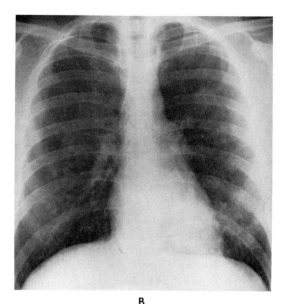

B

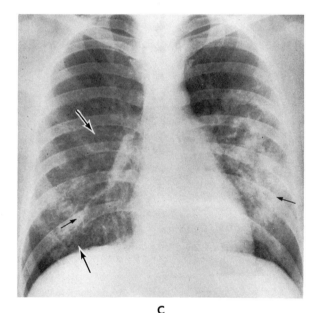

C

Figure 9–97 **Nitrogen Oxide Inhalation: Remission and Late Relapse.**

A, Chest film made after exposure discloses patches of alveolar infiltrates mainly in the left lung. There were minimal respiratory symptoms.

B, Spontaneous clinical improvement within two days was paralleled by disappearance of the pulmonary infiltrates.

C, Six weeks later there was a sudden onset of fever, cyanosis, and cough. Extensive confluent densities appeared in the left lung and right base *(small black arrows).* Many smaller nodules were also present *(large arrows),* although these are not well demonstrated on the picture. The pulmonary lesions persisted for five weeks, but clinical recovery was complete.

After prolonged exposure, chest findings develop within hours and may appear as a severe pulmonary infiltrate simulating edema, or as a diffuse patchy pneumonitis. The delayed form often appears after apparent recovery from an acute episode, a variable latent period intervening. (Courtesy Drs. R. L. Tse and A. A. Bockman, Philadelphia, Pennsylvania.)

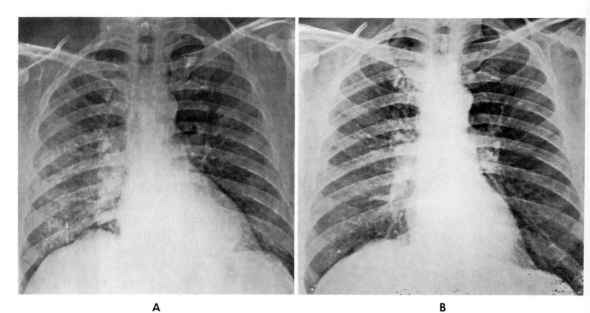

A   B

Figure 9–98 **Nitrogen Oxide Inhalation: Delayed Findings.**

A, Four weeks after moderate exposure to fresh silage there is diffuse nodulation throughout both lung fields due to diffuse chemical obliterating bronchiolitis.

B, One month later nodulation has decreased considerably.

The lesions may clear completely, but sometimes diffuse fibrosis and interstitial emphysema remain permanently.

Mercury Inhalation

Inhalation of mercury vapor can cause extensive pulmonary changes that overshadow the systemic effects resulting from absorption of mercury from the lungs. Generalized necrotizing bronchiolitis with bronchiolar obstruction develops, producing diffuse emphysema intermingled with areas of increased density. Often there is an ill-defined haze over the chest, similar to pulmonary edema. The diaphragms are depressed by the emphysema, and spontaneous pneumothorax frequently occurs. Late complications include fibrosis and secondary bacterial pneumonia. There is a high percentage of fatalities, and even with clinical recovery, interstitial emphysema may persist.

Ingestion of soluble mercury compounds can lead to severe ulcerative and necrotizing inflammation of the entire gastrointestinal tract, with secondary involvement of the central nervous system and kidneys. The roentgenographic changes in the gastrointestinal tract are unknown, because acute episodes generally preclude radiographic examination.[108]

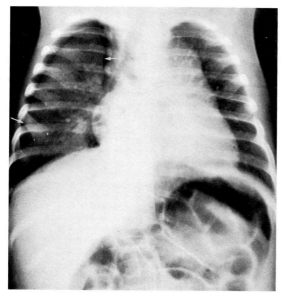

Figure 9–99 **Fatal Acute Mercury Vapor Poisoning.** A 4 month old infant developed diffuse bilateral interstitial and alveolar emphysema one day following exposure. Pneumothorax has developed on the right; pleural air surrounds the entire right lung *(arrows)*. The diaphragms are depressed. The infant died four days later. (Courtesy Dr. J. C. Brennan, Houston, Texas.)

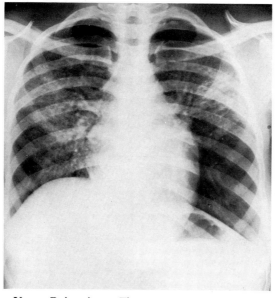

Figure 9–100 **Acute Mercury Vapor Poisoning.** The young mother of the infant whose case is illustrated in Figure 9–99 developed inflammatory edema, which appears as bilateral areas of density extending from the hila. There is apical and basal emphysema, and the left diaphragm is considerably depressed.

Recovery was slow; the patient remained emphysematous months later. (Courtesy Dr. J. C. Brennan, Houston, Texas.)

Smoke, Hydrocarbon, and Zinc Chloride Inhalation

Smoke inhalation is probably the most frequent cause of chemical pneumonitis. The more irritating smoke fumes contain sizable amounts of the oxides of nitrogen and cause radiologic changes similar to those described in *Nitrogen Oxide Inhalation* (p. 522). Acute pulmonary edema and areas of pneumonitis are the usual roentgenologic findings; however, these rarely appear earlier than 24 to 36 hours after exposure and may progress for a few days, a peculiarity that is important for proper diagnosis and management.

Hydrocarbon (kerosene, lighter fluid, and so on) inhalation frequently occurs during accidental ingestion of the liquid. Some investigators believe that the chemical reaches the lungs via the bloodstream after being absorbed from the upper gastrointestinal tract. Basal pulmonary infiltrates appear within hours after ingestion; perihilar infiltrates and pulmonary edema are rare, perhaps because hydrocarbons are a relatively mild irritant. Rarely, pneumatoceles develop. Upper lobe lesions are almost never seen.

Moderate inhalation of zinc chloride smoke can produce extensive pulmonary infiltrates and a pulmonary edema–like picture. The chest film is often negative for the first 24 hours, especially in milder cases. In more severe cases, pulmonary lesions develop earlier, often with acute emphysema and subpleural hemorrhage. Many victims die after severe exposure, but recovery, if it follows, generally is complete without pulmonary sequelae.[109-111]

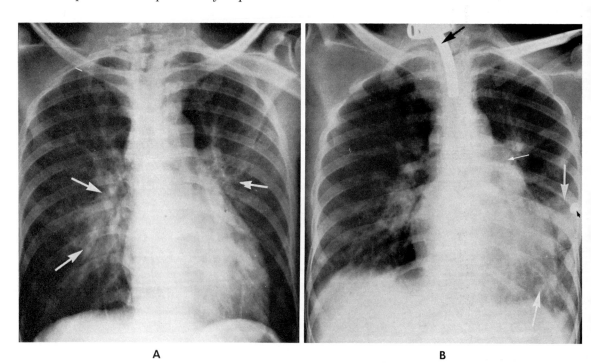

A B

Figure 9–101 **Smoke Inhalation: Chemical Pneumonitis.**

A, Forty-eight hours after the patient was rescued from a burning house there are extensive perihilar alveolar densities *(arrows),* most marked on the right.

B, Two days later the perihilar densities have decreased, but chemical pneumonitis has developed at the left base *(large arrows).* The marked increase in size in the main pulmonary artery segment *(small arrow)* reflects cor pulmonale probably due to the extensive bronchiolitis that has caused obliteration of the capillary bed. The round density on the left *(small black arrow)* is a metallic marker. A tracheostomy tube *(large black arrow)* was inserted because of progressive respiratory difficulty.

As a rule, the findings are negative for the first 24 hours following smoke inhalation.

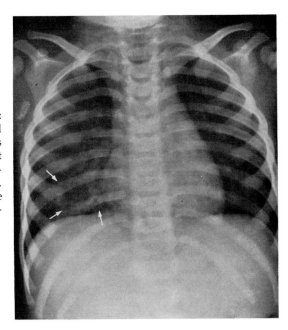

Figure 9–102 **Kerosene Inhalation and Ingestion: Chemical Pneumonitis.** A child had ingested and aspirated kerosene several hours before this film was made. There are soft nodular infiltrates at the right base *(arrows)*. These basal infiltrates are the most common lesions encountered in hydrocarbon aspiration. Perihilar infiltrates are seen less often, and upper lobe lesions are seen least often. Recovery is usually complete within a week.

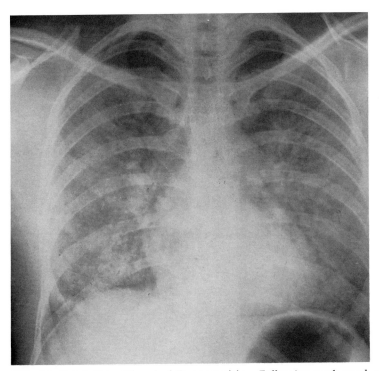

Figure 9–103 **Zinc Chloride Inhalation: Chemical Pneumonitis.** Following prolonged exposure to smoke screen during army training maneuvers, the patient developed in both lungs extensive diffuse infiltrates that resembled pulmonary edema. The main pulmonary artery is unusually prominent. Despite this picture, the physical findings in the chest were minimal, and the patient recovered in about four weeks. (Courtesy Drs. R. B. Stonehill and F. A. Johnson, San Antonio, Texas.)

Acute Carbon Monoxide Poisoning

The most common initial pulmonary finding in acute carbon monoxide poisoning is a ground-glass appearance of interstitial edema, occurring predominantly in the peripheral lung fields.

This ground-glass appearance usually will clear but in severe and more serious intoxication may progress to perihilar haze and even to frank alveolar edema. These findings are usually associated with some cardiac enlargement and a poorer prognosis.

The pulmonary and cardiac changes are probably due to tissue hypoxia.[112]

Pulmonary Oxygen Toxicity

Mechanical ventilation with oxygen in high concentration for prolonged periods can lead to progressive deterioration of pulmonary function. Histologically, the alveoli become filled with exudate, hemorrhage, edema, and finally hyaline membrane formation. Radiographically, the early lesions begin as patchy areas of alveolar consolidation, and within a few days can progress to diffuse bilateral alveolar infiltrates. The latter resemble pulmonary edema but do not maintain a central perihilar distribution.

Although oxygen toxicity should be suspected when these infiltrates develop in a patient receiving a high concentration of oxygen for two or three days, it may be difficult to rule out aspiration pneumonitis, pulmonary edema, diffuse pneumonia, shock lung, or recurrent pulmonary emboli.[113, 114]

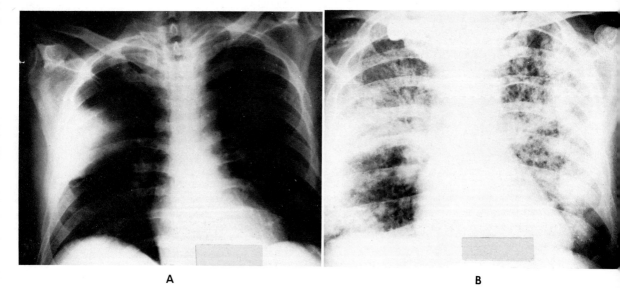

A   B

Figure 9–104 **Pulmonary Oxygen Alveolopathy.**

A, Admission diagnosis of pneumonia in a 47 year old man was confirmed by chest film. Pneumococci were cultured.

B, After three and one half days of 100 per cent oxygen therapy, there are extensive diffuse bilateral alveolar densities, with progressive deterioration of respiratory function. The patient died.

Autopsy disclosed that the lung changes were not due to widespread infection but consisted of hyperplasia of alveolar cells and hyaline membrane formation, characteristic of oxygen toxicity. (Courtesy Drs. N. Joffe and M. Simon, Boston, Massachusetts.)

Radiation Injury to the Lung

The occurrence and severity of changes due to radiation injury of the lung will vary with total dosage, time-dose relationship, and individual susceptibility. A dose greater than 6000 rads will virtually always cause severe and crippling changes. Much smaller doses can produce changes in a susceptible individual.

Active injury becomes apparent within six months after therapy but may occur within one month after intensive treatment. Cough and dyspnea are accompanied by a diffuse haziness in the irradiated area, which is due to edema and exudation. Pleural reaction is unusual. If a lung tumor has been treated, the radiation pneumonitis may be mistaken for extension of the neoplastic process.

These exudative changes will diminish and may disappear completely within a few months, but more often they gradually change into a permanent fibrosis. This may present as coarse strands of fibrotic density or as a fibrotic contracture of a segment or lobe, often with persistent respiratory symptoms. The late changes may simulate lymphangitic tumor extension. (See Fig. 1–11.)[115, 116]

NEOPLASMS OF THE LUNG

Benign Tumors of the Lung

Benign tumors are considerably less frequent than malignant lung tumors. Bronchial adenoma and chondrohamartoma are the most usual benign lesions, although adenomas are actually tumors of low grade malignancy. Other infrequent benign tumors include papilloma, lipoma, neurofibroma, leiomyoma, myoblastoma, and intrabronchial mucocele (bronchocele).[117, 118]

Bronchial Adenoma

Bronchial adenomas constitute 5 to 10 per cent of primary pulmonary neoplasms. About 90 per cent are carcinoids histologically. The other 10 per cent are cylindromas. The adenomas are highly vascular lesions of low grade malignancy, and about 40 per cent of them metastasize to the regional hilar nodes. Very rarely, lytic or blastic bone metastases occur. The cylindroma is the more malignant lesion. Hemoptysis and recurrent pneumonia are the most frequent symptoms; rarely, a carcinoid syndrome develops.

Since the adenoma arises intrabronchially, there may be no radiographic changes until bronchial obstruction leads to atelectasis or pneumonia. Planigraphy may reveal a characteristic "cap" defect obstructing a bronchus. About one third of adenomas present as a mass close to the hilum, with or without parenchymal infiltrates. Bronchography will disclose the intrabronchial defect or bronchial obstruction.

Bronchial adenoma usually cannot be distinguished radiographically from an intrabronchial bronchogenic carcinoma. However, the adenomas occur more frequently in females and in a younger age group than does carcinoma.[119-121]

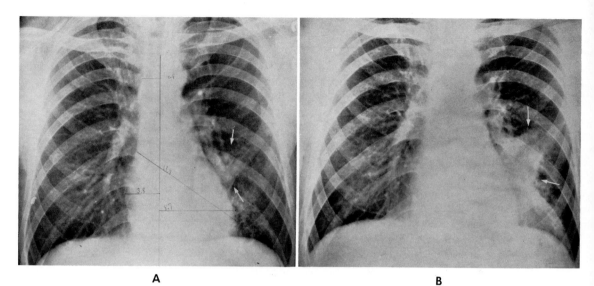

A   B

Figure 9–105 **Bronchial Adenoma: Progressive Lung Changes.**

   A, There is an area of pneumonitis below the left hilum *(arrows)* in 35 year old male who had hemoptysis.

   B, Five months later, pneumonia has progressed and segmental atelectasis extends to the periphery *(arrows).* A small bronchial adenoma was found during surgery.

   Generally, the adenoma itself is not apparent in chest films unless it becomes very large and extends beyond the bronchus.

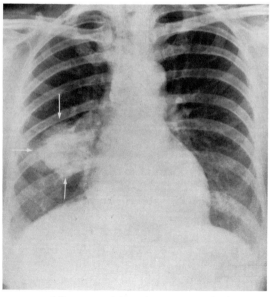

Figure 9–106 **Bronchial Adenoma and Pneumonitis.**   A circumscribed perihilar mass *(arrows)* in a young male who had hemoptysis was found at surgery to be a bronchial adenoma.

Chondrohamartoma (Hamartoma, Chondromatous Adenoma)

The intrapulmonary hamartoma is asymptomatic and grows slowly. It appears as a well-circumscribed, somewhat lobulated mass surrounded by normal lung. Most of these lesions are 1 to 4 cm. in diameter and are located in the lung periphery, close to the parietal or interlobar pleura. Calcification occurs in less than one third of these tumors, and when present it is irregular and scattered throughout the lesion, giving the characteristic "popcorn" appearance. Sometimes there are minute radiolucencies in the periphery of the mass. These are islands of fat imbedded in the cartilage. They may be demonstrable by planigraphy. Very rarely, a hamartoma arises intrabronchially, clinically and radiographically simulating a bronchial adenoma.

Radiographic distinction from a noncalcified granuloma, a malignant tumor, or a calcified granuloma usually cannot be made. Tomographic demonstration of peripheral lucencies would favor hamartoma. The presence of popcorn calcification is highly suggestive of a chondrohamartoma.[122-126]

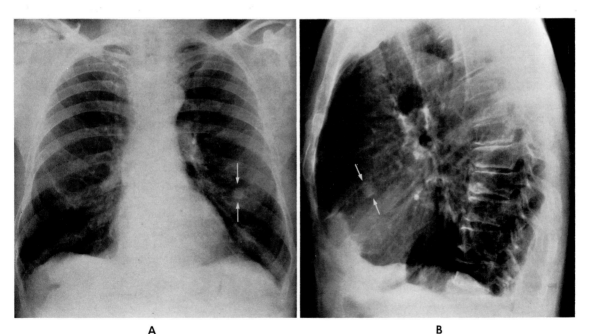

A   **B**

Figure 9–107 **Chondrohamartoma.**

A, Posteroanterior view in a 66 year old asymptomatic male. A small, well-demarcated, smooth-edged nodule is demonstrated in the left lower lung field *(arrows).* No calcification was noted during routine studies or planigraphy.

B, Lateral view indicates location of the nodule in the lingula of the left upper lobe *(arrows).* At surgery this proved to be a hamartoma. In the absence of calcification, the lesion is indistinguishable from noncalcified granuloma or malignant nodule. (Courtesy Dr. Arlyne Shockman, Veterans Administration Hospital, Philadelphia, Pennsylvania.)

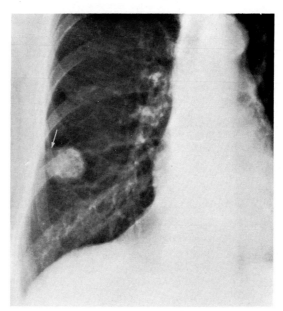

Figure 9-108 **Intrapulmonary Chondrohamartoma with Calcification.** The circumscribed density in the right lower lung field has a slightly lobulated lateral border *(arrow)*. Crescentic mottled calcification is seen in the medial two thirds of the lesion.

Calcification has been emphasized as a diagnostic criterion in hamartoma, but it can be demonstrated in only 10 to 15 per cent of cases. Radiographically, a calcified hamartoma cannot be distinguished from a tuberculoma, and an uncalcified hamartoma is indistinguishable from any coin lesion, including a primary or metastatic malignant nodule or a granuloma. Malignant nodules generally grow much more rapidly than a hamartoma.

The majority of hamartomas are intrapulmonary and produce few or no symptoms. A small percentage are intrabronchial, producing radiographic and clinical evidence of segmental atelectasis.

Bronchial Papilloma

A solitary intrabronchial papilloma, like an adenoma, produces no radiographic changes until bronchial obstruction leads to focal pneumonitis or atelectasis. Bronchography may reveal a nodular intrabronchial defect that is indistinguishable from adenoma or carcinoma.

Diffuse papillomatosis of the bronchial tree is rare and is always associated with papillomatosis of the trachea. These lesions may cause widespread parenchymal inflammatory disease—recurrent pneumonias, bronchiectasis, and fibrosis. The roentgen picture is usually more suggestive of chronic lung inflammatory disease than of neoplastic lesions. Bronchography will disclose the multiple endobronchial defects.[127]

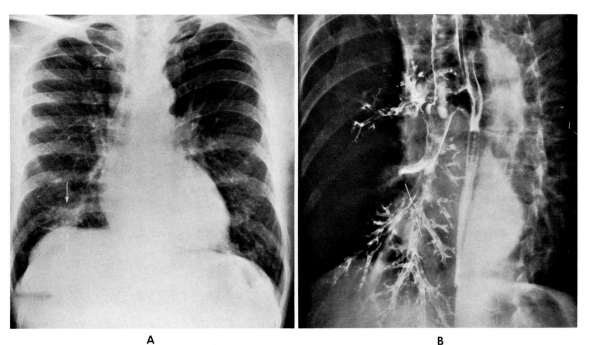

A   B

Figure 9–109 **Bronchial Papilloma.**

A, Posteroanterior view discloses a nonspecific patch of pneumonitis at the right base *(arrows)*.

B, Bronchogram indicates occlusion of the medial basal segment of the right lower lobe *(arrows)*. The lesion was a benign bronchial papilloma.

Bronchography is required for diagnosis of these lesions. They may be single or multiple and diffusely scattered throughout the bronchial tree. (Courtesy Dr. Arlyne Shockman, Veterans Administration Hospital, Philadelphia, Pennsylvania.)

Primary Lymphosarcoma of the Lung

This uncommon lesion arises from the lymphoid tissue of the bronchial mucosa or the larger blood vessels. It lacks a characteristic picture, sometimes resembling an alveolar pneumonia or a discrete solitary mass. Occasionally a miliary or nodular pattern is seen. Rarely is bronchial obstruction, atelectasis, or pleural reaction observed. The usual picture is a pneumonic density, often with air bronchograms, that may enlarge to fill almost the entire lung field. Bilateral involvement often occurs. Linear streaks of lymphangitic spread may extend from the lesion.

When there is a large tumor mass, especially if near the hilum and without evidence of atelectasis, lymphosarcoma should be considered in the differential diagnosis.[128, 129]

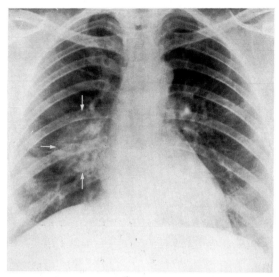

Figure 9–110 **Primary Lymphosarcoma of Lung.** There is a large, poorly demarcated mass *(arrows)* in the middle lobe adjacent to the hilum. The hilar shadow is not enlarged, and there is no evidence of atelectasis. The appearance is nonspecific and could be interpreted as pneumonia or a bronchogenic carcinoma. (Courtesy Dr. Walter M. Whitehouse, Department of Radiology, University of Michigan, Ann Arbor, Michigan.)

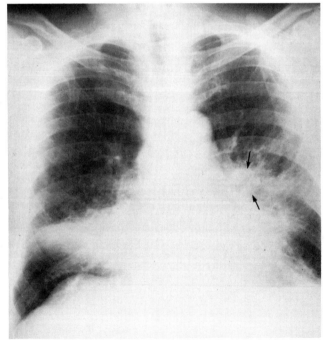

Figure 9–111 **Primary Pulmonary Lymphosarcoma: Bilateral Disease.** The persistent bilateral confluent densities were due to primary pulmonary lymphosarcoma. There is a large node in the left hilum *(arrows)*, and streaky densities extend into the left periphery as a result of lymphatic extension.

Bronchogenic Carcinoma

Bronchogenic carcinoma can produce a variety of changes that depend upon the size and location of the tumor.

About half of the bronchogenic carcinomas arise in the lung periphery, and these are more readily detected radiographically. The carcinoma appears as a nodular density or homogeneous infiltrate with borders that may be sharp and regular or indistinct and irregular. The latter appearance may be mistaken for a pneumonitis. Tumor calcification is extremely rare, and its presence virtually rules out malignancy. Serial film studies to measure growth rate may prove helpful in uncertain cases or when biopsy or surgery cannot be done. Malignant lesions will usually double in volume within three months to a year; benign lesions have a longer doubling time. A doubling time of a few weeks usually indicates an inflammatory process.

The tumors arising in the larger bronchi in or near a hilum may remain undetected for a long time. Serial films over an extended period will reveal slow but steady enlargement of a hilum, and an eventual definite hilar or perihilar mass. Often the earliest findings of a bronchogenic carcinoma are changes secondary to bronchial compromise or occlusion. These changes include recurrent pneumonitis in one area, unresolving pneumonia, obstructive emphysema, atelectasis, and lung abscess. When the tumor arises in a lung apex (Pancoast tumor), it may be overlooked or mistaken for apical pleural thickening. Erosions of the first or second rib often occur before the nature of the apical density is appreciated. Occasionally, necrosis within the tumor causes irregular cavitation, with thick walls and with small masses projecting into the cavity; there may or may not be fluid levels. However, the cavitation may be thin-walled and simulate an inflammatory abscess.

When a lesion is inaccessible to bronchoscopy or when brush biopsy is inconclusive, bronchography may be very helpful and will disclose abrupt bronchial obstruction, annular constriction, narrowing, or bronchial displacement. Planigraphy may disclose significant narrowing of a bronchus and may help distinguish a hilar mass from the hilar vascular shadows. Pulmonary arteriography is of limited value, since bronchogenic carcinomas are supplied by the bronchial arteries. Selective bronchial arteriography is difficult, but it may show abnormal vessels and tumor stain.

Local spread and metastases may produce additional pulmonary nodules, masses, or infiltrates, enlarged hilar or mediastinal masses (nodes), pleural effusion, local rib destruction, or the pulmonary changes of lymphangitic spread. Superior vena cavography may disclose displacement or pressure occlusion of the mediastinal veins. Metastases to bone are frequent. The squamous cell tumors virtually always cause destructive lytic metastases. Small cell (oat cell) lesions and adenocarcinoma not infrequently metastasize diffusely to the bone marrow and may produce a widely distributed osteoblastic change. Cerebral metastases often occur and may clinically simulate a primary brain tumor.

In some cases of bronchogenic carcinoma, hypertrophic pulmonary osteoarthropathy develops. It may be discovered before a bronchogenic lesion is clinically suspected. Osteoarthropathy appears as periosteal reaction and subperiosteal new bone layering, most frequently in the radius, ulna, tibia, and fibula, and less commonly in the metacarpals. Soft tissue clubbing of the fingers may be associated. The periosteal changes often regress after resection of the lung tumor.

Most bronchogenic carcinomas are epidermoid or small cell lesions. The less common adenocarcinoma occurs in younger males and in females. The rarer alveolar cell carcinoma is discussed on page 542.[130-138]

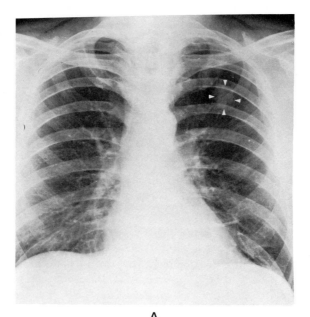

A

B

Figure 9–112   **Growth of Peripheral Bronchogenic Carcinoma: Serial Films.**

   A, There is a poorly defined noncalcified nodule in the left upper lobe *(arrowheads).*
   B, Six months later there is a marked increase in the size of the nodule *(arrowheads).* The diameter of the nodule is 50 per cent greater, indicating tripling of the tumor volume. Doubling of tumor volume within three months to a year suggests a malignant lesion.

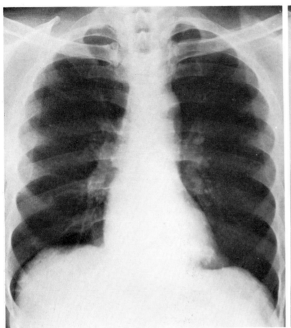

A

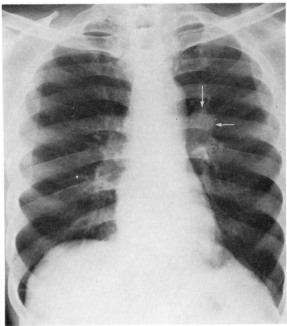

B

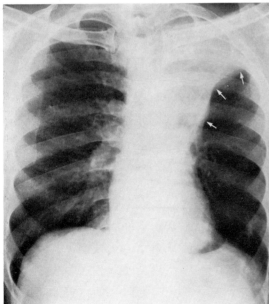

C

Figure 9–113 **Bronchogenic Carcinoma: Progressive Hilar Changes.**

A, Chest film in a 54 year old male who had a chronic cough is apparently normal.

B, One year later there is left hilar enlargement *(arrows)* that is apparent only on comparison with the original film. The enlargement was a bronchogenic carcinoma.

C, Nine months later, atelectasis of the superior division of the left upper lobe has developed. The line formed by the inferior surface of the atelectatic upper lobe *(upper two arrows)* and the bulge of hilar tumor mass *(lower arrow)* form an S-shaped curve. This curve is suggestive of bronchogenic carcinoma. The aerated portions of the left lung are hyperlucent.

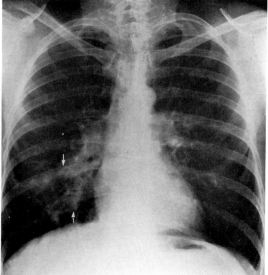

Figure 9-114 **Bronchogenic Carcinoma: Pneumonic Picture.** Pneumonic consolidation *(arrows)* with irregular borders persisted with little change over six weeks. It was a small carcinoma in the segmental bronchus. Persisting, recurring, or unresolved pneumonia, especially in a patient over 40 years of age, should be investigated for bronchogenic carcinoma.

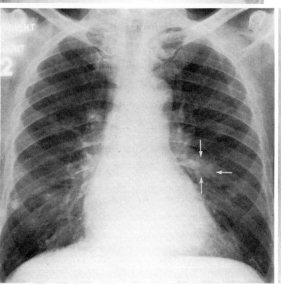

A

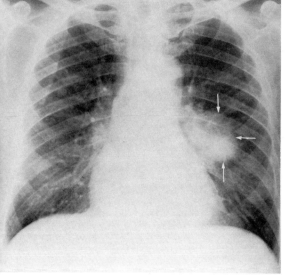

B

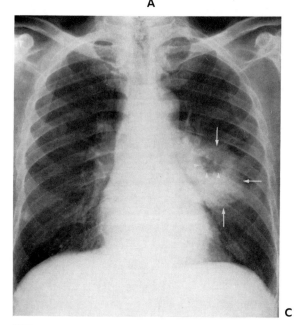

Figure 9-115 **Bronchogenic Carcinoma: Progressive Growth and Cavitation.**

A, There is a small density in the lower left hilum *(arrows).* The patient was asymptomatic and refused surgery.

B, Six months later there is a marked increase in the size of the mass *(arrows).* The mass was lobulated but still well demarcated.

C, Two months later the mass is even larger *(arrows),* and has undergone central necrosis *(arrowheads).* Irregularity of the cavity, with protruding nodular densities, is characteristic of a cavitating bronchogenic carcinoma.

C

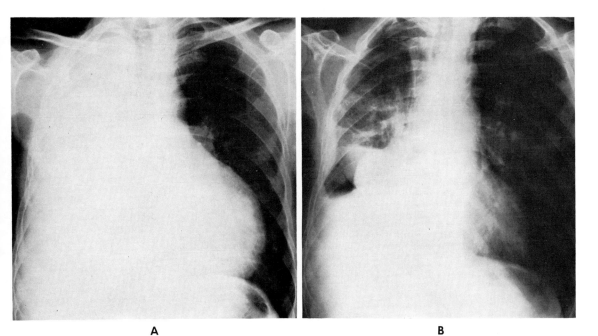

A   B

Figure 9–116 **Bronchogenic Carcinoma: Atelectasis and Pleural Effusion.**

A, There is homogeneous opacification of the entire right lung without significant shift of the mediastinum. The opacification is due to a combination of atelectasis and pleural fluid; each offsets the other's positional effects on the heart and mediastinum.

B, Following withdrawal of fluid, the heart and mediastinum are shifted to the right, and the underlying atelectasis in the right lung becomes apparent.

Large collections of pleural fluid without mediastinal shift to the opposite side, or moderate amounts of fluid with shift to the side of the fluid suggest atelectasis from a bronchogenic carcinoma, as well as fixation of the mediastinum by invasive neoplasm.

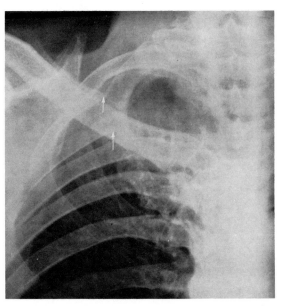

Figure 9–117 **Superior Sulcus Tumor.** The posterior portion of the third and fourth ribs *(arrows)* has been destroyed. The innocuous-appearing density in the right apex is due to an infiltrating anaplastic carcinoma that has destroyed the ribs.

Tumors in the apical sulcus frequently simulate benign pleural thickenings, but rib erosion and the clinical symptoms of intractable shoulder pain and Horner's syndrome should aid in diagnosis. A superior sulcus lesion (Pancoast tumor) may be either a squamous cell bronchogenic tumor or an anaplastic lesion.

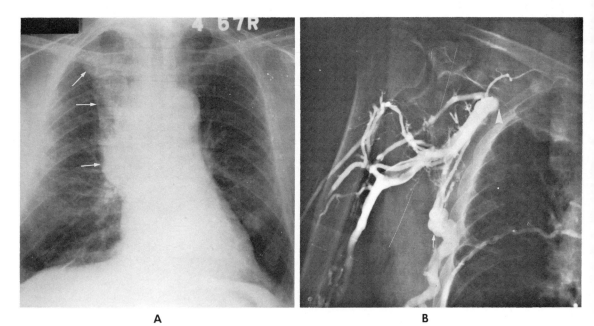

A   **B**

Figure 9–118   **Bronchogenic Carcinoma: Superior Vena Caval Obstruction.**

A, There is enlargement of the right hilum and widening of the mediastinum on the right *(arrows)* due to mediastinal extension of a bronchogenic carcinoma. There was clinical evidence of superior vena caval obstruction.

B, Venogram reveals complete blockage of the subclavian vein *(arrowhead).* There are large, dilated, and anastomotic channels *(arrow)* extending between the superior vena cava and subclavian vein. In the presence of such findings, the lesion is inoperable.

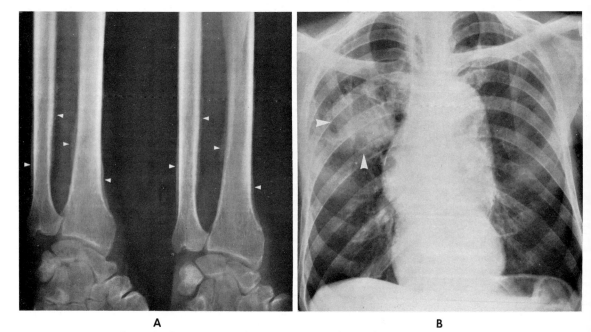

A   **B**

Figure 9–119   **Bronchogenic Carcinoma: Pulmonary Osteoarthropathy.**

A, There are linear irregular areas of subperiosteal new bone formation in the distal radius and ulna *(arrowheads).*

B, Large bronchogenic carcinoma is apparent in the right upper lobe *(arrowheads).*

Following resection, periosteal reaction cleared completely. Evidence of pulmonary osteoarthropathy may occasionally be the first clue to bronchogenic carcinoma.

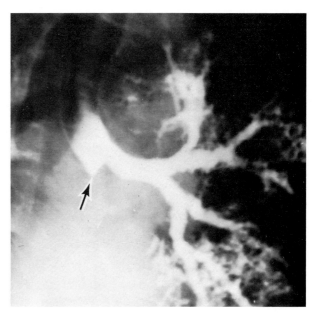

Figure 9–120 **Bronchogenic Carcinoma: Bronchogram.** The dorsal branch of the left lower lobe bronchus is completely occluded (*arrow*) by a bronchogenic carcinoma.

In a patient with a mass suspicious of bronchogenic carcinoma, bronchography will often confirm the diagnosis with evidence of narrowing or occlusion of a bronchus in the vicinity of the mass.

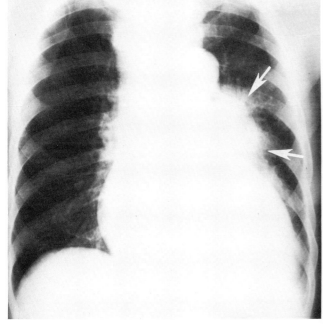

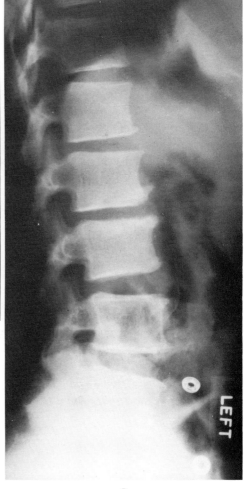

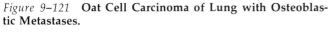

A

Figure 9–121 **Oat Cell Carcinoma of Lung with Osteoblastic Metastases.**

A, The left hilar mass and infiltration (*arrows*) in a 60 year old man proved to be an oat cell carcinoma.

B, The lumbar bodies show a generalized increased density without destructive changes. Osteosclerosis was present in the axial skeleton and ribs and was due to diffuse medullary metastases.

B

Alveolar Cell Carcinoma

This tumor constitutes 2 to 5 per cent of primary lung malignancies. Unlike bronchogenic carcinoma, there is no sex predilection and no apparent relationship to cigarette smoking.

Two distinct forms are encountered. The most frequent type presents in an asymptomatic patient as a nonspecific nodule-like irregular density in the lung periphery, often with some linear streaking. Small lucencies due to unconsolidated alveoli are often present. Cavitation only rarely occurs. The appearance is frequently mistaken for that of an inflammatory infiltrate. Growth is usually slow, and the lesion may remain localized for several years.

The more diffuse symptomatic form has several radiographic patterns. There may be widespread nodular densities of varying sizes with ill-defined margins, irregular pneumonic infiltrates, often with air bronchograms, or a mixture of both. Atelectasis is infrequent. Extensive involvement of one lung is common; less often bilateral lesions are encountered. Radiographic progression may be moderate or slow, but there is usually a steadily downhill clinical course. Late findings include pleural effusion, large mediastinal nodes, and, infrequently, lymphangitic spread characterized by interstitial densities.

The diffuse forms of alveolar cell carcinoma may be mistaken radiographically for sarcoidosis, chronic granulomatous or fungal disease, metastatic lesions, tuberculosis, primary lymphosarcoma, or nonspecific pneumonia.[139-142]

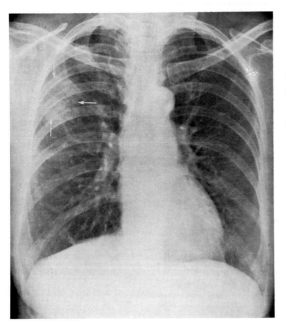

Figure 9-122 **Alveolar Cell Carcinoma.** Chest film in a 55 year old woman demonstrates an ill-defined density with an indistinct lower border *(arrows)* in the right upper lobe. The hilar nodes are not enlarged. This picture simulated pneumonitis but proved to be an alveolar cell carcinoma.

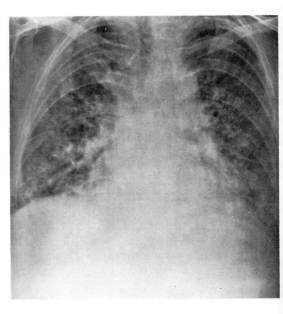

Figure 9-123 **Diffuse Alveolar Cell Carcinoma.** There are diffuse irregular nodular infiltrates in both lung fields similar to the pattern of certain granulomatous diseases. These findings in an elderly female should suggest alveolar cell carcinoma.

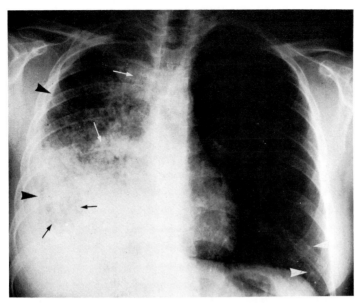

Figure 9–124 **Advanced Alveolar Cell Carcinoma in a 29 Year Old Woman.** The right lung is extensively infiltrated with linear and nodular densities and conglomerate areas in the upper lobe and perihilar area *(white arrows)*. The entire lower lung field is densely infiltrated and contains numerous small lucencies of emphysematous lung *(black arrows)*. The pleural reaction along the lateral wall *(black arrowheads)* is a late finding in alveolar cell carcinoma. A few neoplastic nodules have appeared in the left base *(white arrowheads)*.

The original infiltrate was in the right upper lobe. Progressive spread through an entire lung, simulating a chronic inflammatory process, is relatively common. The youthful age of the patient is unusual, since alveolar cell carcinoma is rarely encountered before the fifth decade.

Metastases to the Lung

Metastases from distant malignancies are brought to the lungs through the bloodstream (hematogenous) or via the lymphatics. Direct extension can occur from neighboring organs, such as breast, esophagus, or stomach, and from primary lung lesions. Most sarcomas, and carcinomas of the breast, kidney, thyroid, colon, and testes, are prone to produce pulmonary metastases.

Hematogenous metastases are usually well-outlined round or oval nodules of varying size, number, and distribution. The sharpness of their borders depends upon their rate of growth. Sarcomatous metastases tend to be large and well rounded, the so-called cannon-ball lesions. A solitary metastasis cannot be differentiated from a primary tumor by radiographic features alone. Metastatic nodules rarely calcify and rarely cause bronchial obstruction, in contrast to granulomas and bronchogenic carcinomas, respectively. Necrosis and cavitation of a metastatic nodule sometimes occur, generally in a squamous cell metastasis. Pleural effusion with or without parenchymal lesions is not uncommon and represents pleural metastases. This is frequent in breast carcinomas.

Metastatic choriocarcinoma nodules are often highly vascular, containing arteriovenous shunts. These may bleed into the adjacent lung, which may give ill-defined irregular borders to the individual lesions. Hemoptysis may be a prominent symptom.

In general, well-developed nodular metastatic disease is readily diagnosed radiographically even before the primary lesion has been uncovered. Occasionally, distinction from nodular granulomas, multiple arteriovenous malformations, pneumoconioses, or nodular forms of vasculitis may be difficult.

Lymphangitic lesions arise from metastatic hilar nodes and appear as interstitial linear densities extending from the hila into the periphery. Small nodular densities may be superimposed on these linear opacities. Frequently the small lymphatics in the peripheral interstitial septa become enlarged with tumor tissue or from central blockage, giving rise to short transverse densities in the periphery — Kerley B lines. Pleural extension often produces pleural effusion. Radiographically, lymphangitic metastases may simulate other interstitial nonneoplastic conditions, but the diagnosis should be considered when radiating linear densities are associated with enlarged hila and with Kerley lines. Carcinoma of the stomach, breast, and pancreas are the tumors most often responsible for lymphangitic metastases.[133, 143, 144]

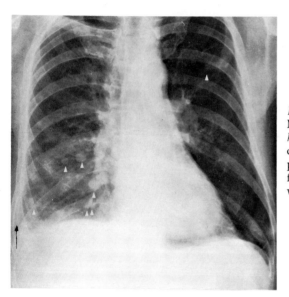

Figure 9–125 **Hematogenous Metastases to the Lung.** Multiple sharply outlined nodular densities *(arrowheads)* in the lungs are due to a metastatic adenocarcinoma from the kidney. The fluid *(arrow)* indicates pleural involvement. Although many nodules are often found in one lobe, a metastatic lesion can develop anywhere within the lung.

Figure 9–126 **Hematogenous Metastases to the Lung.** There are multiple sharply outlined nodules in both lung fields *(arrows)* from a squamous cell carcinoma of the kidney; several nodules have undergone cavitation *(arrowheads).* Breakdown of a metastatic nodule is rare and most frequently occurs in squamous cell metastasis.

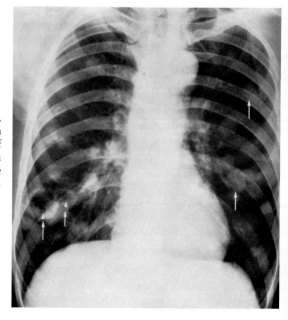

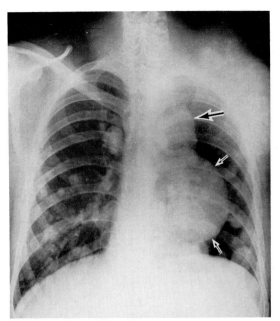

Figure 9-127   **Nodular (Cannon-ball) Metastases from an Osteogenic Sarcoma.**   Nodular masses of vary-ing size are scattered throughout both lower lung fields. A huge metastasis *(small arrows)* obliterates the left cardiac border. The left upper mediastinum is widened by metastases *(large arrow)*. The left clavicle, scapu-la, and entire upper extremity had been resected one year previously because of osteogenic sarcoma.

Lung metastases from primary sarcomas are almost always nodular.

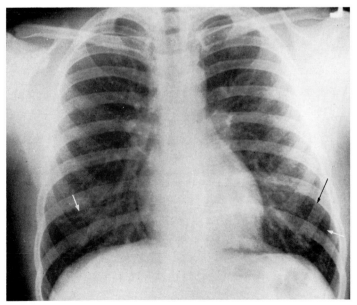

Figure 9-128   **Miliary Metastases from Thyroid Carcinoma in 15 Year Old Boy.**   Both lower lung fields are diffusely studded with very small nodular densities *(arrows)*, which are difficult to appreciate on this reproduction.

Miliary metastases cannot be distinguished from other miliary lung lesions (such as miliary tubercu-losis, pneumoconiosis, hemosiderosis, and so forth).

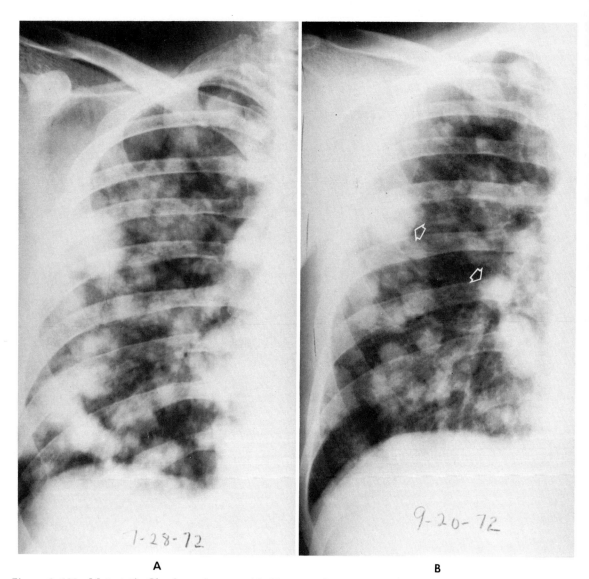

A   **B**

Figure 9–129   **Metastatic Choriocarcinoma with Hemoptysis.**

   A, Chest film of a young man with a two-week history of unexplained hemoptysis reveals bilateral diffuse nodular densities with hazy indistinct borders. These proved to be metastatic choriocarcinoma lesions that caused hemorrhage into the surrounding lung tissue, making the borders of the metastatic lesions blurred and hazy. The primary lesion was found in the testicle.

   B, After almost two months of chemotherapy, the lesions are somewhat fewer in number and now show the usual sharp borders *(arrows)* of metastatic lung lesions. The hemoptysis ceased after a week of chemotherapy.

   Aggressive choriocarcinoma metastases (from uterus or testicle) tend to cause hemorrhage at their periphery and clinical hemoptysis.

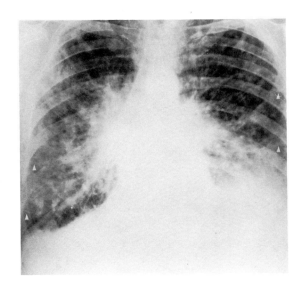

Figure 9-130 **Lymphangitic Metastases to the Lung.** Innumerable irregular linear and nodular densities extend from the hilum to the periphery. The hilar shadows are enlarged due to infiltration of the peribronchial hilar lymph nodes. The engorged, prominent lymphatic channels are most clearly seen in the outer third of the lung fields *(arrowheads),* and pleural changes are seen at the left base.

Tumors of the Trachea

Carcinoma is the most common primary tumor of the trachea but is quite rare. In an adult with upper airway symptoms but with clear lung fields on the radiograph, careful scrutiny of the tracheal air column should be made.

On conventional, well-penetrated chest films there may be evidence of constriction or narrowing of the trachea. An intraluminal mass is sometimes seen. Confirmatory evidence will be found on planigrams or after contrast material is instilled into the trachea, although the latter study is rarely necessary.[145-147]

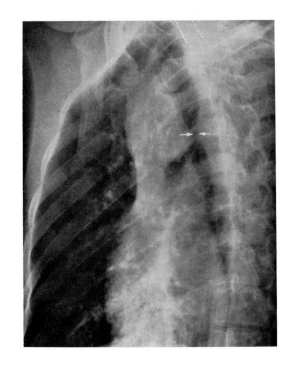

Figure 9-131 **Carcinoma of the Trachea.** The trachea is narrowed *(arrows),* and the surrounding soft tissues are thickened. There is an abrupt shift to a normal lumen inferiorly, which is characteristic of an encircling neoplasm. (Courtesy Dr. Arlyne Shockman, Veterans Administration Hospital, Philadelphia, Pennsylvania.)

MISCELLANEOUS DISORDERS OF THE LUNG

Hughes-Stovin Syndrome

This unusual syndrome is characterized by thromboses in the vena cava or peripheral veins, recurrent pulmonary emboli or thrombi, and formation of one or more aneurysms of major branches of the pulmonary arteries. Young males are predominantly affected. The aneurysms are nonmycotic, and origin of the syndrome is unknown. Some authors consider it a collagen disease.

Radiologically, there is usually evidence of pulmonary thromboembolism (a pleural pneumonia, localized diminished vasculature, and so forth) and later development of one or more rounded densities in the inner lung zones, which prove to be aneurysms on pulmonary angiography. The aneurysms continue to enlarge, and death from sudden rupture usually occurs within a few years after onset.

Thrombosis in the inferior vena cava can be demonstrated by a vena cavagram. Thrombosis of the sagittal sinus is an early finding in some cases and is associated with signs and symptoms of pseudotumor cerebri (see p. 354).[148, 149]

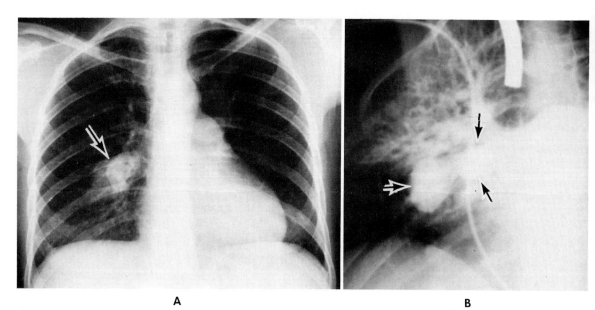

A   B

Figure 9–132 **Hughes-Stovin Syndrome.**

A, Chest film made several months after the first of several episodes of pulmonary thromboembolism shows a mass density *(arrows)* below the right hilum. Note the sparsity of vasculature at both bases.

B, Oblique film of pulmonary angiogram reveals two adjacent contrast-filled aneurysms *(arrows)* and absence of opacified vessels in the lower lobe. Smaller aneurysms were opacified in the right upper lobe and left hilum.

The patient died of sudden massive hemoptysis. No organisms were found in the aneurysms. An old thrombus was found in the inferior vena cava.

Pulmonary Fat Emboli

Fat embolism in the lungs usually follows the release of large quantities of fat globules into the circulation, most often after fractures of the femur or multiple bones. Fortunately, pulmonary fat embolism is quite rare, occurring in less than one per cent of severe fractures. Chest findings appear within hours to several days after the injuries. The usual findings are bilateral alveolar densities, sometimes resembling pulmonary edema but without cardiac enlargement or vascular engorgement. Perihilar and basal densities predominate, and coarse mottling or a somewhat nodular pattern can occur.

The clinical signs of a pulmonary involvement may sometimes be overshadowed by central nervous system symptoms. Radiographic clearing of the chest usually occurs within one week. A similar pulmonary radiographic picture can occur from extensive oil emboli following iodized oil lymphangiography.[150, 151]

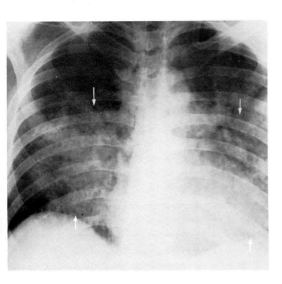

Figure 9–133 **Pulmonary Fat Emboli Following Extensive Fractures.** Extensive bilateral nodular densities in the perihilar and basilar areas *(arrows)* are caused by fat emboli. The absence of cardiomegaly and vascular congestion help differentiate this picture from pulmonary edema and changes caused by congestion. These findings in a patient with a history of extensive fractures suggest fat embolism.

Figure 9–134 **Traumatic Pulmonary Fat Emboli.** Chest film made five days after the patient sustained comminuted fractures of the pelvis and tibia shows coarse patchy alveolar densities diffusely scattered through both lung fields. The chest film made immediately after the accident had been normal. The patient developed fever, cough, dyspnea, and hemoptysis shortly before the film was made. Fat globules were found in his sputum. (Courtesy Dr. W. R. Cole, Perth, Australia, and Journal of College of Radiologists of Australasia.)

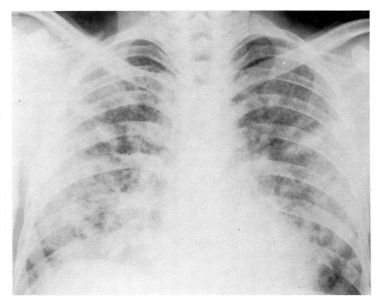

Pulmonary Arteriovenous Fistula (Vascular Malformation)

The lesion generally appears as a lobulated area of homogeneous density, occurring most frequently in the middle and lower lobes, although any segment of the lung may be involved. Multiple lesions are not uncommon. The density is usually connected to the hilum by two enlarged vascular shadows. These afferent and efferent vessels are often difficult to visualize, and planigraphy and pulmonary arteriography may be required for identification. Calcification of the fistula is rare. The mass or masses may pulsate during fluoroscopy and decrease in size as intrathoracic pressure increases (Valsalva's maneuver). Multiple lesions may simulate metastatic disease, especially if the feeding vessels cannot be identified.

For definitive diagnosis, pulmonary angiography is essential, and the procedure may uncover other unsuspected vascular malformations.

Cardiac changes are rare in pulmonary arteriovenous fistula unless there is pronounced shunting. Almost half the pulmonary arteriovenous malformations are part of Rendu-Osler-Weber disease (hereditary hemorrhagic telangiectasia).[152, 153]

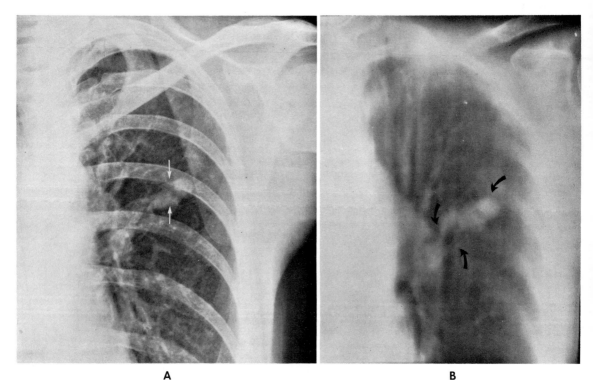

A   **B**

Figure 9–135 **Pulmonary Arteriovenous Fistula: Planigram.**

A, The patient is a 32 year old woman who had hemoptysis. A sharply demarcated density *(arrows)* apparently extending to the hilum is seen in the left upper lobe.

B, Planigram demonstrates the afferent and efferent connecting vessels *(central black arrows),* which are characteristic. The fistula *(upper arrow)* pulsated on fluoroscopy and decreased in size when Valsalva's maneuver was performed.

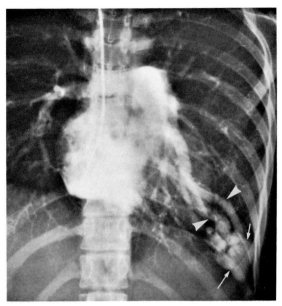

Figure 9–136 **Pulmonary Arteriovenous Fistula: Dextrocardiogram.** At the left base there is opacification of dilated abnormal vascular channels that are characteristic of arteriovenous malformation *(arrows).* The large afferent and efferent vessels are clearly seen *(arrowheads).* Notice that the veins leading from the arteriovenous malformation are tortuous and dilated.

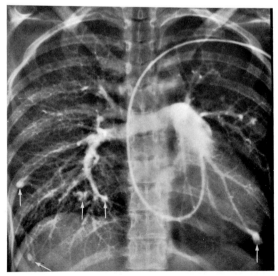

Figure 9–137 **Multiple Pulmonary Arteriovenous Fistulas: Pulmonary Arteriogram.** Multiple small fistulas *(arrows)* are opacified in both lung fields; these were not detectable on chest films. The heart is normal in size. Cardiac dilatation is generally not associated with pulmonary arteriovenous disease, as it is with peripheral arteriovenous fistulas. The patient had Rendu-Osler-Weber disease.

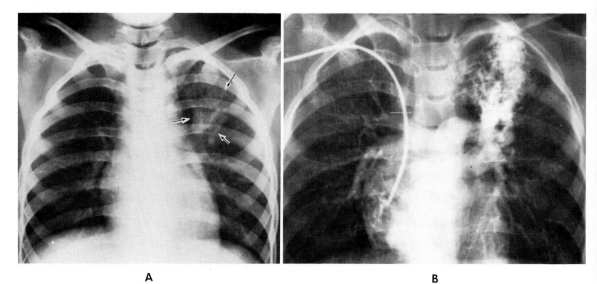

A   **B**

Figure 9–138 **Pulmonary Arteriovenous Malformation.**

A, There is an extensive infiltrate in the right upper lobe *(arrows).* The patient had recurrent episodes of hemoptysis.

B, Angiocardiogram shows that the "infiltrate" is a large, densely opacified vascular malformation, considerably more extensive than the plain film suggested.

Sequestration of the Lung

Sequestration may be intralobar or, less commonly, extralobar.

In intralobar sequestration, a cystic mass of pulmonary tissue, usually in a lower lobe, receives its blood supply from one or more branches of the aorta instead of from the pulmonary artery, but drains into the pulmonary veins. Bronchial communication is usually absent or hypoplastic. The sequestered segment appears as a well-defined oval or round, fairly homogeneous radiographic density in the posterior, inferior, and medial portion of a lower lobe, most often the left lobe. Usually the sequestration replaces the normal posterior basal segment of the lower lobe.

Uncomplicated intralobar sequestration is asymptomatic, but inflammatory complications are not uncommon. Erosion of an adjacent bronchus may produce air-fluid levels within the mass, simulating a local bronchiectatic cavity, abscess, or infected cyst.

Bronchography usually reveals displacement of neighboring bronchi by a noncommunicating mass. In the rare case with bronchial communication, cystic cavities are outlined by contrast material within the mass. Definitive diagnosis requires aortography, which can demonstrate the aortic branch to the sequestration.

Extralobar sequestration is actually accessory tissue, usually beneath a lower lobe and sometimes below the diaphragm. This is also supplied directly from the aorta but drains into the caval circulation. It is much more infrequent than intralobar sequestration and is rarely symptomatic.[154-157]

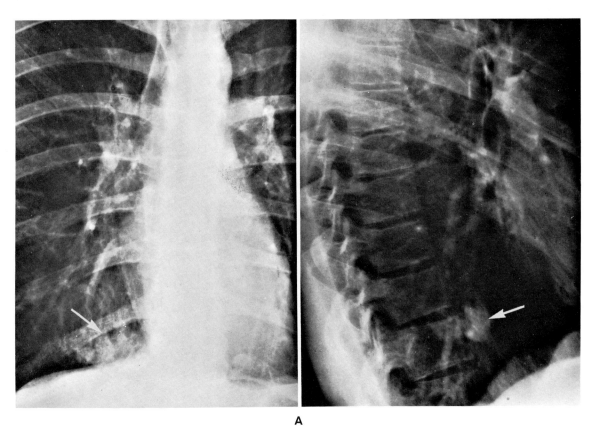

A

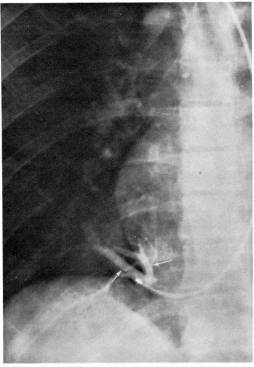

B

Figure 9-139 **Intralobar Sequestration: Angiogram.**

A, Posterior and lateral views demonstrate a round density in the posterior and medial basilar segments of the right lower lung *(arrows),* a typical location for sequestration. The radiographic appearance is nonspecific.

B, Arteriogram demonstrates the separate arterial blood supply of the segment *(arrows)* arising from the aorta, confirming the diagnosis.

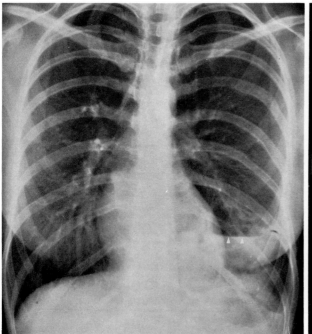
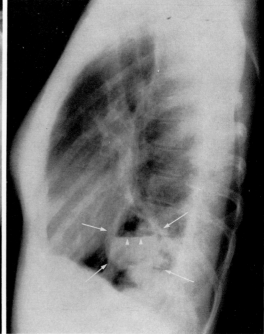

A

Figure 9–140 **Intralobar Sequestration: Bronchogram.**

A, In posteroanterior and lateral views, a cystlike lesion is seen in the posterior medial portion of the left lower lung *(arrows)*. The air-fluid level *(arrowheads)* indicates bronchial communication. This lesion was a sequestered segment of lung that obtained its blood supply from a systemic vessel arising from the aorta.

B, Bronchogram shows contrast material entering the involved segment. Note the fluid level, indicating bronchial communication *(arrow)*.

Most sequestered segments do not communicate with the normal bronchial tree, but in this case the infected segment eroded into the adjacent normal bronchus.

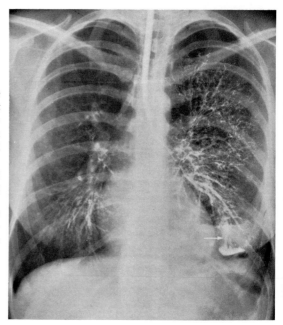

B

Agenesis and Hypogenesis of the Lung

This anomaly may vary from complete absence of one major bronchus and lung (agenesis) to hypoplasia of the bronchus in which there is a rudimentary undeveloped lung. The anomaly is compatible with normal life.

A homogeneous density is seen in the affected hemithorax, which has a smaller volume, narrowed interspaces, and an elevated diaphragm. The heart and mediastinum are shifted to this side. The normal lung shows compensatory hyperinflation, frequently with herniation across the mediastinum to the affected side. Other congenital anomalies, especially hemivertebra, are frequently associated.

Diagnosis requires bronchography and angiography: the former demonstrates absence or hypoplasia of the bronchus and lung; the latter demonstrates complete absence or diminution of the pulmonary artery. All degrees of hypoplasia, including agenesis of an entire lung, can occur. Obstructive atelectasis of an entire lung may produce a similar roentgenologic picture (see p. 470).[44, 158, 159]

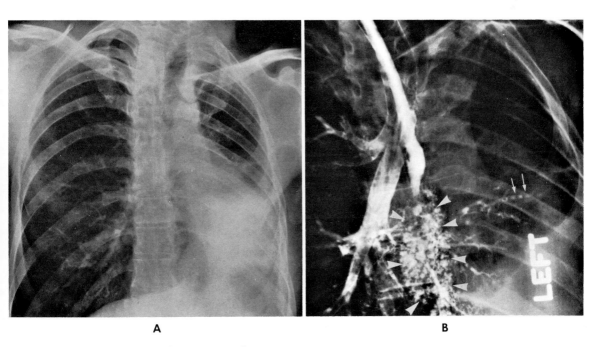

A   B

Figure 9–141 **Hypogenesis of Lung: Bronchogram.**

A, There is asymmetry and marked volume loss in the left thorax. The left diaphragm is elevated, the intercostal spaces are narrowed, and the heart is shifted into the left thorax. What appears to be left upper lobe is actually a herniated segment of the hyperinflated right lung. The left cardiac border is obscured, since adjacent, contrasting, air-containing lung is absent. The picture is characteristic of partial agenesis of the left lung.

B, Bronchogram demonstrates residual left lung parenchyma (*arrowheads*). The bronchi extending to the left lower chest wall arise from the right bronchial tree (*arrows*). Pulmonary angiography would also be diagnostic.

Wilson-Mikity Syndrome (Pulmonary Dysmaturity)

This respiratory distress syndrome, of unknown origin, occurs primarily in premature infants. While this condition may be fatal, recovery may occur over a period of months. The acute stage occurs a few days or weeks after birth and is characterized by a general hyperinflation associated with diffuse bilateral reticulonodular infiltrates interspersed with small lucent areas of focal hyperaeration, giving a "bubbly" appearance. Within a few weeks or months, coarse streaks appear, radiating from the hila, especially into the upper lobes. Focal hyperaeration disappears, but hyperinflation persists. Often, changes of pulmonary hypertension develop. Within 4 to 11 months, the lungs are usually completely clear, but in some infants cardiomegaly and pulmonary hypertension persist.

Hyaline membrane disease, a more common cause of respiratory distress syndrome in premature infants, is readily distinguishable both clinically and radiographically.[160-162]

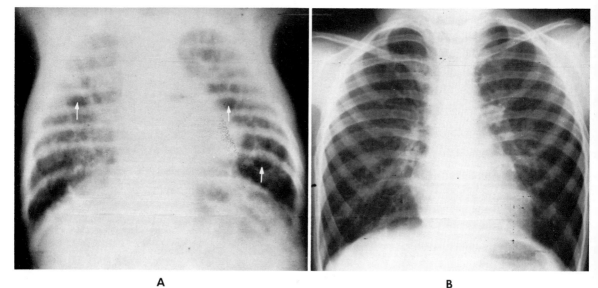

A   B

Figure 9–142 **Wilson Mikity-Syndrome.**

A, Chest film of neonate with respiratory distress symptoms shows diffuse bilateral interstitial reticulonodular densities with both diffuse and focal hyperaeration *(arrows).* The findings are characteristic of pulmonary dysmaturity—Wilson-Mikity syndrome.

B, At 2 years of age, the child is clinically well and the lung fields appear completely normal. This recovery is characteristic of most cases of Wilson-Mikity syndrome.

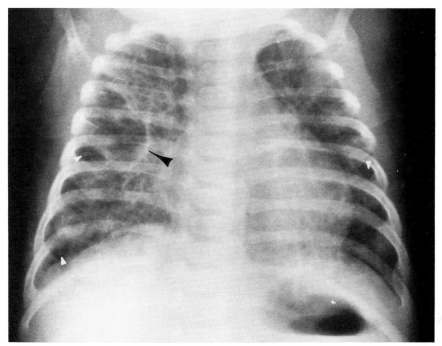

Figure 9–143 **Wilson-Mikity Syndrome in Premature Neonate.** There are diffuse reticulonodular infiltrates, coarse streaks *(black arrowhead),* and areas of focal hyperinflation *(white arrowheads),* giving the lung fields a somewhat characteristic "bubbly" appearance.

This child had respiratory distress symptoms and cyanosis. Recovery occurred within three weeks. The lungs did not return to normal until many months later.

Lymphangiomyomatosis

This unusual syndrome is radiographically characterized by repeated episodes of chylous pleural effusion and/or a diffuse reticular lung pattern. Spontaneous pneumothorax is a frequent complication and may be the presenting symptom. Apparently, only women are affected.

The basic lesion is a hamartomatous lymphangiomyoma, which may be circumscribed or diffuse. The lungs, lymph nodes, mediastinum, and retroperitoneum are the sites of predilection. The mediastinal lymphangiomyoma frequently involves the thoracic duct and may cause a fistulous connection to the pleura, with resultant chylothorax. The interstitial reticular lung pattern is due to diffuse lymphangiomyomatous involvement of the smooth muscles of bronchioles, pulmonary vessels, and lymphatics. Distended peripheral lymphatics from thoracic duct obstruction may also increase the interstitial markings and may lead to Kerley B lines. In advanced disease, a honeycomb lung picture may occur, and progressive respiratory insufficiency is the usual cause of death.

Pulmonary interstitial disease, chylous effusions, and spontaneous pneumothorax in a young or middle-aged female should suggest the diagnosis.[163, 164]

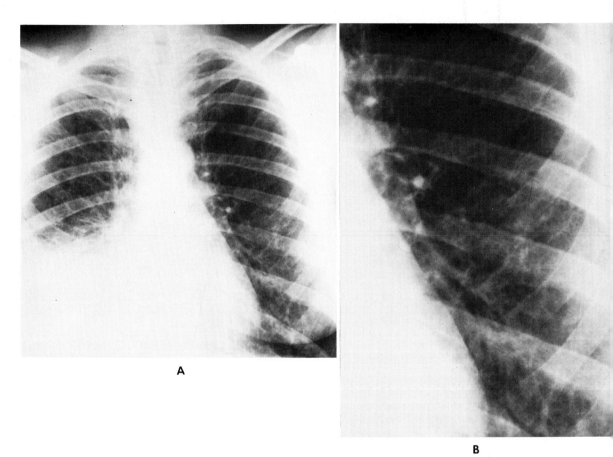

Figure 9–144   **Pulmonary Lymphangiomyomatosis in 56 year old woman.**

A, Diffuse linear interstitial infiltrates, best seen at the left base, are associated with a pleural effusion on the right. The pleural fluid was chylous.

B, Closeup of left lower lung clearly shows the linear reticular interstitial pattern.

A chylous pleural effusion associated with interstitial lung disease is virtually pathognomonic of lymphangiomyomatosis. (Courtesy Dr. W. T. Miller, Philadelphia, Pennsylvania.)

DISEASES OF THE PLEURA

Pleural Effusion

Pleural effusion can result from a great number of local or systemic conditions and diseases. Congestive heart failure, metastatic lesions to the pleura, bacterial and Coxsackie infections, pulmonary infarction, and tuberculous pleuritis are the more common causes. Less frequent causes include pleural mesothelioma, rheumatoid pleural disease, lupus erythematosus, the nephrotic syndrome, pleural trauma, and Meigs' syndrome.

On conventional erect films the earliest finding is blunting of the costophrenic angle, with an upward concave border. Less than 250 ml. of fluid will usually not be detected on posteroanterior projection. Lesser amounts can be seen on lateral projection and on decubitus films. The latter will show a linear density of fluid along the dependent border of the chest. Larger effusions produce a homogeneous density in the lower chest, with a curvilinear upper lateral border. The diaphragm and adjacent heart borders are obscured. Massive effusions may cause widening of the interspaces and contralateral displacement of the heart and mediastinum.

On recumbent films, sizable effusions will produce a haze or density over the entire hemithorax.

The ordinary pleural effusion is easily recognized radiographically, and can be distinguished from pleural thickening by the contour and postural shifting of the fluid density. However, infrapulmonic pleural fluid and loculated collections may be more difficult to diagnose.

A fluid collection between the diaphragm and the inferior surface of a lower lobe (infrapulmonic effusion) may closely simulate an elevated hemidiaphragm. The apex of this "diaphragm" is often more lateral than normally seen. Blunting of the costophrenic angle may be present, a valuable clue. In a right-sided infrapulmonic effusion, the minor fissure may appear closer than normal to the "elevated diaphragm." A left-sided infrapulmonic effusion will increase the distance between the gastric fundus and the left "diaphragm." Lateral decubitus films are necessary to demonstrate the presence and amount of infrapulmonic fluid, and will show the true position of the diaphragm.

Loculation of pleural fluid can occur between the lung and chest wall, or within the interlobar fissures. Loculations along the parietal pleura are most often posterior but can occur against any portion of the chest wall. It appears as a circumscribed homogeneous peripheral density, and simulates a tumor mass; its sharply marginated borders are suggestive of an extrapulmonary location. Fluid in an interlobar fissure may resemble an intrapulmonary tumor, but on lateral view the density conforms to the location and direction of an interlobar fissure. Evidence of other pleural fluid elsewhere, and changes of size and shape on serial studies are helpful diagnostic findings.

Air-fluid levels can appear within free or loculated fluid after thoracocentesis, from a bronchopleural fistula, or from a gas-producing organism.[165-167]

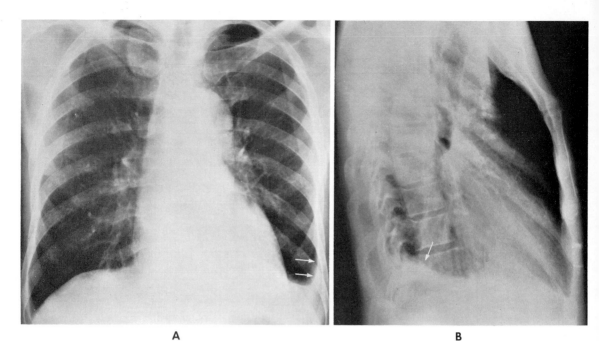

| A | B |

Figure 9–145 **Pleural Effusion.**

A, In posteroanterior view, the left costophrenic sulcus is blunted by a homogeneous curvilinear density extending along the left lateral chest wall *(arrows)*.

B, In lateral view, the posterior portion of the left hemidiaphragm is obliterated by a density with a concave upper border *(arrow)*.

Although the effusion appears small, over 400 ml. of fluid was removed by thoracocentesis. Blunting of the posterior sulcus on lateral view is generally the earliest indication of the presence of fluid.

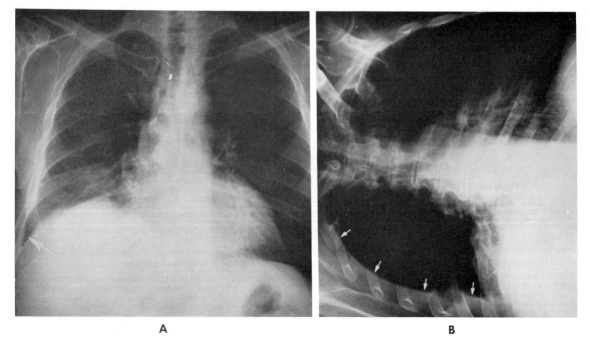

| A | B |

Figure 9–146 **Infrapulmonic Pleural Effusion.**

A, In erect view, the right diaphragm appears to be elevated. The costophrenic angle is clear and sharp *(arrow)*. The density was caused by accumulation of fluid between the diaphragm and the inferior surface of the right lung.

B, Right lateral decubitus view demonstrates the fluid *(arrows)* that has accumulated along the dependent right pleural surface.

An isolated infrapulmonic effusion generally develops on the right side and may resemble an elevated diaphragm. Decubitus views are necessary to demonstrate the presence and amount of fluid, and they are useful for distinguishing pleural fluid from pleural thickening.

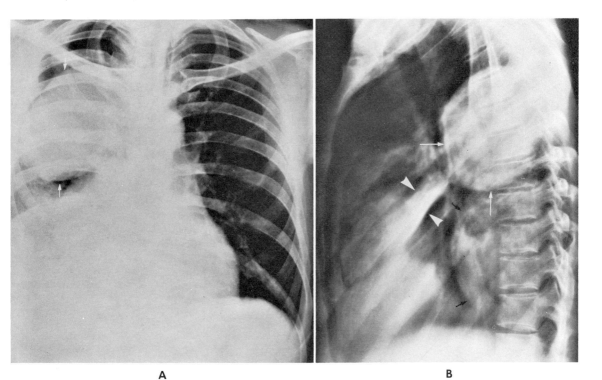

A   B

Figure 9–147 **Loculated Empyema: Postpneumonic.**

    A, In posteroanterior view, a sharply demarcated homogeneous density is seen in the right thorax *(arrows)* obscuring the entire right base. The left lung is clear, and there is no mediastinal shift. This confusing picture is clarified on the lateral film.

    B, In lateral view, three densities are clearly discernible. The upper one *(white arrows)* is a loculate accumulation between the layers of the major fissure. The second density is in the remaining portions of the major fissure *(arrowheads);* this indicates its location in the pleura. The third density *(black arrows)* is due to a large loculate accumulation of fluid adjacent to the posterior pleural surface. These accumulations frequently simulate solid tumors, but evidence of associated pleural changes simplifies the differential diagnosis.

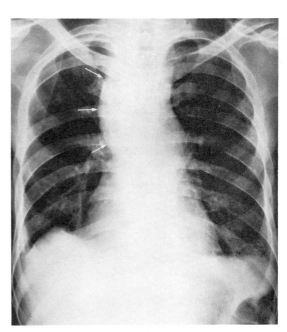

Figure 9–148 **Loculated Mediastinal Pleural Effusion.**
Sharply outlined density in the right upper mediastinum *(arrows)* was a loculate accumulation of fluid. This is seen most frequently following chest or mediastinal surgery. In the present case, it developed secondary to thyroidectomy. The fifth rib had been removed. Loculate fluid between the mediastinum and lung resembles mediastinal tumor.

Tuberculous Pleurisy

Unilateral pleural effusion without parenchymal lesions in a child or young adult is frequently the result of primary tuberculosis. Such an effusion rarely loculates and may persist for long periods. There is no radiographic feature that differentiates this condition from other effusions. A high incidence of subsequent parenchymal tuberculosis has been observed even after the effusion has cleared entirely.

Small pleural effusions occasionally occur with reinfection pulmonary tuberculosis.[168]

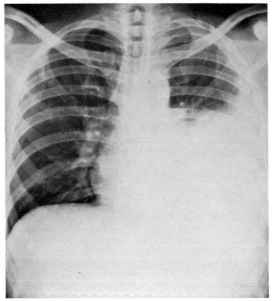

Figure 9–149 **Tuberculous Pleurisy.** A large homogeneous density in the left thorax obliterates the diaphragm and extends along the lateral chest wall; the upper margin is concave. The appearance and the shift of the heart to the right suggest the presence of fluid.

Pleural effusion developing in a young adult who has no parenchymal lesions or other systemic disease suggests tuberculosis.

Empyema and Pyopneumothorax

Suppurative inflammation of the pleura gives rise to a purulent exudate in the pleural space. A dense area of varying size is seen along a pleural surface. Before loculation occurs, empyema is indistinguishable from ordinary pleural effusion. Following loculation, one sees a mass density against the pleural surface, with a sharply defined border, usually concave toward the hilum.

Pyopneumothorax — air in a free or loculated collection — occurs in empyema secondary to pneumothorax, bronchopleural fistula, or after thoracocentesis.

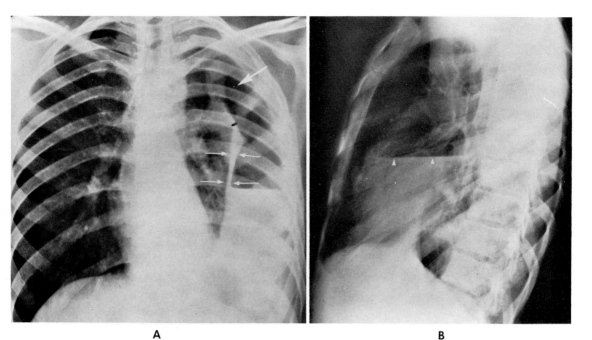

A   B

Figure 9–150 **Pyopneumothorax.**

A, Ten weeks following the onset of staphylococcal pneumonia in a 19 year old male, pneumothorax has occurred in the left lateral chest *(large arrow)*. There is a fluid level at the ninth posterior interspace, and inflammatory thickening of the visceral pleura *(small arrows)*.

B, On lateral view there is a long fluid level anteriorly *(arrowheads)*. The fluid level does not extend across the entire chest, thereby suggesting loculation.

Fibrosis of the Pleura

Any inflammatory pleural reaction or effusion may result in pleural thickening and fibrosis. Irregular hazy opacities are seen between the thoracic cage and the lung; these vary from small plaques to extensive collections surrounding large segments of lung. The densities remain unchanged with alterations in position, and in this way are distinguished from fluid. Loss of volume in the adjacent lung area can ensue. Any pleural surface or interlobar fissure may become fibrotic, and calcification may develop. Generally there are no symptoms due to pleural thickening or calcification.

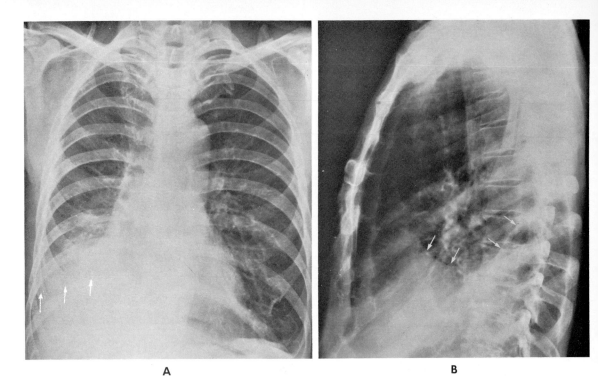

A   B

Figure 9–151 **Fibrosis of the Pleura.**

   A, This is a posteroanterior view in a 61 year old male. The film was taken six months following an episode of bilateral pleurisy. Hazy densities obscure the right diaphragm and obliterate the right costophrenic
angle *(arrows).* These were areas of thickened pleura. Note loss of volume in the right thorax.

   B, Lateral view indicates that involvement was mainly posterior *(arrows).* The pleural density remains
unchanged with alteration in position, and there is little or no radiologic change over extended periods.

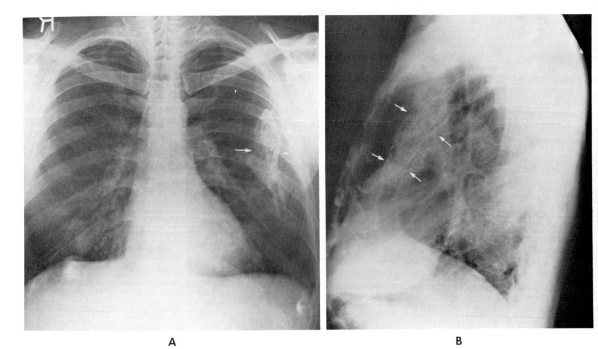

A   B

Figure 9–152 **Calcification of the Pleura.**

   A, Sharply demarcated irregular calcific density overlies the left lateral chest wall *(arrows):* no other abnormalities are seen. The density in the right lower lobe area is an artifact.

   B, Lateral view demonstrates calcification along the anterior and lateral portion of the chest wall
(arrows).

   Plaques of calcification are characteristic and may develop in any pleural segment affected by postin-
flammatory fibrosis.

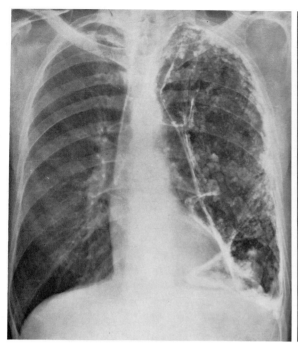 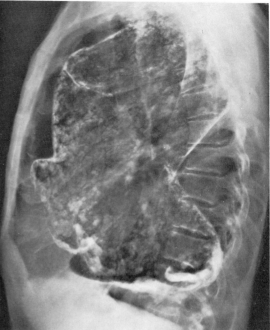

Figure 9–153 **Extensive Calcification of Left Pleura.** Posteroanterior and lateral views in a 44 year old male with history of pneumonia and pleurisy reveals extensive calcification of almost the entire left pleural surface. The left costophrenic sulcus is blunted by old pleural thickening.

Pneumothorax

In younger persons, pneumothorax can result from spontaneous rupture of a subpleural emphysematous bleb, which usually cannot be demonstrated radiographically. In older persons there is generally underlying pulmonary disease with either demonstrable bullae or diffuse interstitial fibrosis and emphysema. Spontaneous pneumothorax is a frequent complication of staphylococcal pneumonia, especially in children. Stab wounds and other penetrating injuries to the chest can tear the pleura or damage the lung and thus cause pneumothorax. Recurrent spontaneous pneumothorax, usually right-sided, can occur during menstrual flow (catamenial pneumothorax). Trauma to the lung apex during insertion of a central venous pressure catheter can often cause pneumothorax.

Pneumothorax appears as a hyperlucent area in which all lung markings are absent. Usually the visceral pleura separating the lung from the pneumothorax is visible. In questionable cases, inspiratory and expiratory films may be helpful; the expiratory film increases the relative density of the lung fields and decreases the lung volume, so that minimal pneumothorax is more clearly demonstrated. Usually some fluid appears in the pleural space within 24 hours.

Tension pneumothorax develops when air continues to enter the pleural space but cannot leave it. The lung on the involved side collapses completely, unless it is diseased or unless adhesions prevent total collapse. The heart and mediastinum are shifted into the normal hemithorax. Thus, cardiac and respiratory function is compromised. The increased intrapleural pressure widens the interspaces and depresses the diaphragm.

Pneumothorax may become chronic, in which case the thickened fibrotic visceral pleura and adhesions prevent the lung from reexpanding even if the air is removed or absorbed.[169]

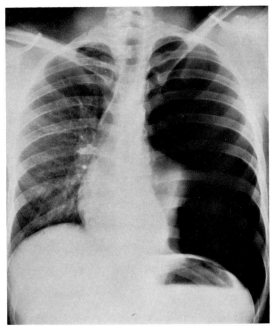

Figure 9–154 **Spontaneous Tension Pneumothorax.**   The entire left lung is collapsed, and the heart and mediastinum are shifted to the right. Note the lucency of the left hemithorax, the depressed diaphragm, and the widened interspaces. Pleural fluid has not yet accumulated, and the costophrenic angle is sharp. Tension pneumothorax may be progressive because of the formation of check valves that allow air to enter but not to leave. Surgical intervention is frequently necessary.

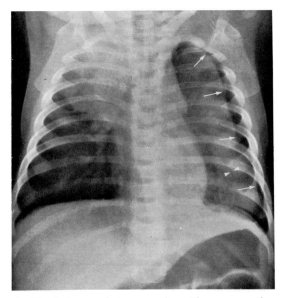

Figure 9–155 **Pneumothorax in Staphylococcal Pneumonia.**   The pneumothorax along the left lateral chest wall *(arrows)* is readily appreciated by contrast with the consolidated left lung. There is consolidation of the right upper lobe due to staphylococcal pneumonia. The presence of an air bronchogram *(arrowheads)* confirms the diagnosis of pulmonary consolidation.

   Pneumothorax is not uncommon in staphylococcal pneumonia, especially in children. It is caused by rupture of pneumatoceles that frequently develop in the lung. The combination of extensive pneumonia and pneumothorax in children strongly suggests staphylococcal pneumonia.

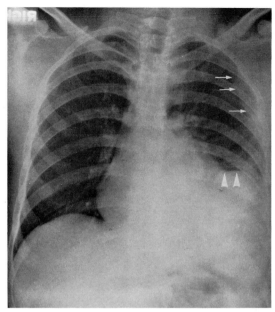

Figure 9–156 **Traumatic Hemopneumothorax.**   A stab wound in the left thorax has caused a pneumo-thorax (*arrows*) with air-fluid levels (*arrowheads*). The fluid was blood, which does not clot in the pleural space. A pleural fluid level always indicates a pneumothorax. Hemopneumothorax cannot be differentiated radiographically from hydropneumothorax. Rarely, hemopneumothorax is spontaneous.

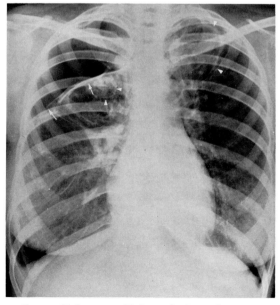

Figure 9–157 **Chronic Pneumothorax: Pulmonary Tuberculosis.**   There is a pneumothorax localized to the right upper lobe. Lung markings are absent. Thickening of the visceral pleura (*arrows*) of the chronically collapsed upper lobe prevents reexpansion. Tuberculous cavitation and infiltration are seen in the left apex and in the right upper lobe *(arrowheads)*.

Mesothelioma

There are two forms of mesothelioma: a benign localized form and a malignant or diffuse type. The benign pleural mesothelioma, sometimes called fibroma, appears as a localized intrathoracic mass often in the lower pleural spaces or in an interlobar fissure. Occasionally it contains calcification. Hypertrophic osteoarthropathy is found in a high percentage of patients with localized or fibrous mesothelioma.

Malignant mesothelioma often is first observed as a localized scalloped density of a pleural surface. Later there is an increase in the nodular pleural thickening, and often extensive pleural effusion develops, obscuring the underlying masses. Withdrawal of fluid and introduction of air will demonstrate the nodularity, which may involve both the parietal and visceral pleura. There may be destruction of the adjacent ribs. The effusion may be massive or quite minimal, but surprisingly little or no mediastinal shift occurs even with larger effusions. The tumor may grow to encase the entire thoracic cavity and sometimes involve the opposite pleural space. Extension into the lung parenchyma, mediastinum, and pericardium is not uncommon. The malignant mesothelioma may begin as a nodular density that may be mistaken for a peripheral parenchymal tumor; its pleural base may be recognized by proper oblique or tangential films.

A high incidence of malignant mesothelioma occurs in patients with asbestosis (see p. 513 and Fig. 9–87). Therefore, the parenchymal fibrosis and pleural calcification of asbestosis may often be associated with scalloped or nodular pleural densities or the effusion of the mesothelioma.[170-173]

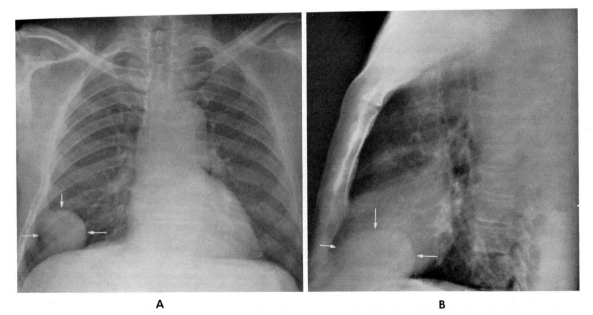

A   **B**

Figure 9–158 **Benign Localized Mesothelioma: Fibroma.**

A, A homogeneous, well-demarcated density *(arrows)* is seen in the lower lung field just above the diaphragm.

B, Lateral view reveals an anterior mass above the diaphragm *(arrows).* The mass was a benign mesothelioma within the major fissure. Radiographically it is indistinguishable from any other intrathoracic mass. Diagnostic pneumothorax might demonstrate its pleural origin and help differentiate it from other intrathoracic masses. (Courtesy Dr. Arlyne Shockman, Veterans Administration Hospital, Philadelphia, Pennsylvania.)

Figure 9–159 **Malignant Mesothelioma.** There is a large pleural effusion in the right lung secondary to a mesothelioma. The nodular pleura is not seen; diagnostic pneumothorax would be needed to demonstrate it. The patient also had pulmonary hypertrophic osteoarthropathy.

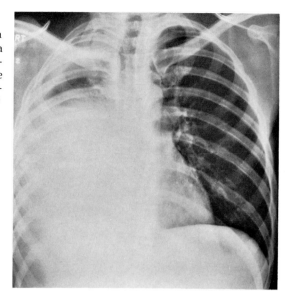

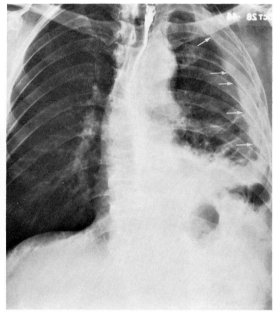

Figure 9–160 **Malignant Mesothelioma.** There is a diffuse nodular thickening of the pleura *(arrows)* due to extensive involvement by a mesothelioma. An infrapulmonic effusion causes apparent elevation of the diaphragm. The patient also had hypertrophic osteoarthropathy in the long bones.

Coxsackie Virus

Pleural and pericardial effusion without parenchymal disease is common in Coxsackie infections. Similar pleural and pericardial findings are seen in lupus erythematosus, tuberculosis, and acute rheumatic fever. (See also p. 145.)

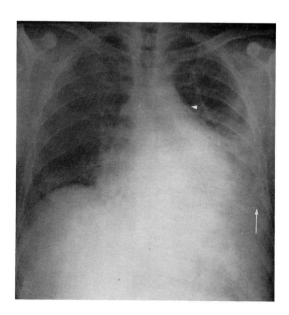

Figure 9–161 **Coxsackie Virus: Pleural and Pericardial Involvement.** Homogeneous density at the left base *(arrow)* is due to a pleural effusion. The cardiac silhouette and the base of the heart *(arrowhead)* are enlarged. This enlargement is due to an associated pericardial effusion. In the absence of clear-cut parenchymal disease, it is difficult to distinguish this picture from that of tuberculosis or lupus erythematosus.

Meigs' Syndrome

This rare syndrome consists of a non-neoplastic pleural effusion and some degree of ascites. These are associated with an ovarian fibroma or other benign ovarian tumors. The pleural effusion is possibly caused by the transfer of ascitic fluid into the pleural space either through a diaphragmatic defect or by way of the diaphragmatic lymphatics, although this has not been verified.

The pleural fluid accumulates preponderantly on one side, most often the right, and reaccumulates rapidly after thoracocentesis. Removal of the diseased ovary leads to rapid disappearance of the pleural and ascitic fluid.

Although a palpable ovarian mass with ascites and pleural effusion could be due to Meigs' syndrome, a malignant ovarian tumor with metastases is overwhelmingly more probable.[174]

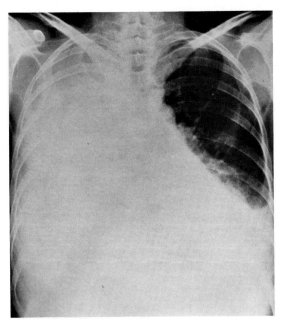

Figure 9–162 **Massive Effusion: Meigs' Syndrome.**   Massive pleural effusion has resulted in development of a homogeneous density in the right thorax. The heart and mediastinum are shifted to the left. A smaller effusion is present at the left base. Such effusion cannot be distinguished radiographically from other types of pleural effusion.

The patient had an ovarian fibroma; the fluid disappeared following surgical removal of the fibroma.

Lupus Erythematosus

The serosal surfaces are frequently involved in lupus erythematosus. Unexplained bilateral pleural thickening or fluid in a woman under the age of 50 should suggest the possibility of lupus. Pleural thickening is seen more frequently than effusion, and enlargement of the cardiac silhouette because of pericardial fluid and myocardial involvement is often an associated roentgen finding. (See also *Systemic Lupus Erythematosus*, p. 43.)

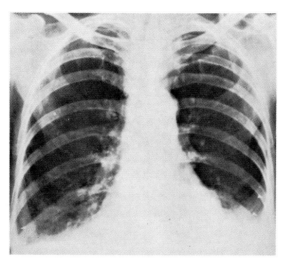

Figure 9–163 **Lupus Erythematosus with Bilateral Pleural Effusions.** The effusion is more pronounced on the left, both costophrenic angles have been obliterated, and there are curvilinear densities extending the length of the lateral chest walls (*arrowheads*). The left diaphragm is almost obscured and the right diaphragm is hazy. Although on radiography the effusion appeared small, over 600 ml. of fluid were removed from the left side. The patient developed systemic symptoms of lupus erythematosus within a few months.

DISEASES OF THE MEDIASTINUM

Acute Mediastinitis

Acute mediastinitis may result from esophageal perforation, from infected mediastinal nodes, or by direct extension of a neck infection.

Diffuse inflammation is characterized by rapidly developing mediastinal widening and ill-defined borders. Mediastinal air will be present if the mediastinitis is due to esophageal or tracheal perforation. Often the inflammation extends into the adjacent lung.

Abscess formation will produce a more sharply bordered mediastinal bulge, and frequently an air-fluid level develops. Adjacent structures, particularly the trachea, may be displaced.[175]

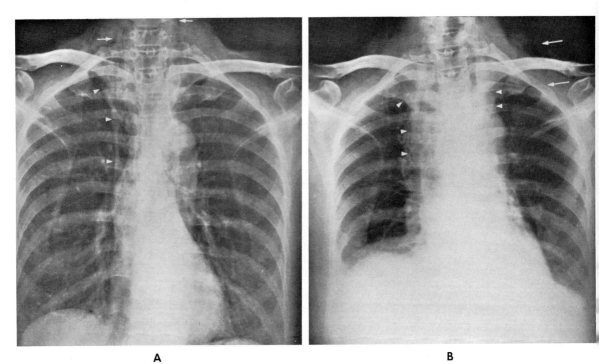

A   **B**

Figure 9–164   **Acute Mediastinitis Following Traumatic Rupture of the Esophagus.**

A, Shortly after esophageal perforation, air is demonstrated in the slightly widened superior mediastinum *(arrowheads)* and neck *(arrows).*

B, Two days later the mediastinum has widened further *(arrowheads)*, and there is residual air in the neck
(arrows). A small right-sided pleural effusion has developed. At surgery, mediastinitis with abscess formation was found.

Chronic Mediastinitis and Idiopathic Mediastinal Fibrosis

Chronic fibrotic mediastinitis may be due to a chronic infectious granuloma, such as histoplasmosis or tuberculosis, but many cases are of unknown origin. The idiopathic form may be related to idiopathic retroperitoneal fibrosis.

Obstruction of the superior vena cava is the most common serious sequela, and about 10 per cent of such obstructions are due to chronic mediastinitis.

Radiographically the mediastinum may appear normal, but when the superior vena cava becomes occluded, there is usually widening either on the right side or bilaterally. The widening is the result of dilated collateral mediastinal veins. Frequently, a mediastinal tumor is suspected from the plain film findings.

Diagnosis is made by opacification studies of the mediastinal veins via injection of contrast material into the veins of the arm. Obstruction of the superior vena cava is seen. The dilated superior vena cava and azygos vein system on the right will correspond to the mediastinal shadow. A dilated tortuous accessory hemiazygos and intercostal vein on the left may also be opacified. These will cause widening of the mediastinum on the left.

Oral barium studies will often disclose multiple intraluminal filling defects in the upper half of the esophagus due to varices. Since the veins of the upper esophagus drain into the azygos system, these veins may become dilated and varicose if there is obstruction of the superior vena cava. Occasionally, the esophagus may be compressed or indented from encircling mediastinal fibrotic tissue.[176-178]

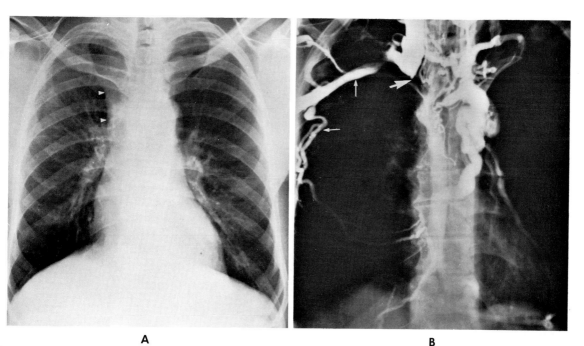

A   **B**

Figure 9–165 **Chronic Mediastinitis with Superior Vena Caval Syndrome Due to Histoplasmosis.**

A, The right superior mediastinum is widened *(arrowheads).*

B, Venography reveals obstruction of the superior vena cava and extensive filling of the collateral circulation *(small arrows).* Note tapering of the obstruction *(large arrow),* which contrasts with the sharply circumscribed obstruction usually caused by malignant mediastinal lesions. The dilated azygos and hemiazygos veins are shown in the mediastinum, descending over the spinal shadow. (Courtesy Dr. Arlyne Shockman, Veterans Administration Hospital, Philadelphia, Pennsylvania.)

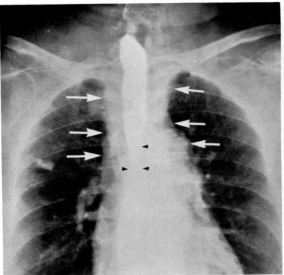

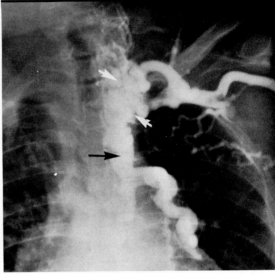

A   B

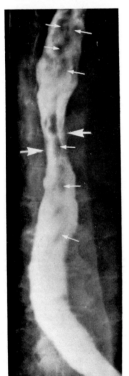

C

Figure 9–166 **Extensive Mediastinal Fibrosis from Tuberculosis: Superior Vena Cava Obstruction and Esophageal Varices.**

This 57 year old man had a superior vena cava clinical syndrome. There had been several episodes of severe gastrointestinal bleeding.

A, The superior mediastinum is irregularly widened bilaterally *(arrows).* The esophagus is persistently narrowed *(arrowheads)* at the T5–T6 level.

B, A left subclavian venogram demonstrates the opacification of a greatly dilated and tortuous left superior intercostal vein *(white arrows),* which drains into a dilated accessory hemiazygos vein *(black arrow).* The flow then extends into the hemiazygos and other collaterals, and it will drain into the inferior vena cava. The superior vena cava was completely blocked.

In superior vena cava block, the dilated superior intercostal vein and dilated accessory hemiazygos frequently produce widening of the left superior mediastinum seen on plain films.

C, Esophagogram demonstrates multiple lucencies due to varices *(small arrows)* in the upper half of the esophagus. They are most prominent in the upper third of the esophagus. The esophageal narrowing is again noted *(large arrows);* the left indentation might be partially due to the aortic knob.

Varices limited to the upper half of the esophagus are almost always due to obstruction of the superior vena cava; this is in contrast to the more common lower esophageal varices, which are usually due to portal hypertension.

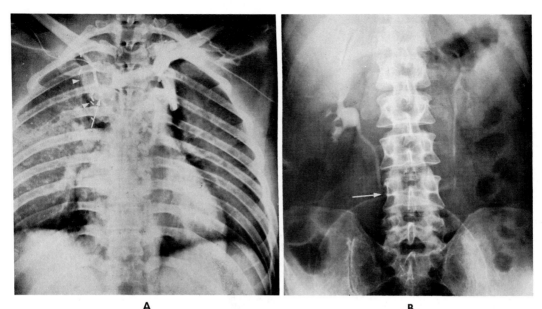

A   B

Figure 9–167 **Idiopathic Mediastinal Fibrosis and Retroperitoneal Fibrosis.**

A, Venography demonstrates both the mediastinal widening *(arrowheads)* and the superior vena cava obstruction *(arrow)* due to idiopathic mediastinal fibrosis. Bilateral parenchymal infiltrates are present.

B, Intravenous urogram shows right midureter displaced toward the midline *(arrow)*, secondary to retroperitoneal fibrosis. There is minimal obstruction of the urinary tract at this time. Idiopathic retroperitoneal fibrosis is occasionally associated with chronic idiopathic mediastinal fibrosis. (Courtesy Dr. Douglas G. Cameron, Montreal, Canada.)

Mediastinal Emphysema (Pneumomediastinum)

Rupture of a pulmonary alveolar bleb may cause dissection along a blood vessel with entrance of air into the mediastinum. This may occur without cause or as a result of severe cough and straining (spontaneous pneumomediastinum). Symptoms are minimal and recovery is spontaneous. Air in the mediastinum may also occur from rupture of a mediastinal air-containing structure (such as the esophagus, trachea, or bronchus), or from spread of retroperitoneal or retropharyngeal air. Pneumomediastinum is also a common finding after chest surgery.

The thin lucent streaks of air are best seen in the superior mediastinum, especially on lateral view, in which air streaks may be seen behind the sternum. There is often evidence of air in the soft tissues of the neck. The mediastinum may be widened. Frequently, air dissects the mediastinal pleura, which then becomes visible as a thin white line. An associated pneumothorax may result from rupture into the pleural space. A primary pneumothorax, however, does not lead to pneumomediastinum. Differentiation of pneumomediastinum from pneumopericardium may require lateral decubitus films.[26, 179, 180]

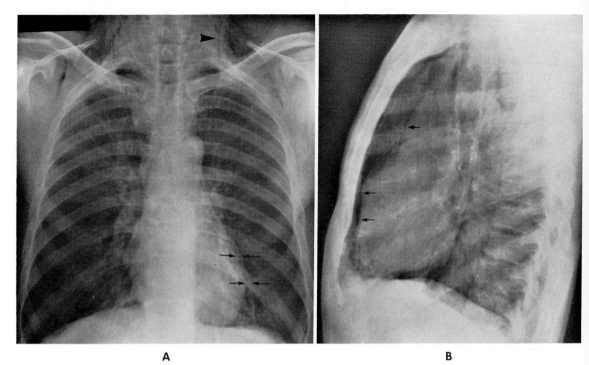

A   B

Figure 9–168 **Mediastinal Emphysema.**

A, Posteroanterior view demonstrates radiolucent streaks of air surrounding the pericardium *(medial arrows).* Notice the thin white line of the mediastinal pleura dissected by air *(lateral arrows).* Thin lucent streaks are also present in the superior mediastinum *(arrowhead).*

B, Lateral view demonstrates the presence of air along the anterior mediastinal pleura *(arrows).* This pneumomediastinum cleared spontaneously and was believed to have been caused by rupture of a bleb.

Herniation Through the Mediastinum

Herniation of a portion of lung through the mediastinum into the opposite thorax most often takes place through the anterior-superior mediastinum, where a potential space exists behind the sternum. A lung may herniate to compensate for marked loss of volume on the opposite side (atelectasis) or as a result of increased pressure on the ipsilateral side, as in a large pleural effusion or tension pneumothorax.

On the posteroanterior film the border of the herniated lung appears as a thin crescentic line superimposed on the other lung, close to the mediastinum and usually in the upper portion of the chest. On lateral view there is an increased lucency in the superior retrosternal area. The posterior-superior and the posterior-inferior portions of the mediastinum are less frequent sites of herniation.

Symptoms result from the primary pulmonary condition and not from the herniation.[181]

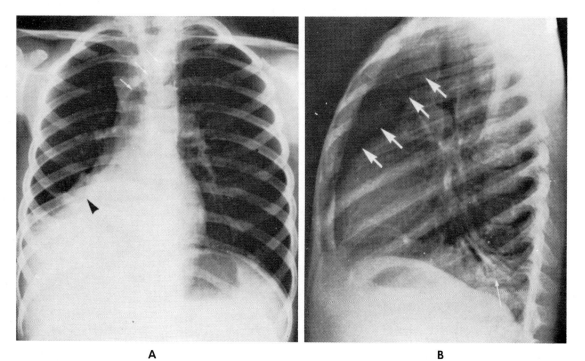

A

B

Figure 9–169 **Mediastinal Herniation Secondary to Atelectasis.**

A, The upper left lung has herniated through the mediastinum into the right hemithorax *(arrows)* and has a crescentic shape. Herniation compensated for atelectasis of the right lower lobe *(arrowhead)*.

B, Increased retrosternal lucency on lateral view represents the herniated left lung *(large arrows)*. The atelectatic right lower lobe *(small arrow)* obscures the right diaphragm.

Mediastinal Mass Lesions

For radiographic localization, the mediastinum can be arbitrarily divided into an *anterior compartment,* which lies anterior to the trachea and to the posterior normal heart border; a *middle compartment,* which extends back to the anterior vertebral bodies; and a *posterior compartment,* consisting of the paravertebral spaces. Each compartment is divided into a superior and an inferior portion.

In the anterior-superior mediastinum the most frequently seen lesions are thymic tumors, teratodermoids, substernal thyroid, lipomas, and lymph node enlargements that are usually lymphomatous. In the middle mediastinum may be found vascular anomalies and abnormalities, enlarged lymph nodes, bronchogenic cysts, esophageal lesions, and hiatal hernias. The posterior mediastinum is the site of neurogenic tumors and occasionally of neurenteric canal and duplication cysts and extramedullary hematopoietic masses. In the middle-superior mediastinum or the posterior-superior mediastinum one sees aneurysm of the aorta and brachiocephalic vessels, substernal thyroid, and, occasionally, lymph nodes. Tumors can arise in any tissue normally found in the mediastinum.

Roentgenologic detection of a mediastinal mass may be obvious or extremely subtle; correct etiologic diagnosis is often impossible and must await surgical biopsy. Some lesions, however, do have readily identifiable radiographic features.

In addition to regular, oblique, and overexposed chest films, body section films and opacification studies of mediastinal vessels can be utilized. Diagnostic pneumomediastinum is rarely employed.

On frontal chest films the mediastinum may merely appear widened, but more often a bulging density is apparent. Lateral view may identify a clear-cut mass. Demonstration of calcification or of displacement of mediastinal structures is a helpful finding.

Body section films (planigraphy) can often more clearly define a mediastinal mass or questionable calcifications. Vascular opacification studies (superior vena cavography, arch aortography) may disclose displacements, narrowing, or occlusion of adjacent vessels. The mass may prove to be of vascular origin, such as an aneurysm, a malformation, or dilated veins.

Nonspecific mediastinal masses may be metastatic nodes, and careful roentgen search for a primary malignancy should be instituted before surgical exploration.[175, 180, 182-184]

Lymphoma and Hodgkin's Disease

Lymphomas in the chest most frequently produce hilar and anterior-superior mediastinal adenopathy, although any portion of the mediastinum is occasionally involved. The changes are usually bilateral. Characteristically the densities are lobular and sharply outlined, although occasionally the border is less clear owing to lymphatic permeation and rapid growth. Lymphatic spread appears as linear densities extending from the mass into the lung fields. Tracheal displacement is unusual.[175, 180, 182, 183]

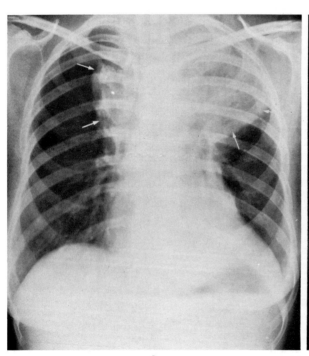 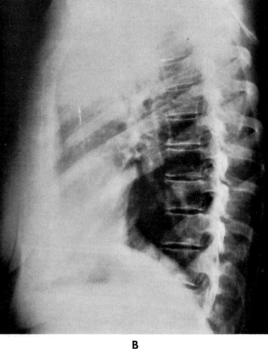

A   B

Figure 9–170   **Hodgkin's Disease in Anterior-Superior Mediastinum.**

A, A 44 year old man had clinical evidence of superior vena cava syndrome. Posteroanterior view demonstrates a large bilateral mass in the anterior-superior mediastinum; the mass extends almost as far as the lateral chest wall on the left *(arrows).* There is little tracheal displacement.

B, Lateral view indicates the predominantly anterior location of the mass *(arrow).*

These changes were attributable to a large mass of conglomerate lymph nodes caused by Hodgkin's disease.

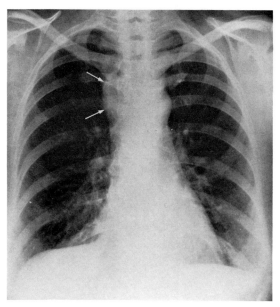

Figure 9–171 **Mediastinal Hodgkin's Disease.** A circumscribed mediastinal mass (*arrows*) is indicated in posteroanterior view without evidence of parenchymal infiltration or tracheal displacement. The appearance of the mass is not diagnostic, but its location suggests the possibility of a lymphoma. Mediastinal tuberculosis and bronchogenic carcinoma may produce a similar appearance.

Thymic Tumors: Thymoma and Thymic Cyst

Thymic tumors arise in the anterior mediastinum at the junction of the great vessels and heart and appear as circumscribed round or lobulated masses. Occasionally they displace the heart and great vessels posteriorly. The tumors may be obscured by the sternum, great vessels, or heart shadow on the posteroanterior film and may be apparent only on the lateral view. An air lucency located where the edge of the mass meets the anterior chest wall (sulcus sign) is frequently apparent on lateral films. The sulcus sign is suggestive of a thymic tumor, since it rarely occurs in an anterior mediastinal teratoma. Planigraphy may be necessary for demonstrating a small thymic tumor.

On frontal view, a unilateral lesion may be benign or malignant, whereas a lesion extending bilaterally is most often malignant. Mottled calcification within the tumor substance or linear streaks in the wall are occasionally seen in either type. Thymic tumors may merge into the heart shadow, so that an angiocardiogram may be needed for differentiation from a cardiac or vascular abnormality. Evidence of invasion seen at surgery is often a better criterion of malignancy than are the histologic findings, which may be equivocal. Myasthenia gravis is frequently associated with thymic tumors.

Thymic cysts occur in the same location as thymomas and are usually smooth and rounded. Calcification of the wall is not uncommon.

The rare thymolipoma may attain huge proportions and often will lie on the anterior aspect of the diaphragms.[182, 184, 185]

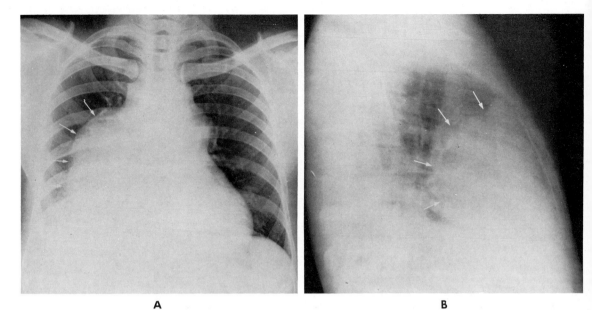

A   **B**

Figure 9–172 **Malignant Thymoma.**

  A, Posteroanterior view reveals a sharply circumscribed mass in the right side of the anterior mediastinum *(arrows).* The right cardiac border is obscured, and there is fluid at the right base.

  B Lateral view indicates that the density is located anteriorly *(arrows).* The right diaphragm is obscured by fluid.

  At surgery a malignant thymoma was found with evidence of metastasis.

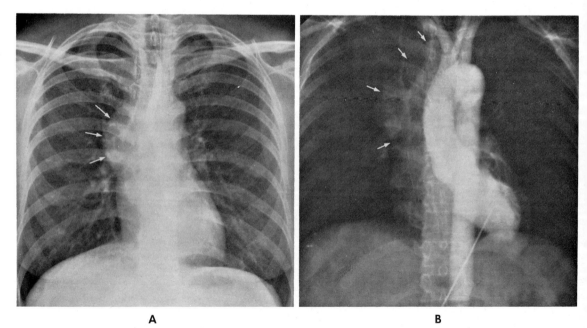

A   **B**

Figure 9–173 **Thymic Cyst: Cardiac Ventriculogram and Arteriogram.**

  A, A circumscribed mass *(arrows)* in the right anterior-superior mediastinum resembles an aortic aneurysm.

  B, Cardiac ventriculography shows the mass *(arrows)* distinct from the opacified aorta.

  A benign anterior mediastinal thymic cyst was found.

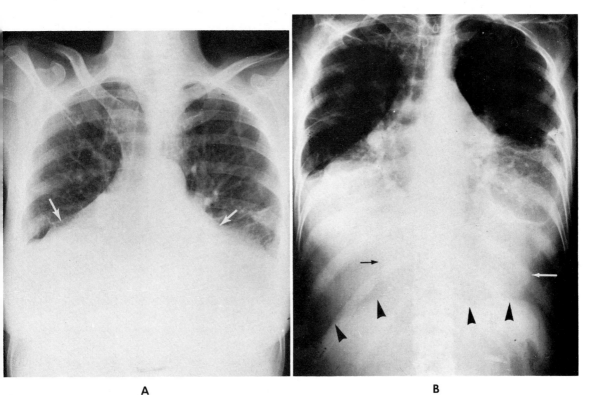

A   **B**

Figure 9–174   **Benign Thymic Tumor: Thymolipoma.**

A, Frontal chest film reveals an apparent huge cardiac shadow and obliteration of the diaphragms (*arrows*).

B, On an overpenetrated film the cardiac (*arrows*) and diaphragmatic shadows (*arrowheads*) can be seen through a mass. The mass was a huge thymolipoma lying directly over the heart anteriorly and on the diaphragms.

A large thymolipoma will obscure the heart and diaphragms, but since the tumor is primarily fatty tissue, a well-penetrated film will allow visualization of the denser heart and diaphragms.

Teratodermoid

Teratodermoids arise in the anterior mediastinum, most frequently at the root of the great vessels. They may be unilateral or bilateral. They are generally circumscribed round or oval masses containing calcified areas. Occasionally, ossifications, including teeth and bones, are seen within the dermoid. There may be sufficient fatty material within the tumor to produce lucent areas on the radiograph. Rarely, perforation may occur into the surrounding structures.[175, 180, 182–184, 186]

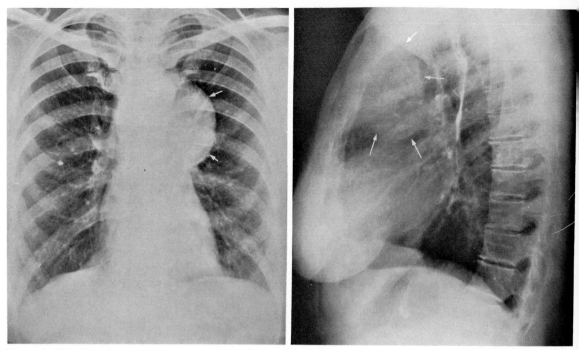

Figure 9–175 **Teratoma.** Posteroanterior and lateral views in an asymptomatic 44 year old woman demonstrate a circumscribed mass in the left anterior mediastinum (*arrows*); the mass is uncalcified. The adjacent structures are not displaced. The radiographic findings are not diagnostic. In the absence of characteristic calcification or areas of fatty lucency, distinction cannot be made between a teratodermoid and a thymic tumor.

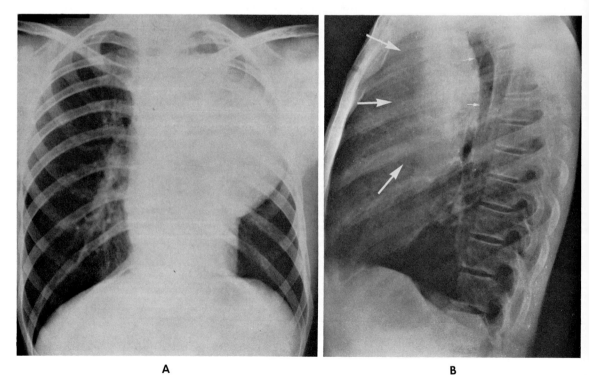

A   B

Figure 9–176 **Dermoid.**

A, There is a large sharply demarcated tumor of the left superior mediastinum.

B, In lateral view, a mass displaces the trachea posteriorly (*small arrows*). The anterior half of the tumor appears lucent (*large arrows*) because of its high fat content.

Dermoids may contain teeth, bones, and large amounts of fat, producing a characteristic picture.

582

Neurogenic Tumors

The schwannoma and ganglioneuroma are the chief neurogenic tumors, and with rare exceptions they arise in the posterior mediastinum. Other tumors include pheochromocytoma, paraganglioma (chemodectoma), neurofibroma, and neurilemoma. They appear as unilateral well-defined round or oval densities arising from or near the intervertebral foramen and usually extending as far back as the gutter. Calcification is infrequent. The tumors often cause erosion and spreading of the ribs, erosion of the pedicles and vertebral bodies, and widening of the intervertebral foramen. On myelography, an intraspinal component may be demonstrated (dumb-bell tumor).

A schwannoma usually appears as a round mass with a short attachment and is seen equally well in posteroanterior and lateral views. A ganglioneuroma is generally elongated and flattened, having indefinite superior and inferior borders, and is sometimes poorly defined on lateral view. Nevertheless, radiologic distinction is usually impossible. Pleural effusions occasionally occur but do not indicate a malignant lesion. Occasionally arteriography is necessary to differentiate a neurogenic tumor from an aneurysm.[175, 180, 187]

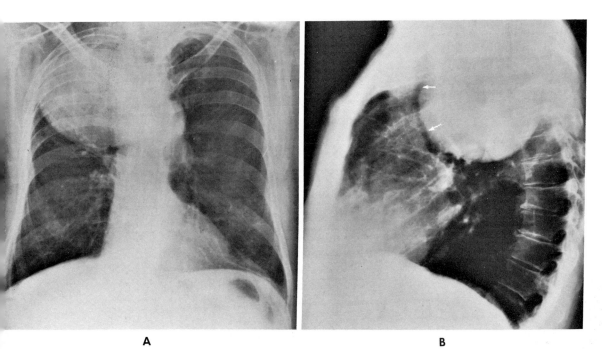

A **B**

Figure 9–177 **Neurogenic Tumor.**

A, A circumscribed density in the right upper thorax is demonstrated in posteroanterior view. There are no bony changes in the ribs or vertebrae.

B, Lateral view demonstrates the density in the posterior mediastinum extending from the posterior gutter anteriorly and displacing the trachea (*arrows*). The location suggests neurogenic tumor.

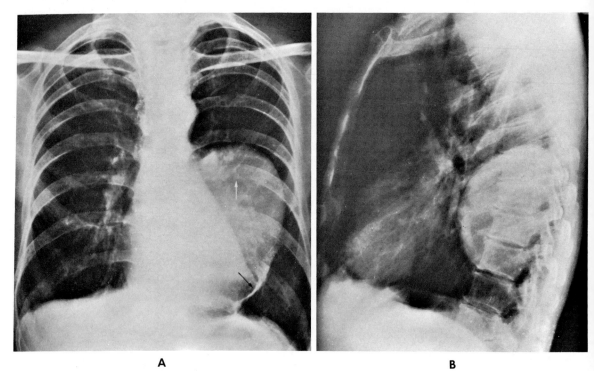

A   **B**

Figure 9–178   **Neurogenic Tumor: Bone Changes.**

   A, A circumscribed density is seen in the left lower thorax. The posterior portion of the eighth rib (*white arrow*) is elevated. The periphery of the tumor is calcified (*black arrow*) — an unusual finding in neurogenic tumors.
   B, Lateral view shows the posterior lesion extending as far as the region of the neural foramen.

Non-Neoplastic Mediastinal Lesions

   These lesions include substernal thyroid, dilatation of the azygos vein, granulomatous lymph nodes, and aneurysm of the innominate artery and aorta.
   A substernal thyroid usually appears as a bilateral, eccentric, sharply defined anterior-superior mediastinal mass that is narrowest on its inferior border. The trachea is almost invariably displaced and is frequently compressed. On fluoroscopy the mass may appear to move during swallowing. Occasionally there is amorphous calcification. Demonstration of extension of the substernal thyroid into the neck confirms the diagnosis. Very rarely a substernal thyroid arises posterior to the trachea. Definitive diagnosis can be made by radioactive iodine scan if the substernal gland is functioning.
   The normal azygos vein does not exceed 6 mm. in diameter on erect views. Dilatation of the vein appears on frontal view as an oval density in the right tracheobronchial angle. Dilatation may be idiopathic or secondary to right heart failure, portal hypertension, obstruction of the inferior or superior vena cava, or absence of the inferior vena cava. Rarely it simulates a mediastinal tumor, but there is a change in size with postural changes and with changes in intrathoracic pressure. Opacification studies of the venous system are diagnostic in doubtful cases.
   Enlarged mediastinal nodes, especially in the paratracheal area, are often seen in sarcoidosis. Usually, hilar adenopathy is associated (see *Sarcoidosis*, p. 107). Other

inflammatory nodes (tuberculosis, histoplasmosis, and so forth) can also involve the mediastinum and simulate a mediastinal tumor.

An aneurysm of the innominate artery appears as a superior mediastinal mass projecting to the right of the midline and continuous with the aortic arch in all projections. It lies partially above the aortic arch and frequently contains linear calcification. It usually displaces the trachea to the left. The aneurysm may be seen to pulsate on fluoroscopic examination, but definitive diagnosis depends on angiographic demonstration of the aneurysm.

Elongation of an arteriosclerotic aorta may cause the innominate artery to buckle. The mediastinal density thus produced is very similar to an aneurysm but generally does not cause tracheal displacement. Angiography may be required to differentiate this condition from a mediastinal tumor or aneurysm.

Less common mediastinal masses include duplication cysts of the esophagus, neuroenteric cysts, intrathoracic meningocele, and masses of extramedullary hematopoietic tissue.[175, 180, 188]

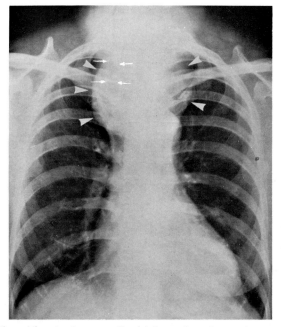

Figure 9-179 **Substernal Thyroid.** A circumscribed bilateral and anterior-superior mediastinal mass (*arrowheads*), which is more prominent on the right, has displaced the trachea to the right (*arrows*). The location, eccentricity, and tracheal displacement are characteristic of substernal thyroid.

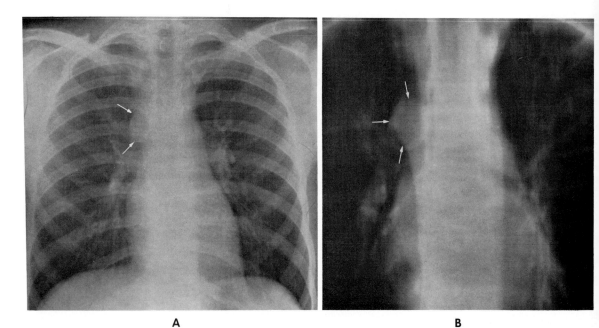

A   **B**

Figure 9–180   **Dilated Azygos Vein.**

A, Posteroanterior view reveals an oval density in the right tracheobronchial angle (*arrows*).

B, Planigram indicates shape and location of the density (*arrows*), which are typical of the azygos vein. At surgery, no cause for the dilatation could be found. (Courtesy Dr. Wilhelm Z. Stern, Montefiore Hospital, New York, New York.)

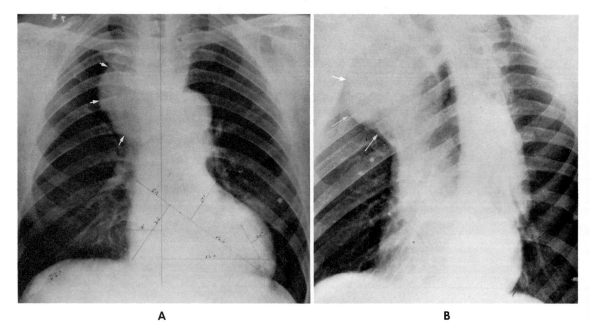

A   **B**

Figure 9–181   **Aneurysm of the Innominate Artery.**

A, In the right superior mediastinum there is a smooth circumscribed density (*arrows*) that extends above the aortic arch.

B, Oblique view indicates anterior location of the mass (*arrows*). It is continuous with the aorta in all projections. Angiography is necessary for definitive diagnosis.

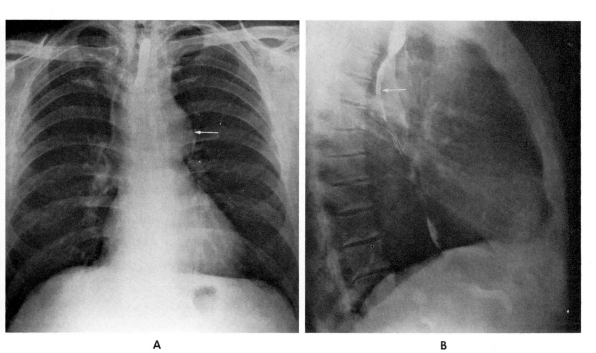

A   B

Figure 9–182   **Aneurysm of the Aorta.**

A, There is a homogeneous mediastinal density just below the aortic knob (*arrow*). The trachea and esophagus are displaced to the right. The mass emerges with the aortic shadow.

B, Lateral view shows posterior displacement of the esophagus (*arrow*) by the mass, which is continuous with the aorta. The trachea is separated from the esophagus by the mass. No vertebral erosion is evident.

Angiography is definitive in the diagnosis of aneurysm. Occasionally a mass lesion develops in the aortic area and is indistinguishable from an aneurysm on plain films.

Aneurysm of the arch of the aorta invariably displaces the trachea and esophagus. Pulsation is usually seen fluoroscopically but is of little value in the differential diagnosis, since pulsations can be transmitted to a lesion adjacent to the aorta.

Benign Lymph Node Hyperplasia

Nonspecific hyperplasia of lymph nodes in the thorax may simulate a variety of tumors. There are no characteristic or distinguishing features.[189]

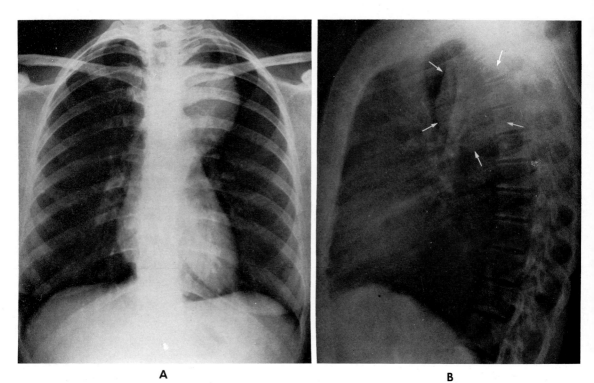

A   **B**

Figure 9–183 **Lymph Node Hyperplasia.**

A, A large smooth density on the left side of the superior mediastinum is seen in posteroanterior view.

B, Lateral view reveals a mass in the posterior mediastinum (*arrows*); the mass displaces the trachea forward and resembles a mediastinal tumor. There are no distinguishing roentgenographic features.

Cystic Lesions of the Mediastinum

Bronchogenic cysts are congenital lesions usually located near the bifurcations of the trachea and major bronchi. The lesion appears as a circumscribed round or oval density that often may displace the adjacent trachea or bronchi. If communication develops with the trachea or bronchi, an air-fluid level may be seen. The cyst may be difficult to visualize on plain films, and displacement of adjacent structures may be the only roentgenologic clue. (See also *Bronchogenic Cysts*, p. 483.)

Esophageal cysts may be acquired or congenital. The acquired type is due to occlusion of the ducts of the secretory glands of the esophagus. Since the congenital cyst is a form of duplication of the esophagus, it may or may not communicate with the esophagus. If the cyst is supplied by an anomalous vessel coming off the aorta, it represents a sequestration. Many duplications do not cause indention of the esophagus, which may appear entirely normal with barium-swallow examination. When the cysts displace the esophagus, they produce a smooth, eccentric, curvilinear impression in the barium-filled esophagus, with a sharp angle at the periphery. The mucosal pattern of the esophagus remains intact. The cyst may become fairly large before symptoms develop.

Thymic cyst is discussed on page 579 and in Figure 9–173.[190, 191]

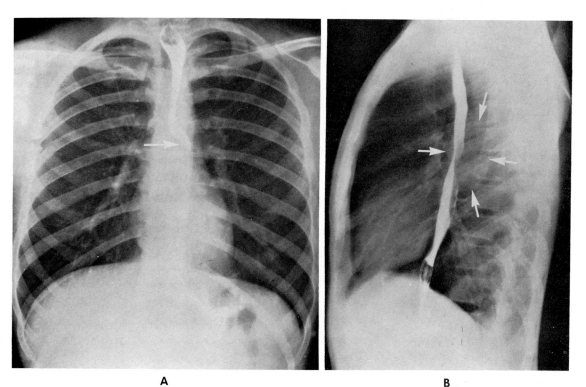

A   B

Figure 9–184 **Bronchogenic Cyst.**

A, In posteroanterior view, the esophagus is displaced to the left (*arrow*), but no other abnormalities are seen.

B, In lateral view, a faint posterior mediastinal mass (*arrows*) is seen indenting the esophagus. It was a bronchogenic cyst.

These lesions are round and well defined and arise most frequently in the region of the carina.

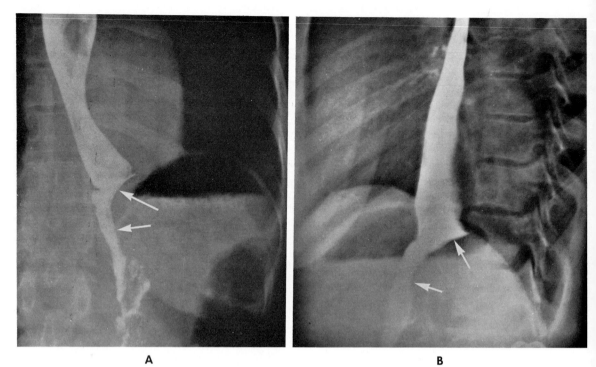

A   **B**

Figure 9–185 **Esophageal Cyst (Duplication Cyst).**

A, In this posteroanterior view, a crescentic impression is seen on the distal esophagus at the cardioesophageal junction adjacent to the stomach gas bubble (*arrows*); the esophageal lumen is narrowed.

B, Lateral view reveals an indentation of the posterior esophagus (*arrow*). The mucosal pattern of the esophagus is intact.

The diagnosis was duplication cyst of the esophagus. Most congenital esophageal cysts arise from the posterior portion of the lower esophagus.

Mediastinal Metastases

Metastases to the mediastinal lymph nodes are usually associated with other pulmonary metastases or with a primary bronchogenic carcinoma, but they may occur without other pulmonary findings. Small mediastinal metastases may be difficult or impossible to detect; larger metastatic masses will cause unilateral or bilateral mediastinal widening. A primary mediastinal tumor or lymphoma will produce a similar roentgen appearance.[175, 180]

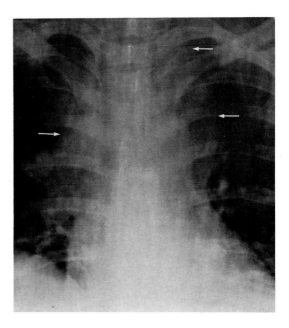

Figure 9–186 **Mediastinal Metastases.** There is diffuse mediastinal widening (*arrows*) due to metastatic lymph node disease. The primary lesion was a small bronchogenic carcinoma in the right hilum.

Subtle focal enlargement of the mediastinum can often be recognized only by serial films.

Mediastinal Extramedullary Hematopoiesis

Circumscribed masses of hematopoietic tissue occasionally are seen in the posterior mediastinum in patients with severe chronic anemia, especially the thalassemic anemias (see p. 1155).

On lateral and well-penetrated frontal chest films, these masses appear as well-demarcated densities, usually multiple and often lobulated. Neither calcification nor adjacent bone erosions are present. The masses are found in the extreme posterior mediastinum, between T_2 and T_{11}.[192, 193]

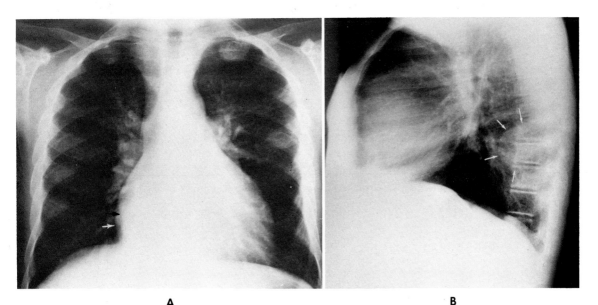

A   B

Figure 9–187 **Extramedullary Hematopoiesis in Thalassemia Major.**

A, There are rib changes characteristic of thalassemia (see Fig. 14–26). The double density (*black arrow*) of the right lower cardiac border was due to a mass of extramedullary hematopoiesis in the posterior mediastinum. White arrow indicates right heart border.

B, Lateral film reveals this mass of extramedullary hematopoiesis (*arrows*) adjacent to the spine.

REFERENCES

1. Campbell, J. A.: The diaphragm in roentgenology of the chest. Radiol. Clin. North Am., *1*:395, 1963.
2. Bartley, O., and Wickbom, I.: Roentgenologic diagnosis of rupture of the diaphragm. Acta Radiol., *53*:33, 1960.
3. Samaan, H. A.: Undiagnosed traumatic diaphragmatic hernia. Br. J. Surg., *58*:257, 1971.
4. Cole, W. R.: Traumatic diaphragmatic hernia. Australas. Radiol., *12*:109, 1968.
5. Reed, J. O., and Lang, E. F.: Diaphragmatic hernia in infancy. Am. J. Roentgenol. Radium Ther. Nucl. Med., *82*:437, 1959.
6. Haines, J. O., and Collins, R. B.: Bochdalek hernia in an adult. Radiology, *95*:277, 1971.
7. Patton, I., and Harris, W. C.: Herniation through the foramen of Morgagni. Br. J. Radiol., *34*:378, 1961.
8. Tamas, A., and Dunbar, J. S.: Eventration of the diaphragm. J. Can. Assoc. Radiol., *8*:1, 1957.
9. Gregg, I., and Trapnell, D. H.: Bronchographic appearance of early chronic bronchitis. Br. J. Radiol., *42*:132, 1969.
10. Knott, J. M. S., and Christie, R. U.: Radiologic diagnosis of emphysema. Lancet, *1*:881, 1951.
11. Friemanis, A. K., and Molnar, W.: Chronic bronchitis and emphysema at bronchography. Radiology, *74*:194, 1960.
12. Brünner, S., and Pock-Steen, O. C.: Localized emphysema. Acta Radiol., *53*:184, 1960.
13. Laws, J. W., and Heard, B. E.: Emphysema and the chest film: a retrospective radiologic and pathologic study. Br. J. Radiol., *35*:750, 1962.
14. Mori, P. A., et al.: The radiologic spectrum of aging and emphysematous lungs. Radiology, *83*:48, 1964.
15. Jacobson, G., et al.: Vascular changes in pulmonary emphysema. Am. J. Roentgenol. Radium Ther. Nucl. Med., *100*:374, 1967.
16. Hepper, N. G., et al.: The prevalence of alpha-1-antitrypsin deficiency in selected groups of patients with chronic obstructive lung disease. Mayo Clin. Proc., *44*:697, 1969.
17. Mittman, C., et al.: Effect of smoking on lung function in alpha-1-antitrypsin heterozygotes. Clin. Res., *17*:553, 1969.
18. Berg, N. O., and Eriksson, S.: Liver disease in adults with alpha-antitrypsin deficiency. N. Engl. J. Med., *287*:1264, 1972.
19. Heitzman, E. R., Markarian, B., and Solomon, J.: Chronic obstructive pulmonary disease. Radiol. Clin. North Am., *11*:49, 1973.
20. Nelson, S. W., and Christoforidis, A.: Reversible bronchiectasis. Radiology, *71*:375, 1958.
21. Burke, E. N.: Laminagraphic appearance of bronchiectasis. Am. J. Roentgenol. Radium Ther. Nucl. Med., *79*:251, 1958.
22. DiRienzo, S., and Weber, H. H.: Radiologische Exploration des Bronchus. Stuttgart, Georg Thieme Verlag, 1960.
23. Ferguson, T. B., and Burford, T. H.: The changing pattern of pulmonary suppuration. Dis. Chest, *53*:396, 1968.
24. Heikel, P. E., and Tarvala, R.: Bronchiectasis in children; high-kilovoltage plain radiography as a diagnostic aid. Ann. Radiol. (Paris), *15*:175, 1972.
25. Fraser, R. G., and Bates, D. V.: Body section roentgenography in the evaluation and differentiation of chronic hypertrophic emphysema and asthma. Am. J. Roentgenol. Radium Ther. Nucl. Med., *82*:39, 1959.
26. Munsell, W. P.: Pneumomediastinum. J.A.M.A., *202*:689, 1967.
27. Lubert, M., and Krause, G. R.: Further observations on lobar collapse. Radiol. Clin. North Am., *1*:331, 1963.
28. Felson, B.: *Fundamentals of Chest Roentgenology*. Philadelphia, W. B. Saunders Company, 1973.
29. Huzly, A.: Middle lobe syndrome. Fortschr. Roentgenstr., *97*:407, 1962.
30. Albo, R. J., and Grimes, O. F.: The middle lobe syndrome. Dis. Chest, *50*:509, 1966.
31. Billig, D. M., and Darling, D. B.: Middle lobe atelectasis in children. Am. J. Dis. Child., *123*:96, 1972.
32. Rapkin, R. H.: The diagnosis of epiglottitis: Simplicity and reliability of radiographs of the neck in the differential diagnosis of the croup syndrome. J. Pediatr., *80*:96, 1972.
33. Boyden, F. M., et al.: Adult epiglottitis. Am. J. Roentgenol. Radium Ther. Nucl. Med., *109*:467, 1970.
34. Bass, J. W., Steele, R. W., and Wiebe, R. A.: Acute epiglottitis. J.A.M.A., *229*:671, 1974.
35. Braman, S. S., and Whitcomb, M. E.: Mucoid impaction of the bronchus. J.A.M.A., *223*:641, 1973.
36. Johnston, R. F.: Tracheobronchomegaly: report of five cases and demonstration of familial occurrence. Am. Rev. Resp. Dis., *91*:35, 1965.
37. Suprenant, E. L., and O'Loughlin, B. J.: Tracheal diverticula and tracheobronchomegaly. Dis. Chest, *49*:345, 1966.
38. Bateson, E. M., and Woo-Ming, M.: Tracheobronchomegaly. Clin. Radiol., *24*:354, 1973.
39. Joress, M. H.: Pulmonary cystic disease: observations in cases treated by exploratory thoracotomy. Dis. Chest, *35*:256, 1959.
40. Price, A. H., and Teplick, J. G.: Progressive bilateral bullous emphysema. Arch. Intern. Med., *77*:132, 1946.

41. Jensen, K. M., Miscall, L., and Steinberg, I.: Angiocardiography in bullous emphysema: its role in selection of the case suitable for surgery. Am. J. Roentgenol. Radium Ther. Nucl. Med., 85:229, 1961.

42. Foreman, S., and Weill, H., et al.: Bullous disease of the lung. Physiologic improvement after surgery. Ann. Intern. Med., 69:757, 1968.

43. Boushy, S. F., and Kohen, R., et al.: Bullous emphysema: Clinical roentgenologic and physiologic study of 49 patients. Dis. Chest, 54:327, 1968.

44. Weir, J. A.: Congenital anomalies of the lung. Ann. Intern. Med., 52:330, 1960.

45. Weinreich, J., and Wolfort, W.: Bronchography in progressive pulmonary dystrophy. Fortschr. Roentgenstr., 95:641, 1961.

46. Ziter, F. M. H., Bramwit, D. N., Holloman, K. R., et al.: Calcified mediastinal bronchogenic cysts. Radiology, 93:1025, 1969.

47. Christoforidis, A. J., Nelson, S. W., and Pratt, P. C.: Bronchiolar dilatation associated with muscular hyperplasia: polycystic lung. Am. J. Roentgenol. Radium Ther. Nucl. Med., 92:513, 1964.

48. Kuisk, H., and Sanchez, J. S.: Desquamative interstitial pneumonia and idiopathic diffuse pulmonary fibrosis. Am. J. Roentgenol. Radium Ther. Nucl. Med., 107:258, 1969.

49. Bower, G. C.: So called bronchiolar emphysema. Am. Rev. Resp. Dis., 96:1049, 1967.

50. Allen, R. P., et al.: Congenital lobar emphysema. Radiology, 86:929, 1961.

51. Roghair, G. D.: Non-operative management of lobar emphysema. Radiology, 102:125, 1972.

52. Weiss, W., and Johnston, D. G.: Pulmonary histiocytosis X. Am. Rev. Tuberc., 75:319, 1951.

53. Crisler, E. C., Durant, J. K., and Parker, J. M.: Pulmonary histiocytosis X. Am. J. Roentgenol. Radium Ther. Nucl. Med., 85:271, 1961.

54. Flaherty, R. A., Keegan, J. M., and Sturtevant, H. N.: Post-pneumonic pulmonary pneumatoceles. Radiology, 74:50, 1960.

55. Rumbaugh, I. F., and Prior, J. A.: Lung abscess: a review of forty-one cases. Ann. Intern. Med., 55:223, 1961.

56. Good, C. A., and Holman, C. B.: Cavitary carcinoma of the lung: features in nineteen cases. Dis. Chest, 37:289, 1960.

57. Margolin, H. N., Rosenberg, L. S., Felson, G., and Baum, G.: Idiopathic hyperlucent lung. Am. J. Roentgenol. Radium Ther. Nucl. Med., 82:63, 1959.

58. Culiner, M. M.: The hyperlucent lung, a problem in differential diagnosis. Dis. Chest, 49:578, 1966.

59. MacPherson, R. I., Cumming, G. R., and Chernick, V.: Unilateral hyperlucent lung. A complication of viral pneumonia. J. Can. Assoc. Radiol., 20:225, 1969.

60. Williams, J. R., and Bonte, F. S.: Pulmonary damage in non-penetrating chest injuries. Radiol. Clin. North Am., 1:439, 1963.

61. Reynolds, J., and Davis, J. T.: Injuries of the chest wall, pleura, pericardium, lungs, bronchi and esophagus. Radiol. Clin. North Am., 4:383, 1966.

62. Hirsch, M., and Bazini, J.: Blast injury of the chest. Clin. Radiol., 20:362, 1969.

63. Wholey, M. H., Good, C. A., and McDonald, J. R.: Disseminated indeterminate pulmonary disease. Radiology, 71:350, 1958.

64. Johnson, T. H., Gajaraj, A., and Feist, J. H.: Patterns of pulmonary interstitial disease. Am. J. Roentgenol. Radium Ther. Nucl. Med., 109:516, 1970.

65. Johnson, T. H., Gajaraj, A., and Feist, J. H.: Vascular key to diagnosis of pulmonary interstitial diseases. Am. J. Roentgenol. Radium Ther. Nucl. Med., 113:518, 1971.

66. Haile, T. S., et al.: Pulmonary edema without cardiomegaly. Am. J. Roentgenol. Radium Ther. Nucl. Med., 103:555, 1968.

67. Stern, W. Z., et al.: Roentgen findings in acute heroin intoxication. Am. J. Roentgenol. Radium Ther. Nucl. Med., 103:522, 1968.

68. Dyck, D. R., and Zylak, C. L.: Acute respiratory distress in adults. Radiology, 106:497, 1973.

69. Robin, E. D., and Carey, L. C., et al.: Capillary leak syndrome with pulmonary edema. Arch. Intern. Med., 130:66, 1972.

70. Hall, R. M., and Margolin, F. R.: Oxygen alveolopathy in adults. Clin. Radiol., 23:11, 1972.

71. Felman, A. H.: Neurogenic pulmonary edema. Am. J. Roentgenol. Radium Ther. Nucl. Med., 112:393, 1971.

72. Lull, G. F., Beyer, J. C., Maier, J. G., and Moss, D. F., Jr.: Pulmonary alveolar proteinosis. Am. J. Roentgenol. Radium Ther. Nucl. Med., 82:76, 1959.

73. Robertson, Hugh E.: Pulmonary alveolar proteinosis. Can. Med. Assoc. J., 93:980, 1965.

74. Farca, A., Maher, G., and Miller, A.: Pulmonary alveolar proteinosis. J.A.M.A., 224:1283, 1973.

75. Olsen, A. M.: Spectrum of aspiration pneumonitis. Ann. Otolaryngol., 79:875, 1970.

76. Cameron, J. L., and Zuidema, G. D.: Aspiration pneumonia. J.A.M.A., 219:1194, 1972.

77. Gooding, C. A., and Gregory, G. A.: Roentgenographic analysis of meconium aspiration of the newborn. Radiology, 100:131, 1971.

78. Liebow, A., Steer, A., and Billingsley, J. G.: Desquamative interstitial pneumonia. Am. J. Med., 39:369, 1965.

79. Klocke, L. A., Augerson, W. S., et al.: Desquamative interstitial pneumonia: a disease with a wide clinical spectrum. Ann. Intern. Med., 66:498, 1967.

80. Lemire, P., and Bettez, P., et al.: Patterns of desquamative interstitial fibrosis. Am. J. Roentgenol. Radium Ther. Nucl. Med., 115:479, 1972.

81. Patchefsky, A. S., Fraimow, W., and Hoch, W. S.: Desquamative interstitial pneumonia. Arch. Intern. Med., *132*:222, 1973.
82. Joffee, N.: Roentgenologic findings in post-shock and post-operative pulmonary insufficiency. Radiology, *94*:369, 1970.
83. De Coster, A., et al.: Concept of shock lung. Lille Med., *19*:424, 1974.
84. Theros, E. G.: An exercise in radiologic-pathologic correlation. Radiology, *89*:524, 1967.
85. Cronstin, M. H., Hooper, G. S., et al.: Congenital pulmonary cystic lymphangiectasis. Am. J. Dis. Child., *114*:330, 1967.
86. Bronson, S. M.: Idiopathic pulmonary hemosiderosis in adults: case and review of literature. Am. J. Roentgenol. Radium Ther. Nucl. Med., *83*:260, 1960.
87. Theros, E. G., et al.: Idiopathic pulmonary hemosiderosis. Radiology, *90*:784, 1968.
88. Cole, W. R.: Pulmonary alveolar microlithiasis. J. Fac. Radiol., *10*:54, 1959.
89. Balikan, J. P., Fuleihan, F. J. D., and Nucho, C. N.: Pulmonary alveolar microlithiasis. Am. J. Roentgenol. Radium Ther. Nucl. Med., *103*:509, 1968.
90. Kino, T., Kohara, Y., and Tsuji, S.: Pulmonary alveolar microlithiasis. Am. Rev. Resp. Dis., *105*:105, 1972.
91. Pendergrass, E. P.: Silicosis and a few of the other pneumoconioses: observations on certain aspects of the problem with emphasis on the role of the radiologist. Am. J. Roentgenol. Radium Ther. Nucl. Med., *80*:1, 1958.
92. Williams, J. L., and Moller, G. A.: Solitary mass in the lungs of coal miners. Am. J. Roentgenol. Radium Ther. Nucl. Med., *117*:765, 1973.
93. Hurwitz, M.: Roentgenologic aspects of asbestoses. Am. J. Roentgenol. Radium Ther. Nucl. Med., *85*:256, 1961.
94. Freundlich, I. M., and Greening, R. R.: Asbestosis and associated medical problems. Radiology, *89*:224, 1967.
95. Fletcher, D. E., and Edge, J. R.: The early radiologic changes in pulmonary and pleural asbestosis. Clin. Radiol., *21*:355, 1970.
96. Meo, G., et al.: Clinical, radiological, and physiopathologic picture in talc workers. Folia Med. (Naples), *46*:893, 1963.
97. Schepers, G. W.: Lung disease caused by inorganic and organic dust (berylliosis, siderosis). Dis. Chest, *44*:133, 1963.
98. Weber, A. L., Stoeckle, J. D., and Hardy, H. L.: Roentgenologic patterns in long-standing beryllium disease. Am. J. Roentgenol. Radium Ther. Nucl. Med., *93*:879, 1965.
99. Sander, O. A.: Berylliosis. Semin. Roentgenol., *2*:306, 1967.
100. Frank, R. G.: Farmer's lung: a form of pneumoconiosis due to organic dusts. Am. J. Roentgenol. Radium Ther. Nucl. Med., *79*:189, 1957.
101. Unger, J. D., Fink, J. N., and Unger, G. F.: Pigeon breeder's disease. Radiology, *90*:683, 1968.
102. Hargreave, F., and Hinson, K. F., et al.: The radiological appearances of allergic alveolitis due to bird sensitivity. Clin. Radiol., *23*:1, 1972.
103. Sahn, S. A., and Richerson, H. B.: Extremes of clinical presentation in parakeet-fanciers lung. Arch. Intern. Med., *130*:913, 1972.
104. Brody, J. S., and Levin, B.: Intralobular septa thickening in lipid pneumonia. Am. J. Roentgenol. Radium Ther. Nucl. Med., *88*:1061, 1962.
105. Weill, H., Ferrans, J., et al.: Early lipoid pneumonia. Am. J. Med., *36*:370, 1964.
106. Cornelius, E. A., and Betlach, E. H.: Silo-filler's disease. Radiology, *74*:232, 1960.
107. Tse, R. L., and Bockman, A. A.: Nitrogen dioxide toxicity. J.A.M.A., *212*:1341, 1970.
108. Teng, C. T., and Brennan, J. C.: Acute mercury vapor poisoning. Radiology, *73*:354, 1959.
109. Bonte, F. J., and Reynolds, J.: Hydrocarbon pneumonitis. Radiology, *71*:391, 1958.
110. Johnson, F. A., and Stonehill, R. B.: Chemical pneumonitis from inhalation of zinc chloride. Dis. Chest, *40*:619, 1961.
111. Campbell, J. B.: Pneumatocele formation following hydrocarbon ingestion. Am. Rev. Resp. Dis., *101*:414, 1970.
112. Sone, S., and Higashihara, T., et al.: Pulmonary manifestations in acute carbon monoxide poisoning. Am. J. Roentgenol. Radium Ther. Nucl. Med., *120*:865, 1974.
113. Joffe, N., and Simon, M.: Pulmonary oxygen toxicity in the adult. Radiology, *93*:460, 1969.
114. Joffe, N.: Roentgenologic findings in post-shock and post-operative pulmonary insufficiency. Radiology, *94*:369, 1970.
115. Smith, J. C.: Radiation pneumonitis. Am. Rev. Resp. Dis., *87*:647, 1963.
116. Libshitz, H. I., and Southard, M. E.: Complications of radiation therapy: The thorax. Semin. Roentgenol., *9*:41, 1974.
117. Talner, L. B., Gmelich, J. T., et al.: The syndrome of bronchial mucocele and regional hyperinflation of the lung. Am. J. Roentgenol. Radium Ther. Nucl. Med., *110*:675, 1970.
118. Lemire, P., Trepaner, A., and Herbert, G.: Bronchocele and blocked bronchiectasis. Am. J. Roentgenol. Radium Ther. Nucl. Med., *110*:687, 1970.
119. Overholt, R. H., Bougas, J. A., and Morse, D. P.: Bronchial adenoma: a study of sixty patients with resections. Am. Rev. Tuberc., *75*:865, 1957.
120. Bower, G.: Bronchial adenoma. Am. Rev. Resp. Dis., *92*:558, 1965.
121. Tolis, G. A., and Fry, W. A., et al.: Bronchial adenomas. Surg., Gynecol. Obstet., *134*:605, 1972.
122. Baleson, E. M., and Abbott, K. E.: Mixed tumors of the lung or hamarto-chondromas. Clin. Radiol., *11*:232, 1960.

123. Metys, R.: Roentgen symptomatology of pulmonary chondrohamartomas. Fortschr. Roentgenstr., *106*:90, 1967.
124. Blair, T. C., and McElvein, R. B.: Hamartoma of the lung. Dis. Chest, *44*:296, 1963.
125. Sagel, S. S., and Ablow, R. C.: Hamartoma; an occasionally rapidly growing tumor of lung. Radiology, *91*:971, 1968.
126. Poirier, T. J., and Van Ordstrand, H. S.: Pulmonary chondromatous hamartomas. Chest, *59*:50, 1971.
127. Singer, D. B., Greenberg, S. D., and Harrison, G. M.: Papillomatosis of the lung. Am. Rev. Resp. Dis., *94*:777, 1966.
128. Bacon, M. G., and Whitehouse, W. M.: Primary lymphosarcoma of the lung. Am. J. Roentgenol. Radium Ther. Nucl. Med., *85*:294, 1961.
129. Dahlgren, S. E., and Ovenfors, C. O.: Primary malignant lymphoma of the lung. Acta Radiol., *8*:401, 1969.
130. Garland, L. H., Beier, R. L., Coulson, W., Heald, J. H., and Stein, R. L.: The apparent sites of origin of carcinomas of the lung. Radiology, *78*:1, 1962.
131. Emerson, G. L., Emerson, M. S., and Sherwood, C. E.: Natural history of carcinoma of the lung. J. Thoracic Surg., *37*:291, 1959.
132. Maruyama, Y., Wilkins, E. W., Jr., and Wyman, S. M.: Evaluation of angiocardiography in pulmonary carcinoma with particular emphasis on prognosis. Radiology, *79*:617, 1962.
133. Nathan, M. H., Collins, V. P., and Adams, R. A.: Differentiation of benign and malignant pulmonary nodules by growth rate. Radiology, *79*:221, 1962.
134. Spratt, J. S., et al.: The detection and growth of intrathoracic neoplasms. Arch. Surg., *86*:283, 1963.
135. Mason, W. E., and Templeton, A. W.: Bronchographic signa useful in the diagnosis of lung cancer. Dis. Chest, *49*:284, 1966.
136. Viamonte, M., Jr.: Angiographic evaluation of lung neoplasms. Radiol. Clin. North Am., *3*:509, 1965.
137. Napoli, L. D., Hansen, H. H., et al.: The incidence of osseous involvement in lung cancer, with special reference to the development of osteoblastic changes. Radiology, *108*:17, 1973.
138. Weiss, W., and Boucot, K. R.: Philadelphia pulmonary neoplasm research project. Early roentgenographic appearance of bronchogenic carcinoma. Arch. Intern. Med., *134*:306, 1974.
139. Kittridge, R. D., and Sherman, R. S.: Roentgen findings in terminal bronchiolar carcinoma. Am. J. Roentgenol. Radium Ther. Nucl. Med., *87*:875, 1962.
140. Watson, W. L., and Farpour, A.: Terminal bronchiolar of "alveolar cell" cancer of the lung. Cancer, *19*:776, 1966.
141. Shapiro, R., Wilson, G. L., et al.: A useful sign in the diagnosis of localized bronchoalveolar carcinoma. Am. J. Roentgenol. Radium Ther. Nucl. Med., *114*:516, 1972.
142. Theros, E. G., and Highman, B.: Radiological-pathological correlation of the month from AFIP: Alveolar Cell Carcinoma. Radiology, *97*:661, 1970.
143. Janower, M. L., and Blennerhassett, J. B.: Lymphangitic spread of metastatic cancer to the lung. Radiology, *101*:267, 1971.
144. Green, J. D., Carden, T. S., et al.: Angiographic demonstration of arteriovenous shunts in pulmonary metastatic choriocarcinoma. Radiology, *108*:67, 1973.
145. Fleming, R. J., Medina, J., and Seaman, W. B.: Roentgenographic aspects of tracheal tumor. Radiology, *79*:629, 1962.
146. Janower, M. L., et al.: Radiological appearances of carcinoma of the trachea. Radiology, *96*:39, 1970.
147. Janower, M. L., et al.: The radiologic appearance of carcinoma of the trachea. Radiology, *96*:39, 1970.
148. Teplick, J. G., Haskin, M. E., and Medwich, A.: The Hughes-Stovin syndrome. Radiology, in press.
149. Wolpert, S. M., Kahn, P. C., and Farbman, M.: Radiology of a Hughes-Stovin Syndrome. Am. J. Roentgenol. Radium Ther. Nucl. Med., *112*:383, 1971.
150. Maruyama, Y., and Little, J. J.: Roentgen manifestations of traumatic pulmonary fat emboli. Radiology, *79*:945, 1962.
151. Acker, S. E., and Greenberg, H. B.: Pulmonary injury from post-traumatic fat embolism. Am. Rev. Resp. Dis., *97*:423, 1968.
152. Sammons, B. P.: Arteriovenous fistula of the lung. Radiology, *72*:710, 1959.
153. Hoffman, R., and Rabens, R.: Evolving pulmonary nodules; multiple pulmonary arteriovenous fistulas. Am. J. Roentgenol. Radium Ther. Nucl. Med., *120*:861, 1974.
154. Salvioni, D., and Golden, R. R.: Intralobar pulmonary sequestration. Dis. Chest, *37*:122, 1960.
155. Sutton, D., and Samuel, R. H.: Thoracic aortography in intra-lobar lung sequestration. Clin. Radiol., *14*:317, 1963.
156. Durnin, R. E., et al.: Bronchopulmonary sequestration. Chest, *57*:454, 1970.
157. Felson, B.: The many faces of pulmonary sequestration. Semin. Roentgenol., *7*:3, 1972.
158. Hülshoff, Th., and Kalvelage, H.: Contribution to the diagnosis and frequency of congenital lung aplasia. Fortschr. Roentgenstr., *91*:725, 1959.
159. Felson, B.: Pulmonary agenesis and related anomalies. Semin. Roentgenol., *7*:17, 1972.
160. Krauss, A. N., Levin, A. R., et al.: Physiologic study on infants with Wilson-Mikity syndrome. J. Pediatr., *77*:27, 1970.
161. Faure, C., Grodemange, M., et al.: Bilateral emphysema in the premature (Wilson-Mikity syndrome). Ann Radiol. (Paris), *9*:731, 1966.

162. Grossman, H., Berdon, W. E., et al.: Neonatal focal hyperaeration of the lungs (Wilson-Mikity syndrome). Radiology, 85:409, 1965.
163. Miller, W. T., Cornog, J. L., and Sullivan, M. A.: Lymphangiomyomatosis. Am. J. Roentgenol. Radium Ther. Nucl. Med., 111:565, 1971.
164. Silverstein, E. F., Ellis, K., et al.: Pulmonary lymphangiomyomatosis. Am. J. Roentgenol. Radium Ther. Nucl. Med., 120:832, 1974.
165. Peterson, J. A.: Recognition of infrapulmonary pleural effusion. Radiology, 74:34, 1960.
166. Fleischner, F. G.: Atypical arrangement of free pleural effusion. Radiol. Clin. North Am., 1:347, 1963.
167. Collins, J. D., et al.: Minimal detectable pleural effusions: Roentgen pathology model. Radiology, 105:51, 1972.
168. Stead, W. W., et al.: The clinical spectrum of primary tuberculosis in adults. Ann. Intern. Med., 68:731, 1968.
169. Lillington, G. A., Mitchell, S. P., and Wood, G. A.: Catamenial pneumothorax. J.A.M.A., 219:1328, 1972.
170. Baume, P., and Monk, I.: Localized pleural mesothelioma. Med. J. Aust., 2:751, 1961.
171. Sleggs, C. A., Marchand, P., and Wagner, J. C.: Diffuse pleural mesotheliomas in South Africa. S. Afr. Med. J., 35:28, 1961.
172. Godwin, M. C.: Diffuse mesothelioma; with comments on their relations to localized fibrous mesothelioma. Cancer, 10:298, 1957.
173. Heller, R. M., Janower, M. L., and Weber, A. L.: The radiologic manifestations of malignant pleural mesothelioma. Am. J. Roentgenol. Radium Ther. Nucl. Med., 108:53, 1970.
174. Lemming, R.: Meigs' syndrome and pathogenesis of pleurisy and polyserositis. Acta Med. Scand., 168:197, 1960.
175. Leigh, T. F., and Weens, H. S.: Roentgen aspects of mediastinal lesions. Semin. Roentgenol., 4:59, 1969.
176. Lull, G. F., Jr., and Winn, D. F., Jr.: Chronic fibrosis mediastinitis due to Histoplasma capsulatum. Radiology, 73:367, 1959.
177. Barrett, N. R.: Idiopathic mediastinal fibrosis. Br. J. Surg., 46:207, 1958.
178. Voog, R., et al.: Idiopathic mediastinal fibrosis and retroperitoneal fibrosis. J. Fr. Med. Chir. Thorac., 24:13, 1970.
179. Cooley, J. C., and Gillespie, J. G.: Mediastinal emphysema. Dis. Chest, 49:104, 1966.
180. Felson, B.: The mediastinum. Semin. Roentgenol., 4:41, 1969.
181. Loden, H.: Mediastinal herniation and displacement studied by transversal tomography. Acta Radiol., 48:337, 1957.
182. Herlitzka, A. J., and Gale, J. W.: Tumors and cysts of the mediastinum. Survey of one hundred seventy four mediastinal tumors treated surgically during the past eighteen years at the University of Wisconsin Hospitals. Arch. Surg., 76:697, 1958.
183. Lyons, H. A.: The diagnosis and classification of mediastinal masses: a study of 782 cases. Ann. Intern. Med., 51:897, 1959.
184. Leigh, T. F.: Mass lesions of the mediastinum. Radiol. Clin. North Am., 1:377, 1963.
185. Ferrané, J., et al.: Abnormal cardiac silhouette and thymic tumors. J. Radiol. Electrol. Med. Nucl., 51:207, 1970.
186. Jungblut, R., and Paschke, K. G.: Roentgen diagnosis of teratoid cysts of the mediastinum. Röntgenblatter, 21.1, 1968.
187. Carey, L. S., Ellis, F. H., Good, C. A., and Woolner, L. B.: Neurogenic tumors of the mediastinum: a clinicopathologic study. Am. J. Roentgenol. Radium Ther. Nucl. Med., 84:189, 1960.
188. Stern, W. Z., and Bloomberg, A. E.: Idiopathic azygous phlebectasia simulating mediastinal tumor. Radiology, 77:622, 1961.
189. Katz, I., and Dziadiw, R.: Localized lymph node hyperplasia. Report of a case simulating posterior mediastinal neurofibroma. Am. J. Roentgenol. Radium Ther. Nucl. Med., 84:206, 1960.
190. Sowerbutts, J. G.: Mediastinal bronchogenic cysts. J. Fac. Radiol., 10:158, 1959.
191. Oldham, H. N., and Sabiston, D. C.: Primary tumors and cysts of the mediastinum. Arch. Surg., 96:71, 1968.
192. Ross, P., and Logan, W.: Roentgen findings in extramedullary hematopoiesis. Am. J. Roentgenol. Radium Ther. Nucl. Med., 106:604, 1969.
193. Korsten, J., Grossman, H., et al.: Extramedullary hematopoiesis in patients with thalassemic anemia. Radiology, 95:257, 1970.

CARDIOVASCULAR DISEASES

RADIOLOGIC FEATURES OF CHAMBER AND VASCULAR ALTERATIONS

Recognition of individual chamber enlargement, of the size of the great vessels, and of the state of the lung vasculature is essential for plain film diagnosis of cardiac disease.

Left ventricular enlargement causes a downward displacement of the cardiac apex on frontal view, and a rounded posterior bulge on lateral and left oblique views.

Right ventricular enlargement leads to rounding and elevation of the apex frontally, to obliteration of the lower retrosternal space on lateral films, and an anterior convexity on right anterior oblique projections.

Left atrial enlargement elevates the left main bronchus and indents or displaces the lower middle third of the esophagus on lateral or right oblique projections. The right border of an enlarged atrium may produce an extra shadow adjacent to the right cardiac border.

Right atrial enlargement is usually difficult to assess. There may be fullness and elongation of the right lower cardiac border.

When multiple chambers are enlarged it is difficult to make an accurate appraisal of individual chamber size and location.

Enlargement of the ascending aorta usually produces a bulge of the right upper border of the silhouette on frontal view. The bulge can be seen best on the left anterior oblique projection.

The size of the aortic arch is usually reflected by the size of the knob. Sometimes the entire arch is clearly visible on left anterior oblique or lateral films. The left border of the descending aorta most often produces a shadow parallel to and to the left of the spine; this is seen on frontal view. If sufficiently opaque or uncoiled, the descending aorta may be evident on lateral and left oblique films.

The bulge of the main pulmonary artery is seen below the knob on frontal view; on the right oblique view it appears as a convexity extending up from the right ventricle. The size of the central pulmonary vessels can be estimated from the hilar and perihilar shadows, which are seen on all projections.

Increased venous pressure is usually manifested by an increased width of the upper lobe veins and by decreased width of the lower lobe veins.

Increased pulmonary arterial flow, as seen in left to right shunts or in high output hearts, causes widening of both the central and peripheral vascular shadows. The differentiation of veins from arteries is often difficult. The descending branch of the right pulmonary artery is usually easily identified and used as an indicator of the arterial tree.

Decreased size of the pulmonary vessels usually can be recognized, but small cardiac chambers ordinarily cannot be diagnosed.

All of the chambers and vessels can be visualized and evaluated most accurately by selective angiocardiography.

CONGESTIVE FAILURE

Cardiac enlargement, increased pulmonary venous pressure, interstitial edema, and later alveolar edema are mirrored in the radiologic findings of congestive left-sided failure.

The degree of cardiac enlargement depends on the severity of the failure and the precongestive heart size.

In incipient or early left heart failure, increased pulmonary venous pressure leads to prominent enlarged superior pulmonary veins and decreased caliber of the inferior veins, apparently due to reflex venous spasm. The hilar shadows, especially the upper portions, enlarge, and the lateral aspects of the hila, normally concave, become convex.

With onset of interstitial edema, the hilar shadows become somewhat hazy, with accentuation of the perihilar vascular markings (hilar haze). The smaller horizontal peripheral interlobular septa become prominent (Kerley B lines), and often lines of accentuated central lymphatics (Kerley A lines) appear. The engorged interlobular septa may be superimposed on the lower lung field, producing a fine reticular pattern (Kerley C lines). Perivascular interstitial edema will cause some fuzziness of the peripheral vascular markings and often will produce a peripheral haze. Stripes of peripheral subpleural edema may appear near the chest wall or at an interlobar fissure.

A further rise in venous pressure may lead to alveolar or pleural transudates. The alveolar transudates of pulmonary edema are characterized by bilateral fan-shaped densities extending from the hila. The lung peripheries, including the apices and bases, remain relatively clear. The patchy density may become more confluent and homogeneous and can simulate areas of pneumonitis. Rarely, the pulmonary edema is unilateral.

The pleural transudates lead to pleural effusions of varying extent. The effusion occurs most often on the right side. If bilateral, the effusion is more marked on the right side. The interlobar pleural lines become thickened and prominent.

If compensation is regained, the heart shadow usually becomes smaller and the vascular abnormalities regress. The pleural effusions disappear. Pleural fluid that is loculated within a fissure tends to resorb more slowly and sometimes may simulate a pulmonary mass, the so-called vanishing tumor.

In right heart failure there is dilatation of the right ventricle and right atrium and often prominence of the right superior mediastinum due to widening of the superior vena cava. The liver becomes enlarged and may cause elevation of the right diaphragm. If right heart failure develops following left heart failure, pulmonary venous congestion from the left-sided failure will diminish.[1-8]

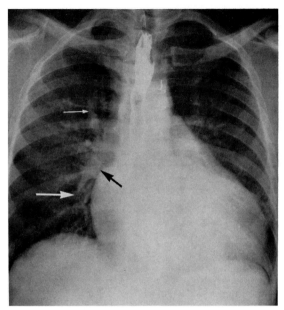

Figure 10-1 **Early Congestive Failure.** The heart is enlarged. There is engorgement of the superior pulmonary veins *(small white arrow)* and diminution of the veins in the lower lung fields *(large white arrow).* The right hilum is broadened, and the clear space between the hilum and the heart has disappeared *(black arrow).*

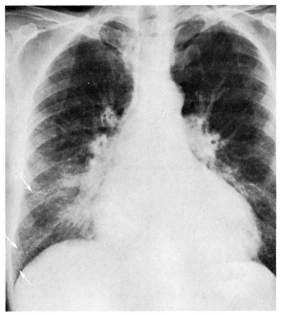

Figure 10-2 **Congestive Failure.** Cardiomegaly, vascular engorgement, and hilar prominence are due to congestive heart failure. There are also interlobular septal densities (Kerley B lines) *(white arrows)* in the right lower lung field that are due to engorgement of the septal venules and lymphatics resulting from increased postcapillary resistance.

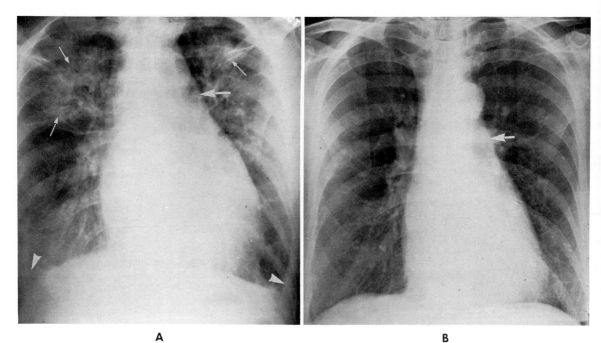

A   **B**

Figure 10-3   **Acute Pulmonary Edema, and Recovery.**

A, The heart is enlarged, and its left border is straightened, the aortic knob is obscured by the dilated main pulmonary artery segment *(large arrow).* There are confluent alveolar densities *(small arrows)* in both upper lung fields, representing areas of pulmonary edema. The upper lobe vessels are dilated, and the lower lobe vasculature is decreased. The densities in both costophrenic angles *(arrowheads)* are due to pleural effusions.

The feathery appearance of the pulmonary edema densities is characteristic of both alveolar exudates and transudates. If the congestive transudate is asymmetric or in an unusual location, radiologic differentiation from a pneumonia may be difficult.

This episode of acute cardiac decompensation was due to a rheumatic myocarditis.

B, One week later, with medical management, the heart is considerably decreased in size; the main pulmonary artery *(arrow)* is smaller and no longer obscures the aortic knob. The pulmonary edema and the pleural fluid have completely cleared, and the lung fields and pulmonary vasculature now appear completely normal.

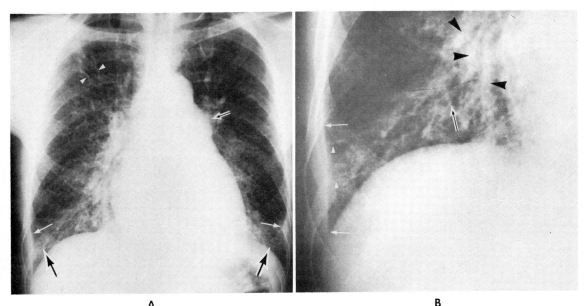

A   B

Figure 10–4   **Congestive Failure: Chronic Interstitial Edema.**

A, The heart is enlarged and the main pulmonary artery is dilated and prominent *(small black-white arrow),* but there is no obvious vascular congestion. There are stripes of density along both lower chest walls *(white arrows)* due to subpleural edema; Kerley B lines *(large black-white arrows),* which are more clearly seen in the enlarged view; Kerley A lines *(arrowheads),* which are curved lines directed toward the hilum; and a somewhat reticular pattern at the bases.

B, An enlarged view of the right lower chest shows more distinctly the Kerley B lines *(white arrowheads)* and the subpleural edematous stripe *(white arrows).* The reticular pattern in the lung field *(black-white arrow)* is due to interstitial tissue edema; it partially obscures the normally sharp borders of the larger vessels *(black arrowheads).*

Interstitial edema may precede, accompany, or follow frank congestive failure.

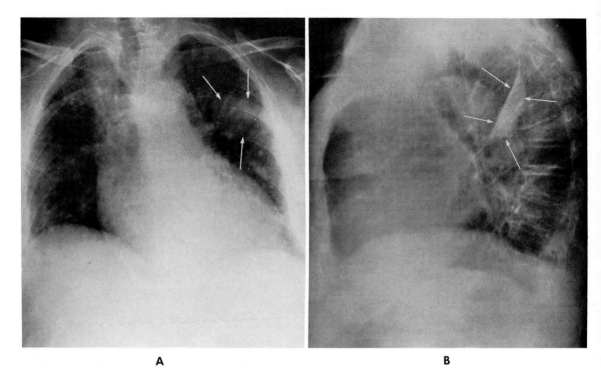

A   B

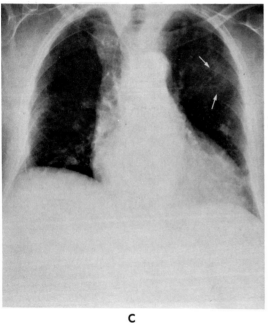

C

Figure 10–5   **Congestive Failure with Loculated Interlobar Effusion (Vanishing Tumor).**

A, The heart is enlarged, and there is an ill-defined
round density in the left upper lung field *(arrows).*

B, Lateral view indicates that the density seen in *A*
is due to a localized collection of interlobar fluid
(arrows) in the upper portion of the left major fissure.

C, Five days later there is almost complete disappearance of the "mass" *(arrows).*

Pleural fluid consequent to congestive failure is
usually free in the pleural space but may become
loculated, simulating a mass; however, its location
within a fissure and its clearing following cardiac compensation simplify the differential diagnosis.

PULMONARY EMBOLISM AND INFARCTION

Pulmonary embolism may occur without infarction of the lung, particularly in younger individuals with adequate collateral circulation. Plain chest films often appear negative, but careful scrutiny may sometimes disclose suggestive changes. Often there will be one or more zones of oligemia, with sparse and thin vascular markings. If marked, the oligemic area may appear hyperlucent (Westermark sign). A major arterial branch may become widened and show a sudden diminution in caliber distally. One or both hilar shadows may increase in size, and acute right heart strain may be manifested by fullness of the main pulmonary artery and dilatation of the right ventricle. These changes are more apparent when a large embolism is in a major arterial branch. Definitive diagnosis, however, requires a pulmonary scan or pulmonary arteriogram.

Infarction of the lung may follow embolism, especially in older individuals whose collateral blood supply is inadequate. The infarct most commonly appears as a parenchymal density that extends to a pleural surface, producing some pleural reaction. Occasionally the pleural reaction and fluid are the predominant findings. In general, the findings resemble a nonspecific pleural pneumonia. However, unlike ordinary pleural pneumonia, infarcts generally heal much more slowly and often leave fibrotic scarring in the lung and persistent pleural thickening. The slow resolution and residual scarring may be helpful diagnostic points in retrospect. Sometimes the infarcted area is not clearly adjacent to a pleural surface and may appear radiographically as a nodular infiltrate without any obvious pleural reaction.

There may be evidence of acute right heart strain. If serial films are available, changes in the main pulmonary artery segment are often observed; this is strong confirmatory evidence.

In a patient with congestive failure or a compromised pulmonary blood supply from chronic lung disease, a pleuropneumonia should be considered as a possible pulmonary infarct. When a parenchymal lesion has developed, the isotope scan is of limited diagnostic value.

Septic pulmonary emboli may occur from right-sided bacterial endocarditis, from septic peripheral thrombophlebitis, or from intravenous drug abuse in addicts. These emboli appear as small scattered areas of consolidation, or as round or wedge-shaped peripheral densities. Rapid enlargement and cavitation occur; thin-walled cavities may simulate pneumatoceles. Larger abscesses may result from coalescence. Pleural involvement may cause empyema, pneumothorax, or bronchopleural fistula.

Definitive roentgen diagnosis of acute or chronic pulmonary embolism is made by pulmonary arteriography. Intraluminal filling defects, cutoffs, and pruning of larger arterial vessels are direct signs. Oligemia, prolongation of the arterial phase, and delay in filling of one or more pulmonary segments are secondary signs of disturbance of blood flow.

In cases of repeated and multiple embolization, cor pulmonale and pulmonary hypertension may be a permanent end result. (See also *Pulmonary Hypertension,* p. 609.)[9-15]

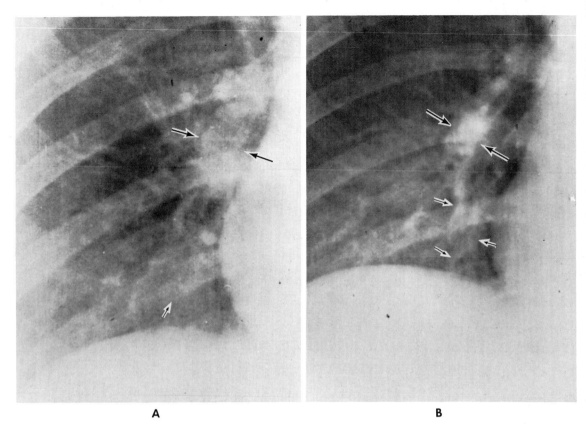

A   **B**

Figure 10–6 **Pulmonary Embolism Without Infarction.**

A, Chest film of a 30 year old man was made within 24 hours after an acute pulmonary embolism that was confirmed by a lung scan. The right pulmonary artery is greatly dilated *(large arrows)* at and below the hilum, and there is a definite paucity of vessels at the right base *(small arrow).*

B, Eight days later the right pulmonary artery has greatly decreased in width *(large arrows)* and the normal vessel shadows have reappeared at the base *(small arrows).*

In younger individuals, the rapid disappearance of a pulmonary embolus and restoration of normal circulation have been corroborated by angiographic studies.

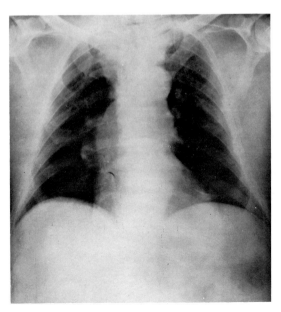

Figure 10–7 **Massive Pulmonary Embolism Without Infarction.**   The patient had had a massive pulmonary embolus in the main pulmonary artery, yet little more than hyperlucent lung fields with relative diminution of the bronchovascular markings is seen. The patient died within an hour after radiography.

The paucity of findings in the presence of massive emboli makes roentgenologic diagnosis extremely difficult; a sudden increase in size of the main pulmonary artery segment may be the sole significant finding.

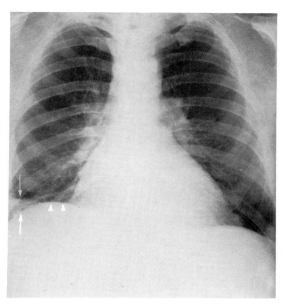

Figure 10–8 **Pulmonary Embolism with Infarction.** Postoperative film in 70 year old man. There is a small area of parenchymal density at the right base, with associated pleural reaction *(white arrows)*. There is slight elevation of the right hemidiaphragm *(arrowheads)*. Note the cone shape of the infarcted area. A truncated cone having its base on the pleura was originally described as the classic appearance of pulmonary infarction, but is rarely seen. Diaphragmatic elevation and fixation caused by pleural involvement is frequently associated.

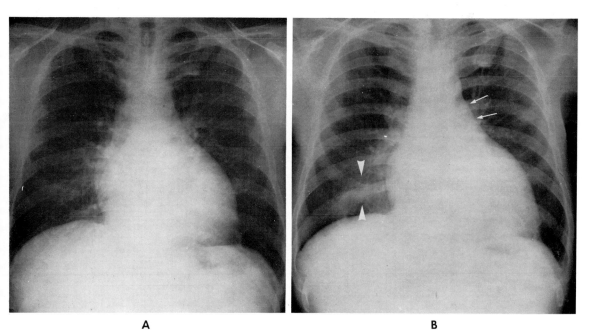

A   B

Figure 10–9 **Pulmonary Embolus with Infarction.**

A, These films in a 42 year old male were made before embolism occurred. There is cardiomegaly, but the chest is otherwise normal.

B, Density at the right base abuts against the posterior pleura *(arrowheads)*. The main pulmonary artery segment *(arrows)* is slightly yet definitely more prominent when compared with that shown in *A.* This is indicative of an acute right heart strain.

Increased prominence of the main pulmonary artery segment in association with pleural pneumonia strongly suggests pulmonary infarction.

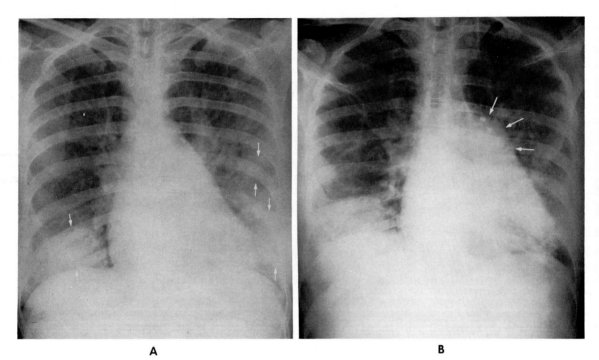

A   **B**

Figure 10–10 **Multiple Pulmonary Emboli.**

A, There are extensive pleuropneumonic opacities at both bases *(arrows)*.

B, Subsequent film shows a marked increase in size of the main pulmonary artery segment *(arrows)*, which suggests pulmonary emboli.

Repeated multiple small pulmonary emboli are a cause of pulmonary hypertension.

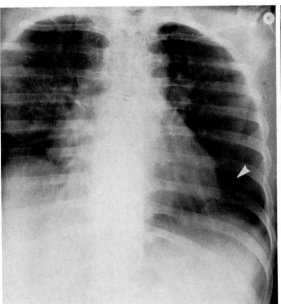

A

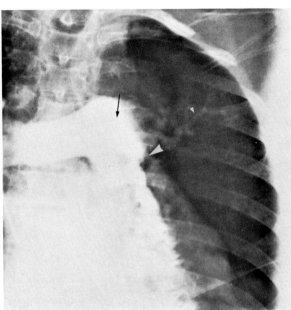

B

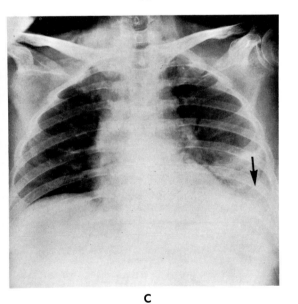

C

Figure 10–11 **Pulmonary Embolism and Infarction: Arteriogram.**

A, Anteroposterior view demonstrates hyperlucency of the left lower lung (Westermark's sign) *(arrowhead).* Pulmonary vessels are lacking.

B, Pulmonary arteriogram demonstrates dilatation of the main pulmonary artery segment *(arrow),* opacification of the superior pulmonary artery *(small arrowhead),* and abrupt occlusion of the vessels in the left lower lobe *(large arrowhead)* by an obstructing pulmonary embolus.

C, One week later there is pleural pneumonia at the left base *(arrow),* which is indicative of the development of pulmonary infarction. (Courtesy Albert Einstein Medical Center, Philadelphia, Pennsylvania.)

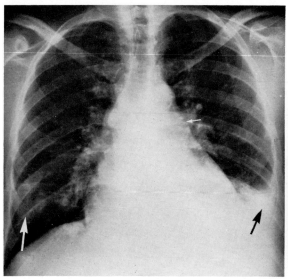
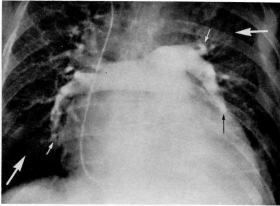

| A | B |

Figure 10–12 **Acute and Chronic Pulmonary Emboli: Pulmonary Arteriogram.**

This 60 year old man was admitted with a history suggestive of recent pulmonary embolism. There had been several similar episodes during the past few years.

A, Chest film shows at the left base a pleural and parenchymal density *(black arrow)* thought to be a pulmonary infarct. A less dense parenchymal density at the right *(large white arrow)* was considered residual from a previous infarct. The heart is greatly enlarged, and the main pulmonary artery segment *(small white arrow)* is full, suggesting cor pulmonale.

B, Pulmonary arteriography shows a sharp cutoff of the main left lower lobe artery *(black arrow),* with no vascular opacification of the left lower lobe. The left upper and right lower lobe arteries are partially cutoff *(small white arrows),* and these lobes show a sparse, thinned blood supply *(large white arrows).* Only the right upper lobe shows good vascularization.

On later films there was better vascularization of the left upper and right lower lobes.

The arterial cutoff and avascularity of the left lower lobe was due to the recent embolus. The early sparsity of vessels in the other lobes, with delay in opacification, is also characteristic of embolization—in this case, of older emboli. Rarely, an embolus itself can be visualized as a defect in the opacified larger arteries.

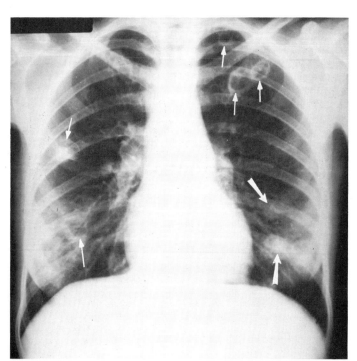

Figure 10–13 **Septic Emboli and Abscess in a Heroin Addict.** The scattered nodular densities *(large arrows),* many of which have undergone cavitation *(small arrows),* are abscesses resulting from a shower of septic emboli.

Septic emboli can occur in septicemia or in bacterial endocarditis, or from intravenous administration of unsterile particulate matter in an addict.

PULMONARY HYPERTENSION

Pulmonary hypertension results from increased resistance in the pulmonary circulation. This can occur secondary to cardiac disease or from intrinsic pulmonary vascular disorders, recurrent pulmonary embolization, obliterative parenchymal pulmonary disease, and hypoventilation.

Longstanding increased venous pressure from chronic congestive failure or mitral stenosis and increased arterial flow from severe left to right cardiac shunts can cause secondary narrowing of the pulmonary arterioles. Reduction of the vascular bed from intrinsic vessel disturbances can occur in vasculitis (collagen diseases, schistosomiasis, recurrent pulmonary emboli, and primary pulmonary hypertension). Pulmonary conditions that can cause obliterative vascular changes include interstitial fibrosis, sarcoidosis, vasculosis, and other causes of alveolar capillary block. A vasospastic form of pulmonary hypertension can result from hypoventilation that may occur in emphysema, chronic enlarged adenoids, obesity, and kyphoscoliosis.

Most cases of chronic pulmonary hypertension are due to congenital heart disease with longstanding large left to right shunts (Eisenmenger's physiology) or to obliterative lung disease. (See *Ventricular Septal Defect,* p. 628, and *Atrial Septal Defect,* p. 624.)

The classic radiographic findings of pulmonary hypertension are enlarged main and central pulmonary arteries with abrupt narrowing of the distal vessels, producing a marked disparity between the central and peripheral vasculature. The lung periphery appears unusually clear and lucent. When pulmonary hypertension develops in a patient with a left to right shunt, these vascular changes often first appear in the left upper lobe. Not infrequently, however, the disparity between the central and peripheral vessels may be minimal or absent. The definitive diagnosis of pulmonary hypertension will depend upon catheterization pressure studies. Pulmonary arteriography will usually best demonstrate the disparity between the central and peripheral vessels. Calcification of one or more major pulmonary arteries may occur in longstanding severe pulmonary hypertension.

Right ventricular hypertrophy invariably occurs in chronic pulmonary hypertension, but radiographic recognition is difficult until dilatation ensues. Partial obliteration of the retrosternal lung space by the enlarging right ventricle may be an early finding on the lateral chest film.

There may be changes in the lung fields due to the underlying lung disease that is responsible for the pulmonary hypertension. When pulmonary hypertension supervenes on a left to right shunt, the pre-existing hypervascularity diminishes in the lung periphery and the shunt may become balanced or even reversed. If only the main pulmonary artery appears dilated the plain film picture may then suggest an incorrect diagnosis of pulmonic valvular stenosis.

In cases of chronic venous congestion (mitral stenosis, chronic left heart failure), the development of pulmonary hypertension will cause decrease of the venous engorgement and diminution or disappearance of Kerley's lines.[16-21]

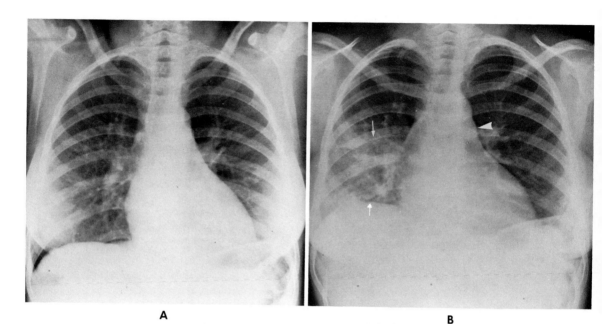

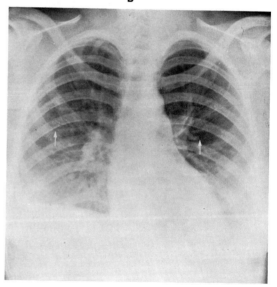

Figure 10–14 **Multiple Pulmonary Infarcts: Pulmonary Hypertension.**

A, The lung fields are clear. The patient had sickle cell disease.

B, Following pulmonary infarction there is extensive pleural pneumonia at the right base *(arrows),* increased fullness in the main pulmonary artery segment *(arrowhead),* and cardiac enlargement.

C, Following several infarctions, the pulmonary artery segment and main pulmonary arteries are very prominent; characteristically, there is an abrupt decrease in arterial size *(arrows),* and relatively clear peripheral lung fields. The picture is typical of pulmonary hypertension.

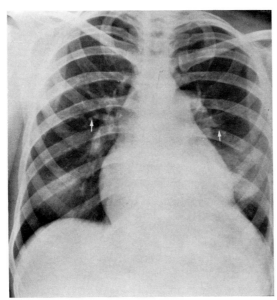

Figure 10–15 **Idiopathic Pulmonary Hypertension.**   There is pulmonary hypertension without evidence of parenchymal pulmonary disease or congenital heart disease in a 30 year old female. Note dilatation of the central vessels with an abrupt change in caliber near the hilum *(arrows)*. The main pulmonary artery segment and pulmonary arteries are prominent, and there is right ventricular enlargement. The patient was considered to have primary pulmonary hypertension.

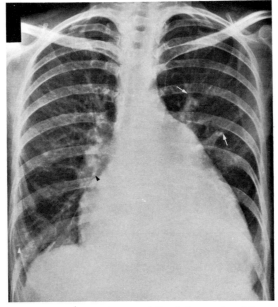

Figure 10–16 **Pulmonary Hypertension Secondary to Mitral Stenosis.**   The cardiac silhouette is typical of mitral disease: there is a small aortic knob, a dilated main pulmonary artery segment, a dilated left atrium (not apparent on this film), and right ventricular enlargement. There is an abrupt change in caliber of the peripheral pulmonary arteries *(arrows)* that is due to pulmonary hypertension. Note selective clearing in the left lung field. Pulmonary hypertension frequently begins in the left upper lobe. Signs of increased pulmonary venous pressure from mitral stenosis are still present but less marked than prior to development of pulmonary hypertension; these include the B lines of Kerley *(white arrowhead)*, and filling of the clear space between the hilar shadows and the left atrium *(black arrowhead)*.

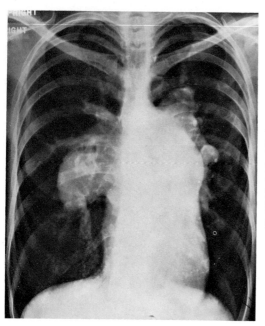

Figure 10–17 **Pulmonary Hypertension Secondary to Atrial Septal Defect.** Extreme dilatation of the pulmonary artery segment and hilar vessels contrasts with barely visible peripheral vessels. Clarity of the peripheral lung fields is attributable to narrowing of the peripheral vessels. Pulmonary hypertension commonly develops in longstanding congenital heart disease in which there are large left to right shunts, and is known as Eisenmenger's physiology.

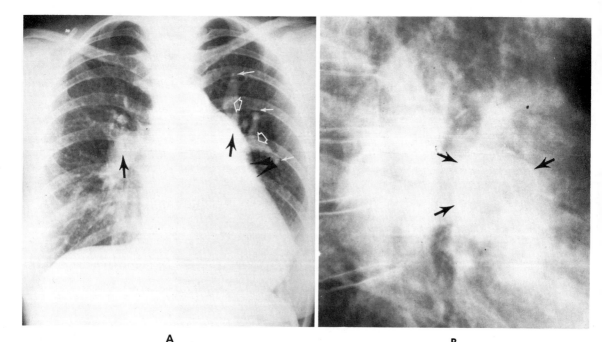

A   B

Figure 10–18   **Pulmonary Hypertension in Ventricular Septal Defect: Pulmonary Artery Calcification.**

A, The dilated main and central pulmonary arteries *(black arrows)* and the dilated proximal vessels *(open arrows)*, which abruptly narrow *(white arrows)*, are characteristic findings of longstanding severe pulmonary hypertension.

An arc of calcification *(black arrowheads)* is faintly visible below the main pulmonary artery.

B, Close-up lateral view shows calcification in the wall of the left pulmonary artery *(arrows)*.

Longstanding severe pulmonary hypertension can cause calcification of the pulmonary arteries.

COR PULMONALE

Cor pulmonale refers to cardiac changes due to pulmonary arterial hypertension that is caused by intrinsic lung disease or hypoventilation. It may also be caused by hypoventilation due to extrapulmonary causes such as kyphoscoliosis, rheumatoid spondylitis, extreme obesity, and various neuromuscular disorders.

Fullness and increased convexity of the main pulmonary artery segment and prominence of the main branches of the pulmonary artery are the constant and significant findings. The abrupt decrease in caliber of the peripheral arteries, which is indicative of pulmonary hypertension, may or may not be apparent, but some disparity between the peripheral and the hilar vessels is generally seen.

Initially, the right ventricular hypertrophy causes no apparent increase in cardiac size. Eventually dilatation of the right heart leads to cardiac enlargement. With progressive right ventricular enlargement, there may be elevation and rounding of the cardiac apex and increased convexity of the lower anterior cardiac outline in the right anterior oblique projection. With marked right ventricular enlargement, there may be a decrease in the retrosternal space on the lateral view. Generally there is no evidence of left atrial or left ventricular enlargement. The aorta is usually not remarkable since the left heart is not involved.

With right heart failure there is further cardiac enlargement. The superior mediastinum may appear widened due to distention of the superior vena cava. Frequently the right hemidiaphragm appears elevated due to congestive hepatomegaly. Pleural transudates may occur.

Changes in the lung fields may indicate the primary condition, such as emphysema, pneumoconiosis, sarcoidosis, or diffuse pulmonary fibrosis.[17-20]

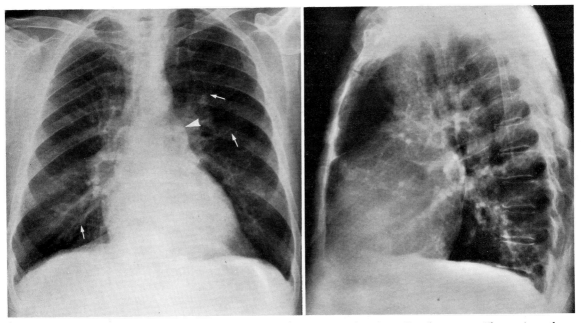

Figure 10-19 **Cor Pulmonale and Pulmonary Hypertension: Pulmonary Emphysema.** The main pulmonary artery segment *(arrowhead)* is dilated, and the peripheral vessels *(arrows)* are narrowed in a patient with longstanding emphysema. Depressed diaphragms and unequal distribution of the peripheral vessels are indicative of emphysema. Enlargement of the right heart is seen most clearly on lateral view.

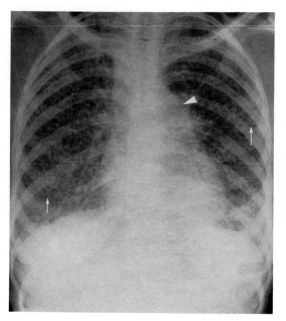

Figure 10–20 **Cor Pulmonale and Pulmonary Hypertension: Sarcoid Fibrosis.** The main pulmonary artery segment *(arrowhead)* is dilated, and the right heart is enlarged; these changes are secondary to pulmonary hypertension from diffuse interstitial fibrosis *(arrows)* of sarcoidosis. The extensive densities in the lung field prevent identification of the vasculature. Cor pulmonale can occur in any condition producing interstitial fibrosis.

CONGENITAL HEART DISEASE

The basic approach to the radiographic diagnosis of congenital heart disease is the correlation of anatomic and hemodynamic information gleaned from the chest films with clinical and electrocardiographic data. For definitive diagnosis, especially when corrective surgery is contemplated, catheterization studies and contrast angiocardiography are essential. Unsuspected multiple cardiac defects may be uncovered by these studies.

Analysis of the radiographs generally begins with evaluation of the pulmonary vasculature, since vascular dynamics usually reflect the basic anatomic abnormality. A morphologic evaluation is then made of the overall heart size, the size and location of the aorta, and the size of the main pulmonary artery segments. The size of the heart chambers is then assessed. When there are complex anomalies or when the chambers are not in their usual location it is difficult or impossible to gain chamber size information from the plain films.

A high percentage of sternal abnormalities occur in congenital heart disease. Premature fusion of the sternal segments is the most common abnormality. Abnormal sternal bowing (pectus carinatum) is also frequent, especially in cyanotic disease.

In congenital *cyanotic* heart disease, clubbing of the fingers and toes is a fairly constant finding, but only 15 per cent of patients have some degree of hypertrophic osteoarthropathy. About 30 per cent have some medullary widening in the long bones, which is due to increased hematopoiesis with marrow hyperplasia. These bone changes are seen only in the more severe cases.

Asplenia or polysplenia is sometimes associated with congenital heart disease and other gut and pulmonary anomalies.

In the following table, congenital heart disease is first divided into the cyano-

CLASSIFICATION OF CONGENITAL HEART DISEASE

Decreased Pulmonary Vasculature	**Cyanotic Heart Disease** Pulmonary Hypertension (Dilated proximal vessels, constricted peripheral vessels)	Increased Pulmonary Vasculature
Right to Left Shunt Tetralogy of Fallot Tricuspid atresia Ebstein's anomaly Pulmonary stenosis with atrial septal defect Pulmonary stenosis with transposition	*Reversed Shunt* Pulmonary hypertension with atrial septal defect ventricular septal defect patent ductus arteriosus (Eisenmenger's physiology)	*Mixed Lesions with Major Malformations* Complete transposition Truncus arteriosus Single ventricle Bilocular heart Total anomalous venous return

Normal Pulmonary Vasculature	**Acyanotic Heart Disease** Increased Pulmonary Vasculature (left to right shunt)	
Stenotic Lesions Coarctation of aorta Aortic stenosis Subaortic stenosis Pulmonary stenosis *Nonstenotic Lesion* Endocardial fibroelastosis	*Intracardiac Shunt* Atrial septal defect Ventricular septal defect	*Extracardiac Shunt* Patent ductus arteriosus Aorticopulmonary defect Partial anomalous venous return

tic and acyanotic groups. Further subdivision is made according to the appearance of the pulmonary vasculature. The diagnostic possibilities in any given case can be considerably reduced by applying these two criteria. The roentgen picture may be drastically altered if multiple defects coexist.[22-25]

Acyanotic Defects

Congenital Aortic Valvular Stenosis

In over half the cases, congenital aortic stenosis is due to a valvular abnormality. The remaining cases are the result of subvalvular or supravalvular defects.

Valvular stenosis is caused by deformed cusps, often associated with bicuspid valves. On plain films there is usually left ventricular enlargement and frequently poststenotic dilatation of the ascending aorta; these findings are quite similar to those of acquired aortic stenosis. The left atrium is usually normal but is somewhat enlarged in a minority of cases. Valvular calcification increases with aging; it is usually present by age 20 and almost invariably present by age 30. When present, it usually indicates a severe stenosis. Left ventricular angiocardiography reveals thickened, irregular, dome-shaped aortic valves, with a jet stream through the stenotic orifice. Left ventricular enlargement and poststenotic aortic dilatation will be apparent.[26-30]

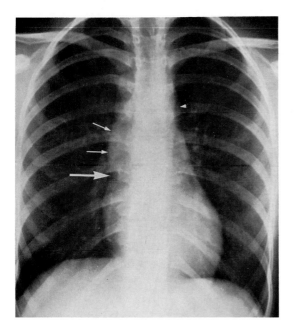

Figure 10–21 **Congenital Aortic Valvular Stenosis.** The heart is not enlarged, since the hypertrophied left ventricle is not dilated. The ascending aorta *(small arrows)* is bulging and prominent due to poststenotic dilatation. The aortic knob is normal *(arrowhead).* The aortic valve is not calcified.

The aortic bulge accentuates the notch *(large arrow)* that separates the ascending aorta from the right atrial shadow.

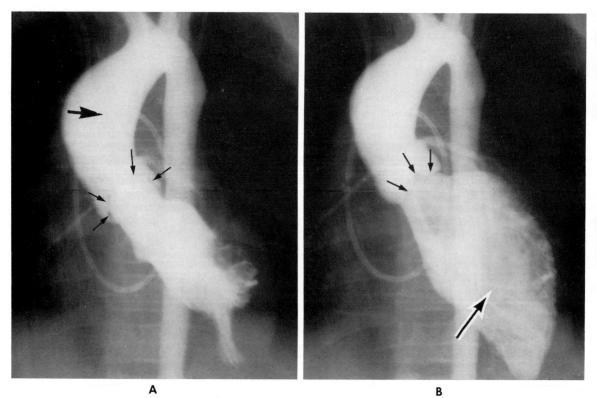

A   B

Figure 10–22 **Congenital Aortic Valvular Stenosis: Angiocardiogram.**

A, In systole, the aortic valves *(small arrows)* are characteristically dome-shaped and thickened. There is poststenotic dilatation of the ascending aorta *(large arrow).*

B, In diastole, the valves are less domed, and their irregular thickening is evident *(black arrows).* The left ventricular chamber *(black-white arrow)* is quite enlarged. (Courtesy Dr. O. W. Kincaid, Mayo Clinic, Rochester, Minnesota.)

Subvalvular Stenosis (Idiopathic Hypertrophic Subaortic Stenosis)

Narrowing of the left ventricular outflow tract from subvalvular stenosis can occur in two forms. In the congenital discrete type, a thin membrane or fibrotic (or fibromuscular) ring is present just below the aortic valve. In the more common idiopathic muscular hypertrophic stenosis, the outflow area is narrowed by a thickened musculature, usually in the interventricular septum.

Radiographically there is generally some degree of left ventricular enlargement. Left atrial enlargement is seen in about 25 per cent of cases and is due to associated mitral insufficiency. Dilatation of the ascending aorta occurs in 25 to 30 per cent of cases but is less prominent than that in valvular stenosis.

Angiography will show the outflow tract eccentrically narrowed by septal hypertrophy, best seen during systole. In the rarer discrete form, the narrowing is closer to the valve, more localized, and more symmetric.[31, 32, 34]

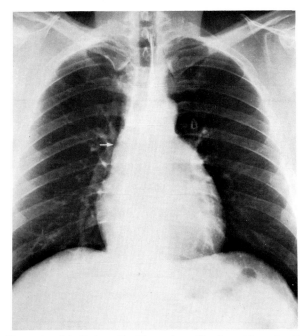

Figure 10-23 **Idiopathic Hypertrophic Subaortic Stenosis.** The heart is only slightly enlarged, and the ascending aorta *(arrow)* is not unduly prominent. Contrast left ventriculography is necessary to make a definitive diagnosis.

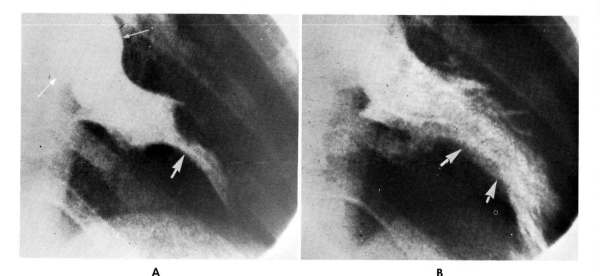

A	B

Figure 10–24   **Idiopathic Hypertrophic Subaortic Stenosis.**

A, Enlargement of a cineradiographic frame obtained during catheter ventriculography in systole demonstrates protrusion of the myocardium *(large arrow)* in the subvalvular area, which has caused narrowing of the left ventricular outflow tract. The root of the aorta *(small arrows)* is normal. In this case the septum was hypertrophied.

B, Cineradiographic frame in the diastolic phase shows the opacified ventricle. Continued bulging of the septum *(arrows)* suggests localized hypertrophy. Notice the change in caliber of the subaortic area in systole and in diastole.

Supravalvular Stenosis

Supravalvular stenosis usually begins at the upper margins of the sinuses of Valsalva and may be due to a fibrous thickening or an encircling fibrous ring. It is frequently associated with idiopathic hypercalcemia of infancy. On plain films the heart is normal or enlarged, the left atrium is normal, and the aorta is normal or small. Angiography will disclose the exact site and length of the stenosis.[34, 35]

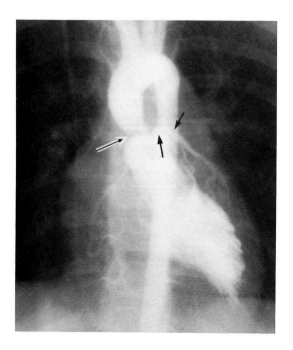

Figure 10–25   **Supravalvular Stenosis in Six Year Old Child: Selective Angiocardiogram.** There is a diaphragm-like stricture *(black-white arrow)* of the ascending aorta above the origin of the coronary arteries *(black arrows).* Note the absence of significant poststenotic dilatation of the ascending aorta. (Courtesy Dr. O. W. Kincaid, Mayo Clinic, Rochester, Minnesota.)

Coarctation of the Aorta

In the adult, coarctation of the aorta often produces characteristic roentgenologic changes in the aorta and notching of the ribs. The cardiac alterations are minimal.

The ascending aorta is often dilated and prominent. The aortic knob, however, appears small in about half the cases. In virtually every instance the narrowed aortic segment appears as a notch in the contour of the descending aorta just below the knob. Poststenotic dilatation immediately below the notch can produce the lower bulge of the characteristic figure-of-three appearance of the descending aorta. The tortuous dilated left subclavian artery often produces a clearly defined bulge above the aortic knob.

Cardiac enlargement occurs in most cases, but is rarely marked; the enlargement is predominantly of the left ventricle, due to increased resistance to left heart outflow.

Erosion and scalloping of the inferior margins of several ribs (rib notching) is the most diagnostic finding. The rib erosions result from pulsatile pressure from the dilated intercostal arteries, which serve as collateral pathways to the poststenotic portion of the aorta. The notching is most frequent in the fourth through the eighth ribs and is usually bilateral but not necessarily symmetric. The first and second ribs do not become notched, since their intercostals do not serve as collaterals. In children, rib notching rarely appears before the sixth or seventh year, and recognition of poststenotic dilatation of the descending aorta may be the only radiographic clue. Other causes of rib notching have been reported, but these are most often unilateral and localized and do not seriously enter into the differential diagnosis.

In some cases the dilated internal mammary artery is visualized on the lateral film as a soft tissue density behind and parallel to the sternum. This may occasionally be seen in children before the rib notching appears.

There are several anatomic variants of coarctation, but the most common form arises at or shortly beyond the ligamentum arteriosum, and the narrowed segment is quite short. Bicuspid aortic valves are a frequently associated anomaly.

In the infantile type of this defect, also known as the hypoplastic aortic arch syndrome, the coarctation extends over a long segment of the aortic arch and may involve the orifice of the left subclavian artery. Opacification studies of the aorta are often employed for definitive diagnosis and for anatomic detail of the coarctation.

Dissecting aneurysm is the most frequent complication of coarctation. A relatively high incidence of coarctation occurs in females with Turner's syndrome.[35-37]

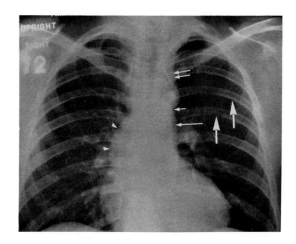

Figure 10–26 **Coarctation of Aorta.** In posteroanterior view, indentation of the descending aorta below the knob *(short arrow)* is apparent due to constriction of the aortic segment. Slight bulging below the indentation is due to poststenotic dilatation *(long arrow)*. Fullness in the left superior mediastinum *(double arrow)* is from dilatation of the left subclavian artery. Large arrows indicate notching on the inferior aspect of the ribs. The ascending aorta is dilated and prominent *(arrowheads)*.

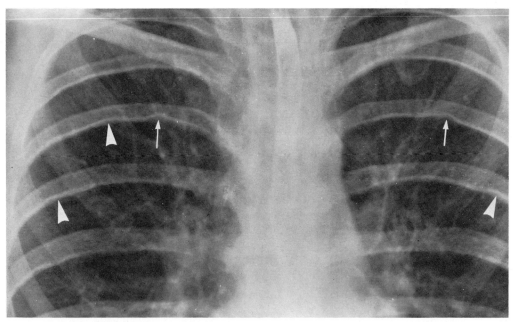

Figure 10–27 **Coarctation of Aorta: Rib Notching.** Magnified view of chest and upper ribs reveals clear-cut and unmistakable notching of both fifth ribs *(arrows)*. Slight and more subtle notching is seen in other ribs *(arrowheads)*: minimal upward displacement of the cortical rib line is apparent on close inspection. There were no other obvious changes in the heart or aorta that are characteristic of coarctation.

    Rib notching may be the only diagnostic finding on a chest film.

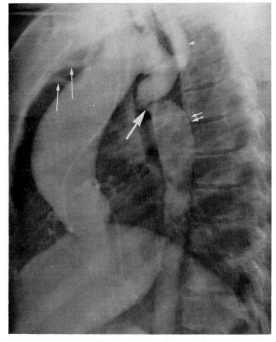

Figure 10–28 **Coarctation of Aorta: Retrograde Aortogram.** A lateral aortogram demonstrates narrowing of the aorta at the site of coarctation *(large arrow),* with apparent pulling forward of the aorta at the level of the ligamentum arteriosum. The proximal subclavian artery is tortuous and dilated *(arrowhead).*

    There is moderate localized poststenotic dilatation of the aorta *(double arrow)*. The ascending aorta is dilated, while the descending aorta beyond the coarctation is hypoplastic. The opacified internal mammary arteries *(long arrows)* are dilated.

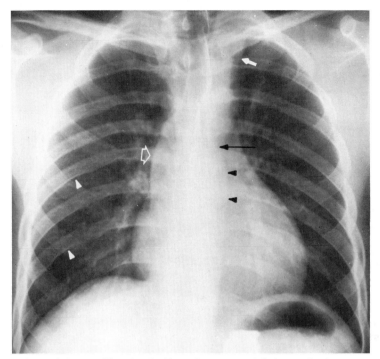

Figure 10–29 **Coarctation of Aorta: Classic Findings in Young Adults.** The heart is somewhat enlarged, with a prominent ascending aorta *(open arrow).* There is an indentation of the descending aortic shadow *(black arrow)* and a poststenotic dilatation *(black arrowheads),* producing the figure-of-three appearance. The bulge in the left superior mediastinum *(white arrow)* is due to the dilated left subclavian vessels. Rib notching *(white arrowheads)* is present but not marked.

Pseudocoarctation of the Aorta

This uncommon congenital anomaly consists of buckling or kinking of the aortic arch in the region of the ligamentum arteriosum. Although anatomically similar to true coarctation, there is no obstruction or hemodynamic abnormality, and neither rib notching nor a hypoplastic descending aorta is associated.

On plain frontal films there are two bulges in the region of the aortic knob, similar to the configuration in coarctation. These bulges represent the somewhat dilated portions of the aorta just proximal and distal to the kink. With physiologic uncoiling as a result of advancing age, these bulges become more prominent. Either or both of the bulges can indent the esophagus.

The upper bulge may simulate a mediastinal tumor. Normal hemodynamic findings in the lower extremities and absence of rib notching rule out true coarctation. Occasionally, the buckling can be identified on oblique or lateral chest films, especially with planigraphy. However, aortic angiography is usually necessary for ruling out mediastinal tumor and for correct diagnosis; a kinked aorta without a pressure gradient across the kink is the definitive angiographic finding.

Pseudocoarctation is asymptomatic, but there may be other associated cardiovascular anomalies.[38–42]

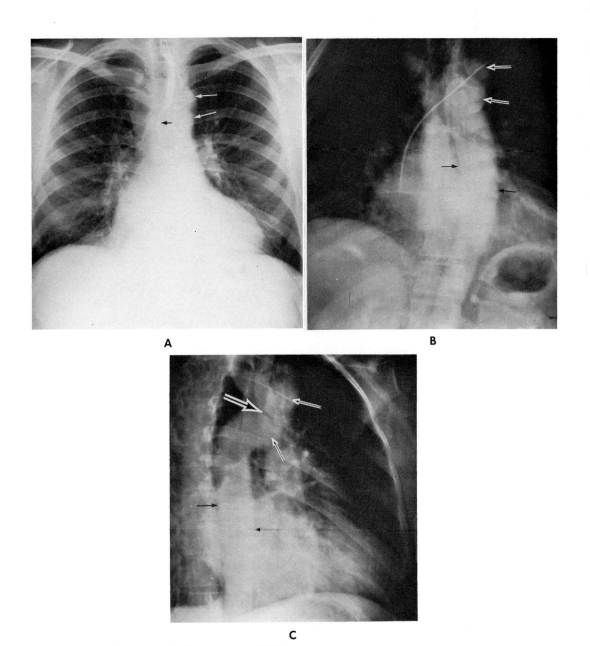

Figure 10–30 **Pseudocoarctation in a 40 Year Old Man.**

A, On frontal chest film the aortic knob appears notched, and two distinct bulges *(white arrows)* can be identified. The esophagus is indented to the right *(black arrow)* at the level of the lower bulge.

The aortic configuration is similar to true coarctation, but the absence of rib notching and of dilatation of the ascending aorta suggests pseudocoarctation.

B and *C,* On thoracic aortograms, the aortic bulges *(small black-white arrows)* are opacified above and below the kink. The oblique view best demonstrates the localized tortuosity and kinking *(large black-white arrow)* of the aortic arch in the region of the ligamentum arteriosum. The descending aorta *(black arrows)* is of normal caliber. This is in contrast to true coarctation, in which the aorta becomes abnormally small distal to a short segment of poststenotic dilatation.

Pulmonic Stenosis

Valvular pulmonic stenosis is a relatively frequent anomaly. The right ventricle has an increased workload and undergoes muscular hypertrophy, but there is little or no change in heart size; in more severe cases right ventricular enlargement may be recognized. The main pulmonary artery is usually prominent because of poststenotic dilatation. The aorta remains normal. The pulmonary vasculature is usually normal. However, in severe stenosis it may be decreased. The left pulmonary artery frequently is more prominent and pulsatile than the right, and the left lung vasculature often appears more prominent. Valvular calcification is extremely rare. In the majority of cases the chest findings are entirely normal except for prominence of the main pulmonary artery.

Isolated infundibular pulmonic stenosis is much less common. There is no poststenotic dilatation of the main pulmonary artery, which is generally less prominent than usual. The pulmonary vasculature and the heart size most often appear normal.

Right heart angiography in valvular stenosis usually demonstrates a thickened pulmonary valve that is dome-shaped and that remains fixed during systole. The poststenotic dilatation is usually apparent.[43-45]

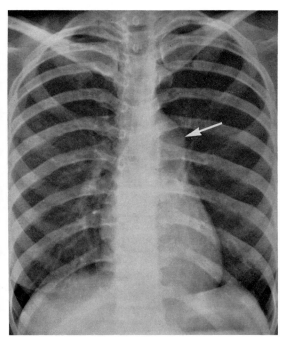

Figure 10–31 **Isolated Pulmonic Valvular Stenosis.** The pulmonary artery segment is usually prominent *(arrow),* and the pulmonary vasculature is normal or slightly diminished. The heart and aorta are of normal size.

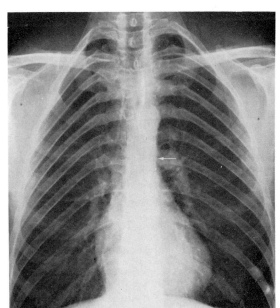

Figure 10–32 **Pulmonic Infundibular Stenosis.** Concavity of the pulmonary artery segment *(arrow)* is due to absence of poststenotic dilatation. The heart, aorta, and chest are usually normal. Catheterization studies and angiography are required for diagnosis.

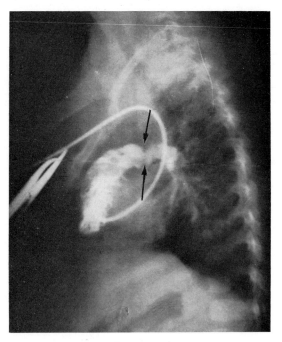

Figure 10–33 **Pulmonic Valvular Stenosis: Selective Angiocardiogram.** Lateral projection shows the catheter in the right ventricle. The right ventricle and the pulmonary artery are partially opacified. There is narrowing of the pulmonary valve *(arrows)*. Characteristically the pulmonic valve is unchanged during systole; and doming of the pulmonic valve, when seen, is diagnostic. Note the trabeculations in the opacified right ventricle, indicative of hypertrophy. Opacification studies are seldom needed in pulmonic valvular stenosis; in this case angiocardiography was done because of associated congenital lesions.

Atrial Septal Defect

Atrial septal defect (ASD) is by far the most common congenital cardiac lesion. It can occur alone or in association with a multitude of other cardiac anomalies. Any portion of the septum may be deficient.

Uncomplicated ASD, if small, will produce few or no radiographic changes. Larger defects cause a hemodynamically significant left to right shunt, thereby increasing pulmonary blood flow. This is evidenced radiographically by prominent dilated pulmonary vessels (arteries) that extend to the outer third of the lung fields. The shunted flow into the right heart also causes enlargement of the pulmonary outflow tract, including the main pulmonary artery segment and the hilar vessels; in addition, there is dilatation of the right ventricle and, often, of the right atrium. The ascending aorta and knob remain small, in contrast to its enlargement in patent ductus. The left atrium and left ventricle do not enlarge. This feature helps to distinguish ASD from ventricular septal defect and patent ductus arteriosus.

Complicating pulmonary hypertension can develop in ASD or in any severe left to right shunt (Eisenmenger's physiology). This is characterized by a relatively abrupt narrowing of the peripheral vessels and increased fullness of the main and central pulmonary arteries, causing a marked disparity between the central and peripheral vessels. The lung periphery becomes more lucent. The dilatation of the main pulmonary artery may assume aneurysmal proportions. The elevated pulmonary pressure may balance or reverse the shunt. The right ventricle may then become more hypertrophied but less dilated, and the actual heart size may decrease.

When ASD is associated with other defects, the clinical and radiologic picture is usually very different. However, when ASD is associated with congenital mitral stenosis (Lutembacher's syndrome) or with mild pulmonary stenosis, the clinical and plain film findings are virtually identical with uncomplicated ASD.

Angiocardiography is rarely necessary for diagnosis of uncomplicated ASD, but it is useful for demonstrating the size and location of the septal defect if surgery is contemplated. The defect is best seen in the left anterior oblique position, which brings the septum into profile. Angiocardiography may also uncover unsuspected associated defects.[46, 47]

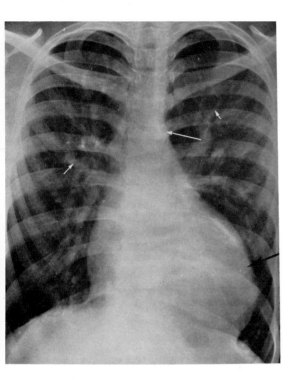

Figure 10–34 **Uncomplicated Atrial Septal Defect.** The patient was a 30 year old male. Prominent dilated pulmonary arteries *(short white arrows)* extend to the periphery of the lung. The small aortic knob *(long white arrow)* reflects decreased left ventricular output, and the elevated cardiac apex *(black arrow)* suggests right ventricular enlargement.

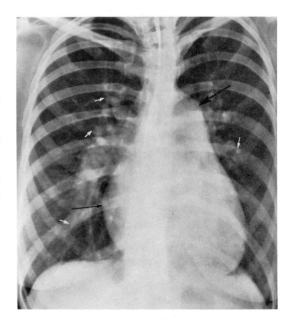

Figure 10–35 **Atrial Septal Defect: Pulmonary Hypertension.** The prominent main pulmonary artery segment *(large black arrow)* and dilated main pulmonary artery branches are in sharp contrast to the abrupt vascular narrowing toward the periphery *(small white arrows)*. This sudden narrowing is the significant finding in pulmonary hypertension and gives rise to the apparent lucency of the peripheral lung fields. The prominent right atrium *(small black arrow)* and the inconspicuous aortic knob are characteristically associated with atrial septal defect.

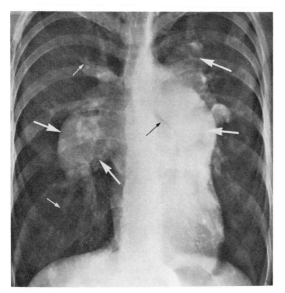

Figure 10–36   **Atrial Septal Defect: Advanced Pulmonary Hypertension.**   There is aneurysmal dilatation of the main and central pulmonary arteries *(large white arrows)* and abrupt narrowing of the branch arteries *(small white arrows),* causing an apparent paucity of peripheral vasculature in the outer third of the lung fields. The left main bronchus *(black arrow)* appears narrow, probably because of pressure from the huge left pulmonary artery.

The relatively small size of the heart is due to the diminished pulmonary circulatory volume. Progressive pulmonary hypertension is often associated with a decrease in heart size.

Actually, there was marked hypertrophy of the right ventricular wall—again illustrating that ventricular hypertrophy without dilatation is not apparent on a radiograph.

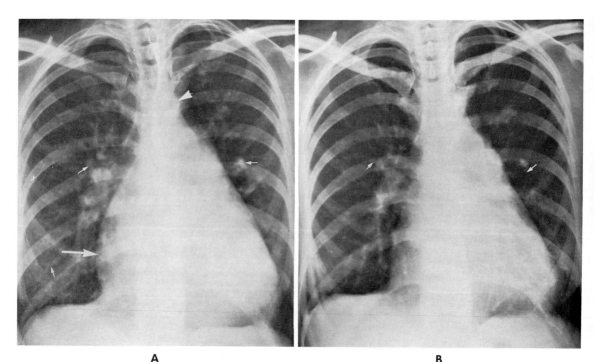

A   **B**

Figure 10–37   **Atrial Septal Defect: Before and After Surgical Closure of Defect.**

A, Preoperative posteroanterior view demonstrates the typical findings in atrial septal defect: a prominent pulmonary artery segment, engorgement of the pulmonary vasculature *(short white arrows),* a small aortic knob *(arrowhead),* and a prominent right atrium *(large white arrow).*

B, Three months after surgical correction the heart appears smaller, and the vasculature *(arrows)* has returned to normal. The right atrium is no longer prominent.

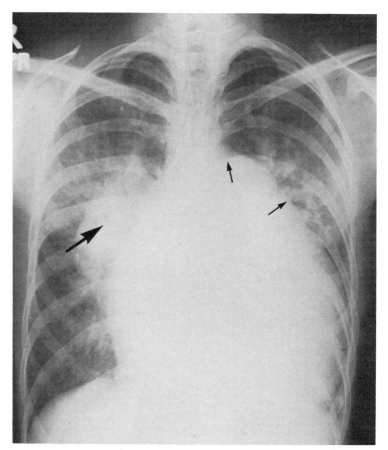

Figure 10–38 **Lutembacher's Syndrome in Young Male.** The increased vasculature, the greatly enlarged right ventricle and right atrium, and the small aorta are consistent with a large atrial septal defect. The unusually large hila (*large arrow*) and the huge main pulmonary artery (*small arrows*) are due to the associated mitral stenosis, which causes greatly increased shunting into the right atrium during atrial systole. There was no significant left atrial enlargement.

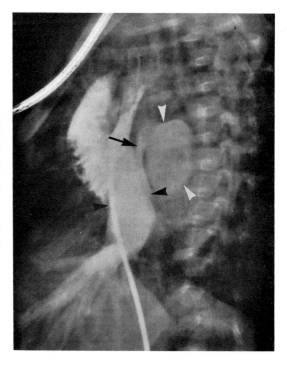

Figure 10–39 **Atrial Septal Defect: Angiocardiogram.** A lateral film shows filling of the right atrium (*black arrowheads*) and simultaneous opacification of the left atrium (*white arrowheads*). The shunting of the contrast medium across the septum between the two chambers is clearly visualized (*black arrow*). Angiocardiography is not usually necessary for diagnosis of atrial septal defect and is generally done to evaluate other intracardiac anomalies.

Injection of contrast medium into the right heart ordinarily does not allow demonstration of the defect, since the shunt is from left to right.

Ventricular Septal Defect

In this disorder the prominence of the pulmonary arterial vasculature and the pulmonary outflow tract depends on the size of the defect and the difference in pressure between the two ventricles. In addition, the larger the shunt, the larger the left atrium and left ventricle and, also, the smaller the aortic knob.

The roentgenologic findings vary from a normal appearing heart with normal vasculature, to an increase in pulmonary vascularity and a prominent main pulmonary artery segment, with an enlarged left atrium and left ventricle. The aortic knob is small. The right ventricle is sometimes enlarged. Selective left ventricular angiocardiography demonstrates the size of the defect, but the degree of shunt is best estimated by catheterization studies.

If pulmonary hypertension (Eisenmenger's physiology) develops, there is abrupt narrowing of the vessels in the middle third of the lung field, with increased lucency of the peripheral lung. Right ventricular hypertrophy increases, and the left ventricle does less work; this may produce an apparent decrease in heart size, since hypertrophy without dilatation does not produce enlargement roentgenographically.[48, 49]

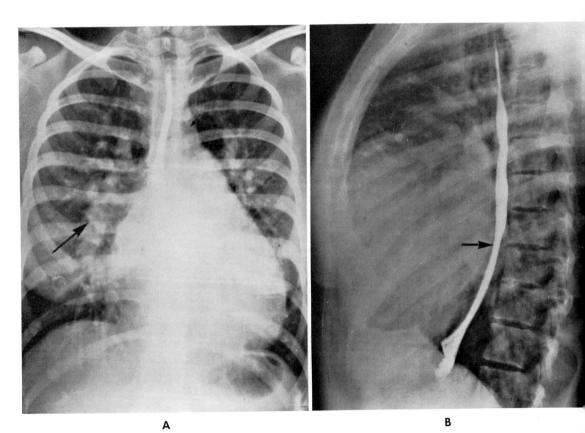

A   **B**

Figure 10–40 **Large Ventricular Septal Defect.**

A, Engorgement of the pulmonary vasculature *(large arrow)* indicates left to right shunt through the septal defect. The aortic knob is small *(small arrow),* and the main pulmonary artery segment and arteries are prominent. Cardiac enlargement appears to be predominantly left ventricular.

B, Lateral view indicates impression on the esophagus *(arrow)* owing to left atrial enlargement.

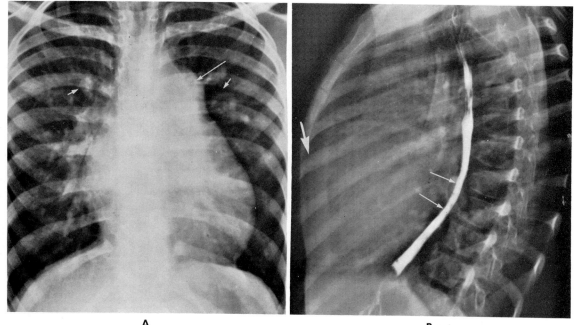

A   B

Figure 10–41   **Ventricular Septal Defect: Pulmonary Hypertension.**

A, In addition to cardiac enlargement, dilatation of the main pulmonary artery segment *(long arrow),* and increased central pulmonary vascularity due to the shunt, there is abrupt narrowing of the peripheral pulmonary vessels *(short arrows)* owing to pulmonary hypertension.

B, In deep oblique view, the enlarged left atrium presses upon the esophagus *(small arrows),* and the enlarged right ventricle encroaches on the retrosternal space *(large arrow).*

Reversal of a left to right shunt is due to development of pulmonary hypertension (Eisenmenger's physiology) and may cause cyanosis. Eisenmenger's physiology can also occur in cases of atrial septal defect and patent ductus arteriosus.

High Ventricular Septal Defect with Aortic Insufficiency

A defect high in the ventricular septum can cause prolapse of one aortic cusp, usually the noncoronary cusp, with resultant aortic insufficiency. There is left ventricular dilatation and moderate dilatation of the ascending aorta. The shunt results in pulmonary vascular engorgement. Thoracic aortography demonstrates the prolapsed cusp and aortic regurgitation.[48, 50]

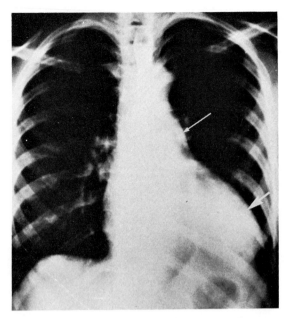

Figure 10–42 **Ventricular Septal Defect and Aortic Insufficiency.** Enlargement of the left ventricle *(large arrow)* is due to aortic insufficiency, which dominates in this case. The main pulmonary artery segment is prominent *(small arrow).* Increased vascularity was present but is not shown well on this film.

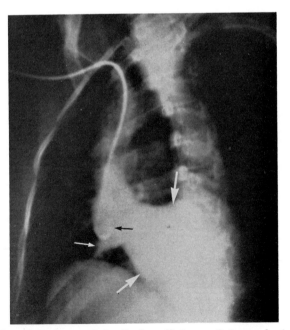

Figure 10–43 **Ventricular Septal Defect and Aortic Insufficiency: Retrograde Aortogram in Left Anterior Oblique Projection.** The dilated left ventricle is opacified *(large white arrows)* refluxly. The aortic cusp *(black arrow)* has prolapsed into the ventricular septal defect and has caused aortic insufficiency. Opacification *(small white arrow)* beneath the cusp is brought about by opaque material entering the septal defect. The interventricular septum *(lower large arrow)* is clearly defined, permitting good delineation of the septal defect.

Patent Ductus Arteriosus

A patent ductus is an extracardiac left to right shunt between the aorta and the pulmonary artery. The increase in pulmonary vascularity and the prominence of the pulmonary artery segment vary with the degree of shunting. The central pulmonary arteries are enlarged. Often the right hilar shadow is larger than the left, due to directional greater shunting into the right pulmonary artery. The increased pulmonary flow to the left atrium results in left atrial enlargement and subsequent left ventricular enlargement. Moderate cardiomegaly due to left ventricular enlargement is seen in almost 75 per cent of cases. Right ventricular enlargement occurs only if pulmonary hypertension develops.

The aorta proximal to the shunt carries a greater volume of blood than normal and may become dilated, with a prominent knob. Aortic enlargement is rare before 2 years of age but is found in 75 per cent of patients above 10 years of age. A convex bulge may be seen on the left border of the aorta just below the knob. This is the site of origin of the ductus and is called the infundibulum of the ductus. This infundibulum sign is seen on posteroanterior films in approximately one third of patients. Sometimes, a small area of calcification is evident in the region of the ductus.

In aorticopulmonic window or fenestration, the shunt is in the ascending aorta proximal to the knob; consequently, the knob does not enlarge.

The diagnosis of patent ductus should be considered in a young person who manifests a large aortic knob and increased pulmonary vascularity. The aortic knob continues to enlarge as the patient becomes older. A retrograde aortogram outlines the ductus and demonstrates the shunt into the pulmonary artery, but this procedure is rarely necessary. On plain films, in the absence of a large aortic knob, patent ductus cannot be differentiated from ventricular septal defect.

Pulmonary hypertension (Eisenmenger's physiology) is a frequent late development.[51, 52]

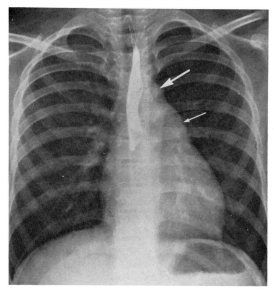

Figure 10–44 **Patent Ductus Arteriosus.** The aortic knob *(large arrow)* is rather large, and the main pulmonary artery segment *(small arrow)* is prominent. The vasculature is not significantly increased. A prominent aortic knob in a young person is abnormal and should suggest a cardiac abnormality.

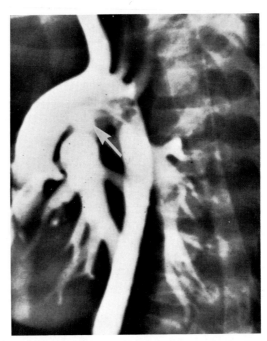

Figure 10–45 **Patent Ductus Arteriosus: Retrograde Aortogram.** Lateral projection of retrograde aortogram, in which the catheter was introduced through the brachial artery, reveals simultaneous opacification of the aorta and the pulmonary arteries through a patent ductus *(arrow)*. This is the diagnostic finding in patent ductus. It is also the diagnostic finding in aorticopulmonary fenestration, but in the latter case the communication is in the ascending portion of the aorta.

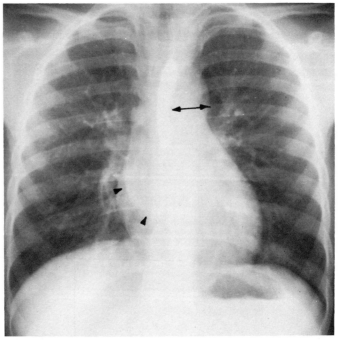

Figure 10–46 **Patent Ductus Arteriosus in Young Adult.** The pulmonary vasculature is moderately increased, the aortic knob shadow is enlarged *(double-headed arrow),* and although the overall heart size is not enlarged, the left atrium *(arrowheads)* seems prominent and large.

    In a younger individual with clinical or radiographic suspicion of a left to right shunt, a prominent aortic knob is highly suggestive of patent ductus arteriosus.

Congenital Coronary Fistula

In this defect the coronary arteries arise normally from the aortic root, but a communication exists between one coronary vessel and a cardiac chamber or a pulmonary trunk. Blood may be shunted into the right side (a left to right shunt) or to the left atrium or ventricle. The right coronary artery is more commonly involved and terminates, in order of frequency, in the right ventricle, right atrium, the coronary sinus, or the pulmonary trunk. The involved coronary is tortuous, dilated, and may be calcified.

Plain films reveal a nonspecific left to right shunt with dilated pulmonary vessels and a prominent main pulmonary artery segment The aortic knob is normal in size.

Diagnosis is made by coronary arteriography, which demonstrates the dilated tortuous coronary artery and its abnormal communication.[53-56]

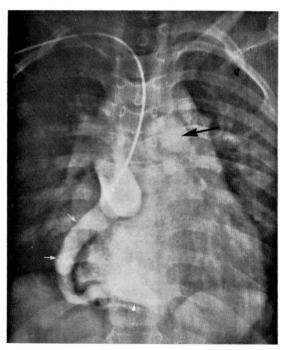

Figure 10–47 **Congenital Coronary Fistula: Retrograde Aortogram.** Vessel *(white arrows)* arising from the aortic root is a dilated anomalous right coronary artery that enters and opacifies the right ventricle *(arrowhead)* and pulmonary arteries *(black arrow)*.

The plain film findings were indistinguishable from those of any left to right shunt. (Courtesy Dr. F. Silverman, Children's Hospital, Cincinnati, Ohio.)

Anomalous Left Coronary Artery: Left Coronary Artery Originating from Pulmonary Artery

This is the most common of the anomalies of the coronary circulation. Blood flows from the normal right coronary artery through intercoronary anastomoses into the anomalous left coronary, and then retrograde into the pulmonary artery. If these intercoronary anastomoses are extensive, there is virtually a small left to right heart shunt.

The left heart ischemia causes impaired contractility and enlargement (dilatation) of the left ventricle and, frequently, of the left atrium. The vasculature remains normal unless congestive failure supervenes. The clinical and radiographic picture is similar to that of idiopathic myocardiopathy or fibroelastosis.

Coronary arteriography is necessary for definitive diagnosis. Only the right coronary will be opacified from the aorta, and the left coronary fills by retrograde flow from intercoronary anastomoses, with subsequent opacification of the pulmonary artery. The anomalous left coronary artery may also become opacified during pulmonary arteriography if the pulmonary arterial pressure is sufficiently high.[53, 57-59]

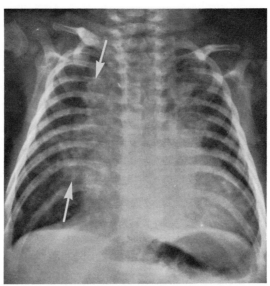

Figure 10–48 **Anomalous Left Coronary Artery in an Infant.** Cardiac enlargement and vascular engorgement *(arrows),* especially in the upper lung fields, are indicative of congestive heart failure. Anomalous coronary artery, fibroelastosis, or glycogen storage disease should be considered in infants who have generalized cardiomegaly.

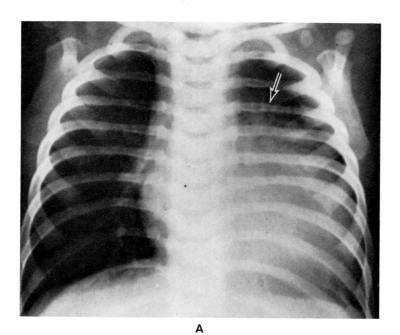

A

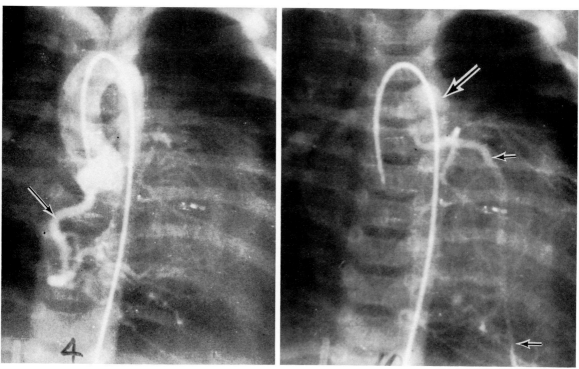

B   C

Figure 10-49   **Anomalous Origin of Left Coronary Artery from Pulmonary Artery.**

A, The heart is greatly enlarged, and the upper lobe veins *(arrow)* are dilated (early congestive failure).
B, Root aortography shows opacification of the right coronary artery *(arrow)* and its branches.
C, Moments later, the left coronary artery *(small arrows)* is opacified by retrograde flow from the right coronary circulation. The left coronary can be seen emptying into the pulmonary artery *(large arrow).*

Endocardial Cushion Defects (Atrioventricularis Communis)

An atrial septal defect at the site of the ostium primum may be associated with a ventricular septal defect, both defects adjacent to the mitral and tricuspid valves. These valves are absent or severely deformed, so that there is a persistent single atrioventricular canal. A single valve may separate the atria from the ventricles. These defects are most frequently encountered in children with Down's syndrome.

Endocardial cushion defects may be classified as follows:

(1) Incomplete form. Ostium primum defect, usually with a cleft mitral valve, and occasionally with a cleft tricuspid valve.

(2) Complete atrioventricular canal. Large defect between atria and ventricles with a common valve and often with ventricular outflow tract abnormalities.

(3) Intermediate forms. (a) Ostium primum with partial fusion of the atrioventricular valve rings. (b) Cleft mitral valve attached to a deficient ventricular septum. (c) Cleft tricuspid valve with septal defect.

Conventional films usually demonstrate cardiomegaly with enlargement of the right heart chambers, prominence of the pulmonary artery, and increased pulmonary vascularity, findings indicative of a nonspecific left to right shunt. Pulmonary hypertension often develops in adults.

Angiocardiography usually reveals virtually immediate opacification of all four cardiac chambers and simultaneous opacification of the aorta and pulmonary artery. Selective left ventriculography will demonstrate a scalloped medial left ventricular border and a goose neck deformity of the subaortic outflow segment of the left ventricle, which is due to the cleft mitral valve. This is a characteristic and virtually diagnostic finding.

Pulmonic stenosis and other cardiac anomalies are often associated with atrioventricularis communis.[60-62]

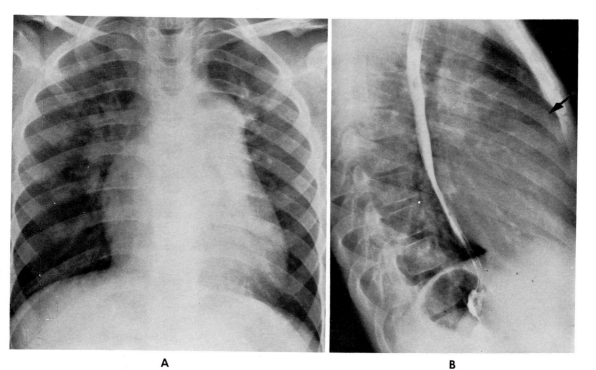

A   B

Figure 10–50 **Endocardial Cushion Defect (Atrioventricularis Communis).**

A, Posteroanterior view reveals a prominent pulmonary artery segment, prominence of the pulmonary vasculature, and a small aortic knob—all characteristic of a left to right shunt. Cardiomegaly is usually more pronounced than it is in this case.

B, Lateral film indicates prominence of the anterior heart border, which is due to right ventricle enlargement *(arrow).* Selective angiocardiograms are needed for definitive demonstration and classification of the defects.

Ruptured Aortic Sinus (Sinus of Valsalva) Aneurysm

The radiologic features of aneurysm of the aortic sinuses (sinuses of Valsalva) are discussed elsewhere (p. 737).

Rupture of a sinus aneurysm is usually into the right atrium or right ventricle. This produces an acute left to right intracardiac shunt. Radiographically there will be an increase in heart size, an increase in the pulmonary vasculature, and prominence of the main pulmonary artery segment, all of which are dramatic changes when contrasted with the prior, relatively normal appearance.

Retrograde aortography in oblique projections demonstrates the flow of contrast material through the aneurysm into the right heart chambers.[63, 64]

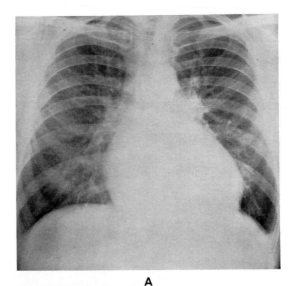

A

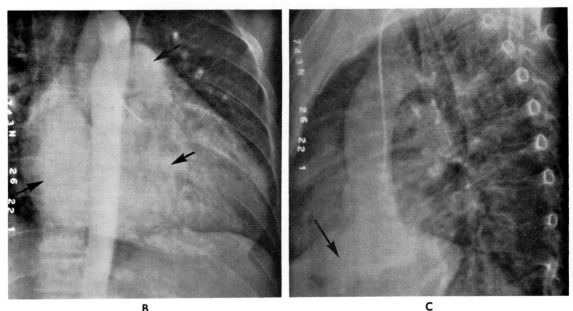

B   C

Figure 10-51   **Ruptured Aortic Sinus Aneurysm: Retrograde Aortogram.**

   A, There is cardiomegaly and increased vascularity in a patient whose heart had been of normal size previously. Rapid enlargement of the heart was due to a left to right shunt secondary to rupture of an aortic sinus aneurysm into the right heart.

   B, In anteroposterior projection of retrograde aortogram, a catheter is seen in the ascending aorta. The right ventricle and pulmonary artery *(arrows)* are opacified simultaneously with the aorta, because of the ruptured aortic sinus. The actual aneurysm is not clearly demonstrated.

   C, Lateral film shows opacification of the right ventricle *(arrow)* after aortic injection. Opacification of the right heart from the aorta can also occur in certain congenital lesions, but sudden onset suggests traumatic or inflammatory rupture of an aortic sinus aneurysm.

Corrected Transposition

   In this condition there is a transposition of the great vessels, but this is hemodynamically corrected by inversion of the ventricles. The anatomic smooth-walled left ventricle with its mitral valve functions as the right ventricle; the thick-walled trabeculated right ventricle with its tricuspid valve gives rise to the aorta and functions as a left ventricle but lies to the left of and anterior to the other ventricle. The aorta arises more anteriorly than normal and forms the upper segment of the left heart border. Uncomplicated corrected transposition is extremely uncommon; the majority of cases have one or more associated defects, most often ventricular septal defect, a single ventricle, or pulmonic stenosis. Deformity of the "tricuspid" valve in the functional left ventricle is common and can cause a valvular insufficiency.

The radiologic findings usually reflect the reversed positions of the great vessels; the aorta originates above, anterior to, and to the left of the pulmonary artery, which now has a more medial origin and position. A smooth convexity or bulge of the upper left cardiac border due to the left-sided ascending aorta replaces the double bulge of the aortic knob and main pulmonary artery. The main and left pulmonary arteries are more medial and lower than normal and may cause an unusual indentation on the esophagus. Cardiac enlargement and increased vasculature will occur in patients with septal defect or single ventricle. If significant pulmonary stenosis is present, the heart will be normal or only slightly enlarged, and the vasculature will be normal or decreased.

Angiographic findings are diagnostic. They consist of a low medial origin of the pulmonary artery and a high, lateral, and anterior aorta. The associated anomalies can be demonstrated by properly selective techniques.[65-68]

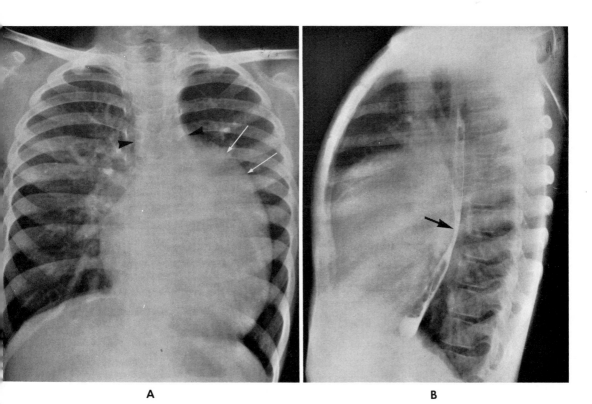

A   B

Figure 10–52   **Corrected Transposition: Characteristic Findings.**

A, The left heart border is prominent, particularly in the upper segment, and has an unusual globular shape. A single bulge of the left-sided ascending aorta *(white arrows)* replaces the usual double bulge of the aortic knob and main pulmonary artery. Pulmonary vascularity is increased due to an associated septal defect. Superimposition of the aorta and pulmonary artery narrows the shadow of the great vessels *(black arrowheads).*

B, Lateral view shows indentation of the esophagus *(arrow)* by the posteriorly displaced pulmonary artery.

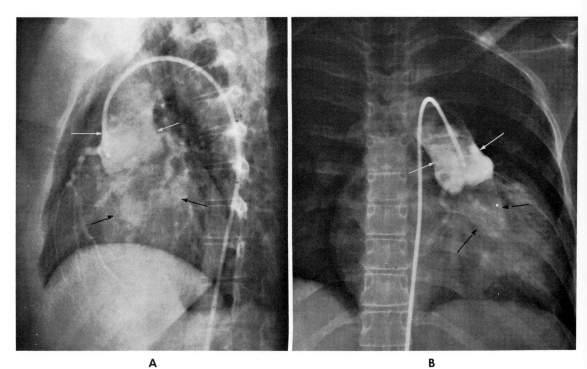

A   **B**

Figure 10–53   **Corrected Transposition: Retrograde Aortogram.**

A, Lateral projection shows the high anterior position of the root of the aorta *(white arrows)*, a characteristic finding in corrected transposition.

Angiocardiogram would outline the abnormal position of the pulmonary artery.

B, Anteroposterior projection demonstrates the abnormal and high position of the ascending aorta *(white arrows)* on the left side: this causes an abnormal convexity of the left heart border on plain films.

A minor degree of aortic insufficiency is responsible for left ventricular opacification below the aortic valve, as shown in both films *(black arrows)*.

Cor Triatriatum Sinistrum

The anomaly is characterized by a transverse incomplete fibromuscular membrane dividing the left atrium into two chambers; the mitral orifice is in the lower chamber and the pulmonary veins are in the upper chamber.

The degree of impedance of venous flow into the left heart will depend on the size and number of openings in the membrane. The hemodynamic effect and radiographic findings are similar to those of mitral stenosis. The main pulmonary artery segment is prominent, the hilar vessels are enlarged, and there is evidence of pulmonary venous hypertension with prominent upper lobe veins and Kerley B lines. Right ventricular enlargement usually is present, but left atrial enlargement may be minimal or absent. In about one third of adult cases an atrial septal defect or anomalous drainage of the pulmonary veins, or both, are associated. These allow the venous blood to bypass the obstruction in the left atrium.

This radiographic picture of pulmonary venous hypertension in cor triatriatum may be difficult to distinguish from that of left ventricular failure, mitral stenosis, left atrial myxoma, ball valve thrombus in left atrium, obstruction of the pulmonary veins by tumor, thrombosis, or mediastinal fibrosis, and total anomalous venous return.

Pulmonary angiography will usually identify the double chambered left atrium and the obstructing membrane. Surgical correction may be lifesaving in infants.[69-71]

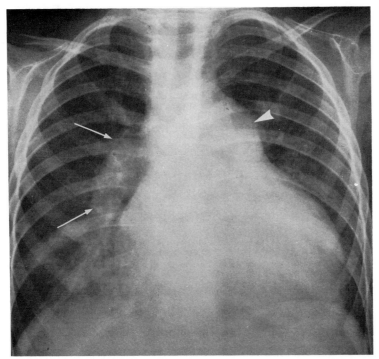

Figure 10-54 **Cor Triatriatum Sinistrum.** The heart is enlarged and the pulmonary vessels *(arrows)* are dilated. The main pulmonary artery segment is prominent *(arrowhead)*. Right atrial and right ventricular enlargement are present, but the left atrium is not enlarged. There are no associated defects. The roentgenologic findings are nonspecific, and angiocardiography is necessary for diagnosis. (Courtesy Drs. C. H. Chang and J. V. Rogers, Atlanta, Georgia.)

Fibroelastosis: Subendocardial Sclerosis

The endocardial thickening may occur alone (primary fibroelastosis) or may be associated with other congenital cardiac lesions (secondary fibroelastosis), particularly aortic stenosis and coarctation of the aorta. The contractility and distensibility of cardiac muscle are hindered by the endocardial thickening.

In the primary disease there is always cardiomegaly; often the cardiac shape is globular. Left ventricular enlargement is always present, and moderate to severe mitral insufficiency with left atrial enlargement is frequent. Right-sided enlargement is often also associated. The aortic knob is relatively small because of decreased left ventricular output. The pulmonary vasculature is normal unless heart failure supervenes. The roentgen picture is similar to that of glycogen storage disease, anomalous coronary artery origin, and other cardiomyopathies.

Angiocardiography discloses a dilated and poorly contracting left ventricle with a thickened wall. Usually the left atrium also is dilated and contrasts poorly; mitral insufficiency is frequent. The right ventricle may pulsate normally. Mild to moderate aortic insufficiency is often seen.

In secondary fibroelastosis, similar roentgen findings are present, but there may be additional changes due to associated hemodynamically significant defects.[72, 73]

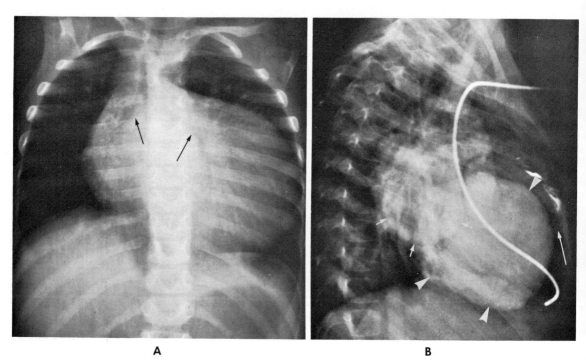

A   B

Figure 10–55 **Fibroelastosis.**

A, Shown is a large globular heart with an inconspicuous aortic knob and normal pulmonary vasculature in a child. The enlarged left atrium *(arrows)* has caused bronchial elevation.

B, Angiocardiography by means of direct left ventricular puncture reveals a large dilated left ventricle *(arrowheads).* The clear space between the ventricle and anterior chest wall *(long arrow)* is due to extreme thickening of the ventricular wall.

The enlarged left atrium *(short arrows)* is also opacified because of associated relative mitral insufficiency. These chambers remained unchanged on multiple films during successive cardiac cycles.

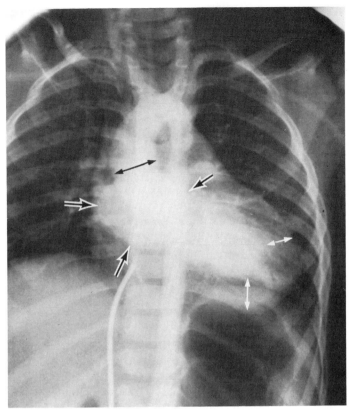

Figure 10-56 **Fibroelastosis in 3 Year Old Child: Angiocardiogram.** Cardiomegaly is present but not pronounced. The film, made five seconds after injection of contrast material into the right atrium, reveals a dilated left atrium *(black-white arrows).* The left ventricle is enlarged and shows greatly thickened walls *(white arrows).* Both of these chambers remained opacified longer than usual, which is a characteristic finding.

The dilatation of the ascending aorta *(black arrow)* is due to an associated aortic valvular stenosis; this classifies the anomaly as secondary fibroelastosis.

Partial Anomalous Pulmonary Venous Return

Partial anomalous venous return is more common and usually asymptomatic. The chest films may appear normal or may show hypervascularity, depending on the amount of blood diverted back into the right side. Selective angiocardiography demonstrates the anomalous veins and their drainage connections.

In the scimitar syndrome there is anomalous venous drainage of the right lung by a large vein emptying into the inferior vena cava, associated with hypoplasia of the right lung and an anomalous arterial supply to the right lower lobe from the aorta. Radiographically the anomalous vein appears as a band of density curving through the right lower lung field toward the midline (scimitar sign), in the direction of the inferior vena cava. The heart is somewhat displaced into the right thorax, and the righthpulmonary artery is hypoplastic, producing a small right hilar shadow. The abnormal vein and the small hilar shadow are often best demonstrated by laminography. Angiography is diagnostic.[74-77]

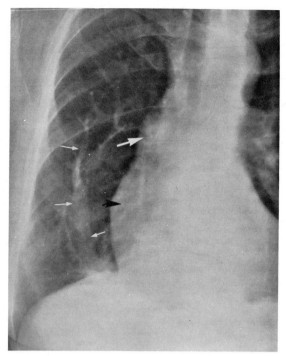

Figure 10–57 **Partial Anomalous Venous Return: Scimitar Syndrome.**   In posteroanterior view of right lung, a crescent shaped density *(small white arrows)* extends toward the medial portion of the diaphragm. This is an anomalous pulmonary vein draining into the inferior vena cava. Note the small size of the right hilum *(large white arrow)* and dextroposition of the heart. Prominence of the right atrium *(black arrow)* is due to the increased volume of venous return.

The anomalous right pulmonary vein associated with a hypoplastic right pulmonary artery and dextroposition of the heart is known as the scimitar syndrome.

Defects Characterized by Cyanosis

Total Anomalous Pulmonary Venous Return

In total anomalous venous return, all the pulmonary veins empty into the right heart, usually into a persistent left superior vena cava, which then empties into the right superior vena cava. The prominent right and left venae cavae and the enlarged right heart produce the characteristic figure-of-eight, or snowman, appearance. The aorta is hypoplastic, and the pulmonary vessels are very prominent, producing a shelf-like density below the small aortic knob. A right to left shunt through an atrial defect is necessary to maintain life. Cyanosis is persistently present.

Selective angiocardiography demonstrates drainage of the pulmonary veins into a persistent left superior vena cava, which, in turn, drains into the right superior vena cava and thence into the right atrium.

Many variations of total anomalous venous return can occur, including emptying into the coronary sinus or into a vein in the abdomen. In the latter condition, especially if the pulmonary veins empty into the portal vein or ductus venosus, pulmonary venous obstruction may occur and lead to pulmonary venous distention and edema, but with a normal-sized heart. This form occurs only in infancy; a normal-sized heart and congestive vascular changes without cardiac murmur in a cyanotic male infant are highly suggestive. Angiography demonstrates the common pulmonary vein extending below the diaphragm.[74–77]

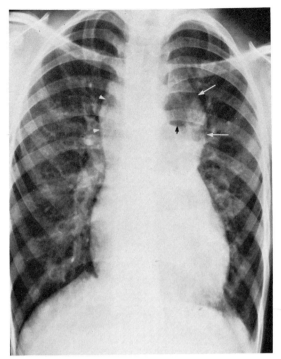

Figure 10–58 **Total Anomalous Venous Return to Superior Vena Cava.** Posteroanterior film shows the characteristic figure-of-eight, or snowman, configuration of the cardiovascular shadow. The abnormal upper left density *(white arrows)* is a persistent left superior vena cava. The upper right density *(arrowheads)* is a distended right superior vena cava. The shelf-like density *(black arrow)* represents the upper border of the distended pulmonary artery.

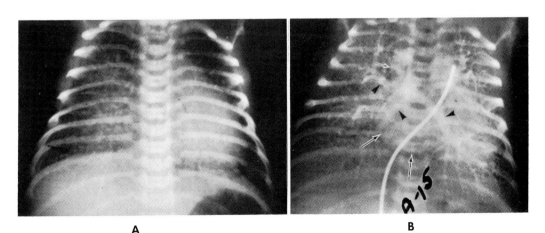

A   B

Figure 10–59 **Total Anomalous Venous Return to Superior Vena Cava.**

A, Chest film of a cyanotic 6 day old infant shows an enlarged heart and severe venous congestion throughout the lung fields. The engorged veins obscure the right cardiac border.

B, Venous phase of pulmonary arteriography demonstrates the pulmonary veins *(arrowheads)* converging and emptying into the superior vena cava *(small arrow)*. The right atrium *(large arrows)* is greatly enlarged.

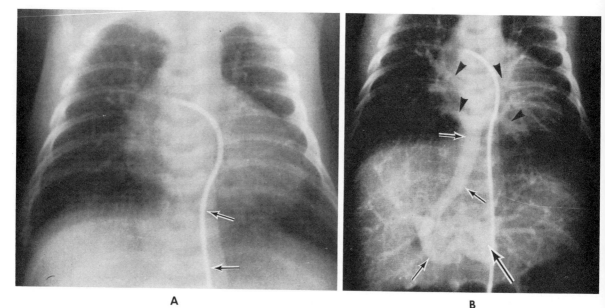

Figure 10–60   **Total Anomalous Venous Return to Portal Vein.**

A, The heart of this 8 month old infant is enlarged. The right atrial border is unusually long and high, and the inferior vena cava, identified by the catheter (*arrows*), is left of midline. The catheter tip is in the pulmonary artery. The upper lung fields are congested, but the hilar shadows are unusually small.

B, The pulmonary arterial tree was normal, but the pulmonary veins (*arrowheads*) join to form a single large accessory vein (*small arrows*) that passes below the diaphragm and drains into the portal vein (*long arrow*), opacifying the portal circulation of the liver.

Transposition of Great Vessels (Uncorrected Complete Transposition)

This is the most frequent form of congenital heart disease associated with cyanosis present at birth.

The relationships of the aorta and the pulmonary artery are reversed. The aorta arises anteriorly from the right ventricle, and the pulmonary artery originates posteriorly from the left ventricle. One or more shunts must be present for survival, ventricular septal defect or patent ductus, or both, being the most common.

The heart is enlarged and assumes a characteristic egg-on-side appearance. Usually the right border is quite prominent due to right atrial enlargement. Frequently, the vascular pedicle appears narrow in the frontal projection; this results from superimposition of the aorta on the pulmonary artery and the absence of a thymic shadow. On oblique and lateral projections, the vascular pedicle appears widened.

Vascularity of the lungs is always considerably increased unless transposition is associated with pulmonary stenosis (which occurs in about 15 per cent of patients). There is often a discrepancy between the pulmonary plethora and the hilar and main pulmonary artery shadows on frontal projection; this is due to the midline position of the main pulmonary artery.

Angiocardiography is diagnostic, demonstrating the abnormal position of the great vessels and the right ventricular origin of the aorta. Demonstration of the associated shunts may require selective angiocardiography.[78, 79]

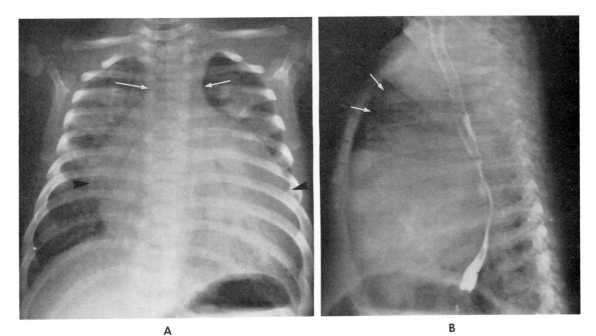

A   B

Figure 10–61   **Transposition of Great Vessels.**

A, Posteroanterior view in a cyanotic infant shows a large globular heart *(arrowheads)* extending to the lateral chest wall. The right border is prominent and shows a long, high convexity. The base of the heart *(arrows)* is narrow, and the pulmonary vasculature is prominent.

B, Lateral view shows the extreme anterior position of the aorta *(arrows).* The superior mediastinal vascular shadow appears widened because the pulmonary artery lies directly behind the aorta. This finding, associated with the narrow vascular pedicle seen on the posteroanterior projection, is characteristic of transposition, although the narrow vascular pedicle is found in less than half the cases.

Figure 10–62   **Transposition of Great Vessels: Angiocardiogram in Anteroposterior Projection.**   There is simultaneous opacification of the right and the left heart because of an associated ventricular septal defect. The aorta (AO) arises from the right ventricle and is anterior to, and to the right of, the pulmonary arteries. The less densely opacified pulmonary artery (PA) arises from the partially filled left ventricle. The opacified right atrium, right ventricle, and left ventricle are indicated with the letters RA, RV, and LV. The relationship of the great vessels is often more clearly seen on lateral views.

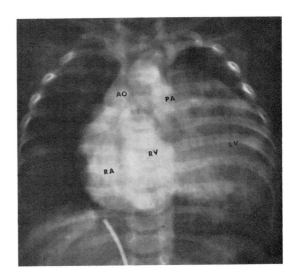

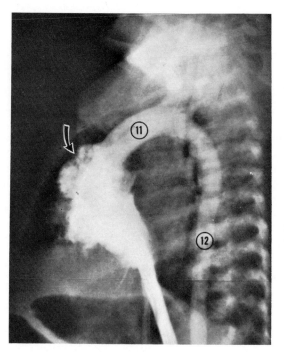

Figure 10–63 **Transposition of Great Vessels: Lateral View of Venous Angiocardiogram.** One half second after injection, there is opacification of the right heart chambers and of the aortic arch (11) and descending aorta (12), which arise from the right ventricle. The right atrium and right ventricle are superimposed; the right atrial appendage can be identified *(arrow)*. The origin of the aorta is more anterior than it is normally—a characteristic finding in complete transposition.

Truncus Arteriosus

In this anomaly, a single artery (the truncus) receives the outflow of both ventricles. The aorta is large and is continuous with the truncus. The pulmonary arteries arise in one of several patterns from the ascending portion of the truncus (types I, II, and III). A ventricular septal defect is usually present. In type IV truncus, the pulmonary arteries are absent; the pulmonary circulation is supplied by bronchial arteries arising in the descending aorta.

Characteristically there is right ventricular enlargement with a concavity in the region of the pulmonary artery. The cardiac appearance may resemble a tetralogy, but the right ventricular size is generally greater in truncus, and the lungs are hypervascular. The left pulmonary artery appears abnormally high and may seem separated from the mediastinum. The aortic arch is high and prominent and is right sided in 20 to 30 per cent of cases.

If hypoplasia of one or both pulmonary arteries is associated, there may be normal or decreased vasculature in one or both lungs.

Angiocardiography indicates filling of the truncus from the right ventricle with subsequent rapid filling of the pulmonary vessels and the aorta. It also demonstrates the site of origin of the pulmonary artery.

Pseudotruncus arteriosus is the term used to designate pulmonary valve atresia and a ventricular septal defect. It is a form of tetralogy. The cardiac silhouette is similar to that found in truncus, but the pulmonary vasculature is decreased or normal and often appears as a peculiar reticular nodular pattern, which represents bronchial artery collateral blood supply. Pseudotruncus may be difficult to distinguish from a type IV truncus.[78–84]

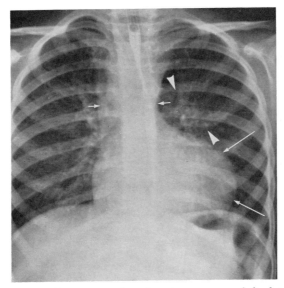

Figure 10–64 **Truncus Arteriosus.** Posteroanterior view in a young adult shows enlargement of the right ventricle with elevation of the cardiac apex *(long arrows)*. The heart and the aortic knob are enlarged and resemble the configuration seen in tetralogy, but the pulmonary vasculature is more prominent than in tetralogy. The waist of the heart is narrow, and the ascending "aorta" is prominent *(short arrows)*. The left pulmonary artery is elevated, has an abnormal appearance, and appears separated from the main cardiac shadow *(arrowheads)*.

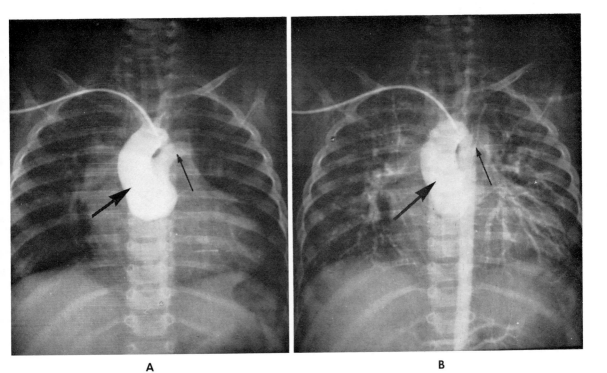

A   B

Figure 10–65 **Truncus Arteriosus: Retrograde Aortogram.**

A, Early phase of an aortogram shows filling of the truncus arteriosus *(large arrow)* and the abnormal pulmonary artery *(small arrow)* arising from the truncus.

B, Later phase of the aortogram shows more extensive opacification of the thoracic aorta and the pulmonary arteries. The various types of truncus arteriosus can be determined with the opacification study.

Single Ventricle (Cor Triloculare Biatriatum)

In this rare and complex anomaly, a combination of several developmental abnormalities results in a single functional ventricle. Other anomalies are usually associated, the most frequent being pulmonary artery stenosis, both valvular and subvalvular. If stenosis is severe, the pulmonary circulation may be provided by the bronchial arteries.

The plain film findings are neither specific nor diagnostic. There is usually some degree of cardiomegaly and, in about half the cases, an enlarged left atrium. The pulmonary vasculature is increased unless severe pulmonary stenosis is associated. The appearance often suggests a ventricular septal defect. Dextrocardia, sometimes with situs inversus, occurs in up to 25 per cent of cases.

Angiocardiography may show both atrioventricular valves opening into a common ventricle, and may demonstrate either a rudimentary right ventricle from which the pulmonary artery arises, or a single ventricle with no separate outlet chamber for the pulmonary artery. Often the associated cardiac anomalies can be detected. The complexity of the angiographic findings often makes exact diagnosis difficult.[85-87]

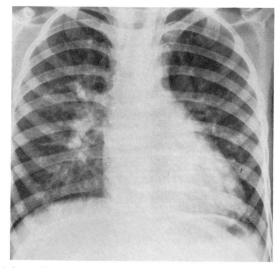

Figure 10–66 **Single Ventricle.** There are no characteristic findings. The heart is moderately enlarged in the posteroanterior projection, and the pulmonary vasculature is prominent, suggesting a left to right shunt; however, the clinical finding of cyanosis indicates that the shunt is really right to left or bidirectional, thus suggesting a single ventricle.

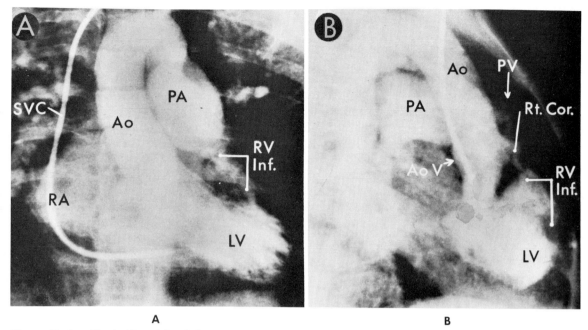

Figure 10–67 **Single Ventricle: Selective Angiocardiogram.**

A, Frontal view: catheter has entered the common ventricle (LV) via the superior vena cava (SVC) and the right atrium (RA). The opacified ventricle (LV) is large and coarsely trabeculated. Both great vessels arise from this single ventricle, the aorta (Ao) directly and the pulmonary artery (PA) from a ventricular pouch (RV Inf.).

B, Lateral projection: there is opacification of the single ventricle (LV), which contains a pouch of rudimentary right ventricle (RV Inf.) from which the pulmonary artery (PA) arises. The pulmonary valves (PV) lie above the aortic valves (AoV). The spatial relationship of the aorta and pulmonary artery is normal. The right coronary artery (Rt. Cor.) is opacified.

Selective angiography affords clearest demonstration of the single ventricle and its associated malformations. (Courtesy Dr. R. Van Praagh, Baltimore, Maryland.)

Tetralogy of Fallot

The primary changes in tetralogy of Fallot are infundibular pulmonic stenosis and a high ventricular septal defect. Overriding of the aorta and right ventricular hypertrophy are secondary structural defects.

The roentgenologic findings on plain films reflect the degree of pulmonic stenosis and the size of the septal defect. Right ventricular hypertrophy is the most common finding. In milder cases this may be evidenced only by fullness of the anterior cardiac mass on lateral view. Greater right ventricular enlargement produces elevation and rounding of the apex on frontal view—the coeur en sabot appearance. Heart size is usually normal or slightly increased.

Because blood enters the aorta from both ventricles there may be considerable enlargement of the aorta, particularly if the pulmonic stenosis is severe. In about one fourth of cases, a right-sided aortic arch is seen.

Pulmonic stenosis is usually infundibular, though valvular stenosis may coexist; poststenotic dilatation of the pulmonary artery is therefore rare. The pulmonary artery is usually inconspicuous. Consequently, the waist of the heart appears narrow. The pulmonary vasculature is usually diminished. In severe pulmonic stenosis, additional pulmonary blood is supplied by dilated bronchial arteries, which produces a reticular vascular lung pattern.

Any or all of these roentgen features may be absent or minimal, and sometimes the silhouette appears completely normal. Classically, in about two thirds of cases, the heart is boot-shaped, the cardiac waist is narrow, the ascending aorta and the aortic arch are large, and the pulmonary vasculature is decreased. The presence of a right-sided aortic arch is a suggestive finding.

Angiocardiography can demonstrate the infundibular stenosis, the simultaneous filling of the aorta and pulmonic vessels from the right ventricle, and the right ventricular enlargement. In about 25 per cent of cases, angiocardiography may also disclose other associated cardiovascular anomalies, most often stenosis of the main pulmonary artery and/or its major branches.

After surgical correction of a tetralogy by subclavian-pulmonary artery anastomosis (Blalock-Taussig shunt), rib notching may develop on the side of the shunt. A complicating pulmonary hypertension may cause postoperative enlargement of the central pulmonary arteries.[88-91]

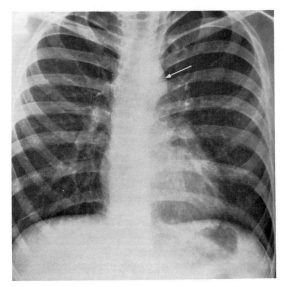

Figure 10–68 **Tetralogy of Fallot: Minimal Findings.**   The aortic knob is moderately enlarged *(arrow)*; the configuration is otherwise normal, and the pulmonary vasculature appears normal or slightly decreased.

In a young person, the combination of a prominent aortic knob and an otherwise normal appearing heart should suggest tetralogy, especially if a congenital heart disease is suspected clinically.

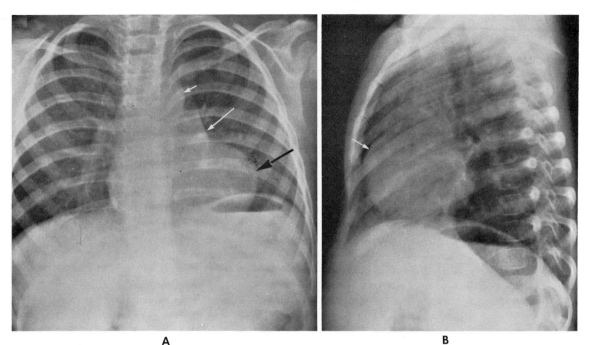

A	**B**

Figure 10–69 **Tetralogy of Fallot: Classic Findings.**

    A, Coeur en sabot with characteristic elevation of the apex is seen *(black arrow)*. The cardiac waist is narrow *(long white arrow)* because of the abscence of poststenotic pulmonary artery dilatation. The pulmonary vasculature is decreased, and the aortic knob is prominent *(short white arrow)*.

    B, Lateral view demonstrates the prominent right ventricle *(arrow)*. The increased size of the aorta is not evident on this film.

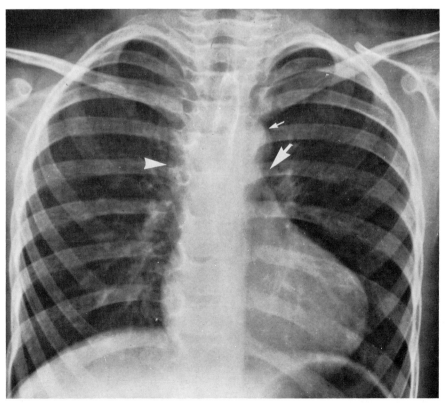

Figure 10–70 **Tetralogy of Fallot: Classic Findings.** The heart has a coeur en sabot configuration, characteristic of right ventricular enlargement. The concavity of the cardiac waist *(large arrow)* is due to the small size of the main pulmonary artery.

    The aortic knob is prominent and on the left *(small arrow)*; the ascending aorta is prominent and bulging *(arrowhead)*. The hilar shadows are small, and the pulmonary vasculature is definitely decreased.

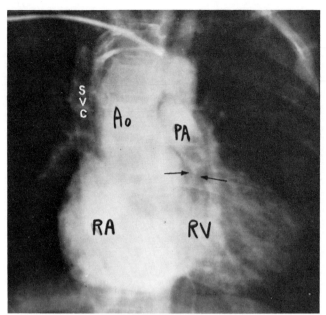

Figure 10–71 **Tetralogy of Fallot: Angiocardiographic Findings.** Frontal film demonstrates opacification of the right atrium (RA) and the right ventricle (RV), but the left ventricle is still unopacified. Simultaneous filling of the aorta (Ao) and the pulmonary artery (PA) indicates that at least a part of the aortic root arises from the right ventricle. Infundibular stenosis of the pulmonary artery *(arrows)* is apparent. Note that the aorta is unusually large and that its root originates further to the right than normal: The superior vena cava (SVC) is still partially opacified.

Simultaneous filling of the aorta and pulmonary artery before left ventricular opacification, and demonstration of infundibular pulmonary stenosis are highly characteristic of tetralogy of Fallot.

Figure 10–72 **Tetralogy of Fallot: Angiocardiogram in Left Anterior Oblique Position.** The contrast material introduced by catheter into the right ventricle (RV) passes through the high ventricular septal defect (VSD) and partially fills the left ventricle (LV).

There is also simultaneous filling of the large aorta (Ao) and the pulmonary artery (PA). The markedly narrowed infundibulum is faintly evident *(arrows)*. The root of the aorta is clearly seen arising from both the right and the left ventricles where they are joined by the large ventricular septal defect.

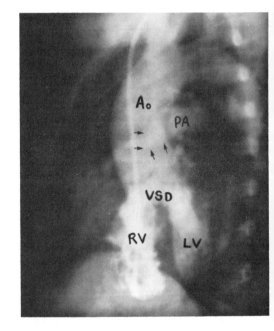

Tricuspid Atresia

Tricuspid atresia is always associated with an atrial septal defect, permitting a right to left shunt. A ventricular septal defect is usually present. Pulmonary vascularity is decreased unless there is associated complete transposition of the great vessels without pulmonic stenosis. The right ventricle is small, atretic, and essentially nonfunctioning. The left ventricle and right atrium are enlarged.

In "pure" tricuspid atresia, the heart size is normal or moderately enlarged. The left border is rounded and elevated, simulating right ventricular enlargement. Right atrial enlargement usually occurs, the degree depending inversely on the size of the atrial shunt. Vena caval enlargement is also common. The right cardiac border is often flattened, the result of the atretic right ventricle. The aorta is prominent, but the pulmonary artery is small, and the cardiac waist is narrow. The cardiac configuration may often mimic tetralogy of Fallot. The oligemic lungs may receive additional blood from the bronchial arteries, which produces a reticular lung pattern.

Transposition of the great vessels, which is present in about half the cases, alters the cardiac findings. The aorta is small, and often the abnormally located right auricular appendage produces a peculiar bulge in the left upper cardiac border. The pulmonary vasculature is usually increased.

Angiocardiograms demonstrate a typical sequence of opacification – the enlarged right atrium, the left atrium, the enlarged left ventricle – and also the "right ventricular window" (due to the atretic right ventricle), which is a triangular clear zone just below and medial to the lower margin of the right atrium. This filling defect, or window, is almost pathognomonic. Dilatation of the inferior vena cava is a frequent finding.

The combination of decreased pulmonary vasculature and left axis deviation in a cyanotic individual strongly suggests the diagnosis of tricuspid atresia.[92-94]

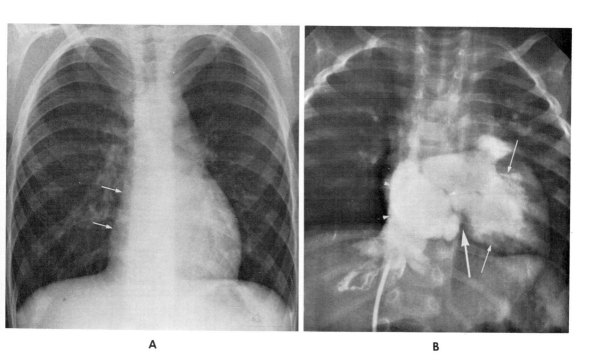

A   B

Figure 10–73 **Tricuspid Atresia: Posteroanterior View: Angiocardiogram.**

A, The right heart border is flattened *(arrows),* and the left heart border is round. The pulmonary vasculature appears diminished; the aorta is relatively large. These findings are similar to those of tetralogy of Fallot.

B, A representative film from an angiocardiogram shows a large right atrium *(arrowheads)* and almost simultaneous opacification of the left ventricle *(small arrows).* The blood is shunted directly from the right atrium into the left atrium and thence into the left ventricle; the right ventricle is not opacified. The characteristic and diagnostic filling defect representing the atretic right ventricle *(large arrow)* is seen adjacent to the right atrium. The aorta is not yet opacified.

Ebstein's Anomaly

The basic abnormality is a downward displacement of the tricuspid valve, in basket-like fashion, into the right ventricle. The valve is usually incompetent. There is always a right to left shunt through an atrial septal defect.

Characteristically, the heart is considerably enlarged and has a globular or box-like configuration. The "squaring" of the right border is due to great enlargement of the right atrium. A bulge that squares the left border is often seen just below the hilum, simulating the bulge of left atrial enlargement. It is caused by a dilated and left-displaced right ventricular outflow segment (infundibulum). Usually the pulmonary artery segment is concave, the hilar shadows are small, and the pulmonary vasculature is decreased. In some cases, however, the cardiac silhouette may appear completely normal.

Angiography reveals a massive right atrium and appendage, an enlarged bulging right ventricular infundibulum, and a small right ventricle that is usually obscured by the dilated right atrium. Prolonged contrast filling of these structures is characteristic of Ebstein's anomaly. There is early filling of the left heart and aorta via the atrial shunt. The enlarged infundibulum of the right ventricle expands markedly during diastole, and it may rise above the pulmonary artery. The displaced tricuspid valve can sometimes be identified.[95-98]

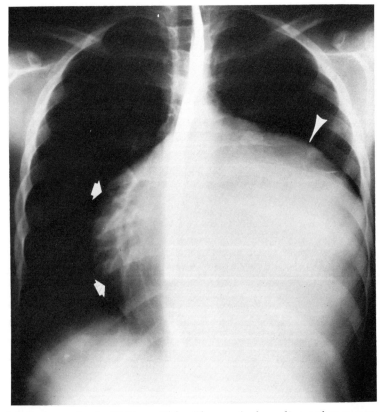

Figure 10-74 **Ebstein's Anomaly in 15 Year Old.** The marked cardiac enlargement, the massive right atrium *(arrows)*, the high bulge on the left *(arrowhead)* due to the dilated and displaced right ventricular outflow infundibulum, and the decreased hilar and parenchymal vascularity are characteristic of Ebstein's anomaly.

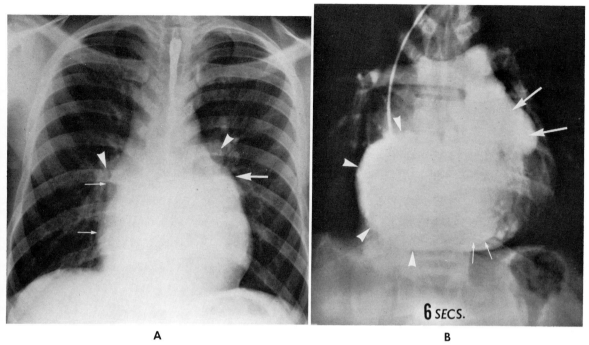

A **B**

Figure 10–75 **Ebstein's Anomaly: Angiocardiogram.**

A, The heart is enlarged and has a squared-off or box-like configuration due to the enlarged right atrium *(small arrows)* and the bulge of the dilated infundibulum *(large arrow)*. The hilar shadows *(arrowheads)* are small, and the pulmonary vasculature is decreased.

These findings are characteristic of Ebstein's anomaly.

B, Frontal angiocardiogram at six seconds shows the massive right atrium still opacified *(arrowheads)*. The opacified small right ventricle merges with and is partially obscured by the huge right atrium. The greatly dilated right ventricular infundibulum *(large arrows)* and the pulmonary arteries also showed persistent opacification. The displaced and deformed tricuspid valve *(small arrows)* can be identified.

Taussig-Bing Complex: Biventricular Origin of Pulmonary Artery

In this rare variant of incomplete corrected transposition of the great vessels, the pulmonary artery overrides a ventricular septal defect and becomes enlarged because it receives blood from both ventricles. The aorta arises solely from the right ventricle. The heart is enlarged, the cardiac base is prominent, and there is increased pulmonary vascularity. Angiocardiography is necessary for diagnosis and may demonstrate (1) opacification of both great vessels from the right ventricle, (2) pulmonary and aortic valves at the same level, (3) partial transposition of the aorta on lateral view, (4) a filling defect at the base of the right ventricle representing the crista supraventricularis, which forms the two outflow tracts, and (5) a ventricular septal defect just below the crista.[99, 100]

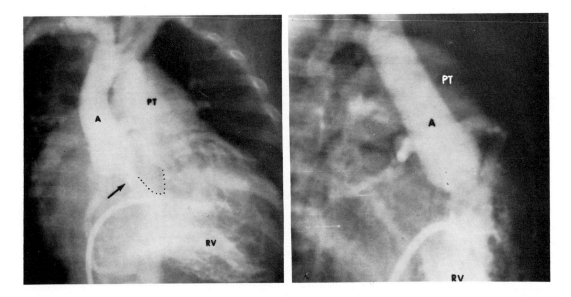

Figure 10–76   **Taussig-Bing Complex: Selective Angiocardiogram.**

A, In anteroposterior projection, a catheter is shown in the right ventricle. With injection of the contrast medium, there is opacification of the right ventricle (RV) and simultaneous filling of the aorta (A) and the larger pulmonary artery (PT).

The dotted line indicates the tongue-like septum (crista supraventricularis) that forms the outflow tract for the small aorta and the overriding pulmonary artery. Arrow points to the narrow outflow tract to the aorta.

B, High right anterior oblique projection shows the aorta's characteristic anterior position and its overlapping of the pulmonary artery. There is simultaneous filling of the aorta (A) and of the enlarged pulmonary trunk (PT) from the right ventricle (RV). (Courtesy Dr. L. S. Carey, St. Paul, Minnesota.)

Miscellaneous Heart Disease

Dextrocardia

There are many types of cardiac malposition and malrotation, and classification is difficult because congenital cardiac anomalies are frequently associated. A roentgen classification has been devised (Elliot, Jue, and Amplatz, 1966) to help identify a great number of cardiac malpositions.

In this classification, the term *situs solitus* means that the descending aorta and gastric fundus are in their normal left-sided position. *Situs inversus* refers to an abnormal right-sided descending aorta and right-sided gastric fundus. *Dextroversion* indicates a right-sided cardiac apex with a left-sided aorta and gastric fundus; *levoversion* refers to a left-sided cardiac apex with a right-sided descending aorta and gastric fundus. Dextro- and levoversion are always associated with axis rotation on the electrocardiogram.

The four basic cardiac positions are as follows:

(1) Situs solitus without dextroversion, which represents the normal heart position.

(2) Situs inversus without levoversion. This is also called "mirror image" dextrocardia, since the chest film when reversed appears perfectly normal. There is usually no associated heart disease; if a congenital cardiac abnormality is present, it is most frequently a corrected transposition of the great vessels.

(3) Situs solitus with dextroversion (right-sided apex, left-sided descending aorta and stomach), which is usually associated with congenital heart disease: most commonly, corrected transposition of the great vessels.

(4) Situs inversus with levoversion (left-sided apex, right-sided descending aorta and stomach). This is invariably assocated with an abnormal heart. The defects in descending order of frequency include corrected transposition of the aorta, ventricular septal defect with pulmonic stenosis, and tricuspid atresia.

In a small number of cases, a left-sided aorta is associated with a right-sided gastric fundus. In this group, interruption of the inferior vena cava, with azygos continuation, is almost always present.

These basic positions of the heart can be recognized on plain chest films. It should be remembered that cardiac malposition does not always imply a hemodynamically abnormal heart.

Agenesis of the spleen is always associated with multiple and bizarre cardiac abnormalities. These lesions are not easily classified.

A simplified and probably more pragmatic modification of the above classification would exclude the terms *dextroversion* and *levoversion*. The four basic positions in order of frequency would be as follows:

(1) Levocardia with situs solitus—the normal heart.

(2) Dextrocardia with situs inversus—the asymptomatic mirror image heart.

(3) Dextrocardia with situs solitus—relatively uncommon; high probability of some form of chamber transposition, particularly corrected transposition.

(4) Levocardia with situs inversus—extremely rare, virtually always associated with cardiac defect.

Angiocardiography may be needed in many cases of malposition and malrotation in order to determine specific chamber location and the associated congenital cardiac abnormalities.

Cardiac malposition may have *extracardiac* causes (false dextrocardia). This is best described as dextroposition or levoposition and must be distinguished from the group of true dextrocardias. The extracardiac causes include agenesis or hypoplasia of the pulmonary artery or lung, fibrothorax, retraction of the heart or mediastinum due to inflammatory processes, and large pleural effusions.[101-103]

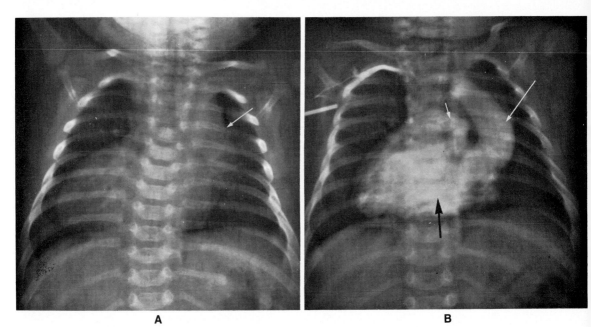

A	**B**

Figure 10–77 **Dextrocardia: Situs Solitus with Dextroversion: Single Ventricle and Corrected Transposition.**

A, In situs solitus, the aorta *(arrow)* and gastric fundus are on the left. The heart is enlarged and is within the right chest. The cardiac apex is displaced to the right.

B, A large common ventricle *(black arrow)* and corrected transposition of the aorta are seen by angiocardiogram. There is simultaneous opacification of the pulmonary artery *(short white arrow)* and the aorta *(long white arrow),* which arises to the left of and anterior to the pulmonary artery.

Situs solitus with dextroversion is usually associated with intracardiac anomalies. Angiocardiography is needed to determine the location of the cardiac chambers and the presence of associated anomalies.

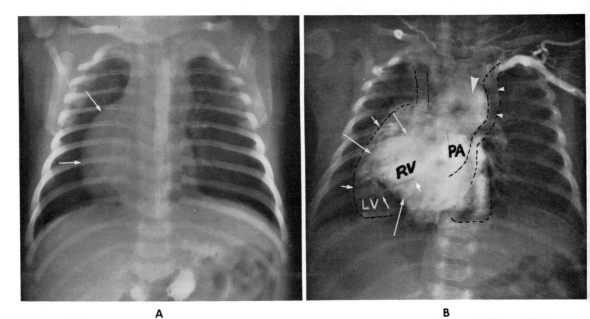

A	**B**

Figure 10–78 **Dextrocardia: Situs Solitus with Dextroversion: Corrected Transposition.**

A, Dextroversion is indicated by location of the cardiac apex in the right chest *(arrows),* and situs solitus by location of the aorta and stomach on the left.

B, On angiocardiogram, the right ventricle (RV) *(long arrows)* and the left ventricular chamber (LV) *(short arrows)* are right-sided and anterior. Both atrial chambers are posterior. The superior vena cava *(small arrowheads)* lies to the left of the aorta. The aorta *(large arrowhead)* arises on the left and anterior to the pulmonary artery (PA)—the characteristic location of a corrected transposition.

660

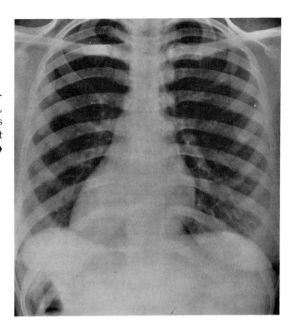

Figure 10–79 **Mirror Image Dextrocardia: Situs Inversus without Levoversion.** The aorta, the stomach, and the splenic flexure are on the right; the liver is under the left diaphragm. The heart shadow and great vessels appear normal if the film is reversed. ▶

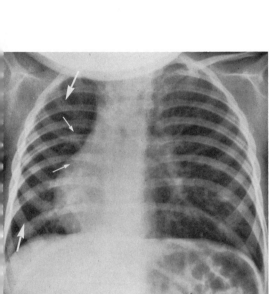

Figure 10–80 **Dextroposition (False Dextrocardia) Due to Hypoplasia of Lung.** The heart is displaced into the right thoracic cavity to compensate for the decreased right thoracic volume secondary to hypoplasia of the right lung *(large arrows)* and pulmonary artery. Note the small size of the right hilum and the paucity of the pulmonary branches *(small arrows)*. The vasculature of the left lung appears increased. ◀

Aortic Arch Malformations

There are numerous anomalies of the aortic arch, many of which are clinically insignificant. Some, however, may produce symptomatic narrowing of the esophagus and trachea and require surgical intervention. In addition, some anomalies are associated with underlying congenital heart disease.

The normal left aortic arch produces a pressure indentation on the left side of the esophagus on frontal view, and some anterior indentation on lateral view, but pressure defect on the posterior esophagus is not normally seen. Deviation of the tracheal air shadow is also frequently observed on frontal view.

The more common anomalies include a right-sided arch with an aberrant left subclavian artery and persistent left ligamentum arteriosum; a right-sided arch with mirror branching of the major arteries (the innominate artery arises on the left); an anomalous right subclavian artery, which is the most frequent anomaly; and a double arch. Indentation of the posterior esophagus occurs in all but the right-sided arch with mirror branches. The latter anomaly is almost always associated with congenital cyanotic heart disease, usually with tetralogy of Fallot, and less frequently with truncus arteriosus and transposition of the great vessels.

The nature of the aortic arch malformation can frequently be surmised from the pattern of esophageal indentation with barium swallow. Right-sided arch produces a right indentation; if associated with an aberrant left subclavian artery, a prominent posterior indentation will also be seen. An aberrant right subclavian artery causes an obliquely upward left indentation and a posterior defect. The double arch produces a left, right, and posterior indentation. Tracheal compression, if present, is readily recognized on well-exposed films. The descending aorta may be right sided or normally left sided.

The arch anomalies may simulate a mediastinal mass or an aneurysm. Opacification studies differentiate the anomaly from mediastinal mass and are also of great value for preoperative evaluation of the precise anatomy of the malformation. (See also *Aberrant Left Pulmonary Artery*, p. 667.)[104-108]

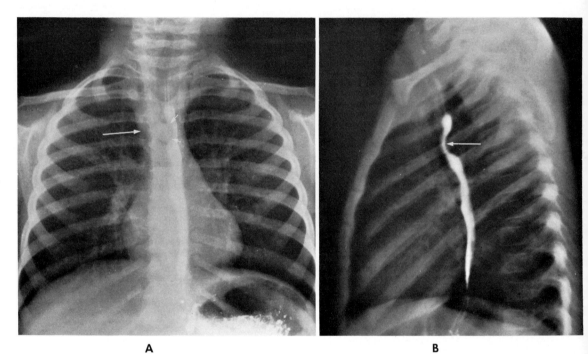

A   **B**

Figure 10–81 **Right-Sided Aortic Arch.**

A, In posteroanterior view the right side of the barium-filled esophagus *(short arrows)* is indented by the right-sided aortic arch. The aortic knob is seen on the right *(long arrow).*

B, Lateral view shows a characteristic posterior indentation of the esophagus by the right-sided arch *(arrows),* which extends behind the esophagus.

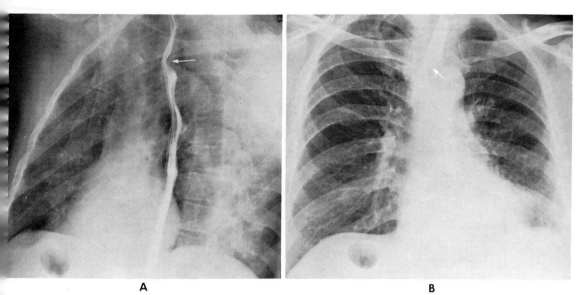

A   B

Figure 10–82   **Anomalous Origin of Right Subclavian Artery.**

A, The right subclavian artery arises from the descending aorta in the left hemithorax. As it crosses the esophagus posteriorly, it produces an oblique pressure defect *(arrow),* which is best seen in the oblique view.

B, Posteroanterior view reveals a small pressure defect *(arrow)* on the left side of the esophagus; the defect resembles the normal aortic indentation but is located somewhat higher. This, too, is due to pressure by the anomalous right subclavian artery.

The pressure on the esophagus may be severe enough to produce dysphagia (dysphagia lusoria).

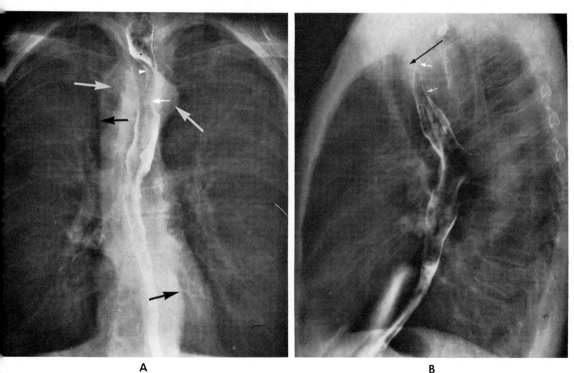

A   B

Figure 10–83   **Double Aortic Arch.**

A, In the posteroanterior view there is an esophageal pressure defect on the right *(arrowhead)* due to a right-sided aortic arch *(large white arrow),* and the left arch *(large white arrow)* produces a similar defect *(small white arrow)* on the left side of the esophagus. The descending aorta is on the right of the thoracic spine *(upper black arrow)* and curves to the left just above the diaphragm *(lower black arrow).* Tracheal compression cannot be appreciated in this film.

B, In the lateral view, marked indentation is apparent on the posterior aspect of the esophagus *(white arrows),* and the trachea is displaced anteriorly *(black arrow)* but is not narrowed. Compression of the trachea by the anterior portion of a fully developed aortic ring can cause significant respiratory symptoms. (Courtesy Dr. Arlyne Shockman, Veterans Administration Hospital, Philadelphia, Pennsylvania.)

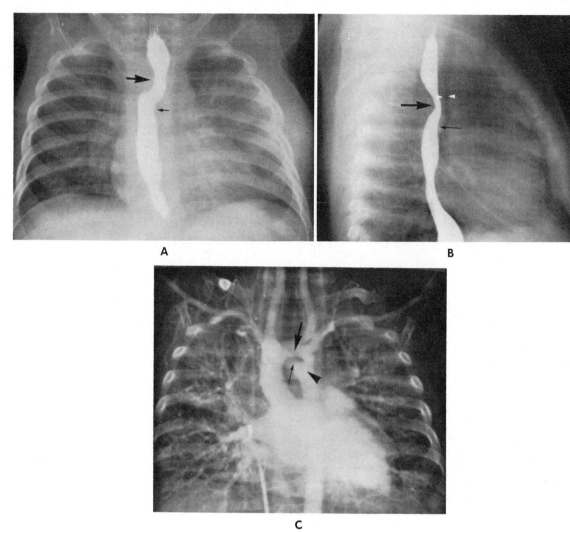

Figure 10–84   **Double Aortic Arch with Tracheal Narrowing: Plain Films and Angiocardiogram.**

A, Posteroanterior chest film of infant with mild respiratory symptoms shows a large right-sided indentation of the esophagus *(large arrow)* at the aortic arch level, and a second smaller indentation on the left side *(small arrow)* at a lower level. These findings are highly suggestive of a double aortic arch.

B, Lateral film discloses the large posterior indentation *(large arrow)* from the posterior arch, and the smaller lower indentation from the anterior arch *(small arrow).* The trachea, at the arch level, is narrowed *(arrowheads).*

C, Late phase of angiocardiogram demonstrates the double arch. The higher posterior arch *(large arrow)* and the lower anterior arch *(small arrow)* join *(arrowhead)* to form the descending aorta.

The trachea and esophagus lie within the ring produced by the double arch.

Localized Constriction of Pulmonary Arteries (Coarctation of the Pulmonary Arteries)

This congenital condition is an isolated lesion in 40 per cent of cases. In 60 per cent of patients, other congenital anomalies are associated, the most frequent being pulmonary valvular stenosis and tetralogy of Fallot. It is also frequently found in the congenital rubella syndrome.

Single or multiple constrictions may occur in the main pulmonary branches or in the peripheral vessels. Pulmonary hypertension frequently ensues. When a main pulmonary artery is involved, the corresponding hilar vessels are small and less distinct. The vessels immediately distal to the narrowing are prominent because of poststenotic dilatation. This unusual combination of small hilar arteries and large posthilar central branches suggests the diagnosis. Involvement and roentgen findings can be unilateral or bilateral. The right ventricle is often prominent. When a more peripheral artery is constricted, the narrowing is sometimes recognized on the chest film. Definitive diagnosis is made by selective pulmonary arteriography. Associated pulmonary valvular stenosis is found in over half the cases.[109-111]

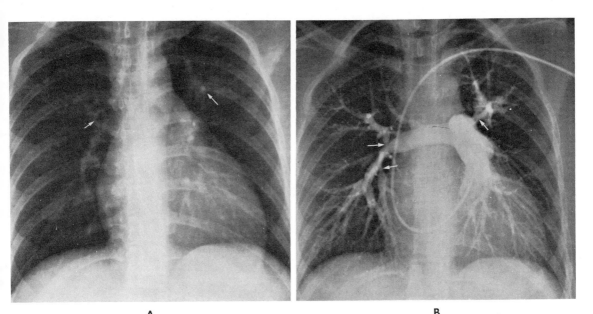

A   **B**

Figure 10–85 **Multiple Constrictions of Larger Pulmonary Arteries: Selective Pulmonary Angiogram.**

A, There are small central hilar shadows and prominent peripheral vessels *(arrows)*. The heart is moderately enlarged.

B, Pulmonary angiogram demonstrates local constrictions of the left upper pulmonary artery and both right pulmonary arteries *(arrows)*. Note the poststenotic dilatation.

Agenesis or Hypoplasia of a Pulmonary Artery

Absence or hypoplasia of a main pulmonary artery is an uncommon anomaly. It occurs more often on the right side. In hypoplasia, the roentgen changes are identical to, but less marked than, those of total agenesis of the artery.

The ipsilateral hemithorax is small, and the lung is either more lucent owing to a decreased number of vascular markings or may show decreased lucency from a heavy bronchial collateral circulation. In virtually every case, the vascularity of each lung is dissimilar. The hilar shadow is small. Blood is supplied by the bronchial arteries, and this produces a fine reticular lung pattern. The plethoric bronchial circulation may anastomose and enlarge one or more intercostal arteries, causing unilateral rib notching. The heart and mediastinum are rotated to the affected side; right-sided agenesis of the pulmonary artery is a cause of dextroposition of the heart. The diaphragm is usually somewhat elevated but moves normally. Compensatory hyperinflation is seen on the normal side.

Bronchography may be normal, or it may show absence of one or more major bronchi, which is indicative of partial lung agenesis.

Angiography demonstrates the absence or hypoplasia of the pulmonary artery and may also show collateral bronchial branches arising from the aorta.[111-114]

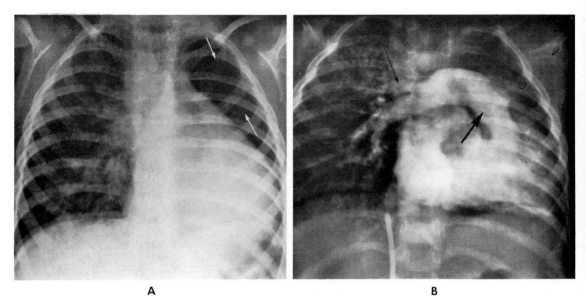

A   B

Figure 10–86 **Agenesis of Left Pulmonary Artery: Angiocardiogram.**

A, Posteroanterior view shows the heart and mediastinum slightly shifted to the left. The left lung is small and appears hyperlucent *(arrows)* owing to the absence of normal pulmonary vasculature.

B, Angiocardiography reveals a huge main pulmonary artery *(large arrow)* with large right-sided branches *(small arrow)*. The left pulmonary artery is absent.

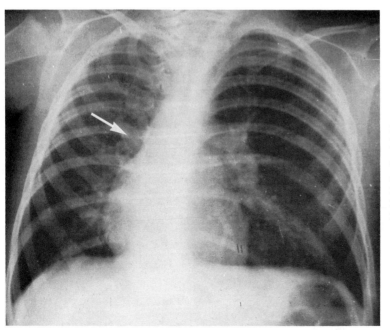

Figure 10–87 **Hypoplastic Right Pulmonary Artery.** The shift of the heart to the right, the very small right hilar shadow *(arrow)*, and the decreased volume of the right hemithorax are characteristic findings of a hypoplastic right pulmonary artery.

Aberrant Left Pulmonary Artery

This rare anomaly, also called *pulmonary sling,* can cause respiratory stridor and cyanosis similar to that seen in the vascular rings (See p. 661).

The aberrant pulmonary artery takes an abnormal turn to the right and courses behind the trachea before entering the left lung. It lies between the trachea and the esophagus. Radiographically, on the lateral chest film, the aberrant artery may cause a posterior impression on the trachea just above the carina and an anterior impression on the barium-filled esophagus. These findings are highly characteristic. Pulmonary angiography will confirm the diagnosis.[115]

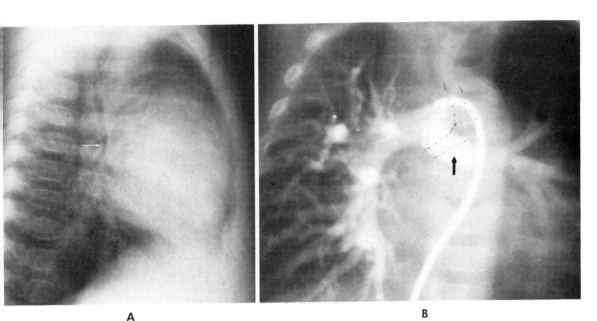

A   B

Figure 10–88   **Aberrant Left Pulmonary Artery.**

A, Lateral chest film shows a forward displacement of the lower trachea *(arrow).* The esophagus, if opacified, would be separated from this portion of the trachea by the aberrant artery.

B, Pulmonary angiogram shows the left pulmonary artery *(arrow)* arising on the right side. It courses between the trachea and esophagus to reach the left lung. Dotted lines indicate displaced trachea.

The Straight Back Syndrome

In this syndrome there are clinical signs that mimic organic heart disease. However, these are due to displacement of the heart and great vessels caused by a developmental absence of the normal thoracic kyphosis and by a narrowed anteroposterior diameter of the chest. The syndrome is found mostly in young adults.

Radiographically the heart may appear normal, but often it has a pancake appearance simulating cardiomegaly on frontal view. In some cases the heart is also displaced to the left, and the main pulmonary artery is prominent. On lateral film the thoracic spine is unusually straight; the anteroposterior diameter of the chest is decreased both absolutely and in ratio to the transverse diameter, and the heart is flattened against the sternum.

A systolic ejection murmur is the usual clinical finding. Mild pulmonary stenosis, atrial septal defect, or idiopathic pulmonary artery dilatation may be suspected, but catheterization and angiocardiographic studies will be normal.

Recognition of the radiographic changes in a patient with a systolic ejection murmur will obviate the need for further cardiac investigations.[116]

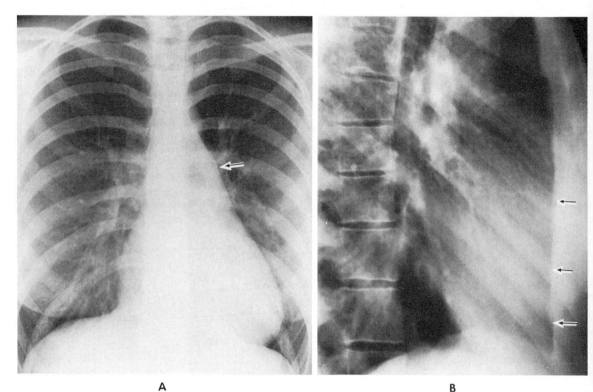

A   B

Figure 10–89 **Straight Back Syndrome.**

A, Chest film of a young woman with a systolic ejection murmur reveals a moderately "enlarged" heart and a prominent main pulmonary artery segment *(arrow).* The vasculature in the right hilum and medial portion of the right base appears prominent.

B, Lateral film reveals a narrow anteroposterior diameter and a straight spine with absence of normal thoracic kyphosis. The heart does not appear enlarged, but it is flattened *(arrows)* against the anterior chest wall.

There was no organic heart disease. The prominence of the right vascular shadows on the frontal film was due to slight displacement of the heart to the left. (Courtesy Dr. H. L. Twigg, Washington, D.C.)

ACQUIRED VALVULAR HEART DISEASE

Acute Rheumatic Heart Disease

The principal change roentgenographically is cardiac enlargement, with or without evidence of pulmonary congestion. The cardiac enlargement may result from myocarditis with or without pericarditis, or from an acute incompetence of the mitral valve. With heart failure, pulmonary edema can develop, and it may be difficult to distinguish this from rheumatic pneumonitis. A decrease in cardiac pulsation amplitude is seen on fluoroscopy. Rheumatic myocarditis was a common cause of cardiomegaly in children but is currently a much more infrequent cause.[117]

Figure 10-90 **Acute Rheumatic Myocarditis: Congestion.** There is both left and right cardiac enlargement with fullness of the base of the heart. There is considerable engorgement of the hilar and perihilar vessels (arrows).

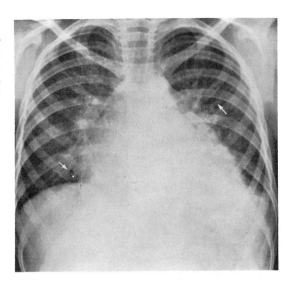

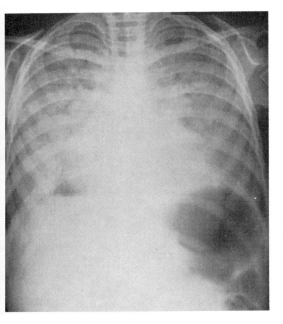

Figure 10-91 **Acute Rheumatic Myocarditis: Pulmonary Edema.** The heart is greatly enlarged, and there is a "butterfly wing" distribution of confluent alveolar densities in both lung fields, which is characteristic of extensive pulmonary edema. (Courtesy Dr. David Goldring, St. Louis, Missouri.)

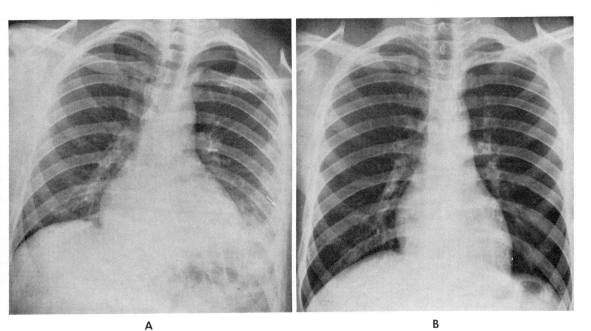

A   B

Figure 10-92 **Recovery Following Rheumatic Myocarditis.**

A, There is diffuse cardiomegaly with predominantly left ventricular enlargement during an acute phase of rheumatic fever. There is enlargement of the hilar vasculature and an effusion at the left base (arrow).

B, Six weeks later the heart has returned to normal size, the vasculature is normal, and the effusion has disappeared.

Mitral Stenosis

Acquired mitral stenosis is the most common rheumatic valvular lesion of clinical significance. There is obstruction of flow from the left atrium into the left ventricle during diastole, resulting in increased pressure and eventual enlargement of the left atrium. This increased pressure is transmitted to the pulmonary veins and eventually extends to the pulmonary arteries and the right heart.

The roentgen changes in the heart and pulmonary vasculature reflect the hemodynamic alterations, which depend on the degree of mitral obstruction. In a hemodynamically insignificant stenosis the heart and vasculature appear normal, in spite of a typical murmur.

The characteristic heart changes include the following:

(1) Left atrial enlargement that causes elevation of the left main bronchus, a double density within the frontal heart shadow, and indentation or displacement of the barium-filled esophagus, which is seen on lateral or right anterior oblique films. Atrial enlargement is the earliest and most characteristic finding of pure mitral stenosis. The left atrial appendage is often displaced by the enlarged atrium and may appear as a bulge on the left border below the pulmonary artery.

(2) A prominent main pulmonary artery segment and enlarged hilar vessels.

(3) A small aortic knob due to decreased left ventricular output. If a prominent ascending aorta is seen, there is probably coexisting aortic valvular disease.

(4) Right ventricular enlargement, with a normal-sized left ventricle.

The left cardiac border appears straightened because of the small aortic knob and the enlarged main pulmonary artery filling the cardiac waist. Overall cardiac enlargement is minimal until dilatation and failure occur. In a few longstanding cases, calcification of the mitral valve or the left atrial wall develops.

The typical vascular features are as follows:

(1) Distention of upper lobe veins and decreased caliber of the lower lobe veins. These changes are indicative of increased postcapillary resistance (increased venous pressure).

(2) Septal or Kerley B lines, which are interlobular septa that have been thickened and made visible by lymphatic distention secondary to increased venous pressure. These are best seen in the outer portion of the lower lung fields.

(3) Dilatation of the central pulmonary arteries and narrowing of the peripheral arteries, if pulmonary arterial hypertension of moderate severity develops.

Parenchymal lung lesions that can develop in longstanding mitral stenosis include the following:

(1) Fine granular opacities in the mid and lower lung fields, due to hemosiderosis.

(2) Isolated areas of bone formation, a rare but diagnostically important finding.

(3) Nonspecific fibrosis, hyperinflation, or pleural thickening chiefly in the lung bases, probably due to episodes of pulmonary infarction.

Following successful corrective surgery, hemodynamic improvement will be evidenced by decreased size of the left atrium, main pulmonary artery, and right descending pulmonary artery. The latter is easily identified on the film.[118-120]

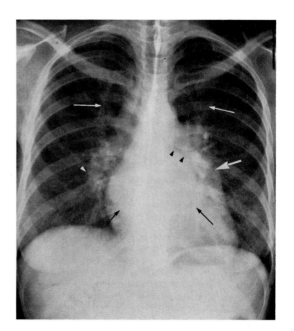

Figure 10-93 **Mitral Stenosis: Typical Findings in Heart and Lung.** The aortic knob appears small. The main pulmonary artery segment is prominent; this prominence, in conjunction with bulging of the left atrial appendage *(large white arrow),* produces mitralization of the left heart border. The enlarged left atrium can be seen as a double density *(black arrows)* through the heart shadow; it also produces elevation of the left main bronchus *(black arrowheads).* The hilar and pulmonary vasculature *(white arrowhead)* is prominent. The superior pulmonary veins *(small white arrows)* are distended, and the inferior pulmonary veins are inconspicuous. These venous changes reflect the increased postcapillary venous pressure.

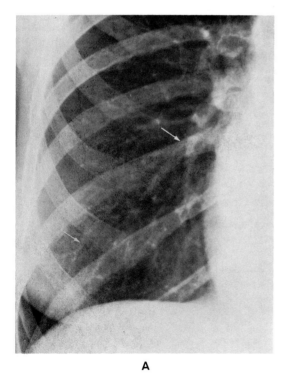

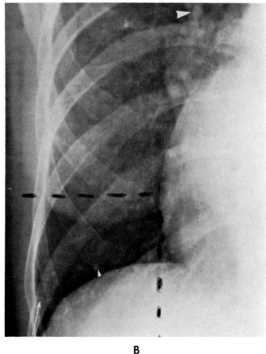

A   B

Figure 10-94 **Mitral Stenosis: Vascular Changes.**

A, Enlarged view of the lung in a normal subject demonstrates the gradual tapering of the vascular channels from the hilum to the periphery *(arrows).* The hilar shadow is of normal size.

B, Corresponding area in a patient with advanced mitral stenosis is shown. There is attenuation of the venous channels in the lower lung field and prominence of the superior pulmonary veins *(large arrowhead).* There are characteristic fine transverse septal lines—Kerley B lines *(arrow).* Small round bone densities are seen in the lower lung fields *(small arrowhead).* These bone densities are rare and are almost specific for mitral stenosis.

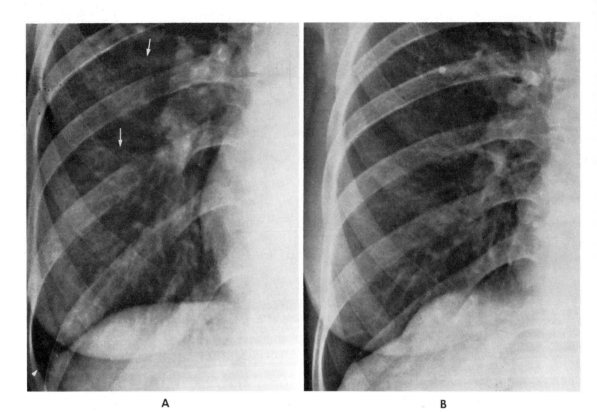

A   B

Figure 10–95 **Mitral Stenosis: Vasculature Before and After Commissurotomy.**

   A, Before surgery there was a generalized venous engorgement *(arrows),* especially in the mid and upper lung fields. The right hilar shadow is quite large. Kerley B lines are seen at the extreme right base *(arrowhead).*

   B, Ten months after commissurotomy there is marked regression of the dilated venous channels, and there is clearing of the lung fields particularly in the mid and upper areas. The Kerley B lines have disappeared, and the right hilar shadow has decreased in size.

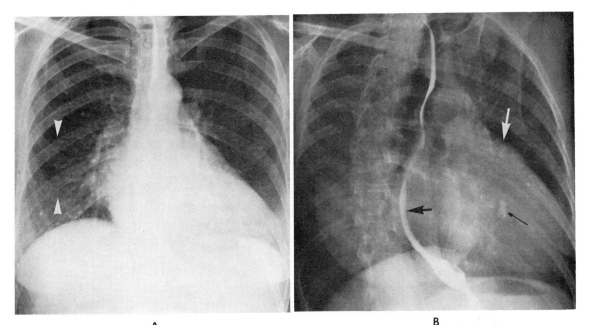

A   B

Figure 10–96 **Hemosiderosis in Mitral Valvular Disease.**

A, On posteroanterior view, granular opacities and a reticular pattern *(arrowheads)* due to hemosiderosis are seen in the mid and lower lung fields. The patient had longstanding combined mitral valvular disease.

B, Right anterior oblique view reveals marked enlargement of the left atrium *(large black arrow)* displacing the esophagus. The mitral valve *(small black arrow)* is calcified, and the pulmonary outflow tract is markedly enlarged *(white arrow)*.

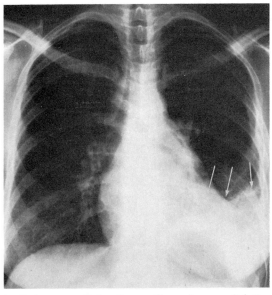

Figure 10–97 **Mitral Stenosis: Pulmonary Infarction.**   There is consolidation and pleural reaction at the left base *(arrows)* owing to pulmonary infarction. The cardiac silhouette is typical of mitral stenosis. A pleuropneumonic process associated with mitral valvular disease suggests a pulmonary infarction. Pulmonary infarction is a common complication of mitral valvular disease, being caused by emboli originating either from peripheral thrombosis because of poor cardiac output, or from right atrial thrombosis secondary to atrial fibrillation.

The fracture of the left fourth rib is a complication of surgery.

Mitral Insufficiency

Rheumatic valvulitis is the most common cause, but mitral regurgitation may also be due to functional dilatation or distortion of the mitral ring from other cardiac diseases, from rupture of the chordae tendineae, or from papillary muscle dysfunction. The roentgen findings reflect the hemodynamic changes.

In pure uncomplicated mitral insufficiency, regurgitation of blood during ventricular systole leads to overfilling and dilatation of the left atrium. In turn, the increased volume of flow from the overfilled atrium during ventricular diastole causes dilatation and enlargement of the left ventricle. The left atrium is usually compliant and enlarges without much pressure elevation. Therefore, the pulmonary venous pressures remain near normal. Radiographically, left atrial enlargement is the most constant and significant finding. The left atrium can sometimes become enormous, occupying the entire enlarged heart diameter in frontal projection. The left ventricle is enlarged, while the aortic knob is normal or small in cases of rheumatic origin. The pulmonary vasculature is usually normal, and no Kerley B lines appear. In these cases the enlarged left ventricle and the normal vasculature will distinguish mitral regurgitation from mitral stenosis.

If the atrium is less compliant or if regurgitation is very severe, the pressure in the left atrium and pulmonary veins increases sufficiently to produce dilatation of the superior veins and diminution of the inferior veins, findings very similar to those in mitral stenosis. The main pulmonary artery segment is prominent. Kerley B lines, however, are less frequent and less prominent than in mitral stenosis. Nevertheless, in these cases the roentgen appearance of the vasculature is virtually indistinguishable from that of mitral stenosis. In many patients both valvular conditions coexist. An enormous left atrium and marked cardiac enlargement are more suggestive of insufficiency. Calcification of the mitral valve is rare in pure insufficiency, but it is somewhat more frequent if stenosis is also present. Calcification of the walls of the enlarged left atrium is a striking but uncommon finding in mitral insufficiency.

Mitral disease is often associated with aortic disease or tricuspid disease, or both; this alters the cardiac silhouette and vasculature patterns. In such cases accurate diagnosis requires catheterization studies or angiocardiography.[121-123]

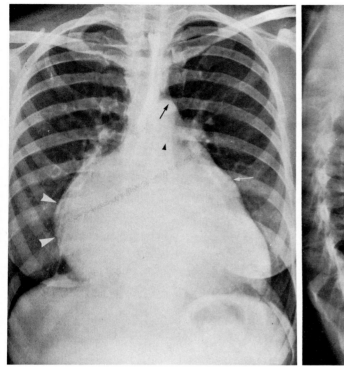
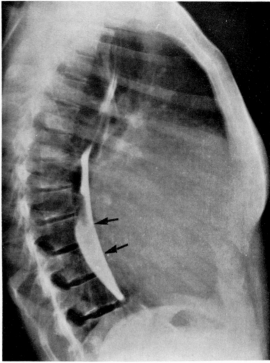

A   B

Figure 10–98 **Mitral Insufficiency: Giant Left Atrium.**

A, Posteroanterior view shows pronounced enlargement of the cardiac silhouette with straightening of the heart border. The aortic knob is small *(black arrow).* The huge left atrium has caused elevation of the left main stem bronchus *(black arrowhead)* and is also responsible for the bulging of the right heart border *(white arrowheads).* The bulging of the left border *(white arrow)* is caused by dilatation of the left atrial appendage. There is both left and right ventricular enlargement. The pulmonary vasculature is only slightly increased.

B, Lateral film indicates marked pressure indentation and posterior displacement of the barium-filled esophagus by the enlarged left atrium *(arrows).*

Figure 10–99 **Mitral Insufficiency: Retrograde Brachial Artery Catheterization.**   In lateral view there is marked regurgitation of contrast medium from the left ventricle *(small arrows)* through the incompetent mitral valve into the large left atrium *(arrowheads).* Note filling of the pulmonary veins *(large arrows).* There is also excellent visualization of the entire thoracic aorta. Estimation of the degree of mitral insufficiency is possible with opacification studies and may be important in making decisions regarding surgery.

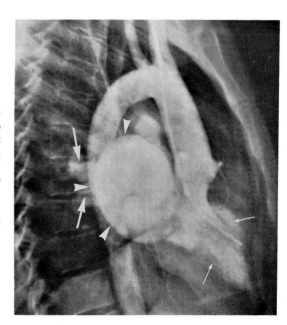

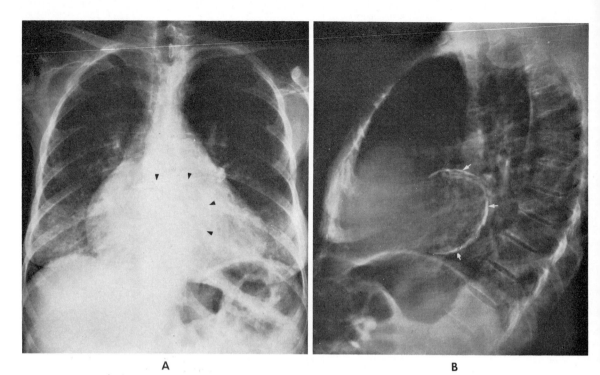

A	B

Figure 10–100 **Mitral Insufficiency: Calcification of Left Atrial Wall.**

A, Posteroanterior view demonstrates the enlarged left atrium. The wall is calcified *(arrowheads)* and outlines over half the chamber. The findings in mitral insufficiency include a small aortic knob, elevation of the left main bronchus, and enlargement of the left ventricle and left atrium.

B, Lateral view shows the calcification of the left atrial wall *(arrows)* more clearly.

Atrial calcification occurs in a very small number of patients with mitral valvular disease, and although diagnostic of rheumatic involvement, it does not distinguish the type of lesion.

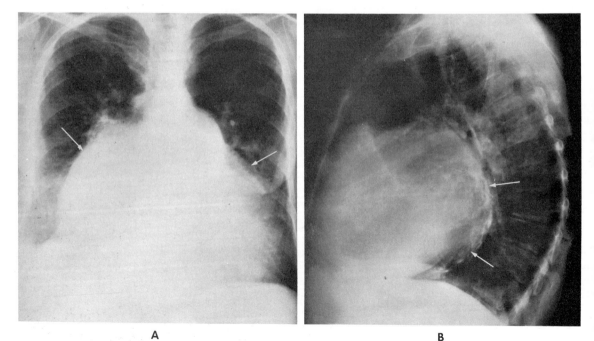

A	B

Figure 10–101 **Mitral Insufficiency: Giant Left Atrium and Calcification of Atrial Wall.**

A, Pronounced enlargement of the cardiac silhouette is caused mainly by the extreme enlargement of the left atrium *(arrows)*. The aortic knob is small; the main pulmonary artery segment is extremely prominent. A small pleural effusion is seen at the right base.

B, On lateral view the posterior border of the huge left atrium is outlined by its calcified wall *(arrows)*. Mitral insufficiency causes greater enlargement of the left atrium than almost any other condition.

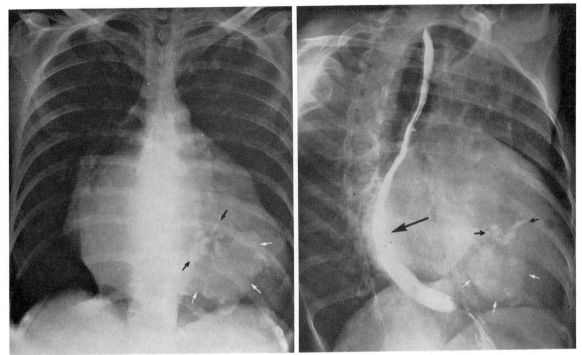

Figure 10–102 **Mitral Valvular Disease: Calcified Valve and Calcified Thrombus in Left Ventricle.** Posteroanterior and right anterior oblique projections of the chest demonstrate the typical cardiac configuration of mitral valvular disease. The enlarged left atrium displaces the esophagus *(large black arrow)* in the right anterior oblique view. Both films show a calcific density in the left ventricle *(white arrows)*; this was a calcified thrombus. The thrombus probably arose in the left atrium and subsequently prolapsed into the left ventricle. Calcification of the mitral valve is also well demonstrated *(black arrows)*.

Figure 10–103 **Mitral Insufficiency and Mitral Stenosis.** Enlargement of the left ventricle *(large arrowheads)* is attributable to mitral insufficiency. The small aortic knob, the prominent main pulmonary artery *(white arrow)*, and the enlargement of the left atrium are attributable to both mitral insufficiency and mitral stenosis. Prominence of the superior pulmonary veins *(small arrowhead)*, diminution of the inferior pulmonary veins, and Kerley B lines *(black arrow)* result from the increased postcapillary resistance caused by mitral stenosis.

In combined mitral valvular lesions, changes in the cardiac silhouette and the vasculature depend on which of the two lesions is predominant.

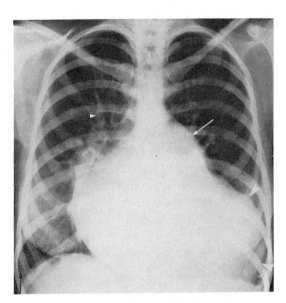

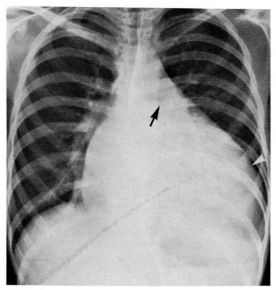

Figure 10–104   **Mitral Insufficiency and Aortic Insufficiency.**   This frequent combination can produce characteristic effects on the configuration of the heart. Often the effects of each lesion are not evident radiographically, and the configuration of the heart conforms to the predominating lesion.

The left ventricle is greatly enlarged *(arrowhead).* Left atrial enlargement is evidenced by elevation of the left main bronchus *(black arrow).* The vasculature is apparently normal. The film strongly suggests mitral insufficiency, and the aortic valvular lesion was not suspected from the radiographic findings.

It is often impossible to diagnose the combined lesion on plain films, and catheterization and opacification studies are usually necessary. The combination of mitral stenosis and aortic stenosis may produce an overall cardiac configuration that is nearly normal.

Aortic Stenosis

Typically the heart is normal size or only slightly enlarged. The increased workload imposed on the left ventricle by the stenosis will lead to concentric hypertrophy. There is little or no roentgenographic enlargement, but the cardiac contour may be altered, usually by a convex bulging of the left lower border. If congestive failure develops, dilatation and enlargement of the left ventricle may be seen. Left atrial dilatation may develop and persist after the failure clears.

Poststenotic dilatation occurs in the ascending aorta and is often best seen on left anterior oblique projections. Increased aortic pulsations may be noted fluoroscopically. The aortic knob remains normal or small. Calcification of the diseased aortic valve can be demonstrated in over 85 per cent of cases; the incidence increases with age. The degree of valve calcification is the only parameter that seems to correlate with the severity of the stenosis. Fluoroscopy and laminography aid in its detection.

Angiocardiography shows decreased mobility, as well as thickening and irregularity, of the aortic cusps. There may be an eccentric jet through the valvular orifice due to the asymmetric and incomplete opening of the aortic valve. Doming of the aortic valve in systole is more frequent in congenital aortic stenosis.

Isolated aortic stenosis on a rheumatic basis is relatively uncommon; it is more often associated with other valvular involvement. In a young individual, pure aortic stenosis is most commonly due to a congenital bicuspid valve.[124, 125]

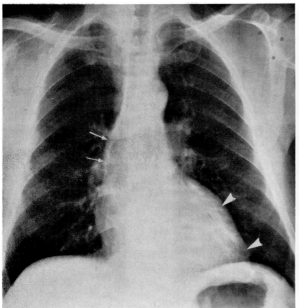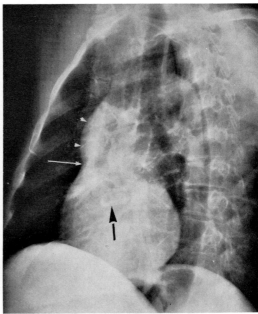

Figure 10-105 **Aortic Stenosis: Acquired Form.**

A, Plain film demonstrates the characteristic heart configuration in aortic stenosis. The left heart border is rounded *(arrowheads),* and the apex is displaced downward owing to left ventricular hypertrophy. The ascending aorta is dilated and bulges slightly *(arrows),* but the aortic knob is not enlarged. The pulmonary vasculature is normal.

B, Left anterior oblique projection demonstrates the characteristic prominence of the ascending aorta because of poststenotic dilatation *(arrowheads)* and a notch *(white arrow)* between the ascending aorta and the heart border. There is calcification of the aortic valve *(black arrow).*

In a young person with heart murmur, a dilated ascending aorta with or without left ventricular enlargement should suggest the possibility of aortic stenosis. Calcification of the valve is confirmatory.

Figure 10-106 **Aortic Stenosis: Valvular Calcification.** A left anterior oblique film shows calcification of the aortic valve *(arrows).* The dilated ascending aorta is also well demonstrated *(arrowheads).* Calcification of the aortic valve is virtually diagnostic of aortic stenosis.

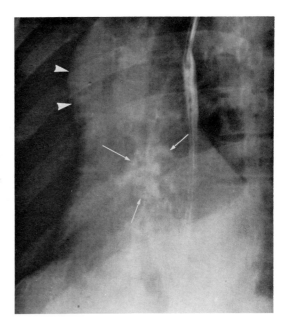

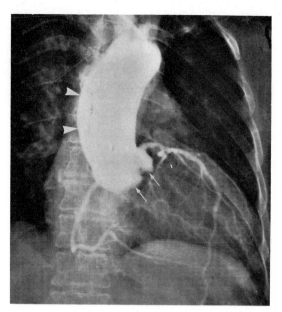

Figure 10–107 **Aortic Stenosis: Aortogram.** The dilated ascending aorta *(arrowheads)* and asymmetric and deformed aortic cusps *(arrows)* are demonstrated. Serial films indicated limitation of valvular motion.

Another change found on aortography is the eccentric jet of contrast material through the valvular orifice; this is seen during systole and results from the asymmetric and incomplete opening of the aortic valve.

The coronary arteries appear normal, although the patient had angina from inadequate blood supply to the hypertrophied myocardium.

Aortic Insufficiency

The reflux of aortic blood during ventricular diastole causes a dilated left ventricle. There is a downward, lateral, and posterior displacement of the cardiac apex, and the apex may appear to project below the diaphragm on the routine posteroanterior film. The enlarged left ventricle and the normal main pulmonary artery segment lead to a relative concavity of the waist of the cardiac silhouette. The ascending aorta may be dilated but usually not to the degree seen in aortic stenosis. If there is marked dilatation, combined aortic stenosis and insufficiency should be suspected. Little change is seen in the aortic knob.

With recurrent episodes of heart failure, left atrial enlargement and relative mitral regurgitation may develop.

The degree of aortic valvular insufficiency can be estimated by the amount of reflux into the left ventricle during contrast aortography.

The most common cause of acquired aortic regurgitation is rheumatic valvulitis. Other causes include luetic aortitis, rheumatoid spondylitis, dissecting aneurysm, and Marfan's syndrome. These are discussed elsewhere (see pages 732, 740, and 1424).

Figure 10–108 **Uncomplicated Rheumatic Aortic Insufficiency.** The left ventricle is dilated and enlarged. The apex is displaced downward and outward *(large arrow)*. The ascending aorta is only slightly prominent *(small arrows),* and the waist of the heart is fairly narrow. The degree of aortic dilatation and a qualitative estimation of the aortic insufficiency can be made during contrast studies. In luetic aortic insufficiency, aortic dilatation may be considerably more pronounced than in the rheumatic form.

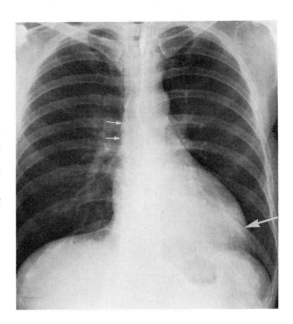

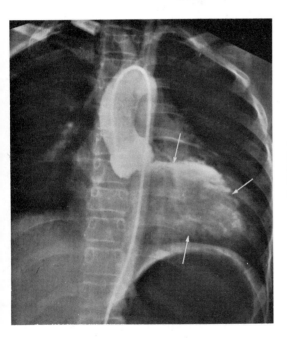

Figure 10–109 **Aortic Insufficiency: Retrograde Aortogram.** The contrast material introduced into the ascending aorta through a retrograde catheter refluxes into the dilated left ventricle *(arrows).* This procedure allows qualitative estimation of the reflux flow.

Combined Aortic Insufficiency and Stenosis

This combined disorder is more common than pure stenosis or pure regurgitation. The enlarged left ventricle and the dilated ascending aorta are characteristic features, and the dilatation is often more prominent than it is in single lesions. Often there is left atrial enlargement. The roentgenologic appearance is sometimes indistinguishable from that of pure aortic stenosis. In the absence of heart failure the pulmonary vasculature remains normal.[125, 126]

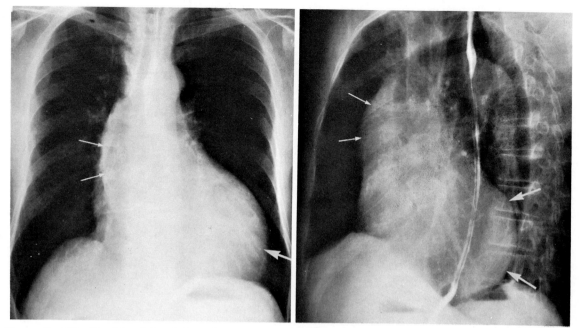

Figure 10–110 **Combined Aortic Insufficiency and Stenosis.** There is left ventricular enlargement *(large arrows)* on both the posteroanterior and left anterior oblique films. Dilatation of the ascending aorta is also seen on both projections *(small arrows).* The aortic knob is of normal size, and the pulmonary vasculature is normal.

Pronounced dilatation of the ascending aorta associated with dilatation of the left ventricle should suggest a combined lesion. In pure aortic insufficiency there is little aortic dilatation; in pure aortic stenosis, the left ventricle, in the absence of failure, is usually not greatly enlarged.

Tricuspid Valve Disease

Acquired tricuspid lesions are usually of rheumatic origin and are almost always associated with mitral or aortic valvular disease, most often mitral stenosis. The rare solitary tricuspid stenosis may occur in carcinoid syndrome, lupus erythematosus, or endomyocardial fibrosis.

Tricuspid insufficiency and stenosis produce almost identical roentgenologic findings. The pooling of blood in the right atrium causes fullness and rounding of the right cardiac border. The superior vena cava is often prominent, and the congested liver may elevate the right diaphragm. Overall cardiac enlargement is usual but is due mainly to the other coexisting valvular lesions. The valve does not calcify.

When a tricuspid lesion develops after pre-existing mitral stenosis, the pulmonary venous congestion and Kerley B lines often decrease or disappear.

Angiocardiography shows prolonged opacification of a greatly enlarged right atrium. The study also rules out myxoma or clot in the right atrium, lesions that can cause right atrial enlargement.[127]

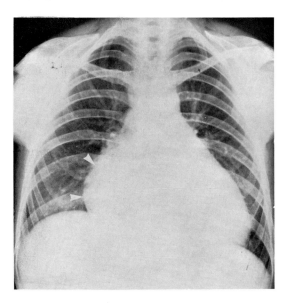

Figure 10–111 **Tricuspid Insufficiency and Mitral Stenosis.** The heart is enlarged and globular. The enlarged right atrium causes the right cardiac border to bulge (arrowheads). Fullness of the superior vena cava is not well seen. The pulmonary vasculature is normal despite the mitral valve disease. Elevation of the right hemidiaphragm is caused by congestive enlargement of the liver.

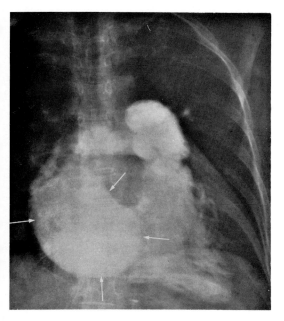

Figure 10–112 **Tricuspid Insufficiency: Angiocardiogram.** The right atrium is greatly enlarged (arrows) and empties slowly—a characteristic finding in tricuspid valve disease.

Figure 10–113 **Tricuspid Stenosis, Aortic Stenosis, and Mitral Stenosis.** Enlargement of the right atrium has produced the right-sided bulge (large arrows) in the posteroanterior view. Enlargement of the left side of the heart is caused by associated valvular disease. Note the decrease in the pulmonary vasculature. On conventional films it is practically impossible to distinguish tricuspid stenosis from tricuspid insufficiency. Notice that the cardiac configuration is almost identical to that shown in Figure 10–111.

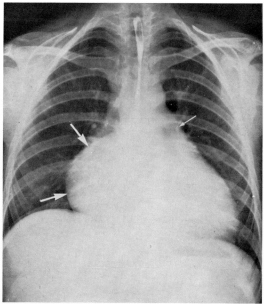

Pulmonary Valvular Lesions

Acquired pulmonary valvular lesions are rarely isolated or severe. Pulmonary valve insufficiency may appear after a pyogenic endocarditis or severe pulmonary hypertension, or during the carcinoid syndrome. No definite roentgen changes ensue. Pulmonary angiography is necessary for definitive diagnosis.[128]

Bacterial Endocarditis

The cardiac shadow may be normal, it may show evidence of previous valvular heart disease, or it may be dilated owing to congestive failure; hence, radiography is usually of little value in the diagnosis of bacterial endocarditis. Splenomegaly may be present and detectable radiographically. The combination of an abnormal cardiac silhouette or history of murmur and splenomegaly in a febrile patient should suggest bacterial endocarditis.

The aortic cusp may perforate or rupture, so that sudden aortic insufficiency with rapid cardiac dilatation and failure ensue. If the rupture occurs into the right heart, a left to right shunt is seen radiographically.

Although emboli are frequent and often extensive, they are usually small and produce few, if any, radiographic findings unless a vital artery such as the mesenteric is involved. Small infarctions of the kidney are frequent complications (see Fig. 11–114).

Septic pulmonary emboli can occur from a right-sided bacterial endocarditis. The emboli appear as one or more patches of density which enlarge and usually cavitate.[129]

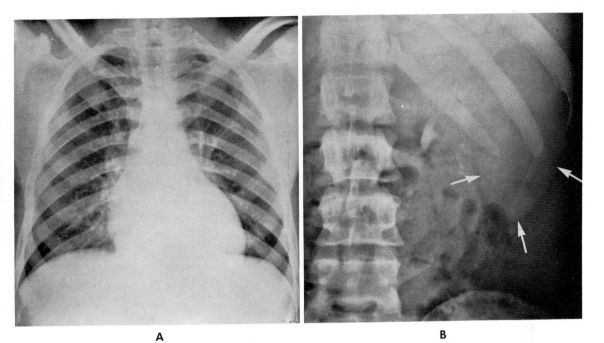

A   B

Figure 10–114 **Subacute Bacterial Endocarditis.**

    A, The chest film is within normal limits. Clinically, the patient had fever, anemia, and significant cardiac murmurs.

    B, Abdominal film demonstrates considerable enlargement of the spleen *(arrows),* which, in the presence of fever and significant murmurs, suggests subacute bacterial endocarditis.

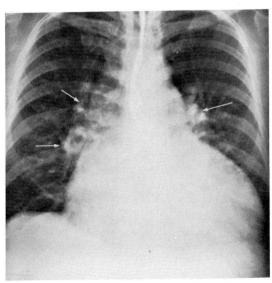

Figure 10–115 **Subacute Bacterial Endocarditis: Ruptured Aortic Cusp.** A patient who had subacute bacterial endocarditis and who previously had aortic valvular disease rapidly developed massive cardiomegaly, prominence of the main pulmonary artery, and prominence of the pulmonary arterial vasculature *(arrows).* These sudden changes were caused by rupture of the aortic root into the right atrium, which produced a left to right shunt.

ARTERIAL HYPERTENSION

Hypertensive Heart

The essential cardiac changes attributable to chronic hypertension are the result of the increased workload of the left ventricle. At first this ventricle undergoes concentric hypertrophy, which generally produces few or no changes in the cardiac silhouette. Then, continued strain leads to dilatation and the typical picture of left ventricular enlargement. The ventricle enlarges to the left and downward, with the apex often projecting below the left diaphragm. The aortic knob enlarges concomitantly, giving the cardiac waist a somewhat concave appearance; this combination produces the aortic configuration of the heart. Pulmonary vascular changes are seen only when decompensation supervenes.[130]

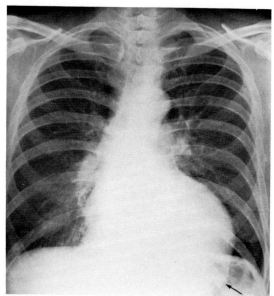

Figure 10–116 **Hypertensive Heart.** There is a prominent enlarged left cardiac border with a low apex *(black arrow)*. The aortic knob is also enlarged. This configuration is typical of prolonged uncomplicated hypertension or aortic valvular disease.

Renovascular Hypertension

A stenotic lesion in one or both renal arteries is found in 3 to 10 per cent of hypertensive patients. Definitive diagnosis is made by renal arteriography, but several other radiographic studies may provide evidence of unilateral renal ischemia.

Early sequential pyelographic films made after rapid injection of the contrast material may show a decreased density of the nephrogram and an initial delay in the appearance of contrast material in the collecting system. This may be the most reliable single finding. This kidney is usually smaller than the other; a difference of 1.5 cm. or more in the lengths of the kidneys is significant. Often the calices of the affected kidney are thin and spidery. Later films may show a greater than normal density of the calices and pelvis due to decreased filtration and slower passage of the filtrate through the tubules, thereby allowing greater concentration of the contrast material. Fairly severe unilateral renal ischemia can occur, however, without positive pyelographic findings. If the renal artery stenosis is severe, a notching due to compression by collateral arteries is sometimes seen in the upper ureter.

A washout test, performed when pyelographic density is well established, consists of rapid infusion of a hypertonic solution of mannitol or urea. The contrast density is washed out earlier and more completely in the kidney with a normal blood supply. This test is meaningful only if there is no interference with drainage as a result of other renal or extrarenal causes. This examination is now rarely employed.

The most frequent renal artery lesion disclosed by arteriography is atherosclerotic narrowing, usually in the proximal portion of the vessel, close to its aortic origin. It may be better demonstrated on oblique films. Most often, atherosclerotic narrowing is unilateral and found in older patients; other vessels may also show atherosclerotic changes. Hemodynamically significant stenotic lesions are usually accompanied by poststenotic dilatation, by some delay in arterial emptying, and by filling of proximal collaterals.

Fibromuscular hyperplasia of a renal artery is a less common finding, appearing in about one third of cases of renal vascular hypertension. This lesion occurs most often in females between the ages of 30 and 50 and is bilateral in about half the cases. Radiographically, the usual form presents with the appearance of a string of beads, involving and narrowing the middle or distal third of the artery. The beads represent microaneurysms. A less common form appears as a relatively long stricture with smooth borders. Both types may coexist. In up to 10 per cent of patients, other arteries are also affected, particularly the mesenteric, iliac, and celiac branches, and the carotids. "Congenital" strictures of the renal arteries may be a form of fibromuscular hyperplasia. There is some evidence that this dysplasia can affect any coat of a smaller artery.

Atherosclerotic narrowing or fibromuscular hyperplasia of a renal artery may also occur in normotensive individuals.[131–136]

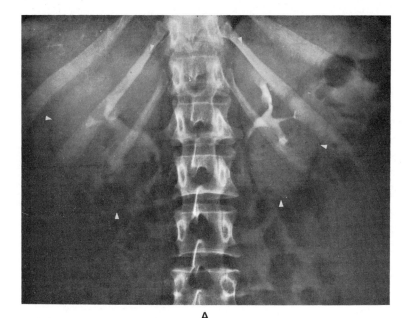

A

Figure 10–117   **Renovascular Hypertension: Pyelogram: Urea Washout and Abdominal Arteriogram.**

A, Excretory urogram shows the left kidney to be considerably smaller than the right *(arrowheads).* Increased density in the pyelogram of the left kidney suggests impaired arterial flow, which allows greater concentration of excreted contrast medium in the collecting system.

Illustration continued on the opposite page

B

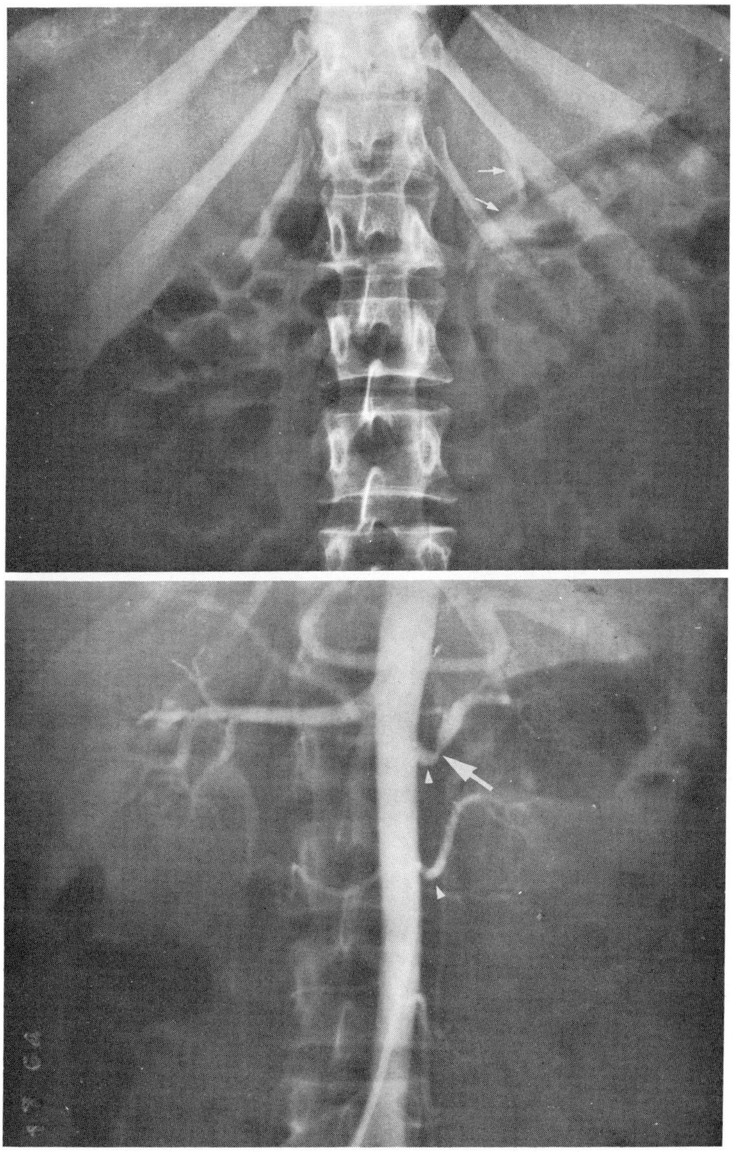

C

Figure 10–117 Continued.

B, In the late phase of a urea washout study there is still considerable contrast density in the upper calices of the left kidney *(arrows),* while the density in the normal right kidney and in the inferior calices of the left kidney has faded almost completely. This suggests ischemia of the upper portion of the left kidney.

C, Arteriogram shows two left renal arteries *(arrowheads)* with an area of stenosis in the upper artery *(arrow).* Note the poststenotic dilatation.

Figure 10–118   **Renovascular Hypertension: Pyelogram: Urea Washout and Abdominal Arteriogram.**

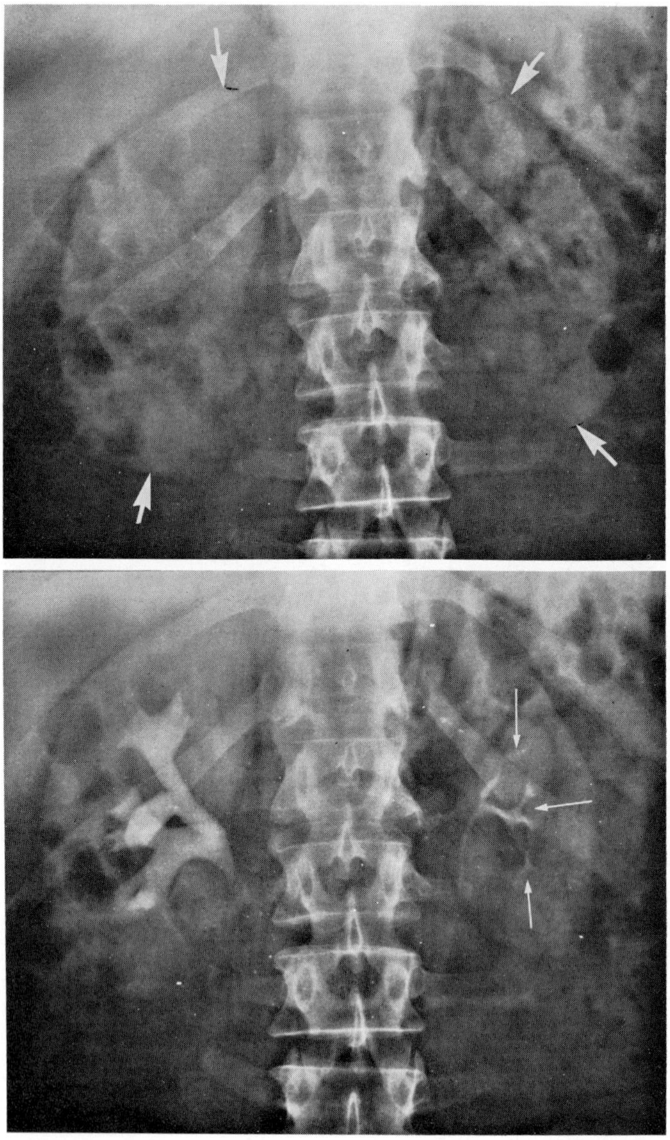

A, The left kidney is significantly smaller than the right and shows a less-dense immediate nephrogram *(arrows).*

B, Excretory urogram confirms the disparity in the sizes of the kidneys and in the sizes of the collecting systems. The thin spidery calices on the left *(arrows)* are due to underfilling caused by slow filtration.

Illustration continued on the opposite page

C

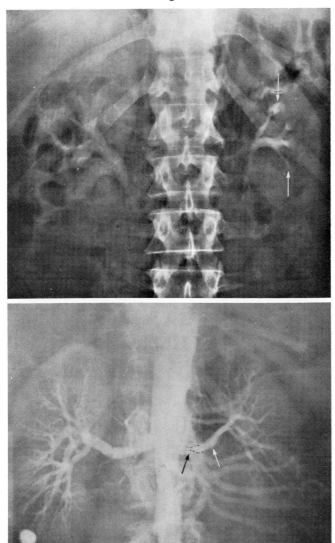

D

Figure 10–118 Continued.

C, Film made after urea washout reveals abnormally prolonged opacification of the collecting system of the left kidney (*arrows*), compared with the almost complete washout on the right side. These findings suggest ischemia on the left side.

D, Arteriogram demonstrates narrowed segment (*black arrow*) of the left renal artery with poststenotic dilatation (*white arrow*). (The density at lower right is an artifact.)

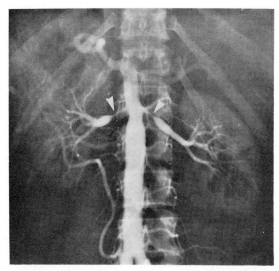

Figure 10–119   **Renovascular Hypertension in Young Adult: Stricture of Both Renal Arteries.**   A short stricture of the left renal artery and a long stricture of the right renal artery (*arrowheads*) are both accompanied by poststenotic dilatation. Strictures and short localized narrowings of renal arteries in younger individuals are currently thought to be forms of fibromuscular hyperplasia, which has a wider spectrum of radiographic changes than was formerly thought.

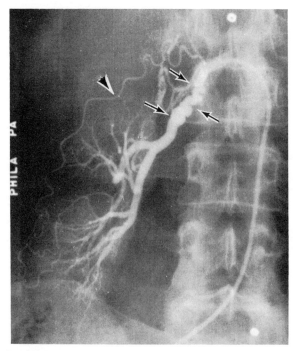

Figure 10–120   **Renovascular Hypertension: Fibromuscular Hyperplasia of the Renal Artery.**   Selective right renal arteriogram of a 39 year old woman with hypertension discloses a rather long corkscrew segment in the middle third of the main renal artery (*arrows*). This is the classic appearance of fibromuscular hyperplasia involving the medial coat of the artery. The capsular artery (*arrowhead*) is opacified and supplies some collateral branches to the lower renal parenchyma, confirming decreased flow through the diseased main renal artery.

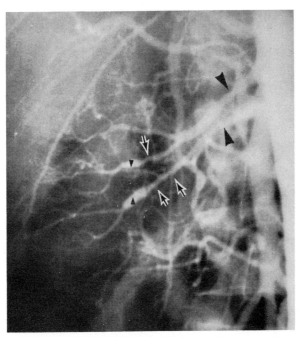

Figure 10–121 **Extensive Fibromuscular Hyperplasia with Severe Hypertension in 39 Year Old Man.** Segments of irregular beaded narrowing are present in the midportions of the two main renal arteries (*large arrowheads*) and in several larger intrarenal branches (*arrows*). Poststenotic dilatation (*small arrowheads*) and opacification of many collaterals confirm the hemodynamic significance of the narrowed segments.

Nephrosclerosis and Hypertension

In hypertension associated with benign arteriosclerosis, the renal arteriogram may be normal. More advanced cases show some reductions in the caliber of the major intrarenal arteries, and there may be a widening of the angle of the more peripheral bifurcations. Often the interlobar arteries are moderately dilated, and an increased resistance to flow is apparent. The smaller arteries sometimes become beaded and tortuous. Kidney size and function remain normal.

In progressive malignant nephrosclerosis and hypertension the kidneys become contracted, and reduction of function is reflected by a diminished nephrographic density. The intrarenal arteries appear pruned because of disproportionately narrow, sparse, and tortuous branches. Circulation time is prolonged. Cortical vessels are sparse or entirely absent. The main renal artery and its primary branches at first may be somewhat dilated. But eventually these narrow, and there is tapering of the main renal arteries. The angiographic appearance in advanced malignant nephrosclerosis is often indistinguishable from that of chronic glomerulonephritis.[137, 138]

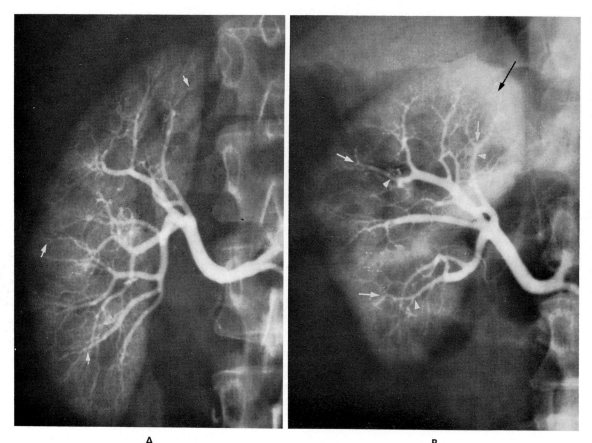

A   B

Figure 10–122 **Essential Hypertension: Benign Arteriolar Nephrosclerosis in Two Patients: Renal Arteriogram.**

A, The small peripheral arteries have a somewhat tortuous and beaded pattern *(arrows)*. These subtle early changes are best appreciated when compared with the vascular tree seen in a normal arteriogram. The arteriogram is normal in other respects, and the kidney size is normal.

B, In the second patient, who had more severe disease, the interlobar arteries are considerably reduced in diameter *(arrowheads),* and the bifurcation angles are increased *(white arrows)*. The more peripheral vessels are somewhat tortuous and beaded. The primary branches of the renal artery are normal. Note the normal nephrographic density of the cortex *(black arrow),* an indication of preserved renal function. (Courtesy Dr. Wade H. Shuford, Atlanta, Georgia.)

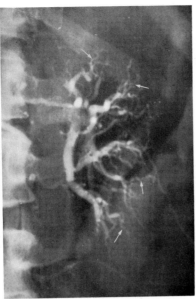 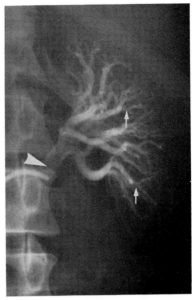

A   B

Figure 10–123   **Malignant Nephrosclerosis in Two Patients: Renal Arteriogram.**

A, The primary branches of the renal artery are dilated. The interlobar arteries are small, short, and somewhat tortuous *(arrows),* and many end abruptly. The cortical vessels are greatly reduced in number. The arterial tree appears bare and pruned. The kidney is small, and the nephrographic density is faint.

B, Renal arteriography in the second patient reveals dilatation of the main *(arrowhead)* and primary branches of the renal artery. The interlobar arteries are abruptly narrowed and shortened, and they terminate suddenly without filling the cortical arteries. The peripheral vasculature is sparse and shortened. The kidney is small, the cortex is thin, and the subsequent nephrogram is very faint. (Courtesy Drs. Richard S. Foster, Wade H. Shuford, and H. Stephen Weens, Atlanta, Georgia.)

Primary Reninism and Juxtaglomerular Cell Adenoma

A syndrome of hypertension, hyperreninemia, and secondary aldosteronism due to release of renin from a renal cortical hemangiopericytoma is a definite clinical entity. Complete cure follows surgical removal of the benign tumor.

Renal vein renin studies will show an abnormally high renin value on one side. Since the tumor may be quite small, selective renal arteriograms may appear grossly normal. However, careful scrutiny may disclose a small cortical area containing a few dilated and tortuous vessels. This area may prove to be somewhat lucent in the nephrogram, and there may be a slight bulging of the cortex. Search for these subtle angiographic findings should be made in cases of hypertension associated with biochemical findings suggestive of primary hyperaldosteronism, but with a unilaterally elevated renal vein renin level.[139, 140]

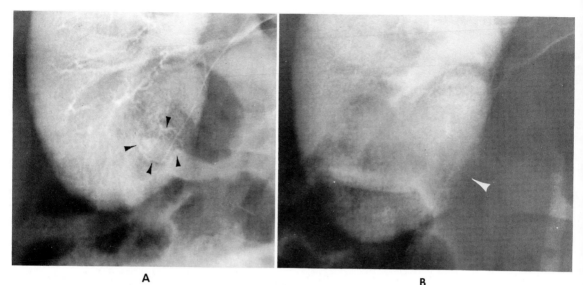

A   **B**

Figure 10–124 **Primary Reninism: Juxtaglomerular Cell Adenoma.**

A, Renal arteriogram of an 18 year old boy with a five-year history of severe hypertension, hypokalemia, and elevated plasma renin and aldosterone reveals vessels of abnormal course and size (*arrowheads*) in the lower kidney.

B, Nephrographic phase shows a lucency and slight renal bulge in this area of abnormal vasculature (*arrowhead*).

A 1.5 cm. juxtaglomerular adenoma was found in the area of these subtle vascular changes. After nephrectomy, complete clinical and biochemical recovery occurred. (Courtesy Dr. J. J. Bookstein, Ann Arbor, Michigan.)

CORONARY ARTERY DISEASE

Coronary Artery Calcification and Atherosclerosis

Coronary artery calcification is only rarely visualized on routine chest films, although it is a frequent autopsy finding. With image-amplification fluoroscopy, coronary calcification has been visualized in about 15 per cent of individuals over age 40, and in about 50 per cent of patients with clinical coronary artery disease.

The calcifications may appear as punctate densities, as patchy densities, or as parallel flecks. The proximal arterial trunks, particularly at the origin of the anterior descending branch of the left coronary artery, are favorite sites. Coronary calcification is practically always within an intimal atherosclerotic plaque.

Contrast coronary arteriography is employed for definitive demonstration of plaques, narrowing, or occlusion. In over 90 per cent of patients with clinical coronary artery disease, multiple sites of obstructive atherosclerotic involvement can be demonstrated. Usually two main branches are involved with obstructive lesions. Coronary arteriography can also demonstrate congenital vascular anomalies (such as aneurysm, fistula, or malformation). It may help rule out coronary artery disease in some patients with angina due to aortic valvular disease.[141–146]

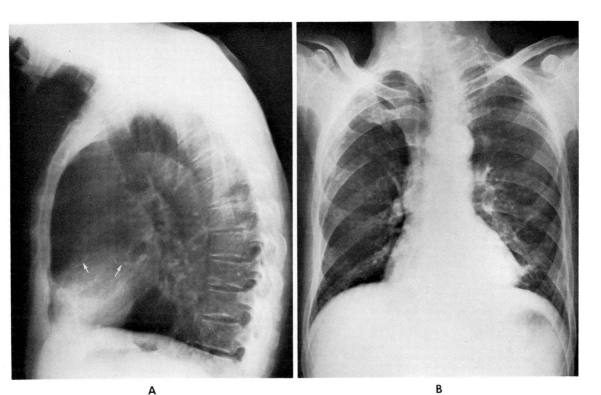

A   **B**

Figure 10–125 **Calcified Coronary Arteries.**

A, Lateral film shows curvilinear calcifications *(arrows)* within the heart shadow; they represent calcific plaques in the left and right coronary arteries.

B, The calcifications cannot be seen in the routine posteroanterior film.

Most areas of coronary artery calcification can be demonstrated only by cineradiography or image-amplification fluoroscopy. (Courtesy Dr. Bernard Pastor, Veterans Administration Hospital, Philadelphia, Pennsylvania.)

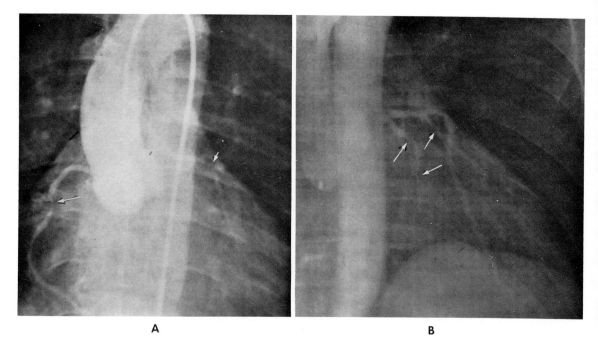

A   B

Figure 10-126   **Arteriosclerosis of Coronary Vessels: Retrograde Aortogram.**

A, There is arteriosclerotic narrowing of the right coronary artery *(long arrow)*, as well as narrowing of the left main coronary artery just above the bifurcation *(short arrow)*.

B, Multiple areas of narrowing of the left coronary artery branches are demonstrated *(arrows)*.

Selective arteriography of a single coronary vessel provides the clearest delineation and is more informative in respect to collateral circulation than is bilateral coronary opacification from the aorta.

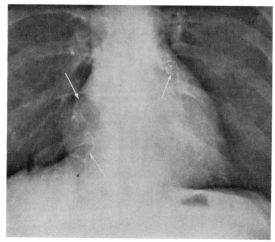

Figure 10-127   **Calcification of Aneurysm of Coronary Arteries.**   Three ring-calcifications *(arrows)* within the heart shadow proved to be calcified congenital aneurysms of the coronary arteries. Aneurysms rarely develop in coronary arteries, and it is even more rare for them to calcify.

Coronary Ischemia and Bypass Surgery

Atherosclerotic involvement is seen most often in the proximal portion of the anterior descending branch of the left coronary artery or the main right coronary artery, or both. The posterior descending or circumflex arteries are only infrequently involved. Preoperative arteriography will disclose the site or sites of major artery narrowing, the state of the small vessels, and the extent of collateral circulation.

Postoperative aortic and selective arteriography of the graft and the other coronary artery will evaluate the patency and functional efficiency of the graft. If a poorly functioning bypass graft is opacified from aortic or selective injection, there will be slow flow, delayed washout of the contrast material, and poor filling of the artery. If the graft cannot be directly filled, its impaired function will be evidenced by filling of the nongrafted coronary artery and by retrograde filling and poor washout of the graft.

A poorly functioning graft is due to stenosis of the aortic or coronary anastomosis or to a disproportionately wide-lumened venous graft. A variable percentage of the grafts become occluded either from thrombi, atheromatous plaques, or intimal fibrosis.[147-149]

Acute Myocardial Infarction and Postmyocardial Infarction Syndrome

As a rule, chest films made after an acute myocardial infarction are unrevealing. Careful fluoroscopy may show a localized area of decreased myocardial contractility, or even paradoxical movement.

If decompensation supervenes, there will be some cardiac dilatation and pulmonary vascular congestion. Acute pulmonary edema sometimes develops rapidly after the cardiac insult.

In about 20 per cent of cases, linear densities appear in the lung bases within a day or two after hospitalization. These are probably plates of atelectasis from shallow respiration.

Myocardial aneurysm may be an early or late complication (see p. 701). Infrequently, the myocardial scar calcifies and can be detected roentgenologically.

The postmyocardial infarction syndrome (Dressler's syndrome) can occur within a few days or up to two months following an acute infarction. Pericardial effusion, pleural effusion, and basal pneumonitis, occurring alone or in combination, are the characteristic roentgen findings. Pleural effusions are seen in over 80 per cent of cases; in nearly half of these it is bilateral, although usually greater on the left side. The pulmonary infiltrates are ill-defined patches, often with linear zones of atelectasis. These lesions either are confined to the left base alone or are bilateral. Infiltrates occur in nearly 60 per cent of cases. Pericardial effusion is manifested by a rapid increase in the heart size and is apparent in 70 per cent of cases. Pleural effusion or infiltrates can occur alone, but pericardial effusion is always accompanied by either pleural effusion or pulmonary infiltrates, or both. The roentgen findings often help distinguish the Dressler syndrome from pulmonary embolization or recurrent myocardial infarction.

Clinical and roentgen clearing may take days to months. Response to steroid therapy is often striking.[150-154]

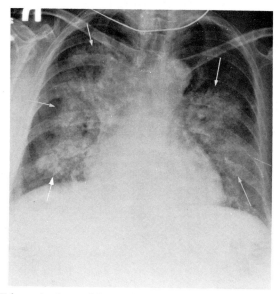

Figure 10–128 **Pulmonary Edema in Acute Myocardial Infarction.** Extensive bilateral confluent densities are shown (*arrows*); however, the periphery of the lungs remains relatively uninvolved. The heart is moderately enlarged, a characteristic finding in pulmonary edema due to left heart failure. The densities may develop rapidly and may also disappear quickly with compensation.

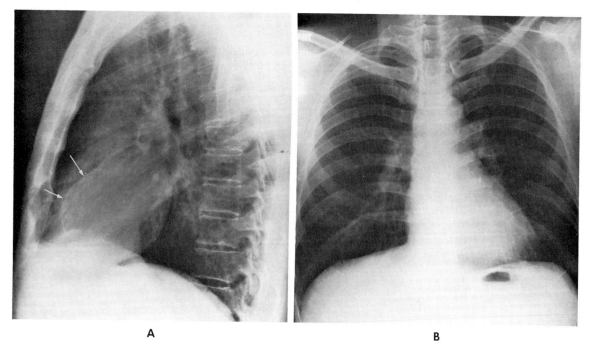

A   B

Figure 10–129 **Myocardial Scar Calcification.**

A, Lateral film demonstrates calcification of the myocardial scar (*arrows*) subsequent to an infarction one year previously.

B, The calcification cannot be seen on the posteroanterior film.

Pericardial calcification is always seen at the cardiac border, whereas myocardial calcification usually lies beneath the roentgen border of the heart, but distinction may be difficult or impossible.

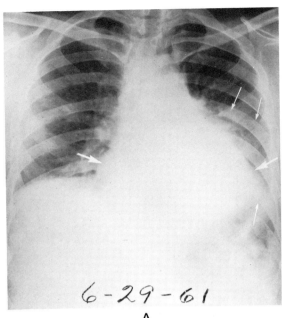

6-29-61

A

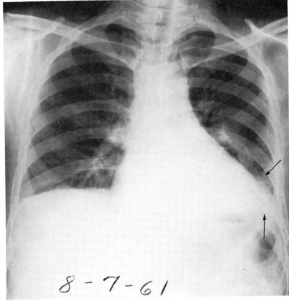

8-7-61

B

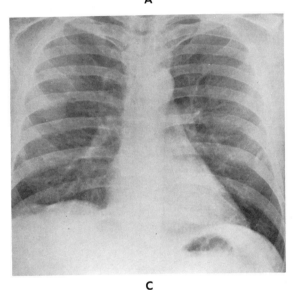

C

Figure 10–130 **Postmyocardial Infarction Syndrome (Dressler's Syndrome).**

A, Four to five weeks after myocardial infarction there is cardiac enlargement *(large arrows)* and a pleuropneumonic process at the left lung base *(small arrows).*

B, Six weeks later the heart size is reduced, and there is a marked subsidence in the pleuropneumonic process *(arrows).*

C, After three months the chest film is normal, the lung fields are clear, and the heart has returned to normal size.

Myocardial Aneurysm

This complication of myocardial infarction may develop early in the softened infarcted area, or much later in the fibrous myocardial scar.

On plain films there may be a bulge or an unusual prominence in the left ventricular border; oblique or lateral films may be needed for adequate demonstration. Paradoxical or limited pulsation is observed in the bulge; cineradiography is preferable to fluoroscopy for detecting paradoxical pulsations. Calcification in the aneurysmal wall is a highly suggestive but infrequent finding. Aneurysm of the septum produces no findings on plain films.

Opacification studies of the ventricles may be necessary for diagnosis and for differentiation from cardiac tumors, adjacent mediastinal masses, pericardial cysts, or localized pericardial effusions. Opacification findings include unusual thinning or stiffening of the wall, increased residual blood in the ventricle, paradoxical bulging during systole, and a local aneurysmal chamber.[2, 155–157]

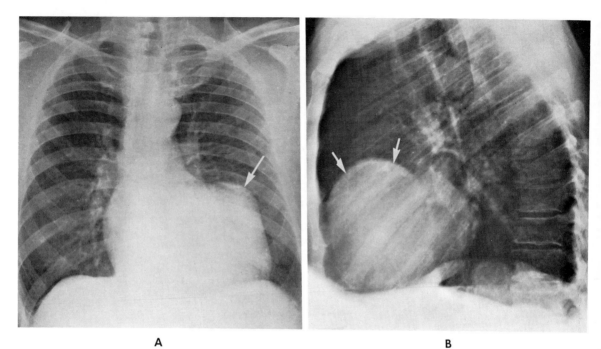

A   B

Figure 10–131 **Ventricular Aneurysm: Postmyocardial Infarction.**

A, There is characteristic bulging of the left heart border *(arrow),* which showed paradoxical pulsation on fluoroscopy. Notice the "squared off" appearance of the left heart.

B, The aneurysmal bulge *(arrows)* is clearly seen in the lateral film.

This left ventricular aneurysm was discovered one year after a myocardial infarction.

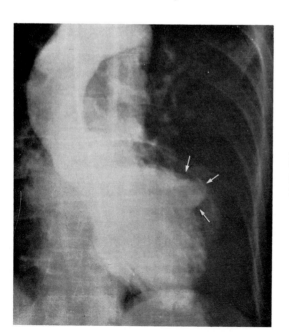

Figure 10–132 **Ventricular Aneurysm: Opacification Study.** Film made during ventricular systole shows the contrast-filled aneurysmal bulge *(arrows).* This may not be apparent during diastole.

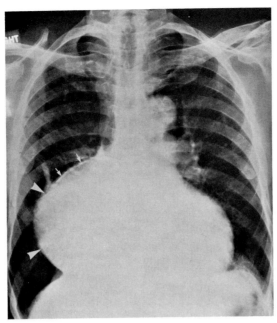

Figure 10-133 **Calcified Right Ventricular Aneurysm.** An unusual huge right ventricular aneurysm causes bulging on the right side of the heart *(arrowheads).* It developed several years after a myocardial infarction. The wall of the aneurysm is calcified *(small arrows).*

CARDIAC ARRHYTHMIA

Chest films of a patient with arrhythmia may give information about the underlying cardiac abnormality. In auricular fibrillation, for example, there may be radiographic evidence of a mitral heart, of calcification of the mitral valve or annulus, of a high output heart of hyperthyroidism, or of calcification of a coronary artery. Calcifications of the upper portion of the interventricular septum are occasionally seen in patients with heart block; presumably these are calcified infarcts.

In patients with an implanted pacemaker, radiographic studies can determine the position of the pacemaker and its electrodes, detect breaks in the wires, and assess the condition of the batteries.[158-160]

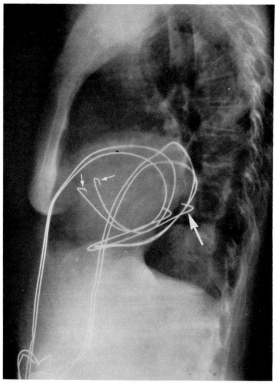

Figure 10–134 **Pacemaker in Heart: Broken Leads.**   Lateral view shows pacemaker terminals attached to the myocardium. These leads have broken *(small arrows)* because of myocardial pulsations, and a new set of wires has been inserted *(large arrow)*. Such breakage is not uncommon and is readily detected on x-ray examination.

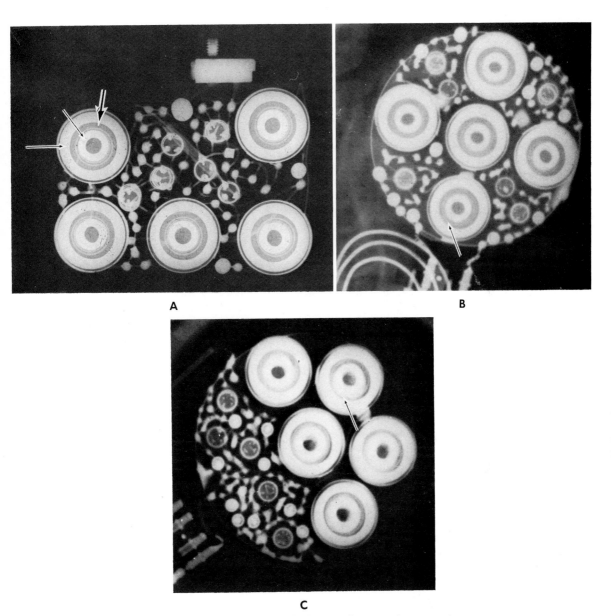

A

B

C

Figure 10–135 **Radiographic Evaluation of State of Charge of Pacemaker Batteries.**

A, In a fully charged mercury battery, the lucent ring of electrolytes *(large arrow)* is of uniform width and is sharply demarcated from the surrounding opaque rings *(small arrows)* of metallic material.

B, In a partially run-down battery, the lucent ring *(arrow)* has become thinner, and its internal border has become somewhat fuzzy and irregular due to metallic deposits.

C, In a nearly exhausted battery, opaque metallic deposits have greatly narrowed the lucent ring *(arrow)*; portions of the ring have been obliterated. The opaque inner ring has increased in width.

To visualize the battery rings without distortion, the batteries must be directly perpendicular to the film.

PERICARDITIS

Pericarditis with Effusion

There are many causes of pericarditis with effusion, but the roentgenologic findings are similar and nonspecific. Usually the cardiac silhouette is enlarged, the degree depending on the amount of fluid. At least 200 ml. of pericardial fluid is necessary before cardiac enlargement is apparent. A rapid change in the cardiac silhouette in the absence of pulmonary vascular engorgement is highly suggestive. Pericardial effusion should be considered if cardiac pulsations are diminished or almost absent and if changes in the cardiac contour occur with postural alterations. Large effusions tend to obliterate normal cardiac markings and alter the cardiohepatic angle; in small or moderate effusions the angle becomes more acute, and in massive effusions it may be completely obliterated. On lateral view, a posteriorly displaced epicardial fat pad lying within the cardiac silhouette is strong evidence of pericardial effusion, but this fatty line is usually difficult to identify. Radiograms made after pericardial tap followed by air injection indicate the degree of effusion and the actual size of the heart.

Intravenous carbon dioxide injection, done with the patient lying on his left side, causes gas to accumulate in the right atrium, so that the thickness of the right cardiac wall and pericardium can be estimated. Similarly, estimation can be made by opacification of the heart chambers during angiocardiography. These methods are not accurate in the presence of significant right pleural effusion.

Distinction between pericardial thickening and pericardial fluid can often be made by cineradiography following carbon dioxide injection. With pericardial effusion, the distance between the gas bubble in the right atrium and the cardiac outline changes with cardiac contraction, whereas with pericardial thickening there is no change. A radioisotope scan of the heart blood volume can be superimposed on the roentgen cardiac density. A marked disparity is indicative of pericardial effusion or thickening.

Nonspecific cardiomegaly without pulmonary congestion should raise the possiblity of pericardial effusion, particularly if enlargement of the cardiac silhouette is of recent origin.[161-163]

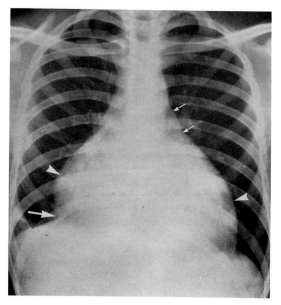

Figure 10–136 **Large Pericardial Effusion.** Generalized symmetric enlargement of the cardiac silhouette *(arrowheads)* is caused by a large pericardial effusion. Normal bulging of the aorta and the pulmonary segments is decreased by the effusion *(small arrows)*. Bulging of the right border causes obliteration of the acute cardiophrenic angle *(large arrow)*. The absence of pulmonary congestion is characteristic.

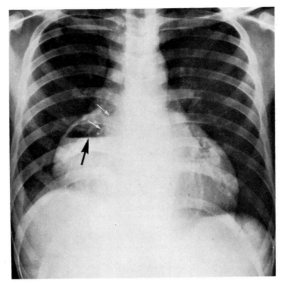

Figure 10–137 **Pericardial Effusion: Diagnostic Pneumopericardium.** Following pericardial tap and air injection, an air-fluid level *(large arrow)* occupies the pericardial space and delineates its size. The true cardiac borders *(small arrows)* are visible through the air bubble.

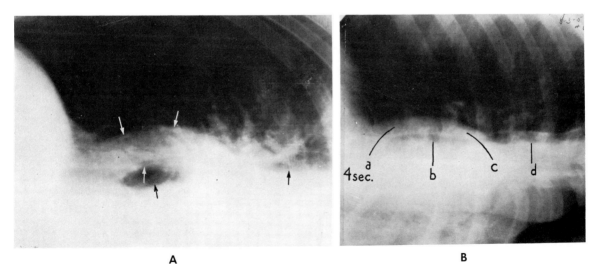

A   B

Figure 10–138 **Pericardial Effusion: Intravenous Carbon Dioxide Study in Two Patients.**

A, Following intravenous carbon dioxide injection, with the patient in left lateral decubitus position, the right atrium is filled with gas, thereby demonstrating the widened pericardial space *(between white arrows)*. In the normal pericardium this thickness does not exceed 4 mm. Black arrows point to accumulation of carbon dioxide in the right atrium and the superior vena cava.

B, Study done in left lateral decubitus position following injection of carbon dioxide into the antecubital vein reveals accumulation of gas in the superior vena cava (d) and the right atrium (b). The pericardial space (a,c) is widened, indicating either pericardial effusion or a thickened pericardium. (Courtesy Dr. James H. Scatliff, New Haven, Connecticut.)

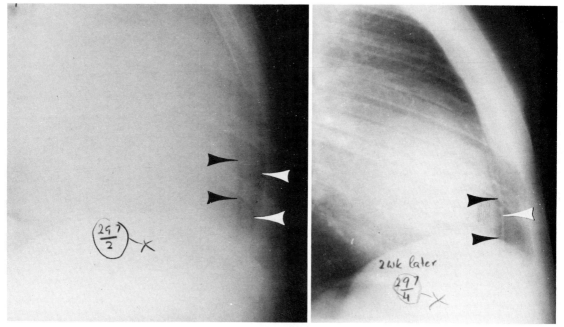

Figure 10-139   **Pericardial Effusion: Pericardial-Epicardial Fat Line Separation.**

A, The epicardial fat line *(black arrowheads)* is separated from the pericardial fat line *(white arrowheads)* by pericardial fluid.

Normally, the two fat lines, if visible on the lateral chest film, are virtually contiguous.

B, Two weeks later, the effusion has almost completely disappeared, and the two fat lines *(arrowheads)* are much closer together; they are separated only by a thin line of density.

Pericardial-epicardial fat line separation is diagnostic of pericardial effusion; unfortunately, these two fat lines cannot be identified on the majority of lateral chest films.

Cardiac Tamponade

Tamponade occurs when pericardial fluid collections prevent adequate venous return to the right atrium. This leads to diminished cardiac output, elevated systemic venous pressure, hypotension, and circulatory shock. Chest pain and dyspnea are usually experienced.

Acute tamponade is usually by traumatic origin, generally from a stab wound of the heart. The rapid accumulation of fluid may lead to tamponade with minimal changes in the cardiac silhouette. Diminished cardiac pulsation may be the only apparent roentgen finding.

Subacute tamponade may occur from chronic uremic effusions; inflammatory pericarditis, especially tuberculosis; or from neoplastic pericardial masses. Radiographically, there is usually enlargement of the cardiac silhouette, diminished cardiac pulsations with clear lung fields, some decrease in pulmonary vasculature, and usually prominence of the superior vena cava shadow.

Venous angiocardiography generally discloses evidence of pericardial effusion with a concavity of the right atrial border in contrast to the straightened flat border seen in constrictive pericarditis or the normal convex border seen in uncomplicated pericardial effusions. Vena cava emptying is prolonged.[164-166]

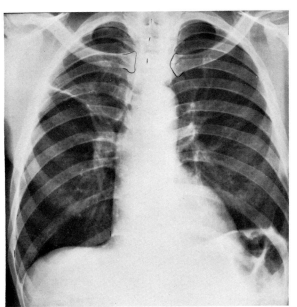

A

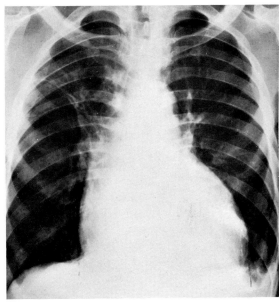

B

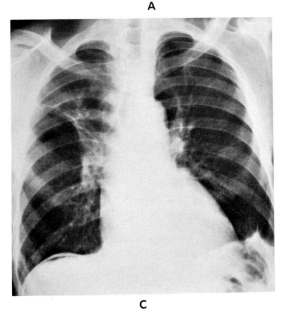

C

Figure 10–140 **Cardiac Tamponade: Serial Films Following Stab Wound.**

A, Cardiac silhouette is normal shortly after a stab injury to the epigastrium. Fibrosis of the right upper lobe was an incidental finding.

B, A few days later the patient developed circulatory embarrassment. There is cardiac enlargement with minimal decrease in the pulmonary vasculature. The enlargement was caused by seepage of blood into the pericardium.

C, Following evacuation of the pericardial blood, the heart is of normal size. The free air under the right diaphragm is secondary to recent surgery.

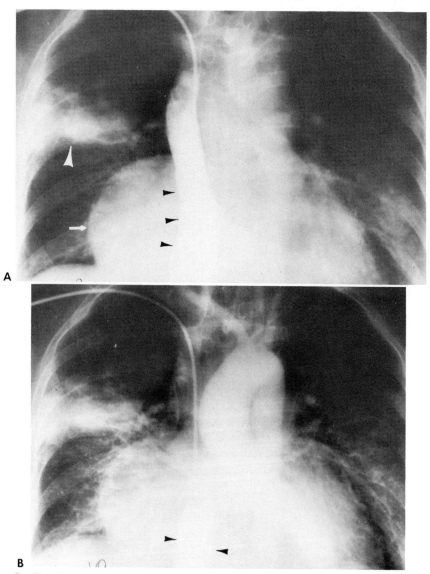

A

B

Figure 10–141   **Cardiac Tamponade from Suppurative Nocardial Pericarditis.**

A, This patient with pulmonary nocardiosis *(white arrowhead)* developed a rounded bulging right cardiac border *(arrow)* and decreased pulmonary vasculature. The neck veins were distended, especially on inspiration.

The venous angiocardiogram shows a greatly increased distance between the right cardiac border and the lateral border of the right atrium *(black arrowheads).* This border is concave owing to compression; normally, the right atrial border is somewhat convex.

B, A few seconds later, most of the contrast material is in the aorta, but there is still contrast material in the inferior vena cava *(arrowheads),* which is partially occluded by the suppurative pericardial collection.

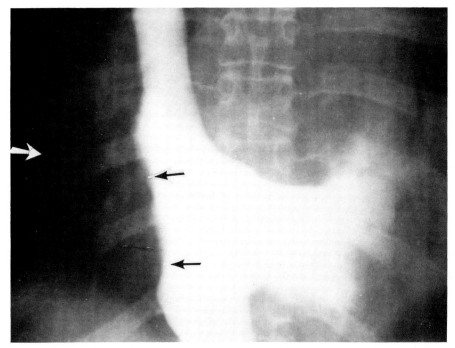

Figure 10–142 **Cardiac Tamponade from Tuberculous Pericarditis: Angiocardiogram.** The right cardiac border (*white arrow*) is separated from the right atrium by a large pericardial collection. The lateral border of the right atrium has been compressed into a concave shape (*black-white arrows*). There was prolonged opacification of the superior and inferior vena cava. (Courtesy Drs. H. B. Spitz and J. C. Holmes, Cincinnati, Ohio.)

Constrictive Pericarditis

Mild to moderate enlargement of the cardiac shadow on plain films is most often seen, although the presence of a normal-sized heart is not rare. The configuration, however, is rarely normal. Diminished cardiac pulsations can be recognized on fluoroscopy in almost every case.

Pericardial calcification is the most striking and suggestive finding, and occurs in over half the cases. However, pericardial calcification can be present without clinical constrictive pericarditis. The ventricular borders and the anterior and posterior atrioventricular grooves are the most common sites of calcification in symptomatic constrictive pericarditis.

The shape of the cardiac silhouette is variable, but in about three quarters of cases the waist of the heart is filled in by prominence of the main pulmonary artery or by localized pericardial thickening. Left atrial enlargement is seen in over half the cases; fullness of the superior vena cava is found in almost half the patients. Pulmonary venous congestion and pleural effusion, either alone or in combination, are common. Often the cardiac outline resembles that of mitral stenosis.

Definitive diagnosis requires special roentgen studies. Angiocardiographic or intravenous carbon dioxide studies demonstrate flattening and rigidity of the lateral border of the right atrium during the cardiac cycle (see *Pericarditis with Effusion*, p. 706). The atrial pericardium is usually thickened (greater than 4 mm.). The dilatation of the superior vena cava and pulmonary artery are demonstrated during angiography. Abrupt restriction of diastolic expansion of the ventricles is said to be a specific finding during cineangiocardiography.

In rare cases, chronic constrictive pericarditis leads to a protein-losing enteropathy associated with jejunal lymphangiectasia. The latter may produce thickening of the jejunal mucosa, seen on barium studies.[167–171]

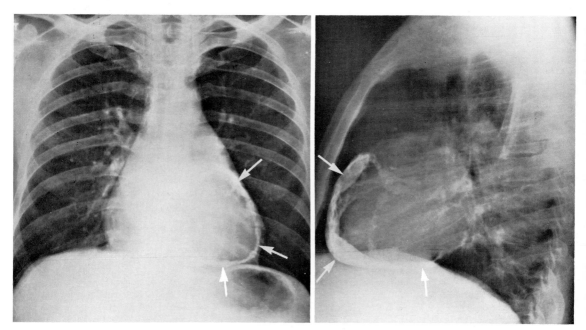

Figure 10–143 **Constrictive Pericarditis with Calcification.** The left and right ventricles *(arrows)* are extensively calcified; calcification around the right ventricle is most clearly demonstrated in lateral view. The heart is moderately enlarged, the conus is prominent, and pulsations were decreased on fluoroscopy. In older persons, pericardial calcifications may be seen in the absence of constrictive changes.

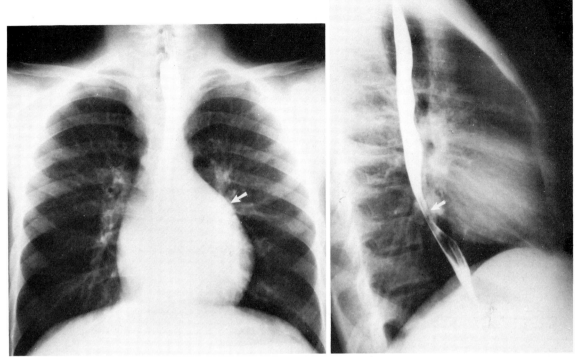

Figure 10–144 **Constrictive Pericarditis.** The heart is somewhat enlarged, and there is evidence of enlargement of the left atrium and its appendage *(arrows)*. The upper lobe vessels are prominent, and the entire roentgen picture is suggestive of mitral disease.

The patient, a 13 year old boy, had no cardiac murmurs but did have a small pulse pressure and attacks of syncope. Constrictive pericarditis without calcification was diagnosed by angiocardiography and confirmed surgically.

In a proper clinical setting, roentgen evidence of left atrial enlargement without a murmur of mitral disease should suggest possible constrictive pericarditis.

Pericardial Cyst

The majority of pericardial cysts occur anteriorly in the right cardiophrenic angle. Less than one fourth arise on the left side. They are usually coelomic cysts and do not communicate with the pericardial cavity. The rare communicating cyst is more correctly a pericardial diverticulum. Pericardial cysts are smooth and regular with sharp margins, and cause bulging in the cardiophrenic angle when they extend to the diaphragm. The density cannot be separated from the cardiac shadow in any projection—a highly suggestive finding. There may be some alteration of shape with postural changes or with respiratory excursions.

Radiographically, the diagnosis is usually not difficult. Occasionally the cardiophrenic density of the cyst can simulate an omental hernia, a cardiac or pericardial tumor, a large pericardial fat pad, or aneurysm of the sinus of Valsalva. Angiocardiography, diagnostic pneumoperitoneum, or direct needling of the cyst and injection of contrast material can be used for more definitive diagnosis.[172, 173]

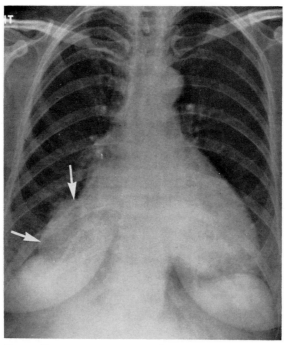

Figure 10–145 **Pericardial Cyst.** Posteroanterior view demonstrates typical sharply delineated bulge obliterating the right cardiophrenic angle *(arrows).*

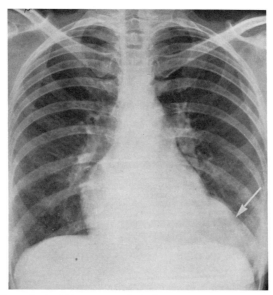

Figure 10–146 **Pericardial Cyst: Left Side.** Smooth bulging density *(arrow)* adjacent to the cardiac apex was due to a pericardial cyst. Most pericardial cysts are located on the right side.

Congenital Absence of the Pericardium

The left pericardium may be totally or partially absent. Only rarely is the right pericardium absent.

In complete absence of the left pericardium, the heart is displaced to the left (levoposition) without tracheal deviation. The main pulmonary artery segment is prominent. A lucent zone is seen between the aorta and pulmonary artery, due to interposition of lung. Diagnosis can be confirmed by an artificial left pneumothorax; air will be seen on the right side of the heart.

Partial absence of the left pericardium occurs in the region of the left atrium. Radiographically, there will be a bulge of the left cardiac border in the region of the left atrium, below the main pulmonary artery segment. Cineangiography will show pulsatile expansion of the left atrium through the defect.

Levoposition of the heart, in the absence of a thoracic pectus and cardiac disease, is suggestive of complete absence of the left pericardium. An atrial bulge on the left cardiac border in an otherwise normal heart is suspicious of partial absence of the left pericardium.[174, 175]

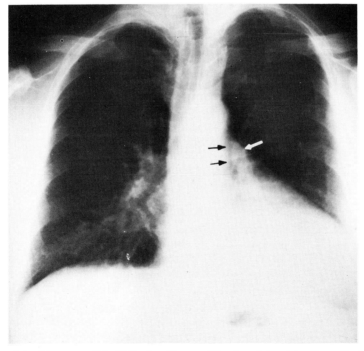

Figure 10–147 **Complete Absence of Left Pericardium.** The heart is virtually entirely in the left thorax. The right border is obscured by the spine. The main pulmonary artery segment is prominent *(white arrow)*, and there is a lucent stripe *(black arrows)* between the aorta and pulmonary artery due to the presence of a piece of lung in the space normally occupied by the pericardium.

The absence of the entire left pericardium was confirmed during coronary bypass surgery.

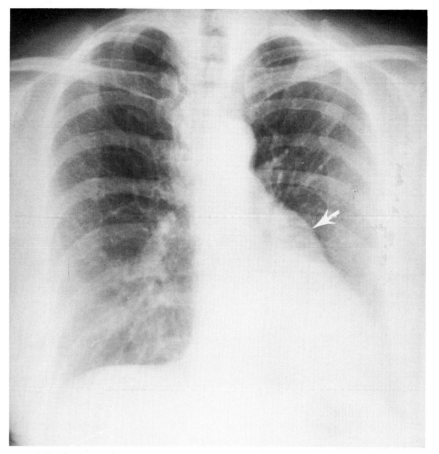

Figure 10–148 **Partial Absence of Left Pericardium.**   There is a marked bulge on the left cardiac border *(arrow)* in the region of the left atrial appendage. Oblique and lateral films made with barium in the esophagus showed no left atrial enlargement.

There was partial absence of the pericardium in the region of the left atrial appendage. This is the most common site of partial absence of the pericardium, and the bulge of the atrial appendage is characteristic of this defect.

DISEASE OF THE MYOCARDIUM

Cardiomyopathy (Cardiopathy)

Although there are numerous causes of cardiomyopathy, the resultant hemo-dynamic changes are identical, and similar roentgenologic changes are found in most cases. Poor contractility and decreased ejection volume from the left ventricle are the basic physiologic alterations. The usual chest finding is diffuse cardiac enlargement of various degrees, with predominant left ventricular enlargement. Left atrial enlargement is almost always associated. The aorta and its knob are usually small because of the decreased cardiac output. However, in older patients the aorta may be enlarged as a result of pre-existing aortic atherosclerosis. Some signs of left ventricular failure are usually present, varying from evidence of pulmonary venous hypertension (disproportionately large upper lobe vessels and hilar fullness) to Kerley B lines, interstitial and intra-alveolar edema, and pleural fluid.

Right-sided changes can appear, including some enlargement of the main pulmonary artery, right atrium, and right ventricle, and fullness of the superior vena cava.

The combination of left ventricular and left atrial enlargement, a small aorta, and some radiographic evidence of failure are highly suggestive of cardiomyopathy, especially if clinically significant murmurs are absent. However, with marked cardiomegaly, confusing murmurs due to functional incompetence of the mitral and tricuspid valves can develop. A complicating pulmonary embolism is not uncommon, confusing the radiographic picture.

Selective angiocardiography will disclose an enlarged, dilated, poorly contracting left ventricle with decreased ejection fraction and increased end-diastolic volume.

Cardiomyopathies are often listed by subgroups, but this classification is not entirely satisfactory. The idiopathic group due to primary myocardial disease includes endocardial fibroelastosis, idiopathic hypertrophic subaortic stenosis, postpartum cardiopathy, the alcohol heart, thiamine or tryptophan deficiency, and others. Myocarditis due to infection can occur in Coxsackie infections, trichinosis, Chagas' disease, diphtheria, and toxoplasmosis. The infiltrative group includes amyloidosis and glycogen storage disease. Endocrine cardiopathies include hypothyroidism, acromegaly, and Cushing's disease. Cardiopathies associated with neurologic disorders are found in the muscle dystrophies and Friedreich's ataxia.[176-180]

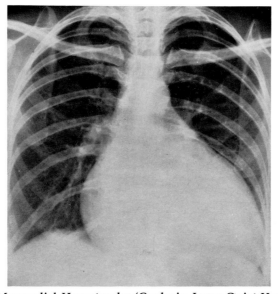

Figure 10–149 **Idiopathic Myocardial Hypertrophy (Cardosis, Large Quiet Heart, Cardiomyopathy).** The heart is greatly enlarged, with left ventricular dilatation predominating. Left atrial enlargement is not well demonstrated in posteroanterior view. The aorta is small, and the pulmonary artery is full. No congestive changes are seen. When this picture occurs in a young person who is not hypertensive and has no significant murmur, it suggests a nonspecific cardiomyopathy.

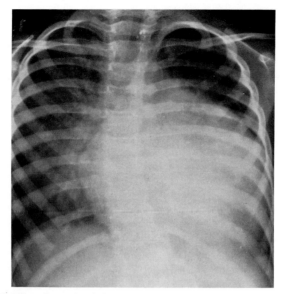

Figure 10–150 **Endocardial Fibroelastosis (Idiopathic Cardiomyopathy).** Marked cardiomegaly is seen with increased pulmonary vasculature due to congestive failure. The latter is a frequent complication of cardiomyopathy. When no cause can be determined for cardiac enlargement and congestive failure in an infant, fibroelastosis should be considered.

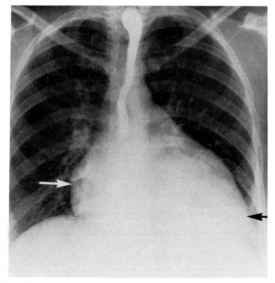

Figure 10–151 **Postpartum Cardiomyopathy.** Cardiomegaly *(arrows)* is attributable to predominantly left ventricular enlargement with a relatively small aorta and only slight prominence of the pulmonary artery segment. The pulmonary vasculature is normal. The cause of postpartum cardiomyopathy is uncertain. It strongly resembles the other cardiomyopathies.

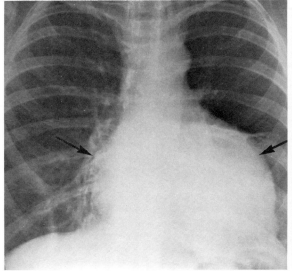

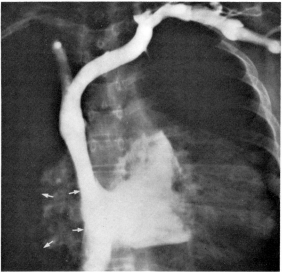

A

B

Figure 10–152 **Myxedema Heart: Endocrine Cardiomyopathy.**

A, The heart is considerably enlarged (arrows) with left ventricular prominence. The aorta is relatively small. There is no pulmonary vascular congestion.

B, Dextrophase of angiocardiogram shows increased distance between the right heart border and the opacified lumen of the right atrium (arrows).

C, Levophase shows opacification of the left ventricle (large arrow) with a widened space between the chamber and the cardiac border (small arrows). Thickened walls on the right and left are due to a combination of myocardial thickening and pericardial effusion, the latter being present in almost every case of myxedema heart. (Courtesy Drs. R. D. Kittredge, E. J. Arida, and N. Finby, New York.)

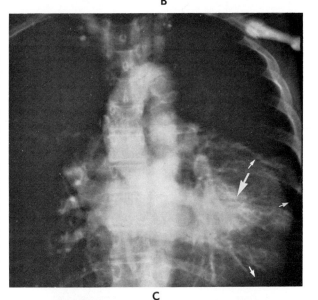

C

Figure 10–153 **Cardiomyopathy in Friedreich's Ataxia.** Nonspecific cardiomegaly is attributable to cardiomyopathy, chiefly left ventricular enlargement. The aortic knob is small and the pulmonary vasculature is normal, since there is no cardiac failure. The high dorsal scoliosis (arrow) is a frequent occurrence in Friedreich's ataxia.

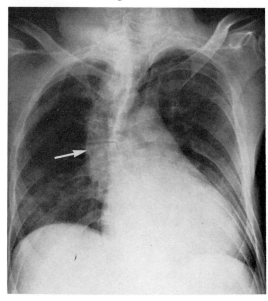

Tumors of the Heart and Pericardium

About 80 per cent of primary tumors of the heart are benign. Malignant primary tumors are all sarcomas.

Myxoma is the most common benign tumor. It arises from the atrial septum, usually in the left atrium, and often has a pedicle. Radiographically the heart and vasculature may be normal, with or without some left atrial enlargement. A large myxoma may lead to cardiac and vasculature alterations indistinguishable from those seen in mitral stenosis. Calcification of a myxoma can occur, but this is rare. Diagnosis is made by angiocardiography, which demonstrates a well-defined filling defect in the opacified atrium. This finding, however, is often indistinguishable from a blood clot. Prolapse of the tumor into the ventricle may produce a ventricular filling defect.

In malignant primary sarcomas of the myocardium the roentgenologic findings frequently are those of unexplained heart failure. The heart contour may be altered, and bizarre configurations may result. Abnormalities of pulsation may be detected fluoroscopically. Occasionally, pericardial effusion develops rapidly and may be sanguineous.

Metastatic disease to the myocardium and pericardium is the most common form of malignant involvement and is most often from bronchogenic or breast carcinoma and sometimes from lymphoma or leukemia. Autopsy information indicates that as many as 50 per cent of patients with bronchogenic carcinoma may develop cardiac metastases; however, in only a small percentage of cases are abnormalities apparent on roentgenogram or electrocardiogram. Cardiac enlargement is the most frequent abnormality and is usually due to a malignant pericardial effusion. Unexplained cardiomegaly or congestive failure in a patient with bronchogenic or breast carcinoma should suggest myocardial or pericardial metastases.[181-186]

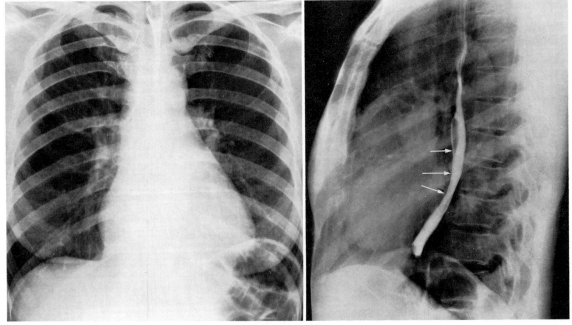

Figure 10–154 **Myxoma of Left Atrium.** There is barium in the esophagus. Moderate left atrial enlargement *(arrows)* is demonstrated most clearly in lateral view; otherwise, the cardiac contour is within normal limits. Enlargement was caused by a myxoma of the left atrium. In this condition, clinical murmurs often simulate mitral valvular disease; myxoma may be suspected if the mitral murmurs show marked variation with change in the patient's position.

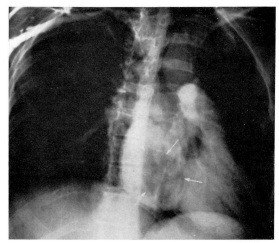

Figure 10–155 **Myxoma of Right Atrium: Angiocardiogram.** Dextrophase at four seconds reveals a large filling defect (*arrows*) in the region of the right atrium; the defect was due to an atrial myxoma. Demonstration of a filling defect on contrast angiocardiography is required for diagnosis of an intracardiac tumor. A blood clot within the heart can produce an identical filling defect.

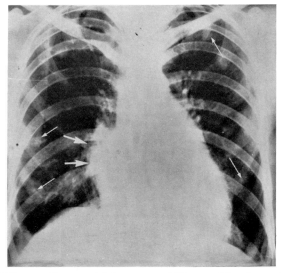

Figure 10–156 **Primary Malignant Hemangioblastoma of Heart.** The heart is enlarged, and there is a localized irregular density on the right side (*large arrows*) below the hilum; this is a tumor mass attached to the heart muscle. Note the metastatic nodular densities in both lung fields (*small arrows*). A metastatic neoplasm to the heart sometimes produces a similar picture. Although irregular and bizarre cardiac configurations might suggest primary cardiac tumor, diagnosis is rarely made during life. (Courtesy Dr. R. Bergonzini, Mondena, Italy.)

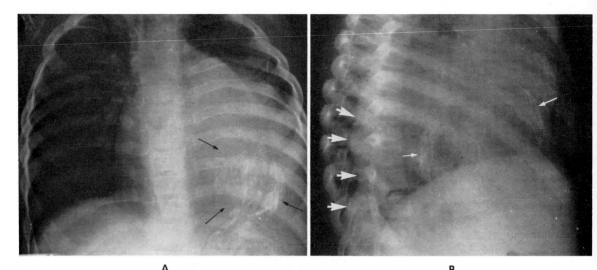

A   **B**

Figure 10–157   **Huge Primary Rhabdomyosarcoma of Heart in an Infant.**

A, The cardiac silhouette is huge, but there is no vascular congestion. A large collection of linear and curvilinear calcifications *(arrows)* is located in the left lower heart density.

B, On lateral view the calcifications are seen to occupy a large segment *(small arrows)* of the cardiac area. The posterior border of the heart mass *(large arrows)* is lobulated, suggesting a tumor mass.

In earlier films, before the calcifications had appeared, a diagnosis of congenital heart disease with cardiomegaly was made.

A huge inoperable rhabdomyosarcoma arising from the ventricular muscle was found at surgery. Palliative radiation was employed to alleviate the severe dyspnea.

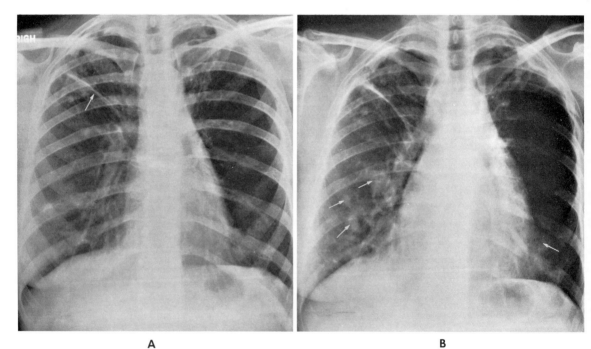

A   **B**

Figure 10–158   **Neoplastic Pericarditis: Breast Carcinoma.**

A, This 38 year old woman had a left mastectomy for carcinoma. Film made a few months later demonstrates old fibrotic disease in both upper lobes, with apical pleural thickening. The short fissure *(small arrow)* and the right hilum are pulled upward by right apical fibrosis. Absence of the left breast shadow is responsible for increased lucency of the left lung field. The heart is normal in size and configuration.

B, Three years later there are multiple metastatic nodules *(arrows),* mainly in the right lung. The overall heart size is greatly increased, but there is no congestion. The pericardium contained fluid and manifested metastatic disease.

In patients with known malignant disease, especially of the breast or lung, a gradual increase in heart size may be indicative of effusion caused by metastatic disease of the pericardium. If congestive failure ensues, myocardial metastases should be suspected.

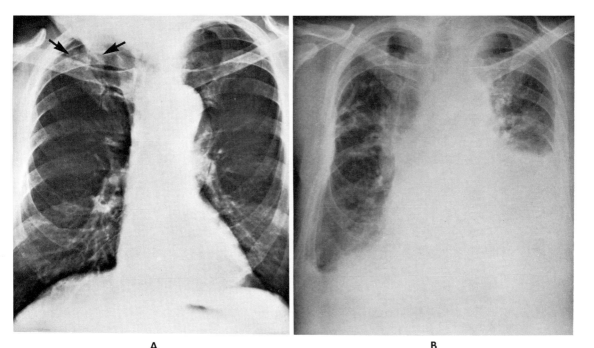

A   **B**

Figure 10–159 **Metastatic Tumor of Heart.**

A, Preoperative film indicates density in the right apex *(arrows)*; this represents a bronchogenic carcinoma. Note the normal cardiac shadow and pulmonary vasculature.

B, Six months postoperatively there is marked cardiac enlargement caused by a malignant pericardial effusion and metastatic pericardial and myocardial involvement. Congestive heart failure with bilateral pleural effusions is present.

In a patient with known bronchogenic carcinoma, late development of cardiac enlargement and pericardial effusion, either alone or in combination, is frequently a result of metastases to the heart or to the pericardium, or to both.

DISEASES OF THE AORTA

Atherosclerosis of the Aorta

Localized atherosclerotic change in the thoracic aorta is often marked by a linear calcific plaque. This is seen most often in the aortic knob, but can occur in any portion; sometimes almost the entire aorta is opacified by extensive calcification in its wall. Similar calcifications can occur in the major vessels of the arch, but this is less common.

Diffuse atherosclerotic involvement of the thoracic aorta may lead to elongation and later to tortuosity of this vessel. On the posteroanterior chest film the elongated ascending and descending aorta bulges laterally, and the knob is high and prominent. The uncoiled aorta is usually best seen on a lateral film if the aortic shadow is sufficiently tortuous and becomes partially surrounded by the air-containing lung. The barium-filled esophagus is adherent to the descending aorta and will follow its elongated or tortuous path. (See also *Peripheral Arteriosclerosis*, p. 746.)

Calcific atherosclerotic plaques are also commonly seen in the abdominal aorta. The best visualization occurs on lateral films of the abdomen.[187]

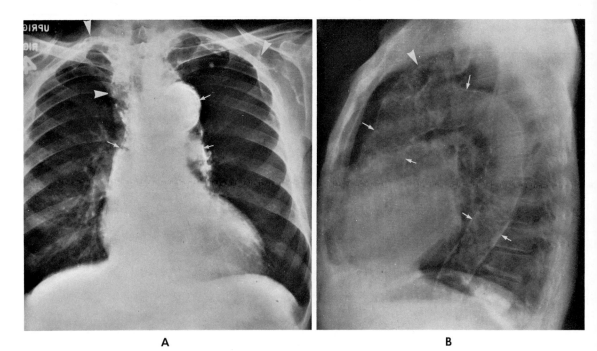

A   **B**

Figure 10–160   **Extensive Calcification of Thoracic Aorta.**

A, In posteroanterior film dense calcified plaques are seen in the wall of the ascending and descending aorta and the arch *(arrows)*. The lateral position of the calcified descending aorta is due to its tortuosity and elongation. The brachiocephalic vessels are also calcified *(arrowheads)*.

B, In lateral view, calcification is seen throughout the thoracic aorta *(arrows)*. Note the pronounced elevation of the aortic arch; this is also due to elongation. Calcification of the innominate artery is clearly demonstrated *(arrowhead)*.

Elongation and Kinking of the Aorta

Progressive aortic medial sclerosis and decreased elasticity often lead to elongation (lengthening, uncoiling) of the aorta in aging individuals. An uncoiled aorta extends outside the usual mediastinal borders and often can be identified and evaluated on lateral and oblique films.

Elongation of the ascending aorta produces a lateral bulging of its right border on frontal films. Its increased convexity is also appreciated on left anterior oblique and lateral films.

An elongated arch leads to a prominent and elevated knob on frontal view; the knob may reach the suprasternal notch. The elevation of the arch may also lead to buckling of the innominate artery, which produces a density in the right superior mediastinum.

The left paraspinal shadow of an elongated and tortuous descending aorta becomes broadened and convex. Oblique views may best demonstrate the tortuosity. The lower esophagus is intimately related to the descending aorta, and esophagogram will show esophageal displacement and curving that correspond to the curves of the descending aorta. When the patient is in erect posture, a delay of lower esophageal emptying may result from pressure caused by an atherosclerotic aorta.

A severely elongated or kinked aortic segment may simulate an aneurysm or a tumor. In the region of the ligamentum arteriosum, kinking of an elongated aorta can produce a configuration similar to that of congenital pseudocoarctation (see p. 621). The kinked areas can sometimes be recognized on oblique or lateral views; planigraphy may give clearer delineation. Aortography will conclusively resolve any uncertainties.

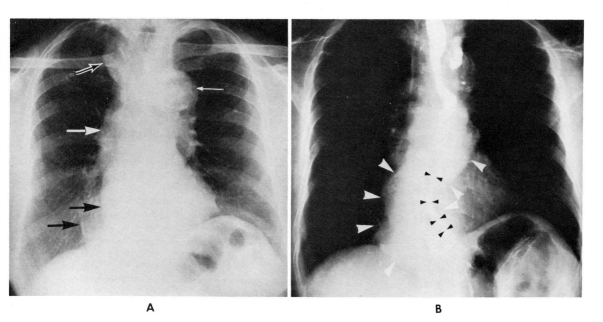

A   B

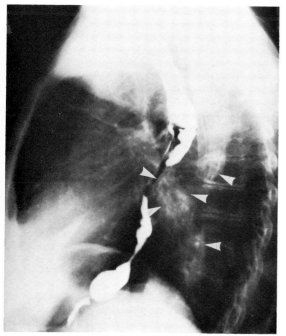

Figure 10–161 **Elongation and Tortuosity of Thoracic Aorta.**

A, Posteroanterior film of a 77 year old man reveals a prominent convex ascending aorta *(large white arrow),* a large, highly situated aortic arch *(small white arrow),* and a double shadow on the right side of the heart *(black arrows).* The lateral border of the double shadow is the aorta; the medial density is the heart border. The density in the right upper mediastinum *(black-white arrow)* is the buckled innominate artery.

B, Penetrated Bucky film with barium in the esophagus discloses the outline of the tortuous aorta *(white arrowheads),* which swings to the right of the heart before entering the diaphragm. Note the conformity of the lower esophagus *(black arrowheads)* to the aortic curves.

C, Lateral film shows the anterior curve of the mid descending aorta *(arrowheads)* pushing the esophagus forward. An elongated descending aorta often curves posteriorly; the retracted esophagus may then simulate displacement by an enlarged left ventricle or atrium.

C

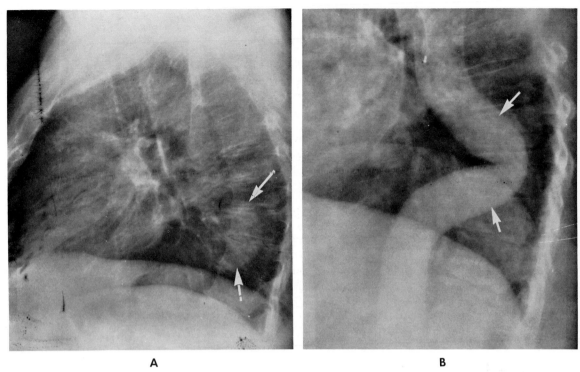

A   B

Figure 10–162   **Atherosclerotic Kinking of Aorta Simulating Aneurysm.**

A, On routine lateral view a density that simulates an aneurysm or a mass lesion is seen *(arrows)*.

B, With opacification of the aorta, lateral view demonstrates a sharp kinking of the lower descending aorta *(arrows)*, which is responsible for the local density.

Aneurysm of the Aorta

Aneurysmal masses can arise in any portion of the aorta but are most frequent in the abdominal aorta. Atherosclerosis is the most common cause; a syphilitic etiology, though now uncommon, most often affects the ascending aorta and arch. Dissecting and post-traumatic aneurysms are discussed elsewhere (pp. 732 and 734).

Radiographically the thoracic aneurysm produces a sharply marginated saccular or fusiform mass of homogeneous density. The mass cannot be separated from the aorta in any projection. Curvilinear calcification in the aneurysmal wall is a frequent and significant finding. Displacement of the adjacent trachea and esophagus, especially by aneurysms of the ascending aorta and aortic arch, is a characteristic finding. Pulsatile pressure frequently erodes adjacent bones; sternal erosions result from ascending aortic lesions, and erosions of the anterior vertebral borders result from aneurysms of the descending thoracic and abdominal aorta. The presence or absence of pulsations in the mass is not very helpful in diagnosis, since solid tumors adjacent to the aorta can often show transmitted pulsations. Aneurysmal enlargement is slow, and months or years may elapse before significant changes in size are seen radiographically. Rupture or leakage produces haziness of the aneurysmal border, and a contiguous density may appear in the lung fields. Later, evidence of pleural or pericardial effusion will develop.

The presence of curvilinear wall calcification and characteristic bone erosion, either alone or in combination, is an almost pathognomonic plain film finding. However, distinction from a mediastinal tumor is often difficult or impossible without contrast aortography.

Aneurysm of the abdominal aorta can sometimes be identified by calcification of its walls, which delineates the enlarged aortic dimensions. Lateral view more clearly demonstrates the calcification of both walls and may show vertebral erosion. The mass is often clinically palpable. Abdominal aortography will usually make the diagnosis, but ultrasound examination is the definite diagnostic study. If rupture occurs, there will be roentgen evidence of abdominal fluid, often a paralytic ileus bowel pattern, and obliteration of the shadow of one or both psoas muscles.[188, 189]

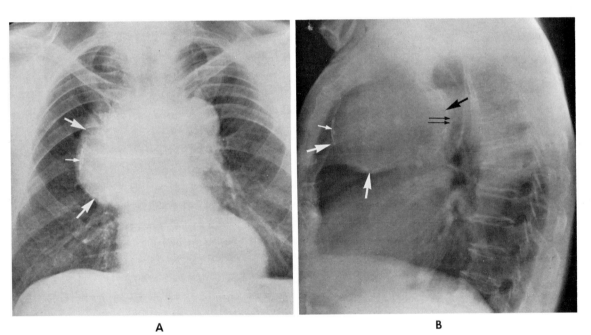

A   B

Figure 10–163   **Saccular Aneurysm of Ascending Aorta: Luetic Aneurysm.**

A, In posteroanterior view there is a saccular mass in the right mediastinum adjacent to and inseparable from the ascending aorta *(large arrows).* There is a curvilinear calcification in its lateral wall *(small arrow),* which strongly suggests an aneurysm.

B, On lateral view, the mass *(large black and white arrows)* is located anteriorly, near the sternum, and is displacing the trachea posteriorly *(double black arrows).* Calcification is seen in the anterior wall of the aneurysm *(small white arrow).* No erosion of the sternum is evident.

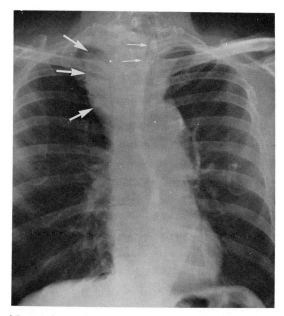

Figure 10–164 **Aneurysm of Innominate Artery.**   A circumscribed mass in the right upper mediastinum *(large arrows)* displaces the esophagus and trachea *(small arrows)* to the left. The appearance simulates a substernal thyroid, but the lesion was an innominate artery aneurysm.

   Opacification studies are often necessary to distinguish aneurysm of the innominate artery from other mediastinal masses. Atherosclerotic buckling of the innominate artery can produce a similar density but generally does not cause displacement of the trachea or esophagus.

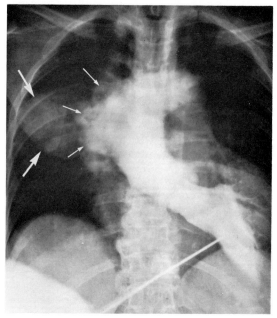

Figure 10–165 **Aneurysm of Ascending Aorta: Ventricular Opacification by Direct Puncture.**   Aneurysm *(large arrows)* of the ascending aorta is only partially opacified *(small arrows),* since the larger portion of the aneurysmal sac is filled with blood clot. This clot also causes the opacified borders to appear irregular.

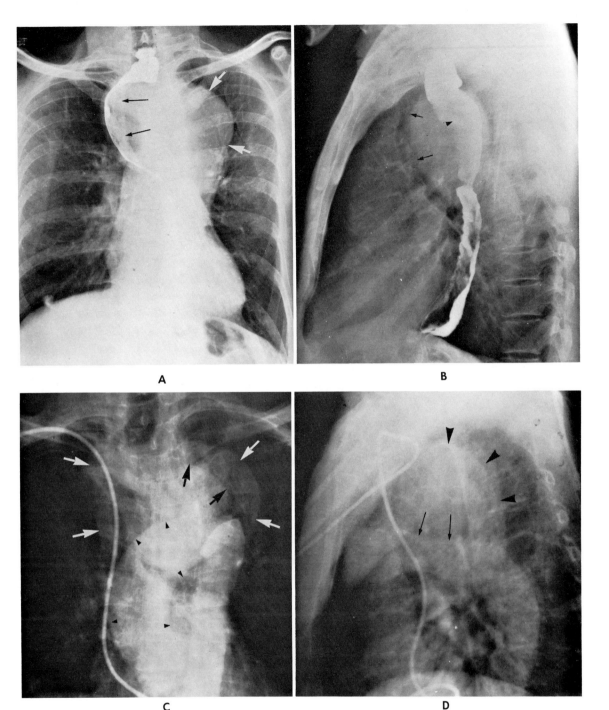

Figure 10–166 Aneurysm of Aortic Arch: Levophase of Angiocardiogram.

A, A huge upper mediastinal mass *(white arrows)* displaces the esophagus and the trachea *(black arrows)* to the right. The aortic arch is obscured. The mass was a large saccular aneurysm of the aortic arch.

B, In lateral view, taken after the patient had swallowed barium, there is anterior displacement of the trachea *(arrows),* and posterior displacement of the esophagus *(arrowhead)* by the aneurysmal mass. There is no vertebral erosion.

Characteristically, aortic arch aneurysms displace the trachea and the esophagus in the fashion illustrated; however, opacification study is required for confirmattion. Note incidental tertiary contractions of the esophagus.

C, Opacification of the aorta in levophase outlines the lumen *(black arrows)* of the aortic arch aneurysm. The opacified portion *(black arrows)* appears much smaller than the entire mass *(white arrows)* because of blood clots along the walls *(unopacified space between black and white arrows).* Outline of the displaced ascending aorta is indicated by small arrowheads.

D, Lateral view clearly shows marked depression *(arrows)* of the aortic arch due to the aneurysm *(arrowheads).*

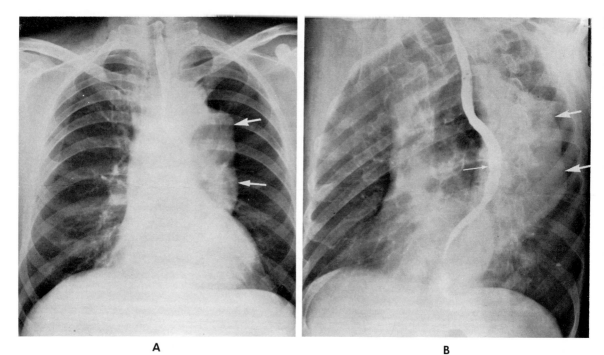

A   B

Figure 10–167   **Fusiform Aneurysm of Descending Aorta: Atherosclerosis.**

   A, The aneurysm is seen as a lobulated mass density *(arrows)* on the left side of the heart.
   B, On the anterior oblique view the fusiform aneurysm *(large arrows)* is continuous with the descending aorta. The barium-filled esophagus *(small arrow)* follows the tortuous course of the aorta.

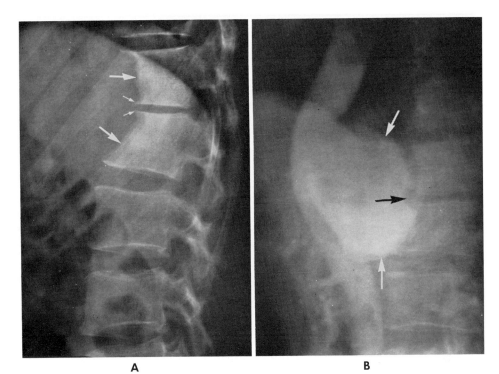

A   B

Figure 10–168 **Saccular Aneurysm of Upper Abdominal Aorta: Erosion of Vertebral Bodies and Aortogram.**

A, In enlarged lateral view of the lumbar spine there is characteristic erosion *(large arrows)* of the vertebral bodies due to pulsations of an aneurysm. Note that the area is smoothly eroded except for the protruding articular disc *(small arrows),* which typically is spared.

B, Aortography reveals the saccular aneurysm *(arrows)* arising from the posterior aspect of the aorta. Note relationship of the aneurysm to the vertebral erosion. Bulging of the intervertebral cartilage, which resists erosion, causes indentation of the aneurysmal sac *(black arrow).*

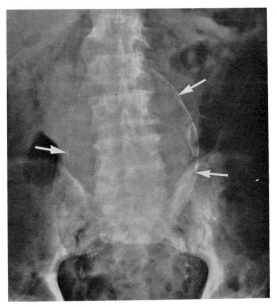

Figure 10–169 **Aneurysm of Abdominal Aorta: Extensive Calcification.** A huge soft tissue mass in the midabdomen is outlined by linear calcifications *(arrows).* An abdominal aneurysm usually bulges to the left side and is frequently identified by such calcification, since the mass itself is often difficult to visualize. The aneurysm is best demonstrated on a lateral abdominal film, which usually depicts the posterior location and the anterior bulging of the calcified aneurysmal wall.

Dissecting Aneurysm

Dissection arises near the root of the aorta in 70 per cent of cases, and 25 per cent arise just distal to the left subclavian artery. Hypertension is frequently associated. A greater than normal incidence is also found in Marfan's syndrome, in coarctation of the aorta, and in Cushing's disease. The significant roentgen finding is progressive widening of the aortic shadow. The increased width of the aortic wall may sometimes be surmised from the increased distance between the outer aortic wall and pre-existing intimal calcifications. The thickness of the normal aortic wall is not more than 3 to 4 mm.

With rupture, the sharp outline of the aortic wall may be obliterated. Later, a bloody pleural or pericardial effusion may occur.

Definitive diagnosis requires thoracic aortography, preferably retrograde. The most consistent aortographic finding is the narrowing of the true lumen by the false channel, frequently with irregularity on the affected side. The false channel produces an extraluminal soft tissue density. Variable degrees of opacification of the false channel may occur, and a thin radiolucent septum may be seen separating the two channels; this septum, if present, is diagnostic. Intercostal and other arterial branches that arise from the false channel will not be opacified. The site of intimal tear may sometimes be identified as an ulcer-like projection of contrast material from the true lumen. If the tear extends to the aortic valve, there may be acute aortic regurgitation, with reflux of contrast into the left ventricle.

The dissection will often extend into the abdominal aorta and may occlude one or both renal arteries.[189-193]

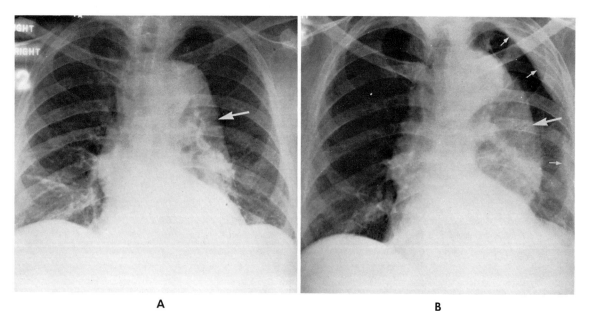

A   B

Figure 10–170 **Dissecting Aneurysm: Progressive Changes.**

A, Posteroanterior view demonstrates an enlarged, elongated, tortuous, atherosclerotic descending aorta *(arrow),* and moderate left ventricular enlargement.

B, Two weeks later the patient developed chest pain. There was a considerable increase in the width of the descending aorta *(large arrow);* this is strongly suggestive of dissecting aneurysm. A left pleural effusion *(small arrows),* probably containing blood, also developed. Hemothorax is a frequent occurrence in dissecting aneurysm.

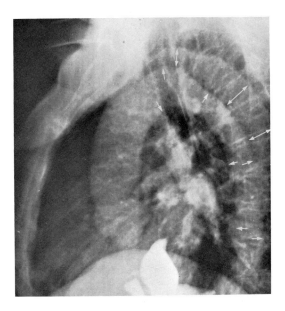

Figure 10–171 **Dissecting Aneurysm: Levophase of Venous Angiocardiogram.** Lateral projection of a venous angiocardiogram demonstrates the characteristic narrowing and irregularity of the true channel *(small arrows).* The lucent septum between the opacified open channel and the unopacified false channel *(double-headed arrows)* is pathognomonic. The density of this unopacified segment is the normal density of the aorta. There is residual barium in the stomach from a previous study.

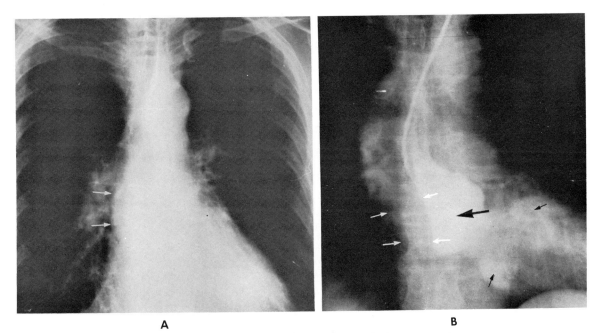

A B

Figure 10–172 **Acute Aortic Insufficiency Due to Dissecting Aortic Aneurysm: Aortogram.**

A, A 58 year old man suddenly developed chest pain and symptoms of aortic insufficiency. The heart is enlarged, and the left ventricle is prominent. The ascending aorta *(arrows)* is somewhat prominent.

B, A retrograde aortogram reveals opacification of the medial portion of the root of the aorta *(large black arrow);* the nonopacified lateral portion, having a distinct border *(between white arrows),* is characteristic of a dissecting aneurysm which in this case has extended to the aortic valves. The contrast medium has entered the left ventricle by reflux *(small black arrows),* indicating aortic insufficiency.

Occasionally the contrast material enters the dissection slowly, so that delayed opacification of the aneurysm occurs.

Aortic insufficiency that develops abruptly can result from rupture of an aortic leaflet due to bacterial endocarditis or from a dissecting aneurysm involving the aortic valve.

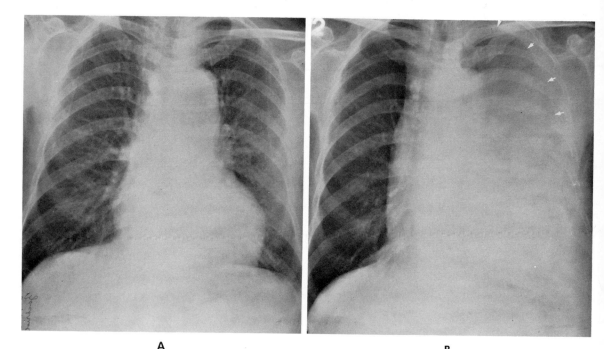

A   **B**

Figure 10–173 **Dissecting Aortic Aneurysm: Rupture.**

A, In early film the aorta is somewhat prominent, and there is some left ventricular enlargement. The patient was symptomatic.

B, Five days later there is evidence of fluid in the left chest (hemothorax) *(arrows),* due to dissection of and leakage from the aorta. The aortic outline has been obliterated by the blood, as is characteristic in ruptured aneurysm or leaking dissection.

Traumatic Aneurysm

Traumatic aneurysm of the aorta generally occurs after closed thoracic trauma, such as blast, compression, or rapid deceleration. The aortic tear occurs most commonly just distal to the origin of the left subclavian artery.

In the acute phase, there may be few or no localizing symptoms of the aortic hematoma, unless bleeding into the pleural cavity has occurred. There is generally increased width of the mediastinum, deviation of the trachea to the right, and loss of the aortic knob shadow.

Within several weeks, the widening becomes a mass or bulge in the region of the aortic knob. In unsuspected cases, this evidence of aneurysm may not be detected for several years after the accident. Slowly progressive enlargement in the region of the aortic knob may be noted on long-interval films. The wall of the aneurysm may eventually become calcified. Angiocardiography confirms the diagnosis.

Following severe chest trauma, interval films made over several months should be carefully inspected for mediastinal widening in the vicinity of the aortic knob.[194, 195]

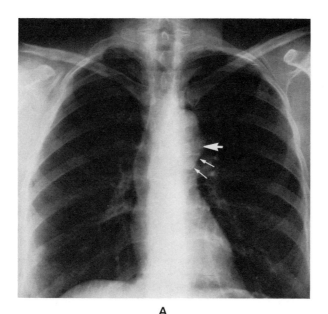

A

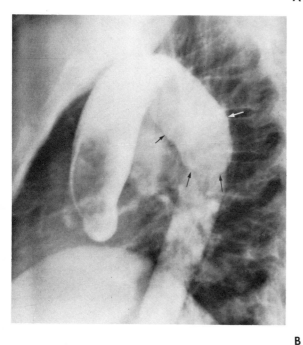

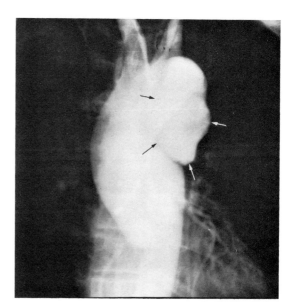

B

Figure 10–174 **Post-Traumatic Aortic Aneurysm.**

A, Chest film of a 44 year old woman discloses a bulging density *(large arrow)* just below the aortic knob. Linear calcification *(small arrows)* is seen in the lower portion of the density. Lateral chest film was unrevealing.

A history of severe crush injury to the chest in an automobile accident six years previously was elicited.

B, Posteroanterior and lateral contrast aortography confirms the presence of an aneurysm just below the aortic knob *(arrows).*

The location and appearance of the aneurysm and the history of trauma are typical. Almost all traumatic aortic aneurysms develop near the ligamentum arteriosum. Wall calcification usually develops months to years after the injury.

Mycotic Aneurysm

Mycotic aneurysms are usually a complication of bacterial endocarditis, septicemia, or neighboring abscess; approximately a third of them develop in the aorta. The infected wall of the involved artery is weakened, and aneurysmal dilatation subsequently occurs. Occasionally, mycotic aneurysm may erode into the adjacent vein, producing an arteriovenous fistula. Opacification studies are necessary for diagnosis.

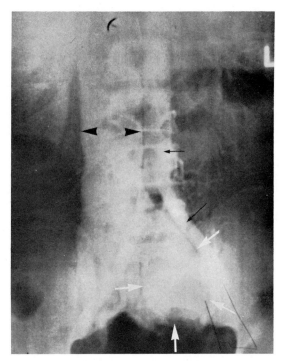

Figure 10–175 **Mycotic Aneurysm in Subacute Bacterial Endocarditis: Development of Arteriovenous Fistula: Arteriogram.** There is simultaneous opacification of the abdominal aorta, iliac vessels *(black arrows),* and the dilated inferior vena cava *(black arrowheads).* The large mycotic aneurysm *(white arrows)* arises from the left common iliac artery, and much blood is shunted into the vena cava through a fistulous connection. (Courtesy Dr. Howard Warner, Temple University School of Medicine, Philadelphia, Pennsylvania.)

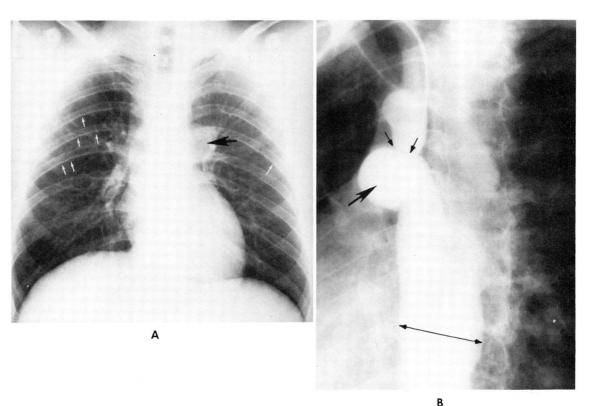

Figure 10-176 **Mycotic Aortic Aneurysm Following Subacute Bacterial Endocarditis.** This young male with coarctation of the aorta recovered from subacute bacterial endocarditis a year prior to this hospital admission.

A, Chest film discloses a left hilar mass *(black arrow)* that obliterated the shadow of the descending aorta. Rib notching *(white arrows)* was present but not marked.

B, Aortogram discloses an aneurysmal mass *(large arrow)* arising from the aorta immediately below the coarctated segment *(small arrows)*, which had been the seat of the endocarditis. Poststenotic dilatation from the coarctation is apparent *(double-headed arrow)*.

Aneurysms of the Sinuses of Valsalva (Aortic Sinuses)

Congenital aneurysms of the aortic sinuses are usually small and asymptomatic. The majority occur in the right aortic sinus and may be associated with Marfan's disease. Syphilitic aortitis is the principal cause of the acquired lesions, although endocarditis and atherosclerosis are occasionally responsible. The acquired aneurysms also occur most often in the right sinus but can become quite large.

Usually there are no suggestive plain film findings, since most lesions are small and intracardiac. There may be left ventricular enlargement from associated aortic insufficiency. The larger aneurysms, usually luetic, may produce a smooth large bulge of the right cardiac border. The rarer left sinus aneurysm bulges into the left supracardiac area. There may be calcification of the aneurysmal wall, especially in the larger lesions. Diagnosis can be made by opacification of the aneurysm during thoracic aortography.

Rupture of an aneurysm of the sinus of Valsalva occurs most often into the right atrium or the right ventricle and will cause an acute left to right intracardiac shunt. The pulmonary vasculature, the main pulmonary artery segment, and the heart size are increased (see Fig. 10–51). The roentgen appearance is in striking contrast to the relatively normal appearance prior to rupture. Retrograde aortography will demonstrate opacification of the right heart chambers and may also identify the aneurysm and a jet of contrast material entering a right heart chamber.[196-199]

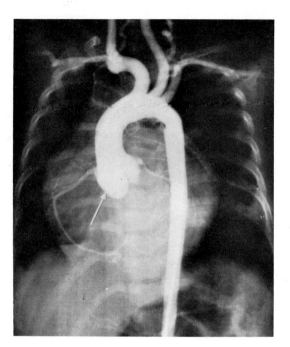

Figure 10–177 **Aneurysm of Sinus of Valsalva: Marfan's Disease.** In a child with Marfan's disease, contrast aortography demonstrates dilatation of the sinuses of Valsalva on the right side (*arrow*). There is dilatation of the entire ascending aorta, which is a common finding in Marfan's disease. In many cases all the aortic sinuses are dilated. (Courtesy Drs. A. C. Papaioannou, M. H. Agustsson, and B. M. Gasul, Athens, Greece.)

Figure 10–178 **Aneurysm of Sinus of Valsalva: Lues.** Aortography demonstrates an aneurysmal dilatation of the sinus of Valsalva (*large arrows*) in an adult male with syphilis. The other sinuses are normal (*small arrows*). In contrast to the picture in Marfan's syndrome, usually only one sinus is involved in syphilis or bacterial endocarditis.

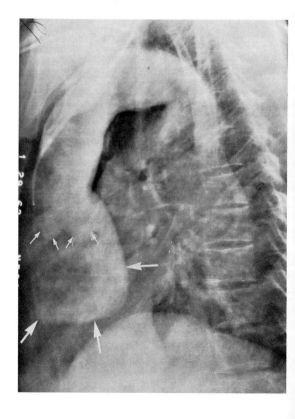

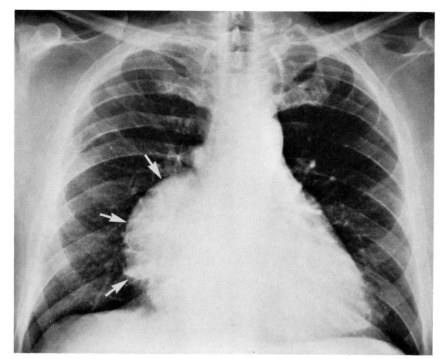

Figure 10–179 **Aneurysm of Sinus of Valsalva: Lues.** Chest film of a 60 year old man with positive serology discloses a large bulge *(arrows)* in the region of the right atrium. The "mass" could not be separated from the heart shadow and was anteriorly located. No calcification was detected.

A chest film made 10 years previously had disclosed a normally shaped but somewhat enlarged heart, which was attributed to his hypertension.

Contrast cardiac studies disclosed that the right-sided "mass" was a huge aneurysm arising from the right sinus of Valsalva.

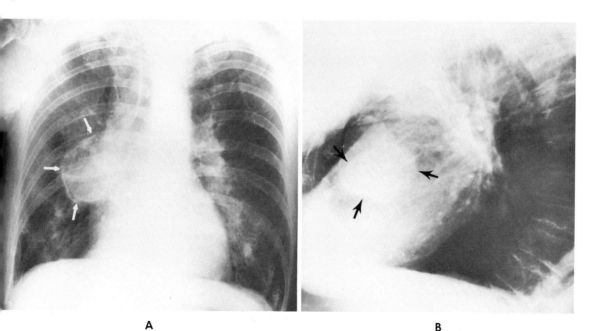

A   B

Figures 10–180 **Calcified Aneurysm of Sinus of Valsalva: Lues.**

A, The mass projecting from the right hilar area *(arrows)* has a calcified rim; its base in the heart shadow is uncalcified.

B, On lateral view, the mass *(arrows)* is situated in the region of the aortic root. This aneurysm was due to luetic involvement of the ascending aorta and its root.

Most aneurysms of the aortic root arise from the right sinus and become visible only after they have grown large enough to project outside of the heart shadow.

Syphilis of the Aorta

Luetic aortitis primarily affects the ascending aorta. It may produce dilatation and prominence of this segment. Sometimes the border of a markedly dilated ascending aorta extends more laterally than the right atrial border. Linear calcification of the ascending aortic wall is a frequent and significant finding, occurring in almost half the cases. Atherosclerosis can cause similar dilatation and calcification, but usually other portions of the aorta are similarly involved; luetic aortitis is limited to the ascending aorta.

Frequently, luetic aortitis involves the aortic valvular ring, and aortic insufficiency is a common complication. The resulting cardiac configuration is quite similar to that of rheumatic aortic insufficiency (see p. 680), but in the latter the ascending aorta is less dilated, and calcification of the aortic wall does not occur. Luetic aortic regurgitation can develop without the other findings of aortitis.

Coronary ostial narrowing, with or without symptoms, occurs in up to 35 per cent of patients with luetic aortitis. Coronary arteriography is necessary for definitive diagnosis.[200, 201]

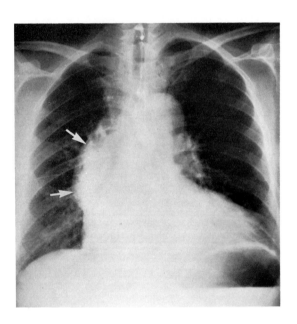

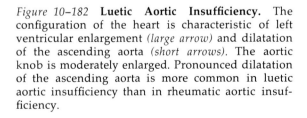

Figure 10–181 **Luetic Aortitis.** The ascending aorta is widened and extends lateral to the right atrium. Faint calcification can be seen *(arrows)* within its wall. Enlargement of the heart is caused by associated luetic aortic regurgitation.

Figure 10–182 **Luetic Aortic Insufficiency.** The configuration of the heart is characteristic of left ventricular enlargement *(large arrow)* and dilatation of the ascending aorta *(short arrows)*. The aortic knob is moderately enlarged. Pronounced dilatation of the ascending aorta is more common in luetic aortic insufficiency than in rheumatic aortic insufficiency.

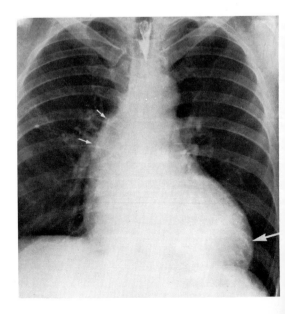

Thrombosis and Embolism of the Terminal Aorta

The thrombosis usually develops in an atherosclerotic aorta. There may be sudden symptoms due to rapid occlusion of the aorta, or more gradual ones, giving rise to the Leriche syndrome (ischemic symptoms in lower extremities and buttocks, and impotence if in males). The aorta is most often occluded between the renal arteries and its bifurcation.

On contrast aortography the aortic outline usually appears irregular owing to the presence of atherosclerotic plaques. Contrast aortography demonstrates the size and the extent of occlusion, particularly in relation to the renal arteries. This procedure also discloses collaterals and associated vascular lesions.

On aortography, embolism of the terminal aorta is indistinguishable from thrombosis. However, an abnormal configuration of the heart on chest film and the absence of collateral vessels suggest embolism in cases of sudden occlusion of the aorta.[202]

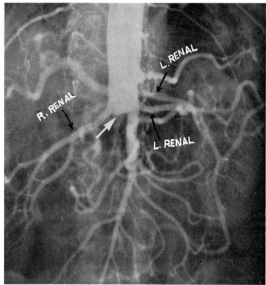

Figure 10–183 **Thrombosis of Terminal Aorta: Leriche Syndrome.** The aorta *(white arrow)* is entirely occluded just below the origin of the renal arteries *(black arrows)*. Obstruction at this location results from a large thrombus extending from the aortic bifurcation. The vessels above the occlusion are well filled and provide a collateral blood supply distally.

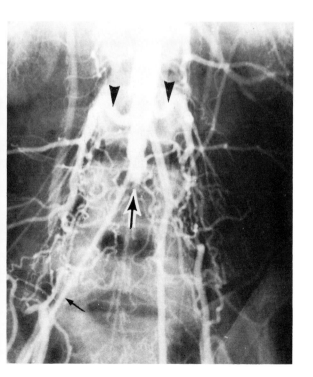

Figure 10–184 **Leriche Syndrome.** The aorta is completely occluded *(black-white arrow)* at its bifurcation, although the catheter *(black arrow)* in the right iliac artery did manage to enter the aorta. The circulation to the pelvis and lower limbs is partially maintained through collaterals that arise mainly from enlarged lumbar branches *(arrowheads)*.

Aortic Arch Arteritis (Takayasu's Disease)

Originally described as "pulseless" disease in young females, this disorder is now recognized as a systemic condition that can affect almost any portion or major branch of the aorta. The most common sites are the carotid and subclavian arteries, the aortic arch, the iliac arteries, and the abdominal aorta. However, segmental occlusions of the pulmonary arteries are not uncommon. The disease is more common in younger women but can occur in either sex up to the sixth decade. It is postulated to be an auto-immune condition.

During the symptomatic systemic phase, thoracic aortitis may lead to dilatation of the ascending aorta and to segmental or generalized dilatation of the descending aorta. These changes are most clearly recognized by serial films; contrast aortography best demonstrates the dilatation. Calcification of the aorta or of the involved branches may occur.

In the later occlusive phase (pulseless phase, if the subclavian is involved), aortography will demonstrate segmental narrowing or occlusion of one or more major arterial trunks of the thoracic aorta (or abdominal aorta). There may be poststenotic dilatation, and in older patients it may be difficult to exlude an atherosclerotic cause. The occlusion in Takayasu's arteritis is often tapered, a feature rarely seen in atherosclerotic occlusion.

Dilatation of the ascending aorta and occlusive disease of several major aortic branches in younger women with febrile and systemic symptoms should suggest Takayasu's arteritis. Angiography is necessary for diagnosis.[203-206]

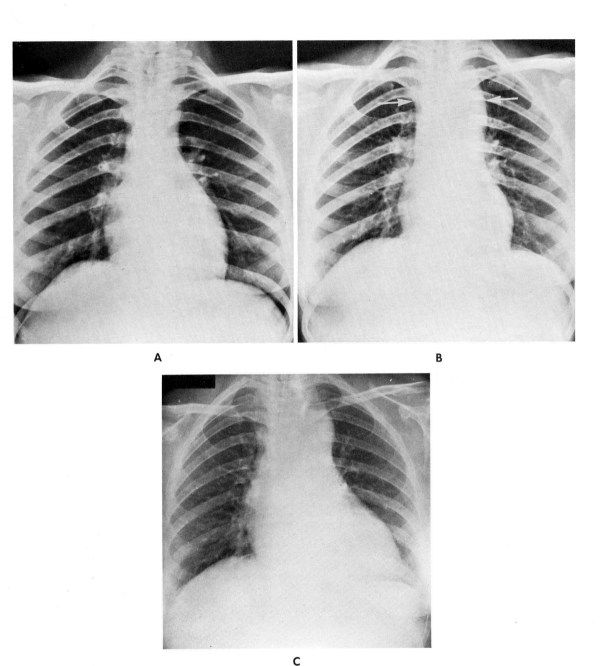

A

B

C

Figure 10–185 **Takayasu's Disease: Active Phase in Young Female.**

A, Chest film of 22 year old girl with symptoms of fever and cough was negative.

B, Chest film one year later shows broadening of the superior mediastinum *(arrows)* originally considered to be caused by enlargement of mediastinal nodes but later found to be due to dilatation of the innominate and carotid arteries.

C, Ten years later there is definite dilatation of the aortic arch and enlargement of the left ventricle. An aortogram at this time (not shown) revealed a dilated aorta, a dilated right innominate artery, and marked narrowing of the left axillary artery.

Progressive enlargement of the ascending aorta or aortic arch in a young female with undiagnosed symptoms is suggestive of Takayasu's disease. (Courtesy L. Gillanders and R. Strachan: Clin. Radiol., *16*:119, 1965.)

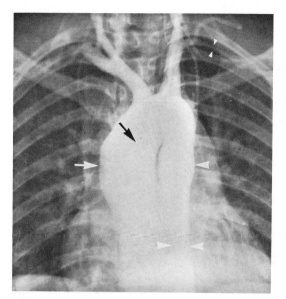

Figure 10–186 **Takayasu's Disease: Aortogram in 19 Year Old Female.**   Symptoms included fatigue, cough, and fever. The aortogram reveals a dilated ascending aorta *(black arrow)* that is irregular and flattened *(white arrow),* and fusiform dilatation and narrowing of the descending aorta *(large arrowheads).* The left subclavian artery is narrowed *(small arrowheads).*

  Although the thoracic aorta is the usual site, other arteries, especially the abdominal aorta and its branches, can be involved. The left superficial femoral artery was narrowed. (Courtesy L. Gillanders and R. Strachan: Clin. Radiol., *16*:119, 1965.)

Figure 10–187 **Takayasu's Disease: Abdominal Aortogram in Late Occlusive Phase.**   The pedal and arm pulses were absent in this 47 year old woman. Fifteen years previously she had suffered from pericarditis and a leg ulcer, and her sedimentation rate was elevated.

  The aortic lumen *(black arrows)* is irregular, and the superior mesenteric artery *(white arrows)* is irregularly narrowed throughout; the renal arteries are narrowed, but this is not clearly demonstrated. Both iliac vessels and the right carotid artery (not shown) were also involved.

  Although similar findings are seen in atherosclerosis, the combination described is consistent with Takayasu's disease. (Courtesy L. Gillanders and R. Strachan: Clin. Radiol., *16*:119, 1965.)

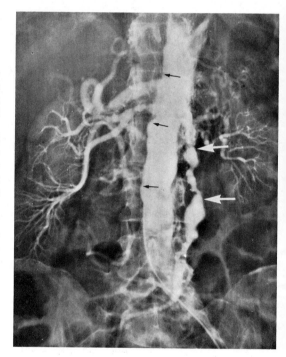

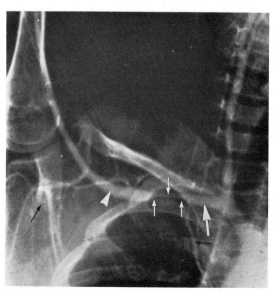

Figure 10–188 **Takayasu's Syndrome: Occlusive or Pulseless Phase.**   Contrast aortography reveals a smooth narrowed segment *(small white arrows)* of the right subclavian artery *(large arrow),* with numerous collaterals *(black arrows)* filling the distal subclavian artery *(arrowhead).* Note the slight poststenotic dilatation. The right radial pulse was barely palpable.

  In some cases the involved artery or arteries are completely occluded. (Courtesy Dr. D. Cooney, Chicago, Illinois.)

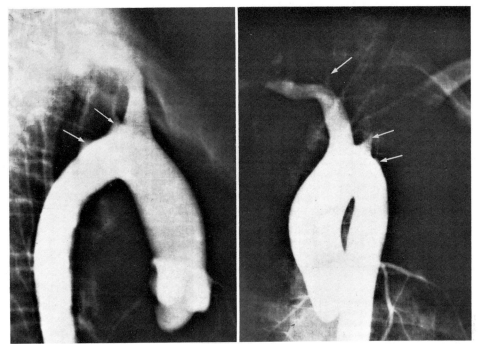

Figure 10–189 **Takayasu's Disease: Aortic Arch Syndrome.** Complete obstruction of the right and left common carotid arteries and the left subclavian artery is seen in anteroposterior and lateral aortogram. The innominate artery and the right subclavian artery are patent. These branches of the aortic arch are occluded close to their origin *(arrows)*. Note tapering of the occluded vessels. Aortitis of the ascending aorta, which commonly develops in the aortic arch syndrome, has caused dilatation of the aorta.

Rheumatoid Aortitis

Aortic valvular insufficiency resulting from a rheumatoid aortitis occasionally develops in adults with rheumatoid spondylitis. The roentgen findings are identical to those of rheumatic aortic insufficiency; the ascending aorta may be slightly enlarged or dilated, and the left ventricle is enlarged (see p. 680). These findings in a patient with rheumatoid spondylitis suggest the diagnosis.[2, 207]

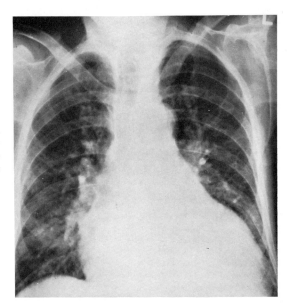

Figure 10–190 **Rheumatoid Aortitis.** In a male with longstanding rheumatoid spondylitis, there is cardiomegaly with left ventricular enlargement due to aortic insufficiency. Widening of the aorta is due to rheumatoid aortitis.

DISEASES OF THE PERIPHERAL VESSELS

Peripheral Arteriosclerosis

Atherosclerotic plaques give rise to irregular linear or curvilinear patches of calcification scattered throughout the peripheral arteries. In contrast, medial or Mönckeberg's sclerosis usually presents as more uniform calcification of an entire segment. It is seen chiefly in peripheral arteries that have a muscular coat, and it does not cause occlusive arterial disease, although it frequently coexists with intimal arteriosclerosis.

Calcification in the smaller vessels, particularly the posterior tibial and dorsalis pedis, is frequently seen in patients with early and accelerated atherosclerosis; most often this occurs in diabetics.

Accurate estimation of the extent of atherosclerosis can be made by contrast arteriography, which may demonstrate diffuse narrowing, irregular lumen, filling defects, or patchy variations of density in the opaque column, the last caused by plaques impinging on the contrast column. There may be complete obstruction. An estimation of the degree of collateral circulation and a preoperative evaluation of the circulation distal to the obstruction (runoff) are best made via arteriography.[208–211]

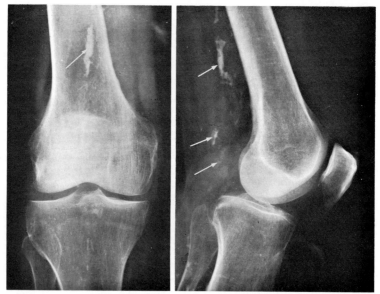

Figure 10–191 **Atherosclerotic Calcification.** Scattered patches of calcification are seen along the course of the popliteal artery *(arrows)*. These are calcified intimal plaques and may indicate ischemic vascular disease.

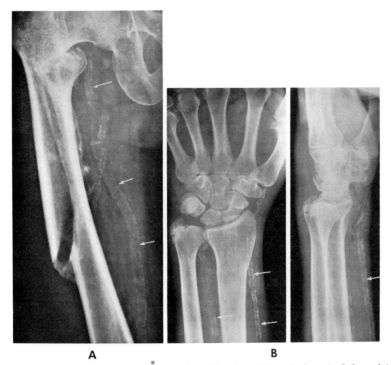

A   B

Figure 10–192 **Medical Sclerosis in Two Patients: Calcification (Mönckeberg's Sclerosis).**

A, Uniform calcification of the wall of the arteries of the thigh is demonstrated *(arrows)* in a patient with a fractured femur.

B, There is extensive calcification caused by Mönckeberg's sclerosis of the radial artery and ulnar artery just above the wrist *(arrows)* in second patient.

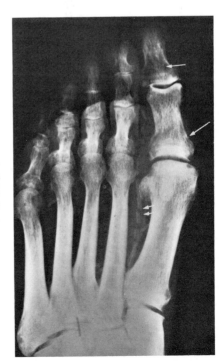

Figure 10–193 **Arteriosclerosis Associated with Diabetes: Gangrene of Foot.** There is extensive calcification of the interosseous arteries *(double arrows)*, associated with vascular insufficiency and gangrene. Spotty demineralization of the toes *(single arrows)* is due to disuse and ischemia and is frequently seen in diabetic patients.

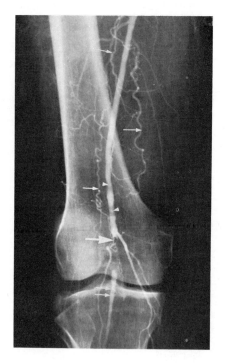

Figure 10–194 **Arteriosclerotic Occlusion of Popliteal Artery.**
Femoral arteriography demonstrates complete occlusion of the
popliteal artery *(large arrow)*. In the arterial segment above the
occlusion there are filling defects along the vessel wall, variations in
diameter, and variations in density of the contrast medium *(arrowheads)*. Note the extensive smaller tortuous collateral vessels *(cork
screw appearance)* *(small arrows)* that supply the popliteal artery
distal to the obstructed area (runoff) *(double arrows)*.

Thromboangiitis Obliterans (Buerger's Disease)

This disease is characterized by a segmental arteritis and panphlebitis of medium and small vessels of the extremities and, occasionally, the viscera. It is seen
predominantly in young nondiabetic males. Thrombotic occlusion is a frequent occurrence. Osteoporosis of the bones of the foot and the lower leg may occur. If
tissue gangrene develops, there may be evidence of superimposed osteomyelitis.
Peripheral arteriography is confirmatory and helps to differentiate Buerger's disease from other vascular disorders. The small or medium-sized arteries of the lower
extremities are almost always affected, and the vessels of the upper extremities are
also involved in a large percentage of cases.

The angiographic findings are highly suggestive. There is usually no evidence
of atherosclerotic plaques, and involvement is segmental. The arteries appear normal up to the point of narrowing or occlusion. Multiple fine collaterals are often
seen near the area of obstruction, and these vessels tend to run parallel to the
main occluded artery. The runoff may be poor or absent. Sometimes the diseased
portions of a larger artery display a corrugated appearance.[210, 212–214]

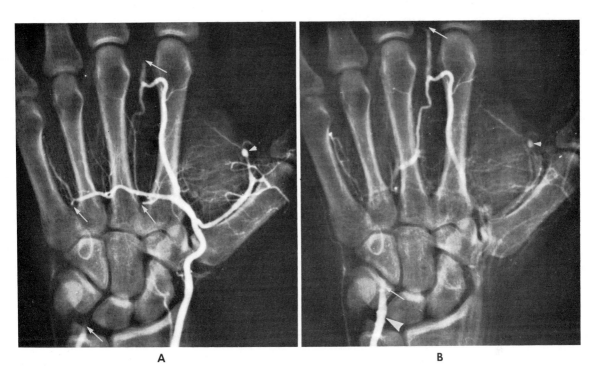

A   B

Figure 10–195   **Thromboangiitis Obliterans (Buerger's Disease): Arteriogram.**

A, There are multiple arterial occlusions involving the ulnar artery at the wrist and the branches of the radial artery *(arrows).* Some of the involved arteries taper gradually. One of the interosseous arteries shows an aneurysmal dilatation *(arrowhead).* Atheromatous plaques are absent. Note the fine collaterals in the interosseous spaces.

B, A later phase of the arteriogram demonstrates somewhat greater penetration by the contrast medium of the occluded channels *(arrows)* and prolonged stasis of contrast medium. The aneurysmal dilatation is still opacified *(small arrowhead).* The corrugated appearance *(large arrowhead)* of the late-filled segment of the ulnar artery is a recognized but infrequent pattern in this disease.

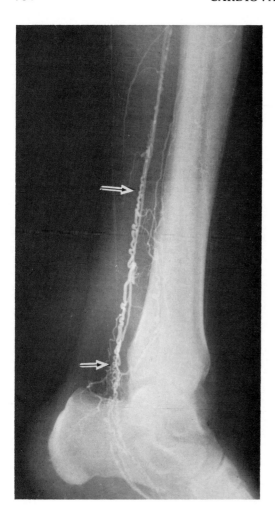

Figure 10–196 **Buerger's Disease: Femoral Arteriogram.** The main arteries óf thé leg aie uccluded and not opacified. Long corkscrew collaterals *(arrows)* are coursing parallel to the occluded arteries. There is no evidence of atherosclerotic change or vascular calcification. (Courtesy Dr. Paul Roy, Montreal, Canada.)

Arterial Embolism

Emboli in the peripheral arteries usually arise from intracardiac thrombi, which can occur in endocarditis, auricular fibrillation, or myocardial infarction. Atherosclerotic plaques may break off and form emboli.

Definitive localization requires contrast arteriography. In the occluded artery the contrast medium terminates abruptly, and the convex upper border of the embolus may project into the lumen. Collaterals are relatively few and narrow in the early postembolic period. This is due to vasospasm of the proximal branches. An incompletely obstructing embolus may present as an ovoid defect in the contrast column.[208, 210]

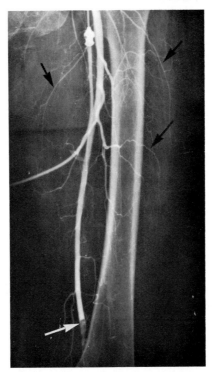

Figure 10–197 **Embolus in Superficial Femoral Artery: Arteriogram.** The blood supply to the superficial femoral artery *(white arrow)* is abruptly cut off by an embolus. Slight seepage of contrast medium around the embolus partially outlines the lumen in the obstructed area, and demonstrates the typical appearance of the acute embolus, with the concavity directed upward. In chronic thrombosis the concavity is more likely to be directed distally.

Few collateral arteries are filled, and the branch vessels opacified proximal to the embolus are small and spastic *(black arrows)*.

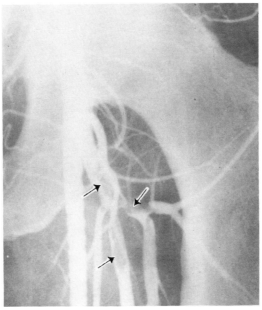

Figure 10–198 **Emboli in Deep Femoral Artery: Arteriogram.** The emboli are visualized as intraluminal defects *(arrows)*, with contrast material coursing around them. The obstruction is incomplete.

Raynaud's Phenomenon

Raynaud's phenomenon occurs almost exclusively in females. Although usually a solitary condition of unknown origin (Raynaud's disease), it is a frequent finding in scleroderma and an occasional complication of rheumatoid arthritis. It can occur after trauma.

In prolonged cases, roentgen changes can be seen in the terminal phalanges. Trophic osteolysis and absorption of the ungual tufts give a tapered, shortened appearance to the terminal phalanges. Soft tissue atrophy occurs in the finger tips. In scleroderma there may be discrete calcium deposits in the finger tips.

The clinical and roentgen findings are sufficiently characteristic for diagnosis. Arteriography may be normal or may show occlusion of a digital artery.[214-216]

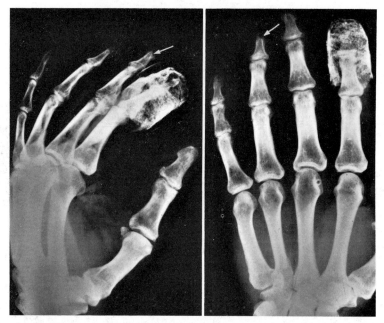

Figure 10–199 **Raynaud's Phenomenon.** The tufts of the terminal phalanges of the middle and fourth digits have been resorbed *(arrows),* as often happens in severe cases. (The bandaged finger is infected.) Soft tissue atrophy is present but is not apparent in this film. The Raynaud phenomenon is found in a variety of diseases in which vasoconstriction is inherent, especially scleroderma.

Ergotism

This condition, once common and due to ingestion of contaminated rye, is now sporadic and usually due to overdose of or sensitivity to the ergotamine drugs. In addition to gastrointestinal and central nervous system symptoms, ischemic symptoms in the extremities due to vasoconstriction are usually present. Gangrene of the digits may occur. The lower extremities are more frequently affected.

Angiography will reveal bilateral and symmetric arterial spasm, often with thread-like narrowing and tapering and sometimes with total occlusion. The femoral and popliteal arteries are most often affected. In more chronic cases, extensive collateral vessels are seen. Rarely, the carotid, coronary, and renal arteries may also be involved. The appearance of the spastic renal arteries may then be indistinguishable from fibromuscular hyperplasia. The peripheral and visceral arterial changes are usually entirely reversible with discontinuance of the drug. Occasionally thrombi develop, presenting as intraluminal filling defects.[217-219]

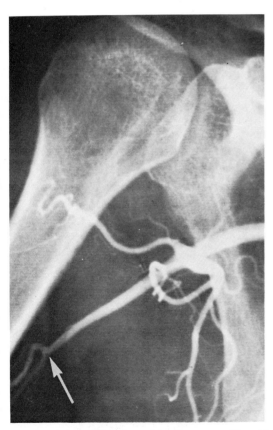 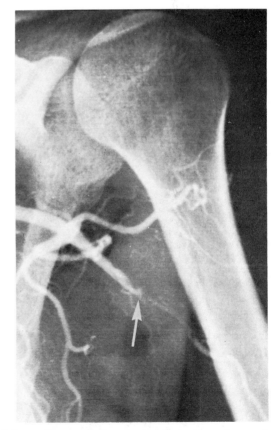

Figure 10-200 **Ergotism: Arteriogram.** Opacification of the axillary arteries demonstrates tapering and complete obstruction *(arrows)*, with moderate filling of collaterals. These changes developed following prolonged use of an ergot derivative for migraine headaches. This bilateral symmetric involvement is characteristic of ergotism. (Courtesy Drs. R. A. Kramer, S. P. Hecker, and B. I. Lewis, Palo Alto, California.)

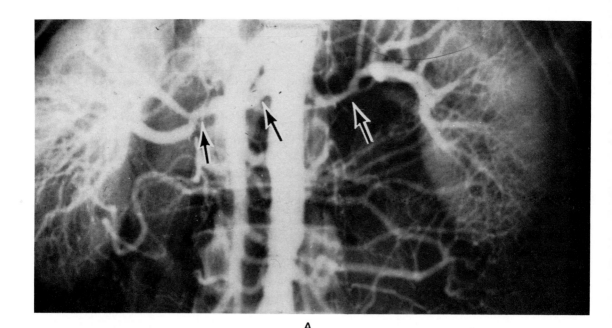

A

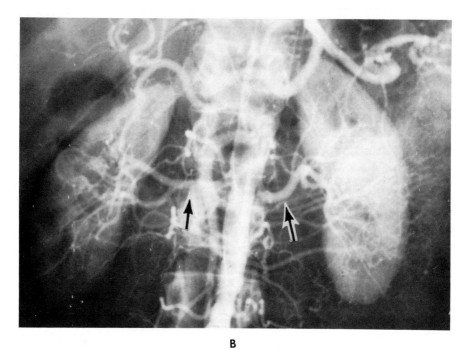

B

Figure 10–201 **Ergotism: Renal Artery Spasm: Aortogram.**

   A, Both renal arteries are irregularly narrowed *(arrows),* with an appearance suggestive of fibromuscular hyperplasia.

   B, Angiographic study 10 days after the ergotamine was discontinued shows completely normal renal arteries *(arrows).* (Courtesy Dr. M. S. Fedotin, Durham, North Carolina.)

Arteriovenous Fistula

Abnormal arteriovenous communications in the systemic circulation are either due to congenital malformations or to acquired fistula, usually post-traumatic.

There are usually no significant findings on plain films, although sometimes there will be overgrowth of bone or soft tissues if the local lesion has been present prior to epiphyseal closure. Sometimes phleboliths may be seen within the lesion. If the arterial shunting is extensive, there may be ischemia distal to the shunt, and the shunted venous return can cause a high output heart (enlarged heart, increased arterial vasculature).

Arteriography of a malformation will usually indicate an enlarged feeding artery with opacification of the many abnormal vessels and rapid filling of numerous irregular veins during the arterial phase. In the post-traumatic fistula, there is generally direct filling of an adjacent vein from an arterial branch. The arterial tree distal to the fistulous connection is often very poorly opacified, since most of the contrast material passes directly into the vein.

Venous malformations, such as phlebectasia or venous angiomas, are not demonstrated by arterial filling, but require phlebography for opacification.[220, 221]

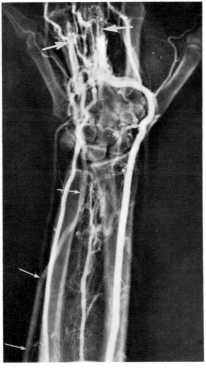

Figure 10–202 **Congenital Arteriovenous Fistula of Hand: Arteriogram.** The site of communication is indicated by the numerous vessels in the region of the third and fourth metacarpals *(large arrows)*. There is simultaneous filling of the venous channels *(small arrows)*. The feeding arteries are somewhat enlarged. This shunting caused ischemia of the distal phalanges.

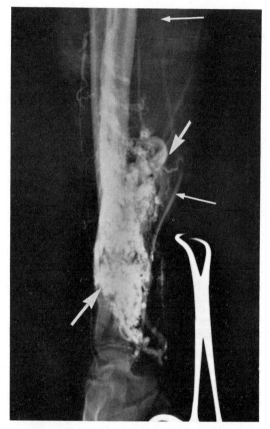

Figure 10–203 **Congenital Arteriovenous Fistula of Forearm: Arteriogram.** Numerous abnormal vessels are clearly demonstrated *(large arrows)*. Absence of vascular filling distal to the fistula led to clinical symptoms of ischemia of the hand. Early venous filling *(small arrows)* is characteristic of an arteriovenous fistula.

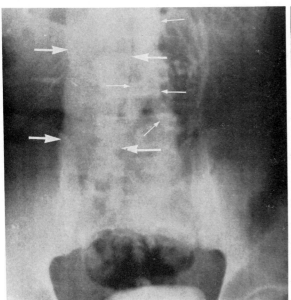

A

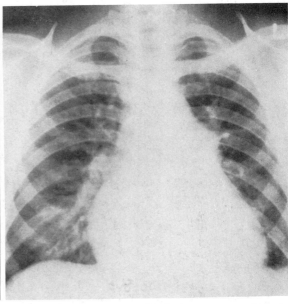

B

Figure 10-204 **Traumatic Arteriovenous Fistula with High Output Heart.**

A, Abnormal communication between the right common iliac artery and vena cava is due to a traumatic arteriovenous fistula. The arteriogram demonstrates simultaneous filling of the aorta and its iliac branches *(small arrows)* and the vena cava *(large arrows).* Because of the rapid shunting, opacification is less dense, and there is no opacification below the iliac arteries.

B, Enlarged high output heart is attributable to the extensive peripheral shunt. The pulmonary vasculature is prominent. The main pulmonary artery is dilated, and there is marked dilatation of the heart.

C, Following surgical correction of the fistula, the heart and vasculature are of normal proportions. (Courtesy Dr. Howard Warner, Temple University School of Medicine, Philadelphia, Pennsylvania.)

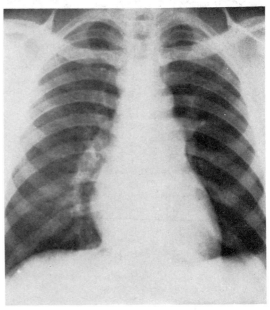

C

Glomangioma or Glomus Tumor

These small arteriovenous tumors are seen in both the upper and lower extremities. They are very painful when under a fingernail or toenail.

Films of the digits may show an indentation or small osseous defect of the dorsal aspect of the distal phalanx due to pressure from the subungual tumor. However, most of these tumors are quite small and produce no osseous erosion.[222]

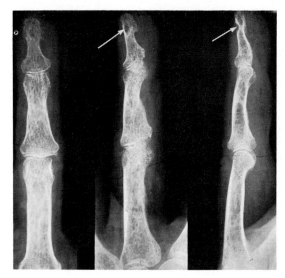

Figure 10–205 **Glomus Tumor.** Small bony defect *(arrows)* on the dorsal aspect of the distal phalanx is the result of pressure exerted by the glomus tumor beneath the nail. The bony cortex is intact at the site of the lesion. (Courtesy Dr. W. R. Harris, Toronto, Canada.)

Varicose Veins

In varicosities with venous stasis, small thrombi may develop, some of which calcify (phleboliths). These appear as rounded calcifications, sometimes with lucent centers. In severe cases of stasis there may be a periosteal reaction in the adjacent long bones. Rarely, calcifications are deposited in the chronically congested soft tissues.

Venography demonstrates the varicosities and indicates the state of the communicating and deep veins. The varicosities appear as tortuous dilated venous segments. When a tourniquet is applied to the superficial veins before injection, the deep veins will be opacified if they are patent. The communicating veins, if incompetent, will also be opacified. In patients with varicosities, venography is performed primarily to prove the patency of the deep veins before surgical removal of the varicosities.[223–225]

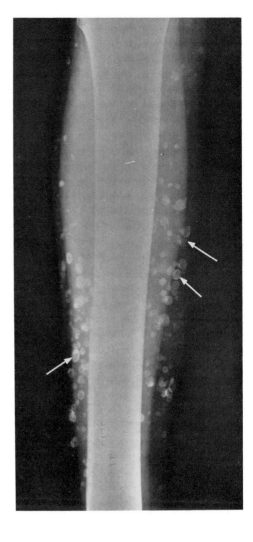

Figure 10–206 **Varicose Veins: Phleboliths.** Numerous areas of round and oval calcification have developed in the soft tissues of a patient with longstanding varicose veins. These are calcified thrombi in the varicosities. Note the lucent centers *(arrows)* of some of the phleboliths.

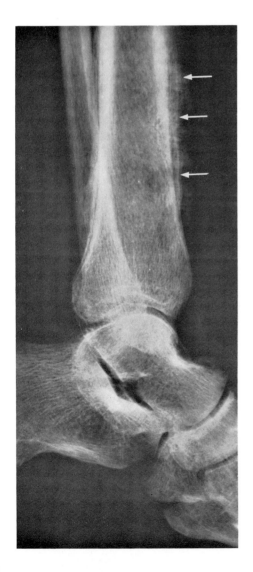

Figure 10–207 **Varicose Veins: Stasis Periostitis.** There is extensive periosteal bone formation on the anterior aspect of the tibial shaft *(arrows)* caused by longstanding vascular stasis from varicose veins.

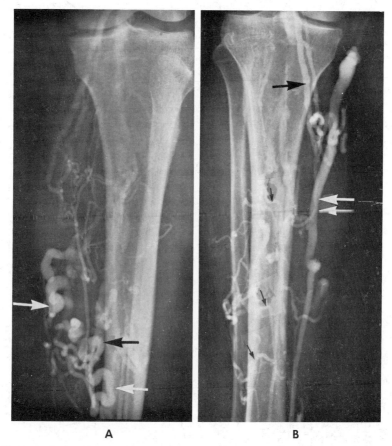

Figure 10–208   **Varicose Veins: Venogram.**

A, The tortuous dilated veins are clearly demonstrated *(arrows).* The deep veins are also filled.

B, The venogram demonstrates only a few varicosities, but the communicating veins are filled *(small arrows),* and there is abnormal filling of the saphenous vein *(white arrows)* from the deep system through the incompetent communicating veins.

A competent communicating vein *(large black arrow)* returns saphenous blood to the popliteal system.

Thrombophlebitis

The origin of thrombophlebitis or thrombosis of the veins of the extremities is frequently uncertain. Among the more specific causes or associations are trauma, stasis, infection, prolonged birth control medication, abdominal surgery, strenuous efforts (effort thrombophlebitis of the axillary vein), the Hughes-Stovin syndrome (see p. 548), Buerger's disease, and pancreatic or bronchogenic carcinoma. Indeed, a thrombophlebitis may be the earliest symptom of an unsuspected lung or pancreatic malignancy. Conversely, pulmonary embolization is often the first manifestation of asymptomatic peripheral venous thrombosis.

Meticulous peripheral venography can disclose the location, extent, and number of thrombi. Defects within the lumen, nonfilling of deep veins, and collateral venous circulation will be seen.

When life-threatening recurrent pulmonary emboli occur in a patient with venous thromboses of the extremities, instead of ligation or plication of the vena cava, an umbrella filter can be installed, using fluoroscopic guidance, into the vena cava below the renal veins.

Phleboliths, which are small calcified thrombi, are commonly observed in the pelvic soft tissues. These round or oval calcifications with lucent centers are seen with increasing frequency in older individuals and are asymptomatic. They may be mistaken for calculi in a lower ureter. Phleboliths also occur in soft tissue hemangiomas and in longstanding venous stasis, such as varicose veins.[224, 226-229]

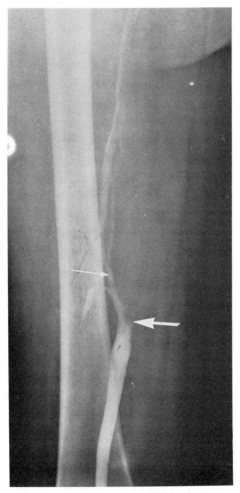

Figure 10–209 **Thrombotic Occlusion of Femoral Vein.** The femoral vein in the midthigh is occluded by thrombosis *(large arrow);* a collateral vein transports the contrast medium cephalad *(small arrow).*

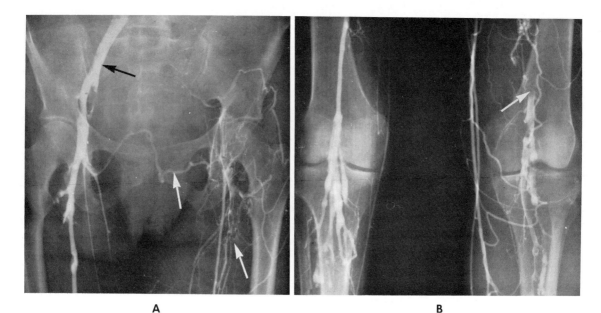

A   B

Figure 10–210 **Thrombosis of Femoral Vein: Venogram.**

A, The femoral and iliac veins are normally filled on the right side *(black arrow),* but on the left there is no opacification of these veins. The network of collaterals *(white arrows)* from left to right is diagnostic of venous obstruction on the left.

B, The popliteal and femoral veins are normal on the right side; while on the left, the femoral vein is obstructed by thrombosis *(arrow),* and there is collateral runoff to the upper thigh.

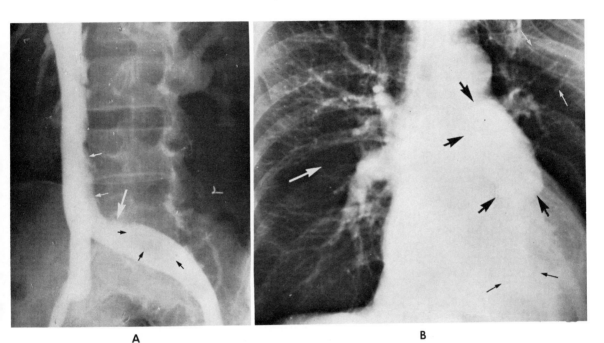

A B

Figure 10-211 **Thrombosis of Iliac Vein and Vena Cava: Multiple Pulmonary Infarcts: Vena Cavagram and Pulmonary Arteriogram.**

A, A vena cavagram was made after injection of opaque material simultaneously into both femoral veins. The patient had had several episodes of pulmonary embolism.

There is an ill-defined filling defect in the left iliac vein *(black arrows)* and an irregular defect on the lateral wall of the vein *(large white arrow)*. The defect extends upward into the left side of the inferior vena cava *(small white arrows)*. The cause was a large thrombus that had originated in the iliac vein and extended into the vena cava.

B, Pulmonary arteriogram indicates absence of arterial filling of the left lower lobe *(small black arrows)*, paucity of vessels in the left upper lobe *(small white arrows)*, and an avascular area in the superior segment of the right lower lobe *(large white arrow)*. These findings are characteristic of multiple pulmonary emboli.

The greatly enlarged main pulmonary artery *(large black arrows)* reflects increased pulmonary arterial pressure.

Opacification studies of the deep iliac veins and the inferior vena cava often disclose the site of a thrombus that is responsible for the development of pulmonary emboli.

Lymphedema

Primary lymphedema occurs in several rare diseases primarily affecting the peripheral lymphatics (congenital lymphedema, lymphedema praecox, Milroy's disease). Secondary lymphedema can result from obstruction of lymphatic nodes or channels in such conditions as filariasis or neoplastic diseases, or from tissue fluid production in excess of lymphatic drainage capabilities, as occurs in venous obstruction. Not infrequently, the condition cannot be classified as primary or secondary, since infection may be superimposed on an area of focal lymphatic deficiency.

In the primary diseases, lymphangiography may show a marked paucity or virtual absence of lymphatic channels.

In secondary occlusive lymphatic conditions there is an early increase in the number of opacified channels; later, these become dilated and tortuous, with nonfilling of the more cephalad lymphatics. In the severer cases, complete obstruction leads to opacification of the numerous smaller superficial transverse channels below the obstruction, producing the characteristic picture of dermal backflow. In cases where excess production of tissue fluid occurs, the overworked lymphatic channels become dilated and increase in number.[230-233]

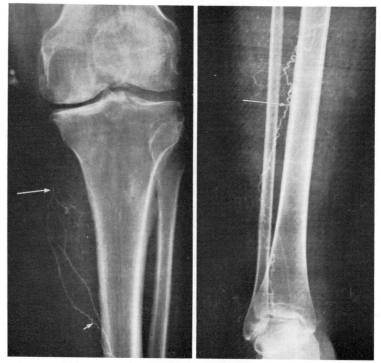

Figure 10–212 **Secondary Lymphedema.** Lymphangiogram of the lower extremities shows markedly tortuous lymphatics *(arrows);* there is no filling beyond the upper tibia.

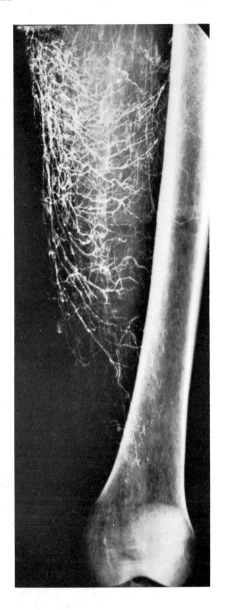

Figure 10-213 **Obstructive Lymphedema with Dermal Back-flow.** Characteristic transverse lymphatics are shown on lymphangiography in the area of the thigh. These represent superficial lymphatics in the skin filled because of occlusion of the normal channels (dermal backflow). (Courtesy Dr. Burton Schaeffer, Philadelphia, Pennsylvania.)

Vena Cava Obstruction

Vena cava obstruction may present clinically as dilatation or varicosities of the peripheral veins. It is generally due to extrinsic causes, although a primary thrombosis may also occur.

Superior vena cava obstruction can be caused by mediastinal neoplasms, especially bronchogenic carcinoma, or by inflammatory or idiopathic mediastinitis (see Fig. 9-118 and p. 573). Cavography, via a vein of the upper extremity, will demonstrate the site of the superior vena cava occlusion and the collateral circulation through the azygos and hemiazygos system veins. Varices of the upper esophagus frequently develop and can be demonstrated on a contrast esophagogram.

Inferior vena cava obstruction occurs principally in the middle third of the vein; a retroperitoneal tumor, usually renal malignancy, is the most frequent cause, although idiopathic thrombosis can also occur. Inferior vena cavography will demonstrate the site of obstruction and may also show tumor masses indenting the cava or even growing into its lumen.

In acute inferior vena cava thrombotic obstruction, the distended collateral veins may cause multiple pressure deformities and medial displacement of the lower ureters and bladder. The latter may become pear-shaped, similar to its appearance in pelvic lipomatosis (see p. 1239).[234-236]

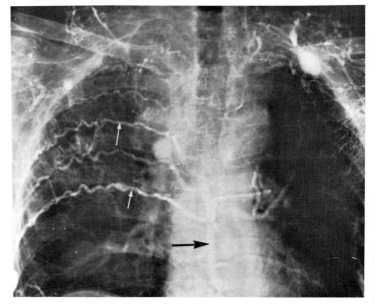

A

Figure 10–214 **Occlusion of Superior Vena Cava: Bronchogenic Carcinoma.**

A, Venogram of both upper extremities shows blockage of both subclavian veins *(large arrows)* due to occlusion of the superior vena cava and the subclavian veins by a large mediastinal extension of a bronchogenic carcinoma *(small arrows).* Note the extensive collaterals around the neck and shoulders.

B, A later film shows retrograde filling of the tortuous intercostal veins *(white arrows)* from the vessels of the chest wall. These intercostals drain into the azygos vein *(black arrow).*

B

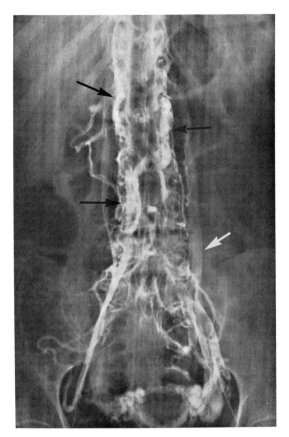

Figure 10-215 **Thrombosis of Inferior Vena Cava: Venogram.** Injection of opaque medium into both femoral veins fails to produce visualization of the inferior vena cava, which is obstructed. Numerous paravertebral collateral veins *(black arrows)* transport blood cephalad. The large spermatic veins *(white arrow)* are an essential part of the collateral circulation.

REFERENCES

1. Gould, S. E.: *Pathology of the Heart.* Springfield, Charles C Thomas, 1960.
2. Friedberg, C. K.: *Diseases of the Heart,* 3rd ed. Philadelphia, W. B. Saunders Company, 1966.
3. Schinz, H. R., Glauner, R., and Uehlinger, E.: *Roentgen Diagnostics,* Progress Volume I, New York, Grune & Stratton, Inc., 1958.
4. Longue, R. B., Roger, I. V., Jr., and Gay, B. B., Jr.: Subtle roentgenographic signs of left heart failure. Am. Heart J., *65*:464, 1963.
5. Gould, D. M., and Torrance, D. J.: Pulmonary edema. Am. J. Roentgenol. Radium Ther. Nucl. Med., *73*:366, 1955.
6. Ormond, R. S., and Eyler, W. R.: Roentgen diagnosis of cardiac competence. Radiology, *79*:378, 1962.
7. Heitzman, R., and Ziter, F. M., Jr.: Acute interstitial pulmonary edema. Am. J. Roentgenol. Radium Ther. Nucl. Med., *98*:291, 1961.
8. Meszaros, W. T.: Lung changes in left heart failure. Circulation, *47*:859, 1973.
9. Fleisner, F. G.: Unilateral pulmonary embolism. Radiology, *73*:591, 1959.
10. Melnick, G. S.: Angiographic demonstration of acute pulmonary emboli. Angiology, *14*:491, 1963.
11. Teplick, J. G., Haskin, M. E., and Steinberg, S. B.: Changes in the main pulmonary artery segment following pulmonary embolism. Am. J. Roentgenol. Radium Ther. Nucl. Med., *92*:557, 1964.
12. Stein, P. D., et al.: Angiographic diagnosis of acute pulmonary embolism. Am. Heart J., *73*:730, 1967.
13. Heitzman, E. R., Markarian, B., and Dailey, E. T.: Pulmonary thromboembolic disease. Radiology, *103*:529, 1972.
14. Kerr, I. H., and Sutton, F. G.: The value of the plain radiograph in acute massive pulmonary embolism. Br. J. Radiol., *44*:751, 1971.
15. Jaffe, R. B., and Koschmann, E. B.: Septic pulmonary emboli. Radiology, *96*:527, 1970.
16. Simon, M.: The pulmonary vessels: their hemodynamic evaluation using routine radiographs. Radiol. Clin. North Am., *1*:363, 1963.
17. Evans, W.: The less common forms of pulmonary hypertension. Br. Heart J., *21*:197, 1959.
18. Schuster, B., Imm, C. W., Yavuz, F., and Johnson, G. F.: Pulmonary arteriolar changes in congeni-

tal heart disease, as demonstrated by pre-wedge pulmonary arteriography. Angiology, 15:239, 1964.

19. Goodwin, J. F.: The nature of pulmonary hypertension. Br. J. Radiol., 31:174, 1958.
20. Steiner, R. E.: Radiological appearances of the pulmonary vessels in pulmonary hypertension. Br. J. Radiol., 31:188, 1958.
21. Anderson, R. E., Baltaxe, H. A., and Amplatz, K.: Eisenmenger's complex mimicking pulmonary stenosis on plain films. Radiology, 98:381, 1971.
22. Taussig, H. B.: Congenital Malformations of the Heart, 2nd ed. New York, Commonwealth Fund, Harvard University Press, 1960.
23. Singh, H., and Parkash, A., et al.: Bone changes in congenital cyanotic heart disease. Br. Heart J., 34:412, 1972.
24. Fischer, K. C., and White, R. I., et al.: Sternal abnormalities in patients with congenital heart disease. Am. J. Roentgenol. Radium Ther. Nucl. Med., 119:530, 1973.
25. Randall, P. A., Moller, J. H., and Amplatz, K.: The spleen and congenital heart disease. Am. J. Roentgenol. Radium Ther. Nucl. Med., 119:551, 1973.
26. Bristow, J. D.: Symposium. Left ventricular outflow tract obstruction. Recognition of left ventricular outflow obstruction. Circulation, 31:600, 1965.
27. Freimanis, A. K., Wooley, C. F., Meckstroth, C. V., and Molnar, W.: Roentgenographic aspects of congenital left ventricular outflow tract obstruction. Am. J. Roentgenol. Radium Ther. Nucl. Med., 95:573, 1965.
28. Takekawa, S. D., et al.: Congenital aortic stenosis. Am. J. Roentgenol. Radium Ther. Nucl. Med., 98:800, 1966.
29. Roberts, W. C., and Elliott, L. P.: Lesions complicating the congenitally bicuspid aortic valve. Radiol. Clin. North Am., 6:409, 1968.
30. Batson, G. A., Urquhart, W., and Sideris, D.: Radiologic features in aortic stenosis. Clin. Radiol., 23:140, 1972.
31. Simon, A. L.: Angiographic diagnosis of idiopathic hypertrophic subaortic stenosis. Radiol. Clin. North Am., 6:423, 1968.
32. Deutsch, V., et al.: Subaortic stenosis. Radiology, 101:275, 1971.
33. Kurlander, G. J., et al.: Supravalvular aortic stenosis. Am. J. Roentgenol. Radium Ther. Nucl. Med., 80:782, 1966.
34. Thelen, M., et al.: Subvalvular aortic stenosis. Fortschr. Roentgenstr., 111:481, 1969.
35. Kjellberg, S. R., Mannheimer, E., Rudhe, U., and Jonsson, B.: Diagnosis of Congenital Heart Disease. Chicago, Year Book Medical Publishers, Inc., 1955.
36. Hirst, A. E., Jr., Varner, J. J., and Kime, S. W.: Dissecting aneurysm of the aorta. Medicine, 37:217, 1968.
37. Nikardoh, H., et al.: Aortic rupture in children as a complication of coarctation of the aorta. Arch. Surg., 107:838, 1973.
38. Griffin, J. F.: Congenital kinking of the aorta (pseudocoarctation). N. Engl. J. Med., 271:726, 1964.
39. Steinberg, I.: Anomalies (pseudocoarctation) of the aortic arch. Am. J. Roentgenol. Radium Ther. Nucl. Med., 88:73, 1962.
40. Shapiro, I. L., et al.: Pseudocoarctation of the aorta. Arch. Intern. Med., 122:345, 1968.
41. Cheng, T. O.: Pseudocoarctation of aorta. Am. J. Med., 49:551, 1970.
42. Steinberg, I., et al.: Pseudocoarctation of the aorta associated with congenital heart disease. Am. J. Roentgenol. Radium Ther. Nucl. Med., 106:1, 1969.
43. Rudhe, U.: Angiocardiography in pulmonary stenosis. Radiol. Clin. North Am., 2:395, 1964.
44. Desilets, D. T., et al.: Severe pulmonary valve stenosis with atresia. Radiol. Clin. North Am., 6:367, 1968.
45. Chen, J. T. T., et al.: Uneven distribution of pulmonary blood flow between left and right lungs in isolated valvular pulmonary stenosis. Am. J. Roentgenol. Radium Ther. Nucl. Med., 107:343, 1969.
46. Spitz, H. B.: The roentgenology of atrial septal defect in the adult. Semin. Roentgenol., 1:67, 1966.
47. Silbiger, M. L., et al.: Correlation of chest roentgenograms and hemodynamic findings following surgical closure of secundum atrial septal defects. Radiology, 91:742, 1968.
48. Rosenbaum, H. D., Lieber, A., Hanson, D. J., and Bernard, J. D.: Roentgen findings in ventricular septal defect. Semin. Roentgenol., 1:47, 1966.
49. Schwarz, E. D., and Dorst, J. P., et al.: Reliability of roentgenographic evaluation of ventricular septal defects in children. Johns Hopkins Med. J., 127:164, 1970.
50. Deutsch, V., and Blieden, L. C., et al.: Ventricular septal defect associated with aortic insufficiency. Am. J. Roentgenol. Radium Ther. Nucl. Med., 106:32, 1969.
51. Klatte, E. C., and Burko, H.: The roentgen diagnosis of patent ductus arteriosus. Semin. Roentgenol., 1:87, 1966.
52. Currarino, G., and Jackson, J. H.: Calcification of ductus arteriosus and ligamentum Botalli. Radiology, 94:139, 1970.
53. Elliot, L. D.: Other forms of left to right shunt. Semin. Roentgenol., 1:120, 1966.
54. Currarino, G., Silverman, F. N., and Landing, B. H.: Abnormal congenital fistulous communications of the coronary arteries. Am. J. Roentgenol. Radium Ther. Nucl. Med., 82:392, 1959.
55. de Nef, J. J. E., et al.: Congenital coronary artery fistula. Br. Heart J., 33:857, 1971.

56. Steinberg, I., and Holswadi, G. R.: Coronary arterio-venous fistula. Am. J. Roentgenol. Radium Ther. Nucl. Med., *116*:82, 1972.

57. Lundquist, C., and Amplatz, K.: Anomalous origin of the left coronary artery from the pulmonary artery. Am. J. Roentgenol. Radium Ther. Nucl. Med., *95*:611, 1965.

58. Friedenberg, M. J., et al.: Opacification from the pulmonary artery of an anomalous left coronary artery. Radiology, *80*:806, 1963.

59. Ogden, J. A.: Congenital anomalies of the coronary arteries. Am. J. Cardiol., *25*:474, 1970.

60. Rubinstein, B. M., Young, D., Pinals, D., and Jacobson, H. G.: The roentgen spectrum in persistent common atrioventricular canal. Radiology, *86*:860, 1966.

61. Baron, M. G.: Endocardial cushion defects. Radiol. Clin. North Am., *6*:343, 1968.

62. Fellows, K., and Henschel, W. G., et al.: Left ventricular angiocardiography in endocardial cushion defects. Ann. Radiol., *15*:223, 1972.

63. Harris, E. J.: Aneurysm of sinus of Valsalva. Am. J. Roentgenol. Radium Ther. Nucl. Med., *76*:767, 1956.

64. Elliot, L. P.: Other forms of left to right shunt. Semin. Roentgenol., *6*:120, 1966.

65. Lester, R. G., Anderson, R. C., Amplatz, K., and Adams, P.: Roentgen diagnosis of congenitally corrected transposition of the great vessels. Am. J. Roentgenol. Radium Ther. Nucl. Med., *83*:985, 1960.

66. Carey, L. S., and Ruttenberg, H. D.: Roentgenographic features of congenital corrected transposition of the great vessels. Am. J. Roentgenol. Radium Ther. Nucl. Med., *92*:623, 1964.

67. Friedberg, D. A., and Nadas, A. S.: Congenital corrected transposition of the great arteries. N. Engl. J. Med., *282*:1053, 1970.

68. Pilapil, V. R., and Bennett, K. R., et al.: Corrected transposition of the great arteries. Am. J. Med., *51*:482, 1971.

69. Chang, C. H., and Rogers, J. V.: Cor triatriatum sinistrum. Am. J. Roentgenol. Radium Ther. Nucl. Med., *80*:405, 1958.

70. Robinson, A. E., et al.: Left sided obstructive disease of the heart and great vessels. Semin. Roentgenol., *31*:410, 1968.

71. McLoughlin, N. J.: Cor triatriatum sinister. Clin. Radiol., *21*:287, 1970.

72. Johnsrude, I. S., and Carey, L. S.: Roentgenographic manifestations of endocardial fibroelastosis. Am. J. Roentgenol. Radium Ther. Nucl. Med., *94*:109, 1965.

73. Ainger, L. E.: Mitral and aortic valve incompetence in endocardial fibroelastosis. Am. J. Cardiol., *28*:309, 1971.

74. Lester, R. G., Mauck, H. P., and Grubb, W. L.: Anomalous pulmonary venous return to the right side of the heart. Semin. Roentgenol., *1*:102, 1966.

75. Darling, R. C., Rothney, W. B., and Craig, J. M.: Total pulmonary venous drainage into the right side of the heart. J. Exp. Mech. Pathol., *6*:10, 1957.

76. Roehm, J. O. F., Jue, K. L., and Amplatz, K.: Radiologic features of the scimitar syndrome. Radiology, *86*:856, 1966.

77. Lester, R. G., et al.: Anomalous pulmonary venous return to the right heart. Semin. Roentgenol., *1*:102, 1966.

78. Carey, L. S., and Elliott, L. P.: Transposition of great vessels. Am. J. Roentgenol. Radium Ther. Nucl. Med., *91*:529, 1964.

79. Grainger, R. G.: Transposition of the great arteries and of the pulmonary veins. Clin. Radiol., *21*:335, 1970.

80. Abrams, H. L., (ed.): *Arteriography.* Boston, Little, Brown and Company, 1961.

81. Lafargue, R. T., Vogel, J. H. K., et al.: Pseudotruncus arteriosus. Am. J. Cardiol., *19*:239, 1967.

82. Collett, R. W., and Edwards, J. E.: Persistent truncus arteriosus: a classification according to anatomical types. Surg. Clin. North Am., *29*:1245, 1949.

83. Dalinka, M. K., Rubinstein, B. M., and Lopez, F.: The roentgen findings in truncus arteriosus. J. Can. Assoc. Radiol., *21*:85, 1970.

84. Hallermann, F. J., et al.: Persistent truncus arteriosus: A radiographic and angiographic study. Am. J. Roentgenol. Radium Ther. Nucl. Med., *107*:827, 1969.

85. Carey, L. S., and Edwards, J. E.: Origin of great vessels from the right ventricle. Am. J. Roentgenol. Radium Ther. Nucl. Med., *93*:296, 1965.

86. Von Praagh, R., von Praagh, S., Vlad, P., and Keith, J. D.: Diagnosis of the anatomic types of single or common ventricle. Am. J. Cardiol., *15*:345, 1965.

87. Kozuka, T., and Sato, K., et al.: Roentgenographic diagnosis of single ventricle. Am. J. Roentgenol. Radium Ther. Nucl. Med., *119*:512, 1973.

88. Lester, R. G., Robinson, A. E., and Osteen, R. T.: Tetralogy of Fallot. A detailed angiographic study. Am. J. Roentgenol. Radium Ther. Nucl. Med., *94*:92, 1965.

89. Johnson, C.: Fallot's tetralogy — a review of the radiological appearances in thirty-three cases. Clin. Radiol., *16*:199, 1965.

90. Reicher-Reiss, H., and Rubenstein, B., et al.: Pulmonary vasculature in tetralogy of Fallot. Clin. Radiol., *22*:341, 1971.

91. Marr, K., Giargiana, F. A., and White, R. I.: Radiographic diagnosis of pulmonary hypertension following Blalock-Taussig shunts in patients with tetralogy of Fallot. Am. J. Roentgenol. Radium Ther. Nucl. Med., *122*:125, 1974.

92. Kieffer, S. A., and Carey, L. S.: Tricuspid atresia with normal aortic root. Roentgen. anatomical correlation. Radiology, *80*:605, 1963.

93. Guller, B., et al.: Angiographic findings in tricuspid atresia. Radiology, *93*:531, 1969.

94. Elliott, L. P., et al.: The roentgenology of tricuspid atresia. Semin. Roentgenol., *3*:399, 1968.

95. Amplatz, K., Lester, R. G., Schiebler, G. L., Adams, P., and Anderson, R. C.: Roentgen features of Ebstein's anomaly. Am. J. Roentgenol. Radium Ther. Nucl. Med., *81*:788, 1959.

96. Tori, G., and Garusi, G. F.: Angiocardiographic diagnosis of Ebstein's anomaly of the tricuspid valve. Radiol. Med. (Tor.), *47*:673, 1961.

97. Elliott, L. P., and Hartmann, A. F.: The right ventricular infundibulum in Ebstein's anomaly of the tricuspid valve. Am. J. Roentgenol. Radium Ther. Nucl. Med., *89*:694, 1967.

98. Biolostozky, D., Horwitz, S., and Espino-Vela, J.: Ebstein's malformation of the tricuspid valve. A review of 65 cases. Am. J. Cardiol., *29*:826, 1972.

99. Carey, L. S., and Edwards, J. E.: Roentgenographic features in cases with origin of both great vessels from right ventricle without pulmonary stenosis. Am. J. Roentgenol. Radium Ther. Nucl. Med., *93*:269, 1965.

100. Klatte, E. C., and Burko, H.: Differential diagnosis of cyanotic congenital heart disease. Semin. Roentgenol., *3*:358, 1968.

101. Elliott, L. P., Jue, K. L., and Amplatz, K.: Roentgen classification of cardiac malpositions. Invest. Radiol., *1*:17, 1966.

102. Steinberg, I., and Ayres, S. M.: Roentgen features of a dextrorotation of the heart. Report on 49 cases. Am. J. Roentgenol. Radium Ther. Nucl. Med., *91*:340, 1964.

103. Cooley, R. N.: Congenital dextrocardia and the general radiologist. Am. J. Roentgenol. Radium Ther. Nucl. Med., *116*:211, 1972.

104. Steinberg, I.: Anomalies of the arch of the aorta. Am. J. Roentgenol. Radium Ther. Nucl. Med., *88*:73, 1962.

105. Felson, B., Cohen, S., Courter, S., and McGuire, J.: Anomalous right subclavian artery. Radiology, *54*:340, 1950.

106. Stewart, J. R., Kincaid, O. W., and Titus, J. L.: Right aortic arch: plain film diagnosis and significance. Am. J. Roentgenol. Radium Ther. Nucl. Med., *97*:377, 1966.

107. Baron, M. G.: Right aortic arch. Circulation, *44*:1137, 1971.

108. Shuford, W. H., Sybers, R. G., and Edwards, F. K.: The three types of right aortic arch. Am. J. Roentgenol., *109*:67, 1970.

109. Gay, B. B., French, R. H., Shuford, W. E., and Rogers, J. V.: The roentgenologic features of single and multiple coarctations of the pulmonary artery and branches. Am. J. Roentgenol. Radium Ther. Nucl. Med., *90*:599, 1963.

110. Kieffer, S. A., Amplatz, K., Anderson, R., and Lillehei, W.: Proximal interruptions of a pulmonary artery. Am. J. Roentgenol. Radium Ther. Nucl. Med., *95*:592, 1965.

111. Tang, J. S., Kauffman, S. L., and Lynfield, J.: Hypoplasia of the pulmonary arteries in infants with congenital rubella. Am. J. Cardiol., *27*:491, 1971.

112. Paul, L. W., and Juhl, J. H.: *The Essentials of Roentgen Interpretations,* New York, Hoeber Medical Division. Harper & Row, Publishers, 1965.

113. Sawa, T. I., and Taplin, G. V.: Unilateral pulmonary artery agenesis, stenosis and hypoplasia. Radiology, *99*:605, 1971.

114. Sage, M. R., and Brown, J. H.: Congenital unilateral absence of a pulmonary artery. Aust. Radiol., *16*:228, 1972.

115. Philp, T., and Sumerling, M. D., et al.: Aberrant left pulmonary artery. Clin. Radiol., *23*:153, 1972.

116. Twigg, H. L,, et al.: Straight back syndrome: radiographic manifestations. Radiology, *88*:274, 1967.

117. Goldring, D., Behrer, M. F., Brown, G., and Elliott, G.: Rheumatic pneumonitis. J. Pediatr., *53*:547, 1958.

118. Kerley, P.: Lung changes in acquired heart disease. Am. J. Roentgenol. Radium Ther. Nucl. Med., *80*:256, 1958.

119. Simon, G.: The value of radiology in critical mitral stenosis – an amendment. Clin. Radiol., *23*:145, 1972.

120. Seningen, R. P., and Chen, J. T., et al.: Roentgen interpretation of postoperative changes in pure mitral stenosis. Am. J. Roentgenol. Radium Ther. Nucl. Med., *113*:693, 1971.

121. Steiner, R. E., Jacobson, G., Dinsmore, R., and Parizel, G.: Mitral regurgitation. Clin. Radiol., *14*:113, 1963.

122. Uricchio, F. J., Lehman, J. S., Lemmon, W. M., Fitch, E. A., Boyer, R. A., and Likoff, W.: Newer techniques helpful in the study of regurgitant lesions of the cardiac valves. Am. J. Cardiol., *4*:696, 1959.

123. Wexler, L., and Silverman, W. L., et al.: Angiographic features of rheumatic and non-rheumatic mitral regurgitation. Circulation, *44*:1080, 1971.

124. Lehman, J. S., Florence, H., Schimert, A. P., and Evans, G. C.: Acquired aortic valvular stenosis. Radiology, *81*:24, 1963.

125. Batson, G. A., Urquhart, W., and Sideris, D. A.: Radiologic features in aortic stenosis. Clin. Radiol., *23*:140, 1972.

126. Klatte, E. C., Tampos, J. P., Campbell, J. A., and Lauri, P. R.: Aortic stenosis – aortic insuf-

ficiency; roentgenographic manifestation. Am. J. Roentgenol. Radium Ther. Nucl. Med., *88*:51, 1962.

127. Tillolson, P. M., and Steinberg, I.: Roentgen features of rheumatic tricuspid stenosis. Am. J. Roentgenol. Radium Ther. Nucl. Med., *87*:948, 1962.

128. Holmes, J. C., Fowler, N. O., and Kaplan, S.: Pulmonary valvular insufficiency. Am. J. Med., *44*:851, 1968.

129. Jaffe, R. B., and Koschmann, E. B.: Septic pulmonary emboli. Radiology, *96*:527, 1970.

130. Sannerstedt, R., Paulin, S., and Varnauskas, E.: Correlation between radiological chest findings and systemic hemodynamics in human arterial hypertension. Br. Heart H., *32*:477, 1970.

131. Bookstein, J. J., and Stewart, B. H.: The current status of renal arteriography. Radiol., Clin. North Am., *2*:461, 1964.

132. Denni, J. M., Wolfel, D. A., and Young, J. D.: Diagnosis of reno-vascular hypertension: role of the radiologist. Radiol. Clin. North Am., *1*:61, 1963.

133. Sutton, D., et al.: Fibromuscular, fibrous and non-atheromatous renal artery stenosis and hypertension. Clin. Radiol., *14*:381, 1963.

134. Kincaid, O. W., et al.: Fibromuscular dysplasia of the renal arteries. Am. J. Roentgenol. Radium Ther. Nucl. Med., *104*:271, 1968.

135. Foster, J. H., Oates, J. A., et al.: Detection and treatment of patients with renovascular hypertension. Surgery, *60*:240, 1966.

136. Bookstein, J. J., and Abrams, H. L., et al.: Radiologic aspects of renovascular hypertension: A symposium. J.A.M.A., *220*:1209; *220*:1218; *221*:368, 1972.

137. Foster, R. S., Shuford, W. H., and Weens, S.: Selective renal arteriography in medical diseases of the kidney. Am. J. Roentgenol. Radium Ther. Nucl. Med., *95*:291, 1965.

138. Gill, W. M., and Pudvan, W. R.: Arteriographic diagnosis of renal parenchymal diseases. Radiology, *96*:81, 1970.

139. Conn, J. W., and Cohen, E. L., et al.: Primary reninism. Arch. Intern. Med., *130*:682, 1972.

140. Conn, J. W., Bookstein, J. J., and Cohen, E. L.: Renin-secreting juxtaglomerular cell adenoma. Radiology, *106*:543, 1973.

141. Jorgens, J., Boardman, W. J., Damberg, S. W., Kinney, W. N., and Kundel, R. R.: The significance of coronary calcification. Am. J. Roentgenol. Radium Ther. Nucl. Med., *95*:667, 1965.

142. Lehman, J. S., Novack, P., Kasparian, H., Likoff, W., and Perlmutter, H. I.: Selective coronary arteriography. Radiology, *83*:846, 1964.

143. Jorgens, J., et al.: Significance of coronary artery calcification. Am. J. Roentgenol. Radium Ther. Nucl. Med., *95*:667, 1965.

144. Gensini, G. G., and Buonanno, C.: Coronary arteriography. Dis. Chest, *54*:90, 1968.

145. Adams, D. F., Abrams, H. L., and Ruttley, M.: The roentgen pathology of coronary artery disease. Semin. Roentgenol., *7*:319, 1972.

146. Fowler, N. O.: Coronary arteriography; a perspective for radiologists. Semin. Roentgenol., *7*:352, 1972.

147. Rosch, J., and Judkins, M. P., et al.: Aortocoronary venous by-pass grafts. Radiology, *102*:567, 1972.

148. Rizk, G., and Cueto, L., et al.: Abnormal angiographic finding in patent aortocoronary by-pass graft. Radiology, *106*:43, 1973.

149. Baltaxe, H. A., and Levin, D. C.: Angiographic demonstration of complications related to the saphenous aorto-coronary bypass procedures. Am. J. Roentgenol. Radium Ther. Nucl. Med., *119*:484, 1973.

150. Dressler, W.: The post-myocardial infarction syndrome. Arch. Intern. Med., *103*:28, 1959.

151. Arnold, H. R.: Post-myocardial infarction syndrome. Am. J. Roentgenol. Radium Ther. Nucl. Med., *90*:628, 1963.

152. Levin, E. J., and Bryk, D.: Dressler syndrome. Radiology, *87*:731, 1966.

153. Sanders, I., and Woessner, M. E.: Cardiac blur sign of post-infarction myocardial scar. Am. J. Roentgenol. Radium Ther. Nucl. Med., *113*:703, 1971.

154. Tudor, J., and Maurer, B. J., et al.: Lung shadows after acute myocardial infarction. Clin. Radiol., *24*:365, 1973.

155. Steinberg, I.: Angiographic findings in ventricular aneurysm due to arteriosclerotic myocardial infarction. Am. J. Roentgenol. Radium Ther. Nucl. Med., *97*:321, 1966.

156. Bjork, L.: Roentgen diagnosis of left ventricular aneurysm. Am. J. Roentgenol. Radium Ther. Nucl. Med., *97*:338, 1966.

157. Baron, M. G.: Postinfarction aneurysm of the left ventricle. Circulation, *43*:762, 1971.

158. Rosenbaum, H. D.: Roentgen demonstration of broken cardiac pacemaker wires. Radiology, *84*:933, 1965.

159. Björk, J.: Isolated calcifications in the interventricular septum. Acta Radiol. [Diag.], *3*:430, 1965.

160. Lillehei, C. W., Cruz, A. B., et al.: A new method of assessing the state of charge of implanted cardiac pacemaker batteries. Am. J. Cardiol., *16*:717, 1965.

161. Steinberg, I., vonGal, H., and Finby, N.: Roentgen diagnosis of pericardial effusion. Am. J. Roentgenol. Radium Ther. Nucl. Med., *79*:321, 1958.

162. Scatliff, J. H., Kummer, A. J., and Jansen, A. H.: Diagnosis of pericardial effusion with intracardiac carbon dioxide. Radiology, *73*:817, 1959.

163. Baron, M. G.: Pericardial effusion. Circulation, *44* :294, 1971.
164. Marikas, G., Samartzis, M., and Marretos, S.: Massive cardiac tamponade in uremic pericarditis with complete recovery. N. Engl. J. Med., *266*:1089, 1962.
165. Spitz, H. B., and Holmes, J. C.: Right atrial contour in cardiac tamponade. Radiology, *103*:69, 1972.
166. Stein, L., Shubin, H., and Weil, N. H.: Recognition and management of pericardial tamponade. J.A.M.A., *225*:503, 1973.
167. Plum, G. E., Brewer, A. G., and Clayett, O. T.: Chronic constrictive pericarditis; roentgenologic findings in 35 surgically proved cases. Mayo Clin. Proc., *32*:555, 1957.
168. Wilkinson, P., Pinto, B., and Senior, J. R.: Reversible protein-losing enteropathy with intestinal lymphangiectasia secondary to chronic constrictive pericarditis. N. Engl. J. Med., *273*:1178, 1965.
169. Shawdon, H. H., and Dinsmore, R. E.: Pericardial calcification: radiologic features and clinical significance in twenty-six patients. Clin. Radiol., *18*:205, 1967.
170. Cornell, S. H., and Rossi, N. P.: Roentgenologic findings in constrictive pericarditis. Am. J. Roentgenol. Radium Ther. Nucl. Med., *102*:301, 1968.
171. Steinberg, I., and Hagstrom, W. C.: Angiocardiography in diagnosis of effusive-restrictive pericarditis. Am. J. Roentgenol. Radium Ther. Nucl. Med., *102*:305, 1968.
172. Dragoni, G., and Romanini, A.: Differential diagnosis of coelomatic cysts of the pericardium. Minerva Med., *52*:1175, 1961.
173. Klatte, E. C., and Yune, H. Y.: Diagnosis and treatment of pericardial cysts. Radiology, *104*:541, 1972.
174. Glover, L. B., Barcia, A., and Reeves, T. J.: Congenital absence of the pericardium. Am. J. Roentgenol. Radium Ther. Nucl. Med., *106*:542, 1969.
175. Nasser, W. K., and Helmen, C., et al.: Congenital absence of the left pericardium. Circulation, *41*:469, 1970.
176. Davies, J. N. P.: Review of obscure disease affecting the mural endocardium. Am. Heart J., *59*:4, 1960.
177. Becker, E. F., and Taube, H.: Myocarditis of obscure etiology associated with pregnancy. N. Engl. J. Med., *266*:62, 1962.
178. Battersby, E. J., and Glenner, G. G.: Familial cardiopathy. Am. J. Med., *30*:282, 1961.
179. Kittredge, R. D., Arida, E. J., and Finby, N.: The role of angiocardiography in myxedema heart disease. Radiology, *80*:430, 1963.
180. Gotsman, M. S., Van der Horst, R. L., and Winship, W. S.: The chest radiograph in primary myocardial disease. Radiology, *99*:1, 1971.
181. Belle, M. S.: Right atrial myxoma. Circulation, *19*:910, 1959.
182. Elke, M., Ludin, H., Wobman, E., and Hartman, G.: Zur Diagnose intrakavitaerer Herztumoren. Fortschr. Roentgenstr., *101*:265, 1964.
183. Bergonzini, R., and Calzavara, M.: Contributo casistico e rassegna della letteratura sui tumori maligni primitivi del cuore. Radiol. Med. (Tor.), Milan, *45*:865, 1959.
184. Mauri, C., and Fontanini, F.: Sulla diagnosi di tumore maligno del cuore. Cardiol. Prat., *10*:21, 1959.
185. Abrams, H. L., Adams, D. F., and Grant, H. A.: The radiology of tumors of the heart. Radiol. Clin. North Am., *9*:299, 1971.
186. Castleman, B., et al.: Case records of Massachusetts General Hospital. N. Engl. J. Med., *287*:555, 1972.
187. Burhenne, J., and Strasser, E.: A simple radiographic method and epidemiologic study of atherosclerosis. Radiology, *97*:180, 1970.
188. Steinberg, I., and Stein, H. L.: Visualization of abdominal aortic aneurysm. Am. J. Roentgenol. Radium Ther. Nucl. Med., *95*:684, 1965.
189. Greenspan, R. H., et al.: Aortography in diagnosis of dissecting aneurysm. Radiology, *78*:665, 1962.
190. Shuford, W. H., et al.: Problems in the aortographic diagnosis of dissecting aneurysms of the aorta. N. Engl. J. Med., *280*:225, 1969.
191. Hartel, M.: Roentgen diagnosis of dissecting aneurysm of the aorta. Schweiz. Med. Wochenschr., *101*:965, 1971.
192. Baron, M. G.: Dissecting aneurysm of the aorta. Circulation, *43*:933, 1971.
193. Beachley, M. C., Ranninger, K., and Roth, F. J.: Roentgenologic evaluation of dissecting aneurysms of the aorta. Am. J. Roentgenol. Radium Ther. Nucl. Med., *121*:617, 1974.
194. Steinberg, I.: Traumatic aneurysm of the thoracic aorta. Am. J. Roentgenol. Radium Ther. Nucl. Med., *91*:1295, 1964.
195. Flaherty, T. T., and Wegner, G. P., et al.: Non-penetrating injuries to the thoracic aorta. Radiology, *92*:541, 1969.
196. Davidsen, G., Petersen, O., and Thomsen, G.: Roentgenologic findings of five cases of congenital aneurysm of aortic sinuses. Acta Radiol., *49*:205, 1958.
197. Harris, E. J.: Aneurysms of the sinus of Valsalva. Am. J. Roentgenol. Radium Ther. Nucl. Med., *76*:767, 1956.
198. Rowley, J. C., and Holmes, R. B.: Sinus of Valsalva aneurysms. J. Can. Assoc. Radiol., *16*:254, 1965.

199. Keene, R. J., and Steiner, R. E., et al.: Aortic root aneurysm; radiographic and pathologic features. Clin. Radiol., *22*:330, 1971.

200. Smith, W. G., and Leonard, J. G.: Radiologic features of syphilitic aortic incompetence. Br. Heart J., *21*:162, 1959.

201. Macleod, C. A., et al.: Angiographic demonstration of coronary ostial stenosis. Am. J. Cardiol., *22*:122, 1968.

202. Szilagyi, D. E.: Lumbar and peripheral arteriography: clinical aspects. Radiology, *69*:177, 1957.

203. Gillanders, L. A., and Strachan, R. W.: Role of radiology in diagnosis of Takayasu's arteriopathy (pulseless disease). Clin. Radiol., *16*:119, 1965.

204. Mengis, C. L., et al.: Aortic arch syndrome of Takayasu. Am. Heart J., *55*:435, 1958.

205. Lande, A., and Gross, A.: Total aortography in the diagnosis of Takayasu's arteritis. Am. J. Roentgenol. Radium Ther. Nucl. Med., *116*:165, 1972.

206. Deutsch, W., Wexler, L., and Deutsch, H.: Takayasu's arteritis. Am. J. Roentgenol. Radium Ther. Nucl. Med., *122*:13, 1974.

207. Spangler, R. D., MacCallister, B. D., and McGoon, D. C.: Aortic valve replacement in patients with severe aortic valve incompetence associated with rheumatoid spondylitis. Am. J. Cardiol., *26*:130, 1970.

208. Lindbom, A.: Arteriosclerosis and arterial thrombosis in lower limb. Acta Radiol. (Supplement 80), 1957.

209. Keats, T. E.: Trends in peripheral arteriography. Radiol. Clin. North Am., *2*:483, 1964.

210. Roy, P.: Peripheral angiography in ischemic arterial disease of the limbs. Radiol. Clin. North Am., *5*:467, 1967.

211. Carlin, R. A., and Amplatz, K.: Downstream aortography. Am. J. Roentgenol. Radium Ther. Nucl. Med., *109*:536, 1970.

212. McKusick, V. A., Harris, W. S., Ottesen, O. E., Goodman, R. M., Shelley, W. M., and Bloodwell, R. D.: Buerger's disease: a distinct clinical and pathologic entity. J.A.M.A., *181*:5, 1962.

213. McKusick, V. A., and Harris, W. S.: The Buerger syndrome in the Orient. Bull. Johns Hopkins Hosp., *90*:241, 1961.

214. Grollman, J. H., Lecky, J. W., and Rösch, J.: Miscellaneous diseases of arteries. Semin. Roentgenol., *5*:306, 1970.

215. Laws, J., W., et al.: Angiographic appearances in rheumatoid arthritis and other disorders. Br. J. Radiol., *36*:477, 1963.

216. Yune, H. Y., Vix, V. A., and Klatte, E. C.: Early fingertip changes in scleroderma. J.A.M.A., *215*:1113, 1971.

217. Kramer, R. A., Hecker, S. P., and Lewis, B. I.: Ergotism: report of a case studied arteriographically. Radiology, *84*:308, 1965.

218. Bagby, R. J., and Cooper, R. D.: Angiography in ergotism. Am. J. Roentgenol. Radium Ther. Nucl. Med., *116*:179, 1972.

219. Fedotin, M. S., and Hartman, C.: Ergotamine poisoning producing renal arterial spasm. N. Engl. J. Med., *283*:518, 1970.

220. Steinberg, I., Tillotson, P. M., and Halpern, M.: Roentgenography of systemic (congenital and traumatic) arteriovenous fistulas. Am. J. Roentgenol. Radium Ther. Nucl. Med., *89*:343, 1963.

221. Thomas, M. L., and Andress, M. R.: Angiography and venous dysplasia. Am. J. Roentgenol. Radium Ther. Nucl. Med., *113*:722, 1971.

222. Harris, W. R.: Erosion of bone produced by glomus tumor. Can. Med. Assoc. J., *70*:684, 1954.

223. Richardson, J. A., Jr., and Diddams, A. C.: Maffucci's syndrome. Arch. Intern. Med., *109*:186, 1962.

224. Hayt, D. B.: Roentgenographic signs of thrombosis of the superior vena cava and tributaries in neoplastic disease. Am. J. Roentgenol. Radium Ther. Nucl. Med., *93*:87, 1965.

225. Halliday, R.: Distribution of radiopaque dye during phlebography of the lower limb. Aust. Radiol., *13*:296, 1969.

226. Byrd, R. B., et al.: Bronchogenic carcinoma and thromboembolic disease. J.A.M.A., *202*:107, 1967.

227. Brodelius, A., Loring, P., and Nylander, G.: Phlebographic techniques in the diagnosis of acute deep venous thrombosis of the lower limbs. Am. J. Roentgenol. Radium Ther. Nucl. Med., *111*:794, 1971.

228. Thomas, M. L., McCallister, V., and Tonge, K.: The radiologic appearance of deep venous thrombosis. Clin. Radiol., *22*:495, 1971.

229. Mobi-Uddin, K., and Callard, G. M., et al.: Transvenous caval interruption with umbrella filter. N. Engl. J. Med., *286*:55, 1972.

230. Buoncore, E., and Young, J. R.: Lymphographic evaluation of lymphedema and lymphatic flow. Am. J. Roentgenol. Radium Ther. Nucl. Med., *95*:751, 1965.

231. Jacobson, S., and Johansson, S.: Lymphangiography in lymphedema. Acta Radiol., *57*:81, 1962.

232. Wallace, S., Jackson, L., Schaffer, B., Gould, J., Greening, R. R., Weiss, A., and Kramer, S.: Lymphangiograms: their diagnostic and therapeutic potential. Radiology, *76*:179, 1961.

233. Craig, O.: Primary lymphedema and lymphatica porosa. Radiology, *92*:1216, 1969.

234. Leiter, E.: Inferior vena caval thrombosis in malignant renal lesions. J.A.M.A., *198*:1167, 1966.

235. Hall, R., and Jenkins, J. D.: Idiopathic intravenous pyelography in acute idiopathic inferior vena caval thrombosis. Br. J. Radiol., *43*:781, 1970.

236. Amoe, H. E., and Lewis, R. E.: Urographic and barium enema appearances in inferior vena cava obstruction. Radiology, *108*:307, 1973.

ROENTGEN SIGNS AND ASSOCIATED DISEASES

In this section, diseases and disorders are alphabetically listed under categories of roentgen signs or appearance. With a few exceptions, the roentgen signs are broad in scope and fairly frequently encountered. Highly specific or pathognomonic roentgen findings have been intentionally omitted.

For purposes of differential diagnosis, the list of conditions under each roentgen category is reasonably complete but not exhaustive. Disorders which are extremely infrequent or those in which the designated roentgen sign is seen only rarely may be absent from the list. Many conditions appear in several lists, which may help narrow the differential diagnostic possibilities.

ROENTGEN SIGNS OF JOINTS

ROENTGEN SIGNS OF KIDNEYS AND URINARY TRACTS

ROENTGEN SIGNS OF OSSEOUS SYSTEM

MISCELLANEOUS ROENTGEN SIGNS

I. CARDIOVASCULAR SYSTEM

ROENTGEN SIGNS OF CARDIOVASCULAR SYSTEM

(1) Increased Pulmonary Vascularity

Predominantly Venous	Predominantly Arterial
Congestive failure	Atrial septal defect
Cor triatriatum sinistrum	Congenital coronary fistula
Left atrial ball valve thrombus	Endocardial cushion defects
Left atrial myxoma	High output heart
Mitral insufficiency	Lutembacher's syndrome
Mitral stenosis	Partial anomalous venous return
Pulmonary venous obstruction by tumor, thrombus, or mediastinal fibrosis	Patent ductus arteriosus
Total anomalous venous return to below diaphragm	Pickwickian obesity
	Primary polycythemia
	Single ventricle
	Taussig-Bing complex
	Transposition of great vessels
	Truncus arteriosus
	Ventricular septal defect

(2) Pulmonary Hypertension and Cor Pulmonale

Chronic Hypoxia	Diffuse Pulmonary Arteritis
Chest deformities	Idiopathic
Chronic airway obstruction (adenoids)	Left to right shunts (Eisenmenger's physiology)
High altitude dwelling	Mitral stenosis (longstanding)
Pickwickian obesity	Schistosomiasis

Diffuse Interstitial Lung Disease	Thromboembolism
Chronic bronchitis	Recurrent pulmonary emboli
Dermatomyositis	Sickle cell anemia (pulmonary thrombi)
Emphysema	
Hamman-Rich syndrome	**Hypervolemia of Right Heart**
Idiopathic fibrosis	
Lupus erythematosus	Ventriculoatrial shunt for hydrocephalus
Lymphangitic metastases	
Mucoviscidosis	
Periarteritis nodosa	
Pneumoconioses	
Scleroderma	
Tuberculosis with extensive fibrosis	

(3) High Output Heart

Arteriovenous fistula (extrapulmonary)
Beriberi
Hypervolemia (transfusions, fluid overload)
Paget's disease
Pickwickian obesity
Pregnancy
Primary polycythemia
Pyrexia
Severe anemia
Sickle cell anemia
Thyrotoxicosis

ROENTGEN SIGNS OF CARDIOVASCULA R SYSTEM *(Continued)*

(4) Prominent or Enlarged Ascending Aorta and/or Aortic Arch

Aneurysm
Ankylosing spondylitis
Aortic insufficiency
Aortic stenosis
Atherosclerosis
Coarctation of aorta
Homocystinuria
Marfan's disease
Patent ductus arteriosus
Pseudocoarctation
Pseudoxanthoma elasticum
Syphilitic aortitis
Takayasu's arteritis
Tetralogy of Fallot
Tricuspid atresia without transposition
Truncus arteriosus (type IV)

(5) Enlarged Left Atrium

Constrictive pericarditis
Heart failure (left sided)
Left to right shunt (excluding atrial septal defect)
   Congenital coronary fistula
   Patent ductus arteriosus
   Ventricular septal defect
Mitral valvular disease
   Insufficiency
   Stenosis
Myocardiopathies
Myxoma of left atrium

(6) Pericardial Effusion

Amebic pericarditis
Congestive failure
Coxsackie pericarditis
Dissecting aneurysm with leakage
Histoplasmosis
Idiopathic
Lupus erythematosus
Mycotic infection
Myxedema
Neoplastic pericarditis
Nephrosis
Periarteritis nodosa
Postcoronary artery bypass surgery
Postmyocardial infarction syndrome
Pyogenic pericarditis
Rheumatic fever
Scleroderma
Syphilis
Toxoplasmosis
Trauma
Tuberculosis
Uremia
Viral infections

(7) Cardiac Calcifications

Aneurysm of coronary artery
Aneurysm of sinus of Valsalva
Atrial clot
Coronary atherosclerosis
Left atrium (mitral disease)
Malignant tumor of heart
Mitral annulus
Myocardial infarct
Myxoma of atrium
Pericardium
Valvular
   Aortic (aortic stenosis)
   Mitral (mitral stenosis)

Table continued on following page.

ROENTGEN SIGNS OF CARDIOVASCULAR SYSTEM *(Continued)*

(8) Vascular Calcifications

Arterial	Venous
Arteriosclerosis	Arteriovenous malformation
Homocystinuria	Hemangioma
Hyperparathyroidism	Maffucci's syndrome
Premature arteriosclerosis	Phleboliths
Familial hyperlipemias	Varicose veins
Progeria	
Secondary hyperlipemias	
Cushing's syndrome	
Diabetes	
Glycogen storage disease	
Hypothyroidism	
Lipodystrophy	
Nephrotic syndrome	
Renal homotransplantation	
Pseudoxanthoma elasticum	
Werner's syndrome	

ROENTGEN SIGNS OF CHEST

(9) Disseminated Alveolar Infiltrates

Infectious	Neoplastic	Other
Bacterial pneumonias (diffuse)	Alveolar cell carcinoma	Alveolar microlithiasis
Giant-cell pneumonia	Hodgkin's disease	Alveolar proteinosis
Mycotic infections	Leukemia	Aspiration pneumonia
Pneumocystis carinii pneumonia	Lymphoma	Blast injury
Tuberculosis		Chemical inhalation
Viral pneumonias (diffuse);		Congenital pulmonary
especially influenzal virus,		lymphangiectasis
measles, varicella		Contusion
		Desquamative interstitial
		pneumonia
		Drug hypersensitivity
		Fat emboli
		Goodpasture's syndrome
		Hemosiderosis
		Hyaline membrane
		disease
		Hydrocarbon pneumonia
		Idiopathic respiratory
		distress syndrome
		Lipoid pneumonia
		Mucoviscidosis
		Oxygen toxicity
		Pulmonary edema
		(see following table)
		Pulmonary hemorrhage
		Rheumatic fever
		pneumonia
		Riley-Day syndrome
		(aspiration)
		Sarcoidosis
		Shock lung

ROENTGEN SIGNS OF CHEST *(Continued)*

(10) Pulmonary Edema

Acute glomerulonephritis
Aspiration
Cardiac failure (left heart)
Chest trauma
Diffuse capillary leak syndrome
Diffuse pulmonary embolism
Drug hypersensitivity
Drug overdose (heroin)
Fluid overload
High altitude
Hypoalbuminemia
Inhalation of toxic agents
Intracranial disease
Malaria (falciparum)
Mediastinal tumor (with venous or lymphatic obstruction)
Near drowning
Oxygen toxicity
Shock lung
Uremia

(11) Disseminated Interstitial (Reticular-Reticulonodular) Infiltrates

Inflammatory	Neoplastic	Other
Diffuse mycotic infections	Hodgkin's disease	Chemical inhalation
Diffuse viral and adenoviral pneumonia	Leukemia	Dermatomyositis
Schistosomiasis	Lymphangitic metastases	Desquamative interstitial pneumonia
Tuberculosis	Lymphomas	Diffuse interstitial fibrosis (idiopathic)
		Drug hypersensitivity
		Goodpasture's syndrome
		Hamman-Rich syndrome
		Hemosiderosis
		Histiocytosis X
		Interstitial pulmonary edema
		Mucoviscidosis
		Neurofibromatosis
		Niemann-Pick disease
		Organic dust inhalation
		Oxygen toxicity
		Pneumoconioses
		Polyarteritis
		Recurrent pulmonary infarction
		Rheumatoid arthritis
		Sarcoidosis
		Scleroderma
		Tuberous sclerosis

Table continued on following page.

ROENTGEN SIGNS OF CHEST *(Continued)*

(12) Pulmonary Fibrosis (Diffuse)

Chemical inhalation (late stage)
Dermatomyositis
Desquamative pneumonia
Hamman-Rich syndrome
Histiocytosis X
Idiopathic pulmonary hemosiderosis
Mucoviscidosis
Mycotic diseases

Neurofibromatosis
Pneumoconioses (organic and inorganic)
Radiation fibrosis
Rheumatoid lung
Sarcoidosis
Scleroderma
Tuberculosis
Tuberous sclerosis

(13) Kerley B Lines

Congestive failure
Diffuse interstitial fibrosis of varied etiology
Hamman-Rich syndrome
Interstitial lower lobe pneumonia
Lipoid pneumonia
Lymphangitic metastases
Mitral stenosis
Pneumoconioses
Pulmonary alveolar proteinosis (unusual)
Sarcoidosis

(14) Miliary and Small Nodular Lesions (Disseminated)

Inflammatory	Neoplastic	Other
Cytomegalovirus pneumonia	Alveolar cell carcinoma	Amyloid disease
Dirofilariasis	Hodgkin's disease	Arteriovenous mal-
Fungal infections	Lymphosarcoma	formation
Listeriosis	Metastatic disease	Caplan's syndrome
Miliary tuberculosis		Gaucher's disease
Paragonimiasis		Hemosiderosis
Septic infarcts		Histiocytosis X
Varicella pneumonia		Niemann-Pick disease
Viral pneumonia		Pneumoconioses
		Polyarteritis and vasculitis
		Pseudoxanthoma
		elasticum
		Rheumatoid nodules
		Sarcoidosis
		Siderosis
		Wegener's granulomatosis

(15) Large Nodular Lesions

Inflammatory	Neoplastic	Other
Abscess (pyogenic and others)	Alveolar cell carcinoma	Amyloidosis
Dirofilariasis	Bronchogenic carcinoma	Anthracosilicosis
Mycotic infections	Hamartoma	Arteriovenous
Paragonimiasis	Hodgkin's disease	malformation
	Lymphosarcoma	Caplan's syndrome
	Metastases	Lipoid pneumonia
	Papilloma	Pulmonary hematoma
	Plasmacytoma	Rheumatoid nodules
		Sequestration
		Wegener's granulomatosis

ROENTGEN SIGNS OF CHEST *(Continued)*

(16) Cavitary Lesions

Inflammatory	Neoplastic	Other
Abscess (pyogenic)	Bronchogenic carcinoma	Bulla (infected)
Amebic abscess	Hodgkin's disease	Cyst
Bronchiectasis	Metastatic lesions	Infarct
Echinococcus disease	(especially squamous cell)	Polyarteritis
Gram-negative infections		Polycystic lung
B. proteus		Rheumatoid nodules
E. coli		Sequestration
Pseudomonas		Traumatic lung cyst
Klebsiella pneumonia		Wegener's granulomatosis
Melioidosis		
Mycotic infections		
Actinomycosis		
Aspergillosis		
Blastomycosis		
Coccidioidomycosis		
Cryptococcosis		
Histoplasmosis		
Mucormycosis		
Nocardiosis		
Sporotrichosis		
Paragonimiasis		
Staphylococcal pneumonia		
Tuberculosis		

(17) Atelectasis (Obstructive)

Inflammatory	Neoplastic	Other
Middle lobe syndrome	Bronchial adenoma	Agammaglobulinemia
Pertussis	Bronchogenic carcinoma	Bronchial asthma
Tuberculosis	Granular cell myoblastoma	Bulbar poliomyelitis
	Invasive mediastinal	Chondromalacia of
	malignancy	bronchus
	Metastases to bronchus	Foreign body
		Lymph node enlargement
		(peribronchial)
		Mucoviscidosis
		Papilloma of bronchus
		Peritoneal dialysis
		Postoperative mucus plug
		Primary polycythemia

Table continued on following page.

II. CHEST

ROENTGEN SIGNS OF CHEST *(Continued)*

(18) Adenopathy—Hilar and/or Mediastinal

Infectious	Neoplastic	Other
Adenoviral infections	Bronchogenic carcinoma	Behçet's disease
Bacterial pneumonias	Extramedullary	Erythema nodosum
(children)	plasmacytoma	Sarcoidosis
Blastomycosis	Follicular lymphoma	Silicosis
Candidiasis	Hodgkin's disease	
Coccidioidomycosis	Leukemia (acute and	
Histoplasmosis	chronic)	
Infectious mononucleosis	Lymphosarcoma	
Measles	Metastatic disease	
Mycoplasma	Reticulum cell sarcoma	
Pertussis		
Plague pneumonia		
Psittacosis		
Sporotrichosis		
Tuberculosis		
Tularemia		
Varicella pneumonia		
Viral pneumonias		

(19) Mediastinal Masses and Widening

Inflammatory	Neoplastic	Other
Acute mediastinitis	Bronchogenic carcinoma	Aneurysm
Chronic mediastinitis	Extramedullary	Cushing's syndrome
Histoplasmosis	plasmacytoma	Myasthenia gravis
Lymph nodes	Follicular lymphoma	Sarcoidosis
Tuberculosis	Hodgkin's disease	Steroid lipomatosis
	Leukemia	Substernal thyroid
	Lymphosarcoma	Superior vena caval
	Metastatic disease	obstruction
	Reticulum cell sarcoma	Thymic enlargement
	Thymic tumors	
	Tumors, benign and	
	malignant	

ROENTGEN SIGNS OF CHEST *(Continued)*

(20) Pleural Fluid and/or Thickening

Inflammatory	Neoplastic	Other
Amebiasis	Bronchogenic carcinoma	Acute pancreatitis
Bacterial pneumonia	Hodgkin's disease	Asbestosis
Coccidioidomycosis	Leukemia	Cirrhosis
Coxsackie infections	Lymphosarcoma	Constrictive pericarditis
Empyema	Malignant mesothelioma	Familial Mediterranean
Hemophilus influenzae pneumonia	Metastatic disease	fever
Mycoplasma pneumonia	Primary macroglobulinemia	Heart failure
Nocardiosis	of Waldenström	Lupus erythematosus
Pneumococcal pneumonia		Meigs' syndrome
Q fever		Nephrotic syndrome
Staphylococcal pneumonia		Peritoneal dialysis
Tuberculosis		Postmyocardial infarction
Tularemia		syndrome
Viral pneumonia		Pulmonary infarction
		Rheumatoid arthritis
		Subdiaphragmatic abscess
		Superior vena caval
		obstruction
		Trauma

(21) Pneumothorax

Bronchopleural fistula
Bullous emphysema
Chemical pneumonitis
Histiocytosis X
Honeycomb lung
Ruptured esophagus
Spontaneous pneumothorax
Staphylococcal pneumonia
Trauma
Tuberculosis

ROENTGEN SIGNS OF GASTROINTESTINAL TRACT AND ABDOMEN

(22) Esophagus—Delayed Emptying and/or Dilatation

From Diffuse Peristaltic Abnormality	From Segmental Peristaltic Abnormality
Aperistalsis (achalasia)	Chalasia
Chagas' disease	Diffuse lower esophageal spasm
Dermatomyositis	Hypertrophic esophagogastric sphincter
Diabetes	Myasthenia gravis
Moniliasis	Myotonia atrophica
Riley-Day syndrome	Pseudobulbar palsy
Scleroderma	

Table continued on following page.

ROENTGEN SIGNS OF GASTROINTESTINAL TRACT AND ABDOMEN
(Continued)

From Intrinsic Obstructive Lesion	*From Extrinsic Obstructive Lesion*
Benign tumors	Aneurysm
Carcinoma and other malignant tumors	Cricopharyngeal muscle hypertrophy
Congenital strictures	Mediastinal tumors and masses
Duplication of esophagus	Vascular ring (aortic)
Foreign body	
Peptic or reflux esophagitis	
Peptic ulcer of esophagus	
Schatzki ring	
Sideropenic web	
Stricture (inflammatory or chemical)	
Zenker's diverticulum	

(23) Distal Esophagus — Narrowing and/or Deformity

Achalasia
Carcinoma of esophagus
Carcinoma of gastric cardia
Chagas' disease
Diffuse lower esophageal spasm
Duplication cyst
Hypertrophic esophagogastric sphincter
Leiomyoma
Peptic esophagitis
Reflux esophagitis
Stricture (chemical, inflammatory)
Trauma (intubation and instrumentation)

(24) Stomach — Dilatation

Acute hemorrhagic gastritis	Obstructive pyloroduodenal lesions
Anticholinergic drugs in gastric ulcer disease	Inflammatory
Bezoar	Neoplastic
Diabetes	Postoperative gastric dilatation
Duodenal ulcer with obstruction	Pyloric stenosis
Generalized ileus	Small bowel obstruction
Hypokalemia	Uremia and other toxic states
	Volvulus

(25) Stomach — Intraluminal Filling Defects

Aberrant pancreas	Carcinoma
Benign tumors	Duplication of stomach
Angioma	Eosinophilic granuloma
Carcinoid	Foreign body
Fibroma	Leiomyosarcoma
Leiomyoma	Lymphosarcoma
Lipoma	Polyposis (multiple)
Polyp	Varices (multiple)
Bezoar	Villous adenoma
Blood clots	

ROENTGEN SIGNS OF GASTROINTESTINAL TRACT AND ABDOMEN
(Continued)

(26) Stomach—Antropyloric Narrowing

Amyloidosis
Antral gastritis
Antral or pyloric ulcer
Carcinoma of antrum
Eosinophilic gastroenteritis
Foreign body
Granulomatous enteritis with antral involvement
Hypertrophic pyloric stenosis (infantile, adult)
Inflammatory and chemical stricture
Pancreatic tumors
Perforation and abscess
Syphilis (rare)
Tuberculosis (rare)

(27) Mesenteric Small Intestine—Mucosal Thickening Usually With Thickened Wall

Acute radiation injury
Amyloidosis
Eosinophilic gastroenteritis
Giardiasis
Hodgkin's disease
Hypoalbuminemia (bowel edema)
   Cirrhosis of liver
   Nephrotic syndrome
Intestinal lymphangiectasia

Intramural hemorrhage
   Anticoagulants
   Hemophilia
   Purpuras
Lymphosarcoma
Mastocytosis
Primary macroglobulinemia of Waldenström
Reticulum cell sarcoma
Scleroderma
Vascular occlusion
Whipple's disease

(28) Mesenteric Small Intestine—Multiple Intraluminal Defects

Carcinoids
Cronkhite-Canada syndrome
Dysgammaglobulinemia
Hodgkin's disease
Lymphosarcoma
Metastatic nodules (usually malignant melanoma)
Nodular lymphoid hyperplasia
Peutz-Jeghers syndrome
Regional enteritis
Reticulum cell sarcoma
Submucosal hemorrhages
   Anticoagulants
   Hemophilia
   Purpuras
   Vascular occlusion
Typhoid fever

Table continued on following page.

ROENTGEN SIGNS OF GASTROINTESTINAL TRACT AND ABDOMEN
(Continued)

(29) Mesenteric Small Intestine—Inflammatory-Like Appearance
(mucosal thickening or deformity, thickened walls, segmental narrowing)

Actinomycosis
Carcinoids
Enteritis necroticans
Eosinophilic gastroenteritis
Granulomatous enteritis
Hodgkin's disease
Parasitic infections
Potassium thiazide enteritis
Radiation injury
Retractile mesenteritis
Strongyloidiasis
Tuberculosis
Vascular occlusion
Zollinger-Ellison syndrome

(30) Mesenteric Small Intestine—Deficiency Pattern

Carcinoid syndrome
Celiac disease
Chronic mesenteric vascular insufficiency
Emotional states
Familial Mediterranean fever
γ-A heavy chain disease
Gastrointestinal allergy
Hyperthyroidism

Hypoparathyroidism
Kwashiorkor
Lymphosarcoma
Mucoviscidosis
Nontropical sprue
Pancreatic insufficiency
Parasitic infections
Tropical sprue

(31) Colon—Inflammatory-Like Appearance
(mucosal changes, ulcerations, irritability, segmental narrowing)

Actinomycosis
Amebiasis
Antibiotic-induced enterocolitis
Bacillary dysentery
Diverticulitis
Granulomatous colitis
Ischemic colitis

Lymphogranuloma inguinale
Lymphoma (diffuse)
Postirradiation colitis
Schistosomiasis
Trichuriasis
Tuberculosis
Ulcerative colitis

(32) Colon—Solitary Mass Defect

Adenomatous polyp
Ameboma
Benign tumors (leiomyoma, fibroma, etc.)
Carcinoid
Carcinoma
Endometrioma
Fecal impaction
Lipoma
Lymphosarcoma
Schistosomiasis
Villous adenoma

ROENTGEN SIGNS OF GASTROINTESTINAL TRACT AND ABDOMEN
(Continued)

(33) Colon—Multiple Filling Defects

Amebomas
Amyloidosis
Ischemic colitis
Lymphosarcoma
Nodular lymphoid hyperplasia
Pneumatosis cystoides intestinalis
Polyps
  Cronkhite-Canada syndrome

Familial polyposis
Gardner's syndrome
Nonfamilial polyposis
Peutz-Jeghers syndrome
Pseudopolyps in ulcerative colitis
Pseudopolyps in granulomatous colitis
Retained fecal contents

(34) Megacolon

Chagas' disease
Chronic constipation
Diabetes
Familial Mediterranean fever
Generalized peritonitis
Granulomatous colitis
Hirschsprung's disease
Lead poisoning
Low mechanical obstruction
Parkinsonism
Porphyria
Toxic ileus
Ulcerative colitis (toxic megacolon)

(35) Intra-abdominal Calcifications

Inflammatory	Neoplastic	Other
Amebic abscess	Dermoid cyst	Adrenal
Echinococcus cyst	Hemangioma	Aneurysm
Healed splenic	Hepatoma	Arteriosclerotic vessels
histoplasmosis	Mucus-producing	Atherosclerotic aorta
Meconium peritonitis	adenocarcinoma	Epiploic appendages
Mesenteric lymph nodes	Neuroblastoma	Fecolith (appendix, colon)
Schistosomiasis	Ovarian carcinoma	Gallbladder wall
Tuberculous peritonitis	Peritoneal metastases	Gallstones
	Pheochromocytoma	Hematoma
	Renal cyst or tumor	Mesenteric cyst
	Retroperitoneal sarcoma	Milk of calcium bile
	Uterine fibroids	Nephrocalcinosis (see
		Table 44)
		Pancreatic calculi
		Phleboliths
		Prostatic calculi
		Pseudocyst of pancreas
		Renal (see Tables 42–45)
		Retroperitoneal
		hematoma
		Rib cartilage (extra-
		abdominal)
		Seminal vesicles
		Urinary tract calculi
		Vas deferens

Table continued on following page.

ROENTGEN SIGNS OF GASTROINTESTINAL TRACT AND ABDOMEN
(Continued)

(36) Splenomegaly

Inflammatory	Neoplastic	Other
Cytomegalic inclusion disease	Hodgkin's disease	Acute infarct
Echinococcus cyst	Leukemia	Adrenocortical insuffi-
Histoplasmosis	Lymphosarcoma	ciency
Infectious mononucleosis	Sarcoma	Amyloidosis
Malaria		Cirrhosis of liver
Miliary tuberculosis		Congestive splenomegaly
Subacute bacterial endocarditis		Cooley's anemia
Viral hepatitis		Dysgammaglobulinemia
		Felty's syndrome
		Gaucher's disease
		Hemochromatosis
		Hemolytic anemia
		Hurler's syndrome
		Juvenile rheumatoid arthritis
		Lupus erythematosus
		Mastocytosis (urticaria pigmentosa)
		Mucoviscidosis
		Myelofibrosis
		Myelophthisic anemia
		Niemann-Pick disease
		Osteopetrosis
		Portal hypertension
		Primary polycythemia
		Sickle cell anemia (S-C)
		Splenic cyst
		Splenic rupture
		Trauma

(37) Liver Calcifications

Abscess (pyogenic or amebic)	Gumma
Alveolar hydatid disease	Hemangioma; A-V malformation
Aneurysms of hepatic artery	Hepatic duct calculi
Congenital cysts	Metastases (ovary, breast, colon, mesothelioma, neuroblastoma)
Cystic hydatid disease	Portal vein thrombus
Granulomas (tuberculosis, histoplasmosis, brucellosis)	Primary neoplasm (malignant hepatoma, cholangioma)

ROENTGEN SIGNS OF JOINTS

(38) Rheumatoid and Rheumatoid-Like Arthritis

Agammaglobulinemia arthritis
Gout
Hemochromatosis (rare)
Juvenile rheumatoid arthritis
Lupus erythematosus
Polychondritis
Psoriatic arthritis
Reiter's syndrome
Sjögren's syndrome
Ulcerative colitis arthritis

(39) Osteoarthritis—Degenerative Joint Disease

Acromegaly
Aseptic necrosis (late)
Ehlers-Danlos syndrome
Familial Mediterranean fever
Frostbite
Gout
Hemochromatosis (rare)
Hemophilia
Hepatolenticular degeneration
Late result of joint inflammation or trauma
Ochronosis
Osteoarthritis and spondylosis of aging
Trauma

(40) Neuropathic Joint (Charcot Joint)

Diabetes
Leprosy
Riley-Day syndrome
Spinal cord disease
Steroid therapy (Charcot-like)
Syphilis (tabes dorsalis)
Syringomyelia

(41) Periarticular Calcifications

Dermatomyositis
Hyperparathyroidism
Hypervitaminosis A
Myositis ossificans
Neurotrophic joint
Pseudogout
Pyogenic arthritis
Scleroderma
Synovioma
Trauma
Tuberculosis

ROENTGEN SIGNS OF KIDNEYS AND URINARY TRACTS

(42) Enlargement

Unilateral	Bilateral
Acute pyelonephritis	Acromegaly
Acute transplant rejection	Acute glomerulonephritis
Compensatory hypertrophy	Acute renal failure
Cyst and cystic disease	Amyloidosis
Duplicated kidney	Bilateral acute pyelonephritis
Echinococcus disease	Bilateral metastatic disease
Hematoma	Bilateral obstructive hydronephrosis
Multicystic kidney	Idiopathic lipoid nephrosis
Neoplastic lesions	Leukemia
(primary and secondary)	Lymphosarcoma
Obstructive hydronephrosis	Polycystic kidneys
Renal vein thrombosis	Renal cortical necrosis
	Renal vein thrombosis
	Tuberous sclerosis (bilateral angiomyolipomas)

(43) Decreased Size

Unilateral	Bilateral
Chronic pyelonephritis	Balkan nephritis
Congenital hypoplastic kidney	Bilateral renal artery stenosis
Obstructive atrophy	Chronic glomerulonephritis
Renal artery stenosis	Chronic gouty nephritis
Renal infarction (late)	Chronic interstitial nephritis
	Chronic pyelonephritis
	Diffuse fibromuscular hyperplasia
	Hereditary nephritis (Alport's disease)
	Kimmelstein-Wilson disease
	Medullary cystic disease
	Scleroderma
	Systemic lupus erythematosus

(44) Nephrocalcinosis

Intrinsic Renal Disease	Systemic Disturbance
Medullary sponge kidney	Carcinomatosis of bone
Papillary necrosis	Cortisone therapy
Renal tubular acidosis	Cretinism
Tuberculosis	Cushing's disease
	Hyperoxaluria
	Hyperparathyroidism
	Hypervitaminosis D
	Idiopathic hypercalcinuria
	Idiopathic hypercalcemia
	Milk-alkali ingestion
	Multiple myeloma
	Sarcoidosis
	Sjögren's syndrome
	Sulfonamide therapy

ROENTGEN SIGNS OF KIDNEYS AND URINARY TRACTS *(Continued)*

(45) Focal Calcifications

Abscess
Aneurysm (intrarenal artery)
Benign tumors
Chronic glomerulonephritis (cortical calcification)
Cyst
Cystinuria lithiasis
Echinococcus disease
Gouty lithiasis
Hematoma
Hydronephrotic cavity
Infarct
Lithiasis (idiopathic)
Malignant tumors
Ochronosis lithiasis
Pyonephrosis
Renal cortical necrosis (tramway calcification)
Tuberculosis and any condition causing nephrocalcinosis
   (see preceding table)
Vascular malformation

ROENTGEN SIGNS OF OSSEOUS SYSTEM

(46) Demineralization and/or Osteoporosis

Acromegaly	Hypothyroidism (cretinism)	Postmenopausal
Acro-osteolysis	Idiopathic hypercalcemia (late)	Primary macroglobulinemia
Ankylosing spondylitis	Immobilization	of Waldenström
Cooley's anemia	Iron deficiency anemia	Protein deficiency
Cortisone therapy	(severe, in children)	Renal osteodystrophy
Cushing's disease	Juvenile rheumatoid arthritis	Renal tubular acidosis
Dermatomyositis	Leukemia (acute)	Rheumatoid arthritis
Familial Mediterranean	Malabsorption	Rickets
fever	Multiple myeloma	Scurvy
Fanconi's syndrome	Muscular dystrophy	Senility
Frostbite	Niemann-Pick disease	Shoulder-hand syndrome
Gaucher's disease	Ochronosis	Sickle cell anemia
Hemochromatosis (systemic)	Osteitis fibrosa cystica	Sudeck's atrophy
Hemophilia	Osteogenesis imperfecta	Thyrotoxicosis
Hepatic rickets	Osteomalacia	Turner's syndrome
Hepatolenticular	Osteoporosis circumscripta	
degeneration	Oxalosis	
Homocystinuria	Paget's disease	
Hyperparathyroidism	Phenylketonuria	
Hyperthyroidism		
Hypoparathyroidism with		
steatorrhea		

Table continued on following page.

ROENTGEN SIGNS OF OSSEOUS SYSTEM *(Continued)*

(47) Osteosclerosis

Inflammatory (Focal)	Neoplastic (Usually Focal, Occasionally Diffuse)	Other (Usually Diffuse)
Brodie's abscess	Hodgkin's disease	*Engelmann's disease
Cytomegalic inclusion disease	Leukemia (chronic)	Fibrous dysplasia
Garre's osteomyelitis	Lymphoma	Fluorosis
Osteomyelitis (chronic and healed)	Metastatic carcinoma	Hyperparathyroidism
	Multiple myeloma (rare)	Hypoparathyroidism
		Hypothyroidism
		Idiopathic hypercalcemia
		Lipoatrophic diabetes mellitus
		Mastocytosis
		Melorheostosis
		Myeloid metaplasia
		Osteopetrosis
		Paget's disease
		Pycnodysostosis
		Renal osteodystrophy
		Tuberous sclerosis
		Vitamin D intoxication

(48) Periosteal Reaction

Caffey's disease	Pachydermoperiostosis
Congenital syphilis	Reiter's syndrome
Fluorosis	Rickets (healing)
Hypertrophic pulmonary osteoarthropathy	Scurvy
Juvenile rheumatoid arthritis	Thyroid acropachy
Leukemia (acute)	Trauma
Malignant bone tumors	Tropical ulcer
Metastatic neuroblastoma	Tuberous sclerosis
Osteoid osteoma	Varicose ulcers
Osteomyelitis (acute)	Vitamin A intoxication

(49) Rickets and Rachitic-Like Changes

Atresia of common duct (hepatic rickets)
Fanconi's syndrome
Hepatolenticular degeneration
Homocystinuria
Hypophosphatasia
Malabsorption
Oxalosis
Phenylketonuria
Renal tubular acidosis
Uremic osteodystrophy
Vitamin D deficiency

ROENTGEN SIGNS OF OSSEOUS SYSTEM *(Continued)*

(50) Rib Notching or Erosion

Aplasia of pulmonary artery
Bulbar poliomyelitis
Coarctation of aorta
Increased pulse pressure
Neurofibromatosis
Vascular malformation (intercostal artery)
Vascular obstruction (vena caval, subclavian, or
 innominate artery or vein)
Vascular surgery, for correction of tetralogy of Fallot

(51) Aseptic Necrosis

Caisson disease
Cushing's syndrome
Gaucher's disease
Hemophilia joint disease
Hypothyroidism
Idiopathic
Lupus erythematosus
Sickle cell anemia
Steroid therapy
Trauma

(52) Bone Infarcts

Arteriosclerosis
Caisson disease
Idiopathic
Pancreatitis (acute and chronic)
Sickle cell anemia

(53) Acro-osteolysis

Angiomatous malformation
Epidermolysis bullosa
Familial acro-osteolysis
Hyperparathyroidism
Occupational (vinyl chloride)
Progeria
Raynaud's phenomenon
Scleroderma

(54) Advanced Bone Age and Skeletal Maturation

Adrenogenital syndrome (adrenal hyperplasia or adenoma)
Constitutional
Exogenous obesity
Juvenile rheumatoid arthritis
Lipoatrophic diabetes mellitus
Ovarian endocrine tumors
Pinealoma
Polyostatic fibrous dysplasia (Albright's syndrome)
Primary hyperaldosteronism
Sato's syndrome (pituitary gigantism)

Table continued on following page.

ROENTGEN SIGNS OF OSSEOUS SYSTEM *(Continued)*

(55) Retarded Bone Age and Skeletal Maturation

Congenital hyperuricosuria
Cushing's syndrome
Homocystinuria
Hypogonadism
Hypopituitarism
Hypothyroidism
Malnutrition
Phenylketonuria
Turner's syndrome

(56) Skull Tables—Thickening and/or Sclerosis

Acromegaly
Anemias of childhood (severe: Cooley's, sickle cell,
   spherocytosis, iron deficiency)
Cerebral atrophy (childhood)
Craniostenosis
Engelmann's disease
Fibrous dysplasia
Hyperostosis frontalis interna
Idiopathic
Meningioma
Microcephaly
Myotonia atrophica
Neuroblastoma (metastatic)
Osteoma
Osteopetrosis
Paget's disease

VII. MISCELLANEOUS

MISCELLANEOUS ROENTGEN SIGNS

(57) Soft Tissue Calcifications in Extremities (Excluding Focal Soft Tissue Disease)

Metabolic and Collagen	Parasitic	Vascular
Chronic renal disease (amorphous, periarticular)	Cysticercosis (oval, in muscle planes)	Atherosclerosis
Dermatomyositis (amorphous, subcutaneous sheets)	Dracunculosis (coiled, tubelike)	Medial sclerosis (Mönckeberg's)
Ehlers-Danlos syndrome (ring calcifications)	Filariasis (coiled, linear)	Phleboliths
Fluorosis (ligamentous)	Leprosy (linear, in nerve sheaths)	Venous stasis
Gout (amorphous, periarticular)		
Hyperparathyroidism, primary and secondary (amorphous, periarticular; more common in secondary disease)		
Hypervitaminosis A (ligaments)		
Hypervitaminosis D (flocculent periarticular masses)		
Hypoparathyroidism (arterial, periarticular)		
Idiopathic calcinosis (amorphous, plaques)		
Idiopathic hypercalcemia (flocculent periarticular masses)		
Milk-alkali syndrome (amorphous, periarticular)		
Ochronosis (ligaments, cartilage)		
Paraplegia (periarticular)		
Poliomyelitis (periarticular, disc cartilage)		
Progressive myositis ossificans (massive, subcutaneous, bone formation)		
Pseudogout (synovial, cartilage)		
Pseudohypoparathyroidism (small subcutaneous)		
Pseudopseudohypoparathyroidism (small subcutaneous)		
Pseudoxanthoma elasticum (faint subcutaneous sheets)		
Scleroderma (amorphous, subcutaneous sheets, periarticular in hands)		
Werner's syndrome (amorphous, vascular, ligaments, periarticular)		

VII. MISCELLANEOUS

MISCELLANEOUS ROENTGEN SIGNS (*Continued*)

(58) Cartilage Calcifications

Adrenal insufficiency (ear cartilage)
Chondrocalcinosis (joint cartilage)
Chronic respiratory poliomyelitis
Gout (joint cartilage)
Hypercalcemia (ear cartilage)
Hyperparathyroidism (joint and ear cartilage)
Ochronosis (intervertebral discs, ear cartilage)
Osteoarthritis (joint cartilage)
Physiologic (costal and laryngeal cartilage)
Polychondritis (joint, ear and other cartilage)
Senility
Tietze's syndrome (late)
Trauma
Vitamin D intoxication (joint cartilage)

(59) Intracranial Calcifications

Asymptomatic-Physiologic	Inflammatory	Neoplastic	Other
Choroid plexus	Abscess	Craniopharyngioma	Aneurysm
Dura	Cysticercosis	Gliomas	Arteriosclerotic vessels
Falx	Cytomegalic inclusion disease	Hemangioma	Arteriovenous malformation
Habenula	Paragonimiasis	Lipoma of corpus callosum	Familial cerebral calcification
Internal carotid artery (parasellar)	Torulosis (rare)	Meningioma	Hemorrhage
Petroclinoid ligament	Toxoplasmosis	Metastatic disease (rare)	Idiopathic basal ganglia calcification
Pineal	Trichinosis (rare)	Metastatic retinoblastoma	Primary hypo-parathyroidism
Tentorium	Tuberculoma	Pinealoma	Pseudohypopara-thyroidism
	Tuberculous meningitis	Pituitary adenoma (rare)	Pseudopseudo-hypopara-thyroidism
			Sturge-Weber disease
			Subdural hema-toma
			Tuberous sclerosis

INDEX

Colon (*Continued*)
  shoulder defect of, A31
  stricture of, in inferior mesenteric artery
    occlusion, **1038**
  thumbprinting defects in, A34
  transverse, granulomatous colitis of, **1012**
    in Morgagni hernia, 452
  urticaria of, **943**
Colon cutoff sign, in acute mesenteric
    vascular insufficiency, 1032
  in acute pancreatitis, 983, **985**
Coma, drug, aspiration pneumonia during, **501**
Common duct
  calculi in, 1088, **1088, 1089, 1090**
  carcinoma of, **1098**
  congenital atresia of, 1090, **1091**
  cystic dilatation of, congenital, 1091, **1092**
Communicating hydrocephalus, 390, **391, 392**
Computerized axial tomography (CAT). See
  CT scan.
Computerized tomography (CT). See *CT scan.*
Computerized transaxial tomography (CTT).
  See *CT scan.*
Congenital agammaglobulinemia, 25, **26, 27**
Congenital hypogammaglobulinemia, 25
Congenital multicystic kidney, 847, 848
Congenital rubella, 140, **141**
Congestion, pulmonary, vs. pulmonary fat
  embolism, **549**
Congestive failure, 598–602, **599, 600, 601,
  602**
  infarction of lung in, 603
Connective tissue, diseases of, 35–50
Consolidation, lobar, A16
Constipation, chronic, 912, **912, 913**
  in Parkinson's disease, 296
Contusion, pulmonary, closed chest trauma
  and, 494
Cooley's anemia, 1155, **1155, 1156, 1157**
Cor pulmonale, 613, **613, 614**
  diseases associated with, B4
  in chronic bronchitis, 457
  in cystic fibrosis, **997**
  in pulmonary emphysema, 459, **461**
  in schistosomiasis, 270, **271**
Cor triatriatum sinistrum, 640, **641**
Cor triloculare biatriatum, 650, **650, 651**
Cord bladder, in paraplegia, **371**
Corkscrew esophagus, 884, **885, 886**
Coronary artery, anomalous left, 633, **634, 635**
  calcification of, and atherosclerosis, 696,
    697, 698
  left, from pulmonary artery, 633, **634, 635**
Coronary artery disease, 696–703
Coronary fistula, congenital, 633, **633**
Coronary ischemia, and bypass surgery, 699
Corpus callosum, agenesis of, 381, **382, 383**
  tumors of, 339, **341**
Cortical irregularity, avulsive, 1389, **1390**
Cortical thinning, renal, A39
Cortisone, and hypercalcemic nephropathy,
  822
Cortisone therapy, effects of, 1
Costal chondritis, 104, **104**
"Cotton wool" appearance, in osteitis
  deformans, 1359, **1363**

Coxa valga, in Niemann-Pick disease, 1198
  in progeria, **1436**
Coxsackie disease, 145, **146, 147**
  pleural effusion in, 569, **569**
Cranial arteritis, 113, **113**
Cranial sutures, premature closure of, 385,
  386, 387
Craniopharyngioma, and hypopituitarism,
  1254
  calcification in, 319
  suprasellar, 347, **348, 349**
Craniostenosis, 385, **386, 387**
Craniosynostosis, 385, **386, 387**
  and microcrania, 379
Creeping eruption, 290, **290**
Cretinism, 1263, **1263, 1264, 1265**
Cricopharyngeal dysphagia, 882, **883**
Crohn's disease, 1006
  sacroiliac changes in, 78
Cronkhite-Canada syndrome, 1076, **1076**
Croup, acute epiglottitis and, 474, **475**
Cryptococcosis, 231, **232, 233**
CT scan, 293, **294, 295**
  in acoustic neurinoma, 343, **345**
  in agenesis of cerebellum, 384, **385**
  in Arnold-Chiari malformation, 356
  in arteriovenous malformation of brain,
    306, **307**
  in brain atrophy, 377, **378, 379**
  in brain infarcts, 297, **299**
  in cerebellar hemangioblastoma, **330**
  in cerebellar medulloblastoma, **329**
  in cerebellar tumors, 325, **329, 330**
  in cerebral aneurysm, 304
  in cerebral hypoplasia, 380, **381**
  in communicating hydrocephalus, 391
  in "empty" sella turcica, 404
  in focal epilepsy, 405
  in frontal lobe tumors, 330, **332**
  in glioblastoma, **341**
  in glioma, **329, 342**
  in Huntington's chorea, 400, **400**
  in internal hydrocephalus, 357, **361, 362**
  in intracranial abscess, 313, **315**
  in intracranial hemorrhage, 302, **304**
  in intracranial tumors, 319
  in lipoma of corpus callosum, 381, **383**
  in lymphosarcoma, **321**
  in meningioma, 321, **321, 324, 325**
  in metastatic brain tumors, 350, **353**
  in microcephaly, 379
  in neurolemmoma, **320**
  in occipital lobe tumors, 337, **338**
  in orbital sarcoma, 320
  in orbital tumors, 320, **320, 321**
  in parietal lobe tumors, 332, **335**
  in Parkinson's disease, **296**
  in pinealoma, 1259, **1261**
  in pituitary adenoma, 346, **347**
  in porencephaly, 383, **384**
  in pseudohypoparathyroidism, **1314**
  in pseudotumor cerebri, 355
  in skull fractures, 362
  in spinocerebellar ataxias, 393
  in Sturge-Weber disease, 394
  in subdural hematoma, 309, **311**